STUDENT

SOLUTIONS MANUAL

John Garlow

Tarrant County College

BEGINNING & INTERMEDIATE ALGEBRA

THIRD EDITION

K. Elayn Martin-Gay

PEARSON

Prentice Hall

Upper Saddle River, NJ 07458

Editor-in-Chief: Chris Hoag
Senior Acquisitions Editor: Paul Murphy
Project Manager: Mary Beckwith
Assistant Editor: Christina Simoneau
Vice President of Production & Manufacturing: David W. Riccardi
Executive Managing Editor: Kathleen Schiaparelli
Assistant Managing Editor: Nicole Jackson
Production Editor: Nicole Jackson
Supplement Cover Designer: Joanne Alexandris
Supplement Cover Manager: Paul Gourhan
Manufacturing Buyer: Ilene Kahn

© 2005 Pearson Education, Inc.
Pearson Prentice Hall
Pearson Education, Inc.
Upper Saddle River, NJ 07458

Printed in the United States of America

10 9 8 7 6 5 4 3

ISBN 0-13-149357-4

Pearson Education Ltd., *London*
Pearson Education Australia Pty. Ltd., *Sydney*
Pearson Education Singapore, Pte. Ltd.
Pearson Education North Asia Ltd., *Hong Kong*
Pearson Education Canada, Inc., *Toronto*
Pearson Educación de Mexico, S.A. de C.V.
Pearson Education—Japan, *Tokyo*
Pearson Education Malaysia, Pte. Ltd.

Table of Contents

Chapter 1

Exercise Set 1.1

Answers will vary on Exercises 1-19.

Exercise Set 1.2

1. $7 > 3$

3. $6.26 = 6.26$

5. $0 < 7$

7. $-2 < 2$

9. $32 < 212$

11. $44,300 > 34,611$

13. True, since $11 = 11$.

15. False, since 10 is to the left of 11 on the number line.

17. False, since 11 is to the left of 24 on the number line.

19. True, since 7 is to the right of 0 on the number line.

21. $30 \le 45$

23. $8 < 12$

25. $5 \ge 4$

27. $15 \ne -2$

29. 535 represents an altitude of 535 feet.
-8 represents 8 feet below sea level.

31. $-34,841$ represents a population decrease of 34,841.

33. 350 represents a deposit of $350.
-126 represents a withdrawal of $126.

35. 1993

37. 1993, 1994

39. $827 \ge 818$

41. The number 0 belongs to the sets of: whole numbers, integers, rational numbers, and real numbers.

43. The number -2 belongs to the sets of: integers, rational numbers, and real numbers.

45. The number 6 belongs to the sets of: natural numbers, whole numbers, integers, rational numbers, and real numbers.

47. The number 2/3 belongs to the sets of: rational numbers and real numbers.

49. The number $-\sqrt{5}$ belongs to the sets of: irrational numbers and real numbers.

51. False. Rational numbers may be non-integers.

53. True

55. True

57. True

59. False. An irrational number may not be written as a fraction

59. False. Irrational numbers are real.

61. $-10 > -100$

63. $32 > 5.2$

65. $\dfrac{18}{3} < \dfrac{24}{3}$

67. $-51 < -50$

69. $|-5| > -4$ since $5 > -4$

71. $|-1| = |1|$ since $1 = 1$

73. $|-2| < |-3|$ since $2 < 3$

75. $|0| < |-8|$ since $8 < 8$

77. $-0.04 > -26.7$

79. The sun is brighter since $-26.7 < -0.04$.

81. The sun is the brightest since -26.7 is to the left of all other numbers listed.

83. $20 \le 25$

85. $6 > 0$

87. $-12 < -10$

89. Answers may vary.

Section 1.3

Mental Math

1. $\dfrac{3}{8}$

2. $\dfrac{1}{4}$

3. $\dfrac{5}{7}$

4. $\dfrac{2}{5}$

5. numerator, denominator

6. $\dfrac{11}{7}$

Exercise Set 1.3

1. $33 = 3 \cdot 11$

3. $98 = 2 \cdot 7 \cdot 7$

5. $20 = 2 \cdot 2 \cdot 5$

7. $75 = 3 \cdot 5 \cdot 5$

9. $45 = 3 \cdot 3 \cdot 5$

11. $\dfrac{2}{4} = \dfrac{2}{2 \cdot 2} = \dfrac{1}{2}$

13. $\dfrac{10}{15} = \dfrac{2 \cdot 5}{3 \cdot 5} = \dfrac{2}{3}$

15. $\dfrac{3}{7} = \dfrac{3}{7}$

17. $\dfrac{18}{30} = \dfrac{2 \cdot 3 \cdot 3}{2 \cdot 3 \cdot 5} = \dfrac{3}{5}$

19. $\dfrac{1}{2} \cdot \dfrac{3}{4} = \dfrac{1 \cdot 3}{2 \cdot 2 \cdot 2} = \dfrac{3}{8}$

21. $\dfrac{2}{3} \cdot \dfrac{3}{4} = \dfrac{2 \cdot 3}{3 \cdot 2 \cdot 2} = \dfrac{1}{2}$

23. $\dfrac{1}{2} \div \dfrac{7}{12} = \dfrac{1}{2} \cdot \dfrac{12}{7} = \dfrac{1 \cdot 2 \cdot 2 \cdot 3}{2 \cdot 7} = \dfrac{2 \cdot 3}{7} = \dfrac{6}{7}$

25. $\dfrac{3}{4} \div \dfrac{1}{20} = \dfrac{3}{4} \cdot \dfrac{20}{1} = \dfrac{3 \cdot 2 \cdot 2 \cdot 5}{2 \cdot 2} = \dfrac{3 \cdot 5}{1} = 15$

27. $\dfrac{7}{10} \cdot \dfrac{5}{21} = \dfrac{7 \cdot 5}{2 \cdot 5 \cdot 3 \cdot 7} = \dfrac{1}{2 \cdot 3} = \dfrac{1}{6}$

29. $2\dfrac{7}{9} \cdot \dfrac{1}{3} = \dfrac{25}{9} \cdot \dfrac{1}{3} = \dfrac{5 \cdot 5 \cdot 1}{3 \cdot 3 \cdot 3} = \dfrac{25}{27}$

31. Area $= \dfrac{11}{12} \cdot \dfrac{3}{5} = \dfrac{11 \cdot 3}{2 \cdot 2 \cdot 3 \cdot 5} = \dfrac{11}{2 \cdot 2 \cdot 5}$

            $= \dfrac{11}{20}$ sq. mi

33. $\dfrac{4}{5} - \dfrac{1}{5} = \dfrac{4-1}{5} = \dfrac{3}{5}$

35. $\dfrac{4}{5} + \dfrac{1}{5} = \dfrac{4+1}{5} = \dfrac{5}{5} = 1$

37. $\dfrac{17}{21} - \dfrac{10}{21} = \dfrac{17-10}{21} = \dfrac{7}{21} = \dfrac{7}{3 \cdot 7} = \dfrac{1}{3}$

39. $\dfrac{23}{105} + \dfrac{4}{105} = \dfrac{23+4}{105} = \dfrac{27}{105} = \dfrac{3 \cdot 3 \cdot 3}{3 \cdot 5 \cdot 7}$

            $= \dfrac{3 \cdot 3}{5 \cdot 7} = \dfrac{9}{35}$

41. $\dfrac{7}{10} = \dfrac{7 \cdot 3}{10 \cdot 3} = \dfrac{21}{30}$

43. $\dfrac{2}{9} = \dfrac{2 \cdot 2}{9 \cdot 2} = \dfrac{4}{18}$

45. $\dfrac{4}{5} = \dfrac{4 \cdot 4}{5 \cdot 4} = \dfrac{16}{20}$

47. $\dfrac{2}{3} + \dfrac{3}{7} = \dfrac{2 \cdot 7}{3 \cdot 7} + \dfrac{3 \cdot 3}{7 \cdot 3} = \dfrac{14}{21} + \dfrac{9}{21} = \dfrac{23}{21}$

49. $2\dfrac{13}{15} - 1\dfrac{1}{5} = \dfrac{43}{15} - \dfrac{6}{5} = \dfrac{43}{15} - \dfrac{6 \cdot 3}{5 \cdot 3}$

            $= \dfrac{43-18}{15} = \dfrac{25}{15} = 1\dfrac{2}{3}$

51. $\dfrac{5}{22} - \dfrac{5}{33} = \dfrac{5 \cdot 3}{22 \cdot 3} - \dfrac{5 \cdot 2}{33 \cdot 2} = \dfrac{15}{66} - \dfrac{10}{66} = \dfrac{5}{66}$

53. $\dfrac{12}{5} - 1 = \dfrac{12}{5} - \dfrac{5}{5} = \dfrac{12-5}{5} = \dfrac{7}{5}$

55. $1 - \dfrac{3}{10} - \dfrac{5}{10} = \dfrac{10}{10} - \dfrac{3}{10} - \dfrac{5}{10} = \dfrac{10 - 3 - 5}{10}$

$\qquad = \dfrac{2}{10} = \dfrac{2}{2 \cdot 5} = \dfrac{1}{5}$

The unknown part is $\dfrac{1}{5}$

57. $1 - \dfrac{1}{4} - \dfrac{3}{8} = \dfrac{8}{8} - \dfrac{1 \cdot 2}{4 \cdot 2} - \dfrac{3}{8} = \dfrac{8 - 2 - 3}{8} = \dfrac{3}{8}$

The unknown part is $\dfrac{3}{8}$

59. $1 - \dfrac{1}{2} - \dfrac{1}{6} - \dfrac{2}{9} = \dfrac{18}{18} - \dfrac{1 \cdot 9}{2 \cdot 9} - \dfrac{1 \cdot 3}{6 \cdot 3} - \dfrac{2 \cdot 2}{9 \cdot 2}$

$\qquad = \dfrac{18 - 9 - 3 - 4}{18} = \dfrac{2}{18} = \dfrac{1}{9}$

The unknown part is $\dfrac{1}{9}$

61. $\dfrac{10}{21} + \dfrac{5}{21} = \dfrac{10 + 5}{21} = \dfrac{15}{21} = \dfrac{3 \cdot 5}{3 \cdot 7} = \dfrac{5}{7}$

63. $\dfrac{10}{3} - \dfrac{5}{21} = \dfrac{10 \cdot 7}{3 \cdot 7} - \dfrac{5}{3 \cdot 7} = \dfrac{70}{21} - \dfrac{5}{21} = \dfrac{65}{21}$

65. $\dfrac{2}{3} \cdot \dfrac{3}{5} = \dfrac{2 \cdot 3}{3 \cdot 5} = \dfrac{2}{5}$

67. $\dfrac{3}{4} \div \dfrac{7}{12} = \dfrac{3}{4} \cdot \dfrac{12}{7} = \dfrac{3 \cdot 3 \cdot 4}{4 \cdot 7} = \dfrac{9}{7}$

69. $\dfrac{5}{12} + \dfrac{4}{12} = \dfrac{5 + 4}{12} = \dfrac{9}{12} = \dfrac{3 \cdot 3}{3 \cdot 4} = \dfrac{3}{4}$

71. $5 + \dfrac{2}{3} = \dfrac{15}{3} + \dfrac{2}{3} = \dfrac{15 + 2}{3} = \dfrac{17}{3}$

73. $\dfrac{7}{8} \div 3\dfrac{1}{4} = \dfrac{7}{8} \div \dfrac{13}{4} = \dfrac{7}{8} \cdot \dfrac{4}{13} = \dfrac{7 \cdot 4}{2 \cdot 4 \cdot 13} = \dfrac{7}{26}$

75. $\dfrac{7}{18} \div \dfrac{14}{36} = \dfrac{7}{18} \cdot \dfrac{36}{14} = \dfrac{7 \cdot 2 \cdot 18}{18 \cdot 2 \cdot 7} = 1$

77. $\dfrac{23}{105} - \dfrac{2}{105} = \dfrac{23 - 2}{105} = \dfrac{21}{105} = \dfrac{21}{21 \cdot 5} = \dfrac{1}{5}$

79. $1\dfrac{1}{2} + 3\dfrac{2}{3} = \dfrac{3}{2} + \dfrac{11}{3} = \dfrac{3 \cdot 3}{2 \cdot 3} + \dfrac{11 \cdot 2}{3 \cdot 2} = \dfrac{9 + 22}{6}$

$\qquad = \dfrac{31}{6} = 5\dfrac{1}{6}$

81. $\dfrac{2}{3} - \dfrac{5}{9} + \dfrac{5}{6} = \dfrac{2 \cdot 2 \cdot 3}{3 \cdot 2 \cdot 3} - \dfrac{5 \cdot 2}{9 \cdot 2} + \dfrac{5 \cdot 3}{6 \cdot 3}$

$\qquad = \dfrac{12 - 10 + 15}{18} = \dfrac{17}{18}$

83. $5 + 4\dfrac{1}{8} + 4\dfrac{1}{8} + 15\dfrac{3}{4} + 15\dfrac{3}{4} + 10\dfrac{1}{2}$

$\qquad = \dfrac{40}{8} + \dfrac{33}{8} + \dfrac{33}{8} + \dfrac{126}{8} + \dfrac{126}{8} + \dfrac{84}{8}$

$\qquad = \dfrac{40 + 33 + 33 + 126 + 126 + 84}{8}$

$\qquad = \dfrac{442}{8}$

$\qquad = 55\dfrac{1}{4}$ feet

85. $4\dfrac{3}{4} + 1\dfrac{2}{5} = \dfrac{19}{4} + \dfrac{7}{5} = \dfrac{19 \cdot 5}{4 \cdot 5} + \dfrac{7 \cdot 4}{5 \cdot 4} = \dfrac{95 + 28}{20}$

$\qquad = \dfrac{123}{20} = 6\dfrac{3}{20}$ meters

87. Answers may vary.

89. $5\frac{1}{2} - 2\frac{1}{8} = \frac{11}{2} - \frac{17}{8} = \frac{11 \cdot 4}{2 \cdot 4} - \frac{17}{8} = \frac{44 - 17}{8}$

$= \frac{27}{8} = 3\frac{3}{8}$ miles

91. $\frac{7}{50}$ are in the physical sciences

93. $1 - \frac{4}{25} - \frac{7}{50} - \frac{7}{50} - \frac{7}{100} - \frac{21}{100} - \frac{3}{100}$

$= \frac{100}{100} - \frac{4 \cdot 4}{25 \cdot 4} - \frac{7 \cdot 2}{50 \cdot 2} - \frac{7 \cdot 2}{50 \cdot 2} - \frac{7}{100}$

$- \frac{21}{100} - \frac{3}{100}$

$= \frac{100 - 16 - 14 - 14 - 7 - 21 - 3}{100}$

$= \frac{25}{100} = \frac{1}{4}$

$\frac{1}{4}$ are in the biological and

agricultural sciences

95. $\frac{666}{3678} = \frac{111}{613}$ were Old Navy stores.

97. Area $= \frac{1}{2} \cdot \frac{7}{8} \cdot \frac{4}{9} = \frac{7 \cdot 4}{2 \cdot 2 \cdot 4 \cdot 9} = \frac{7}{36}$ sq ft

Section 1.4

Calculator Explorations

1. $5^4 = 625$

3. $9^5 = 59,049$

5. $2(20 - 5) = 30$

7. $24(862 - 455) + 89 = 9857$

9. $\frac{4623 + 129}{36 - 34} = 2376$

Mental Math

1. multiply

2. Add

3. subtract

4. divide

Exercise Set 1.4

1. $3^5 = 3 \cdot 3 \cdot 3 \cdot 3 \cdot 3 = 243$

3. $3^3 = 3 \cdot 3 \cdot 3 = 27$

5. $1^5 = 1 \cdot 1 \cdot 1 \cdot 1 \cdot 1 = 1$

7. $5^1 = 5$

9. $\left(\frac{1}{5}\right)^3 = \left(\frac{1}{5}\right)\left(\frac{1}{5}\right)\left(\frac{1}{5}\right) = \frac{1 \cdot 1 \cdot 1}{5 \cdot 5 \cdot 5} = \frac{1}{125}$

11. $\left(\frac{2}{3}\right)^4 = \left(\frac{2}{3}\right)\left(\frac{2}{3}\right)\left(\frac{2}{3}\right)\left(\frac{2}{3}\right) = \frac{2 \cdot 2 \cdot 2 \cdot 2}{3 \cdot 3 \cdot 3 \cdot 3} = \frac{16}{81}$

13. $7^2 = 7 \cdot 7 = 49$

15. $(4)^2 = (4) \cdot (4) = 16$

17. $(1.2)^2 = (1.2) \cdot (1.2) = 1.44$

19. $5 + 6 \cdot 2 = 5 + 12 = 17$

21. $4 \cdot 8 - 6 \cdot 2 = 32 - 12 = 20$

23. $2(8 - 3) = 2(5) = 10$

25. $2 + (5 - 2) + 4^2 = 2 + 3 + 4^2 = 2 + 3 + 16 = 21$

27. $5 \cdot 3^2 = 5 \cdot 9 = 45$

29. $\dfrac{1}{4} \cdot \dfrac{2}{3} - \dfrac{1}{6} = \dfrac{2}{12} - \dfrac{1}{6} = \dfrac{1}{6} - \dfrac{1}{6} = 0$

31. $\dfrac{6 - 4}{9 - 2} = \dfrac{2}{7}$

33. $2[5 + 2(8 - 3)] = 2[5 + 2(5)] = 2[5 + 10]$
$= 2[15] = 30$

35. $\dfrac{19 - 3 \cdot 5}{6 - 4} = \dfrac{19 - 15}{6 - 4} = \dfrac{4}{2} = 2$

37. $\dfrac{|6 - 2| + 3}{8 + 2 \cdot 5} = \dfrac{|4| + 3}{8 + 2 \cdot 5} = \dfrac{4 + 3}{8 + 2 \cdot 5} = \dfrac{4 + 3}{8 + 10}$
$= \dfrac{7}{18}$

39. $\dfrac{3 + 3(5 + 3)}{3^2 + 1} = \dfrac{3 + 3(8)}{3^2 + 1} = \dfrac{3 + 3(8)}{9 + 1}$
$= \dfrac{3 + 24}{9 + 1} = \dfrac{27}{10}$

41. $\dfrac{6 + |8 - 2| + 3^2}{18 - 3} = \dfrac{6 + |6| + 3^2}{18 - 3} = \dfrac{6 + 6 + 3^2}{18 - 3}$
$= \dfrac{6 + 6 + 9}{18 - 3} = \dfrac{21}{15} = \dfrac{3 \cdot 7}{3 \cdot 5}$
$= \dfrac{7}{5}$

43. No; since in the absence of grouping symbols we always perform multiplications or divisions before additions or subtractions in any expression.

45. **a.** $(6 + 2) \cdot (5 + 3) = 8 \cdot 8 = 64$

    b. $(6 + 2) \cdot 5 + 3 = 8 \cdot 5 + 3 = 40 + 3$
            $= 43$

    c. $6 + 2 \cdot 5 + 3 = 6 + 10 + 3 = 19$

    d. $6 + 2 \cdot (5 + 3) = 6 + 2 \cdot 8 = 6 + 16$
            $= 22$

47. Let $y = 3$
$3y = 3(3) = 9$

49. Let $x = 1$ and $z = 5$
$\dfrac{z}{5x} = \dfrac{5}{5(1)} = \dfrac{5}{5} = 1$

51. Let $x = 1$
$3x - 2 = 3(1) - 2 = 3 - 2 = 1$

53. Let $x = 1$ and $y = 3$
$|2x + 3y| = |2(1) + 3(3)| = |2 + 9| = |11| = 11$

55. Let $y = 3$
$5y^2 = 5(3)^2 = 5(9) = 45$

57. Let $x = 12$, $y = 8$ and $z = 4$
$\dfrac{x}{z} + 3y = \dfrac{12}{4} + 3(8) = 3 + 24 = 27$

59.   Let $x = 12$ and $y = 8$.

$$x^2 - 3y + x = (12)^2 - 3(8) + 12$$
$$= 144 - 24 + 12$$
$$= 132$$

61.   Let $x = 12$, $y = 8$ and $z = 4$

$$\frac{x^2 + z}{y^2 + 2z} = \frac{(12)^2 + 4}{(8)^2 + 2(4)} = \frac{144 + 4}{64 + 8}$$
$$= \frac{148}{72} = \frac{37}{18}$$

63. Evaluate $16t^2$ for each value of t.

$t = 1;\ 16(1)^2 = 16(1) = 16$

$t = 2;\ 16(2)^2 = 16(4) = 64$

$t = 3;\ 16(3)^2 = 16(9) = 144$

$t = 4;\ 16(4)^2 = 16(16) = 256$

Time t (in seconds)	Distance $16t^2$ (in feet)
1	16
2	64
3	144
4	256

65. Let $x = 5$

$3x - 6 = 9$

$3(5) - 6 \overset{?}{=} 9$

$15 - 6 \overset{?}{=} 9$

$9 = 9$, true

5 is a solution of the equation.

67. Let $x = 0$

$2x + 6 = 5x - 1$

$2(0) + 6 \overset{?}{=} 5(0) - 1$

$0 + 6 \overset{?}{=} 0 - 1$

$6 = -1$, false

0 is not a solution of the equation.

69. Let $x = 8$

$2x - 5 = 5$

$2(8) - 5 \overset{?}{=} 5$

$16 - 5 \overset{?}{=} 5$

$9 = 5$, false

8 is not a solution of the equation.

71. Let $x = 2$

$x + 6 = x + 6$

$2 + 6 \overset{?}{=} 2 + 6$

$8 = 8$, true

2 is a solution of the equation.

73. Let $x = 0$

$x = 5x + 15$

$(0) \overset{?}{=} 5(0) + 15$

$0 \overset{?}{=} 0 + 15$

$0 = 15$, false

0 is not a solution of the equation.

75. $x + 15$

77. $x - 5$

79. $3x + 22$

81. $1 + 2 = 9 \div 3$

83. $3 \neq 4 \div 2$

85. $5 + x = 20$

87. $13 - 3x = 13$

89. $\dfrac{12}{x} = \dfrac{1}{2}$

91. Answers may vary.

93. $(20 - 4) \cdot 4 \div 2 = (16) \cdot 4 \div 2 = 64 \div 2 = 32$

95. Let $l = 8$ and $w = 6$
$2l + 2w = 2(8) + 2(6) = 16 + 12 = 28$ m.

97. Let $l = 120$ and $w = 100$
$lw = (120)(100) = 12,000$ sq.ft.

99. Let $P = 650$, $T = 3$, and $I = 126.75$
$\dfrac{I}{PT} = \dfrac{126.75}{(650)(3)} = \dfrac{126.75}{1950} = 0.065 = 6.5\%$

101. Let $m = 228$.
$$4.00 + 0.07m = 4.00 + 0.07(228)$$
$$= 4.00 + 15.96$$
$$= \$19.96$$

Section 1.5

Mental Math

1. negative

2. positive

3. 0

4. negative

5. negative

6. 0

Exercise Set 1.5

1. $6 + 3 = 9$

3. $-6 + (-8) = -14$

5. $8 + (-7) = 1$

7. $-14 + 2 = -12$

9. $-2 + (-3) = -5$

11. $-9 + (-3) = -12$

13. $-7 + 3 = -4$

15. $10 + (-3) = 7$

17. $5 + (-7) = -2$

19. $-16 + 16 = 0$

21. $27 + (-46) = -19$

23. $-18 + 49 = 31$

25. $-33 + (-14) = -47$

27. $6.3 + (-8.4) = -2.1$

29. $|-8| + (-16) = 8 + (-16) = -8$

31. $117 + (-79) = 38$

33. $-9.6 + (-3.5) = -13.1$

35. $-\dfrac{3}{8} + \dfrac{5}{8} = \dfrac{2}{8} = \dfrac{1}{4}$

37. $-\dfrac{7}{16} + \dfrac{1}{4} = -\dfrac{7}{16} + \dfrac{1 \cdot 4}{4 \cdot 4} = -\dfrac{7}{16} + \dfrac{4}{16} = -\dfrac{3}{16}$

39. $-\dfrac{7}{10} + \left(-\dfrac{3}{5}\right) = -\dfrac{7}{10} + \left(-\dfrac{3 \cdot 2}{5 \cdot 2}\right)$
$= -\dfrac{7}{10} + \left(-\dfrac{6}{10}\right) = -\dfrac{13}{10}$

41. $-15 + 9 + (-2) = -6 + (-2) = -8$

43. $-21 + (-16) + (-22) = -37 + (-22) = -59$

45. $-23 + 16 + (-2) = -7 + (-2) = -9$

47. $|5 + (-10)| = |-5| = 5$

49. $6 + (-4) + 9 = 2 + 9 = 11$

51. $[-17 + (-4)] + [-12 + 15] = [-21] + [3]$
$= -18$

53. $|9 + (-12)| + |-16| = |-3| + 16 = 3 + 16 = 19$

55. $-1.3 + [0.5 + (-0.3) + 0.4]$
$= -1.3 + [0.2 + 0.4]$
$= -1.3 + [0.6]$
$= -0.7$

57. $-15 + 9 = -6$
The high temperature in Anoka was $-6°$.

59. $-1312 + 658 = -654$
You are 654 feet below sea level.

61. $(-1411) + (-567) + (-149)$
$= (-1978) + (-149)$
$= -2127$
The total net income
was $-\$2127$ million.

63. $5 + (-5) + (-2) = 0 + (-2) = -2$
Her score was 2 under par.

65. The opposite of 6 is -6.

67. The opposite of -2 is 2.

69. The opposite of 0 is 0.

71. Since $|-6|$ is 6, the opposite of $|-6|$ is -6.

73. Answers may vary

75. $-|-2| = -2$

77. $-|0| = -0 = 0$

79. $-\left|-\dfrac{2}{3}\right| = -\dfrac{2}{3}$

81. Answers may vary

83. Let $x = -4$

$$x + 9 = 5$$

$$(-4) + 9 \overset{?}{=} 5$$

$$5 = 5, \text{ true}$$

-4 is a solution of the equation.

85. Let $y = -1$

$$y + (-3) = -7$$

$$(-1) + (-3) \overset{?}{=} -7$$

$$-4 = -7, \text{ false}$$

-1 is not a solution of the equation.

87. July

89. October

91. $\left[(-9.1) + 14.4 + 8.8\right] \div 3 = \left[5.3 + 8.8\right] \div 3$

$\quad = \left[14.1\right] \div 3 = 4.7$

The average was $4.7°$ F.

93. $-a$ is a <u>negative</u> number.

95. $a + a$ is a <u>positive</u> number.

Exercise Set 1.6

1. $-6 - 4 = -6 + (-4) = -10$

3. $4 - 9 = 4 + (-9) = -5$

5. $16 - (-3) = 16 + 3 = 19$

7. $\dfrac{1}{2} - \dfrac{1}{3} = \dfrac{1}{2} + \left(-\dfrac{1}{3}\right) = \dfrac{1 \cdot 3}{2 \cdot 3} + \left(-\dfrac{1 \cdot 2}{3 \cdot 2}\right)$

$\quad = \dfrac{3}{6} + \left(-\dfrac{2}{6}\right) = \dfrac{1}{6}$

9. $-16 - (-18) = -16 + 18 = 2$

11. $-6 - 5 = -6 + (-5) = -11$

13. $7 - (-4) = 7 + 4 = 11$

15. $-6 - (-11) = -6 + 11 = 5$

17. $16 - (-21) = 16 + 21 = 37$

19. $9.7 - 16.1 = 9.7 + (-16.1) = -6.4$

21. $-44 - 27 = -44 + (-27) = -71$

23. $-21 - (-21) = -21 + 21 = 0$

25. $-2.6 - (-6.7) = -2.6 + 6.7 = 4.1$

27. $-\dfrac{3}{11} - \left(-\dfrac{5}{11}\right) = -\dfrac{3}{11} + \dfrac{5}{11} = \dfrac{2}{11}$

29. $-\dfrac{1}{6} - \dfrac{3}{4} = -\dfrac{1}{6} + \left(-\dfrac{3}{4}\right) = -\dfrac{1 \cdot 2}{6 \cdot 2} + \left(-\dfrac{3 \cdot 3}{4 \cdot 3}\right)$

$\quad = -\dfrac{2}{12} + \left(-\dfrac{9}{12}\right) = -\dfrac{11}{12}$

31. $8.3 - (-0.62) = 8.3 + 0.62 = 8.92$

33. $8 - (-5) = 8 + 5 = 13$

35. $-6-(-1)=-6+1=-5$

37. $7-8=7+(-8)=-1$

39. $-8-15=-8+(-15)=-23$

41. Answers may vary

43. $-10-(-8)+(-4)-20$
$\quad = -10+8+(-4)+(-20)$
$\quad = -2+(-4)+(-20)=-6+(-20)=-26$

45. $5-9+(-4)-8-8$
$\quad = 5+(-9)+(-4)+(-8)+(-8)$
$\quad = -4+(-4)+(-8)+(-8)$
$\quad = -8+(-8)+(-8)=-16+(-8)=-24$

47. $-6-(2-11)=-6-(-9)=-6+9=3$

49. $3^3-8\cdot9=27-8\cdot9$
$\quad = 27-72=27+(-72)=-45$

51. $2-3(8-6)=2-3(2)=2-6=2+(-6)=-4$

53. $(3-6)+4^2=\left[3+(-6)\right]+4^2=\left[-3\right]+4^2$
$\quad = \left[-3\right]+16=13$

55. $-2+\left[(8-11)-(-2-9)\right]$
$\quad = -2+\left[(8+(-11))-(-2+(-9))\right]$
$\quad = -2+\left[(-3)-(-11)\right]=-2+\left[(-3)+11\right]$
$\quad = -2+\left[8\right]=6$

57. $|-3|+2^2+\left[-4-(-6)\right]=3+2^2+\left[-4+6\right]$
$\quad = 3+2^2+\left[2\right]=3+4+\left[2\right]=7+\left[2\right]=9$

59. Let $x=-5$ and $y=4$.
$\quad x-y=-5-4=-5+(-4)=-9$

61. Let $x=-5$, $y=4$, and $t=10$.
$\quad |x|+2t-8y=|-5|+2(10)-8(4)$
$\quad = 5+2(10)-8(4)=5+20-32$
$\quad = 25-32=25+(-32)=-7$

63. Let $x=-5$ and $y=4$.
$$\frac{9-x}{y+6}=\frac{9-(-5)}{4+6}=\frac{9+5}{4+6}=\frac{14}{10}=\frac{2\cdot7}{2\cdot5}=\frac{7}{5}$$

65. Let $x=-5$ and $y=4$.
$\quad y^2-x=4^2-(-5)=16+5=21$

67. Let $x=-5$ and $t=10$.
$$\frac{|x-(-10)|}{2t}=\frac{|-5-(-10)|}{2(10)}=\frac{|-5+10|}{2(10)}$$
$$=\frac{|5|}{2(10)}=\frac{5}{20}=\frac{5}{4\cdot5}=\frac{1}{4}$$

69. The change in temperature is the difference between the last temperature and the first temperature.
$\quad -56-44=-56+(-44)=-100$
The temperature dropped $100°$.

71. Gains: $+2$

Losses: $-5, -20$

$2 + (-5) + (-20) = -3 + (-20) = -23$

Total loss of 23 yards

73. $-475 - 94 = -475 + (-94) = -569$

He was born in 569 B.C.

75. Rises: $+120$

Drops: $-250, -178$

$120 + (-250) + (-178)$

$= -130 + (-178) = -308$

The overall vertical change was a drop of 308 feet.

77. $19,340 - (-512) = 19,340 + 512$

$\qquad\qquad\qquad = 19,852$

19,852 feet higher

79. $y = 180 - 50 = 180 + (-50) = 130$

The supplementary angle is $130°$

81. $x = 90 - 60 = 90 + (-60) = 30$

The complementary angle is $30°$

83. Let $x = -4$

$x - 9 = 5$

$-4 - 9 \overset{?}{=} 5$

$-13 = 5,$ false

-4 is not a solution of the equation.

85. Let $x = -2$

$-x + 6 = -x - 1$

$-(-2) + 6 \overset{?}{=} -(-2) - 1$

$2 + 6 \overset{?}{=} 2 + (-1)$

$8 = 1,$ false

-2 is not a solution of the equation.

87. Let $x = 2$

$-x - 13 = -15$

$-2 - 13 \overset{?}{=} -15$

$-2 + (-13) \overset{?}{=} -15$

$-15 = -15,$ true

2 is a solution of the equation.

89. The change in temperature is the difference between the given month's temperature and the previous month's.

F: $-23.7 - (-19.3) = -23.7 + 19.3 = -4.4°$

Mr: $-21.1 - (-23.7) = -21.1 + 23.7 = 2.6°$

Ap: $-9.1 - (-21.1) = -9.1 + 21.1 = 12°$

Ma: $14.4 - (-9.1) = 14.4 + 9.1 = 23.5°$

Jn: $29.7 - 14.4 = 29.7 + (-14.4) = 15.3°$

Jy: $33.6 - 29.7 = 33.6 + (-29.7) = 3.9°$

Au: $33.3 - 33.6 = 33.3 + (-33.6) = -0.3°$

S: $27.0 - 33.3 = 27.0 + (-33.3) = -6.3°$

O: $8.8 - 27.0 = 8.8 + (-27.0) = -18.2°$

N: $-6.9 - 8.8 = -6.9 + (-8.8) = -15.7°$

D: $-17.2 - (-6.9) = -17.2 + 6.9 = -10.3°$

91. October

93. True: answers may vary.

95. True: answers may vary.

97. $4.362 - 7.0086 = -2.6466$

Section 1.7

Calculator Explorations

1. $-38(26 - 27) = 38$

3. $134 + 25(68 - 91) = -441$

5. $\dfrac{-50(294)}{175 - 265} = 163.\overline{3}$

7. $9^5 - 4550 = 54,499$

9. $(-125)^2 = 15,625$

Mental Math

1. positive

2. positive

3. negative

4. negative

5. positive

6. negative

Exercise Set 1.7

1. $-6(4) = -24$

3. $2(-1) = -2$

5. $-5(-10) = 50$

7. $-3 \cdot 4 = -12$

9. $-7 \cdot 0 = 0$

11. $2(-9) = -18$

13. $-\dfrac{1}{2}\left(-\dfrac{3}{5}\right) = \dfrac{1 \cdot 3}{2 \cdot 5} = \dfrac{3}{10}$

15. $-\dfrac{3}{4}\left(-\dfrac{8}{9}\right) = \dfrac{3 \cdot 8}{4 \cdot 9} = \dfrac{24}{36} = \dfrac{2 \cdot 12}{3 \cdot 12} = \dfrac{2}{3}$

17. $5(-1.4) = -7.0$

19. $-0.2(-0.7) = 0.14$

21. $-10(80) = -800$

23. $4(-7) = -28$

25. $(-5)(-5) = 25$

27. $\dfrac{2}{3}\left(-\dfrac{4}{9}\right) = -\dfrac{2 \cdot 4}{3 \cdot 9} = -\dfrac{8}{27}$

29. $-11(11) = -121$

31. $-\dfrac{20}{25}\left(\dfrac{5}{16}\right) = -\dfrac{20\cdot 5}{25\cdot 16} = -\dfrac{100}{400} = -\dfrac{1}{4}$

33. $(-1)(2)(-3)(-5)$
$= -2(-3)(-5) = 6(-5) = -30$

35. $(-2)(5) - (-11)(3) = -10 - (-33)$
$\qquad\qquad\qquad = -10 + 33$
$\qquad\qquad\qquad = 23$

37. $(-6)(-1)(-2) - (-5) = -12 + 5 = -7$

39. True

41. False

43. $(-2)^4 = (-2)(-2)(-2)(-2) = 4(-2)(-2)$
$\qquad\quad = -8(-2) = 16$

45. $-1^5 = -(1)(1)(1)(1)(1) = -1$

47. $(-5)^2 = (-5)(-5) = 25$

49. $-7^2 = -(7)(7) = -49$

51. Reciprocal of 9 is $\dfrac{1}{9}$ since $9\cdot\dfrac{1}{9} = 1$

53. Reciprocal of $\dfrac{2}{3}$ is $\dfrac{3}{2}$ since $\dfrac{2}{3}\cdot\dfrac{3}{2} = 1$

55. Reciprocal of -14 is $-\dfrac{1}{14}$

since $-14\cdot-\dfrac{1}{14} = 1$

57. Reciprocal of $-\dfrac{3}{11}$ is $-\dfrac{11}{3}$

since $-\dfrac{3}{11}\cdot-\dfrac{11}{3} = 1$

59. Reciprocal of 0.2 is $\dfrac{1}{0.2}$

since $0.2\cdot\dfrac{1}{0.2} = 1$

61. Reciprocal of $\dfrac{1}{-6.3}$ is -6.3

since $\dfrac{1}{-6.3}\cdot-6.3 = 1$

63. $\dfrac{18}{-2} = 18\cdot-\dfrac{1}{2} = -9$

65. $\dfrac{-16}{-4} = -16\cdot-\dfrac{1}{4} = 4$

67. $\dfrac{-48}{12} = -48\cdot\dfrac{1}{12} = -4$

69. $\dfrac{0}{-4} = 0\cdot-\dfrac{1}{4} = 0$

71. $-\dfrac{15}{3} = -15\cdot\dfrac{1}{3} = -5$

73. $\dfrac{5}{0}$ is undefined

75. $\dfrac{-12}{-4} = -12\cdot-\dfrac{1}{4} = 3$

77. $\dfrac{30}{-2} = 30\cdot-\dfrac{1}{2} = -15$

79. $\dfrac{6}{7} \div -\dfrac{1}{3} = \dfrac{6}{7} \cdot \left(-\dfrac{3}{1}\right) = -\dfrac{6 \cdot 3}{7 \cdot 1} = -\dfrac{18}{7}$

81. $-\dfrac{5}{9} \div \left(-\dfrac{3}{4}\right) = -\dfrac{5}{9} \cdot \left(-\dfrac{4}{3}\right) = \dfrac{5 \cdot 4}{9 \cdot 3} = \dfrac{20}{27}$

83. $-\dfrac{4}{9} \div \dfrac{4}{9} = -\dfrac{4}{9} \cdot \dfrac{9}{4} = -1$

85. $\dfrac{-9(-3)}{-6} = \dfrac{27}{-6} = -\dfrac{9}{2}$

87. $\dfrac{12}{9-12} = \dfrac{12}{-3} = -4$

89. $\dfrac{-6^2 + 4}{-2} = \dfrac{-36+4}{-2} = \dfrac{-32}{-2} = 16$

91. $\dfrac{8 + (-4)^2}{4-12} = \dfrac{8+16}{4-12} = \dfrac{24}{-8} = -3$

93. $\dfrac{22 + (3)(-2)}{-5-2} = \dfrac{22 + (-6)}{-5-2} = \dfrac{16}{-7}$

95. $\dfrac{-3 - 5^2}{2(-7)} = \dfrac{-3-25}{2(-7)} = \dfrac{-3 + (-25)}{-14} = \dfrac{-28}{-14} = 2$

97. $\dfrac{6 - 2(-3)}{4 - 3(-2)} = \dfrac{6 - (-6)}{4 - (-6)} = \dfrac{6+6}{4+6} = \dfrac{12}{10} = \dfrac{6}{5}$

99. $\dfrac{-3 - 2(-9)}{-15 - 3(-4)} = \dfrac{-3 - (-18)}{-15 - (-12)}$

$\qquad = \dfrac{-3 + 18}{-15 + 12} = \dfrac{15}{-3} = -5$

101. $\dfrac{|5-9| + |10-15|}{|2(-3)|} = \dfrac{|-4| + |-5|}{|-6|}$

$\qquad = \dfrac{4+5}{6} = \dfrac{9}{6} = \dfrac{3}{2}$

103. Let $x = -5$ and $y = -3$.

$\qquad 3x + 2y = 3(-5) + 2(-3)$

$\qquad\qquad = -15 + (-6)$

$\qquad\qquad = -21$

105. Let $x = -5$ and $y = -3$.

$\qquad 2x^2 - y^2 = 2(-5)^2 - (-3)^2 = 2(25) - 9$

$\qquad\qquad = 50 + (-9) = 41$

107. Let $x = -5$ and $y = -3$.

$\qquad x^3 + 3y = (-5)^3 + 3(-3)$

$\qquad\qquad = -125 + (-9) = -134$

109. Let $x = -5$ and $y = -3$.

$\qquad \dfrac{2x - 5}{y - 2} = \dfrac{2(-5) - 5}{-3 - 2} = \dfrac{-10 - 5}{-3 - 2} = \dfrac{-15}{-5} = 3$

111. Let $x = -5$ and $y = -3$.

$\qquad \dfrac{-3 - y}{x - 4} = \dfrac{-3 - (-3)}{-5 - 4} = \dfrac{-3 + 3}{-5 - 4} = \dfrac{0}{-9} = 0$

113. $4(-1,272) = -5,088$

The net income will be

$-\$5,088$ million.

115. Let $x = 7$

$-5x = -35$

$-5(7) \overset{?}{=} -35$

$-35 = -35$, true

7 is a solution of the equation.

117. Let $x = -20$

$\dfrac{x}{10} = 2$

$\dfrac{-20}{10} \overset{?}{=} 2$

$-2 = 2$, false

-20 is not a solution of the equation.

119. Let $x = 5$

$-3x - 5 = -20$

$-3(5) - 5 \overset{?}{=} -20$

$-15 - 5 \overset{?}{=} -20$

$-20 = -20$, true

5 is a solution of the equation.

121. Answers may vary

123. -1 and 1 are their own reciprocals.

125. Positive

127. Not possible

129. Negative

131. $-2 + \dfrac{-15}{3} = \dfrac{-2 \cdot 3}{1 \cdot 3} + \dfrac{-15}{3} = \dfrac{-6 + (-15)}{3}$

$= \dfrac{-21}{3} = -7$

133. $2\left[-5 + (-3)\right] = 2(-8) = -16$

Integrated Review

1. positive

2. positive

3. negative

4. negative

5. positive

6. negative

7. negative

8. positive

9. $5(-7) = -35$

10. $-3(-10) = 30$

11. $\dfrac{-20}{-4} = 5$

12. $\dfrac{30}{-6} = -5$

13. $7 - (-3) = 7 + 3 = 10$

14. $-8 - 10 = -8 + (-10) = -18$

15. $-14 - (-12) = -14 + 12 = -2$

16. $-3 - (-1) = -3 + 1 = -2$

17. $-\dfrac{1}{2}\left(-\dfrac{3}{4}\right) = \dfrac{1 \cdot 3}{2 \cdot 4} = \dfrac{3}{8}$

18. $-\dfrac{2}{7}\left(\dfrac{11}{12}\right) = -\dfrac{2 \cdot 11}{7 \cdot 12} = -\dfrac{22}{84} = -\dfrac{11}{42}$

19. $\dfrac{-12}{0.2} = -60$

20. $\dfrac{-3.8}{-2} = 1.9$

21. $-19 + (-23) = -42$

22. $18 + (-25) = -7$

23. $-15 + 17 = 2$

24. $-2 + (-37) = -39$

25. $(-8)^2 = (-8)(-8) = 64$

26. $-9^2 = -(9)(9) = -81$

27. $-3^2 = -(3)(3)(3) = -27$

28. $(-2)^4 = (-2)(-2)(-2)(-2) = 16$

29. $-1^{10} = -(1)(1)(1)(1)(1)(1)(1)(1)(1)(1) = -1$

30. $(-1)^{10} = (-1)(-1)(-1)(-1)(-1)(-1)$
$\qquad\qquad\quad \cdot (-1)(-1)(-1)(-1)$
$\qquad = 1$

31. $(-2)^5 = (-2)(-2)(-2)(-2)(-2) = -32$

32. $-2^5 = -(2)(2)(2)(2)(2) = -32$

33. $(2)(-8)(-3) = (-16)(-3) = 48$

34. $3(-2)(5) = (-6)(5) = -30$

35. $-6(2) + 20 \div 2 - 4 = -12 + 10 - 4 = -6$

36. $-4(-3) + 9 \div 3 - 6 = 12 + 3 - 6 = 9$

37. $-3^2 - \left[6 + 5|-2 - 1|\right] = -9 - \left[6 + 5|-3|\right]$
$\qquad = -9 - \left[6 + 5(3)\right] = -9 - \left[6 + 15\right] = -9 - 21$
$\qquad = -30$

38. $-5^2 - \left[4 + 3|-3 - 2|\right] = -25 - \left[4 + 3|-5|\right]$
$\qquad = -25 - \left[4 + 3(5)\right] = -25 - \left[4 + 15\right]$
$\qquad = -25 - 19 = -44$

39. $2(19 - 17)^3 - 3(7 - 9)^2$
$\qquad = 2\left[19 + (-17)\right]^3 - 3\left[7 + (-9)\right]^2$
$\qquad = 2[2]^3 - 3[-2]^2 = 2[8] - 3[4]$
$\qquad = 16 - 12 = 16 + (-12) = 4$

40. $3(10 - 9)^2 - 6(20 - 19)^3$
$\qquad = 3\left[10 + (-9)\right]^3 - 6\left[20 + (-19)\right]^2$
$\qquad = 3[1]^3 - 6[1]^2 = 3[1] - 6[1]$
$\qquad = 3 - 6 = 3 + (-6) = -3$

41. $\dfrac{19 - 25}{3(-1)} = \dfrac{19 + (-25)}{-3} = \dfrac{-6}{-3} = 2$

42. $\dfrac{8(-4)}{-2} = \dfrac{-32}{-2} = 16$

43. $\dfrac{-2(3-6) - 6(10-9)}{-6 - (-5)}$

$= \dfrac{-2\big[3 + (-6)\big] - 6\big[10 + (-9)\big]}{-6 - (-5)}$

$= \dfrac{-2[-3] - 6[1]}{-6 - (-5)} = \dfrac{6 - 6}{-6 - (-5)} = \dfrac{6 + (-6)}{-6 + 5}$

$= \dfrac{0}{-1} = 0$

44. $\dfrac{5(7-9) - 3(100-97)}{4-5}$

$= \dfrac{5\big[7 + (-9)\big] - 3\big[100 + (-97)\big]}{4-5}$

$= \dfrac{5[-2] - 3[3]}{4-5} = \dfrac{-10 - 9}{4-5} = \dfrac{-10 + (-9)}{4 + (-5)}$

$= \dfrac{-19}{-1} = 19$

45. $\dfrac{-4(8-10)^3}{-2-1-12} = \dfrac{-4\big[8 + (-10)\big]^3}{-2-1-12}$

$= \dfrac{-4[-2]^3}{-2-1-12} = \dfrac{-4[-8]}{-2-1-12}$

$= \dfrac{32}{-2 + (-1) + (-12)} = \dfrac{32}{-15} = -\dfrac{32}{15}$

46. $\dfrac{6(7-10)^2}{6-(-1)-2} = \dfrac{6\big[7 + (-10)\big]^2}{6-(-1)-2}$

$= \dfrac{6\big[-3\big]^2}{6-(-1)-2} = \dfrac{6[9]}{6-(-1)-2}$

$= \dfrac{54}{6 + 1 + (-2)} = \dfrac{54}{5}$

Exercise Set 1.8

1. $x + 16 = 16 + x$

3. $-4 \cdot y = y \cdot (-4)$

5. $xy = yx$

7. $2x + 13 = 13 + 2x$

9. $(xy) \cdot z = x \cdot (yz)$

11. $2 + (a + b) = (2 + a) + b$

13. $4 \cdot (ab) = 4a \cdot (b)$

15. $(a + b) + c = a + (b + c)$

17. $8 + (9 + b) = (8 + 9) + b = 17 + b$

19. $4(6y) = (4 \cdot 6)y = 24y$

21. $\dfrac{1}{5}(5y) = \left(\dfrac{1}{5} \cdot 5\right)y = 1 \cdot y = y$

23. $(13 + a) + 13 = (a + 13) + 13 = a + (13 + 13)$

$= a + 26$

25. $-9(8x) = (-9 \cdot 8)x = -72x$

27. $\dfrac{3}{4}\left(\dfrac{4}{3}s\right) = \left(\dfrac{3}{4} \cdot \dfrac{4}{3}\right)s = 1s = s$

29. Answers may vary

31. $4(x+y) = 4x + 4y$

33. $9(x-6) = 9x - 9 \cdot 6 = 9x - 54$

35. $2(3x+5) = 2(3x) + 2(5) = 6x + 10$

37. $7(4x-3) = 7(4x) - 7(3) = 28x - 21$

39. $3(6+x) = 3(6) + 3x = 18 + 3x$

41. $-2(y-z) = -2y - (-2)z = -2y + 2z$

43. $-7(3y+5) = -7(3y) + (-7)(5) = -21y - 35$

45. $5(x+4m+2) = 5x + 5(4m) + 5(2)$
$$= 5x + 20m + 10$$

47. $-4(1-2m+n) = -4(1) - (-4)(2m) + (-4)n$
$$= -4 + 8m - 4n$$

49. $-(5x+2) = -1(5x+2) = -1(5x) + (-1)(2)$
$$= -5x - 2$$

51. $-(r-3-7p) = -1(r-3-7p)$
$$= -1r - (-1)(3) - (-1)(7p)$$
$$= -r + 3 + 7p$$

53. $\dfrac{1}{2}(6x+8) = \dfrac{1}{2}(6x) + \dfrac{1}{2}(8)$
$$= \left(\dfrac{1}{2} \cdot 6\right)x + \left(\dfrac{1}{2} \cdot 8\right) = 3x + 4$$

55. $-\dfrac{1}{3}(3x-9y) = -\dfrac{1}{3}(3x) - \left(-\dfrac{1}{3}\right)(9y)$
$$= \left(-\dfrac{1}{3} \cdot 3\right)x - \left(-\dfrac{1}{3} \cdot 9\right)y = -1 \cdot x + 3 \cdot y$$
$$= -x + 3y$$

57. $3(2r+5) - 7 = 3(2r) + 3(5) - 7$
$$= 6r + 15 + (-7) = 6r + 8$$

59. $-9(4x+8) + 2 = -9(4x) + (-9)(8) + 2$
$$= -36x - 72 + 2 = -36x - 70$$

61. $-4(4x+5) - 5 = -4(4x) + (-4)(5) - 5$
$$= -16x + (-20) + (-5) = -16x - 25$$

63. $4 \cdot 1 + 4 \cdot y = 4(1+y)$

65. $11x + 11y = 11(x+y)$

67. $(-1) \cdot 5 + (-1) \cdot x = -1(5+x) = -(5+x)$

69. $30a + 30b = 30(a+b)$

71. Commutative property of multiplication

73. Associative property of addition

75. Distributive property

77. Associative property of multiplication

79. Identity element of addition

81. Distributive property

83. Associative and commutative properties of multiplication

85.

Expression	Opposite	Reciprocal
8	-8	$\dfrac{1}{8}$

87.

Expression	Opposite	Reciprocal
x	$-x$	$\dfrac{1}{x}$

89.

Expression	Opposite	Reciprocal
$2x$	$-2x$	$\dfrac{1}{2x}$

91. No

93. Yes

95. Answers may vary

Chapter 1 Review

1. $8 < 10$

2. $7 > 2$

3. $-4 > -5$

4. $\dfrac{12}{2} > -8$

5. $|-7| < |-8|$

6. $|-9| > -9$

7. $-|-1| = -1$

8. $|-14| = -(-14)$

9. $1.2 > 1.02$

10. $-\dfrac{3}{2} < -\dfrac{3}{4}$

11. $4 \geq -3$

12. $6 \neq 5$

13. $0.03 < 0.3$

14. $50 > 40$

15. **a.** The natural numbers are 1 and 3.
     b. The whole numbers are 0, 1, and 3.
     c. The integers are -6, 0, 1, and 3.
     d. The rational numbers are $-6, 0, 1,$ $1\dfrac{1}{2}, 3,$ and 9.62.
     e. The irrational number is π.
     f. The real numbers are all numbers in the given set.

16. a. The natural numbers are 2 and 5.

b. The whole numbers are 2 and 5.

c. The integers are -3, 2, and 5.

d. The rational numbers are $-3, -1.6,$

$2, 5, \dfrac{11}{2}$, and 15.1.

e. The irrational numbers are $\sqrt{5}$ and 2π.

f. The real numbers are all numbers
   in the given set.

17. Friday

18. Wednesday

19. $36 = 2 \cdot 2 \cdot 3 \cdot 3$

20. $120 = 2 \cdot 2 \cdot 2 \cdot 3 \cdot 5$

21. $\dfrac{8}{15} \cdot \dfrac{27}{30} = \dfrac{2 \cdot 4 \cdot 3 \cdot 3 \cdot 3}{3 \cdot 5 \cdot 2 \cdot 3 \cdot 5} = \dfrac{12}{25}$

22. $\dfrac{7}{8} \div \dfrac{21}{32} = \dfrac{7}{8} \cdot \dfrac{32}{21} = \dfrac{7 \cdot 8 \cdot 4}{8 \cdot 3 \cdot 7} = \dfrac{4}{3}$

23. $\dfrac{7}{15} + \dfrac{5}{6} = \dfrac{7 \cdot 2}{15 \cdot 2} + \dfrac{5 \cdot 5}{6 \cdot 5} = \dfrac{14}{30} + \dfrac{25}{30}$

$= \dfrac{14 + 25}{30} = \dfrac{39}{30} = \dfrac{3 \cdot 13}{3 \cdot 10} = \dfrac{13}{10}$

24. $\dfrac{3}{4} - \dfrac{3}{20} = \dfrac{3 \cdot 5}{4 \cdot 5} - \dfrac{3}{20} = \dfrac{15}{20} - \dfrac{3}{20} = \dfrac{15 - 3}{20}$

$= \dfrac{12}{20} = \dfrac{3 \cdot 4}{5 \cdot 4} = \dfrac{3}{5}$

25. $2\dfrac{3}{4} + 6\dfrac{5}{8} = \dfrac{11}{4} + \dfrac{53}{8} = \dfrac{11 \cdot 2}{4 \cdot 2} + \dfrac{53}{8}$

$= \dfrac{22 + 53}{8} = \dfrac{75}{8} = 9\dfrac{3}{8}$

26. $7\dfrac{1}{6} - 2\dfrac{2}{3} = \dfrac{43}{6} - \dfrac{8}{3} = \dfrac{43}{6} - \dfrac{8 \cdot 2}{3 \cdot 2}$

$= \dfrac{43 - 16}{6} = \dfrac{27}{6} = \dfrac{9 \cdot 3}{2 \cdot 3} = \dfrac{9}{2} = 4\dfrac{1}{2}$

27. $5 \div \dfrac{1}{3} = 5 \cdot \dfrac{3}{1} = 15$

28. $2 \cdot 8\dfrac{3}{4} = 2 \cdot \dfrac{35}{4} = \dfrac{2 \cdot 35}{2 \cdot 2} = \dfrac{35}{2} = 17\dfrac{1}{2}$

29. $1 - \dfrac{112}{6} - \dfrac{1}{4} = \dfrac{12}{12} - \dfrac{1 \cdot 2}{6 \cdot 2} - \dfrac{1 \cdot 3}{4 \cdot 3}$

$= \dfrac{12 - 2 - 3}{12} = \dfrac{7}{12}$

The unknown part is $\dfrac{7}{12}$

30. $P = 2l + 2w$

$P = 2\left(1\dfrac{1}{3}\right) + 2\left(\dfrac{7}{8}\right) = \dfrac{2}{1} \cdot \dfrac{4}{3} + \dfrac{2}{1} \cdot \dfrac{7}{8}$

$= \dfrac{8}{3} + \dfrac{14}{8} = \dfrac{8 \cdot 8}{3 \cdot 8} + \dfrac{14 \cdot 3}{8 \cdot 3} = \dfrac{64 + 42}{24}$

$= \dfrac{106}{24} = 4\dfrac{10}{24} = 4\dfrac{5}{12}$ meters

$A = lw$

$A = 1\dfrac{1}{3} \cdot \dfrac{7}{8} = \dfrac{4}{3} \cdot \dfrac{7}{8} = \dfrac{4 \cdot 7}{3 \cdot 2 \cdot 4}$

$= \dfrac{7}{6} = 1\dfrac{1}{6}$ sq. meters

31. $P =$ the sum of the lengths of the sides

$$P = \frac{5}{11} + \frac{8}{11} + \frac{3}{11} + \frac{3}{11} + \frac{2}{11} + \frac{5}{11} = \frac{26}{11}$$

$$= 2\frac{4}{11} \text{ in.}$$

$A =$ the sum of the two areas, each
given by lw

$$A = \frac{5}{11} \cdot \frac{5}{11} + \frac{3}{11} \cdot \frac{3}{11} = \frac{25}{121} + \frac{9}{121}$$

$$= \frac{34}{121} \text{ sq. in.}$$

32. $7\frac{1}{2} - 6\frac{1}{8} = \frac{15}{2} - \frac{49}{8} = \frac{15 \cdot 4}{2 \cdot 4} - \frac{49}{8}$

$$= \frac{60 - 49}{8} = \frac{11}{8} = 1\frac{3}{8} \text{ ft}$$

33. $1\frac{1}{8} + 1\frac{13}{16} = \frac{9}{8} + \frac{29}{16} = \frac{9 \cdot 2}{8 \cdot 2} + \frac{29}{16}$

$$= \frac{18 + 29}{16} = \frac{47}{16} = 2\frac{15}{16} \text{ lb.}$$

34. $1\frac{1}{2} + 1\frac{11}{16} + 1\frac{3}{4} + 1\frac{5}{8} + 1\frac{1}{8} + \frac{11}{16}$

$$= \frac{3}{2} + \frac{27}{16} + \frac{7}{4} + \frac{13}{8} + \frac{9}{8} + \frac{11}{16}$$

$$= \frac{3 \cdot 8}{2 \cdot 8} + \frac{27}{16} + \frac{7 \cdot 4}{4 \cdot 4} + \frac{13 \cdot 2}{8 \cdot 2} + \frac{9 \cdot 2}{8 \cdot 2} + \frac{11}{16}$$

$$= \frac{24 + 27 + 28 + 26 + 18 + 11}{16}$$

$$= \frac{134}{16} = 8\frac{3}{8} \text{ lb}$$

35. Total weight = weight of girls
                + weight of boys

$$8\frac{3}{8} + 2\frac{15}{16} = \frac{67}{8} + \frac{47}{16} = \frac{67 \cdot 2}{8 \cdot 2} + \frac{47}{16}$$

$$= \frac{134 + 47}{16} = \frac{181}{16} = 11\frac{5}{16} \text{ lb.}$$

36. Jioke

37. Odera

38. $1\frac{13}{16} - \frac{11}{16} = \frac{29}{16} - \frac{11}{16} = \frac{29 - 11}{16}$

$$= \frac{18}{16} = 1\frac{2}{6} = 1\frac{1}{8} \text{ lb}$$

39. $5\frac{1}{2} - 1\frac{5}{8} = \frac{11}{2} - \frac{13}{8} = \frac{11 \cdot 4}{2 \cdot 4} - \frac{13}{8}$

$$= \frac{44 - 13}{8} = \frac{31}{8} = 3\frac{7}{8} \text{ lb}$$

40. $4\frac{5}{32} - 1\frac{1}{8} = \frac{133}{32} - \frac{9}{8} = \frac{133}{32} - \frac{9 \cdot 4}{8 \cdot 4}$

$$= \frac{133 - 36}{32} = \frac{97}{32} = 3\frac{1}{32} \text{ lb}$$

41. $2^4 = 2 \cdot 2 \cdot 2 \cdot 2 = 16$

42. $5^2 = 5 \cdot 5 = 25$

43. $\left(\frac{2}{7}\right)^2 = \frac{2}{7} \cdot \frac{2}{7} = \frac{4}{49}$

44. $\left(\frac{3}{4}\right)^3 = \frac{3}{4} \cdot \frac{3}{4} \cdot \frac{3}{4} = \frac{27}{64}$

45. $6 \cdot 3^2 + 2 \cdot 8 = 6 \cdot 9 + 2 \cdot 8 = 54 + 16 = 70$

46. $68 - 5 \cdot 2^3 = 68 - 5 \cdot 8 = 68 - 40 = 28$

47. $3(1 + 2 \cdot 5) + 4 = 3(1 + 10) + 4 = 3(11) + 4$
$$= 33 + 4 = 37$$

48. $8 + 3(2 \cdot 6 - 1) = 8 + 3(12 - 1) = 8 + 3(11)$
$$= 8 + 33 = 41$$

49. $\dfrac{4 + |6 - 2| + 8^2}{4 + 6 \cdot 4} = \dfrac{4 + |4| + 64}{4 + 24} = \dfrac{4 + 4 + 64}{4 + 24}$
$$= \dfrac{72}{28} = \dfrac{4 \cdot 18}{4 \cdot 7} = \dfrac{18}{7}$$

50. $5\left[3(2 + 5) - 5\right] = 5\left[3(7) - 5\right] = 5\left[21 - 5\right]$
$$= 5\left[16\right] = 80$$

51. $20 - 12 = 2 \cdot 4$

52. $\dfrac{9}{2} > -5$

53. Let $x = 6$ and $y = 2$.
$2x + 3y = 2(6) + 3(2) = 12 + 6 = 18$

54. Let $x = 6$, $y = 2$, and $z = 8$.
$x(y + 2z) = 6\left[2 + 2(8)\right] = 6\left[2 + 16\right]$
$$= 6\left[18\right] = 108$$

55. Let $x = 6$, $y = 2$, and $z = 8$.
$\dfrac{x}{y} + \dfrac{z}{2y} = \dfrac{6}{2} + \dfrac{8}{2(2)} = \dfrac{6}{2} + \dfrac{8}{4} = 3 + 2 = 5$

56. Let $x = 6$ and $y = 2$.
$x^2 - 3y^2 = (6)^2 - 3(2)^2 = 36 - 3(4)$
$$= 36 - 12 = 36 + (-12) = 24$$

57. Let $a = 37$ and $b = 80$.
$180 - a - b = 180 - 37 - 80$
$$= 180 + (-37) + (-80) = 143 + (-80) = 63°$$

58. Let $x = 3$.
$$7x - 3 = 18$$
$$7(3) - 3 \overset{?}{=} 18$$
$$21 - 3 \overset{?}{=} 18$$
$$18 = 18, \quad \text{true}$$
3 is a solution to the equation.

59. Let $x = 1$.
$$3x^2 + 4 = x - 1$$
$$3(1)^2 + 4 \overset{?}{=} 1 - 1$$
$$3 + 7 \overset{?}{=} 0$$
$$10 = 0, \quad \text{false}$$
1 is not a solution to the equation.

60. The additive inverse of -9 is 9.

61. The additive inverse of $\dfrac{2}{3}$ is $-\dfrac{2}{3}$.

62. The additive inverse of $|-2|$ is -2
since $|-2| = 2$.

63. The additive inverse of $-|-7|$ is 7
since $-|-7| = -7$.

64. $-15 + 4 = -11$

65. $-6 + (-11) = -17$

66. $\dfrac{1}{16} + \left(-\dfrac{1}{4}\right) = \dfrac{1}{16} + \left(-\dfrac{1 \cdot 4}{4 \cdot 4}\right)$

$\qquad = \dfrac{1}{16} + \left(-\dfrac{4}{16}\right) = -\dfrac{3}{16}$

67. $-8 + |-3| = -8 + 3 = -5$

68. $-4.6 + (-9.3) = -13.9$

69. $-2.8 + 6.7 = 3.9$

70. $-282 + 728 = 446$ feet

71. $6 - 20 = 6 + (-20) = -14$

72. $-3.1 - 8.4 = -3.1 + (-8.4) = -11.5$

73. $-6 - (-11) = -6 + 11 = 5$

74. $4 - 15 = 4 + (-15) = -11$

75. $-21 - 16 + 3(8 - 2)$

$\qquad = -21 + (-16) + 3\left[8 + (-2)\right]$

$\qquad = -21 + (-16) + 3[6] = -21 + (-16) + 18$

$\qquad = -37 + 18 = -19$

76. $\dfrac{11 - (-9) + 6(8 - 2)}{2 + 3 \cdot 4} = \dfrac{11 + 9 + 6\left[8 + (-2)\right]}{2 + 3 \cdot 4}$

$\qquad = \dfrac{11 + 9 + 6[6]}{2 + 3 \cdot 4} = \dfrac{11 + 9 + 36}{2 + 12} = \dfrac{56}{14} = 4$

77. Let $x = 3$, $y = -6$, and $z = -9$.

$2x^2 - y + z = 2(3)^2 - (-6) + (-9)$

$\qquad = 2(9) + 6 + (-9) = 18 + 6 + (-9)$

$\qquad = 24 + (-9) = 15$

78. Let $x = 3$ and $y = -6$.

$\dfrac{y - x + 5x}{2x} = \dfrac{y + 4x}{2x} = \dfrac{-6 + 4(3)}{2(3)}$

$\qquad = \dfrac{-6 + 12}{6} = \dfrac{6}{6} = 1$

79. The multiplicative inverse of -6 is $-\dfrac{1}{6}$

since $-6 \cdot -\dfrac{1}{6} = 1$.

80. The multiplicative inverse of $\dfrac{3}{5}$ is $\dfrac{5}{3}$

since $\dfrac{3}{5} \cdot \dfrac{5}{3} = 1$.

81. $6(-8) = -48$

82. $(-2)(-14) = 28$

83. $\dfrac{-18}{-6} = 3$

84. $\dfrac{42}{-3} = -14$

85. $\dfrac{4 \cdot (-3) + (-8)}{2 + (-2)} = \dfrac{-12 + (-8)}{2 + (-2)} = \dfrac{-20}{0}$

The expression is undefined.

86. $\dfrac{3(-2)^2 - 5}{-14} = \dfrac{3(4)-5}{-14} = \dfrac{12-5}{-14} = \dfrac{7}{-14} = -\dfrac{1}{2}$

87. $\dfrac{-6}{0}$ is undefined

88. $\dfrac{0}{-2} = 0$

89. $-4^2 - (-3+5) \div (-1) \cdot 2$

$= -16 - (2) \div (-1) \cdot 2 = -16 + 2 \cdot 2$

$= -16 + 4 = -12$

90. $-5^2 - (2-20) \div (-3) \cdot 3$

$= -25 - (-18) \div (-3) \cdot 3 = -25 - 6 \cdot 3$

$= -25 - 18 = -43$

91. Let $x = -5$ and $y = -2$.

$x^2 - y^4 = (-5)^2 - (-2)^4 = 25 - 16 = 9$

92. Let $x = -5$ and $y = -2$.

$x^2 - y^3 = (-5)^2 - (-2)^3 = 25 - (-8)$

$= 25 + 8 = 33$

93. $\dfrac{-9 + (-7) + 1}{3} = \dfrac{-15}{3} = -5$

Her average score per round
was 5 under par.

94. $\dfrac{-1 + 0 + (-3) + 0}{4} = \dfrac{-4}{4} = -1$

His average score per round
was 1 under par.

95. Commutative property of addition

96. Multiplicative identity property

97. Distributive property

98. Additive inverse property

99. Associative property of addition

100. Commutative property of multiplication

101. Distributive property

102. Associative property of multiplication

103. Multiplicative inverse property

104. Additive identity property

105. Commutative property of addition

Chapter 1 Test

1. $|-7| > 5$

2. $(9+5) \ge 4$

3. $-13 + 8 = -5$

4. $-13 - (-2) = -13 + 2 = -11$

5. $12 \div 4 \cdot 3 - 6 \cdot 2 = 3 \cdot 3 - 12 = 9 - 12 = -3$

6. $(13)(-3) = -39$

7. $(-6)(-2) = 12$

8. $\dfrac{|-16|}{-8} = \dfrac{16}{-8} = -2$

9. $\dfrac{-8}{0}$ is undefined

10. $\dfrac{|-6|+2}{5-6} = \dfrac{6+2}{5+(-6)} = \dfrac{8}{-1} = -8$

11. $\dfrac{1}{2} - \dfrac{5}{6} = \dfrac{1\cdot 3}{2\cdot 3} - \dfrac{5}{6} = \dfrac{3-5}{6} = \dfrac{-2}{6} = -\dfrac{1}{3}$

12. $-1\dfrac{1}{8} + 5\dfrac{3}{4} = -\dfrac{9}{8} + \dfrac{23}{4} = -\dfrac{9}{8} + \dfrac{2\cdot 23}{2\cdot 4}$

$\qquad = \dfrac{-9+46}{8} = \dfrac{37}{8} = 4\dfrac{5}{8}$

13. $(2-6) \div \dfrac{-2-6}{-3-1} - \dfrac{1}{2} = (2-6) \div \dfrac{-8}{-4} - \dfrac{1}{2}$

$\qquad = -4 \div 2 - \dfrac{1}{2} = -2 - \dfrac{1}{2} = -2\dfrac{1}{2}$

14. $3(-4)^2 - 80 = 3(16) - 80 = 48 + (-80) = -32$

15. $6\big[5 + 2(3-8) - 3\big]$

$\qquad = 6\{5 + 2[3 + (-8)] + (-3)\}$

$\qquad = 6\{5 + 2[-5] + (-3)\}$

$\qquad = 6\{5 + (-10) + (-3)\}$

$\qquad = 6\{-5 + (-3)\} = 6\{-8\} = -48$

16. $\dfrac{-12 + 3\cdot 8}{4} = \dfrac{-12+24}{4} = \dfrac{12}{4} = 3$

17. $\dfrac{(-2)(0)(-3)}{-6} = \dfrac{0(-3)}{-6} = \dfrac{0}{-6} = 0$

18. $-3 > -7$

19. $4 > -8$

20. $2 < |-3|$

21. $|-2| = -1 - (-3)$

22. $2221 < 10{,}993$

23. a. The natural numbers are 1 and 7.

     b. The whole numbers are 0, 1 and 7.

     c. The integers are -5, -1, $0, 1$, and 7.

     d. The rational numbers are

        -5, -1, $\dfrac{1}{4}$, $0, 1$, 7, and 11.6.

     e. The irrational numbers are $\sqrt{7}$ and 3π.

     f. The real numbers are all numbers

       in the given set.

24.   Let $x = 6$ and $y = -2$.

$\qquad x^2 + y^2 = (6)^2 + (-2)^2$

$\qquad\qquad\quad = 36 + 4$

$\qquad\qquad\quad = 40$

25.   Let $x = 6$, $y = -2$ and $z = -3$.

$\qquad x + yz = 6 + (-2)(-3)$

$\qquad\qquad = 6 + 6$

$\qquad\qquad = 12$

26.   Let $x = 6$ and $y = -2$.

$\qquad 2 + 3x - y = 2 + 3(6) - (-2)$

$\qquad\qquad\qquad = 2 + 18 + 2$

$\qquad\qquad\qquad = 20 + 2$

$\qquad\qquad\qquad = 22$

27.   Let $x = 6$, $y = -2$ and $z = -3$.

$$\frac{y + z - 1}{x} = \frac{-2 + (-3) - 1}{6}$$

$$= \frac{-5 + (-1)}{6}$$

$$= \frac{-6}{6}$$

$$= -1$$

28. Associative property of addition

29. Commutative property of multiplication

30. Distributive property

31. Multiplicative inverse property

32. The opposite of -9 is 9.

33. The reciprocal of $-\dfrac{1}{3}$ is -3.

34. Second down

35. Gains: 5, 29

Losses: $-10, -2$

Total gain or loss

$= 5 + (-10) + (-2) + 29 = (-5) + (-2) + 29$

$= -7 + 29 = 22$ yards gained.

Yes, they scored a touchdown.

36. Since $-14 + 31 = 17$, the temperature

at noon was $17°$

37. $356 + 460 + (-166) = 650$

The net income was $650 million.

38. Change in value per share $= -1.50$

Change in total value $= 28(-1.50) = -420$

Total loss of $420

Chapter 2

Section 2.1

Mental Math

1. -7

2. 3

3. 1

4. -1

5. 17

6. 1.2

7. $\dfrac{1}{8}$

8. $-\dfrac{5}{3}$

9. $-\dfrac{2}{3}$

10. Like terms

11. Unlike terms

12. Unlike terms

13. Like terms

14. Like terms

15. Unlike terms

Exercise Set 2.1

1. $7y + 8y = (7 + 8)\,y = 15y$

3. $-9n - 6n = (-9 - 6)\,n = -15n$

5. $3.5t - 4.5t = (3.5 - 4.5)\,t = -1t \text{ or } -t$

7. $8w - w + 6w = (8 - 1 + 6)\,w = 13w$

9. $3b - 5 - 10b - 4 = 3b - 10b - 5 - 4$
$$= (3 - 10)\,b - 9 = -7b - 9$$

11. $m - 4m + 2m - 6 = (1 - 4 + 2)\,m - 6$
$$= -m - 6$$

13. $5(y - 4) = 5(y) - 5(4) = 5y - 20$

15. $7(d - 3) + 10 = 7d - 21 + 10 = 7d - 11$

17. $-(3x - 2y + 1) = -3x + 2y - 1$

19. $5(x + 2) - (3x - 4) = 5x + 10 - 3x + 4$
$$= 2x + 14$$

21. Answers may vary

23. $(4x - 10) + (6x + 7) = 4x - 10 + 6x + 7$
$$= 10x - 3$$

25. $(3x - 8) - (7x + 1) = 3x - 8 - 7x - 1 = -4x - 9$

27. $7x^2 + 8x^2 - 10x^2 = (7 + 8 - 10)\,x^2 = 5x^2$

29. $6x - 5x + x - 3 + 2x = 6x - 5x + x + 2x - 3$

$\quad = (6 - 5 + 1 + 2)x - 3 = 4x - 3$

31. $-5 + 8(x - 6) = -5 + 8x - 48 = 8x - 53$

33. $6.2x - 4 + x - 1.2 = 6.2x + x - 4 - 1.2$

$\quad = (6.2 + 1)x - 5.2 = 7.2x - 5.2$

35. $2k - k - 6 = (2 - 1)k - 6 = k - 6$

37. $0.5(m + 2) + 0.4m = 0.5m + 1 - 0.4m$

$\quad\quad\quad\quad = 0.9m + 1$

39. $-4(3y - 4) = -12y + 16$

41. $3(2x - 5) - 5(x - 4) = 6x - 15 - 5x + 20$

$\quad\quad\quad\quad = x + 5$

43. $6x + 0.5 - 4.3x - 0.4x + 3$

$\quad = 6x - 4.3x - 0.4x + 0.5 + 3$

$\quad = (6 - 4.3 - 0.4)x + 3.5$

$\quad = 1.3x + 3.5$

45. $-2(3x - 4) + 7x - 6 = -6x + 8 + 7x - 6$

$\quad\quad\quad\quad = x + 2$

47. $-9x + 4x + 18 - 10x = -9x + 4x - 10x + 18$

$\quad = (-9 + 4 - 10)x + 18 = -15x + 18$

49. $5k - (3k - 10) = 5k - 3k + 10 = 2k + 10$

51. $(3x + 4) - (6x - 1) = 3x + 4 - 6x + 1$

$\quad\quad\quad\quad = -3x + 5$

53.

twice a number	decreased by	4
↓	↓	↓
$2x$	$-$	4

55.

three-fourths of a number	increased by	12
↓	↓	↓
$\dfrac{3}{4}x$	$+$	12

57.

5 times a number	added to	-2	added to	7 times a number
↓	↓	↓	↓	↓
$5x$	$+$	-2	$+$	$7x$

$5x + (-2) + 7x = 12x - 2$

59. $(m - 9) - (5m - 6) = m - 9 - 5m + 6 = -4m - 3$

61.

8 times	the sum of a number and 6
↓	↓
8	$(x + 6)$

$8(x + 6) = 8x + 48$

63.

double a number	minus	the sum of the number and 10
↓	↓	↓
$2x$	$-$	$(x + 10)$

$2x - (x + 10) = 2x - x - 10 = x - 10$

65.

7	multiplied by	the quotient of a number and 6
↓	↓	↓
7	×	$\dfrac{x}{6}$

$$7\left(\frac{x}{6}\right) = \frac{7x}{6}$$

67.

2	added to	3 times a number	added to	-9	added to	4 times a number
↓	↓	↓	↓	↓	↓	↓
2	+	$3x$	+	-9	+	$4x$

$$2 + 3x + (-9) + 4x = 7x - 7$$

69. $5x + (4x - 1) + 5x + (4x - 1)$
$$= 5x + 4x - 1 + 5x + 4x - 1$$
$$= (18x - 2) \text{ ft.}$$

71. $y - x^2 = 3 - (-1)^2 = 3 - 1 = 2$

73. $a - b^2 = 2 - (-5)^2 = 2 - 25 = -23$

75. $yz - y^2 = (-5)(0) - (-5)^2 = 0 - 25 = -25$

77. 1 cone + 1 cylinder $\overset{?}{=}$ 3 cubes

1 cube + 2 cubes $\overset{?}{=}$ 3 cubes

3 cubes = 3 cubes: Balanced

79. 2 cylinders + 1 cube $\overset{?}{=}$ 3 cones + 2 cubes

$2 \cdot 2$ cubes + 1 cube $\overset{?}{=}$ 3 cubes + 2 cubes

4 cubes + 1 cube $\overset{?}{=}$ 3 cubes + 2 cubes

5 cubes = 5 cubes: Balanced

81. $12(x + 2) + (3x - 1) = 12x + 24 + 3x - 1$
$$= 15x + 23$$

The total length is $(15x + 23)$ inches

83. $5b^2c^3 + 8b^3c^2 - 7b^3c^2 = 5b^2c^3 + b^3c^2$

85. $3x - (2x^2 - 6x) + 7x^2$
$$= 3x - 2x^2 + 6x + 7x^2$$
$$= 5x^2 + 9x$$

87. $-(2x^2y + 3z) + 3z - 5x^2y$
$$= -2x^2y - 3z + 3z - 5x^2y$$
$$= -7x^2y$$

Section 2.2

Mental Math

1. 2, **2.** 3, **3.** 12, **4.** 18, **5.** 17, **6.** 21

7. 9, **8.** 6, **9.** 2, **10.** 2, **11.** −5, **12.** −8

Exercise Set 2.2

1. $x + 7 = 10$

$x + 7 - 7 = 10 - 7$

$x = 3$

Check: $x + 7 = 10$

$3 + 7 \overset{?}{=} 10$

$10 = 10$

The solution is 3.

3. $x - 2 = -4$

$x - 2 + 2 = -4 + 2$

$x = -2$

Check: $x - 2 = -4$

$-2 - 2 \overset{?}{=} -4$

$-4 = -4$

The solution is -2.

5. $3 + x = -11$

$3 - 3 + x = -11 - 3$

$x = -14$

Check: $3 + x = -11$

$3 + (-14) \overset{?}{=} -11$

$-11 = -11$ True

The solution is -14.

7. $r - 8.6 = -8.1$

$r - 8.6 + 8.6 = -8.1 + 8.6$

$r = 0.5$

Check: $x - 8.6 = -8.1$

$0.5 - 8.6 \overset{?}{=} -8.1$

$-8.1 = -8.1$

The solution is 0.5.

9. $8x = 7x - 3$

$8x - 7x = 7x - 7x - 3$

$x = -3$

Check: $8x = 7x - 3$

$8(-3) \overset{?}{=} 7(-3) - 3$

$-24 \overset{?}{=} -21 - 3$

$-24 = -24$ True

The solution is -3.

11. $5b - 0.7 = 6b$

$5b - 5b - 0.7 = 6b - 5b$

$-0.7 = b$

Check: $5b - 0.7 = 6b$

$5(-0.7) - 0.7 \overset{?}{=} 6(-0.7)$

$-3.5 - 0.7 \overset{?}{=} -4.2$

$-4.2 = -4.2$

The solution is -0.7.

13. $7x - 3 = 6x$

$7x - 6x - 3 = 6x - 6x$

$x - 3 = 0$

$x - 3 + 3 = 0 + 3$

$x = 3$

Check: $7x - 3 = 6x$

$7(3) - 3 \overset{?}{=} 6(3)$

$21 - 3 \overset{?}{=} 18$

$18 = 18$

The solution is 3.

15.
$$3x - 6 = 2x + 5$$
$$3x - 2x - 6 = 2x - 2x + 5$$
$$x - 6 = 5$$
$$x - 6 + 6 = 5 + 6$$
$$x = 11$$

Check: $3x - 6 = 2x + 5$
$$3(11) - 6 \overset{?}{=} 2(11) + 5$$
$$33 - 6 \overset{?}{=} 22 + 5$$
$$27 = 27$$
The solution is 11.

17. $3t - t - 7 = t - 7$
$$2t - 7 = t - 7$$
$$2t - t - 7 = t - t - 7$$
$$t - 7 = -7$$
$$t - 7 + 7 = -7 + 7$$
$$t = 0$$

Check: $3t - t - 7 = t - 7$
$$3(0) - 0 - 7 \overset{?}{=} 0 - 7$$
$$-7 = -7$$
The solution is 0.

19. $7x + 2x = 8x - 3$
$$9x = 8x - 3$$
$$9x - 8x = 8x - 8x - 3$$
$$x = -3$$

Check: $7x + 2x = 8x - 3$
$$7(-3) + 2(-3) \overset{?}{=} 8(-3) - 3$$
$$-21 - 6 \overset{?}{=} -24 - 3$$
$$-27 = -27$$
The solution is -3.

21. $-2(x + 1) + 3x = 14$
$$-2x - 2 + 3x = 14$$
$$x - 2 = 14$$
$$x - 2 + 2 = 14 + 2$$
$$x = 16$$

Check: $-2(x + 1) + 3x = 14$
$$-2(16 + 1) + 3(16) \overset{?}{=} 14$$
$$-34 + 48 \overset{?}{=} 14$$
$$14 = 14$$
The solution is 16.

23. $-5x = 20$
$$\frac{-5x}{-5} = \frac{20}{-5}$$
$$x = -4$$

Check: $-5x = 20$
$$-5(-4) \overset{?}{=} 20$$
$$20 = 20$$
The solution is -4.

25. $3x = 0$
$$\frac{3x}{3} = \frac{0}{3}$$
$$x = 0$$

Check: $3x = 0$
$$3(0) \overset{?}{=} 0$$
$$0 = 0$$
The solution is 0.

27. $-x = -12$

$$\frac{-x}{-1} = \frac{-12}{-1}$$

$$x = 12$$

Check: $-x = -12$

$$-(12) \overset{?}{=} -12$$

$$-12 = -12$$

The solution is 12.

29. $3x + 2x = 50$

$$5x = 50$$

$$\frac{5x}{5} = \frac{50}{5}$$

$$x = 10$$

Check: $3x + 2x = 50$

$$3(10) + 2(10) \overset{?}{=} 50$$

$$30 + 20 \overset{?}{=} 50$$

$$50 = 50$$

The solution is 10.

31. $\dfrac{2}{3}x = -8$

$$\frac{3}{2}\left(\frac{2}{3}x\right) = \frac{3}{2}(-8)$$

$$x = -12$$

Check: $\dfrac{2}{3}x = -8$

$$\frac{2}{3}(-12) \overset{?}{=} -8$$

$$-8 = -8$$

The solution is -12.

33. $\dfrac{1}{6}d = \dfrac{1}{2}$

$$6\left(\frac{1}{6}d\right) = 6\left(\frac{1}{2}\right)$$

$$d = 3$$

Check: $\dfrac{1}{6}d = \dfrac{1}{2}$

$$\frac{1}{6}(3) \overset{?}{=} \frac{1}{2}$$

$$\frac{1}{2} = \frac{1}{2}$$

The solution is 3.

35. $\dfrac{a}{-2} = 1$

$$2\left(\frac{a}{-2}\right) = 2(1)$$

$$a = -2$$

Check: $\dfrac{a}{-2} = 1$

$$\frac{2}{-2} \overset{?}{=} 1$$

$$1 = 1$$

The solution is -2.

37. $\dfrac{k}{-7} = 0$

$$-7\left(\frac{k}{-7}\right) = -7(0)$$

$$k = 0$$

Check: $\dfrac{k}{-7} = 0$

$$\frac{0}{-7} \overset{?}{=} 0$$

$$0 = 0$$

The solution is 0

39. Answers may vary

41.
$$2x - 4 = 16$$
$$2x - 4 + 4 = 16 + 4$$
$$2x = 20$$
$$\frac{2x}{2} = \frac{20}{2}$$
$$x = 10$$

Check: $2x - 4 = 16$

$$2(10) - 4 \overset{?}{=} 16$$
$$20 - 4 \overset{?}{=} 16$$
$$16 = 16$$

The solution is 10.

43.
$$-x + 2 = 22$$
$$-x + 2 - 2 = 22 - 2$$
$$-x = 20$$
$$x = -20$$

Check: $-x + 2 = 22$

$$-(-20) + 2 \overset{?}{=} 22$$
$$20 + 2 \overset{?}{=} 22$$
$$22 = 22$$

The solution is -20.

45.
$$6a + 3 = 3$$
$$6a + 3 - 3 = 3 - 3$$
$$6a = 0$$
$$\frac{6a}{6} = \frac{0}{6}$$
$$a = 0$$

Check: $6a + 3 = 3$

$$6(0) + 3 \overset{?}{=} 3$$
$$0 + 3 \overset{?}{=} 3$$
$$3 = 3$$

The solution is 0.

47.
$$6x + 10 = -20$$
$$6x + 10 - 10 = -20 - 10$$
$$6x = -30$$
$$\frac{6x}{6} = \frac{-30}{6}$$
$$x = -5$$

Check: $6x + 10 = -20$

$$6(-5) + 10 \overset{?}{=} -20$$
$$-30 + 10 \overset{?}{=} -20$$
$$-20 = -20$$

The solution is -5.

49.
$$5 - 0.3k = 5$$
$$5 - 5 - 0.3k = 5 - 5$$
$$-0.3k = 0$$
$$\frac{-0.3k}{-0.3} = \frac{0}{-0.3}$$
$$k = 0$$

Check: $5 - 0.3k = 5$

$$5 - 0.3(0) \overset{?}{=} 5$$
$$5 - 0 \overset{?}{=} 5$$
$$5 = 5$$

The solution is 0.

51.
$$-2x + \frac{1}{2} = \frac{7}{2}$$
$$-2x + \frac{1}{2} - \frac{1}{2} = \frac{7}{2} - \frac{1}{2}$$
$$-2x = \frac{6}{2}$$
$$-2x = 3$$
$$\frac{-2x}{-2} = \frac{3}{-2}$$
$$x = -\frac{3}{2}$$

Check: $-2x + \frac{1}{2} = \frac{7}{2}$
$$-2\left(-\frac{3}{2}\right) + \frac{1}{2} \stackrel{?}{=} \frac{7}{2}$$
$$3 + \frac{1}{2} \stackrel{?}{=} \frac{7}{2}$$
$$\frac{7}{2} = \frac{7}{2} \quad \text{True}$$

The solution is $-\frac{3}{2}$.

53.
$$\frac{x}{3} + 2 = -5$$
$$\frac{x}{3} + 2 - 2 = -5 - 2$$
$$\frac{x}{3} = -7$$
$$3\left(\frac{x}{3}\right) = 3(-7)$$
$$x = -21$$

Check: $\frac{x}{3} + 2 = -5$
$$\frac{-21}{3} + 2 \stackrel{?}{=} -5$$
$$-7 + 2 \stackrel{?}{=} -5$$
$$-5 = -5$$

The solution is -21.

55.
$$10 = 2x - 1$$
$$10 + 1 = 2x - 1 + 1$$
$$11 = 2x$$
$$\frac{11}{2} = \frac{2x}{2}$$
$$\frac{11}{2} = x$$

Check: $10 = 2x - 1$
$$10 \stackrel{?}{=} 2\left(\frac{11}{2}\right) - 1$$
$$10 \stackrel{?}{=} 11 - 1$$
$$10 = 10$$

The solution is $\frac{11}{2}$.

57. $6z - 8 - z + 3 = 0$
$$5z - 5 = 0$$
$$5z - 5 + 5 = 0 + 5$$
$$5z = 5$$
$$\frac{5z}{5} = \frac{5}{5}$$
$$z = 1$$

Check: $6z - 8 - z + 3 = 0$

$6(1) - 8 - (1) + 3 \overset{?}{=} 0$

$6 - 8 - 1 + 3 \overset{?}{=} 0$

$0 = 0$

The solution is 1.

59. $10 - 3x - 6 - 9x = 7$

$4 - 12x = 7$

$4 - 4 - 12x = 7 - 4$

$-12x = 3$

$\dfrac{-12x}{-12} = \dfrac{3}{-12}$

$x = -\dfrac{1}{4}$

Check: $10 - 3x - 6 - 9x = 7$

$10 - 3\left(-\dfrac{1}{4}\right) - 6 - 9\left(-\dfrac{1}{4}\right) \overset{?}{=} 7$

$10 + \dfrac{3}{4} - 6 + \dfrac{9}{4} \overset{?}{=} 7$

$4 + \dfrac{12}{4} \overset{?}{=} 7$

$4 + 3 \overset{?}{=} 7$

$7 = 7$

The solution is $-\dfrac{1}{4}$.

61. $\dfrac{5}{6}x = 10$

$\dfrac{6}{5}\left(\dfrac{5}{6}x\right) = \dfrac{6}{5}(10)$

$x = 12$

Check: $\dfrac{5}{6}x = 10$

$\dfrac{5}{6}(12) \overset{?}{=} 10$

$10 = 10$

The solution is 12.

63. $1 = 0.4x - 0.6x - 5$

$1 = -0.2x - 5$

$1 + 5 = -0.2x - 5 + 5$

$6 = -0.2x$

$\dfrac{6}{-0.2} = \dfrac{-0.2x}{-0.2}$

$-30 = x$

Check: $1 = 0.4x - 0.6x - 5$

$1 \overset{?}{=} 0.4(-30) - 0.6(-30) - 5$

$1 \overset{?}{=} -12 + 18 - 5$

$1 = 1$

The solution is -30.

65. $z - 5z = 7z - 9 - z$

$-4z = 6z - 9$

$-4z - 6z = 6z - 6z - 9$

$-10z = -9$

$\dfrac{-10z}{-10} = \dfrac{-9}{-10}$

$z = \dfrac{9}{10}$

Check: $z - 5z = 7z - 9 - z$

$$\frac{9}{10} - 5\left(\frac{9}{10}\right) \overset{?}{=} 7\left(\frac{9}{10}\right) - 9 - \frac{9}{10}$$

$$\frac{9}{10} - \frac{45}{10} \overset{?}{=} \frac{63}{10} - 9 - \frac{9}{10}$$

$$-\frac{36}{10} \overset{?}{=} \frac{54}{10} - \frac{90}{10}$$

$$-\frac{36}{10} = -\frac{36}{10}$$

The solution is $\dfrac{9}{10}$.

67. $0.4x - 0.6x - 5 = 1$

$$-0.2x - 5 = 1$$

$$-0.2x - 5 + 5 = 1 + 5$$

$$-0.2x = 6$$

$$\frac{-0.2x}{-0.2} = \frac{6}{-0.2}$$

$$x = -30$$

Check: $0.4x - 0.6x - 5 = 1$

$$0.4(-30) - 0.6(-30) - 5 \overset{?}{=} 1$$

$$-12 + 18 - 5 \overset{?}{=} 1$$

$$1 = 1$$

The solution is -30.

69. $6 - 2x + 8 = 10$

$$-2x + 14 = 10$$

$$-2x + 14 - 14 = 10 - 14$$

$$-2x = -4$$

$$\frac{-2x}{-2} = \frac{-4}{-2}$$

$$x = 2$$

Check: $6 - 2x + 8 = 10$

$$6 - 2(2) + 8 \overset{?}{=} 10$$

$$6 - 4 + 8 \overset{?}{=} 10$$

$$10 = 10$$

The solution is 2.

71. $-3a + 6 + 5a = 7a - 8a$

$$2a + 6 = -a$$

$$2a + a + 6 = -a + a$$

$$3a + 6 = 0$$

$$3a + 6 - 6 = -6$$

$$3a = -6$$

$$\frac{3a}{3} = \frac{-6}{3}$$

$$a = -2$$

Check: $-3a + 6 + 5a = 7a - 8a$

$$-3(-2) + 6 + 5(-2) \overset{?}{=} 7(-2) - 8(-2)$$

$$6 + 6 - 10 \overset{?}{=} -14 + 16$$

$$2 = 2$$

The solution is -2.

73. $20 = -3(2x + 1) + 7x$

$$20 = -6x - 3 + 7x$$

$$20 = -3 + x$$

$$20 + 3 = -3 + 3 + x$$

$$23 = x$$

Check: $20 = -3(2x+1)+7x$

$20 \overset{?}{=} -3\left[2(23)+1\right]+7(23)$

$20 \overset{?}{=} -3(46+1)+161$

$20 \overset{?}{=} -141+161$

$20 = 20$

The solution is 23.

75. The other number is $20 - p$.

77. The length of the other piece is $(10-x)$ feet

79. The supplement of the angle $x°$ is $(180-x)°$.

81. Catarella received $(n+284)$ votes.

83. The length of the Golden Gate Bridge is $(m-60)$ ft.

85. Ortiz received $(n+47,628)$ votes.

87. The area of the Sahara Desert is $7x$ square miles.

89. Sum = first integer + second integer.
Sum $= x+(x+2)$
$\quad = x+x+2 = 2x+2$

91. Sum = first integer + third integer.
Sum $= x+(x+2)$
$\quad = x+x+2 = 2x+2$

93. $5x+2(x-6) = 5x+2x-12 = 7x-12$

95. $-(x-1)+x = -x+1+x = 1$

97. $(-3)^2 > -3^2$

99. $(-2)^3 = -2^3$

101. $180-\left[x+(2x+7)\right] = 180-\left[x+2x+7\right]$
$= 180-\left[3x+7\right] = 180-3x-7 = 173-3x$
The third angle is $(173-3x)°$.

103. $9x = 2100$
$\dfrac{9x}{9} - \dfrac{2100}{9}$
$x = \dfrac{700}{3}$
Each dose should be $\dfrac{700}{3}$ mg

105. $200+150+400+x = 1000$
$750+x = 1000$
$750-750+x = 1000-750$
$x = 250$
The fluid needed by the patient is 250 ml

107. Answers may vary

109. $4.95y = -31.185$
$\dfrac{4.95y}{4.95} = \dfrac{-31.185}{4.95}$
$y = -6.3$

111.
$$0.06y + 2.63 = 2.5562$$
$$0.06y + 2.63 - 2.63 = 2.5562 - 2.63$$
$$0.06y = -0.0738$$
$$\frac{0.06y}{0.06} = \frac{-0.0738}{0.06}$$
$$y = -1.23$$

Section 2.3

Calculator Explorations

1.
$$-2(3x - 4) = 2x$$
$$-6x + 8 = 2x$$
$$-6x - 2x + 8 = 2x - 2x$$
$$-8x + 8 = 0$$
$$-8x + 8 - 8 = 0 - 8$$
$$-8x = -8$$
$$\frac{-8x}{-8} = \frac{-8}{-8}$$
$$x = 1$$

3.
$$4(2n - 1) = (6n + 4) + 1$$
$$8n - 4 = 6x + 4 + 1$$
$$8n - 4 = 6n + 5$$
$$8n - 6n - 4 = 6n - 6n + 5$$
$$2n - 4 = 5$$
$$2n - 4 + 4 = 5 + 4$$
$$2n = 9$$
$$\frac{2n}{2} = \frac{9}{2}$$
$$n = \frac{9}{2}$$

5.
$$3(4y + 2) = 2(1 + 5y) + 8$$
$$12y + 6 = 2 + 10y + 8$$
$$12y + 6 = 10y + 10$$
$$12y - 10y + 6 = 10y - 10y + 10$$
$$2y + 6 = 10$$
$$2y + 6 - 6 = 10 - 6$$
$$2y = 4$$
$$\frac{2y}{2} = \frac{4}{2}$$
$$y = 2$$

Exercise Set 2.3

1.
$$-2(3x - 4) = 2x$$
$$-6x + 8 = 2x$$
$$-6x + 6x + 8 = 2x + 6x$$
$$8 = 8x$$
$$\frac{8}{8} = \frac{8x}{8}$$
$$1 = x$$

3.
$$4(2n - 1) = (6n + 4) + 1$$
$$8n - 4 = 6n + 4 + 1$$
$$8n - 4 = 6n + 5$$
$$8n - 6n - 4 = 6n - 6n + 5$$
$$2n - 4 = 5$$
$$2n - 4 + 4 = 5 + 4$$
$$2n = 9$$
$$\frac{2n}{2} = \frac{9}{2}$$
$$n = \frac{9}{2}$$

5.
$$5(2x-1)-2(3x)=1$$
$$10x-5-6x=1$$
$$4x-5=1$$
$$4x-5+5=1+5$$
$$4x=6$$
$$\frac{4x}{4}=\frac{6}{4}$$
$$x=\frac{3}{2}$$

7.
$$6(x-3)+10=-8$$
$$6x-18+10=-8$$
$$6x-8=-8$$
$$6x-8+8=-8+8$$
$$6x=0$$
$$\frac{6x}{6}=\frac{0}{6}$$
$$x=0$$

9.
$$\frac{3}{4}x-\frac{1}{2}=1$$
$$4\left(\frac{3}{4}x-\frac{1}{2}\right)=4(1)$$
$$3x-2=4$$
$$3x-2+2=4+2$$
$$3x=6$$
$$\frac{3x}{3}=\frac{6}{3}$$
$$x=2$$

11.
$$x+\frac{5}{4}=\frac{3}{4}x$$
$$4\left(x+\frac{5}{4}\right)=4\left(\frac{3}{4}x\right)$$
$$4x+5=3x$$
$$4x-3x+5=3x-3x$$
$$x+5=0$$
$$x+5-5=0-5$$
$$x=-5$$

13.
$$\frac{x}{2}-1=\frac{x}{5}+2$$
$$10\left(\frac{x}{2}-1\right)=10\left(\frac{x}{5}+2\right)$$
$$5x-10=2x+20$$
$$5x-2x-10=2x-2x+20$$
$$3x-10=20$$
$$3x-10+10=20+10$$
$$3x=30$$
$$\frac{3x}{3}=\frac{30}{3}$$
$$x=10$$

15.
$$\frac{6(3-z)}{5}=-z$$
$$5\left[\frac{6(3-z)}{5}\right]=5(-z)$$
$$6(3-z)=-5z$$
$$18-6z=-5z$$
$$18-6z+6z=-5z+6z$$
$$18=z$$

17.
$$\frac{2(x+1)}{4} = 3x - 2$$
$$4\left[\frac{2(x+1)}{4}\right] = 4(3x-2)$$
$$2(x+1) = 12x - 8$$
$$2x + 2 = 12x - 8$$
$$2x - 12x + 2 = 12x - 12x - 8$$
$$-10x + 2 = -8$$
$$-10x + 2 - 2 = -8 - 2$$
$$-10x = -10$$
$$\frac{-10x}{-10} = \frac{-10}{-10}$$
$$x = 1$$

19.
$$0.50x + 0.15(70) = 0.25(142)$$
$$100\left[0.50x + 0.15(70)\right] = 100\left[0.25(142)\right]$$
$$50x + 15(70) = 25(142)$$
$$50x + 1050 = 3550$$
$$50x + 1050 - 1050 = 3550 - 1050$$
$$50x = 2500$$
$$\frac{50x}{50} = \frac{2500}{50}$$
$$x = 50$$

21.
$$0.12(y-6) + 0.06y = 0.08y - 0.07(10)$$
$$100\left[0.12(y-6) + 0.06y\right] = 100\left[0.08y - 0.07(10)\right]$$
$$12(y-6) + 6y = 8y - 7(10)$$
$$12y - 72 + 6y = 8y - 70$$
$$18y - 72 = 8y - 70$$
$$18y - 8y - 72 = 8y - 8y - 70$$
$$10y - 72 = -70$$
$$10y - 72 + 72 = -70 + 72$$
$$10y = 2$$
$$\frac{10y}{10} = \frac{2}{10}$$
$$y = \frac{1}{5} = 0.2$$

23.
$$5x - 5 = 2(x+1) + 3x - 7$$
$$5x - 5 = 2x + 2 + 3x - 7$$
$$5x - 5 = 5x - 5$$
$$5x - 5x - 5 = 5x - 5x - 5$$
$$-5 = -5$$

Every real number is a solution.

25.
$$\frac{x}{4} + 1 = \frac{x}{4}$$
$$4\left(\frac{x}{4} + 1\right) = 4\left(\frac{x}{4}\right)$$
$$x + 4 = x$$
$$x - x + 4 = x - x$$
$$4 = 0$$

There is no solution.

27. $3x - 7 = 3(x + 1)$

$3x - 7 = 3x + 3$

$3x - 3x - 7 = 3x - 3x + 3$

$-7 = 3$

There is no solution.

29. Answers may vary

31. Answers may vary

33. $4x + 3 = 2x + 11$

$4x - 2x + 3 = 2x - 2x + 11$

$2x + 3 = 11$

$2x + 3 - 3 = 11 - 3$

$2x = 8$

$\dfrac{2x}{2} = \dfrac{8}{2}$

$x = 4$

35. $-2y - 10 = 5y + 18$

$-2y - 5y - 10 = 5y - 5y + 18$

$-7y - 10 = 18$

$-7y - 10 + 10 = 18 + 10$

$-7y = 28$

$\dfrac{-7y}{-7} = \dfrac{28}{-7}$

$y = -4$

37. $0.6x - 0.1 = 0.5x + .2$

$0.6x - 0.5x - 0.1 = 0.5x - 0.5x + 0.2$

$0.1x - 0.1 = 0.2$

$0.1x - 0.1 + 0.1 = 0.2 + 0.1$

$0.1x = 0.3$

$\dfrac{0.1x}{0.1} = \dfrac{0.3}{0.1}$

$x = 3$

39. $2y + 2 = y$

$2y - y + 2 = y - y$

$y + 2 = 0$

$y + 2 - 2 = -2$

$y = -2$

41. $3(5c - 1) - 2 = 13c + 3$

$15c - 3 - 2 = 13c + 3$

$15c - 5 = 13c + 3$

$15c - 13c - 5 = 13c - 13c + 3$

$2c - 5 = 3$

$2c - 5 + 5 = 3 + 5$

$2c = 8$

$\dfrac{2c}{2} = \dfrac{8}{2}$

$c = 4$

43. $x + \dfrac{7}{6} = 2x - \dfrac{7}{6}$

$6\left(x + \dfrac{7}{6}\right) = 6\left(2x - \dfrac{7}{6}\right)$

$6x + 7 = 12x - 7$

$6x - 12x + 7 = 12x - 12x - 7$

$-6x + 7 = -7$

$-6x + 7 - 7 = -7 - 7$

$-6x = -14$

$\dfrac{-6x}{-6} = \dfrac{-14}{-6}$

$x = \dfrac{14}{6}$

$x = \dfrac{7}{3}$

45.
$$2(x-5) = 7 + 2x$$
$$2x - 10 = 7 + 2x$$
$$2x - 2x - 10 = 7 + 2x - 2x$$
$$-10 = 7$$
There is no solution.

47.
$$\frac{2(z+3)}{3} = 5 - z$$
$$3\left[\frac{2(z+3)}{3}\right] = 3(5-z)$$
$$2z + 6 = 15 - 3z$$
$$2z + 3z + 6 = 15 - 3z + 3z$$
$$5z + 6 = 15$$
$$5z + 6 - 6 = 15 - 6$$
$$5z = 9$$
$$\frac{5z}{5} = \frac{9}{5}$$
$$z = \frac{9}{5}$$

49.
$$\frac{4(y-1)}{5} = -3y$$
$$5\left[\frac{4(y-1)}{5}\right] = 5(-3y)$$
$$4y - 4 = -15y$$
$$4y + 15y - 4 = -15y + 15y$$
$$19y - 4 = 0$$
$$19y - 4 + 4 = 4$$
$$19y = 4$$
$$\frac{19y}{19} = \frac{4}{19}$$
$$y = \frac{4}{19}$$

51.
$$8 - 2(a-1) = 7 + a$$
$$8 - 2a + 2 = 7 + a$$
$$-2a + 10 = 7 + a$$
$$-2a + 2a + 10 = 7 + a + 2a$$
$$10 = 7 + 3a$$
$$10 - 7 = 7 - 7 + 3a$$
$$3 = 3a$$
$$\frac{3}{3} = \frac{3a}{3}$$
$$1 = a$$

53.
$$2(x+3) - 5 = 5x - 3(1+x)$$
$$2x + 6 - 5 = 5x - 3 - 3x$$
$$2x + 1 = 2x - 3$$
$$2x - 2x + 1 = 2x - 2x - 3$$
$$1 = -3$$
There is no solution.

55.
$$\frac{5x-7}{3} = x$$
$$3\left[\frac{5x-7}{3}\right] = 3(x)$$
$$5x - 7 = 3x$$
$$5x - 3x - 7 = 3x - 3x$$
$$2x - 7 = 0$$
$$2x - 7 + 7 = 7$$
$$2x = 7$$
$$\frac{2x}{2} = \frac{7}{2}$$
$$x = \frac{7}{2}$$

57.
$$\frac{9+5v}{2} = 2v - 4$$
$$2\left(\frac{9+5v}{2}\right) = 2(2v - 4)$$
$$9 + 5v = 4v - 8$$
$$9 + 5v - 4v = 4v - 4v - 8$$
$$9 + v = -8$$
$$9 - 9 + v = -8 - 9$$
$$v = -17$$

59.
$$-3(t - 5) + 2t = 5t - 4$$
$$-3t + 15 + 2t = 5t - 4$$
$$-t + 15 = 5t - 4$$
$$-t - 5t + 15 = 5t - 5t - 4$$
$$-6t + 15 = -4$$
$$-6t + 15 - 15 = -4 - 15$$
$$-6t = -19$$
$$\frac{-6t}{-6} = \frac{-19}{-6}$$
$$t = \frac{19}{6}$$

61.
$$0.02(6t - 3) = 0.12(t - 2) + 0.18$$
$$0.12t - 0.06 = 0.12t - 0.24 + 0.18$$
$$0.12t - 0.06 = 0.12t - 0.06$$
$$0.12t - 0.12t - 0.06 = 0.12t - 0.12t - 0.06$$
$$-0.06 = -0.06$$
Every real number is a solution.

63.
$$0.06 - 0.01(x + 1) = -0.02(2 - x)$$
$$100[0.06 - 0.01(x + 1)] = 100[-0.02(2 - x)]$$
$$6 - (x + 1) = -2(2 - x)$$
$$6 - x - 1 = -4 + 2x$$
$$5 - x = -4 + 2x$$
$$5 - x - 2x = -4 + 2x - 2x$$
$$5 - 3x = -4$$
$$5 - 5 - 3x = -4 - 5$$
$$-3x = -9$$
$$\frac{-3x}{-3} = \frac{-9}{-3}$$
$$x = 3$$

65.
$$\frac{3(x - 5)}{2} = \frac{2(x + 5)}{3}$$
$$6\left[\frac{3(x - 5)}{2}\right] = 6\left[\frac{2(x + 5)}{3}\right]$$
$$9(x - 5) = 4(x + 5)$$
$$9x - 45 = 4x + 20$$
$$9x - 4x - 45 = 4x - 4x + 20$$
$$5x - 45 = 20$$
$$5x - 45 + 45 = 20 + 45$$
$$5x = 65$$
$$\frac{5x}{5} = \frac{65}{5}$$
$$x = 13$$

67.
$$2x + 7 = x + 6$$
$$2x - x + 7 = x - x + 6$$
$$x + 7 = 6$$
$$x + 7 - 7 = 6 - 7$$
$$x = -1$$
The number is -1.

69. $3x - 6 = 2x + 8$

$3x - 2x - 6 = 2x - 2x + 8$

$x - 6 = 8$

$x - 6 + 6 = 8 + 6$

$x = 14$

The number is 14.

71. $\dfrac{1}{3}x = \dfrac{5}{6}$

$6\left(\dfrac{1}{3}x\right) = 6\left(\dfrac{5}{6}\right)$

$2x = 5$

$\dfrac{2x}{2} = \dfrac{5}{2}$

$x = \dfrac{5}{2}$

The number is $\dfrac{5}{2}$.

73. $\dfrac{x}{4} + \dfrac{1}{2} = \dfrac{3}{4}$

$4\left(\dfrac{x}{4} + \dfrac{1}{2}\right) = 4\left(\dfrac{3}{4}\right)$

$x + 2 = 3$

$x + 2 - 2 = 3 - 2$

$x = 1$

The number is 1.

75. $10 - 5x = 3x$

$10 - 5x + 5x = 3x + 5x$

$10 = 8x$

$\dfrac{10}{8} = \dfrac{8x}{8}$

$\dfrac{5}{4} = x$

The number is $\dfrac{5}{4}$

77. Since the perimeter is the sum of the lengths of the sides,

$x + x + x + 2x + 2x = 28$

$7x = 28$

$\dfrac{7x}{7} = \dfrac{28}{7}$

$x = 4$

$2x = 2(4) = 8$

The lengths are 4 cm and 8 cm.

79. $\left|2^3 - 3^2\right| - \left|5 - 7\right|$

$= \left|8 - 9\right| - \left|-2\right|$

$= \left|-1\right| - 2$

$= 1 - 2$

$= -1$

81. $\dfrac{5}{4 + 3 \cdot 7} = \dfrac{5}{4 + 21} = \dfrac{5}{25} = \dfrac{1}{5}$

83. $x + (2x - 3) + (3x - 5) = x + 2x - 3 + 3x - 5$

$= 6x - 8$

The perimeter is $(6x - 8)$ meters

85. Fairview

87.
$$65 + x = 2x - 90$$
$$65 + x - x = 2x - x - 90$$
$$65 = x - 90$$
$$65 + 90 = x - 90 + 90$$
$$155 = x$$

The number of places named
Five Points is 155.

89.
$$1000(7x - 10) = 50(412 + 100x)$$
$$7000x - 10,000 = 20,600 + 5000x$$
$$7000x - 5000x - 10,000$$
$$\qquad = 20,600 + 5000x - 5000x$$
$$2000x - 10,000 = 20,600$$
$$2000x - 10,000 + 10,000$$
$$\qquad = 20,600 + 10,000$$
$$2000x = 30,600$$
$$\frac{2000x}{2000} = \frac{30,600}{2000}$$
$$x = 15.3$$

91.
$$0.035x + 5.112 = 0.010x + 5.107$$
$$1000(0.035x + 5.112) = 1000(0.010x + 5.107)$$
$$35x + 5112 = 10x + 5107$$
$$35x - 10x + 5112 = 10x - 10x + 5107$$
$$25x + 5112 = 5107$$
$$25x + 5112 - 5112 = 5107 - 5112$$
$$25x = -5$$
$$\frac{25x}{25} = \frac{-5}{25}$$
$$x = -\frac{1}{5} = -0.2$$

93.
$$x(x - 3) = x^2 + 5x + 7$$
$$x^2 - 3x = x^2 + 5x + 7$$
$$x^2 - x^2 - 3x = x^2 - x^2 + 5x + 7$$
$$-3x = 5x + 7$$
$$-3x - 5x = 5x - 5x + 7$$
$$-8x = 7$$
$$\frac{-8x}{-8} = \frac{7}{-8}$$
$$x = -\frac{7}{8}$$

95.
$$2z(z + 6) = 2z^2 + 12z - 8$$
$$2z^2 + 12z = 2z^2 + 12z - 8$$
$$2z^2 - 2z^2 + 12z = 2z^2 - 2z^2 + 12z - 8$$
$$12z = 12z - 8$$
$$12z - 12z = 12z - 12z - 8$$
$$0 = -8$$

There is no solution.

Integrated Review

1.
$$x - 10 = -4$$
$$x - 10 + 10 = -4 + 10$$
$$x = 6$$

2.
$$y + 14 = -3$$
$$y + 14 - 14 = -3 - 14$$
$$y = -17$$

3. $9y = 108$
$$\frac{9y}{9} = \frac{108}{9}$$
$$y = 12$$

4. $-3x = 78$

$$\frac{-3x}{-3} = \frac{78}{-3}$$

$$x = -26$$

5. $-6x + 7 = 25$

$$-6x + 7 - 7 = 25 - 7$$

$$-6x = 18$$

$$\frac{-6x}{-6} = \frac{18}{-6}$$

$$x = -3$$

6. $5y - 42 = -47$

$$5y - 42 + 42 = -47 + 42$$

$$5y = -5$$

$$\frac{5y}{5} = \frac{-5}{5}$$

$$y = -1$$

7. $\frac{2}{3}x = 9$

$$\frac{3}{2}\left(\frac{2}{3}x\right) = \frac{3}{2}(9)$$

$$x = \frac{27}{2} = 13.5$$

8. $\frac{4}{5}z = 10$

$$\frac{5}{4}\left(\frac{4}{5}z\right) = \frac{5}{4}(10)$$

$$z = \frac{25}{2} = 12.5$$

9. $\frac{r}{-4} = -2$

$$-4\left(\frac{r}{-4}\right) = -4(-2)$$

$$r = 8$$

10. $\frac{y}{-8} = 8$

$$-8\left(\frac{y}{-8}\right) = -8(8)$$

$$y = -64$$

11. $6 - 2x + 8 = 10$

$$-2x + 14 = 10$$

$$-2x + 14 - 14 = 10 - 14$$

$$-2x = -4$$

$$\frac{-2x}{-2} = \frac{-4}{-2}$$

$$x = 2$$

12. $-5 - 6y + 6 = 19$

$$-6y + 1 = 19$$

$$-6y + 1 - 1 = 19 - 1$$

$$-6y = 18$$

$$\frac{-6y}{-6} = \frac{18}{-6}$$

$$y = -3$$

13. $2x - 7 = 2x - 27$

$$2x - 2x - 7 = 2x - 2x - 27$$

$$-7 = -27 \text{ a contradiction}$$

There is no solution

14. $3 + 8y = 8y - 2$

$3 + 8y - 8y = 8y - 8y - 2$

$3 = -2$ a contradiction

There is no solution

15. $-3a + 6 + 5a = 7a - 8a$

$2a + 6 = -a$

$2a - 2a + 6 = -a - 2a$

$6 = -3a$

$\dfrac{6}{-3} = \dfrac{-3a}{-3}$

$-2 = a$

16. $4b - 8 - b = 10b - 3b$

$3b - 8 = 7b$

$3b - 3b - 8 = 7b - 3b$

$-8 = 4b$

$\dfrac{-8}{4} = \dfrac{4b}{4}$

$-2 = b$

17. $-\dfrac{2}{3}x = \dfrac{5}{9}$

$-\dfrac{3}{2}\left(-\dfrac{2}{3}x\right) = -\dfrac{3}{2}\left(\dfrac{5}{9}\right)$

$x = -\dfrac{5}{6}$

18. $-\dfrac{3}{8}y = -\dfrac{1}{16}$

$-\dfrac{8}{3}\left(-\dfrac{3}{8}y\right) = -\dfrac{8}{3}\left(-\dfrac{1}{16}\right)$

$y = \dfrac{1}{6}$

19. $10 = -6n + 16$

$10 - 16 = -6n + 16 - 16$

$-6 = -6n$

$\dfrac{-6}{-6} = \dfrac{-6n}{-6}$

$1 = n$

20. $-5 = -2m + 7$

$-5 - 7 = -2m + 7 - 7$

$-12 = -2m$

$\dfrac{-12}{-2} = \dfrac{-2m}{-2}$

$6 = m$

21. $3(5c - 1) - 2 = 13c + 3$

$15c - 3 - 2 = 13c + 3$

$15c - 5 = 13c + 3$

$15c - 13c - 5 = 13c - 13c + 3$

$2c - 5 = 3$

$2c - 5 + 5 = 3 + 5$

$2c = 8$

$\dfrac{2c}{2} = \dfrac{8}{2}$

$c = 4$

22. $4(3t + 4) - 20 = 3 + 5t$

$12t + 16 - 20 = 3 + 5t$

$12t - 4 = 3 + 5t$

$12t - 5t - 4 = 3 + 5t - 5t$

$7t - 4 = 3$

$7t - 4 + 4 = 3 + 4$

$7t = 7$

$\dfrac{7t}{7} = \dfrac{7}{7}$

$t = 1$

23.
$$\frac{2(z+3)}{3} = 5 - z$$
$$3\left[\frac{2(z+3)}{3}\right] = 3(5-z)$$
$$2z + 6 = 15 - 3z$$
$$2z + 3z + 6 = 15 - 3z + 3z$$
$$5z + 6 = 15$$
$$5z + 6 - 6 = 15 - 6$$
$$5z = 9$$
$$\frac{5z}{5} = \frac{9}{5}$$
$$z = \frac{9}{5}$$

24.
$$\frac{3(w+2)}{4} = 2w + 3$$
$$4\left[\frac{3(w+2)}{4}\right] = 4(2w+3)$$
$$3w + 6 = 8w + 12$$
$$3w - 8w + 6 = 8w - 8w + 12$$
$$-5w + 6 = 12$$
$$-5w + 6 - 6 = 12 - 6$$
$$-5w = 6$$
$$\frac{-5w}{-5} = \frac{6}{-5}$$
$$w = -\frac{6}{5}$$

25.
$$-2(2x-5) = -3x + 7 - x + 3$$
$$-4x + 10 = -4x + 10$$
$$-4x + 4x + 10 = -4x + 4x + 10$$
$$10 = 10$$
Every real number is a solution.

26.
$$-4(5x-2) = -12x + 4 - 8x + 4$$
$$-20x + 8 = -20x + 8$$
$$-20x + 20x + 8 = -20x + 20x + 8$$
$$8 = 8$$
Every real number is a solution.

27.
$$0.02(6t-3) = 0.04(t-2) + 0.02$$
$$100\left[0.02(6t-3)\right] = 100\left[0.04(t-2) + 0.02\right]$$
$$2(6t-3) = 4(t-2) + 2$$
$$12t - 6 = 4t - 8 + 2$$
$$12t - 6 = 4t - 6$$
$$12t - 4t - 6 = 4t - 4t - 6$$
$$8t - 6 = -6$$
$$8t - 6 + 6 = -6 + 6$$
$$8t = 0$$
$$\frac{8t}{8} = \frac{0}{8}$$
$$t = 0$$

28.
$$0.03(m+7) = 0.02(5-m) + 0.03$$
$$100\left[0.03(m+7)\right] = 100\left[0.02(5-m) + 0.03\right]$$
$$3(m+7) = 2(5-m) + 3$$
$$3m + 21 = 10 - 2m + 3$$
$$3m + 21 = 13 - 2m$$
$$3m + 2m + 21 = 13 - 2m + 2m$$
$$5m + 21 = 13$$
$$5m + 21 - 21 = 13 - 21$$
$$5m = -8$$
$$\frac{5m}{5} = \frac{-8}{5}$$
$$m = -\frac{8}{5} = -1.6$$

29. $\quad -3y = \dfrac{4(y-1)}{5}$

$$5(-3y) = 5\left[\dfrac{4(y-1)}{5}\right]$$

$$-15y = 4y - 4$$

$$-15y - 4y = 4y - 4y - 4$$

$$-19y = -4$$

$$\dfrac{-19y}{-19} = \dfrac{-4}{-19}$$

$$y = \dfrac{4}{19}$$

30. $\quad -4x = \dfrac{5(1-x)}{6}$

$$6(-4x) = 6\left[\dfrac{5(1-x)}{6}\right]$$

$$-24x = 5 - 5x$$

$$-24x + 5x = 5 - 5x + 5x$$

$$-19x = 5$$

$$\dfrac{-19x}{-19} = \dfrac{5}{-19}$$

$$x = -\dfrac{5}{19}$$

31. $\quad \dfrac{5}{3}x - \dfrac{7}{3} = x$

$$3\left(\dfrac{5}{3}x - \dfrac{7}{3}\right) = 3(x)$$

$$5x - 7 = 3x$$

$$5x - 5x - 7 = 3x - 5x$$

$$-7 = -2x$$

$$\dfrac{-7}{-2} = \dfrac{-2x}{-2}$$

$$\dfrac{7}{2} = x$$

32. $\quad \dfrac{7}{5}n + \dfrac{3}{5} = -n$

$$5\left(\dfrac{7}{5}n + \dfrac{3}{5}\right) = 5(-n)$$

$$7n + 3 = -5n$$

$$7n - 7n + 3 = -5n - 7n$$

$$3 = -12n$$

$$\dfrac{3}{-12} = \dfrac{-12n}{-12}$$

$$-\dfrac{1}{4} = n$$

Exercise Set 2.4

1. $\qquad$ Let $x =$ the number.

$$2(x-8) = 3(x+3)$$

$$2x - 16 = 3x + 9$$

$$2x - 2x - 16 = 3x - 2x + 9$$

$$-16 = x + 9$$

$$-16 - 9 = x + 9 - 9$$

$$-25 = x$$

The number $= -25$

3.   Let x = the number.

$$2x(3) = 5x - \frac{3}{4}$$

$$6x = 5x - \frac{3}{4}$$

$$6x - 5x = 5x - 5x - \frac{3}{4}$$

$$x = -\frac{3}{4}$$

The number $= -\frac{3}{4}$

5.   Let x = the number of the left page
and $x + 1 =$ the number of the right page.

$$x + x + 1 = 469$$

$$2x + 1 = 469$$

$$2x + 1 - 1 = 469 - 1$$

$$2x = 468$$

$$\frac{2x}{2} = \frac{468}{2}$$

$$x = 234$$

$$x + 1 = 235$$

The page numbers are 234 and 235.

7.   Let x = the code for Belgium,
$x + 1 =$ the code for France,
$x + 2 =$ the code for Spain.

$$x + x + 1 + x + 2 = 99$$

$$3x + 3 = 99$$

$$3x + 3 - 3 = 99 - 3$$

$$3x = 96$$

$$\frac{3x}{3} = \frac{96}{3}$$

$$x = 32$$

$$x + 1 = 33$$

$$x + 2 = 34$$

The codes are;

Belgium:32; France:33; Spain:34

9.   Let x = length of the first piece,
$2x =$ length of the second piece,
and $5x =$ length of the third piece.

$$x + 2x + 5x = 40$$

$$8x = 40$$

$$\frac{8x}{8} = \frac{40}{8}$$

$$x = 5$$

$$2x = 2(5) = 10$$

$$5x = 5(5) = 25$$

The lengths are 5, 10, and 25 inches

11.   Let x = the salary of the governor of
Nebraska and $2x$ = the salary of
the governor of Washington.

$$x + 2x = 195,000$$

$$3x = 195,000$$

$$\frac{3x}{3} = \frac{195,000}{3}$$

$$x = 65,000$$

$$2x = 2(65,000) = 130,000$$

The governor of Nebraska makes $65,000
and the governor of Washington makes
$130,000.

13.   Let x = the number of miles driven,
0.29 = the charge per mile, and
24.95 = the charge per day.

$$2(24.95) + 0.29x = 100$$
$$49.90 + 0.29x = 100$$
$$49.90 - 49.90 + 0.29x = 100 - 49.90$$
$$0.29x = 50.10$$
$$\frac{0.29x}{0.29} = \frac{50.10}{0.29}$$
$$x = 172$$

172 miles were driven

15. Let x = the number of miles driven,
    0.80 = the charge per mile,
    3.00 = the taxi charge
    4.50 = the toll charges.
    $$0.80x + 3.00 + 4.50 = 27.50$$
    $$0.80x + 7.50 = 27.50$$
    $$0.80x + 7.50 - 7.50 = 27.50 - 7.50$$

    $$0.80x = 20.00$$
    $$\frac{0.80x}{0.80} = \frac{20.00}{0.80}$$
    $$x = 25$$

    You can travel 25 miles

17. Let x = the measure of each of the two
    equal angle, and $2x + 30$ = the measure
    of the third.
    $$x + x + 2x + 30 = 180$$
    $$4x + 30 = 180$$
    $$4x + 30 - 30 = 180 - 30$$
    $$4x = 150$$
    $$\frac{4x}{4} = \frac{150}{4}$$
    $$x = 37.5$$
    $$2x + 30 = 2(37.5) + 30 = 105$$

    The 3 angles are $37.5°, 37.5°, 105°$

19. Let x = the measure of each of the two
    equal angles A and D, and $2x$ = the
    measure of each of the other two
    equal angles C and B.
    $$x + x + 2x + 2x = 360$$
    $$6x = 360$$
    $$\frac{6x}{6} = \frac{360}{6}$$
    $$x = 60$$
    $$2x = 2(60) = 120$$

    The angles are $A = 60°, D = 60°$
    $C = 120°, B = 120°$

21. Let x = length of the shorter piece
    and $2x + 2$ = length of the longer piece.
    $$x + 2x + 2 = 17$$
    $$3x + 2 = 17$$
    $$3x + 2 - 2 = 17 - 2$$
    $$3x = 15$$
    $$\frac{3x}{3} = \frac{15}{3}$$
    $$x = 5$$
    $$2x + 2 = 2(5) + 2 = 12$$

    The shorter piece = 5 ft. and the
    longer piece = 12 ft

23. Let x = the number of millions of
    prescriptions in 1997 and $x + 5.5$ = the
    number in 2001.
    $$x + x + 5.5 = 35.7$$
    $$2x + 5.5 = 35.7$$
    $$2x + 5.5 - 5.5 = 35.7 - 5.5$$

$$2x = 30.2$$
$$\frac{2x}{2} = \frac{30.2}{2}$$
$$x = 15.1$$
$$x + 5.5 = 15.1 + 5.5 = 20.6$$

15.1 million prescriptions were written in 1997 and 20.6 were written in 2001

25. Let x = the measure of the smaller angle and $3x$ = the measure of the other.

$$x + 3x = 180$$
$$4x = 180$$
$$\frac{4x}{4} = \frac{180}{4}$$
$$x = 45$$
$$3x = 3(45) = 135$$

The 2 angles are $45°, 135°$.

27. Let x = the measure of the smallest angle, $x + 2$ = the measure of the second, and $x + 4$ = the measure of the third.

$$x + x + 2 + x + 4 = 180$$
$$3x + 6 = 180$$
$$3x + 6 - 6 = 180 - 6$$
$$3x = 174$$
$$\frac{3x}{3} = \frac{174}{3}$$
$$x = 58$$
$$x + 2 = 58 + 2 = 60$$
$$x + 4 = 58 + 4 = 62$$

The 3 angles are $58°, 60°, 62°$

29. Let x = the number of miles driven, 0.20 = the charge per mile, and 39 = the charge per day.

$$39 + 0.20x = 95$$
$$39 - 39 + 0.20x = 95 - 39$$
$$0.20x = 56$$
$$\frac{0.20x}{0.20} = \frac{56}{0.20}$$
$$x = 280$$

280 miles were driven

31. Let x = the Oakland score and $x + 27$ = the Tampa Bay score.

$$x + x + 27 = 69$$
$$2x + 27 = 69$$
$$2x + 27 - 27 = 69 - 27$$
$$2x = 42$$
$$\frac{2x}{2} = \frac{42}{2}$$
$$x = 21$$
$$x + 27 = 21 + 27 = 48$$

The score was
Tampa Bay 48, Oakland 21.

33. Let x = number of medals won by Germany, $x + 1$ = number of medals won by Australia, and $x + 2$ = number of medals won by China.

$$x + x + 1 + x + 2 = 174$$
$$3x + 3 = 174$$
$$3x + 3 - 3 = 174 - 3$$
$$3x = 171$$
$$\frac{3x}{3} = \frac{171}{3}$$
$$x = 57$$

$x+1 = 57+1 = 58$, $x+2 = 57+2 = 59$

Germany won 57 medals.

Australia won 58 medals.

China won 59 medals.

35. Let x = the number of electoral votes for
for Texas and $x + 21$ = the number
for California.

$$x + x + 21 = 89$$
$$2x + 21 = 89$$
$$2x + 21 - 21 = 89 - 21$$
$$2x = 68$$
$$\frac{2x}{2} = \frac{68}{2}$$
$$x = 34$$

$x + 21 = 34 + 21 = 55$

Texas will have 34 votes.

California will have 55 votes.

37. Let x = number of moons won of
Neptune, $x + 13$ = number of moons
of Uranus, and $2x + 2$ = number
of moons of Saturn.

$$x + x + 13 + 2x + 2 = 47$$
$$4x + 15 = 47$$
$$4x + 15 - 15 = 47 - 15$$
$$4x = 32$$
$$\frac{4x}{4} = \frac{32}{4}$$
$$x = 8$$

$x + 13 = 8 + 13 = 21$, $2x + 2 = 2(8) + 2 = 18$

Neptune: 8 moons; Uranus: 21 moons;

Saturn: 18 moons

39.   Let x = the number.

$$3(x+5) = 2x - 1$$
$$3x + 15 = 2x - 1$$
$$3x - 2x + 15 = 2x - 2x - 1$$
$$x + 15 = -1$$
$$x + 15 - 15 = -1 - 15$$
$$x = -16$$

The number $= -16$

41. Let x = the number of votes for
Randall and $x + 13{,}288$ = the number
for Brown.

$$x + x + 13{,}288 = 119{,}436$$
$$2x + 13{,}288 = 119{,}436$$
$$2x + 13{,}288 - 13{,}288 = 119{,}436 - 13{,}288$$
$$2x = 106{,}148$$
$$\frac{2x}{2} = \frac{106{,}148}{2}$$
$$x = 53{,}074$$

$x + 13{,}288 = 53{,}074 + 13{,}288 = 66{,}362$

Randall had 53,074 votes.

Brown had 66,362 votes.

43. Illinois

45. Let x = the amount spent in millions by
Florida and $x + 2.2$ = the
amount spent in millions by Texas.

$$x + x + 2.2 = 56.6$$
$$2x + 2.2 = 56.6$$
$$2x + 2.2 - 2.2 = 56.6 - 2.2$$

$$2x = 54.4$$

$$\frac{2x}{2} = \frac{54.4}{2}$$

$$x = 27.2$$

$$x + 2.2 = 27.2 + 2.2 = 29.4$$

Florida spent \$27.2 million.

Texas spent \$29.4 million.

47. Answers may vary

49. $-2 + (-8) = -10$

51. $-11 + 2 = -9$

53. $-12 - 3 = -12 + (-3) = -15$

55. Let x = the measure of the smallest angle, $2x$ = the measure of the second, and $3x$ = the measure of the third.

$$x + 2x + 3x = 180$$

$$6x = 180$$

$$\frac{6x}{6} = \frac{180}{6}$$

$$x = 30$$

$$2x = 2(30) = 60$$

$$3x = 3(30) = 90$$

The 3 angles are $30°, 60°, 90°$

57. Answers may vary

59. Answers may vary

61. C

Exercise Set 2.5

1. Let $A = 45$ and $b = 15$

$$A = bh$$

$$45 = 15h$$

$$\frac{45}{15} = \frac{15h}{15}$$

$$3 = h$$

3. Let $S = 102$, $l = 7$, and $w = 3$

$$S = 4lw + 2wh$$

$$102 = 4(7)(3) + 2(3)h$$

$$102 = 84 + 6h$$

$$102 - 84 = 84 - 84 + 6h$$

$$18 = 6h$$

$$\frac{18}{6} = \frac{6h}{6}$$

$$3 = h$$

5. Let $A = 180$, $B = 11$, and $b = 7$

$$A = \frac{1}{2}(B + b)h$$

$$180 = \frac{1}{2}(11 + 7)h$$

$$2(180) = 2\left[\frac{1}{2}(18)h\right]$$

$$360 = 18h$$

$$\frac{360}{18} = \frac{18h}{18}$$

$$20 = h$$

7. Let $P = 30$, $a = 8$, and $b = 10$

$$P = a + b + c$$
$$30 = 8 + 10 + c$$
$$30 = 18 + c$$
$$30 - 18 = 18 - 18 + c$$
$$12 = c$$

9. Let $C = 15.7$, and $\pi = 3.14$

$$C = 2\pi r$$
$$15.7 = 2(3.14) r$$
$$15.7 = 6.28r$$
$$\frac{15.7}{6.28} = \frac{6.28r}{6.28}$$
$$2.5 = r$$

11. Let $I = 3750$, $P = 25,000$, and $R = 0.05$

$$I = PRT$$
$$3750 = 25,000(0.05)T$$
$$3750 = 1250T$$
$$\frac{3750}{1250} = \frac{1250T}{1250}$$
$$3 = T$$

13. Let $V = 565.2$, $r = 6$, and $\pi = 3.14$

$$V = \frac{1}{3}\pi r^2 h$$
$$565.2 = \frac{1}{3}(3.14)(6)^2 h$$
$$565.2 = 37.68h$$
$$\frac{565.2}{37.68} = \frac{37.68h}{37.68}$$
$$15 = h$$

15. $f = 5gh$

$$\frac{f}{5g} = \frac{5gh}{5g}$$
$$\frac{f}{5g} = h$$

17. $V = LWH$

$$\frac{V}{LH} = \frac{LWH}{LH}$$
$$\frac{V}{LH} = W$$

19. $3x + y = 7$

$$3x - 3x + y = 7 - 3x$$
$$y = 7 - 3x$$

21. $A = P + PRT$

$$A - P = P - P + PRT$$
$$A - P = PRT$$
$$\frac{A - P}{PT} = \frac{PRT}{PT}$$
$$\frac{A - P}{PT} = R$$

23. $V = \frac{1}{3}Ah$

$$3V = 3\left(\frac{1}{3}Ah\right)$$
$$3V = Ah$$
$$\frac{3V}{h} = \frac{Ah}{h}$$
$$\frac{3V}{h} = A$$

25. $P = a + b + c$

$P - b - c = a + b - b + c - c$

$P - b - c = a$

27. $S = 2\pi rh + 2\pi r^2$

$S - 2\pi r^2 = 2\pi rh + 2\pi r^2 - 2\pi r^2$

$S - 2\pi r^2 = 2\pi rh$

$\dfrac{S - 2\pi r^2}{2\pi r} = \dfrac{2\pi rh}{2\pi r}$

$\dfrac{S - 2\pi r^2}{2\pi r} = h$

29. Let $A = 52,400$ and $l = 400$

$A = lw$

$52,400 = 400w$

$\dfrac{52,400}{400} = \dfrac{400w}{400}$

$131 = w$

The width is 131 ft

31. Let $F = 14$

$14 = \dfrac{9}{5}C + 32$

$5(14) = 5\left(\dfrac{9}{5}\right)C + 5(32)$

$70 = 9C + 160$

$70 - 160 = 9C + 160 - 160$

$-90 = 9C$

$\dfrac{-90}{9} = \dfrac{9C}{9}$

$-10 = C$

The equivalent temperature is $-10°\,C$.

33. Let $d = 25,000$ and $r = 4000$

$d = rt$

$25,000 = 4000t$

$\dfrac{25,000}{4000} = \dfrac{4000t}{4000}$

$6.25 = t$

It will take 6.25 hours

35. Let $P = 260$ and $w = \dfrac{2}{3}l$

$P = 2l + 2w$

$260 = 2l + 2\left(\dfrac{2}{3}l\right)$

$260 = \dfrac{10}{3}l$

$3(260) = 3\left(\dfrac{10}{3}l\right)$

$780 = 10l$

$\dfrac{780}{10} = \dfrac{10l}{10}$

$78 = l$

$w = \dfrac{2}{3}l = \dfrac{2}{3}(78) = 52$

The width is 52 ft and the length is 78 ft.

37. Let $P = 102$, $a =$ the length of the shortest side, $b = 2a$, and $c = a + 30$

$P = a + b + c$

$102 = a + 2a + a + 30$

$102 = 4a + 30$

$102 - 30 = 4a + 30 - 30$

$72 = 4a$

$$\frac{72}{4} = \frac{4a}{4}$$

$$18 = a$$

$$b = 2a = 2(18) = 36$$

$$c = a + 30 = 18 + 30 = 48$$

The lengths are 18 ft, 36 ft, and 48 ft.

39. Let $t = 2.5$ and $r = 55$

$$d = rt$$

$$d = 55(2.5)$$

$$d = 137.5$$

They are 137.5 miles apart.

41. Let $l = 8$, $w = 3$, and $h = 6$

$$V = lwh$$

$$V = 8(3)(6) = 144$$

Let x = number of piranha and volume

per fish $= 1.5$

$$144 = 1.5x$$

$$\frac{144}{1.5} = \frac{1.5x}{1.5}$$

$$96 = x$$

96 piranhas can be placed in the tank.

43. Let $h = 60$, $B = 130$, and $b = 70$

$$A = \frac{1}{2}(B + b)h$$

$$A = \frac{1}{2}(130 + 70)60 = \frac{1}{2}(200)(60) = 6000$$

Let x = number of bags of fertilizer

and the area per bag = 4000.

$$4000x = 6000$$

$$\frac{4000x}{4000} = \frac{6000}{4000}$$

$$x = 1.5$$

Two bags must be purchased.

45. Let $d = 16$, so $r = 8$

$$A = \pi r^2 = \pi(8)^2 = 64\pi$$

Let $d = 10$, so $r = 5$

$$A = 2\pi r^2 = 2\pi(5)^2 = 50\pi$$

One 16 inch pizza has more area and

therefore gives more pizza for the price.

47. Let $d = 42.8$ and

$$r = 552 \frac{\text{km}}{\text{hr}} \times \frac{1 \text{ hr}}{60 \text{ min}} = 9.2 \frac{\text{km}}{\text{min}}$$

$$d = rt$$

$$42.8 = 9.2t$$

$$\frac{42.8}{9.2} = \frac{9.2t}{9.2}$$

$$4.65 = t$$

It would take 4.65 minutes

49. Let x = the length of a side of the

square and $x + 5$ = the length of a

side of the triangle.

$$P(\text{triangle}) = P(\text{square}) + 7$$

$$3(x + 5) = 4x + 7$$

$$3x + 15 = 4x + 7$$

$$3x - 3x + 15 = 4x - 3x + 7$$

$$15 = x + 7$$
$$15 - 7 = x + 7 - 7$$
$$8 = x$$
$$x + 5 = 8 + 5 = 13$$

The side of the triangle is 13 in.

51. Let $d = 135$ and $r = 60$

$$d = rt$$
$$135 = 60t$$
$$\frac{135}{60} = \frac{60t}{60}$$
$$2.25 = t$$

It would take 2.25 hours.

53. Let $A = 1,813,500$ and $w = 150$

$$A = lw$$
$$1,813,500 = l(150)$$
$$\frac{1,813,500}{150} = \frac{150l}{150}$$
$$12,090 = l$$

The length is 12,090 ft

55. Let $F = 122$

$$122 = \frac{9}{5}C + 32$$
$$5(122) = 5\left(\frac{9}{5}\right)C + 5(32)$$
$$610 = 9C + 160$$
$$610 - 160 = 9C + 160 - 160$$
$$450 = 9C$$
$$\frac{450}{9} = \frac{9C}{9}$$
$$50 = C$$

The equivalent temperature is $50°$ C.

57. Let $l = 199$, $w = 78.5$, and $h = 33$

$$V = lwh$$
$$V = 199(78.5)(33) = 515,509.5$$

The volume must be 515,509.5 cu in.

59. Let $\pi = 3.14$ and $d = 9.5$ so $r = 4.75$

$$V = \frac{4}{3}\pi r^3 = \frac{4}{3}(3.14)(4.75)^3 = 449$$

The volume is 449 cu in.

61. Let $C = 167$

$$F = \frac{9}{5}C + 32 = \frac{9}{5}(167) + 32$$
$$= 300.6 + 32 = 332.6$$

The equivalent temperature is $332.6°$F.

63. $\dfrac{9}{x + 5}$

65. $3(x + 4)$

67. $3(x - 12)$

69. Let $C = -78.5$

$$F = \frac{9}{5}C + 32 = \frac{9}{5}(-78.5) + 32$$
$$= -141.3 + 32 = -109.3$$

The equivalent temperature is $-109.3°$ F.

71. Let $d = 93,000,000$ and $r = 186,000$

$$d = rt$$
$$93,000,000 = 186,000t$$
$$\frac{93,000,000}{186,000} = \frac{186,000t}{186,000}$$
$$500 = t$$

It will take 500 seconds or $8\frac{1}{3}$ minutes.

73. Let $t = 365$ and $r = 20$

$d = rt = 20(365) = 7300$ inches

$\dfrac{7300 \text{ inches}}{1} \cdot \dfrac{1 \text{ foot}}{12 \text{ inch}} \approx 608.33$ feet

It moves about 608.33 feet.

75. Let $h = 15$, and $d = 2$ then $r = 1$.

$h = \dfrac{15 \text{ feet}}{1} \cdot \dfrac{12 \text{ inches}}{1 \text{ foot}} = 180$ inches

Use $\pi = 3.14$.

$V = \pi r^2 h$

$V = (3.14)(1)^2(180)$

$V = 565.2$

The volume of the column is 565.2 cu in.

77. Let $F = 78$

$$78 = \frac{9}{5}C + 32$$

$$5(78) = 5\left(\frac{9}{5}\right)C + 5(32)$$

$$390 = 9C + 160$$

$$390 - 160 = 9C + 160 - 160$$

$$230 = 9C$$

$$\frac{230}{9} = \frac{9C}{9}$$

$$25\frac{5}{9} = C$$

The equivalent temperature is $25\frac{5}{9}^{\circ}$ C.

79. The original parallelogram has an area

$V = bh$

The altered box, has a base $2b$,

a height $2h$ and a new area.

$A = 2b(2h)$

$A = 4bh.$

The area is multiplied by 4.

Exercise Set 2.6

1. Let $x =$ the decrease in price.

$x = 0.25(256) = 64$

The decrease in price is \$64.

The sale price is $256 - 64 = \$192$.

3. Let $x =$ the original price and

$0.25x =$ the discount

$x - 0.25x = 78$

$0.75x = 78$

$\dfrac{0.75x}{0.75} = \dfrac{78}{0.75}$

$x = 104$

The original price was \$104

5. Increase $= 380,000 - 220,000 = 160,000$

Let $x =$ the percent increase

$x(220,000) = 160,000$

$\dfrac{220,000x}{220,000} = \dfrac{160,000}{220,000}$

$x = 0.727$

The percent increase was 73%

7. Let x = the time traveled by the jet plane

$$Rate \cdot Time = Distance$$

Jet	500	x	$500x$
Prop	200	$x+2$	$200(x+2)$

$$d = d$$
$$500x = 200(x+2)$$
$$500x = 200x + 400$$
$$500x - 200x = 200x - 200x + 400$$
$$300x = 400$$
$$\frac{300x}{300} = \frac{400}{300}$$
$$x = \frac{4}{3}$$

The jet traveled for $\frac{4}{3}$ hours

$$d = rt$$
$$d = 500\left(\frac{4}{3}\right) = 666\frac{2}{3} \text{ miles}$$

9. Let x = the time to get to Disneyland
and $7.2 - x$ = the time to return

$$Rate \cdot Time = Distance$$

Going	50	x	$50x$
Returning	40	$7.2-x$	$40(7.2-x)$

$$d = d$$
$$50x = 40(7.2 - x)$$
$$50x = 288 - 40x$$
$$50x + 40x = 288 - 40x + 40x$$
$$90x = 288$$
$$\frac{90x}{90} = \frac{288}{90}$$
$$x = 3.2$$

It took 3.2 hours to get to Disneyland

$$d = rt$$
$$d = 50(3.2) = 160 \text{ miles}$$

11. Let x = the amount of pure acid.

$$No. of \ gallons \cdot Strength = Amt \ of \ Acid$$

100%	x	1.00	x
40%	2	0.4	$2(0.4)$
70%	$x+2$	0.7	$0.7(x+2)$

$$x + 2(0.4) = 0.7(x+2)$$
$$x + 0.8 = 0.7x + 1.4$$
$$x - 0.7x + 0.8 = 0.7x - 0.7x + 1.4$$
$$0.3x + 0.8 = 1.4$$
$$0.3x + 0.8 - 0.8 = 1.4 - 0.8$$
$$0.3x = 0.6$$
$$\frac{0.3x}{0.3} = \frac{0.6}{0.3}$$
$$x = 2$$

Mix 2 gallons of pure acid

13. Let x = the pounds of cashew nuts.

$$Cost/lb \cdot No. of \ lbs = Cost$$

Peanuts	3	20	$3(20)$
Cashews	5	x	$5x$
Mix	3.50	$x+20$	$3.50(x+20)$

$$3(20) + 5x = 3.50(x+20)$$
$$60 + 5x = 3.5x + 70$$
$$60 + 5x - 3.5x = 3.5x - 3.5x + 70$$
$$60 + 1.5x = 70$$
$$60 - 60 + 1.5x = 70 - 60$$

$$1.5x = 10$$

$$\frac{1.5x}{1.5} = \frac{10}{1.5}$$

$$x = 6\frac{2}{3}$$

$$x = 2$$

Add $6\frac{2}{3}$ pounds of cashews

15. No. The 30% solution will only dilute the 50% solution, not make it more concentrated.

17. Let x = the amount invested at 9% for one year.

Principal · Rate = Interest

9%	x	.09	$.09x$
8%	$25,000 - x$	.08	$.08(25,000 - x)$
Total	$25,000$		$2,135$

$$.09x + .08(25,000 - x) = 2,135$$

$$.09x + 2,000 - .08x = 2,135$$

$$.01x + 2,000 = 2,135$$

$$.01x + 2,000 - 2,000 = 2,135 - 2,000$$

$$.01x = 135$$

$$\frac{.01x}{.01} = \frac{135}{.01}$$

$$x = 13,500$$

$$25,000 - x = 25,000 - 13,500 = 11,500$$

She invested $11,500 @ 8% and $13,500 @ 9%

19. Let x = the amount invested at 11% for one year.

Principal · Rate = Interest

11%	x	.11	$.11x$
4%	$10,000 - x$	$-.04$	$-.04(10,000 - x)$
Total	$10,000$		650

$$.11x - .04(10,000 - x) = 650$$

$$.11x - 400 + .04x = 650$$

$$.15x - 400 = 650$$

$$.15x - 400 + 400 = 650 + 400$$

$$.15x = 1050$$

$$\frac{.15x}{.15} = \frac{1050}{.15}$$

$$x = 7,000$$

$$10,000 - x = 10,000 - 7,000 = 3,000$$

He invested $7,000 @ 11% and $3,000 @ 4%

21. Let x = the amount paid.

Profit $= .2x$

Amount paid + profit = *s*elling price

$$x + .2x = 4,680$$

$$1.2x = 4,680$$

$$\frac{1.2x}{1.2} = \frac{4,680}{1.2}$$

$$x = 3,900$$

He paid $3,900

23. Let x = the amount invested at 10% for one year.

Principal · Rate = Interest

10%	x	.10	$.10x$
8%	$54,000 - x$	.08	$.08(54,000 - x)$

$$.10x = .08(54,000 - x)$$
$$.10x = 4320 - .08x$$
$$.10x + .08x = 4320 - .08x + .08x$$
$$.18x = 4320$$
$$\frac{.18x}{.18} = \frac{4320}{.18}$$
$$x = 24,000$$
$$54,000 - x = 54,000 - 24,000 = 30,000$$

Invest $30,000 @ 8% and
$24,000 @ 10%

25. Let x = the time they talk

Rate · Time = Distance

Cade	5	x	$5x$
Kathleen	4	x	$4x$
Total			20

$$5x + 4x = 20$$
$$9x = 20$$
$$\frac{9x}{9} = \frac{20}{9}$$
$$x = 2\frac{2}{9}$$

They can talk for $2\frac{2}{9}$ hours

27. Let x = the amount of 20% alloy.

No. of Oz · Strength = Amt of Copper

50%	200	0.5	$200(0.5)$
20%	x	0.2	$0.2x$
Mix	$x + 200$	0.3	$0.3(x + 200)$

$$200(0.5) + 0.2x = 0.3(x + 200)$$
$$100 + 0.2x = 0.3x + 60$$
$$100 + 0.2x - 0.2x = 0.3x - 0.2x + 60$$
$$100 = 0.1x + 60$$
$$100 - 60 = 0.1x + 60 - 60$$
$$40 = 0.1x$$
$$\frac{40}{0.1} = \frac{0.1x}{0.1}$$
$$400 = x$$

Mix with 400 ounces of 20% alloy.

29. Increase = $70 - 40 = 30$

Let x = the percent increase

$$x(40) = 30$$
$$\frac{40x}{40} = \frac{30}{40}$$
$$x = 0.75$$

The percent increase was 75%

31. Let x = the amount invested at 9% for one year.

Principal · Rate = Interest

9%	x	.09	$.09x$
6%	3000	.06	$.06(3000)$
Total			585

$$.09x + .06(3000) = 585$$
$$.09x + 180 = 585$$
$$.09x + 180 - 180 = 585 - 180$$

$$.09x = 405$$

$$\frac{.09x}{.09} = \frac{405}{.09}$$

$$x = 4500$$

Should invest $4500 @ 9%.

33. Let x = the number of cards(in millions) issued in 2001.

Increase = $1.17x$

No. in 2001 + increase = no. in 2003

$$x + 1.17x = 500$$

$$2.17x = 500$$

$$\frac{2.17x}{2.17} = \frac{500}{2.17}$$

$$x = 230$$

230 million cards were issued in 2001

35. Let x = the rate of hiker1

Rate · Time = Distance

Hiker1	x	2	$2x$
Hiker2	$x+1.1$	2	$2(x+1.1)$
Total			11

$$2x + 2(x+1.1) = 11$$

$$2x + 2x + 2.2 = 11$$

$$4x + 2.2 = 11$$

$$4x + 2.2 - 2.2 = 11 - 2.2$$

$$4x = 8.8$$

$$\frac{4x}{4} = \frac{8.8}{4}$$

$$x = 2.2$$

$$x + 1.1 = 2.2 + 1.1 = 3.3$$

Hiker1: 2.2 mph; Hiker2: 3.3 mph.

37. Increase = $21.0 - 20.7 = 0.3$

Let x = the percent increase

$$x(20.7) = 0.3$$

$$\frac{20.7x}{20.7} = \frac{0.3}{20.7}$$

$$x = 0.0145$$

The percent increase was 1.45%

39. Let x = the time it takes them to meet

Rate · Time = Distance

Nedra	3	x	$3x$
Latonya	4	x	$4x$
Total			12

$$3x + 4x = 12$$

$$7x = 12$$

$$\frac{7x}{7} = \frac{12}{7}$$

$$x = 1\frac{5}{7}$$

They meet in $1\frac{5}{7}$ hours

41. $3 + (-7) = -4$

43. $\dfrac{3}{4} - \dfrac{3}{16} = \dfrac{3}{4}\left(\dfrac{4}{4}\right) - \dfrac{3}{16} = \dfrac{12}{16} - \dfrac{3}{16} = \dfrac{9}{16}$

45. $-5 - (-1) = -5 + 1 = -4$

47. $-5 > -7$

49. $\left|-5\right| = -(-5)$

51.
$$R = C$$
$$24x = 100 + 20x$$
$$24x - 20x = 100 + 20x - 20x$$
$$4x = 100$$
$$\frac{4x}{x} = \frac{100}{4}$$
$$x = 25$$

Should sell 25 skateboards to break even.

53.
$$R = C$$
$$7.50x = 4.50x + 2400$$
$$7.50x - 4.50x = 4.50x - 4.50x + 2400$$
$$3x = 2400$$
$$\frac{3x}{3} = \frac{2400}{3}$$
$$x = 800$$

Should sell 800 books to break even.

55. Answers may vary

57. $x(300) = 23$

$x = \dfrac{23}{300} \approx 0.077 = 7.7\%$

This is about 7.7% of the daily value.

59. Let $x =$ percent of calories from fat.

$x(280) = 9(6)$

$x = \dfrac{54}{280} \approx 0.193 = 19.3\%$

One serving contains 19.3% of calories from fat.

61. Answers may vary

Section 2.7

Mental Math

1. $5x > 10$
$x > 2$

2. $4x < 20$
$x < 5$

3. $2x \geq 16$
$x \geq 8$

4. $9x \leq 63$
$x \leq 7$

Exercise Set 2.7

1. $x \geq 2,\ [2, \infty)$

3. $(-\infty, -5),\ x < -5$

5. $x \leq -1,\ (-\infty, -1]$

7. $x < \dfrac{1}{2},\ \left(-\infty, \dfrac{1}{2}\right)$

9. $y \geq 5,\ [5, \infty)$

11. $2x < -6$

$x < -3, \ (-\infty, -3)$

$$\xleftarrow{\hspace{2.5cm}}\overset{)}{\underset{-3}{\vrule height 6pt}}\xrightarrow{\hspace{2.5cm}}$$

13. $x - 2 \geq -7$

$x \geq -5, \ [-5, \infty)$

$$\xleftarrow{\hspace{2.5cm}}\underset{-5}{\vrule height 6pt}\xrightarrow{\hspace{2.5cm}}$$

15. $-8x \leq 16$

$\dfrac{-8x}{-8} \geq \dfrac{16}{-8}$

$x \geq -2, \ [-2, \infty)$

$$\xleftarrow{\hspace{2.5cm}}\underset{-2}{\vrule height 6pt}\xrightarrow{\hspace{2.5cm}}$$

17. $3x - 5 > 2x - 8$

$x - 5 > -8$

$x > -3, \ (-3, \infty)$

$$\xleftarrow{\hspace{2.5cm}}\underset{-3}{(}\xrightarrow{\hspace{2.5cm}}$$

19. $4x - 1 \leq 5x - 2x$

$4x - 1 \leq 3x$

$x - 1 \leq 0$

$x \leq 1, \ (-\infty, 1]$

$$\xleftarrow{\hspace{2.5cm}}\underset{1}{\vrule height 6pt}\xrightarrow{\hspace{2.5cm}}$$

21. $x - 7 < 3(x + 1)$

$x - 7 < 3x + 3$

$-2x - 7 < 3$

$-2x < 10$

$x > -5, \ (-5, \infty)$

$$\xleftarrow{\hspace{2.5cm}}\underset{-5}{(}\xrightarrow{\hspace{2.5cm}}$$

23. $-6x + 2 \leq 2(5 - x)$

$-6x + 2 \leq 10 - 2x$

$-4x + 2 \leq 10$

$-4x \leq 8$

$x \geq -2, \ [-2, \infty)$

$$\xleftarrow{\hspace{2.5cm}}\underset{-2}{\vrule height 6pt}\xrightarrow{\hspace{2.5cm}}$$

25. $4(3x - 1) \leq 5(2x - 4)$

$12x - 4 \leq 10x - 20$

$2x - 4 \leq -20$

$2x \leq -16$

$x \leq -8, \ (-\infty, -8]$

$$\xleftarrow{\hspace{2.5cm}}\underset{-8}{\vrule height 6pt}\xrightarrow{\hspace{2.5cm}}$$

27. $3(x + 2) - 6 > -2(x - 3) + 14$

$3x + 6 - 6 > -2x + 6 + 14$

$3x > -2x + 20$

$5x > 20$

$x > 4, \ (4, \infty)$

$$\xleftarrow{\hspace{2.5cm}}\underset{4}{(}\xrightarrow{\hspace{2.5cm}}$$

29. $-2x \le -40$

$x \ge 20, \; [20, \infty)$

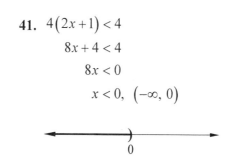

31. $-9 + x > 7$

$x > 16, \; (16, \infty)$

33. $3x - 7 < 6x + 2$

$-3x - 7 < 2$

$-3x < 9$

$x > -3, \; (-3, \infty)$

35. $5x - 7x \ge x + 2$

$-2x \ge x + 2$

$-3x \ge 2$

$x \le -\dfrac{2}{3}, \; \left(-\infty, -\dfrac{2}{3}\right]$

37. $\dfrac{3}{4}x > 2$

$x > \dfrac{8}{3}, \; \left(\dfrac{8}{3}, \infty\right)$

39. $3(x - 5) < 2(2x - 1)$

$3x - 15 < 4x - 2$

$-x - 15 < -2$

$-x < 13$

$x > -13, \; (-13, \infty)$

41. $4(2x + 1) < 4$

$8x + 4 < 4$

$8x < 0$

$x < 0, \; (-\infty, 0)$

43. $-5x + 4 \ge -4(x - 1)$

$-5x + 4 \ge -4x + 4$

$-x + 4 \ge 4$

$-x \ge 0$

$x \le 0, \; (-\infty, 0]$

45. $-2(x - 4) - 3x < -(4x + 1) + 2x$

$-2x + 8 - 3x < -4x - 1 + 2x$

$-5x + 8 < -2x - 1$

$-3x + 8 < -1$

$-3x < -9$

$x > 3, \; (3, \infty)$

Martin-Gay, Beginning and Intermediate Algebra, 3e   **67**

47. $-3x + 6 \geq 2x + 6$

$-5x + 6 \geq 6$

$-5x \geq 0$

$x \leq 0, \ (-\infty, 0]$

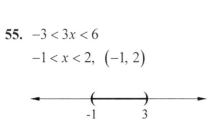

49. Answers may vary

51. $-1 < x < 3, \ (-1, 3)$

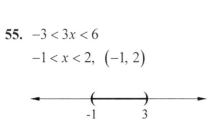

53. $0 \leq y < 2, \ [0, 2)$

55. $-3 < 3x < 6$

$-1 < x < 2, \ (-1, 2)$

57. $2 \leq 3x - 10 \leq 5$

$12 \leq 3x \leq 15$

$4 \leq x \leq 5, \ [4, 5]$

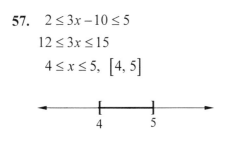

59. $-4 < 2(x - 3) \leq 4$

$-4 < 2x - 6 \leq 4$

$2 < 2x \leq 10$

$1 < x \leq 5, \ (1, 5]$

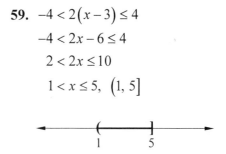

61. $-2 < 3x - 5 < 7$

$3 < 3x < 12$

$1 < x < 4, \ (1, 4)$

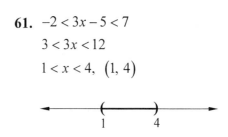

63. $-6 < 3(x - 2) \leq 8$

$-6 < 3x - 6 \leq 8$

$0 < 3x \leq 14$

$0 < x \leq \dfrac{14}{3}, \ \left(0, \dfrac{14}{3}\right]$

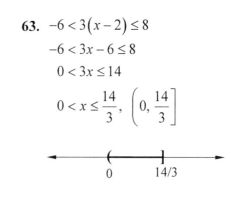

65. Answers may vary

67. $2x + 6 > -14$

$2x > -20$

$x > -10$

69. Let $x =$ the number of people invited.

$34x + 50 \leq 3000$

$34x \leq 2950$

$x \leq 86.8$

They must invite no more than 86 people.

71. Let x = the length.

$$2l + 2w = P$$
$$2x + 2(15) \le 100$$
$$2x + 30 \le 100$$
$$2x \le 70$$
$$x \le 35$$

The length can be no greater than 35 cm..

73. Let x = the rate for $5000 for one year.

Principal · Rate = Interest

11%	10,000	.11	.11(10,000)
?	5000	x	5000x
Total			1600

$$1100 + 5000x \ge 1600$$
$$5000x \ge 500$$
$$x \ge .1$$

Should invest the $5000 at 10% or more.

75. Let x = his score on the third game.

$$\frac{146 + 201 + x}{3} \ge 180$$
$$3\left(\frac{146 + 201 + x}{3}\right) \ge 3(180)$$
$$347 + x \ge 540$$
$$x \ge 193$$

He must score at least 193.

77. $x < 200$　　　　recommended
$200 \le x \le 240$　borderline
$x > 240$　　　　high

79. Let x = the unknown number.

$$-5 < 2x + 1 < 7$$
$$-6 < 2x < 6$$
$$-3 < x < 3$$

All numbers between -3 and 3.

81.
$$C = \frac{5}{9}(F - 32)$$
$$-39 \le \frac{5}{9}(F - 32) \le 45$$
$$-351 \le 5(F - 32) \le 405$$
$$-351 \le 5F - 160 \le 405$$
$$-191 \le 5F \le 565$$
$$-38.2° \le F \le 113°$$

83. $2^3 = (2)(2)(2) = 8$

85. $1^{12} = (1)(1)(1)(1)(1)(1)(1)(1)(1)(1)(1)(1) = 1$

87. $\left(\frac{4}{7}\right)^2 = \left(\frac{4}{7}\right)\left(\frac{4}{7}\right) = \frac{16}{49}$

89. $3400

91. 1986-1990

93.
$$C = 3.14d$$
$$2.9 \le 3.14d \le 3.1$$
$$0.924 \le d \le 0.987$$

The diameter must be between 0.924 cm and 0.987 cm.

95. $x(x+4) > x^2 - 2x + 6$

$x^2 + 4x > x^2 - 2x + 6$

$4x > -2x + 6$

$6x > 6$

$x > 1, \ (1, \infty)$

97. $x^2 + 6x - 10 < x(x-10)$

$x^2 + 6x - 10 < x^2 - 10x$

$6x - 10 < -10x$

$16x - 10 < 0$

$16x < 10$

$x < \dfrac{10}{16} = \dfrac{5}{8}$

$x < \dfrac{5}{8}, \ \left(-\infty, \dfrac{5}{8}\right)$

Chapter 2 Review

1. $5x - x + 2x = 6x$

2. $0.2z - 4.6x - 7.4z = -4.6x - 7.2z$

3. $\dfrac{1}{2}x + 3 + \dfrac{7}{2}x - 5 = \dfrac{8}{2}x - 2 = 4x - 2$

4. $\dfrac{4}{5}y + 1 + \dfrac{6}{5}y + 2 = \dfrac{10}{5}y + 3 = 2y + 3$

5. $2(n-4) + n - 10 = 2n - 8 + n - 10 = 3n - 18$

6. $3(w+2) - (12-w) = 3w + 6 - 12 + w$
$= 4w - 6$

7. $(x+5) - (7x-2) = x + 5 - 7x + 2 = -6x + 7$

8. $(y-0.7) - (1.4y - 3) = y - 0.7 - 1.4y + 3$
$= -0.4y + 2.3$

9. $3x - 7$

10. $3x + 2(x+2.8) = 3x + 2x + 5.6 = 5x + 5.6$

11. $8x + 4 = 9x$
$4 = x$

12. $5y - 3 = 6y$
$-3 = y$

13. $3x - 5 = 4x + 1$
$-5 = x + 1$
$-6 = x$

14. $2x - 6 = x - 6$
$x - 6 = -6$
$x = 0$

15. $4(x+3) = 3(1+x)$
$4x + 12 = 3 + 3x$
$x + 12 = 3$
$x = -9$

16. $6(3+n) = 5(n-1)$
$18 + 6n = 5n - 5$
$18 + n = -5$
$n = -23$

17.
$$x - 5 = 3$$
$$x - 5 + \underline{5} = 3 + \underline{5}$$
$$x = 8$$

18.
$$x + 9 = -2$$
$$x + 9 - \underline{9} = -2 - \underline{9}$$
$$x = -11$$

19. $10 - x$

20. $(x - 5)$ in.

21. $180 - (x + 5) = 180 - x - 5 = (175 - x)^\circ$

22.
$$\frac{3}{4}x = -9$$
$$\frac{4}{3}\left(\frac{3}{4}x\right) = \frac{4}{3}(-9)$$
$$x = -12$$

23.
$$\frac{x}{6} = \frac{2}{3}$$
$$x = 4$$

24.
$$-3x + 1 = 19$$
$$-3x = 18$$
$$x = -6$$

25.
$$5x + 25 = 20$$
$$5x = -5$$
$$x = -1$$

26.
$$5x + x = 9 + 4x - 1 + 6$$
$$6x = 4x + 14$$
$$2x = 14$$
$$x = 7$$

27.
$$-y + 4y = 7 - y - 3 - 8$$
$$3y = -y - 4$$
$$4y = -4$$
$$y = -1$$

28. Let x = the first even integer.
$$\text{Sum } = x + (x + 2) + (x + 4) = 3x + 6$$

29.
$$\frac{2}{7}x - \frac{5}{7} = 1$$
$$7\left(\frac{2}{7}x\right) - 7\left(\frac{5}{7}\right) = 7(1)$$
$$2x - 5 = 7$$
$$2x = 12$$
$$x = 6$$

30.
$$\frac{5}{3}x + 4 = \frac{2}{3}x$$
$$\frac{3}{3}x + 4 = 0$$
$$x = -4$$

31.
$$-(5x + 1) = -7x + 3$$
$$-5x - 1 = -7x + 3$$
$$2x - 1 = 3$$
$$2x = 4$$
$$x = 2$$

32.
$$-4(2x + 1) = -5x + 5$$
$$-8x - 4 = -5x + 5$$
$$-3x - 4 = 5$$
$$-3x = 9$$
$$x = -3$$

33. $-6(2x-5) = -3(9+4x)$

$-12x+30 = -27-12x$

$30 = -27$

There is no solution.

34. $3(8y-1) = 6(5+4y)$

$24y-3 = 30+24y$

$-3 = 30$

There is no solution.

35. $\dfrac{3(2-z)}{5} = z$

$3(2-z) = 5z$

$6-3z = 5z$

$6 = 8z$

$\dfrac{6}{8} = z$

$\dfrac{3}{4} = z$

36. $\dfrac{4(n+2)}{5} = -n$

$4(n+2) = -5n$

$4n+8 = -5n$

$8 = -9n$

$-\dfrac{8}{9} = n$

37. $5(2n-3)-1 = 4(6+2n)$

$10n-15-1 = 24+8n$

$10n-16 = 24+8n$

$2n-16 = 24$

$2n = 40$

$n = 20$

38. $-2(4y-3)+4 = 3(5-y)$

$-8y+6+4 = 15-3y$

$-8y+10 = 15-3y$

$10 = 15+5y$

$-5 = 5y$

$-1 = y$

39. $9z-z+1 = 6(z-1)+7$

$8z+1 = 6z-6+7$

$8z+1 = 6z+1$

$2z+1 = 1$

$2z = 0$

$z = 0$

40. $5t-3-t = 3(t+4)-15$

$4t-3 = 3t+12-15$

$4t-3 = 3t-3$

$t-3 = -3$

$t = 0$

41. $-n+10 = 2(3n-5)$

$-n+10 = 6n-10$

$10 = 7n-10$

$20 = 7n$

$\dfrac{20}{7} = n$

42. $-9-5a = 3(6a-1)$

$-9-5a = 18a-3$

$-9 = 23a-3$

$-6 = 23a$

$-\dfrac{6}{23} = a$

43. $\dfrac{5(c+1)}{6} = 2c - 3$

$5(c+1) = 6(2c-3)$

$5c + 5 = 12c - 18$

$-7c + 5 = -18$

$-7c = -23$

$c = \dfrac{23}{7}$

44. $\dfrac{2(8-a)}{3} = 4 - 4a$

$2(8-a) = 3(4-4a)$

$16 - 2a = 12 - 12a$

$10a + 16 = 12$

$10a = -4$

$a = \dfrac{-4}{10}$

$a = -\dfrac{2}{5}$

45. $200(70x - 3560) = -179(150x - 19,300)$

$14,000x - 712,000 = -26,850x + 3,454,700$

$40,850x - 712,000 = 3,454,700$

$40,850x = 4,166,700$

$x = 102$

46. $1.72y - 0.04y = 0.42$

$1.68y = 0.42$

$y = 0.25$

47. Let x = the unknown number.

$\dfrac{x}{3} = x - 2$

$x = 3(x - 2)$

$x = 3x - 6$

$-2x = -6$

$x = 3$

The number is 3.

48. Let x = the unknown number.

$2(x + 6) = -x$

$2x + 12 = -x$

$12 = -3x$

$-4 = x$

The number is -4.

49. Let x = the length of a side of the base and $3x + 68$ = the height.

$x + 3x + 68 = 1380$

$4x + 68 = 1380$

$4x = 1312$

$x = 328$

$3x + 68 = 3(328) + 68 = 1052$

The height is 1052 ft.

50. Let x = the length of the shorter piece and $2x$ = the length of the other.

$x + 2x = 12$

$3x = 12$

$x = 4$

$2x = 8$

The lengths are 4 feet and 8 feet.

51. Let $x =$ the first area code
and $3x + 34 =$ the other.
$$x + 3x + 34 = 1262$$
$$4x + 34 = 1262$$
$$4x = 1228$$
$$x = 307$$
$$3x + 34 = 3(307) + 34 = 955$$
The codes are 307 and 955.

52. Let $x =$ the first integer, $x + 2 =$ the
second integer, and $x + 4 =$ the third.
$$x + x + 2 + x + 4 = -114$$
$$3x + 6 = -114$$
$$3x = -120$$
$$x = -40$$
$$x + 2 = -38, \text{ and } x + 4 = -36$$
The integers are $-40, -38, -36$.

53. Let $P = 46$ and $l = 14$.
$$P = 2l + 2w$$
$$46 = 2(14) + 2w$$
$$46 = 28 + 2w$$
$$18 = 2w$$
$$9 = w$$

54. Let $V = 192$, $l = 8$, and $w = 6$.
$$V = lwh$$
$$192 = 8(6)h$$
$$192 = 48h$$
$$4 = h$$

55.
$$y = mx + b$$
$$y - b = mx$$
$$\frac{y - b}{x} = m$$

56.
$$r = vst - 9$$
$$r + 9 = vst$$
$$\frac{r + 9}{vt} = s$$

57. $2y - 5x = 7$
$$-5x = -2y + 7$$
$$x = \frac{-2y + 7}{-5} = \frac{2y - 7}{5}$$

58. $3x - 6y = -2$
$$-6y = -3x - 2$$
$$y = \frac{-3x - 2}{-6} = \frac{3x + 2}{6}$$

59. $C = \pi D$
$$\frac{C}{D} = \pi$$

60. $C = 2\pi r$
$$\frac{C}{2r} = \pi$$

61. Let $V = 900$, $l = 20$, and $h = 3$.
$$V = lwh$$
$$900 = 20w(3)$$
$$900 = 60w$$
$$15 = w$$
Width $= 15$ meters

62. Let $F = 104$
$$C = \frac{5}{9}(F - 32) = \frac{5}{9}(104 - 32) = \frac{5}{9}(72)$$
$$= 40$$
The temperature was $40°$ C.

63. Let $d = 10,000$ and $r = 125$

$$d = rt$$
$$10,000 = 125t$$
$$80 = t$$

It will take 1 hour and 20 minutes.

64. Let $x =$ the amount invested at 10.5% for one year.

Principal $\cdot$ *Rate* $=$ *Interest*

10.5%	x	.105	$.105x$
8.5%	$50,000 - x$	.085	$.085(50,000 - x)$
Total	$50,000$		$4,550$

$$.105x + .085(50,000 - x) = 4,550$$
$$.105x + 4,250 - .085x = 4,550$$
$$.02x + 4,250 = 4,550$$
$$.02x = 300$$
$$x = 15,000$$

$50,000 - x = 50,000 - 15,000 = 35,000$

She invested \$35,000 @ 8.5% and \$15,000 @ 10.5%

65. Let $x =$ the number of dimes, $2x =$ the number of quarters, and $500 - x - 2x =$ the number of nickles.

No. of Coins $\cdot$ *Value* $=$ *Amt of Money*

Dimes	x	.1	$.1x$
Quarters	$2x$	.25	$.25(2x)$
Nickles	$500 - 3x$	.05	$.05(500 - 3x)$
Total	500		88

$$.1x + .25(2x) + .05(500 - 3x) = 88$$

$$.1x + .5x + 25 - .15x = 88$$
$$.45x + 25 = 88$$
$$.45x = 63$$
$$x = 140$$

$500 - 3x = 500 - 3(140) = 500 - 420 = 80$

There were 80 nickles in the pay phone.

66. Let $x =$ the time traveled by the Amtrak

Rate $\cdot$ *Time* $=$ *Distance*

Amtrak	60	x	$60x$
Freight	45	$x + 1.5$	$45(x + 1.5)$

$$d = d$$
$$60x = 45(x + 1.5)$$
$$60x = 45x + 67.5$$
$$15x = 67.5$$
$$x = 4.5$$

It will take 4.5 hours

67. Let $x =$ the time to cycle up and $5 - x =$ the time to cycle down

Rate $\cdot$ *Time* $=$ *Distance*

Up	8	x	$8x$
Down	12	$5 - x$	$12(5 - x)$

$$d = d$$
$$8x = 12(5 - x)$$
$$8x = 60 - 12x$$
$$20x = 60$$
$$x = 3$$

It took 3 hours to cycle up.

$$d = rt$$
$$d = 8(3) = 24 \text{ miles up.}$$

Total distance was 48 miles.

68. $x \le -2$, $(-\infty, -2]$

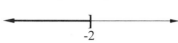

69. $x > 0$, $(0, \infty)$

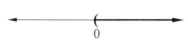

70. $-1 < x < 1$, $(-1, 1)$

71. $0.5 \le y < 1.5$, $[0.5, 1.5)$

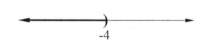

72. $-2x \ge -20$

$$\frac{-2x}{-2} \le \frac{-20}{-2}$$

$$x \le 10, \ (-\infty, 10]$$

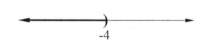

73. $-3x > 12$

$$\frac{-3x}{-3} < \frac{12}{-3}$$

$$x < -4, \ (-\infty, -4)$$

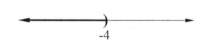

74. $5x - 7 > 8x + 5$

$$-3x - 7 > 5$$

$$-3x > 12$$

$$\frac{-3x}{-3} < \frac{12}{-3}$$

$$x < -4, \ (-\infty, -4)$$

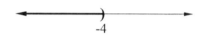

75. $x + 4 \ge 6x - 16$

$$-5x + 4 \ge -16$$

$$-5x \ge -20$$

$$\frac{-5x}{-5} \le \frac{-20}{-5}$$

$$x \le 4, \ (-\infty, 4]$$

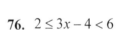

76. $2 \le 3x - 4 < 6$

$$6 \le 3x < 10$$

$$2 \le x < \frac{10}{3}$$

$$2 \le x < \frac{10}{3}, \ \left[2, \frac{10}{3}\right)$$

77. $-3 < 4x - 1 < 2$

$$-2 < 4x < 3$$

$$-\frac{1}{2} < x < \frac{3}{4}, \ \left(-\frac{1}{2}, \frac{3}{4}\right)$$

78. $-2(x-5) > 2(3x-2)$

$$-2x+10 > 6x-4$$
$$-8x+10 > -4$$
$$-8x > -14$$
$$\frac{-8x}{-8} < \frac{-14}{-8}$$
$$x < \frac{7}{4}, \quad \left(-\infty, \frac{7}{4}\right)$$

7/4

79. $4(2x-5) \le 5x-1$

$$8x-20 \le 5x-1$$
$$3x-20 \le -1$$
$$3x \le 19$$
$$x \le \frac{19}{3}, \quad \left(-\infty, \frac{19}{3}\right]$$

19/3

80. Let x = the amount of sales then
$0.05x$ = her commission.
$$175 + 0.05x \ge 300$$
$$0.05x \ge 125$$
$$x \ge 2500$$
Sales must be at least $2500.

81. Let x = her score on the fourth round.
$$\frac{76+82+79+x}{4} < 80$$
$$237 + x < 320$$
$$x < 83$$
Her score must be less than 83.

Chapter 2 Test

1. $2y-6-y-4 = y-10$

2. $2.7x+6.1+3.2x-4.9 = 5.9x+1.2$

3. $4(x-2)-3(2x-6) = 4x-8-6x+18$
$$= -2x+10$$

4. $7+2(5y-3) = 7+10y-6$
$$= 10y+1$$

5. $-\dfrac{4}{5}x = 4$
$$x = -5$$

6. $4(n-5) = -(4-2n)$
$$4n-20 = -4+2n$$
$$2n-20 = -4$$
$$2n = 16$$
$$n = 8$$

7. $5y-7+y = -(y+3y)$
$$6y-7 = -4y$$
$$-7 = -10y$$
$$\frac{7}{10} = y$$

8. $4z+1-z = 1+z$
$$3z+1 = 1+z$$
$$2z+1 = 1$$
$$2z = 0$$
$$z = 0$$

9. $\dfrac{2(x+6)}{3} = x - 5$

$2(x+6) = 3(x-5)$

$2x + 12 = 3x - 15$

$12 = x - 15$

$27 = x$

10. $\dfrac{1}{2} - x + \dfrac{3}{2} = x - 4$

$2\left(\dfrac{1}{2} - x + \dfrac{3}{2}\right) = 2(x-4)$

$1 - 2x + 3 = 2x - 8$

$-2x + 4 = 2x - 8$

$-4x + 4 = -8$

$-4x = -12$

$x = 3$

11. $-0.3(x-4) + x = 0.5(3-x)$

$10\left[-0.3(x-4) + x\right] = 10\left[0.5(3-x)\right]$

$-3(x-4) + 10x = 5(3-x)$

$-3x + 12 + 10x = 15 - 5x$

$7x + 12 = 15 - 5x$

$12x + 12 = 15$

$12x = 3$

$x = \dfrac{3}{12} = \dfrac{1}{4} = 0.25$

12. $-4(a+1) - 3a = -7(2a-3)$

$-4a - 4 - 3a = -14a + 21$

$-7a - 4 = -14a + 21$

$7a - 4 = 21$

$7a = 25$

$a = \dfrac{25}{7}$

13. $-2(x-3) = x + 5 - 3x$

$-2x + 6 = -2x + 5$

$6 = 5$ a contradiction

There is no solution.

14. Let $x =$ the number.

$x + \dfrac{2}{3}x = 35$

$3x + 2x = 105$

$5x = 105$

$x = 21$

The number is 21.

15. Let $l = 35$, and $w = 20$.

$2A = 2lw = 2(35)(20) = 1400$

Let $x =$ the number of gallons needed at 200 square feet per gallon.

$1400 = 200x$

$7 = x$

7 gallons are needed.

16. Let $x =$ the amount invested at 10% for one year.

$Principal \cdot Rate = Interest$

10%	x	.10	$.1x$
12%	$2x$	.12	$.12(2x)$
Total			2890

$.1x + .12(2x) = 2890$

$.1x + .24x = 2890$

$.34x = 2890$

$x = 8500$

$2x = 2(8500) = 17,000$

He invested $8500 @10% and $17,000 @ 12%.

17. Let x = the time they travel

$$Rate \cdot Time = Distance$$

Train1	50	x	$50x$
Train2	64	x	$64x$
Total			285

$$50x + 64x = 285$$
$$114x = 285$$
$$x = 2\frac{1}{2}$$

They must travel for $2\frac{1}{2}$ hours

18. Let $y = -14$, $m = -2$, and $b = -2$.

$$y = mx + b$$
$$-14 = -2x - 2$$
$$-12 = -2x$$
$$6 = x$$

19. $V = \pi r^2 h$

$$\frac{V}{\pi r^2} = \frac{\pi r^2 h}{\pi r^2}$$
$$\frac{V}{\pi r^2} = h$$

20. $3x - 4y = 10$

$$-4y = -3x + 10$$
$$y = \frac{-3x + 10}{-4}$$
$$y = \frac{3x - 10}{4}$$

21. $3x - 5 \geq 7x + 3$

$$-4x - 5 \geq 3$$
$$-4x \geq 8$$
$$\frac{-4x}{-4} \leq \frac{8}{-4}$$
$$x \leq -2, \ \left(-\infty, -2\right]$$

22. $x + 6 > 4x - 6$

$$-3x + 6 > -6$$
$$-3x > -12$$
$$\frac{-3x}{-3} < \frac{-12}{-3}$$
$$x < 4, \ \left(-\infty, 4\right)$$

23. $-2 < 3x + 1 < 8$

$$-3 < 3x < 7$$
$$-1 < x < \frac{7}{3}, \ \left(-1, \frac{7}{3}\right)$$

24. $\dfrac{2(5x + 1)}{3} > 2$

$$2(5x + 1) > 6$$
$$10x + 2 > 6$$
$$10x > 4$$
$$x > \frac{4}{10} = \frac{2}{5}, \ \left(\frac{2}{5}, \infty\right)$$

Chapter 2 Cumulative Review

1. **a.** The natural numbers are 11 and 112.
    b. The whole numbers are 0, 11 and 112.
    c. The integers are $-3, -2, 0, 11$ and 112.
    d. The rational numbers are $-3, -2, 0, \frac{1}{4}, 11$ and 112.
    e. The irrational number is $\sqrt{2}$.
    f. The real numbers are all numbers in the given set.

2. **a.** The natural numbers are 2, 7 and 8.
    b. The whole numbers are 0, 2, 7 and 8.
    c. The integers are $-185, 0, 2, 7$ and 8.
    d. The rational numbers are $-185, -\frac{1}{5}, 0, 2, 7$ and 8.
    e. The irrational number is $\sqrt{3}$.
    f. The real numbers are all numbers in the given set.

3. **a.** $|4| = 4$ **b.** $|-5| = 5$ **c.** $|0| = 0$

4. **a.** $|5| = 5$ **b.** $|-8| = 8$ **c.** $|-2/3| = 2/3$

5. **a.** $40 = 2 \cdot 2 \cdot 2 \cdot 5$ **b.** $63 = 3 \cdot 3 \cdot 7$

6. **a.** $44 = 2 \cdot 2 \cdot 11$ **b.** $90 = 2 \cdot 3 \cdot 3 \cdot 5$

7. $\dfrac{2}{5} = \dfrac{2}{5} \cdot \dfrac{4}{4} = \dfrac{8}{20}$

8. $\dfrac{2}{3} = \dfrac{2}{3} \cdot \dfrac{8}{8} = \dfrac{16}{24}$

9. $\begin{aligned}
3\left[4(5+2)-10\right] &= 3\left[4(7)-10\right] \\
&= 3\left[28-10\right] \\
&= 3\left[18\right] \\
&= 54
\end{aligned}$

10. $\begin{aligned}
5\left[16-4(2+1)\right] &= 5\left[16-4(3)\right] \\
&= 5\left[16-12\right] \\
&= 5\left[4\right] \\
&= 20
\end{aligned}$

11. Let $x = 2$.
$$3x + 10 = 8x$$
$$3(2) + 10 \overset{?}{=} 8(2)$$
$$6 + 10 \overset{?}{=} 16$$
$$16 = 16$$
2 is a solution of the equation.

12. Let $x = 3$.
$$5x - 2 = 4x$$
$$5(3) - 2 \overset{?}{=} 4(3)$$
$$15 - 2 \overset{?}{=} 12$$
$$13 \neq 12$$
3 is not a solution of the equation.

13. $-1 + (-2) = -3$

14. $(-2) + (-8) = -10$

15. $-4 + 6 = 2$

16. $-3 + 10 = 7$

17. a. $-(-10) = 10$ **b.** $-\left(-\dfrac{1}{2}\right) = \dfrac{1}{2}$

     c. $-(-2x) = 2x$ **d.** $-|-6| = -(6) = -6$

18. a. $-(-5) = 5$ **b.** $-\left(-\dfrac{2}{3}\right) = \dfrac{2}{3}$

     c. $-(-a) = a$ **d.** $-|-3| = -(3) = -3$

19. a. $5.3 - (-4.6) = 5.3 + 4.6 = 9.9$

     b. $-\dfrac{3}{10} - \dfrac{5}{10} = -\dfrac{3}{10} + \left(-\dfrac{5}{10}\right) = \dfrac{-3-5}{10}$

$$= -\dfrac{8}{10} = -\dfrac{4}{5}$$

     c. $-\dfrac{2}{3} - \left(-\dfrac{4}{5}\right) = -\dfrac{2}{3} \cdot \dfrac{5}{5} + \dfrac{4}{5} \cdot \dfrac{3}{3}$

$$= -\dfrac{10}{15} + \dfrac{12}{15} = \dfrac{2}{15}$$

20. a. $-2.7 - 8.4 = -2.7 + (-8.4) = -11.1$

     b. $-\dfrac{4}{5} - \left(-\dfrac{3}{5}\right) = -\dfrac{4}{5} + \dfrac{3}{5} = \dfrac{-4+3}{5} = -\dfrac{1}{5}$

     c. $\dfrac{1}{4} - \left(-\dfrac{1}{2}\right) = \dfrac{1}{4} + \dfrac{1}{2} \cdot \dfrac{2}{2} = \dfrac{1}{4} + \dfrac{2}{4} = \dfrac{3}{4}$

21. a. $x = 90 - 38 = 90 + (-38) = 52$

     The complementary angle is $52°$

     b. $y = 180 - 62 = 180 + (-62) = 118$

     The supplementary angle is $118°$

22. a. $x = 90 - 72 = 90 + (-72) = 18$

     The complementary angle is $18°$

     b. $y = 180 - 47 = 180 + (-47) = 133$

     The supplementary angle is $133°$

23. a. $(-1.2)(0.05) = -0.06$

     b. $\dfrac{2}{3}\left(-\dfrac{7}{10}\right) = -\dfrac{2 \cdot 7}{3 \cdot 10} = -\dfrac{14}{30} = -\dfrac{7}{15}$

24. a. $(4.5)(-0.08) = -0.36$

     b. $-\dfrac{3}{4}\left(-\dfrac{8}{17}\right) = \dfrac{3 \cdot 8}{4 \cdot 17} = \dfrac{24}{68} = \dfrac{6}{17}$

25. a. $\dfrac{-24}{-4} = 6$ **b.** $\dfrac{-36}{3} = -12$

     c. $\dfrac{2}{3} \div \left(-\dfrac{5}{4}\right) = \dfrac{2}{3}\left(-\dfrac{4}{5}\right) = -\dfrac{8}{15}$

26. a. $\dfrac{-32}{8} = -4$ **b.** $\dfrac{-108}{-12} = 9$

     c. $-\dfrac{5}{7} \div \left(-\dfrac{9}{2}\right) = -\dfrac{5}{7}\left(-\dfrac{2}{9}\right) = \dfrac{10}{63}$

27. a. $x + 5 = 5 + x$

     b. $3 \cdot x = x \cdot 3$

28. a. $y + 1 = 1 + y$

     b. $y \cdot 4 = 4 \cdot y$

29. a. $8 \cdot 2 + 8 \cdot x = 8(2 + x)$

     b. $7s + 7t = 7(s + t)$

30. a. $4 \cdot y + 4 \cdot \frac{1}{3} = 4\left(y + \frac{1}{3}\right)$

     b. $0.10x + 0.10y = 0.10(x + y)$

31. $(2x - 3) - (4x - 2) = 2x - 3 - 4x + 2$

$$= -2x - 1$$

32. $(-5x + 1) - (10x + 3) = -5x + 1 - 10x - 3$

$$= -15x - 2$$

33.
$$y + 0.6 = -1.0$$
$$y + 0.6 - 0.6 = -1.0 - 0.6$$
$$y = -1.6$$

Check: $y + 0.6 = -1.0$
$$-1.6 + 0.6 \overset{?}{=} -1.0$$
$$-1.0 = -1.0$$
The solution is -1.6.

34.
$$z - 0.4 = 1.5$$
$$z - 0.4 + 0.4 = 1.5 + 0.4$$
$$z = 1.9$$

Check: $z - 0.4 = 1.5$
$$1.9 - 0.4 \overset{?}{=} 1.5$$
$$1.5 = 1.5$$
The solution is 1.9.

35.
$$7 = -5(2a - 1) - (-11a + 6)$$
$$7 = -10a + 5 + 11a - 6$$
$$7 = a - 1$$
$$7 + 1 = a - 1 + 1$$
$$8 = a$$

Check: $7 = -5(2a - 1) - (-11a + 6)$
$$7 \overset{?}{=} -5[2(8) - 1] - [-11(8) + 6]$$
$$7 \overset{?}{=} -5(15) - (-82)$$
$$7 \overset{?}{=} -75 + 82$$
$$7 = 7$$
The solution is 8.

36.
$$-3x + 1 - (-4x - 6) = 10$$
$$-3x + 1 + 4x + 6 = 10$$
$$x + 7 = 10$$
$$x = 3$$

37.
$$\frac{y}{7} = 20$$
$$y = 140$$

38.
$$\frac{x}{4} = 18$$
$$x = 72$$

39.
$$4(2x - 3) + 7 = 3x + 5$$
$$8x - 12 + 7 = 3x + 5$$
$$8x - 5 = 3x + 5$$
$$5x - 5 = 5$$
$$5x = 10$$
$$x = 2$$

40.
$$6x + 5 = 4(x + 4) - 1$$
$$6x + 5 = 4x + 16 - 1$$
$$6x + 5 = 4x + 15$$
$$2x + 5 = 15$$
$$2x = 10$$
$$x = 5$$

41. Let x = a number.
$$2(x + 4) = 4x - 12$$
$$2x + 8 = 4x - 12$$
$$8 = 2x - 12$$
$$20 = 2x$$
$$10 = x$$
The number is 10.

42.　　Let $x =$ a number.

$$x + 4 = 3x - 8$$
$$4 = 2x - 8$$
$$12 = 2x$$
$$6 = x$$

The number is 6.

43.　$V = lwh$

$$\frac{V}{wh} = \frac{lwh}{wh}$$

$$\frac{V}{wh} = l$$

44.　$C = 2\pi r$

$$\frac{C}{2\pi} = \frac{2\pi r}{2\pi}$$

$$\frac{C}{2\pi} = r$$

45.　$x + 4 \le -6$

$$x \le -10, \quad \left(-\infty, -10\right]$$

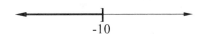

46.　$x - 3 > 2$

$$x > 5, \quad \left(5, \infty\right)$$

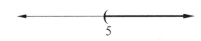

Chapter 3

Section 3.1

Mental Math

1. $x + y = 10$

   Answers may vary; Ex. $(5,5),(7,3)$

2. $x + y = 6$

   Answers may vary; Ex. $(0,6),(6,0)$

3. $x = 3$

   Answers may vary; Ex. $(3,5),(3,0)$

4. $y = -2$

   Answers may vary; Ex. $(0,-2),(1,-2)$

Exercise Set 3.1

1. France

3. France, U.S., Spain

5. 39 million

7. 20

9. 1985

11. 1997

13. 63 million

15. 1900

17. 27 million

19. Answers may vary

Use the following graph for Exercises 21, 23, 25, 27, 29, 31.

21-31.

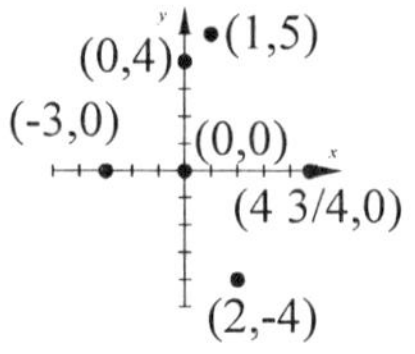

(0,4), •(1,5), (-3,0), (0,0), (4 3/4,0), (2,-4)

21. Point $(1, 5)$ lies in quadrant I.

23. Point $(-3, 0)$ lies on the x-axis.

25. Point $(2, -4)$ lies in quadrant IV.

27. Point $\left(4\frac{3}{4}, 0\right)$ lies on the x-axis.

29. Point $(0, 0)$ lies on the origin.

31. Point $(0, 4)$ lies on the y-axis.

33. $A:(0,0)$, $B:\left(3\frac{1}{2},0\right)$, $C:(3,2)$,
   $D:(-1,3)$, $E:(-2,-2)$, $F:(0,-1)$,
   $G:(2,-1)$

84   Martin-Gay, Beginning and Intermediate Algebra, 3e

35.

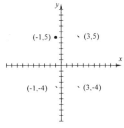

Rectangle is 9 units by 4 units.

Perimeter is $9 + 4 + 9 + 4 = 26$ units.

37.a. $(2313, 2), (2085, 1), (2711, 21), (2869, 39),$
$(2920, 42), (4038, 99), (1783, 0), (2493, 9)$

b.

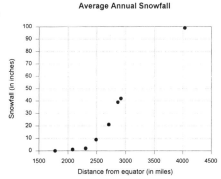

c. The farther from the equator,
the more snowfall.

39.
$$y = -5x$$
$(-1, -5)$
No
$$(-5) \overset{?}{=} -5(-1)$$
$$-5 \neq 5$$

$$y = -5x$$
$(0, 0)$
Yes
$$(0) \overset{?}{=} -5(0)$$
$$0 = 0$$

$$y = -5x$$
$(2, -10)$
Yes
$$(-10) \overset{?}{=} -5(2)$$
$$-10 = -10$$

41.
$$x = 5$$
$(4, 5)$
No
$$(4) \overset{?}{=} 5$$
$$4 \neq 5$$

$$x = 5$$
$(5, 4)$
Yes
$$(5) \overset{?}{=} 5$$
$$5 = 5$$

$$x = 5$$
$(5, 0)$
Yes
$$(5) \overset{?}{=} 5$$
$$5 = 5$$

43.
$$2x - y = 11$$
$(3, -4)$
No
$$2(3) - (-4) \overset{?}{=} 11$$
$$10 \neq 11$$

$$2x - y = 11$$
$(9, 8)$
No
$$2(9) - (8) \overset{?}{=} 11$$
$$10 \neq 11$$

45.
$$x = \frac{1}{3}y$$
$(0, 0)$
Yes
$$0 \overset{?}{=} \frac{1}{3}(0)$$
$$0 = 0$$

$$x = \frac{1}{3}y$$

$(3,9)$ $3 \overset{?}{=} \frac{1}{3}(9)$

Yes

$$3 = 3$$

47. $x - 4y = 4$

$$y = -2, \; x - 4(-2) = 4$$
$$x + 8 = 4$$
$$x = -4; \; (-4, -2)$$
$$x = 4, \; 4 - 4y = 4$$
$$-4y = 0$$
$$y = 0; \; (4, 0)$$

49. $y = -7$

$$x = 11, \; y = -7; \; (11, -7)$$
$$y = -7, \; x = \text{any value}$$

51. $-2x + 7y = -3$

$$y = 1, \; -2x + 7(1) = -3$$
$$x = 5; \; (5, 1)$$
$$x = 1, \; -2(1) + 7y = -3$$
$$y = -\frac{1}{7}; \; \left(1, -\frac{1}{7}\right)$$

53. $2x + 7y = 5$

$$x = 0, \; 2(0) + 7y = 5,$$
$$7y = 5$$
$$y = \frac{5}{7}; \; \left(0, \frac{5}{7}\right)$$

$$y = 0, \; 2x + 7(0) = 5$$
$$2x = 5$$
$$x = \frac{5}{2}; \; \left(\frac{5}{2}, 0\right)$$
$$y = 1, \; 2x + 7(1) = 5,$$
$$2x + 7 = 5$$
$$2x = -2$$
$$x = -1; \; (-1, 1)$$

x	y
0	$\frac{5}{7}$
$\frac{5}{2}$	0
-1	1

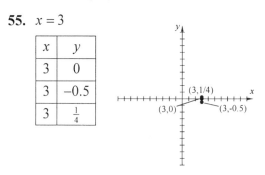

55. $x = 3$

x	y
3	0
3	-0.5
3	$\frac{1}{4}$

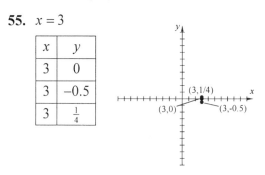

57. $x = -5y$

$$y = 0, \; x = -5(0) = 0$$
$$y = 1, \; x = -5(1) = -5$$
$$x = 10, \; 10 = -5y, \; y = -2$$

x	y
0	0
-5	1
10	-2

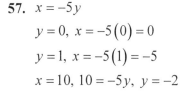

59. Answers may vary.

61. a. $y = 80x + 5000$

$x = 100, \ y = 80(100) + 5000 = 13,000$

$x = 200, \ y = 80(200) + 5000 = 21,000$

$x = 300, \ y = 80(300) + 5000 = 29,000$

x	100	200	300
y	13,000	21,000	29,000

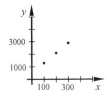

b. Let $y = 8600$

$8600 = 80x + 5000$

$3600 = 80x$

$45 = x$

45 desks can be produced

63. In 1995, there were 670 Target stores.

65. year 6: 66 stores; year 7: 60 stores; year 8: 55 stores

67. When $a = b$.

69. $10x = -5y$

$-2x = y$

71. $x - 3y = 6$

$-3y = -x + 6$

$y = \dfrac{1}{3}x - 2$

73. Quadrant IV

75. Quadrants II or III

77.

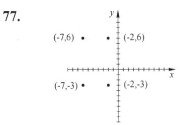

a. The fourth vertex is $(-2, 6)$
The rectangle is 9 units by 5 units.

b. The perimeter is $9 + 5 + 9 + 5 = 28$ units.

c. The area is $9 \times 5 = 45$ square units.

79. 2001, 2002, 2003

81. 46 million

83. $30°$ north, $90°$ west

85. $40°$ north, $104°$ west

Section 3.2

Graphing Calculator Explorations

1. $y = -3x + 7$

3. $y = 2.5x - 7.9$

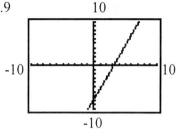

5. $y = -\dfrac{3}{10}x + \dfrac{32}{5}$

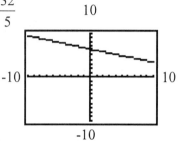

Exercise Set 3.2

1. Yes; It can be written in the form
$Ax + By = C$.

3. Yes; It can be written in the form
$Ax + By = C$.

5. No; x is squared.

7. Yes; It can be written in the form
$Ax + By = C$.

9. $x + y = 4$

x	y
-2	6
0	4
2	2

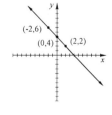

11. $x - y = -2$

x	y
-2	0
0	2
2	4

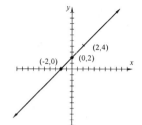

13. $x - 2y = 6$

x	y
-4	-5
0	-3
4	-1

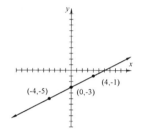

15. $y = 6x + 3$

x	y
-1	-3
0	3
1	9

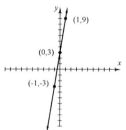

17. $x - 2y = -6$

x	y
-4	1
0	3
4	5

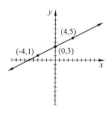

19. $y = 6x$

x	y
-1	-6
0	0
1	6

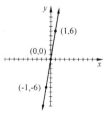

21. $3y - 10 = 5x$

x	y
-4	$-\dfrac{10}{3}$
0	$\dfrac{10}{3}$
1	5

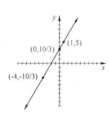

23. $x + 3y = 9$

x	y
-9	6
0	3
3	2

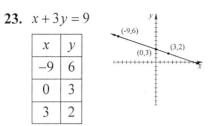

25. $y - x = -1$

x	y
-4	-5
0	-1
4	3

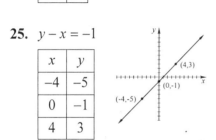

27. $x = -3y$

x	y
-6	2
0	0
6	-2

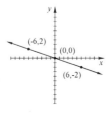

29. $5x - y = 10$

x	y
1	-5
2	0
3	5

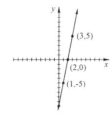

31. $y = \dfrac{1}{2}x + 2$

x	y
-4	0
0	2
4	4

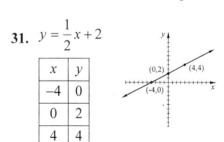

33. $y = 5x$ $y = 5x + 4$

x	y
-1	-5
0	0
1	5

x	y
-1	-1
0	4
1	9

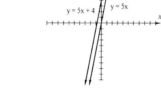

35. $y = -2x$ $y = -2x - 3$

x	y
-2	4
0	0
2	-4

x	y
-2	1
0	-3
2	-7

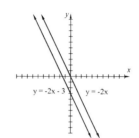

Martin-Gay, Beginning and Intermediate Algebra, 3e   **89**

37. $y = \dfrac{1}{2}x$ $y = \dfrac{1}{2}x + 2$

x	y
-4	-2
0	0
4	2

x	y
-4	0
0	2
4	4

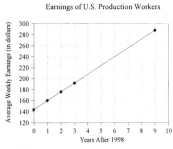

Answers may vary.

39. c

41. d

43. $y = 180x + 450$

In 2009, $x = 13$

$y = 180(13) + 450$

$y = 2790$

The total sales in 2009
should be $2790 million.

45.

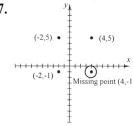

$y = 16x + 144$

The average weekly earnings in 2007
should be $288.

47.

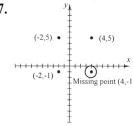

49. $3(x - 2) + 5x = 6x - 16$

$3x - 6 + 5x = 6x - 16$

$8x - 6 = 6x - 16$

$2x - 6 = -16$

$2x = -10$

$x = -5$

51. $3x + \dfrac{2}{5} = \dfrac{1}{10}$

$10(3x) + 10\left(\dfrac{2}{5}\right) = 10\left(\dfrac{1}{10}\right)$

$30x + 4 = 1$

$30x = -3$

$x = -\dfrac{1}{10}$

53. $x - y = -3$

$y = 0, \ x - 0 = -3, \ x = -3$

$x = 0, \ 0 - y = -3, \ y = 3$

x	y
0	3
-3	0

55. $y = 2x$

$y = 0,\ 0 = 2x,\ 0 = x$

$x = 0,\ y = 2(0) = 0$

x	y
0	0
0	0

57. $y = x + 5$

x	y
−3	2
0	5
3	8

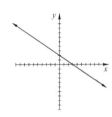

59. $2x + 3y = 6$

x	y
0	2
3	0

61. Answers may vary.

63. If (a, b) is a solution of $x + y = 5$, then (b, a) is also a solution. Explanations may vary.

65. $y = x^2$

x	y
0	0
1	1
−1	1
2	4
−2	4

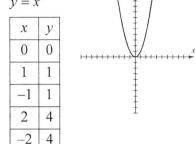

Section 3.3

Graphing Calculator Explorations

1. $x = 3.78y$

$y = \dfrac{x}{3.78}$

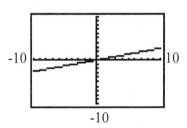

3. $3x + 7y = 21$

$7y = -3x + 21$

$y = -\dfrac{3}{7}x + 3$

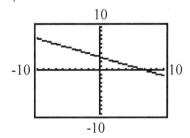

5. $-2.2x + 6.8y = 15.5$

$6.8y = 2.2x + 15.5$

$y = \dfrac{2.2}{6.8}x + \dfrac{15.5}{6.8}$

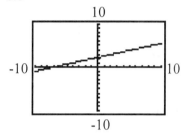

7. Infinite

9. 0

11. $x - y = 3$

$y = 0,\ x - 0 = 3,\ x = 3$

$x = 0,\ 0 - y = 3,\ y = -3$

x-intercept: $(3, 0)$; y-intercept: $(0, -3)$

x	y
3	0
0	−3

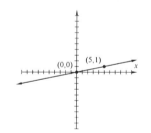

Mental Math

1. False

2. False

3. True

4. True

5. False

6. True

Exercise Set 3.3

1. x-intercept: $(-1, 0)$; y-intercept: $(0, 1)$

3. x-intercept: $(-2, 0)$

5. x-intercepts: $(-1, 0), (1, 0)$

y-intercept: $(0, 1), (0, -2)$

13. $x = 5y$

$y = 0,\ x = 5(0) = 0$

$x = 0,\ 0 = 5y,\ y = 0$

x-intercept: $(0, 0)$; y-intercept: $(0, 0)$

$y = 1,\ x = 5(1) = 5$

x	y
0	0
5	1

15. $-x + 2y = 6$

$y = 0, \; -x + 2(0) = 6, \; x = -6$

$x = 0, \; -0 + 2y = 6, \; y = 3$

x-intercept: $(-6, 0)$; y-intercept: $(0, 3)$

x	y
-6	0
0	3

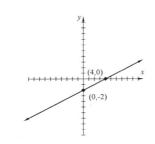

17. $2x - 4y = 8$

$y = 0, \; 2x - 4(0) = 8, \; x = 4$

$x = 0, \; 2(0) - 4y = 8, \; y = -2$

x-intercept: $(4, 0)$; y-intercept: $(0, -2)$

x	y
4	0
0	-2

19. $x = -1$

for all values of y.

21. $y = 0$

for all values of x.

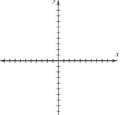

23. $y + 7 = 0$

$\quad\quad y = -7$

$\quad\quad$ for all values of x.

25. $x + 2y = 8$

x-intercept: $(8, 0)$; y-intercept: $(0, 4)$

x	y
0	4
8	0

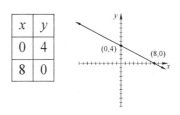

27. $x - 7 = 3y$

x-intercept: $(7, 0)$; y-intercept: $\left(0, -\dfrac{7}{3}\right)$

x	y
0	$-\dfrac{7}{3}$
7	0

29. $x = -3$

for all values of y.

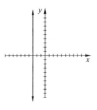

31. $3x + 5y = 7$

x-intercept: $\left(\dfrac{7}{3}, 0\right)$; y-intercept: $\left(0, \dfrac{7}{5}\right)$

x	y
0	$\dfrac{7}{5}$
$\dfrac{7}{3}$	0

33. $x = y$

x-intercept: $(0,0)$; y-intercept: $(0,0)$

Second point $(4,4)$

x	y
4	4
0	0

35. $x + 8y = 8$

x-intercept: $(8,0)$; y-intercept: $(0,1)$

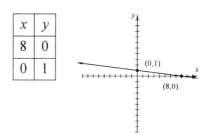

x	y
8	0
0	1

37. $5 = 6x - y$

x-intercept: $\left(\dfrac{5}{6}, 0\right)$; y-intercept: $(0, -5)$

x	y
$\dfrac{5}{6}$	0
0	-5

39. $-x + 10y = 11$

x-intercept: $(-11, 0)$; y-intercept: $\left(0, \dfrac{11}{10}\right)$

x	y
-11	0
0	$\dfrac{11}{10}$

41. $y = 4.5$
for all values of *x*.

43. $y = \dfrac{1}{2}x$

     x-intercept:$(0,0)$; *y*-intercept:$(0,0)$

     Second point, $(6,3)$

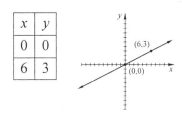

x	y
0	0
6	3

45. $x + 4 = 0$

       $x = -4$

     for all values of *y*.

47. $3x - 4y = -12$

     x-intercept: $(-4,0)$; *y*-intercept: $(0,3)$

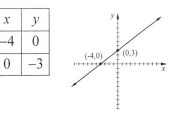

x	y
-4	0
0	-3

49. $2x + 3y = 6$

     x-intercept: $(3,0)$; *y*-intercept: $(0,2)$

x	y
3	0
0	2

51. $y = 3$

     C

53. $x = -1$

     E

55. $y = 2x + 3$

     B

57. $\dfrac{-6-3}{2-8} = \dfrac{-9}{-6} = \dfrac{3}{2}$

59. $\dfrac{-8-(-2)}{-3-(-2)} = \dfrac{-6}{-1} = 6$

61. $\dfrac{0-6}{5-0} = \dfrac{-6}{5} = -\dfrac{6}{5}$

63. $y = 78.1x + 569.9$

     a. $(0, 569.9)$

     b. In 1999, the average price of a digital
        camera was \$569.90.

65. $y = -37.2x + 264.4$

    a. $y = 0,$ $0 = -37.2x + 264.4$

                 $37.2x = 264.4$

                     $x = 7.1$

     $(7.1, 0).$

    b. 7.1 years after 1995 $(2002+)$

        no music cassettes will be shipped.

    c. Answers may vary.

67. $3x + 6y = 1200$

    a. $x = 0,\ 3(0) + 6y = 1200,\ y = 200$

       $(0, 200)$ corresponds to no chairs

         and 200 desks being manufactured.

    b. $y = 0,\ 3x + 6(0) = 1200,\ x = 400$

       $(400, 0)$ corresponds to 400 chairs

         and no desks being manufactured.

    c.

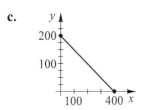

    d. $y = 50,\ 3x + 6(50) = 1200$

              $3x + 300 = 1200$

                  $3x = 900$

                   $x = 300$

       300 chairs can be made.

69. Parallel to $y = -1$ is horizontal.

     y-intercept is $(0, -4)$, so $y = -4$

     for all values of x. $y = -4$

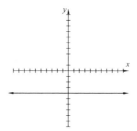

71. Answers may vary

73. Answers may vary

Section 3.4

Graphing Calculator Explorations

1. $y_1 = 3.8x,\ y_2 = 3.8x - 3,\ y_3 = 3.8x + 9$

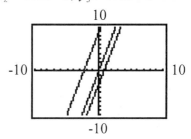

3. $y_1 = \dfrac{1}{4}x,\ y_2 = \dfrac{1}{4}x + 5,\ y_3 = \dfrac{1}{4}x - 8$

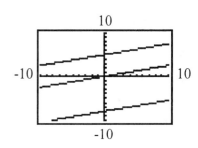

Mental Math

1. Upward

2. Downward

3. Horizontal

4. Vertical

Exercise Set 3.4

1. $(-1, 2)$ and $(2, -2)$

$$m = \frac{y_2 - y_1}{x_2 - x_1} = \frac{-2 - 2}{2 - (-1)} = -\frac{4}{3}$$

3. $(2, -1)$ and $(2, 3)$

$$m = \frac{y_2 - y_1}{x_2 - x_1} = \frac{3 - (-1)}{2 - 2} = \frac{4}{0} \text{ is undefined.}$$

5. $(-3, -2)$ and $(-1, 3)$

$$m = \frac{y_2 - y_1}{x_2 - x_1} = \frac{3 - (-2)}{-1 - (-3)} = \frac{5}{2}$$

7. $(0, 0)$ and $(7, 8)$

$$m = \frac{y_2 - y_1}{x_2 - x_1} = \frac{8 - 0}{7 - 0} = \frac{8}{7}$$

9. $(-1, 5)$ and $(6, -2)$

$$m = \frac{y_2 - y_1}{x_2 - x_1} = \frac{-2 - 5}{6 - (-1)} = -\frac{7}{7} = -1$$

11. $(1, 4)$ and $(5, 3)$

$$m = \frac{y_2 - y_1}{x_2 - x_1} = \frac{3 - 4}{5 - 1} = -\frac{1}{4}$$

13. $(-4, 3)$ and $(-4, 5)$

$$m = \frac{y_2 - y_1}{x_2 - x_1} = \frac{5 - 3}{-4 - (-4)} = \frac{2}{0} \text{ is undefined.}$$

15. $(-2, 8)$ and $(1, 6)$

$$m = \frac{y_2 - y_1}{x_2 - x_1} = \frac{6 - 8}{1 - (-2)} = -\frac{2}{3}$$

17. $(1, 0)$ and $(1, 1)$

$$m = \frac{y_2 - y_1}{x_2 - x_1} = \frac{1 - 0}{1 - 1} = \frac{1}{0} \text{ is undefined.}$$

19. $(5, 1)$ and $(-2, 1)$

$$m = \frac{y_2 - y_1}{x_2 - x_1} = \frac{1 - 1}{-2 - 5} = \frac{0}{-7} = 0$$

21. Line 1

23. Line 2

25. $(0, 0)$ and $(1, 1)$

$$m = \frac{y_2 - y_1}{x_2 - x_1} = \frac{1 - 0}{1 - 0} = 1$$

D

27. A vertical line has undefined slope.

B

29. $(2, 0)$ and $(4, -1)$

$$m = \frac{y_2 - y_1}{x_2 - x_1} = \frac{-1 - 0}{4 - 2} = -\frac{1}{2}$$

E

31. $x = 6$ is a vertical line, so it has an undefined slope.

33. $y = -4$ is a horizontal line, so it has a slope $m = 0$.

35. $x = -3$ is a vertical line, so it has an undefined slope.

37. $y = 0$ is a horizontal line, so it has a slope $m = 0$.

39. $(-3, -3)$ and $(0, 0)$

$$m = \frac{y_2 - y_1}{x_2 - x_1} = \frac{0 - (-3)}{0 - (-3)} = \frac{3}{3} = 1$$

a. $m = 1$

b. $m = -1$

41. $(-8, -4)$ and $(3, 5)$

$$m = \frac{y_2 - y_1}{x_2 - x_1} = \frac{5 - (-4)}{3 - (-8)} = \frac{9}{11}$$

a. $m = \dfrac{9}{11}$

b. $m = -\dfrac{11}{9}$

43. $(0, 6)$ and $(-2, 0)$

$$m_1 = \frac{y_2 - y_1}{x_2 - x_1} = \frac{0 - 6}{-2 - 0} = \frac{-6}{-2} = 3$$

$(0, 5)$ and $(1, 8)$

$$m_2 = \frac{y_2 - y_1}{x_2 - x_1} = \frac{8 - 5}{1 - 0} = \frac{3}{1} = 3$$

$m_1 = m_2$, parallel

45. $(2, 6)$ and $(-2, 8)$

$$m_1 = \frac{y_2 - y_1}{x_2 - x_1} = \frac{8 - 6}{-2 - 2} = \frac{2}{4} = -\frac{1}{2}$$

$(0, 3)$ and $(1, 5)$

$$m_2 = \frac{y_2 - y_1}{x_2 - x_1} = \frac{5 - 3}{1 - 0} = \frac{2}{1} = 2$$

$$m_1 m_2 = \left(-\frac{1}{2}\right)(2) = -1, \text{ perpendicular}$$

47. $(3, 6)$ and $(7, 8)$

$$m_1 = \frac{y_2 - y_1}{x_2 - x_1} = \frac{8 - 6}{7 - 3} = \frac{2}{4} = \frac{1}{2}$$

$(0, 6)$ and $(2, 7)$

$$m_2 = \frac{y_2 - y_1}{x_2 - x_1} = \frac{7 - 6}{2 - 0} = \frac{1}{2}$$

$m_1 = m_2$, parallel

49. $(2, -3)$ and $(6, -5)$

$$m_1 = \frac{y_2 - y_1}{x_2 - x_1} = \frac{-5 - (-3)}{6 - 2} = \frac{-2}{4} = -\frac{1}{2}$$

$(5, -2)$ and $(-3, -4)$

$$m_2 = \frac{y_2 - y_1}{x_2 - x_1} = \frac{-4 - (-2)}{-3 - 5} = \frac{-2}{-8} = \frac{1}{4}$$

$m_1 \neq m_2$ and $m_1 m_2 \neq -1$, neither

51. $(-4, -3)$ and $(-1, 0)$

$$m_1 = \frac{y_2 - y_1}{x_2 - x_1} = \frac{0 - (-3)}{-1 - (-4)} = \frac{3}{3} = 1$$

$(4, -4)$ and $(0, 0)$

$$m_2 = \frac{y_2 - y_1}{x_2 - x_1} = \frac{0 - (-4)}{0 - 4} = \frac{4}{-4} = -1$$

$$m_1 m_2 = (1)(-1) = -1, \text{ perpendicular}$$

53. $(-7, -5)$ and $(-2, -6)$

$$m_1 = \frac{y_2 - y_1}{x_2 - x_1} = \frac{-6 - (-5)}{-2 - (-7)} = \frac{-1}{5} = -\frac{1}{5}$$

parallel; $m_2 = m_1 = -\frac{1}{5}$

55. $(0, 0)$ and $(1, -3)$

$$m_1 = \frac{y_2 - y_1}{x_2 - x_1} = \frac{-3 - 0}{1 - 0} = \frac{-3}{1} = -3$$

perpendicular; $m_2 =$ (negative

reciprocal of m_1) $= \frac{1}{3}$

57. $(3, 3)$ and $(-3, -3)$

$$m_1 = \frac{y_2 - y_1}{x_2 - x_1} = \frac{-3 - 3}{-3 - 3} = \frac{-6}{-6} = 1$$

parallel; $m_2 = m_1 = 1$

59. pitch $= \dfrac{6}{10} = \dfrac{3}{5}$

61. grade $= \dfrac{\text{rise}}{\text{run}} = \dfrac{2}{16} = 0.125 = 12.5\%$

63. grade $= \dfrac{\text{rise}}{\text{run}} = \dfrac{2580}{6450} = 0.40 = 40\%$

65. slope $= \dfrac{\text{rise}}{\text{run}} = \dfrac{0.25}{12} = 0.02$

67. $(2003, 148)$ and $(2007, 168)$

$$m = \frac{y_2 - y_1}{x_2 - x_1} = \frac{168 - 148}{2007 - 2003} = \frac{20}{4} = 5$$

$= 5$ million users per year.

Every year there will be 5 million

more cell phone users.

69. $(5000, 1800)$ and $(20,000, 7200)$

$$m = \frac{y_2 - y_1}{x_2 - x_1} = \frac{7200 - 1800}{20,000 - 5000} = \frac{5400}{15,000}$$

$= 0.36$ dollars per mile

It costs $0.36 per mile to own and

operate a compact car.

71. $x + y = 10$

$\qquad y = -x + 10$

73. $x + 2y = -12$

$\qquad 2y = -x - 12$

$\qquad y = -\dfrac{1}{2}x - 6$

75. $5x - y = 17$

$\qquad y = 5x - 17$

77. 28.3 miles per gallon

79. 1992 the average was 27.6 miles per

gallon.

81. The greatest slope was from 1992 to 1993.

83. pitch $= \dfrac{\text{rise}}{\text{run}}$

$\qquad \dfrac{1}{3} = \dfrac{x}{18}$

$\qquad 3x = 18$

$\qquad x = 6$

85. a. $(1994, 782)$ and $(2001, 1132)$

b. $m = \dfrac{y_2 - y_1}{x_2 - x_1} = \dfrac{1132 - 782}{2001 - 1994} = \dfrac{350}{7} = 50$

c. For the years 1994 through 2001, the price per acre of U.S. farmland rose $50 every year.

87. $(1,1)$, $(-4, 4)$ and $(-3, 0)$

$m_1 = \dfrac{0 - 1}{-3 - 1} = \dfrac{1}{4}$, $m_2 = \dfrac{0 - 4}{-3 - (-4)} = -4$

$m_1 m_2 = -1$, so the sides are perpendicular.

89. $(2.1, 6.7)$ and $(-8.3, 9.3)$

$m = \dfrac{y_2 - y_1}{x_2 - x_1} = \dfrac{9.3 - 6.7}{-8.3 - 2.1} = \dfrac{2.6}{-10.4} = -0.25$

91. $(2.3, 0.2)$ and $(7.9, 5.1)$

$m = \dfrac{y_2 - y_1}{x_2 - x_1} = \dfrac{5.1 - 0.2}{7.9 - 2.3} = \dfrac{4.9}{5.6} = 0.875$

93. $y = -\dfrac{1}{3}x + 2$

$y = -2x + 2$

$y = -4x + 2$

The line becomes steeper.

Integrated Review

1. $(0, 0)$ and $(2, 4)$

$m = \dfrac{y_2 - y_1}{x_2 - x_1} = \dfrac{4 - 0}{2 - 0} = \dfrac{4}{2} = 2$

2. Horizontal line, $m = 0$

3. $(0, 1)$ and $(3, -1)$

$m = \dfrac{y_2 - y_1}{x_2 - x_1} = \dfrac{-1 - 1}{3 - 0} = -\dfrac{2}{3}$

4. Vertical line, slope is undefined.

5. $y = -2x$

$m = -2, b = 0$

6. $x + y = 3$

$y = -x + 3$

$m = -1, b = 3$

7. $x = -1$ for all values of y.

Vertical line

Slope is undefined.

8. $y = 4$ for all values of x

Horizontal line

$m = 0$

9. $x - 2y = 6$

$-2y = -x + 6$

$y = \dfrac{1}{2}x - 3$

$m = \dfrac{1}{2}, b = -3$

10. $y = 3x + 2$

$m = 3, b = 2$

11. $5x + 3y = 15$

x	y
0	5
3	0

12. $2x - 4y = 8$

x	y
0	-2
4	0

13. $(-1, 3)$ and $(1, -3)$

$m_1 = \dfrac{y_2 - y_1}{x_2 - x_1} = \dfrac{-3 - 3}{1 - (-1)} = \dfrac{-6}{2} = -3$

$(2, -1)$ and $(4, -7)$

$m_2 = \dfrac{y_2 - y_1}{x_2 - x_1} = \dfrac{-7 - (-1)}{4 - 2} = \dfrac{-6}{2} = -3$

$m_1 = m_2$, parallel

14. $(-6, -6)$ and $(-1, -2)$

$m_1 = \dfrac{y_2 - y_1}{x_2 - x_1} = \dfrac{-2 - (-6)}{-1 - (-6)} = \dfrac{4}{5}$

$(-4, 3)$ and $(3, -3)$

$m_2 = \dfrac{y_2 - y_1}{x_2 - x_1} = \dfrac{-3 - 3}{3 - (-4)} = \dfrac{-6}{7} = -\dfrac{6}{7}$

$m_1 \neq m_2$ and $m_1 m_2 \neq -1$, neither

15. $y = 110x + 1407$

a. $(0, 1407)$

b. In 2000, there were 1407 million admissions to movie theatres in the U.S.

c. $m = 110$

d. For the years 2000 through 2002, the number of movie theater admissions has increased at a rate of 110 million per year.

Section 3.5

Graphing Calculator Explorations

1. $y_1 = x,\ y_2 = 6x,\ y_3 = -6x$

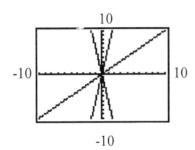

3. $y_1 = \dfrac{1}{2}x + 2,\ y_2 = \dfrac{3}{4}x + 2,\ y_3 = x + 2$

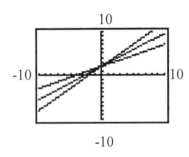

5. $y_1 = -7x + 5,\ y_2 = 7x + 5$

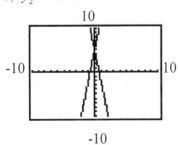

Mental Math

1. $y = 2x - 1$

$m = 2,\ (0, -1)$

2. $y = -7x + 3$

$m = -7,\ (0, 3)$

3. $y = x + \dfrac{1}{3}$

$m = 1,\ \left(0, \dfrac{1}{3}\right)$

4. $y = -x - \dfrac{2}{9}$

$m = -1,\ \left(0, -\dfrac{2}{9}\right)$

5. $y = \dfrac{5}{7}x - 4$

$m = \dfrac{5}{7},\ (0, -4)$

6. $y = -\dfrac{1}{4}x + \dfrac{3}{5}$

$m = -\dfrac{1}{4},\ \left(0, \dfrac{3}{5}\right)$

Exercise Set 3.5

1. $2x + y = 4$
$$y = -2x + 4$$
$$y = mx + b$$
$$m = -2,\ b = 4,\ (0,\ 4)$$

3. $x + 9y = 1$
$$9y = -x + 1$$
$$y = -\frac{1}{9}x + \frac{1}{9}$$
$$y = mx + b$$
$$m = -\frac{1}{9},\ b = \frac{1}{9},\ \left(0,\ \frac{1}{9}\right)$$

5. $4x - 3y = 12$
$$-3y = -4x + 12$$
$$y = \frac{4}{3}x - 4$$
$$y = mx + b$$
$$m = \frac{4}{3},\ b = -4,\ (0,\ -4)$$

7. $x + y = 0$
$$y = -x$$
$$y = mx + b$$
$$m = -1,\ b = 0,\ (0,\ 0)$$

9. $y = -3$
$$y = mx + b$$
$$m = 0,\ b = -3,\ (0,\ -3)$$

11. $-x + 5y = 20$
$$5y = x + 20$$
$$y = \frac{1}{5}x + 4$$
$$y = mx + b$$
$$m = \frac{1}{5},\ b = 4,\ (0,\ 4)$$

13. B

15. D

17. $x - 3y = -6,\ -3y = -x - 6,$
$$y = \frac{1}{3}x + 2,\ m_1 = \frac{1}{3}$$
$$3x - y = 0,\ -y = -3x,\ y = 3x,\ m_2 = 3$$
$$m_1 \neq m_2 \text{ and } m_1 m_2 \neq -1,\ \text{neither}$$

19. $2x - 7y = 1,\ -7y = -2x + 1,$
$$y = \frac{2}{7}x - \frac{1}{7},\ m_1 = \frac{2}{7}$$
$$2y = 7x - 2,\ y = \frac{7}{2}x - 1,\ m_2 = \frac{7}{2}$$
$$m_1 \neq m_2 \text{ and } m_1 m_2 \neq -1,\ \text{neither}$$

21. $10 + 3x = 5y,\ 2 + \frac{3}{5}x = y,\ m_1 = \frac{3}{5}$
$$5x + 3y = 1,\ 3y = -5x + 1,\ y = -\frac{5}{3}x + \frac{1}{3}$$
$$m_2 = -\frac{5}{3}$$
$$m_1 m_2 = \left(\frac{3}{5}\right)\left(-\frac{5}{3}\right) = -1,\ \text{perpendicular}$$

23. $6x = 5y + 1$, $6x - 1 = 5y$, $\dfrac{6}{5}x - \dfrac{1}{5} = y$,

$m_1 = \dfrac{6}{5}$

$-12x + 10y = 1$, $10y = 12x + 1$,

$y = \dfrac{12}{10}x + \dfrac{1}{10}$, $y = \dfrac{6}{5}x + \dfrac{1}{10}$, $m_2 = \dfrac{6}{5}$

$m_1 = m_2$, parallel

25. Answers may vary.

27. $m = -1$, $b = 1$

$y = mx + b$

$y = -x + 1$

29. $m = 2$, $b = \dfrac{3}{4}$

$y = mx + b$

$y = 2x + \dfrac{3}{4}$

31. $m = \dfrac{2}{7}$, $b = 0$

$y = mx + b$

$y = \dfrac{2}{7}x$

33. $y = \dfrac{2}{3}x + 5$

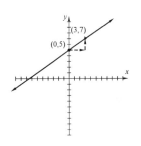

35. $y = -\dfrac{3}{5}x - 2$

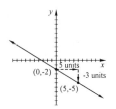

37. $y = 2x + 1$

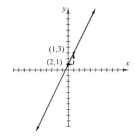

39. $y = -5x$

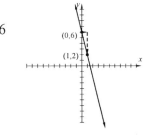

41. $4x + y = 6$

$y = -4x + 6$

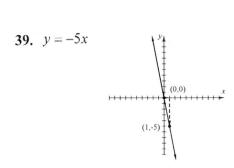

43. $x - y = -2$

$y = x + 2$

45. $3x + 5y = 10$

$5y = -3x + 10$

$y = -\dfrac{3}{5}x + 2$

47. $4x - 7y = -14$

$-7y = -4x - 14$

$y = \dfrac{4}{7}x + 2$

49. $y - (-6) = 2(x - 4)$

$y + 6 = 2x - 8$

$y = 2x - 14$

51. $y - 1 = -6(x - (-2))$

$y - 1 = -6(x + 2)$

$y - 1 = -6x - 12$

$y = -6x - 11$

53. $(0, 0)$ and $(1, 1)$

$m = \dfrac{y_2 - y_1}{x_2 - x_1} = \dfrac{1 - 0}{1 - 0} = 1$

D

55. A vertical line has undefined slope.

B

57. $(2, 0)$ and $(4, -1)$

$m = \dfrac{y_2 - y_1}{x_2 - x_1} = \dfrac{-1 - 0}{4 - 2} = -\dfrac{1}{2}$

E

59. a. $(0, 21)$ and $(22, 45)$

$m = \dfrac{y_2 - y_1}{x_2 - x_1} = \dfrac{45 - 21}{22 - 0} = \dfrac{24}{22} = \dfrac{12}{11}$

b. $m = \dfrac{12}{11}; \ (0, 21)$

$y = mx + b$

$y = \dfrac{12}{11}x + 21$

61. $2y + 4x = 12$

$2y = -4x + 12$

$y = -2x + 6$

$m_1 = -2$

parallel; $m_2 = -2$

point $(0, 5), \ b_2 = 5$

$y = m_2 x + b_2$

$y = -2x + 5$

63. a. The temperature $100°$ Celsius is equivalent to $212°$ Fahrenheit.

b. $68°\,\text{F}$

c. $27°\,\text{C}$

d. $(0, 32)$ and $(100, 212)$

$m = \dfrac{F_2 - F_1}{C_2 - C_1} = \dfrac{212 - 32}{100 - 0} = \dfrac{180}{100} = \dfrac{9}{5}$

$m = \dfrac{9}{5}, \ (0, 32)$

$F = mC + b$

$F = \dfrac{9}{5}C + 32$

Section 3.6

Mental Math

1. $y - 8 = 3(x - 4)$

$m = 3$

Answers may vary. Example: $(4, 8)$

2. $y - 1 = 5(x - 2)$

$m = 5$

Answers may vary. Example: $(2, 1)$

3. $y + 3 = -2(x - 10)$

$m = -2$

Answers may vary. Example: $(10, -3)$

4. $y + 6 = -7(x - 2)$

$m = -7$

Answers may vary. Example: $(2, -6)$

5. $y = \dfrac{2}{5}(x + 1)$

$m = \dfrac{2}{5}$

Answers may vary. Example: $(-1, 0)$

6. $y = \dfrac{3}{7}(x + 4)$

$m = \dfrac{3}{7}$

Answers may vary. Example: $(-4, 0)$

Exercise Set 3.6

1. $m = 6;\ (2, 2)$

$y - y_1 = m(x - x_1)$

$y - 2 = 6(x - 2)$

$y - 2 = 6x - 12$

$6x - y = 10$

3. $m = -8;\ (-1, -5)$

$y - y_1 = m(x - x_1)$

$y - (-5) = -8(x - (-1))$

$y + 5 = -8x - 8$

$8x + y = -13$

5. $m = \dfrac{1}{2};\ (5, -6)$

$y - y_1 = m(x - x_1)$

$y - (-6) = \dfrac{1}{2}(x - 5)$

$2(y + 6) = x - 5$

$2y + 12 = x - 5$

$-x + 2y = -17$

$x - 2y = 17$

7. $(3,2)$ and $(5,6)$

$m = \dfrac{y_2 - y_1}{x_2 - x_1} = \dfrac{6-2}{5-3} = \dfrac{4}{2} = 2$

$m = 2;\ (3,2)$

    $y - y_1 = m(x - x_1)$

    $y - 2 = 2(x - 3)$

    $y - 2 = 2x - 6$

  $-2x + y = -4$

    $2x - y = 4$

9. $(-1,3)$ and $(-2,-5)$

$m = \dfrac{y_2 - y_1}{x_2 - x_1} = \dfrac{-5-3}{-2-(-1)} = \dfrac{-8}{-1} = 8$

$m = 8;\ (-1,3)$

    $y - y_1 = m(x - x_1)$

    $y - 3 = 8(x - (-1))$

    $y - 3 = 8x + 8$

  $-8x + y = 11$

    $8x - y = -11$

11. $(2,3)$ and $(-1,-1)$

$m = \dfrac{y_2 - y_1}{x_2 - x_1} = \dfrac{-1-3}{-1-2} = \dfrac{-4}{-3} = \dfrac{4}{3}$

$m = \dfrac{4}{3};\ (2,3)$

    $y - y_1 = m(x - x_1)$

    $y - 3 = \dfrac{4}{3}(x - 2)$

  $3(y - 3) = 4(x - 2)$

    $3y - 9 = 4x - 8$

  $-4x + 3y = 1$

    $4x - 3y = -1$

13. Vertical line, point $(0,2)$

    $x = c$

    $x = 0$

15. Horizontal line, point $(-1,3)$

    $y = c$

    $y = 3$

17. Vertical line, point $\left(-\dfrac{7}{3}, -\dfrac{2}{5}\right)$

    $x = c$

    $x = -\dfrac{7}{3}$

19. $y = 5$ is horizontal.

Parallel to $y = 5$ is horizontal; $y = c$.

Point $(1,2)$

$y = 2$

21. $x = -3$ is vertical.

Perpendicular to $x = -3$ is horizontal; $y = c$.

Point $(-2,5)$

$y = 5$

23. $x = 0$ is vertical.

Parallel to $x = 0$ is vertical; $x = c$.

Point $(6,-8)$

$x = 6$

25. $m = -\dfrac{1}{2}; \left(0, \dfrac{5}{3}\right)$

$$y = mx + b$$

$$y = -\dfrac{1}{2}x + \dfrac{5}{3}$$

$$6y = -3x + 10$$

$$3x + 6y = 10$$

27. $m = 1; (-7, 9)$

$$y - y_1 = m(x - x_1)$$

$$y - 9 = 1\left[x - (-7)\right]$$

$$y - 9 = x + 7$$

$$x - y = -16$$

29. $(10, 7)$ and $(7, 10)$

$$m = \dfrac{y_2 - y_1}{x_2 - x_1} = \dfrac{10 - 7}{7 - 10} = \dfrac{3}{-3} = -1$$

$$m = -1; (10, 7)$$

$$y - y_1 = m(x - x_1)$$

$$y - 7 = -1(x - 10)$$

$$y - 7 = -x + 10$$

$$x + y = 17$$

31. x-axis is horizontal.

Parallel to y-axis is horizontal; $y = c$.

Point $(6, 7)$

$$y = 7$$

33. $m = -\dfrac{4}{7}; (-1, -2)$

$$y - y_1 = m(x - x_1)$$

$$y - (-2) = -\dfrac{4}{7}\left[x - (-1)\right]$$

$$y + 2 = -\dfrac{4}{7}x - \dfrac{4}{7}$$

$$7y + 14 = -4x - 4$$

$$4x + 7y = -18$$

35. $(-8, 1)$ and $(0, 0)$

$$m = \dfrac{y_2 - y_1}{x_2 - x_1} = \dfrac{0 - 1}{0 - (-8)} = -\dfrac{1}{8}$$

$$m = -\dfrac{1}{8}; (0, 0)$$

$$y - y_1 = m(x - x_1)$$

$$y - 0 = -\dfrac{1}{8}(x - 0)$$

$$8y = -x$$

$$x + 8y = 0$$

37. $m = 3; (0, 0)$

$$y = mx + b$$

$$y = 3x + 0$$

$$3x - y = 0$$

39. $(-6, -6)$ and $(0,0)$

$$m = \frac{y_2 - y_1}{x_2 - x_1} = \frac{0 - (-6)}{0 - (-6)} = \frac{6}{6} = 1$$

$$m = 1;\ (0,0)$$

$$y - y_1 = m(x - x_1)$$

$$y - 0 = 1(x - 0)$$

$$y = x$$

$$x - y = 0$$

41. $m = -5,\ b = 7$

$$y = mx + b$$

$$y = -5x + 7$$

$$5x + y = 7$$

43. $(-1, 5)$ and $(0, -6)$

$$m = \frac{y_2 - y_1}{x_2 - x_1} = \frac{-6 - 5}{0 - (-1)} = \frac{-11}{1} = -11$$

$$m = -11;\ (0, -6)$$

$$y = mx + b$$

$$y = -11x - 6$$

$$11x + y = -6$$

45. Undefined slope is vertical., point $\left(-\frac{3}{4}, 1\right)$

$$x = c$$

$$x = -\frac{3}{4}$$

47. y-axis is vertical.

Perpendicular to y-axis is horizontal; $y = c$.

Point $(-2, -3)$

$$y = -3$$

49. $m = 7;\ (1, 3)$

$$y - y_1 = m(x - x_1)$$

$$y - 3 = 7(x - 1)$$

$$y - 3 = 7x - 7$$

$$7x - y = 4$$

51. a. $(0, 4760)$ and $(3, 6680)$

$$m = \frac{y_2 - y_1}{x_2 - x_1} = \frac{6680 - 4760}{3 - 0} = \frac{1920}{3} = 640$$

$$m = 640;\ (0, 4760)$$

$$y = mx + b$$

$$y = 640x + 4760$$

  b. If $x = 10$,

then $y = 640(10) + 4760 = 11{,}160$

Expect 11,160 vehicles.

53. a. $(1, 32)$ and $(3, 96)$

$$m = \frac{y_2 - y_1}{x_2 - x_1} = \frac{96 - 32}{3 - 1} = \frac{64}{2} = 32$$

$$m = 32;\ (1, 32)$$

$$s - s_1 = m(t - t_1)$$

$$s - 32 = 32(t - 1)$$

$$s - 32 = 32t - 32$$

$$s = 32t$$

  b. If $t = 4$, then $s = 32(4) = 128$ ft/sec.

55. a. $(0, 70.3)$ and $(10, 79.6)$

$$m = \frac{y_2 - y_1}{x_2 - x_1} = \frac{79.6 - 70.3}{10 - 0} - \frac{9.3}{10} = 0.93$$

$m = 0.93; \ (0, 70.3)$

$$y - y_1 = m(x - x_1)$$

$$y - 70.3 = 0.93(x - 0)$$

$$y - 70.3 = 0.93x$$

$$y = 0.93x + 70.3.$$

b. If $x = 17$,

then $y = 0.93(17) + 70.3 = 86.11$

Expect 86.11 person per square mile.

57. a. $(0, 191)$ and $(5, 260)$

b. $m = \dfrac{y_2 - y_1}{x_2 - x_1} = \dfrac{260 - 191}{5 - 0} = \dfrac{69}{5} = 13.8$

$m = 13.8; \ (0, 191)$

$$y - y_1 = m(x - x_1)$$

$$y - 191 = 13.8(x - 0)$$

$$y - 191 = 13.8x$$

$$y = 13.8x + 191$$

c. If $x = 4$,

then $y = 13.8(4) + 191 = 246.2$

Expect $246.2 million in sales.

59. $(10, 63)$ and $(15, 94)$

$$m = \frac{y_2 - y_1}{x_2 - x_1} = \frac{94 - 63}{15 - 10} = \frac{31}{5}$$

$m = \dfrac{31}{5}; \ (10, 63)$

$$y - y_1 = m(x - x_1)$$

$$y - 63 = \frac{31}{5}(x - 10)$$

$$5y - 315 = 31(x - 10)$$

$$5y - 315 = 31x - 310$$

$$31x - 5y = -5$$

61. a. $(3, 10,000)$ and $(5, 8000)$

$$m = \frac{y_2 - y_1}{x_2 - x_1} = \frac{8000 - 10,000}{5 - 3}$$

$$= \frac{-2000}{2} = -1000$$

$m = -1000; \ (5, 8000)$

$$S - S_1 = m(p - p_1)$$

$$S - 8000 = -1000(p - 5)$$

$$S - 8000 = -1000p + 5000$$

$$S = -1000p + 13,000$$

b. If $p = 3.50$,

then $S = -1000(3.5) + 13,000 = 9500$

Expect $9500 in daily sales.

63. If $x = 2$, then

$$x^2 - 3x + 1 = (2)^2 - 3(2) + 1 = 4 - 6 + 1 = -1$$

65. If $x = -1$, then

$$x^2 - 3x + 1 = (-1)^2 - 3(-1) + 1 = 1 + 3 + 1 = 5$$

67. No

69. Yes

71. Answers may vary.

73. $y = 3x - 1$, $m_1 = 3$

    a. Parallel: $m_2 = m_1 = 3$; $(-1, 2)$

$$y - y_1 = m_2 (x - x_1)$$
$$y - 2 = 3(x - (-1))$$
$$y - 2 = 3x + 3$$
$$-3x + y = 5$$
$$3x - y = -5$$

    b. Perpendicular: $m_2 = -\dfrac{1}{m_1} = -\dfrac{1}{3}$;

    $(-1, 2)$

$$y - y_1 = m_2 (x - x_1)$$
$$y - 2 = -\dfrac{1}{3}(x - (-1))$$
$$3(y - 2) = -1(x + 1)$$
$$3y - 6 = -x - 1$$
$$x + 3y = 5$$

75. $3x + 2y = 7$, $y = -\dfrac{3}{2}x + \dfrac{7}{2}$, $m_1 = -\dfrac{3}{2}$

    a. Parallel: $m_2 = m_1 = -\dfrac{3}{2}$; $(3, -5)$

$$y - y_1 = m_2 (x - x_1)$$
$$y - (-5) = -\dfrac{3}{2}(x - 3)$$
$$2(y + 5) = -3(x - 3)$$
$$2y + 10 = -3x + 9$$
$$3x + 2y = -1$$

b. Perpendicular: $m_2 = -\dfrac{1}{m_1} = \dfrac{2}{3}$; $(3, -5)$

$$y - y_1 = m_2 (x - x_1)$$
$$y - (-5) = \dfrac{2}{3}(x - 3)$$
$$3(y + 5) = 2(x - 3)$$
$$3y + 15 = 2x - 6$$
$$2x - 3y = 21$$

Exercise Set 3.7

1. $\{(2, 4), (0, 0), (-7, 10), (10, -7)\}$

    Domain: $\{-7, 0, 2, 10\}$

    Range: $\{-7, 0, 4, 10\}$

3. $\{(0, -2), (1, -2), (5, -2),\}$

    Domain: $\{0, 1, 5\}$

    Range: $\{-2\}$

5. Every point has a unique x-value: it is a function.

7. Two points have the same x-value: it is not a function.

9. No

11. Yes

13. Yes

15. No

17. No

19. Yes

21. Yes; $y = x + 1$ is a non-vertical line.

23. Yes; $y - x = 7$ is a non-vertical line.

25. Yes; $y = 6$ is a non-vertical line.

27. No; $x = -2$ is a vertical line.

29. No; does not pass the vertical line test.

31. 5:20 A.M.

33. Answers may vary

35. $4.75 per hour

37. 1996

39. Yes; answers may vary

41. $f(x) = 2x - 5$

$f(-2) = 2(-2) - 5 = -4 - 5 = -9$

$f(0) = 2(0) - 5 = -5$

$f(3) = 2(3) - 5 = 6 - 5 = 1$

43. $f(x) = x^2 + 2$

$f(-2) = (-2)^2 + 2 = 4 + 2 = 6$

$f(0) = (0)^2 + 2 = 2$

$f(3) = (3)^2 + 2 = 9 + 2 = 11$

45. $f(x) = x^3$

$f(-2) = (-2)^3 = (-8) = -8$

$f(0) = (0)^3 = 0$

$f(3) = (3)^3 = 27$

47. $f(x) = |x|$

$f(-2) = |-2| = 2$

$f(0) = |0| = 0$

$f(3) = |3| = 3$

49. $h(x) = 5x$

$h(-1) = 5(-1) = -5, (-1, -5)$

$h(0) = 5(0) = 0, (0, 0)$

$h(4) = 5(4) = 20, (4, 20)$

51. $h(x) = 2x^2 + 3$

$h(-1) = 2(-1)^2 + 3 = 2 + 3 = 5, (-1, 5)$

$h(0) = 2(0)^2 + 3 = 3, (0, 3)$

$h(4) = 2(4)^2 + 3 = 2 \cdot 16 + 3 = 32 + 3$

$= 35, (4, 35)$

53. $h(x) = -x^2 - 2x + 3$

$h(-1) = -(-1)^2 - 2(-1) + 3 = 4, (-1, 4)$

$h(0) = -(0)^2 - 2(0) + 3 = 3, (0, 3)$

$h(4) = -(4)^2 - 2(4) + 3 = -21, (4, -21)$

55. $h(x) = 6$

$h(-1) = 6, (-1, 6)$

$h(0) = 6, (0, 6)$

$h(4) = 6, (4, 6)$

57. $(-\infty, \infty)$

59. $x + 5 \neq 0 \Rightarrow x \neq -5$, therefore

$(-\infty, -5) \cup (-5, \infty)$

61. $(-\infty, \infty)$

63. D; $(-\infty, \infty)$, R; $x \geq -4$, $[-4, \infty)$

65. D; $(-\infty, \infty)$, R; $(-\infty, \infty)$

67. D; $(-\infty, \infty)$, R; $\{2\}$

69. $(-2, 1)$

71. $(-3, -1)$

73. $H(x) = 2.59x + 47.24$
    a. $H(46) = 2.59(46) + 47.24 = 166.38$ cm
    b. $H(39) = 2.59(39) + 47.24 = 148.25$ cm

75. Answers may vary

77. $y = x + 7$
       $f(x) = x + 7$

79. $g(x) = -3x + 12$
    a. $g(s) = -3(s) + 12 = -3s + 12$
    b. $g(r) = -3(r) + 12 = -3r + 12$

81. $f(x) = x^2 - 12$
    a. $f(12) = (12)^2 - 12 = 132$
    b. $f(a) = (a)^2 - 12 = a^2 - 12$

Chapter 3 Review

1. 128 million

2. 19 million

3. 2001

4. Number is subscribers is increasing

5. Cross country skiing; 181 calories

7. 52 more calories

9.-14.

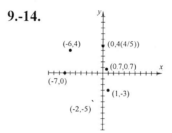

15. A. $(8.00, 1), (7.50, 10), (6.50, 25),$
     $(5.00, 50), (2.00, 100)$

B.

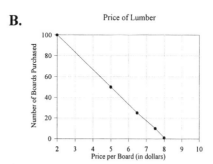

16. a. $(1996, 10.5), (1997, 10), (1998, 9.8),$
      $(1999, 9.9), (2000, 9.6), (2001, 9.8)$

B.

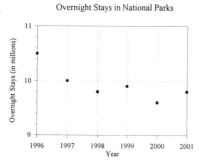

Overnight Stays in National Parks

17. $(0,56)$ No

$$7x - 8y = 56$$
$$7(0) - 8(56) \stackrel{?}{=} 56$$
$$-448 \neq 56$$

$(8,0)$ Yes

$$7x - 8y = 56$$
$$7(8) - 8(0) \stackrel{?}{=} 56$$
$$56 = 56$$

18. $(-5,0)$ Yes

$$-2x + 5y = 10$$
$$-2(-5) + 5(0) \stackrel{?}{=} 10$$
$$10 = 10$$

$(1,1)$ No

$$-2x + 5y = 10$$
$$-2(1) + 5(1) \stackrel{?}{=} 10$$
$$3 \neq 10$$

19. $(13,5)$ Yes

$$x = 13$$
$$(13) \stackrel{?}{=} 13$$
$$13 = 13$$

$(13,13)$ Yes

$$x = 13$$
$$(13) \stackrel{?}{=} 13$$
$$13 = 13$$

20. $(7,2)$ Yes

$$y = 2$$
$$(2) \stackrel{?}{=} 2$$
$$2 = 2$$

$(2,7)$ No

$$y = 2$$
$$(7) \stackrel{?}{=} 2$$
$$7 \neq 2$$

21. $-2 + y = 6x, \quad x = 7$

$$-2 + y = 6(7)$$
$$-2 + y = 42$$
$$y = 44$$
$$(7, 44)$$

22. $y = 3x + 5, \quad y = -8$

$$-8 = 3x + 5$$
$$-13 = 3x$$
$$-\frac{13}{3} = x$$
$$\left(-\frac{13}{3}, -8\right)$$

23. $9 = -3x + 4y$

$y = 0, \ 9 = -3x + 4(0), \ 9 = -3x, \ -3 = x$

$y = 3, \ 9 = -3x + 4(3), \ 9 = -3x + 12$

$\quad -3 = -3x, 1 = x$

$x = 9, \ 9 = -3(9) + 4y, \ 9 = -27 + 4y$

$\quad 36 = 4y, 9 = y$

x	y
-3	0
1	3
9	9

24. $y = 5$ for all values of x.

x	y
7	5
−7	5
0	5

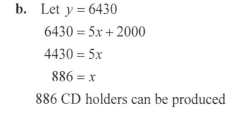

25. $x = 2y$

$y = 0, \; x = 2(0) = 0$

$y = 5, \; x = 2(5) = 10$

$y = -5, \; x = 2(-5) = -10$

x	y
0	0
10	5
−10	−5

26. a. $y = 5x + 2000$

$x = 1, \; y = 5(1) + 2000 = 2005$

$x = 100, \; y = 5(100) + 2000 = 2500$

$x = 1000, \; y = 5(1000) + 2000 = 7000$

x	1	100	1000
y	2005	2500	7000

b. Let $y = 6430$

$6430 = 5x + 2000$

$4430 = 5x$

$886 = x$

886 CD holders can be produced

27. $x - 3y = 12$

x	y
12	0
0	−4

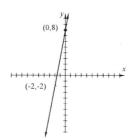

28. $5x - y = -8$

x	y
−2	−2
0	8

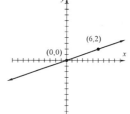

29. $x = 3y$

x	y
0	0
6	2

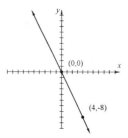

30. $y = -2x$

x	y
0	0
4	−8

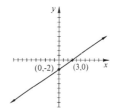

31. $2x - 3y = 6$

x	y
0	−2
3	0

32. $4x - 3y = 12$

x	y
0	-4
3	0

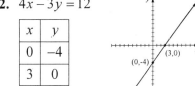

33. $y = 3x + 111$

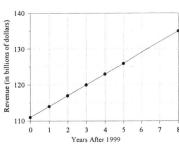

Expect a revenue of \$135 billion in 2007.

34. x-intercept: $(4,0)$

     y-intercept: $(0,-2)$

35. y-intercept: $(0,-3)$

36. x-intercepts: $(-2,0), (2,0)$

     y-intercepts: $(0,2), (0,-2)$

37. x-intercepts: $(-1,0), (2,0), (3, 0)$

     y-intercept: $(0,-2)$

38. $x - 3y = 12$

x	y
0	-4
12	0

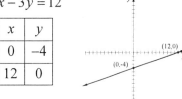

39. $-4x + y = 8$

x	y
0	8
-2	0

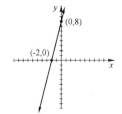

40. $y = -3$ for all x

x	y
0	-3

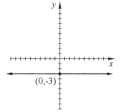

41. $x = 5$ for all y

x	y
5	0

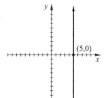

42. $y = -3x$

Find a second point.

x	y
0	0
3	-9

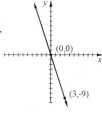

43. $x = 5y$

Find a second point.

x	y
0	0
5	1

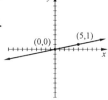

44. $(-1, 2)$, and $(3, -1)$

$$m = \frac{y_2 - y_1}{x_2 - x_1} = \frac{-1 - 2}{3 - (-1)} = -\frac{3}{4}$$

45. $(-2, -2)$, and $(3, -1)$

$$m = \frac{y_2 - y_1}{x_2 - x_1} = \frac{-1 - (-2)}{3 - (-2)} = \frac{1}{5}$$

46. $m = 0$

D

47. $m = -1$

B

48. Slope is undefined.

C

49. $m = 3$

A

50. $m = \dfrac{2}{3}$

E

51. $(2, 5)$, and $(6, 8)$

$$m = \frac{y_2 - y_1}{x_2 - x_1} = \frac{8 - 5}{6 - 2} = \frac{3}{4}$$

52. $(4, 7)$, and $(1, 2)$

$$m = \frac{y_2 - y_1}{x_2 - x_1} = \frac{2 - 7}{1 - 4} = \frac{-5}{-3} = \frac{5}{3}$$

53. $(1, 3)$, and $(-2, -9)$

$$m = \frac{y_2 - y_1}{x_2 - x_1} = \frac{-9 - 3}{-2 - 1} = \frac{-12}{-3} = 4$$

54. $(-4, 1)$, and $(3, -6)$

$$m = \frac{y_2 - y_1}{x_2 - x_1} = \frac{-6 - 1}{3 - (-4)} = \frac{-7}{7} = -1$$

55. Vertical; slope is undefined

56. Horizontal; slope is zero

57. Horizontal; slope is zero

58. Vertical; slope is undefined

59. Every 1 year, monthly day care increases by $17.75.

60. Every 1 year, 7.7 billion more dollars are spent on technology.

61. $3x + y = 7$

$$y = -3x + 7$$
$$y = mx + b$$
$$m = -3, \ y\text{-intercept} = (0, 7)$$

62. $x - 6y = -1$

$$-6y = -x - 1$$
$$y = \frac{1}{6}x + \frac{1}{6}$$
$$y = mx + b$$
$$m = \frac{1}{6}, \ y\text{-intercept} = \left(0, \frac{1}{6}\right)$$

63. $y = 2$

$$y = mx + b$$
$$m = 0, \ y\text{-intercept} = (0, 2)$$

64. $x = -5$

$y = mx + b$

m is undefined,

There is no y-intercept.

65. $x - y = -6, \; -y = -x - 6,$

$y = x + 6, \; m_1 = 1$

$x + y = 3, \; y = -x + 3, \; m_2 = -1$

$m_1 m_2 = (1)(-1) = -1,$ perpendicular

66. $3x + y = 7, \; y = -3x + 7, \; m_1 = -3$

$-3x - y = 10, \; -y = 3x + 10,$

$y = -3x - 10, \; m_2 = -3$

$m_1 = m_2,$ parallel

67. $y = 4x + \dfrac{1}{2}, \; m_1 = 4$

$4x + 2y = 1, \; 2y = -4x + 1,$

$y = -2x + \dfrac{1}{2}, \; m_2 = -2$

$m_1 \neq m_2$ and $m_1 m_2 \neq -1,$ neither

68. $m = -5, \; b = \dfrac{1}{2}$

$y = mx + b$

$y = -5x - \dfrac{1}{2}$

69. $m = \dfrac{2}{3}, \; b = 6$

$y = mx + b$

$y = \dfrac{2}{3}x + 6$

70. $y = -3x$

$y = mx + b$

$m = -3, \; b = 0$

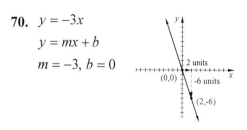

71. $y = 3x - 1$

$y = mx + b$

$m = 3, \; b = -1$

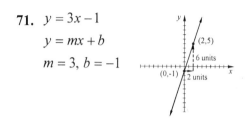

72. $-x + 2y = 8$

$2y = x + 8$

$y = \dfrac{1}{2}x + 4$

$y = mx + b$

$m = \dfrac{1}{2}, \; b = 4$

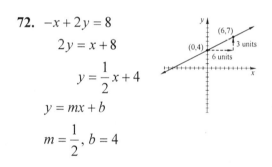

73. $5x - 3y = 15$

$-3y = -5x + 15$

$y = \dfrac{5}{3}x - 5$

$y = mx + b$

$m = \dfrac{5}{3}, \; b = -5$

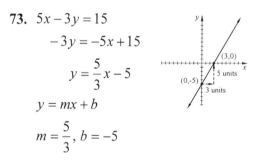

74. $y = -2x + 1$

$m = 2, \; b = 1$

D

75. $y = -4x$

$m = -4, \; b = 0$

C

76. $y = 2x$

$m = 2, b = 0$

A

77. $y = 2x - 1$

$m = 2, b = -1$

B

78. $m = 4; (2, 0)$

$y - y_1 = m(x - x_1)$

$y - 0 = 4(x - 2)$

$y = 4x - 8$

$4x - y = 8$

79. $m = -3; (0, -5)$

$y = mx + b$

$y = -3x - 5$

$3x + y = -5$

80. $m = \dfrac{1}{2}; \left(0, -\dfrac{7}{2}\right)$

$y = mx + b$

$y = \dfrac{1}{2}x - \dfrac{7}{2}$

$2y = x - 7$

$x - 2y = 7$

81. Horizontal line, point $(-2, -3)$

$y = c$

$y = -3$

82. Horizontal line, point $(0, 04)$

$y = c$

$y = 0$

83. $m = -6; (2, -1)$

$y - y_1 = m(x - x_1)$

$y - (-1) = -6(x - 2)$

$y + 1 = -6x + 12$

$6x + y = 11$

84. $m = 12; \left(\dfrac{1}{2}, 5\right)$

$y - y_1 = m(x - x_1)$

$y - 5 = 12\left(x - \dfrac{1}{2}\right)$

$y - 5 = 12x - 6$

$12x - y = 1$

85. $(0, 6)$ and $(6, 0)$

$m = \dfrac{y_2 - y_1}{x_2 - x_1} = \dfrac{0 - 6}{6 - 0} = \dfrac{-6}{6} = -1$

$m = -1; (0, 6)$

$y - y_1 = m(x - x_1)$

$y - 6 = -1(x - 0)$

$y - 6 = -x$

$x + y = 6$

86. $(0, -4)$ and $(-8, 0)$

$m = \dfrac{y_2 - y_1}{x_2 - x_1} = \dfrac{0 - (-4)}{-8 - 0} = \dfrac{4}{-8} = -\dfrac{1}{2}$

$m = -\dfrac{1}{2}; (0, -4)$

$y - y_1 = m(x - x_1)$

$y - (-4) = -\dfrac{1}{2}(x - 0)$

$$y + 4 = -\frac{1}{2}x$$

$$2y + 8 = -x$$

$$x + 2y = -8$$

87. Vertical line, point $(5, 7)$

$$x = c$$

$$x = 5$$

88. Horizontal line, point $(-6, 8)$

$$y = c$$

$$y = 8$$

89. $y = 8$ is horizontal.

Perpendicular to $y = 8$ is vertical; $x = c$.

Point $(6, 0)$

$$x = 6$$

90. $x = -2$ is vertical.

Perpendicular to $x = -2$ is horizontal;

$y = c$, point $(10, 12)$

$$y = 12$$

91. $y = -3x + 7$, $m_1 = -3$

a. Parallel: $m_2 = m_1 = -3$; $(5, 0)$

$$y - y_1 = m_2(x - x_1)$$

$$y - 0 = -3(x - 5)$$

$$y = -3x + 15$$

$$3x + y = 15$$

b. Perpendicular: $m_2 = -\dfrac{1}{m_1} = \dfrac{1}{3}$; $(5, 0)$

$$y - y_1 = m_2(x - x_1)$$

$$y - 0 = \frac{1}{3}(x - 5)$$

$$3(y - 0) = 1(x - 5)$$

$$3y - 0 = x - 5$$

$$x - 3y = 5$$

92. Two points have the same x-value: it is not a function.

93. Every point has a unique x-value: it is a function.

94. Yes; $7x - 6y = 1$ is a non-vertical line.

95. Yes; $y = 7$ is a non-vertical line.

96. No; $x = 2$ is a vertical line.

97. Yes; for each value of x there is only one value of y.

98. No; some values of x give 2 values of y.

99. No

100. Yes

101. $f(x) = -2x + 6$

a. $f(0) = -2(0) + 6 = 6$

b. $f(-2) = -2(-2) + 6 = 4 + 6 = 10$

c. $f\left(\dfrac{1}{2}\right) = -2\left(\dfrac{1}{2}\right) + 6 = -1 + 6 = 5$

102. $h(x) = -5 - 3x$

     a. $h(2) = -5 - 3(2) = -11$

     b. $h(-3) = -5 - 3(-3) = 4$

     c. $h(0) = -5 - 3(0) = -5$

103. $g(x) = x^2 + 12x$

     a. $g(3) = (3)^2 + 12(3) = 45$

     b. $g(-5) = (-5)^2 + 12(-5) = -35$

     c. $g(0) = (0)^2 + 12(0) = 0$

104. $h(x) = 6 - |x|$

     a. $h(-1) = 6 - |-1| = 6 - 1 = 5$

     b. $h(1) = 6 - |1| = 6 - 1 = 5$

     c. $h(-4) = 6 - |-4| = 6 - 4 = 2$

105. $(-\infty, \infty)$

106. $x - 2 \neq 0 \Rightarrow x \neq 2$, therefore

     $(-\infty, 2) \cup (2, \infty)$

107. D; $[-3, 5]$, R; $[-4, 2]$

108. D; $(-\infty, \infty)$, R; $x \geq 0$, $[0, \infty)$

109. D; $\{3\}$, R; $(-\infty, \infty)$

110. D; $(-\infty, \infty)$, R; $x \leq 2$, $(-\infty, 2]$

Chapter 3 Test

1. a. $(1980, 38), (1984, 47), (1988, 51),$

     $(1992, 54), (1996, 59), (2000, 55)$

b.

2. $2x + y = 8$

x	y
4	0
0	8

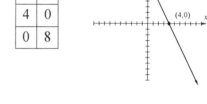

3. $5x - 7y = 10$

x	y
2	0
-5	-5

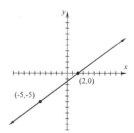

4. $y = -1$

     for all values of x

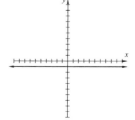

5. $x - 3 = 0$

$x = 3$

for all values of y

(3,5)
(3,-5)

6. $(-1, -1)$, and $(4, 1)$

$m = \dfrac{y_2 - y_1}{x_2 - x_1} = \dfrac{1 - (-1)}{4 - (-1)} = \dfrac{2}{5}$

7. Horizontal line: $m = 0$

8. $(6, -5)$, and $(-1, 2)$

$m = \dfrac{y_2 - y_1}{x_2 - x_1} = \dfrac{2 - (-5)}{-1 - 6} = \dfrac{7}{-7} = -1$

9. $-3x + y = 5$

$y = 3x + 5$

$y = mx + b$

$m = 3$

10. $x = 6$ is a vertical line.

The slope is undefined.

11. $7x - 3y = 2$

$-3y = -7x + 2$

$y = \dfrac{7}{3}x - \dfrac{2}{3}$

$y = mx + b$

$m = \dfrac{7}{3}, b = -\dfrac{2}{3}, \left(0, -\dfrac{2}{3}\right)$

12. $y = 2x - 6, \ m_1 = 2$

$-4x = 2y, \ -2x = y,$

$y = -2x, \ m_2 = -2$

$m_1 \neq m_2$ and $m_1 m_2 \neq -1$, neither

13. $m = -\dfrac{1}{4}; \ (2, 2)$

$y - y_1 = m(x - x_1)$

$y - 2 = -\dfrac{1}{4}(x - 2)$

$4(y - 2) = -(x - 2)$

$4y - 8 = -x + 2$

$x + 4y = 10$

14. $(0, 0)$ and $(6, -7)$

$m = \dfrac{y_2 - y_1}{x_2 - x_1} = \dfrac{-7 - 0}{6 - 0} = -\dfrac{7}{6}$

$m = -\dfrac{7}{6}; \ (0, 0)$

$y - y_1 = m(x - x_1)$

$y - 0 = -\dfrac{7}{6}(x - 0)$

$6y = -7x$

$7x + 6y = 0$

15. $(2, -5)$ and $(1, 3)$

$m = \dfrac{y_2 - y_1}{x_2 - x_1} = \dfrac{3 - (-5)}{1 - 2} = \dfrac{8}{-1} = -8$

$m = -8; \ (1, 3)$

$y - y_1 = m(x - x_1)$

$y - 3 = -8(x - 1)$

$y - 3 = -8x + 8$

$8x + y = 11$

16. $x = 7$ is vertical.

Parallel to $x = 7$ is vertical;

$x = c$, point $(-5, -1)$

$x = -5$

17. $m = \dfrac{1}{8}, b = 12$

$y = mx + b$

$y = \dfrac{1}{8}x + 12$

$8y = x + 96$

$x - 8y = -96$

18. Yes

19. No

20. $h(x) = x^3 - x$

a. $h(-1) = (-1)^3 - (-1) = -1 + 1 = 0$

b. $h(0) = (0)^3 - (0) = 0$

c. $h(4) = (4)^3 - (4) = 64 - 4 = 60$

21. $x + 1 \neq 0 \Rightarrow x \neq -1$, therefore

$(-\infty, -1) \cup (-1, \infty)$

22. $D; (-\infty, \infty), \ R; x \leq 4, \ (-\infty, 4]$

23. $D; (-\infty, \infty), \ R; (-\infty, \infty)$

24. 9 p.m.

25. 4 p.m.

26. January 1st and December 1st

27. June 1st and end of July

28. Yes; it passes the vertical line test.

29. Yes; every location has exactly 1 sunset time per day.

Chapter 3 Cumulative Review

1. a. $2 < 3$ **b.** $7 > 4$ **c.** $72 > 27$

2. $\dfrac{56}{64} = \dfrac{7 \cdot 8}{8 \cdot 8} = \dfrac{7}{8}$

3. $\dfrac{2}{15} \cdot \dfrac{5}{13} = \dfrac{2 \cdot 5}{3 \cdot 5 \cdot 13} = \dfrac{2}{39}$

4. $\dfrac{10}{3} + \dfrac{5}{21} = \dfrac{10 \cdot 7}{3 \cdot 7} + \dfrac{5}{21} = \dfrac{70 + 5}{21} = \dfrac{75}{21}$

$= \dfrac{3 \cdot 25}{3 \cdot 7} = \dfrac{25}{7} = 3\dfrac{4}{7}$

5. $\dfrac{3 + |4 - 3| + 2^2}{6 - 3} = \dfrac{3 + |1| + 2^2}{6 - 3} = \dfrac{3 + 1 + 4}{6 - 3}$

$= \dfrac{8}{3}$

6. $16 - 3 \cdot 3 + 2^4 = 16 - 3 \cdot 3 + 16$

$= 16 - 9 + 16$

$= 23$

7. a. $-8 + (-11) = -19$

b. $-5 + 35 = 30$

c. $0.6 + (-1.1) = -0.5$

d. $-\dfrac{7}{10} + \left(-\dfrac{1}{10}\right) = -\dfrac{8}{10} = -\dfrac{4}{5}$

e. $11.4 + (-4.7) = 6.7$

f. $-\dfrac{3}{8} + \dfrac{2}{5} = -\dfrac{3 \cdot 5}{8 \cdot 5} + \dfrac{2 \cdot 8}{5 \cdot 8} = \dfrac{-15 + 16}{40} = \dfrac{1}{40}$

8. $\left|9+(-20)\right|+\left|-10\right| = \left|-11\right|+\left|-10\right|$

$\qquad = 11+10$

$\qquad = 21$

9. a. $-14-8+10-(-6)$

$\qquad = -14+(-8)+10+6$

$\qquad = -6$

b. $1.6-(-10.3)+(-5.6)$

$\qquad = 1.6+10.3+(-5.6)$

$\qquad = 6.3$

10. $-9-(3-8) = -9-(-5) = -9+5 = -4$

11. Let $x = -2$ and $y = -4$.

a. $5x - y = 5(-2)-(-4) = -10+4$

$\qquad = -6$

b. $x^4 - y^2 = (-2)^4 - (-4)^2 = 16-16 = 0$

c. $\dfrac{3x}{2y} = \dfrac{3(-2)}{2(-4)} = \dfrac{-6}{-8} = \dfrac{3}{4}$

12. $\dfrac{x}{-10} = 2$

Let $x = -20$.

$\dfrac{-20}{-10} \overset{?}{=} 2$

$2 = 2$ True

2 is a solution to the equation.

13. a. $10+(x+12) = 10+x+12 = x+22$

b. $-3(7x) = -21x$

14. $(12+x)-(4x-7) = 12+x-4x+7$

$\qquad = 19-3x$

15. a. -3 **b.** 22 **c.** 1 **d.** -1 **e.** $\dfrac{1}{7}$

16. $-5(x-7) = -5x-(-5)(7) = -5x+35$

17. $y+0.6 = -1.0$

$\qquad y = -1.6$

18. $5(3+z)-(8z+9) = -4$

$15+5z-8z-9 = -4$

$-3z+6 = -4$

$-3z = -10$

$z = \dfrac{10}{3}$

19. $\dfrac{5}{2}x = 15$

$\dfrac{2}{5}\left(\dfrac{5}{2}x\right) = \dfrac{2}{5}(15)$

$x = 6$

20. $\dfrac{x}{4}-1 = -7$

$4\left(\dfrac{x}{4}\right)-4(1) = 4(-7)$

$x-4 = -28$

$x = -24$

21. Sum = first integer + second integer + third integer.

Sum $= x+(x+1)+(x+2)$

$\qquad = x+x+1+x+2$

$\qquad = 3x+3$

22. $\dfrac{x}{3} - 2 = \dfrac{x}{3}$

$$3\left(\dfrac{x}{3}\right) - 3(2) = 3\left(\dfrac{x}{3}\right)$$

$$x - 6 = x$$

$$-6 = 0$$

This is false. There is no solution.

23. $\dfrac{2(a+3)}{3} = 6a + 2$

$$2(a+3) = 18a + 6$$

$$2a + 6 = 18a + 6$$

$$-16a + 6 = 6$$

$$-16a = 0$$

$$a = 0$$

24. $x + 2y = 6$

$$x - x + 2y = 6 - x$$

$$2y = 6 - x$$

$$\dfrac{2y}{2} = \dfrac{6-x}{2}$$

$$y = \dfrac{6-x}{2}$$

25. Let x = the number of Democratic representatives and $x + 15$ = the number of Republican representatives.

$$x + x + 15 = 431$$

$$2x + 15 = 431$$

$$2x+ = 416$$

$$x = 208$$

$$x + 15 = 223$$

There were 208 Democratic representatives and 223 Republican.

26. $5(x + 4) \geq 4(2x + 3)$

$$5x + 20 \geq 8x + 12$$

$$-3x + 20 \geq 12$$

$$-3x \geq -8$$

$$\dfrac{-3x}{-3} \leq \dfrac{-8}{-3}$$

$$x \leq \dfrac{8}{3}, \quad \left(-\infty, \dfrac{8}{3}\right]$$

27. The perimeter of a rectangle is given by the formula $P = 2l + 2w$. Let l = the length of the garden. $P = 2l + 2w$

$$140 = 2l + 2w$$

$$140 = 2l + 2(30)$$

$$140 = 2l + 60$$

$$80 = 2l$$

$$40 = l$$

The length of the garden is 40 feet.

28. $-3 < 4x - 1 \leq 2$

$$-2 < 4x \leq 3$$

$$-\dfrac{1}{2} < x \leq \dfrac{3}{4}, \quad \left(-\dfrac{1}{2}, \dfrac{3}{4}\right]$$

29. $y = mx + b$

$$y - b = mx + b - b$$

$$y - b = mx$$

$$\dfrac{y-b}{m} = \dfrac{mx}{m}$$

$$\dfrac{y-b}{m} = x$$

30. $y = -5x$

x	y
0	0
−1	5
2	−10

31. Let x = the amount of 70% acid.

No. of liters · Strength = Amt of Acid

70%	x	0.7	$0.7x$
40%	$12 - x$	0.4	$0.4(12 - x)$
50%	12	0.5	$0.5(12)$

$0.7x + 0.4(12 - x) = 0.5(12)$

$0.7x + 4.8 - 0.4x = 6$

$0.3x + 4.8 = 6$

$0.3x = 1.2$

$x = 4$

$12 - x = 12 - 4 = 8$

Mix 4 liters of 70% acid

with 8 liters of 40% acid.

32. $y = -3x + 5$

x	y
−1	8
0	0
1	2

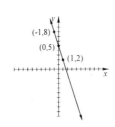

33. $x \geq -1$, $[-1, \infty)$

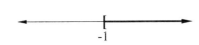

34. $2x + 4y = -8$

x-intercept, $y = 0$

$2x + 4(0) = -8 \Rightarrow x = -4 : (-4, 0)$

y-intercept, $x = 0$

$2(0) + 4y = -8 \Rightarrow y = -2 : (0, -2)$

35. $-1 \leq 2x - 3 < 5$

$2 \leq 2x < 8$

$1 \leq x < 4$, $[1, 4)$

36. $x = 2$

$x = 2$ for all

values of y

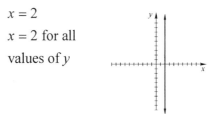

37.a.

$x - 2y = 6$

$(6, 0)$

Yes

$(6) - 2(0) \overset{?}{=} 6$

$6 = 6$

b.

$x - 2y = 6$

$(0, 3)$

No

$(0) - 2(3) \overset{?}{=} 6$

$-6 \neq 6$

$x - 2y = 6$

c. $\left(1, -\dfrac{5}{2}\right)$ $(1) - 2\left(-\dfrac{5}{2}\right) \overset{?}{=} 6$

Yes

$1 + 5 \overset{?}{=} 6$

$6 = 6$

38. $(0, 5)$ and $(-5, 4)$

$$m = \frac{y_2 - y_1}{x_2 - x_1} = \frac{4 - 5}{-5 - 0} = \frac{-1}{-5} = \frac{1}{5}$$

39. a. linear
   b. linear
   c. not linear
   d. linear

40. $x = -10$ is a vertical line.
   The slope is undefined.

41. $y = -1$ is horizontal, slope is 0.

42. $2x - 5y = 10$

$$-5y = -2x + 10$$

$$y = \frac{2}{5}x - 2$$

$$y = mx + b$$

$$m = \frac{2}{5}, \, b = -2$$

The slope is $\frac{2}{5}$.

The y-intercept is $(0, -2)$

43. $(-1, 7)$ and $(2, 2)$

$$m_1 = \frac{y_2 - y_1}{x_2 - x_1} = \frac{2 - 7}{2 - (-1)} = \frac{-5}{3} = -\frac{5}{3}$$

perpendicular; $m_2 =$ (negative

reciprocal of m_1) $= \frac{3}{5}$

44. $(2, 3)$ and $(0, 0)$

$$m = \frac{y_2 - y_1}{x_2 - x_1} = \frac{0 - 3}{0 - 2} = \frac{-3}{-2} = \frac{3}{2}$$

Point: $(0, 0)$

$$y - y_1 = m(x - x_1)$$

$$y - 0 = \frac{3}{2}(x - 0)$$

$$2y = 3x$$

$$3x - 2y = 0$$

Chapter 4

Section 4.1

Graphing Calculator Explorations

1. $\begin{cases} y = -2.68x + 1.21 \\ y = 5.22x - 1.68 \end{cases}$

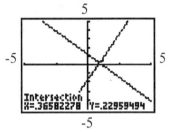

The solution of the system is $(0.37, 0.23)$.

3. $\begin{cases} 4.3x - 2.9y = 5.6 \\ 8.1x + 7.6y = -14.1 \end{cases}$

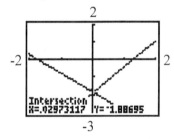

The solution of the system is $(0.03, -1.89)$.

Mental Math

1. One solution, $(-1, 3)$

2. No solution

3. Infinite number of solutions.

4. One solution, $(3, 4)$

5. No solution

6. Infinite number of solutions.

7. One solution, $(3, 2)$

8. One solution, $(0, -3)$

Exercise Set 4.1

1. a. Let $x = 2$ and $y = 4$.

$$x + y = 8 \qquad 3x + 2y = 21$$
$$2 + 4 \overset{?}{=} 8 \qquad 3(2) + 2(4) \overset{?}{=} 21$$
$$6 = 8 \qquad\qquad 6 + 8 \overset{?}{=} 21$$
$$\text{False} \qquad\qquad 14 = 21$$
$$\qquad\qquad\qquad \text{False}$$

$(2, 4)$ is not a solution of the system.

b. Let $x = 5$ and $y = 3$.

$$x + y = 8 \qquad 3x + 2y = 21$$
$$5 + 3 \overset{?}{=} 8 \qquad 3(5) + 2(3) \overset{?}{=} 21$$
$$8 = 8 \qquad\qquad 15 + 6 \overset{?}{=} 21$$
$$\text{True} \qquad\qquad 21 = 21$$
$$\qquad\qquad\qquad \text{True}$$

$(5, 3)$ is a solution of the system.

c. Let $x = 1$ and $y = 9$.

$x + y = 8$ $3x + 2y = 21$

$1 + 9 \overset{?}{=} 8$ $3(1) + 2(9) \overset{?}{=} 21$

$10 = 8$

     False $3 + 18 \overset{?}{=} 21$

                  $21 = 21$

                   True

$(1, 9)$ is not a solution of the system.

3. a. Let $x = 2$ and $y = -1$.

$3x - y = 5$ $x + 2y = 11$

$3(2) - (-1) \overset{?}{=} 5$ $2 + 2(-1) \overset{?}{=} 11$

$6 + 1 \overset{?}{=} 5$ $2 - 2 \overset{?}{=} 11$

$7 = 5$ $0 = 11$

   False   False

$(2, -1)$ is not a solution of the system.

b. Let $x = 3$ and $y = 4$.

$3x - y = 5$ $x + 2y = 11$

$3(3) - 4 \overset{?}{=} 5$ $3 + 2(4) \overset{?}{=} 11$

$9 - 4 \overset{?}{=} 5$ $3 + 8 \overset{?}{=} 11$

$5 = 5$ $11 = 11$

   True   True

$(3, 4)$ is a solution of the system.

c. Let $x = 0$ and $y = -5$.

$3x - y = 5$ $x + 2y = 11$

$3(0) - (-5) \overset{?}{=} 5$ $0 + 2(-5) \overset{?}{=} 11$

$0 + 5 \overset{?}{=} 5$ $0 - 10 \overset{?}{=} 11$

$5 = 5$ $-10 = 11$

   True   False

$(0, -5)$ is not a solution of the system.

5. a. Let $x = -3$ and $y = -6$.

$2y = 4x$ $2x - y = 0$

$2(-6) \overset{?}{=} 4(-3)$ $2(-3) - (-6) \overset{?}{=} 0$

$-12 = -12$ $-6 + 6 \overset{?}{=} 0$

   True $0 = 0$

                 True

$(-3, -6)$ is a solution of the system.

b. Let $x = 0$ and $y = 0$.

$2y = 4x$ $2x - y = 0$

$2(0) \overset{?}{=} 4(0)$ $2(0) - (0) \overset{?}{=} 0$

$0 = 0$ $0 - 0 \overset{?}{=} 0$

   True $0 = 0$

                 True

$(0, 0)$ is a solution of the system.

c. Let $x = 1$ and $y = 2$.

$$2y = 4x \qquad 2x - y = 0$$

$$2(2) \overset{?}{=} 4(1) \qquad 2(1) - (2) \overset{?}{=} 0$$

$$4 = 4 \qquad \qquad 2 - 2 \overset{?}{=} 0$$

$$\text{True} \qquad \qquad 0 = 0$$

$$\qquad \qquad \qquad \text{True}$$

$(1, 2)$ is a solution of the system.

7. Answers may vary

9. $\begin{cases} y = x + 1 \\ y = 2x - 1 \end{cases}$

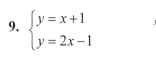

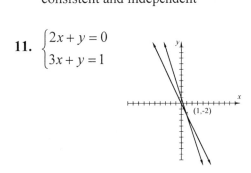

The solution of the system is $(2, 3)$.

consistent and independent

11. $\begin{cases} 2x + y = 0 \\ 3x + y = 1 \end{cases}$

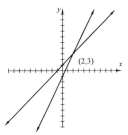

The solution of the system is $(1, -2)$.

consistent and independent

13. $\begin{cases} y = -x - 1 \\ y = 2x + 5 \end{cases}$

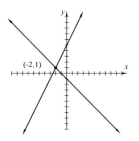

The solution of the system is $(-2, 1)$.

consistent and independent

15. $\begin{cases} 2x - y = 6 \\ y = 2 \end{cases}$

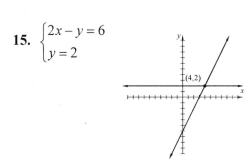

The solution of the system is $(4, 2)$.

consistent and independent

17. $\begin{cases} x + y = 5 \\ x + y = 6 \end{cases}$

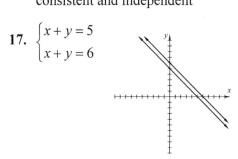

There is no solution.

inconsistent and independent

19. $\begin{cases} y - 3x = -2 \\ 6x - 2y = 4 \end{cases}$

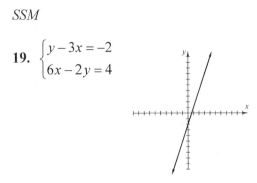

The solution of the system is $(2, 0)$.

consistent and independent

21. $\begin{cases} x - 2y = 2 \\ 3x + 2y = -2 \end{cases}$

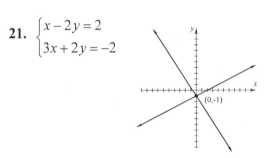

The solution of the system is $(0, -1)$.

consistent and independent

23. $\begin{cases} \dfrac{1}{2}x + y = -1 \\ x = 4 \end{cases}$

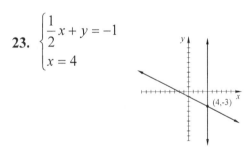

The solution of the system is $(4, -3)$.

consistent and independent

25. $\begin{cases} y = x - 2 \\ y = 2x + 3 \end{cases}$

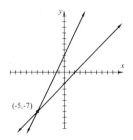

The solution of the system is $(-5, -7)$.

consistent and independent

27. $\begin{cases} x + y = 7 \\ x - y = 3 \end{cases}$

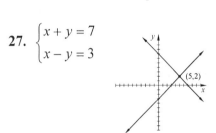

The solution of the system is $(5, 2)$.

consistent and independent

29. Answers may vary

31. Intersecting, one solution

33. Parallel, no solution

35. Identical lines, infinite number of solutions

37. Intersecting, one solution

39. Intersecting, one solution

41. Identical lines, infinite number of solutions

43. Parallel, no solution

45. $5(x-3)+3x=1$

$5x-15+3x=1$

$8x-15=1$

$8x=16$

$x=2$

The solution is 2.

47. $4\left(\dfrac{y+1}{2}\right)+3y=0$

$2(y+1)+3y=0$

$2y+2+3y=0$

$5y+2=0$

$5y=-2$

$y=-\dfrac{2}{5}$

The solution is $-\dfrac{2}{5}$.

49. $8a-2(3a-1)=6$

$8a-6a+2=6$

$2a+2=6$

$2a=4$

$a=2$

The solution is 2.

51. Answers may vary

53. 1984, 1988

55. 1996

57. Answers may vary

59. Answers may vary.

61. a. Each table includes the point $(4,9)$.

Therefore $(4,9)$ is a solution of the system.

b.

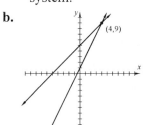

c. Yes

Possible answer

$$\begin{cases} 3x-2y=-2 \\ 6x+4y=4 \end{cases}$$

Section 4.2

Mental Math

1. When solving, you obtain $x=1$. $(1,4)$

2. When solving, you obtain $0=34$.
No solution.

3. When solving, you obtain $0=0$.
Infinite number of solutions.

4. When solving, you obtain $y=0$. $(5,0)$

5. When solving, you obtain $x=0$. $(0,0)$

6. When solving, you obtain $0=0$.
Infinite number of solutions.

Exercise Set 4.2

1. $\begin{cases} x + y = 3 \\ x = 2y \end{cases}$

Substitute $2y$ for x in the first equation.

$2y + y = 3$

$3y = 3$

$y = 1$

Let $y = 1$ in the second equation.

$x = 2(1)$

$x = 2$

The solution is $(2,1)$.

3. $\begin{cases} x + y = 6 \\ y = -3x \end{cases}$

Substitute $-3x$ for y in the first equation.

$x + (-3x) = 6$

$-2x = 6$

$x = -3$

Let $x = -3$ in the second equation.

$y = -3(-3)$

$y = 9$

The solution is $(-3,9)$.

5. $\begin{cases} 3x + 2y = 16 \\ x = 3y - 2 \end{cases}$

Substitute $3y - 2$ for x in the first equation.

$3(3y - 2) + 2y = 16$

$9y - 6 + 2y = 16$

$11y = 22$

$y = 2$

Let $y = 2$ in the second equation.

$x = 3(2) - 2$

$x = 4$

The solution is $(4,2)$.

7. $\begin{cases} 3x - 4y = 10 \\ x = 2y \end{cases}$

Substitute $2y$ for x in the first equation.

$3(2y) - 4y = 10$

$6y - 4y = 10$

$2y = 10$

$y = 5$

Let $y = 5$ in the second equation.

$x = 2(5)$

$x = 10$

The solution is $(10,5)$.

9. $\begin{cases} y = 3x + 1 \\ 4y - 8x = 12 \end{cases}$

Substitute $3x + 1$ for y in the second equation.

$4(3x + 1) - 8x = 12$

$12x + 4 - 8x = 12$

$4x = 8$

$x = 2$

Let $x = 2$ in the first equation.

$y = 3(2) + 1$

$y = 7$

The solution is $(2,7)$.

11. $\begin{cases} x+2y=6 \\ 2x+3y=8 \end{cases}$

Solve the first equation for x.

$x=6-2y$

Substitute $6-2y$ for x in the second equation.

$2(6-2y)+3y=8$

$12-4y+3y=8$

$-y=-4$

$y=4$

Let $y=4$ in $x=6-2y$.

$x=6-2(4)$

$y=-2$

The solution is $(-2,4)$.

13. $\begin{cases} 2x-5y=1 \\ 3x+y=-7 \end{cases}$

Solve the second equation for y.

$y=-7-3x$

Substitute $-7-3x$ for y in the first equation.

$2x-5(-7-3x)=1$

$2x+35+15x=1$

$17x=-34$

$x=-2$

Let $x=-2$ in $y=-7-3x$.

$y=-7-3(-2)$

$y=-1$

The solution is $(-2,-1)$.

15. $\begin{cases} 2y=x+2 \\ 6x-12y=0 \end{cases}$

Solve the first equation for x.

$x=2y-2$

Substitute $2y-2$ for x in the second equation.

$6(2y-2)-12y=0$

$12y-12-12y=0$

$-12=0$

The system has no solution.

17. $\begin{cases} \dfrac{1}{3}x-y=2 \\ x-3y=6 \end{cases}$

Solve the second equation for x.

$x=6+3y$

Substitute $6+3y$ for x in the first equation.

$\dfrac{1}{3}(6+3y)-y=2$

$2+y-y=2$

$2=2$

The equations in the original system are equivalent and there are an infinite number of solutions.

19. $\begin{cases} 4x+y=11 \\ 2x+5y=1 \end{cases}$

Solve the first equation for y.

$y=11-4x$

Substitute $11-4x$ for y in the second equation.

$$2x + 5(11 - 4x) = 1$$
$$2x + 55 - 20x = 1$$
$$-18x = -54$$
$$x = 3$$

Let $x = 3$ in $y = 11 - 4x$.

$$y = 11 - 4(3)$$
$$y = -1$$

The solution is $(3, -1)$.

$$-15y = 5$$
$$y = -\frac{1}{3}$$

Let $y = -\frac{1}{3}$ in $x = -2y$.

$$x = -2\left(-\frac{1}{3}\right) \qquad x = \frac{2}{3}$$

The solution is $\left(\frac{2}{3}, -\frac{1}{3}\right)$.

21. $\begin{cases} 2x - 3y = -9 \\ 3x = y + 4 \end{cases}$

Solve the second equation for y.

$$y = 3x - 4$$

Substitute $3x - 4$ for y in the first equation.

$$2x - 3(3x - 4) = -9$$
$$2x - 9x + 12 = -9$$
$$-7x = -21$$
$$x = 3$$

Let $x = 3$ in $y = 3x - 4$.

$$y = 3(3) - 4$$
$$y = 5$$

The solution is $(3, 5)$.

25. $\begin{cases} 3x - y = 1 \\ 2x - 3y = 10 \end{cases}$

Solve the first equation for y.

$$y = 3x - 1$$

Substitute $3x - 1$ for y in the second equation.

$$2x - 3(3x - 1) = 10$$
$$2x - 9x + 3 = 10$$
$$-7x = 7$$
$$x = -1$$

Let $x = -1$ in $y = 3x - 1$.

$$y = 3(-1) - 1$$
$$y = -4$$

The solution is $(-1, -4)$.

23. $\begin{cases} 6x - 3y = 5 \\ x + 2y = 0 \end{cases}$

Solve the second equation for x.

$$x = -2y$$

Substitute $-2y$ for x in the first equation.

$$6(-2y) - 3y = 5$$
$$-12y - 3y = 5$$

27. $\begin{cases} -x + 2y = 10 \\ -2x + 3y = 18 \end{cases}$

Solve the first equation for x.

$$x = 2y - 10$$

Substitute $2y - 10$ for x in the second equation.

$$-2(2y-10)+3y=18$$
$$-4y+20+3y=18$$
$$-y=-2$$
$$y=2$$

Let $y=2$ in $x=2y-10$.

$$x=2(2)-10$$
$$x=-6$$

The solution is $(-6,2)$.

29. $\begin{cases} 5x+10y=20 \\ 2x+6y=10 \end{cases}$

Solve the first equation for x.

$$x+2y=4$$
$$x=4-2y$$

Substitute $4-2y$ for x in the second equation.

$$2(4-2y)+6y=10$$
$$8-4y+6y=10$$
$$2y=2$$
$$y=1$$

Let $y=1$ in $x=4-2y$.

$$x=4-2(1)$$
$$x=2$$

The solution is $(2,1)$.

31. $\begin{cases} 3x+6y=9 \\ 4x+8y=16 \end{cases}$

Solve the first equation for x.

$$x+2y=3$$
$$x=3-2y$$

Substitute $3-2y$ for x in the second equation.

$$4(3-2y)+8y=16$$
$$12-8y+8y=16$$
$$12=16$$

The system has no solution.

33. $\begin{cases} y=2x+9 \\ y=7x+10 \end{cases}$

Substitute $2x+9$ for y in the second equation.

$$2x+9=7x+10$$
$$-5x=1$$
$$x=-\frac{1}{5}$$

Let $x=-\frac{1}{5}$ in the first equation.

$$y=2\left(-\frac{1}{5}\right)+9 \qquad y=\frac{43}{5}$$

The solution is $\left(-\frac{1}{5},\frac{43}{5}\right)$.

35. Answers may vary.

37. $\begin{cases} -5y+6y=3x+2(x-5)-3x+5 \\ \qquad\quad y=3x+2x-10-3x+5 \\ \qquad\quad y=2x-5 \\ \\ 4(x+y)-x+y=-12 \\ 4x+4y-x+y=-12 \\ \qquad\quad 3x+5y=-12 \end{cases}$

Substitute $2x - 5$ for y in the second equation.

$3x + 5(2x - 5) = -12$

$3x + 10x - 25 = -12$

$13x = 13 \quad x = 1$

Let $x = 1$ in $y = 2x - 5$.

$y = 2(1) - 5$

$y = -3$

The solution is $(1, -3)$.

39. $\quad 3x + 2y = 6$

$-2(3x + 2y) = -2(6)$

$-6x - 4y = -12$

41. $\quad -4x + y = 3$

$3(-4x + y) = 3(3)$

$-12x + 3y = 9$

43. $3n + 6m$

$\underline{2n - 6m}$

$5n$

45. $-5a - 7b$

$\underline{5a - 8b}$

$-15b$

47. a. $\begin{cases} y = 2.5x + 450 \\ y = 11.85x + 337 \end{cases}$

Substitute $11.85x + 337$ for y in the first equation.

$11.85x + 337 = 2.5x + 450$

$9.35x = 113$

$x = 12.09$

Let $x = 12.09$ in $y = 2.5x + 450$

$y = 2.5(12.09) + 450$

$y = 480.225$

The solution is $(12, 480)$.

b. In $1970 + 12 = 1982$, 480,000 men and 480,000 women received bachelor degrees.

c.

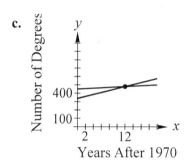

Years After 1970

49. $\begin{cases} y = 5.1x + 14.56 \\ y = -2x - 3.9 \end{cases}$

Substitute $-2x - 3.9$ for y in the first equation.

$-2x - 3.9 = 5.1x + 14.56$

$-7.1x = 18.46$

$x = -2.6$

Let $x = -2.6$ in $y = -2x - 3.9$

$y = -2(-2.6) - 3.9$

$y = 1.3$

The solution is $(-2.6, 1.3)$.

51. $\begin{cases} 3x + 2y = 14.05 \\ 5x + y = 18.5 \end{cases}$

Solve the first equation for $y = -5x + 18.5$

Substitute $-5x + 18.5$ for y in the first equation.

$3x + 2(-5x + 18.5) = 14.05$

$3x - 10x + 37 = 14.05$

$-7x = -22.95$

$x = 3.279$

Let $x = 3.279$ in $y = -5x + 18.5$

$y = -5(3.279) + 18.5$

$y = 2.105$

The solution is $(3.28, 2.11)$.

Exercise Set 4.3

1. $\begin{cases} 3x + y = 5 \\ 6x - y = 4 \end{cases}$

$3x + y = 5$

$\underline{6x - y = 4}$

$9x \quad = 9$

$x \quad = 1$

Let $x = 1$ in the first equation.

$3(1) + y = 5$

$3 + y = 5$

$y = 2$

The solution of the system is $(1, 2)$.

3. $\begin{cases} x - 2y = 8 \\ -x + 5y = -17 \end{cases}$

$x - 2y = \quad 8$

$\underline{-x + 5y = -17}$

$3y = -9$

$y = -3$

Let $y = -3$ in the first equation.

$x - 2(-3) = 8$

$x + 6 = 8$

$x = 2$

The solution of the system is $(2, -3)$.

5. $\begin{cases} x + y = 6 \\ x - y = 6 \end{cases}$

$x + y = 6$

$\underline{x - y = 6}$

$2x \quad = 12$

$x \quad = 6$

Let $x = 6$ in the first equation.

$6 + y = 6$

$y = 0$

The solution of the system is $(6, 0)$.

7. $\begin{cases} 3x + y = 4 \\ 9x + 3y = 6 \end{cases}$

Multiply the first equation by -3.

$-9x - 3y = -12$

$\underline{9x + 3y = \quad 6}$

$0 = -6$

The system has no solution.

9. $\begin{cases} 3x - 2y = 7 \\ 5x + 4y = 8 \end{cases}$

Multiply the first equation by 2.

$6x - 4y = 14$

$\underline{5x + 4y = 8}$

$11x \quad\quad = 22$

$x \quad\quad = 2$

Let $x = 2$ in the first equation.

$3(2) - 2y = 7$

$6 - 2y = 7$

$-2y = 1$

$$y = -\frac{1}{2}$$

The solution of the system is $\left(2, -\dfrac{1}{2}\right)$.

11. $\begin{cases} \dfrac{2}{3}x + 4y = -4 \\ 5x + 6y = 18 \end{cases}$

Multiply the first equation by 3 and the second equation by -2.

$2x + 12y = -12$

$\underline{-10x - 12y = 36}$

$-8x \quad\quad = -48$

$x \quad\quad = 6$

Let $x = 6$ in the first equation.

$\dfrac{2}{3}(6) + 4y = -4$

$4 + 4y = -4$

$4y = -8$

$y = -2$

The solution of the system is $(6, -2)$.

13. $\begin{cases} 4x - 6y = 8 \\ 6x - 9y = 12 \end{cases}$

Multiply the first equation by 3 and the second equation by -2.

$12x - 18y = 24$

$\underline{-12x + 18y = 24}$

$0 = 0$

The equations in the original system are equivalent and there are an infinite number of solutions.

15. $\begin{cases} 3x + y = -11 \\ 6x - 2y = -2 \end{cases}$

Multiply the first equation by 2.

$6x + 2y = -22$

$\underline{6x - 2y = -2}$

$12x \quad\quad = -24$

$x \quad\quad = -2$

Let $x = -2$ in the first equation.

$3(-2) + y = -11$

$-6 + y = -11$

$y = -5$

The solution of the system is $(-2, -5)$.

17. $\begin{cases} 3x + 2y = 11 \\ 5x - 2y = 29 \end{cases}$

$3x + y = 11$

$\underline{5x - 2y = 29}$

$8x \quad\quad = 40$

$x \quad\quad = 5$

Let $x = 5$ in the first equation.

$$3(5) + 2y = 11$$
$$15 + 2y = 11$$
$$2y = -4$$
$$y = -2$$

The solution of the system is $(5, -2)$

19. $\begin{cases} x + 5y = 18 \\ 3x + 2y = -11 \end{cases}$

Multiply the first equation by -3.

$$-3x - 15y = -54$$
$$\underline{3x + 2y = -11}$$
$$-13y = -65$$
$$y = 5$$

Let $y = 5$ in the first equation.

$$x + 5(5) = 18$$
$$x + 25 = 18$$
$$x = -7$$

The solution of the system is $(-7, 5)$.

21. $\begin{cases} 2x - 5y = 4 \\ 3x - 2y = 4 \end{cases}$

Multiply the first equation by -3 and the second equation by 2.

$$-6x + 15y = -12$$
$$\underline{6x - 4y = \quad 8}$$
$$11y = -4$$
$$y = -\frac{4}{11}$$

Let $y = -\dfrac{4}{11}$ in the first equation.

$$2x - 5\left(-\frac{4}{11}\right) = 4$$
$$2x + \frac{20}{11} = 4$$
$$11(2x) + 11\left(\frac{20}{11}\right) = 11(4)$$
$$22x + 20 = 44$$
$$22x = 24$$
$$x = \frac{12}{11}$$

The solution of the system is $\left(\dfrac{12}{11}, -\dfrac{4}{11}\right)$.

23. $\begin{cases} 2x + 3y = 0 \\ 4x + 6y = 3 \end{cases}$

Multiply the first equation by -2.

$$-4x - 6y = 0$$
$$\underline{4x + 6y = 3}$$
$$0 = 3$$

The system has no solution.

25.
$$\begin{cases} \dfrac{x}{3} + \dfrac{y}{6} = 1 \\ \dfrac{x}{2} - \dfrac{y}{4} = 0 \end{cases}$$

Multiply the first equation by 6 and the second equation by 4.

$$\begin{cases} 2x + y = 6 \\ 2x - y = 0 \end{cases} \quad \text{Simplificd system}$$

$$\begin{aligned} 4x &= 6 \\ x &= \frac{3}{2} \end{aligned}$$

Multiply the second equation of the simplified system by -1.

$$\begin{cases} 2x + y = 6 \\ -2x + y = 0 \end{cases}$$

$$\begin{aligned} 2y &= 6 \\ y &= 3 \end{aligned}$$

The solution of the system is $\left(\dfrac{3}{2}, 3 \right)$.

27.
$$\begin{cases} x - \dfrac{y}{3} = -1 \\ -\dfrac{x}{2} + \dfrac{y}{8} = \dfrac{1}{4} \end{cases}$$

Multiply the first equation by 3 and the second equation by 8.

$$\begin{cases} 3x - y = -3 \\ -4x + y = 2 \end{cases} \quad \text{Simplified system}$$

$$\begin{aligned} -x &= -1 \\ x &= 1 \end{aligned}$$

Multiply the first equation of the simplified system by 4 and the second equation by 3.

$$\begin{cases} 12x - 4y = -12 \\ -12x + 3y = 6 \end{cases}$$

$$\begin{aligned} -y &= -6 \\ y &= 6 \end{aligned}$$

The solution of the system is $(1, 6)$.

29.
$$\begin{cases} \dfrac{x}{3} - y = 2 \\ -\dfrac{x}{2} + \dfrac{3y}{2} = -3 \end{cases}$$

Multiply the first equation by 3 and the second equation by 2.

$$\begin{cases} x - 3y = 6 \\ -x + 3y = -6 \end{cases} \quad \text{Simplified system}$$

$$0 = 0$$

The equations in the original system are equivalent and there are an infinite number of solutions.

31.
$$\begin{cases} 8x + 11y = -16 \\ 2x + 3y = -4 \end{cases}$$

Multiply the second equation by -4.

$$\begin{aligned} 8x + 11y &= -16 \\ -8x - 12y &= 16 \end{aligned}$$

$$\begin{aligned} -y &= 0 \\ y &= 0 \end{aligned}$$

Let $y = 0$ in the first equation.

$8x + 11(0) = -16$

$8x = -16$

$x = -2$

The solution of the system is $(-2, 0)$.

33. Answers may vary.

35. $\begin{cases} 2x - 3y = -11 \\ y = 4x - 3 \end{cases}$

Substitute $4x - 3$ for y in the first equation.

$2x - 3(4x - 3) = -11$

$2x - 12x + 9 = -11$

$-10x + 9 = -11$

$-10x = -20$

$x = 2$

Let $x = 2$ in the second equation.

$y = 4(2) - 3$

$y = 5$

The solution of the system is $(2, 5)$.

37. $\begin{cases} x + 2y = 1 \\ 3x + 4y = -1 \end{cases}$

Multiply the first equation by -2.

$-2x - 4y = -2$

$\underline{3x + 4y = -1}$

$x = -3$

Let $x = -3$ in the first equation.

$-3 + 2y = 1$

$2y = 4$

$y = 2$

The solution is $(-3, 2)$.

39. $\begin{cases} 2y = x + 6 \\ 3x - 2y = -6 \end{cases}$

Subtract x from both sides of the first equation.

$\begin{cases} -x + 2y = 6 \\ 3x - 2y = -6 \end{cases}$

$2x \qquad = 0$

$x = 0$

Let $x = 0$ in the first equation.

$2y = 0 + 6$

$y = 3$

The solution of the system is $(0, 3)$.

41. $\begin{cases} y = 2x - 3 \\ y = 5x - 18 \end{cases}$

Substitute $5x - 18$ for y in the first equation.

$5x - 18 = 2x - 3$

$3x - 18 = -3$

$$3x = 15$$
$$x = 5$$

Let $x = 5$ in the second equation.

$$y = 5(5) - 18$$
$$y = 7$$

The solution of the system is $(5, 7)$.

43. $\begin{cases} x + \dfrac{1}{6}y = \dfrac{1}{2} \\ 3x + 2y = 3 \end{cases}$

Multiply the first equation by -12.

$$\begin{cases} -12x - 2y = -6 \\ \underline{3x + 2y = 3} \end{cases}$$
$$-9x = -3$$
$$x = \dfrac{1}{3}$$

Substitute $\dfrac{1}{3}$ for x

in the second equation.

$$3\left(\dfrac{1}{3}\right) + 2y = 3$$
$$1 + 2y = 3$$
$$2y = 2$$
$$y = 1$$

The solution of the system is $\left(\dfrac{1}{3}, 1\right)$.

45. $\begin{cases} \dfrac{x+2}{2} = \dfrac{y+11}{3} \\ \dfrac{x}{2} = \dfrac{2y+16}{6} \end{cases}$

Multiply the first equation by 6
and the second equation by -6.

$\begin{cases} 3(x+2) = 2(y+11) \\ 3x + 6 = 2y + 22 \\ \\ -3x = -2y - 16 \end{cases}$

Add the two equations.

$$6 = 6$$

There are an infinite number of solutions.

47. $\begin{cases} 2x + 3y = 14 \\ 3x - 4y = -69.1 \end{cases}$

Multiply the first equation by 3 and
the second equation by -2.

$$6x + 9y = 42$$
$$\underline{-6x + 8y = 138.2}$$
$$17y = 180.2$$
$$y = 10.6$$

Let $y = 10.6$ in the first equation.

$$2x + 3(10.6) = 14$$
$$2x + 31.8 = 14$$
$$2x = -17.8$$
$$x = -8.9$$

The solution of the system is $(-8.9, 10.6)$.

49. Let $x = $ a number.

$$2x + 6 = x - 3$$

51. Let $x = $ a number.

$$20 - 3x = 2$$

53. Let $n = $ a number.

$$4(n + 6) = 2n$$

55. a. $\begin{cases} 0.35x + y = 9.3 \\ 0.56x - y = -2.5 \end{cases}$

$0.91x \qquad = 6.8$

$\qquad\quad x = 7.47$

The number of skiers will equal the number of snowboarders in $1996 + 8 = 2004$.

b. Let $x = 8$ in the first equation.

$0.35(8) + y = 9.3$

$2.8 + y = 9.3$

$\qquad\quad y = 6.5$

There will be 6.5 million skiers.

Let $x = 8$ in the second equation.

$0.56(8) - y = -2.5$

$4.48 - y = -2.5$

$\qquad\quad y = 6.98$

There will be 6.98 million snowboarders. The numbers differ because of rounding in part a.

57. $\begin{cases} x + y = 5 \\ 3x + 3y = b \end{cases}$

Multiply the first equation by -3.

$-3x - 3y = -15$

$\underline{3x + 3y = \quad b}$

$\qquad\quad 0 = b - 15$

a. The system has an infinite number of solutions if this statement is true.

$b = 15$

b. The system has no solution if this statement is false. $b =$ any real number except 15.

59. a. Answers may vary.

b. Answers may vary

Integrated Review

1. C

2. D

3. A

4. B

5. $\begin{cases} 2x - 3y = -11 \\ y = 4x - 3 \end{cases}$

Substitute $4x - 3$ for y in the first equation.

$2x - 3(4x - 3) = -11$

$2x - 12x + 9 = -11$

$-10x = -20$

$x = 2$

Let $x = 2$ in the second equation.

$y = 4(2) - 3$

$y = 5$

The solution is $(2, 5)$.

6. $\begin{cases} 4x - 5y = 6 \\ y = 3x - 10 \end{cases}$

Substitute $3x - 10$ for y in the first equation.

$4x - 5(3x - 10) = 6$

$4x - 15x + 50 = 6$

$-11x = -44$

$x = 4$

Let $x = 4$ in the second equation.

$y = 3(4) - 10$

$y = 2$

The solution is $(4, 2)$.

7. $\begin{cases} x + y = 3 \\ x - y = 7 \end{cases}$

$2x \quad = 10$

$x \quad = 5$

Let $x = 5$ in the first equation.

$5 + y = 3$

$y = -2$

The solution of the system is $(5, -2)$.

8. $\begin{cases} x - y = 20 \\ x + y = -8 \end{cases}$

$2x \quad = 12$

$x \quad = 6$

Let $x = 6$ in the second equation.

$6 + y = -8$

$y = -14$

The solution of the system is $(6, -14)$.

9. $\begin{cases} x + 2y = 1 \\ 3x + 4y = -1 \end{cases}$

Solve the first equation for x.

$x = 1 - 2y$

Substitute $1 - 2y$ for x in the second equation.

$3(1 - 2y) + 4y = -1$

$3 - 6y + 4y = -1$

$-2y = -4$

$y = 2$

Let $y = 2$ in $x = 1 - 2y$.

$x = 1 - 2(2)$

$x = -3$

The solution is $(-3, 2)$.

10. $\begin{cases} x + 3y = 5 \\ 5x + 6y = -2 \end{cases}$

Solve the first equation for x.

$x = 5 - 3y$

Substitute $5 - 3y$ for x in the second equation.

$5(5 - 3y) + 6y = -2$

$25 - 15y + 6y = -2$

$-9y = -27$

$y = 3$

Let $y = 3$ in $x = 5 - 3y$.

$x = 5 - 3(3)$

$x = -4$

The solution is $(-4, 3)$.

11. $\begin{cases} y = x + 3 \\ 3x - 2y = -6 \end{cases}$

Substitute $x + 3$ for y in the second equation.

$3x - 2(x + 3) = -6$

$3x - 2x - 6 = -6$

$x = 0$

Let $x = 0$ in the first equation.

$y = 0 + 3$

$y = 3$

The solution is $(0, 3)$.

12. $\begin{cases} y = -2x \\ 2x - 3y = -16 \end{cases}$

Substitute $-2x$ for y in the second equation.

$2x - 3(-2x) = -16.$

$2x + 6x = -16$

$8x = -16$

$x = -2$

Let $x = -2$ in the first equation.

$y = -2(-2)$

$y = 4$

The solution is $(-2, 4)$.

13. $\begin{cases} y = 2x - 3 \\ y = 5x - 18 \end{cases}$

Substitute $5x - 18$ for y in the first equation.

$5x - 18 = 2x - 3$

$3x = 15$

$x = 5$

Let $x = 5$ in the second equation.

$y = 5(5) - 18$

$y = 7$

The solution is $(5, 7)$.

14. $\begin{cases} y = 6x - 5 \\ y = 4x - 11 \end{cases}$

Substitute $6x - 5$ for y in the second equation.

$6x - 5 = 4x - 11$

$2x = -6$

$x = -3$

Let $x = -3$ in the first equation.

$y = 6(-3) - 5$

$y = -23$

The solution is $(-3, -23)$.

15. $\begin{cases} x + \dfrac{1}{6}y = \dfrac{1}{2} \\ 3x + 2y = 3 \end{cases}$

Multiply the first equation by 6.

$\begin{cases} 6x + y = 3 \\ 3x + 2y = 3 \end{cases}$ Simplified system

Multiply the first equation of the simplified system by -2.

$\begin{cases} -12x - 2y = -6 \\ \underline{3x + 2y = 3} \end{cases}$

$-9x = -3$

$x = \dfrac{1}{3}$

Multiply the second equation of the simplified system by -2.

$\begin{cases} 6x + y = 3 \\ \underline{-6x - 4y = -6} \end{cases}$

$-3y = -3$

$y = 1$

The solution of the system is $\left(\dfrac{1}{3}, 1\right)$.

16. $\begin{cases} x + \dfrac{1}{3}y = \dfrac{5}{12} \\ 8x + 3y = 4 \end{cases}$

Multiply the first equation by 12.

$\begin{cases} 12x + 4y = 5 \\ 8x + 3y = 4 \end{cases}$ Simplified system

Multiply the first equation of the simplified system by 2 and the second equation by -3.

$\begin{cases} 24x + 8y = 10 \\ \underline{-24x - 9y = -12} \end{cases}$

$-y = -2$

$y = 2$

Multiply the first equation of the simplified system by 3 and the second equation by 4.

$\begin{cases} 36x + 12y = 15 \\ -32x - 12y = -16 \end{cases}$

$4x = -1$

$x = -\dfrac{1}{4}$

The solution of the system is $\left(-\dfrac{1}{4}, 2\right)$.

17. $\begin{cases} x - 5y = 1 \\ -2x + 10y = 3 \end{cases}$

Multiply the first equation by 2.

$2x - 10y = 2$

$\underline{-2x + 10y = 3}$

$0 = 5$

The system has no solution.

18. $\begin{cases} -x + 2y = 3 \\ 3x - 6y = -9 \end{cases}$

Multiply the first equation by 3.

$-3x + 6y = 9$

$\underline{3x - 6y = -9}$

$0 = 0$

The equations in the original system are equivalent and there are an infinite number of solutions.

19. $\begin{cases} 0.2x - 0.3y = -0.95 \\ 0.4x + 0.1y = 0.55 \end{cases}$

Multiply both equations by 10.

$\begin{cases} 2x - 3y = -9.5 \\ 4x + y = 5.5 \end{cases}$ Simplified system

Multiply the first equation of the simplified system by -2.

$\begin{cases} -4x + 6y = 19 \\ 4x + y = 5.5 \end{cases}$

$\qquad 7y = 24.5$

$\qquad y = 3.5$

Multiply the second equation of the simplified system by 3.

$\begin{cases} 2x - 3y = -9.5 \\ 12x + 3y = 16.5 \end{cases}$

$\quad 14x \quad = 7$

$\quad x \quad = 0.5$

The solution of the system is $(0.5, 3.5)$.

20. $\begin{cases} 0.08x - 0.04y = -0.11 \\ 0.02x - 0.06y = -0.09 \end{cases}$

Multiply both equations by 100.

$\begin{cases} 8x - 4y = -11 \\ 2x - 6y = -9 \end{cases}$ Simplified system

Multiply the second equation of the simplified system by -4.

$\begin{cases} 8x - 4y = -11 \\ -8x + 24y = 36 \end{cases}$

$\quad 20y = 25$

$\quad y = 1.25$

Multiply the first equation of the simplified system by -3 and the second equation by 2.

$\begin{cases} -24x + 12y = 33 \\ 4x - 12y = -18 \end{cases}$

$\quad -20x \quad = 15$

$\quad x \quad = -0.75$

The solution of the system is $(-0.75, 1.25)$.

21. $\begin{cases} x = 3y - 7 \\ 2x - 6y = -14 \end{cases}$

Substitute $3y - 7$ for x in the second equation.

$2(3y - 7) - 6y = -14$

$\quad 6y - 14 - 6y = -14$

$\qquad\qquad 0 = 0$

The equations in the original system are equivalent and there are an infinite number of solutions..

22. $\begin{cases} y = \dfrac{x}{2} - 3 \\ 2x - 4y = 0 \end{cases}$

Substitute $\dfrac{x}{2} - 3$ for y in the second equation.

$2x - 4\left(\dfrac{x}{2} - 3 \right) = 0$

$\quad 2x - 2x + 12 = 0$

$\qquad\qquad 12 = 0$

There is no solution.

23-24. Answers may vary.

Exercise Set 4.4

1. $x+y+z=3$ $-x+y+z=5$
     $(-1)+3+1=3$ $-(-1)+3+1=5$
             $3=3$ $5=5$
A is true.   B is true.
     $-x-y+2z=0$ $x+2y-3z=2$
  $-(-1)-3+2(1)=0$ $(-1)+2(3)-3(1)=2$
             $0=0?$ $2=2$
C is false.   D is true.
Therefore, equations A, B, and D.

3. Yes. Answers may vary.

5. $\begin{cases} x-y+z=-4 & (1) \\ 3x+2y-z=5 & (2) \\ -2x+3y-z=15 & (3) \end{cases}$

Add E1 and E2.
$4x+y=1$
Add E1 and E3
$-x+2y=11$
Solve the new system:
$\begin{cases} 4x+y=1 \\ -x+2y=11 \end{cases}$

Multiply the second equation by 4.
$\begin{cases} 4x+y=1 \\ -4x+8y=44 \end{cases}$

Add the equations.
$\quad 4x+y=1$
$\underline{-4x+8y=44}$
$\quad\quad 9y=45$
$\quad\quad\quad y=5$
Replace y with 5 in the equation
$4x+y=1$.
$4x+5=1$
$\quad 4x=-4$
$\quad\quad x=-1$
Replace x with -1 and y with 5 in E1.
$(-1)-(5)+z=-4$
$\quad\quad -6+z=-4$
$\quad\quad\quad\quad z=2$
The solution is $(-1, 5, 2)$.

7. $\begin{cases} x+y \quad\quad =3 & (1) \\ \quad 2y \quad =10 & (2) \\ 3x+2y-3z=1 & (3) \end{cases}$

Solve E2 for y: $y=5$
Replace y with 5 in E1.
$x+5=3$
$\quad x=-2$
Replace x with -1 and y with 5 in E3.
$3(-2)+2(5)-3z=1$
$\quad\quad -6+10-3z=1$
$\quad\quad\quad\quad 4-3z=1$
$\quad\quad\quad\quad\quad -3z=-3$
$\quad\quad\quad\quad\quad\quad z=1$
The solution is $(-2, 5, 12)$.

9. $\begin{cases} 2x+2y+z=1 & (1) \\ -x+y+2z=3 & (2) \\ x+2y+4z=0 & (3) \end{cases}$

Add E2 and E3.
$3y+6z=3$ or $y+2z=1$
Multiply E2 by 2 and add to E1.
$\quad -2x+2y+4z=6$
$\underline{\quad 2x+2y+z=1}$
$\quad\quad\quad 4y+5z=7$
Solve the new system:
$\begin{cases} y+2z=1 \\ 4y+5z=7 \end{cases}$

Multiply the first equation by -4.
$\begin{cases} -4y-8z=-4 \\ \quad 4y+5z=7 \end{cases}$

Add the equations.
$\quad -4y-8z=-4$
$\underline{\quad 4y+5z=7}$
$\quad\quad\quad -3z=3$
$\quad\quad\quad\quad\quad z=-1$
Replace z with -1 in the equation
$y+2z=1$.
$y+2(-1)=1$
$\quad y-2=1$
$\quad\quad y=3$

Replace y with 3 and z with -1 in E3.
$$x + 2(3) + 4(-1) = 0$$
$$x + 6 - 4 = 0$$
$$x + 2 = 0$$
$$x = -2$$
The solution is $(-2, 3, -1)$.

11. $\begin{cases} x - 2y + z = -5 & (1) \\ -3x + 6y - 3z = 15 & (2) \\ 2x - 4y + 2z = -10 & (3) \end{cases}$

Multiply E2 by $-\dfrac{1}{3}$ and E3 by $\dfrac{1}{2}$.

$\begin{cases} x - 2y + z = -5 \\ x - 2y + z = -5 \\ x - 2y + z = -5 \end{cases}$

All three equations are identical. There are infinitely many solutions.
The solution is $\{(x, y, z) \mid x - 2y + z = -5\}$.

13. $\begin{cases} 4x - y + 2z = 5 & (1) \\ 2y + z = 4 & (2) \\ 4x + y + 3z = 10 & (3) \end{cases}$

Multiply E1 by -1 and add to E3.
$$-4x + y - 2z = -5$$
$$\underline{4x + y + 3z = 10}$$
$$2y + z = 5 \quad (4)$$
Multiply E4 by -1 and add to E2.
$$-2y - z = -5$$
$$\underline{2y + z = 4}$$
$$0 = -1 \quad \text{False}$$
Inconsistent system.
The solution is $\varnothing$.

15. $\begin{cases} x + 5z = 0 & (1) \\ 5x + y = 0 & (2) \\ y - 3z = 0 & (3) \end{cases}$

Multiply E3 by -1 and add to E2.
$$-y + 3z = 0$$
$$\underline{5x + y = 0}$$
$$5x + 3z = 0 \quad (4)$$
Multiply E1 by -5 and add to E4.

$$-5x - 25z = 0$$
$$\underline{5x + 3z = 0}$$
$$-22z = 0$$
$$z = 0$$
Replace z with 0 in E4.
$$5x + 3(0) = 0$$
$$5x = 0$$
$$x = 0$$
Replace x with 0 in E2.
$$5(0) + y = 0$$
$$y = 0$$
The solution is $(0, 0, 0)$.

17. $\begin{cases} 6x - 5z = 17 & (1) \\ 5x - y + 3z = -1 & (2) \\ 2x + y = -41 & (3) \end{cases}$

Add E2 and E3.
$$7x + 3z = -42 \quad (4)$$
Multiply E4 by 5, multiply E1 by 3, and add.
$$35x + 15z = -210$$
$$\underline{18x - 15z = 51}$$
$$53x = -159$$
$$x = -3$$
Replace x with -3 in E1.
$$6(-3) - 5z = 17$$
$$-18 - 5z = 17$$
$$-5z = 35$$
$$z = -7$$
Replace x with -3 in E3.
$$2(-3) + y = -41$$
$$-6 + y = -41$$
$$y = -35$$
The solution is $(-3, -35, -7)$.

19. $\begin{cases} x + y + z = 8 & (1) \\ 2x - y - z = -1 & (2) \\ x - 2y - 3z = 22 & (3) \end{cases}$

Add E1 and E2.
$$3x = 18 \quad \text{or} \quad x = 6$$
Add twice E1 to E3.

$$2x + 2y + 2z = 16$$
$$\underline{x - 2y - 3z = 22}$$
$$3x - z = 38$$

Replace x with 6 in this equation.
$$3(6) - z = 38$$
$$18 - z = 38$$
$$-z = 20$$
$$z = -20$$

Replace x with 6 and z with −20 in E1.
$$6 + y + (-20) = 8$$
$$y - 14 = 8$$
$$y = 22$$

The solution is (6, 22, −20).

21. $\begin{cases} x + 2y - z = 5 & (1) \\ 6x + y + z = 7 & (2) \\ 2x + 4y - 2z = 5 & (3) \end{cases}$

Add E1 and E2.
$$7x + 3y = 12 \quad (4)$$

Add twice E2 to E3.
$$12x + 2y + 2z = 14$$
$$\underline{2x + 4y - 2z = 5}$$
$$14x + 6y = 19 \quad (5)$$

Multiply E4 by −2 and add to E5.
$$-14x - 6y = -24$$
$$\underline{14x + 6y = 19}$$
$$0 = -5 \quad \text{False}$$

Inconsistent system.
The solution is $\varnothing$.

23. $\begin{cases} 2x - 3y + z = 2 & (1) \\ x - 5y + 5z = 3 & (2) \\ 3x + y - 3z = 5 & (3) \end{cases}$

Add −2 times E2 to E1.
$$2x - 3y + z = 2$$
$$\underline{-2x + 10y - 10z = -6}$$
$$7y - 9z = -4 \quad (4)$$

Add −3 times E2 to E3.

$$-3x + 15y - 15z = -9$$
$$\underline{3x + y - 3z = 5}$$
$$16y - 18z = -4$$

Solve the new system:
$\begin{cases} 7y - 9z = -4 & (4) \\ 16y - 18z = -4 & (5) \end{cases}$

Multiply E4 by −2 and add to E5.
$$-14y + 18z = 8$$
$$\underline{16y - 18z = -4}$$
$$2y = 4$$
$$y = 2$$

Replace y with 2 in E4.
$$7(2) - 9z = -4$$
$$-9z = -18$$
$$z = 2$$

Replace y with 2 and z with 2 in E1.
$$2x - 3(2) + 2 = 2$$
$$x = 3$$

The solution is (3, 2, 2).

25. $\begin{cases} -2x - 4y + 6z = -8 & (1) \\ x + 2y - 3z = 4 & (2) \\ 4x + 8y - 12z = 16 & (3) \end{cases}$

Add 2 times E2 to E1.
$$2x + 4y - 6z = 8$$
$$\underline{-2x - 4y + 6z = -8}$$
$$0 = 0$$

Add −4 times E2 to E3.
$$-4x - 8y + 12z = -16$$
$$\underline{4x + 8y - 12z = 16}$$
$$0 = 0$$

The system is dependent.
The solution is $\{(x, y, z) \mid x + 2y - 3z = 4\}$.

27. $\begin{cases} 2x + 2y - 3z = 1 & (1) \\ y + 2z = -14 & (2) \\ 3x - 2y = 5 & (3) \end{cases}$

Add E1 to E3.
$$5x - 3z = 0 \quad (4)$$

Add twice E2 to E3.

$$2y + 4z = -28$$
$$\underline{3x - 2y \qquad = -1}$$
$$3x + \qquad 4z = -29 \ (5)$$

Multiply E4 by 4, multiply E5 by 2, and add.

$$20x - 12z = 0$$
$$\underline{9x - 12z = -87}$$
$$29x \qquad = -87$$
$$x = -3$$

Replace x with -3 in E4.
$$5(-3) - 3z = 0$$
$$3z = -15$$
$$z = -5$$

Replace z with -5 in E2.
$$y + 2(-5) = -14$$
$$y - 10 = -14$$
$$y = -4$$

The solution is $(-3, -4, -5)$.

29. $\begin{cases} x + 2y - z = 5 & (1) \\ -3x - 2y - 3z = 11 & (2) \\ 4x + 4y + 5z = -18 & (3) \end{cases}$

Add E1 and E2.
$$-2x - 4z = 16 \ \text{ or } \ x + 2z = -8 \ (4)$$

Add twice E2 to E3.
$$-6x - 4y - 6z = 22$$
$$\underline{4x + 4y + 5z = -18}$$
$$-2x - \qquad z = 4 \ (5)$$

Solve the new system:
$$\begin{cases} x + 2z = -8 & (4) \\ -2x - z = 4 & (5) \end{cases}$$

Add twice E4 to E5.
$$2x + 4z = -16$$
$$\underline{-2x - z = 4}$$
$$3z = -12$$
$$z = -4$$

Replace z with -4 in E4.
$$x + 2(-4) = -8$$
$$x - 8 = -8$$
$$x = 0$$

Replace x with 0 and z with -4 in E1.

$$0 + 2y - (-4) = 5$$
$$2y = 1$$
$$y = \frac{1}{2}$$

The solution is $\left(0, \frac{1}{2}, -4\right)$.

31. $\begin{cases} \frac{3}{4}x - \frac{1}{3}y + \frac{1}{2}z = 9 & (1) \\ \frac{1}{6}x + \frac{1}{3}y - \frac{1}{2}z = 2 & (2) \\ \frac{1}{2}x - y + \frac{1}{2}z = 2 & (3) \end{cases}$

Multiply E1 by 12, multiply E2 by 6, and multiply E3 by 2.
$$\begin{cases} 9x - 4y + 6z = 108 & (4) \\ x + 2y - 3z = 12 & (5) \\ x - 2y + z = 4 & (6) \end{cases}$$

Add twice E5 to E4.
$$2x + 4y - 6z = 24$$
$$\underline{9x - 4y + 6z = 108}$$
$$11x \qquad = 132$$
$$x = 12$$

Add E5 and E6.
$$2x - 2z = 16 \ \text{ or } \ x - z = 8$$

Replace x with 12 in this equation.
$$12 - z = 8$$
$$z = 4$$

Replace x with 12 and z with 4 in E6.
$$12 - 2y + 4 = 4$$
$$12 - 2y = 0$$
$$-2y = -12$$
$$y = 6$$

The solution is $(12, 6, 4)$.

33. Let x = the first number,
then $2x$ = the second number.
$$x + 2x = 45$$
$$3x = 45$$
$$x = 15$$
$$2x = 2(15) = 30$$

The numbers are 15 and 30.

35. $2(x-1)-3x = x-12$
$2x-2-3x = x-12$
$-x-2 = x-12$
$-2x = -10$
$x = 5$
The solution set is $\{5\}$.

37. $-y-5(y+5) = 3y-10$
$-y-5y-25 = 3y-10$
$-6y-25 = 3y-10$
$-9y = 15$
$y = -\dfrac{15}{9} = -\dfrac{5}{3}$
The solution set is $\left\{-\dfrac{5}{3}\right\}$.

39. Answers may vary.

41. Answers may vary.

43. $\begin{cases} x+\ y+\ z = 1 & (1) \\ 2x-\ y+\ z = 0 & (2) \\ -x+2y+2z = -1 & (3) \end{cases}$
Add E1 and E3.
$3y+3z = 0$ or $y+z = 0$ (4)
Add -2 times E1 to E2.
$-2x-2y-2z = -2$
$\underline{2x-\ y+\ z = 0}$
$-3y-\ z = -2$ (5)
Add E4 and E5.
$-2y = -2$
$y = 1$
Replace y with 1 in E4.
$1+z = 0$
$z = -1$
Replace y with 1 and z with -1 in E1.
$x+1+(-1) = 1$
$x = 1$
The solution is $(1, 1, -1)$, and

$\dfrac{x}{8}+\dfrac{y}{4}+\dfrac{z}{3} = \dfrac{1}{8}+\dfrac{1}{4}-\dfrac{1}{3}$
$= \dfrac{3}{24}+\dfrac{6}{24}-\dfrac{8}{24}$
$= \dfrac{1}{24}.$

45. $\begin{cases} x+y\ \ \ \ \ -w = 0 & (1) \\ y+2z+w = 3 & (2) \\ x\ \ \ \ -z\ \ \ \ \ = 1 & (3) \\ 2x-y\ \ \ \ \ -w = -1 & (4) \end{cases}$
Add E1 and E2.
$x+2y+2z = 3$ (4)
Add E2 and E4.
$2x+z = 2$ (5)
Add E3 and E5.
$x-z = 1$
$\underline{2x+z = 2}$
$3x\ \ \ = 3$
$x = 1$
Replace x with 1 in E3.
$1-z = 1$
$z = 0$
Replace x with 1 and z with 0 in E4.
$1+2y+2(0) = 3$
$1+2y = 3$
$2y = 2$
$y = 1$
Replace y with 1, and z with 0 in E2.
$1+2(0)+w = 3$
$1+w = 3$
$w = 2$
The solution is $(1, 1, 0, 2)$.

47. $\begin{cases} x+y+z+w = 5 & (1) \\ 2x+y+z+w = 6 & (2) \\ x+y+z\ \ \ \ \ = 2 & (3) \\ x+y\ \ \ \ \ \ \ \ \ = 0 & (4) \end{cases}$
Add -1 times E4 to E3.
$-x-y\ \ \ \ = 0$
$\underline{x+y+z = 2}$
$z = 2$

Replace z with 2 in E1 and E2.

$$\begin{cases} x + y + w = 3 & (5) \\ 2x + y + w = 4 & (6) \end{cases}$$

Add -1 times E5 to E6.

$$\begin{array}{rcl} -x - y - w &=& -3 \\ 2x + y + w &=& 4 \\ \hline x &=& 1 \end{array}$$

Replace x with 1 in E4.

$$1 + y = 0$$
$$y = -1$$

Replace x with 1, y with -1, and z with 2 in E1.

$$1 + (-1) + 2 + w = 5$$
$$2 + w = 5$$
$$w = 3$$

The solution is $(1, -1, 2, 3)$.

49. Answers may vary.

Exercise Set 4.5

1. Let x = the first number, and y = the second number.

$$\begin{cases} x = y + 2 & (1) \\ 2x = 3y - 4 & (2) \end{cases}$$

Substitute $x = y + 2$ in E2.

$$2(y + 2) = 3y - 4$$
$$2y + 4 = 3y - 4$$
$$y = 8$$

Replace y with 8 in E1.

$$x = 8 + 2 = 10$$

The numbers are 10 and 8.

3. Let x = length of Enterprise, and y = length of Nimitz.

$$\begin{cases} x + y = 2193 & (1) \\ x - y = 9 & (2) \end{cases}$$

Add E1 and E2.

$$2x = 2202$$
$$x = 1101$$

Replace x with 1102 in E1.

$$1101 + y = 2193$$
$$y = 1092$$

The Enterprise is 1101 feet long and the Nimitz is 1092 feet long.

5. Let p = the speed of the plane in still air, and w = the speed of the wind.

$$\begin{cases} p + w = 560 & (1) \\ p - w = 480 & (2) \end{cases}$$

Add E1 and E2.

$$2p = 1040$$
$$p = 520$$

Replace p with 520 in E1.

$$520 + w = 560$$
$$w = 40$$

The speed of the plane in still air is 520 mph and the speed of the wind is 40 mph.

7. Let x = number of quarts of 4% butterfat milk, and
y = number of quarts of 1% butterfat milk.

$$\begin{cases} x + y = 60 & (1) \\ 0.04x + 0.01y = 0.02(60) & (2) \end{cases}$$

Multiply E2 by -100 and add to E1.

$$\begin{array}{rcl} x + y &=& 60 \\ -4x - y &=& -120 \\ \hline -3x &=& -60 \end{array}$$
$$x = 20$$

Replace x with 20 in E1.

$$20 + y = 60$$
$$y = 40$$

There should be 20 quarts of 4% butterfat used and 40 quarts of 1% butterfat used.

9. Let k = number of students studied abroad in the United Kingdom and s = number of students studied abroad in Spain.

$$\begin{cases} k + \quad\ s = 40,012 \quad (1) \\ \quad\quad k = 15,428 + s \ (2) \end{cases}$$

Substitute $k = 15,428 + s$ in E1.

$$(15,428 + s) + s = 40,012$$
$$15,428 + 2s = 40,012$$
$$2s = 24,584$$
$$s = 12,292$$

Replace s with 12,292 in E21.
$$k = 15,428 + 12,292 = 27,720$$

The United Kingdom had 27,720 students and Spain had 12,292 students.

11. Let l = the number of large frames, and s = the number of small frames.

$$\begin{cases} \quad\ l + s = 22 \quad (1) \\ 15l + 8s = 239 \ (2) \end{cases}$$

Multiply E1 by -8 and add to E2.

$$-8l - 8s = -176$$
$$\underline{15l + 8s = 239}$$
$$7l \qquad = 63$$
$$l = 9$$

Replace l with 9 in E1.
$$9 + s = 22$$
$$s = 13$$

She bought 9 large frames and 13 small frames.

13. Let x = the first number, and y = the second number.

$$\begin{cases} \ x = y - 2 \quad (1) \\ 2x = 3y + 4 \ (2) \end{cases}$$

Substitute $x = y - 2$ in E2.

$$2(y - 2) = 3y + 4$$
$$2y - 4 = 3y + 4$$
$$y = -8$$

Replace y with -8 in E1.
$$x = -8 + 2 = -10$$
The numbers are -10 and -8.

15. $\begin{cases} y = 7x + 18.7 \ (1) \\ y = 6x + 27.7 \ (2) \end{cases}$

Substitute $7x + 18.7$ for y in E2.
$$7x + 18.7 = 6x + 27.7$$
$$x = 9$$
$$1996 + 9 = 2005$$
The year would be 2005.

17. Let x = price of each tablet, and y = the price of each pen.

$$\begin{cases} 7x + \ 4y = 6.40 \ (1) \\ 2x + 19y = 5.40 \ (2) \end{cases}$$

Multiply E1 by 2 and E2 by -7 and add.

$$14x + \quad 8y = 12.8$$
$$\underline{-14x - 133y = -37.80}$$
$$-125y = -25$$
$$y = 0.20$$

Replace y with 0.20 in E1.
$$7x + 4(0.20) = 6.40$$
$$7x + 0.80 = 6.40$$
$$7x = 5.60$$
$$x = 0.80$$

Tablets cost \$0.80 each and pens cost \$0.20 each.

19. Let p = the speed of the plane in still air, and w = the speed of the wind.
First note:
$$\frac{2160 \text{ mi}}{3 \text{ hr}} = 720 \text{ mph and}$$
$$\frac{2160 \text{ mi}}{4 \text{ hr}} = 540 \text{ mph}$$

Now,
$$\begin{cases} p + w = 720 \\ p - w = 540 \end{cases}$$

Add E1 and E2.
$$2p = 1260$$
$$p = 630$$

Replace p with 630 in E1.

$630 + w = 720$

$w = 90$

The speed of the plane in still air is 630 mph and the speed of the wind is 90 mph.

21. a. Answers may vary, but notice the slope of each function.

b. $\begin{cases} y = -0.52x + 64.5 & (1) \\ y = 0.02x + 7.2 & (2) \end{cases}$

Substitute $0.02x + 7.2$ for y in E1.

$0.02x + 7.2 = -0.52x + 64.5$

$0.54x = 57.3$

$x = \dfrac{57.3}{0.54} \approx 106.11$

$1990 + 106 = 2006$

They would be the same in 2006.

23. let x = length of shortest two sides, and y = length of the longest side.

$\begin{cases} 2x + y = 93 & (1) \\ y = x + 9 & (2) \end{cases}$

Replace y with $x + 9$ in E1.

$2x + (x + 9) = 93$

$3x + 9 = 93$

$3x = 84$

$x = 28$

Replace x with 28 in E2.

$y = 28 + 9 = 35$

The three sides are 28 cm, 28 cm, and 35 cm.

25. Cost for Hertz: $H(x) = 25 + 0.10x$

Cost for Budget: $B(x) = 20 + 0.25x$

We want to find when $B(x) = 2 \cdot H(x)$

$20 + 0.25x = 2(25 + 0.10x)$

$20 + 0.25x = 50 + 0.20x$

$0.05x = 30$

$x = 600$

The Budget charge will be twice that of the Hertz charge at 600 miles.

27. $\begin{cases} x + y = 180 \\ x = y - 30 \end{cases}$

Replace x with $y - 30$ in E1.

Replace y with $x + 9$ in E1.

$(y - 30) + y = 180$

$2y - 30 = 180$

$2y = 210$

$y = 105$

Replace y with 105 in E2.

$x = 105 - 30 = 75$

The value of x is 75° and the value of y is 105°.

29. $C(x) = 30x + 10,000$

$R(x) = 46x$

$46x = 30x + 10,000$

$16x = 10,000$

$x = 625$

625 units

31. $C(x) = 1.2x + 1500$

$R(x) = 1.7x$

$1.7x = 1.2x + 1500$

$0.5x = 1500$

$x = 3000$ units

3000 units

33. $C(x) = 75x + 160,000$

$R(x) = 200x$

$200x = 75x + 160,000$

$125x = 160,000$

$x = 1280$

1280 units

35. a. $R(x) = 450x$

b. $C(x) = 200x + 6000$

c. $R(x) = C(x)$

$450x = 200x + 6000$

$250x = 6000$

$x = 24$ desks

37. Let x = units of Mix A

y = units of Mix B, and

z = units of Mix C.

$$\begin{cases} 4x+6y+4z=30 & (1) \\ 6x+y+z=16 & (2) \\ 3x+2y+12z=24 & (3) \end{cases}$$

Multiply E2 by –6 and add to E1.

$-36x-6y-6z=-96$

$\underline{4x+6y+4z=30}$

$-32x-2z=-66$ or $16x+z=33$ (4)

Multiply E2 by –2 and add to E3.

$-12x-2y-2z=-32$

$\underline{3x+2y+12z=24}$

$-9x+10z=-8$ (5)

Multiply E4 by –10 and add to E5.

$-160x-10z=-330$

$\underline{-9x+10z=-8}$

$-169x=-338$

$x=2$

Replace x with 2 in E4.

$16(2)+z=33$

$32+z=33$

$z=1$

Replace x with 2 and z with 1 in E2.

$6(2)+y+1=16$

$y=3$

You need 2 units of Mix A, 3 units of Mix B, and 1 unit of Mix C.

39. Let x = length of shortest side,

y = length of longest side, and

z = length of the other two sides

$$\begin{cases} x+y+2z=29 & (1) \\ y=2x & (2) \\ z=x+2 & (3) \end{cases}$$

Substitute $y=2x$ and $z=x+2$ in E1.

$x+(2x)+2(x+2)=29$

$x+2x+2x+4=29$

$5x=25$

$x=5$

Replace x with 5 in E2 and E3.

$-2(5)+y=0 \qquad -(5)+z=2$

$\qquad y=10 \qquad\qquad z=7$

The sides are 5 in., 7 in., 7 in., and 10 in.

41. Let x = the first number

y = the second number, and

z = the third number.

$$\begin{cases} x+y+z=40 \\ x=y+5 \\ x=2z \end{cases}$$

$$\begin{cases} x+y+z=40 & (1) \\ x-y=5 & (2) \\ x-2z=0 & (3) \end{cases}$$

Add E1 and E2.

$2x+z=45$ (4)

Multiply E3 by –2 and add to E4.

$-2x+4z=0$

$\underline{2x+z=45}$

$5z=45$

$z=9$

Replace z with 9 in E3.

$x-2(9)=0$

$x=18$

Replace x with 15 in E2.

$18-y=5$

$y=13$

The numbers are 18, 13, and 9.

43. Let x = number of free throws,

y = number of two-point field goals, and

z = number of three-point field goals

$$\begin{cases} x+2y+3z=698 \\ y=6z-19 \\ x=y-64 \end{cases}$$

$$\begin{cases} x+2y+3z=698 & (1) \\ y-6z=-19 & (2) \\ x-y=-64 & (3) \end{cases}$$

Multiply E3 by –1 and add to E1.

$-x+y=64$

$\underline{x+2y+3z=698}$

$3y+3z=762$ or $y+z=254$ (4)

Multiply E4 by –1 and add to E2.

$$-y - z = -254$$
$$\underline{y - 6z = -19}$$
$$-7z = -273$$
$$z = 39$$

Replace z with 39 in E2.

$$y - 6(39) = -19$$
$$y - 234 = -19$$
$$y = 215$$

Replace y with 215 in E3.

$$x - 215 = -64$$
$$x = 151$$

She made 151 free throws, 215 two-point field goals, and 39 three-point field goals.

45. $\begin{cases} x + y + z = 180 \\ y + 2x + 5 = 180 \\ z + 2x - 5 = 180 \end{cases}$

$\begin{cases} x + y + z = 180 \quad (1) \\ 2x + y \quad\quad\; = 175 \quad (2) \\ 2x + \quad\quad z = 185 \quad (3) \end{cases}$

Multiply E1 by –1 and add to E2.

$$-x - y - z = -180$$
$$\underline{2x + y \quad\quad = 175}$$
$$x \quad\quad - z = -5 \quad (4)$$

Add E3 and E4.

$$3x = 180$$
$$x = 60$$

Replace x with 60 in E4.

$$60 - z = -5$$
$$z = 65$$

Replace x with 60 in E2.

$$2(60) + y = 175$$
$$120 + y = 175$$
$$y = 55$$

$x = 60$, $y = 55$, and $z = 65$

47. $\begin{cases} 3x - y + z = 2 \\ -x + 2y + 3z = 6 \end{cases}$

$$6x - 2y + 2z = 4$$
$$\underline{-x + 2y + 3z = 6}$$
$$5x \quad\quad + 5z = 10$$
$$5x + 5z = 10$$

49. $\begin{cases} x + 2y - z = 0 \\ 3x + y - z = 6 \end{cases}$

$$-3x - 6y + 3z = 0$$
$$\underline{3x + y - z = 2}$$
$$-5y + 2z = 2$$
$$-5y + 2z = 2$$

51. Let x = number filed in 1980 and y = number filed in 2001

$\begin{cases} y = 4x + 200,000 \quad (1) \\ y - x = 1,100,000 \quad (2) \end{cases}$

Substitute $y = 4x + 200,000$ in E2.

$$(4x + 200,000) - x = 1,100,000$$
$$3x + 200,000 = 1,100,000$$
$$3x = 900,000$$
$$x = 300,000$$

Replace x with 300,000 in E1.

$$y = 4(300,000) + 200,000 = 1,400,000$$

There were 300,000 filed in 1980 and 1,400,000 filed in 2001.

53. $y = ax^2 + bx + c$

$(1, 6): \quad 6 = a + b + c \quad (1)$
$(-1, -2): -2 = a - b + c \quad (2)$
$(0, -1): \quad -1 = c \quad\quad\quad\; (3)$

Substitute $c = -1$ in E1 and E2 to obtain

$\begin{cases} a + b = 7 \quad \text{(from E1)} \\ a - b = -1 \text{ (from E2)} \end{cases}$

Add these equations.

$$2a = 6$$
$$a = 3$$

Replace a with 3 in $a + b = 7$.

$3 + b = 7$ so $b = 4$

Therefore, $a = 3$, $b = 4$, and $c = -1$.

55. $y = ax^2 + bx + c$

$(0, 1065) : 1065 = c$ (1)

$(1, 1070) : 1070 = a + b + c$ (2)

$(3, 1175) : 1175 = 9a + 3b + c$ (3)

Substitute $c = 1065$ in E2 and E3 to

obtain $\begin{cases} a + b = 5 \\ 9a + b = 105 \end{cases}$

Multiply the first equation by -1 and add to the second equation.

$-3a - 3b = -15$

$\underline{9a + 3b = 110}$

$6a \quad\quad = 95$

$a = \dfrac{95}{6} = 15\dfrac{5}{6}$

Replace a with $15\dfrac{5}{6}$ in $a + b = 5$.

$\dfrac{95}{6} + b = 5$

$b = -\dfrac{65}{6} = -10\dfrac{5}{6}$

So, $a = 15\dfrac{5}{6}$, $b = -10\dfrac{5}{6}$, and $c = 1065$.

Chapter 4 Review

1. a. Let $x = 12$ and $y = 4$.

$\begin{array}{ll} 2x - 3y = 12 & 3x + 4y = 1 \\ 2(12) - 3(4) \overset{?}{=} 12 & 3(12) + 4(4) \overset{?}{=} 1 \\ 24 - 12 \overset{?}{=} 12 & 36 + 16 \overset{?}{=} 1 \\ 12 = 12 & 52 = 1 \\ \text{True} & \text{False} \end{array}$

$(12, 4)$ is not a solution of the system.

b. Let $x = 3$ and $y = -2$.

$\begin{array}{ll} 2x - 3y = 12 & 3x + 4y = 1 \\ 2(3) - 3(-2) \overset{?}{=} 12 & 3(3) + 4(-2) \overset{?}{=} 1 \\ 6 + 6 \overset{?}{=} 12 & 9 - 8 \overset{?}{=} 1 \\ 2 = 12 & 1 = 1 \\ \text{True} & \text{True} \end{array}$

$(3, -2)$ is a solution of the system

c. Let $x = -3$ and $y = 6$.

$\begin{array}{ll} 2x - 3y = 12 & 3x + 4y = 1 \\ 2(-3) - 3(6) \overset{?}{=} 12 & 3(-3) + 4(6) \overset{?}{=} 1 \\ -6 - 18 \overset{?}{=} 12 & -9 + 24 \overset{?}{=} 1 \\ -24 = 12 & 15 = 1 \\ \text{False} & \text{False} \end{array}$

$(-3, 6)$ is not a solution of the system

2. a. Let $x = \dfrac{3}{4}$ and $y = -3$.

$$4x + y = 0 \qquad\qquad -8x - 5y = 9$$

$$4\left(\frac{3}{4}\right) - 3 \overset{?}{=} 0 \qquad -8\left(\frac{3}{4}\right) - 5(-3) \overset{?}{=} 9$$

$$3 - 3 \overset{?}{=} 0 \qquad\qquad -6 + 15 \overset{?}{=} 9$$

$$0 = 0 \qquad\qquad\qquad 9 = 9$$

$$\text{True} \qquad\qquad\qquad \text{True}$$

$\left(\dfrac{3}{4}, -3\right)$ is a solution of the system.

b. Let $x = -2$ and $y = 8$.

$$4x + y = 0 \qquad\qquad -8x - 5y = 9$$

$$4(-2) + 8 \overset{?}{=} 0 \qquad -8(-2) - 5(8) \overset{?}{=} 9$$

$$-8 + 8 \overset{?}{=} 0 \qquad\qquad 16 - 40 \overset{?}{=} 9$$

$$0 = 0 \qquad\qquad\qquad -24 = 9$$

$$\text{True} \qquad\qquad\qquad \text{False}$$

$(-2, 8)$ is not a solution of the system

c. Let $x = \dfrac{1}{2}$ and $y = -2$.

$$4x + y = 0 \qquad\qquad -8x - 5y = 9$$

$$4\left(\frac{1}{2}\right) - 2 \overset{?}{=} 0 \qquad -8\left(\frac{1}{2}\right) - 5(-2) \overset{?}{=} 9$$

$$2 - 2 \overset{?}{=} 0 \qquad\qquad -4 + 10 \overset{?}{=} 9$$

$$0 = 0 \qquad\qquad\qquad 6 = 9$$

$$\text{True} \qquad\qquad\qquad \text{False}$$

$\left(\dfrac{1}{2}, -2\right)$ is not a solution of the system.

3. a. Let $x = -6$ and $y = -8$.

$$5x - 6y = 18 \qquad\qquad 2y - x = -4$$

$$5(-6) - 6(-8) \overset{?}{=} 18 \quad 2(-8) - (-6) \overset{?}{=} -4$$

$$-30 + 48 \overset{?}{=} 18 \qquad -16 + 6 \overset{?}{=} -4$$

$$18 = 18 \qquad\qquad -10 = -4$$

$$\text{True} \qquad\qquad\qquad \text{False}$$

$(-6, -8)$ is not a solution of the system.

b. Let $x = 3$ and $y = \dfrac{5}{2}$.

$$5x - 6y = 18 \qquad\qquad 2y - x = -4$$

$$5(3) - 6\left(\frac{5}{2}\right) \overset{?}{=} 18 \qquad 2\left(\frac{5}{2}\right) - 3 \overset{?}{=} -4$$

$$15 - 15 \overset{?}{=} 18 \qquad\qquad 5 - 3 \overset{?}{=} -4$$

$$0 = 18 \qquad\qquad\qquad 2 = -4$$

$$\text{False} \qquad\qquad\qquad \text{False}$$

$\left(3, \dfrac{5}{2}\right)$ is not a solution of the system

c. Let $x = 3$ and $y = -\dfrac{1}{2}$.

$$5x - 6y = 18 \qquad\qquad 2y - x = -4$$

$$5(3) - 6\left(-\frac{1}{2}\right) \overset{?}{=} 18 \qquad 2\left(-\frac{1}{2}\right) - 3 \overset{?}{=} -4$$

$$15 + 3 \overset{?}{=} 18 \qquad\qquad -1 - 3 \overset{?}{=} -4$$

$$18 = 18 \qquad\qquad\qquad -4 = -4$$

$$\text{True} \qquad\qquad\qquad \text{True}$$

$\left(3, -\dfrac{1}{2}\right)$ is a solution of the system

4. a. Let $x = 2$ and $y = 2$.

$$2x + 3y = 1 \qquad\qquad 3y - x = 4$$

$$2(2) + 3(2) \overset{?}{=} 1 \qquad 3(2) - (2) \overset{?}{=} 4$$

$$4 + 6 \overset{?}{=} 1 \qquad\qquad 6 - 2 \overset{?}{=} 4$$

$$10 = 1 \qquad\qquad\qquad 4 = 4$$

$$\text{False} \qquad\qquad\qquad \text{True}$$

$(2, 2)$ is not a solution of the system.

b. Let $x = -1$ and $y = 1$.

$$2x + 3y = 1 \qquad\qquad 3y - x = 4$$

$$2(-1) + 3(1) \overset{?}{=} 1 \qquad 3(1) - (-1) \overset{?}{=} 4$$

$$-2 + 3 \overset{?}{=} 1 \qquad\qquad 3 + 1 \overset{?}{=} 4$$

$$1 = 1 \qquad\qquad\qquad 4 = 4$$

$$\text{True} \qquad\qquad\qquad \text{True}$$

$(-1, 1)$ is a solution of the system

c. Let $x = 2$ and $y = -1$.

$$2x + 3y = 1 \qquad\qquad 3y - x = 4$$

$$2(2) + 3(-1) \overset{?}{=} 1 \qquad 3(2) - (-1) \overset{?}{=} 4$$

$$4 - 3 \overset{?}{=} 1 \qquad\qquad 6 + 1 \overset{?}{=} 4$$

$$1 = 1 \qquad\qquad\qquad 5 = 4$$

$$\text{True} \qquad\qquad\qquad \text{False}$$

$(2, -1)$ is not a solution of the system.

5. $\begin{cases} 2x + y = 5 \\ 3y = -x \end{cases}$

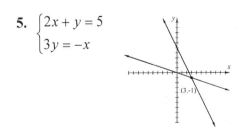

The solution of the system is $(3, -1)$.

6. $\begin{cases} 3x + y = -2 \\ 2x - y = -3 \end{cases}$

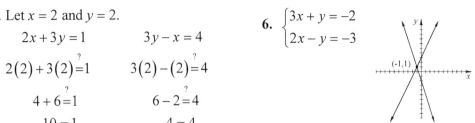

The solution of the system is $(-1, 1)$.

7. $\begin{cases} y - 2x = 4 \\ x + y = -5 \end{cases}$

The solution of the system is $(-3, -2)$.

8. $\begin{cases} y - 3x = 0 \\ 2y - 3 = 6x \end{cases}$

There is no solution of the system.

9. $\begin{cases} 3x + y = 2 \\ 3x - 6 = -9y \end{cases}$

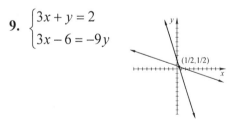

The solution of the system is $\left(\dfrac{1}{2}, \dfrac{1}{2} \right)$.

10. $\begin{cases} 2y + x = 2 \\ x - y = 5 \end{cases}$

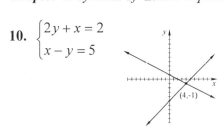

The solution of the system is $(4, -1)$.

11. intersecting, one solution

12. parallel no solution

13. identical,
There is an infinite number of solutions.

14. intersecting, one solution

15. $\begin{cases} y = 2x + 6 \\ 3x - 2y = -11 \end{cases}$

Substitute $2x + 6$ for y in the second equation.

$$3x - 2(2x + 6) = -11$$
$$3x - 4x - 12 = -11$$
$$-x = 1$$
$$x = -1$$

Let $x = -1$ in the first equation.

$$y = 2(-1) + 6$$
$$y = 4$$

The solution is $(-1, 4)$.

16. $\begin{cases} y = 3x - 7 \\ 2x - 3y = 7 \end{cases}$

Substitute $3x - 7$ for y in the second equation.

$$2x - 3(3x - 7) = 7$$

$$2x - 9x + 21 = 7$$
$$-7x = -14$$
$$x = 2$$

Let $x = 2$ in the first equation.

$$y = 3(2) - 7$$
$$y = -1$$

The solution is $(2, -1)$.

17. $\begin{cases} x + 3y = -3 \\ 2x + y = 4 \end{cases}$

Solve the first equation for x.

$$x = -3y - 3$$

Substitute $-3y - 3$ for x in the second equation.

$$2(-3y - 3) + y = 4$$
$$-6y - 6 + y = 4$$
$$-5y = 10$$
$$y = -2$$

Let $y = -2$ in $x = -3y - 3$.

$$x = -3(-2) - 3$$
$$x = 3$$

The solution is $(3, -2)$.

18. $\begin{cases} 3x + y = 11 \\ x + 2y = 12 \end{cases}$

Solve the first equation for y.

$$y = 11 - 3x$$

Substitute $11 - 3x$ for y in the second equation.

$$x + 2(11 - 3x) = 12$$
$$x + 22 - 6x = 12$$

$$-5x = -10$$
$$x = 2$$

Let $x = 2$ in $y = 11 - 3x$.

$$y = 11 - 3(2)$$
$$y = 5$$

The solution is $(2, 5)$.

19. $\begin{cases} 4y = 2x - 3 \\ x - 2y = 4 \end{cases}$

Solve the second equation for x.

$$x = 4 + 2y$$

Substitute $4 + 2y$ for x in the first equation.

$$4y = 2(4 + 2y) - 3$$
$$4y = 8 + 4y - 3$$
$$0 = 5$$

The system has no solution.

20. $\begin{cases} 2x = 3y - 18 \\ x + 4y = 2 \end{cases}$

Solve the second equation for x.

$$x = 2 - 4y$$

Substitute $2 - 4y$ for x in the first equation.

$$2(2 - 4y) = 3y - 18$$
$$4 - 8y = 3y - 18$$
$$-11y = -22$$
$$y = 2$$

Let $y = 2$ in $x = 2 - 4y$.

$$x = 2 - 4(2)$$
$$x = -6$$

The solution is $(-6, 2)$.

21. $\begin{cases} 2(3x - y) = 7x - 5 \\ 6x - 2y = 7x - 5 \\ 5 - 2y = x \\ 3(x - y) = 4x - 6 \\ 3x - 3y = 4x - 6 \\ 6 - 3y = x \end{cases}$

Substitute $5 - 2y$ for x in the second equation.

$$6 - 3y = 5 - 2y$$
$$-y = -1$$
$$y = 1$$

Let $y = 1$ in $5 - 2y = x$.

$$x = 5 - 2(1)$$
$$x = 3$$

The solution is $(3, 1)$.

22. $\begin{cases} 4(x - 3y) = 3x - 1 \\ 4x - 12y = 3x - 1 \\ x = 12y - 1 \\ 3(4y - 3x) = 1 - 8x \\ 12y - 9x = 1 - 8x \\ 12y - 1 = x \end{cases}$

Substitute $12y - 1$ for x in the second equation.

$$12y - 1 = 12y - 1$$
$$0 = 0$$

The system is dependent.

There are an infinite number of solutions.

23. $\begin{cases} \dfrac{3}{4}x + \dfrac{2}{3}y = 2 \\ 3x + y = 18 \end{cases}$

Multiply the first equation by 12.

$\begin{cases} 9x + 8y = 24 \\ 3x + y = 18 \end{cases}$

Solve the second equation of the simplified system for y

$y = 18 - 3x$

Substitute $18 - 3x$ for y in the first equation..

$9x + 8(18 - 3x) = 24$

$9x + 144 - 24x = 24$

$-15x = -120$

$x = 8$

Substitute 8 for x in $y = 18 - 3x$.

$y = 18 - 3(8) = -6$

The solution of the system is $(8, -6)$.

24. $\begin{cases} \dfrac{2}{5}x + \dfrac{3}{4}y = 1 \\ x + 3y = -2 \end{cases}$

Multiply the first equation by 20.

$\begin{cases} 8x + 15y = 20 \\ x + 3y = -2 \end{cases}$

Solve the second equation of the simplified system for x

$x = -3y - 2$

Substitute $-3y - 2$ for x in the first equation..

$8(-3y - 2) + 15y = 20$

$-24y - 16 + 15y = 20$

$-9y = 36$

$y = -4$

Substitute -4 for y in $x = -3y - 2$.

$x = -3(-4) - 2 = 10$

The solution of the system is $(10, -4)$.

25. $\begin{cases} 2x + 3y = -6 \\ \underline{x - 3y = -12} \\ 3x = -18 \\ x = -6 \end{cases}$

Let $x = -6$ in the first equation.

$2(-6) + 3y = -6$

$-12 + 3y = -6$

$3y = 6$

$y = 2$

The solution of the system is $(-6, 2)$

26. $\begin{cases} 4x + y = 15 \\ \underline{-4x + 3y = -19} \\ 4y = -4 \\ y = -1 \end{cases}$

Let $y = -1$ in the first equation.

$4x + (-1) = 15$

$4x - 1 = 15$

$4x = 16$

$x = 4$

The solution of the system is $(4, -1)$.

27. $\begin{cases} 2x - 3y = -15 \\ x + 4y = 31 \end{cases}$

Multiply the second equation by -2.

$2x - 3y = -15$
$\underline{-2x - 8y = -62}$
$\qquad -11y = -77$
$\qquad\qquad y = 7$

Let $y = 7$ in the second equation.

$x + 4(7) = 31$
$x + 28 = 31$
$\qquad x = 3$

The solution of the system is $(3, 7)$.

28. $\begin{cases} x - 5y = -22 \\ 4x + 3y = 4 \end{cases}$

Multiply the first equation by -4.

$-4x + 20y = 88$
$\underline{4x + 3y = 4}$
$\qquad\quad 23y = 92$
$\qquad\qquad y = 4$

Let $y = 4$ in the first equation.

$x - 5(4) = -22$
$x - 20 = -22$
$\qquad x = -2$

The solution of the system is $(-2, 4)$.

29. $\begin{cases} 2x = 6y - 1 \\ \dfrac{1}{3}x - y = \dfrac{-1}{6} \end{cases}$

Rearrange the first equation.

Multiply the second equation by -6.

$2x - 6y = -1$
$\underline{-2x + 6y = 1}$
$\qquad 0 = 0$

The system is dependent.

There are an infinite number of solutions.

30. $\begin{cases} 8x = 3y - 2 \\ \dfrac{4}{7}x - y = \dfrac{-5}{2} \end{cases}$

Rearrange the first equation.

Multiply the second equation by -6.

$8x - 3y = -2$
$\underline{-8x + 14y = 35}$
$\qquad\quad 11y = 33$
$\qquad\qquad y = 3$

Let $y = 3$ in the first equation.

$8x = 3(3) - 2$
$8x = 7$
$\quad x = \dfrac{7}{8}$

The solution of the system is $\left(\dfrac{7}{8}, 3\right)$.

31. $\begin{cases} 5x = 6y + 25 \\ -2y = 7x - 9 \end{cases}$

Rearrange both equations.

$\begin{cases} 5x - 6y = 25 \\ -7x - 2y = -9 \end{cases}$

Multiply the second equation by -3.

$5x - 6y = 25$
$\underline{21x + 6y = 27}$
$26x = 52$
$\qquad x = 2$

Let $x = 2$ in the second equation.

$-2y = 7(2) - 9$

$-2y = 5$

$y = -\dfrac{5}{2}$

The solution of the system is $\left(2, -2\frac{1}{2}\right)$.

32. $\begin{cases} -4x = 8 + 6y \\ -3y = 2x - 3 \end{cases}$

Rearrange both equations.

$\begin{cases} -4x - 6y = 8 \\ -2x - 3y = -3 \end{cases}$

Multiply the second equation by -2.

$\begin{array}{r} -4x - 6y = 8 \\ 4x + 6y = -6 \\ \hline 0 = 2 \end{array}$

The system is inconsistent
and has no solution.

33. $\begin{cases} 3(x - 4) = -2y \\ 3x - 12 = -2y \\ 3x + 2y = 12 \\ \\ 2x = 3(y - 19) \\ 2x = 3y - 57 \\ 2x - 3y = -57 \end{cases}$

Multiply the first equation by 3 and
the second equation by 2.

$\begin{array}{r} 9x + 6y = 36 \\ 4x - 6y = -114 \\ \hline 13x = -78 \\ x = -6 \end{array}$

Let $x = -6$ in the first equation.

$3(-6) + 2y = 12$

$-18 + 2y = 12$

$2y = 30$

$y = 15$

The solution of the system is $(-6, 15)$.

34. $\begin{cases} 4(x + 5) = -3y \\ 4x + 20 = -3y \\ 4x + 3y = -20 \\ \\ 3x - 2(y + 18) = 0 \\ 3x - 2y - 36 = 0 \\ 3x - 2y = 36 \end{cases}$

Multiply the first equation by 2 and
the second equation by 3.

$\begin{array}{r} 8x + 6y = -40 \\ 9x - 6y = 108 \\ \hline 17x = 68 \\ x = 4 \end{array}$

Let $x = 4$ in the first equation.

$4(4) + 3y = -20$

$16 + 3y = -20$

$3y = -36$

$y = -12$

The solution of the system is $(4, -12)$.

35. $\begin{cases} \dfrac{2x+9}{3} = \dfrac{y+1}{2} \\[2mm] \dfrac{x}{3} = \dfrac{y-7}{6} \end{cases}$

Multiply both equations by 6

$\begin{cases} 2(2x+9) = 3(y+1) \\ 4x+18 = 3y+3 \\ 4x-3y = -15 \\[3mm] 2x = y-7 \\ 2x - y = -7 \end{cases}$

Multiply the second equation by -3.

$\begin{aligned} 4x - 3y &= -15 \\ -6x + 3y &= 21 \\ \hline -2x &= 6 \\ x &= -3 \end{aligned}$

Let $x = -3$ in the second equation.

$\begin{aligned} 2(-3) - y &= -7 \\ -6 - y &= -7 \\ -y &= -1 \\ y &= 1 \end{aligned}$

The solution of the system is $(-3, 1)$.

36. $\begin{cases} \dfrac{2-5x}{4} = \dfrac{2y-4}{2} \\[2mm] \dfrac{x+5}{3} = \dfrac{y}{5} \end{cases}$

Multiply the first equation by 4
and the second equation by 15

$\begin{cases} 2 - 5x = 2(2y-4) \\ 2 - 5x = 4y - 8 \\ 5x + 4y = 10 \\[3mm] 5(x+5) = 3y \\ 5x + 25 = 3y \\ 5x - 3y = -25 \end{cases}$

Multiply the second equation by -1.

$\begin{aligned} 5x + 4y &= 10 \\ -5x + 3y &= 25 \\ \hline 7y &= 35 \\ y &= 5 \end{aligned}$

Let $y = 5$ in the second equation.

$\begin{aligned} 5x - 3(5) &= -25 \\ 5x - 15 &= -25 \\ 5x &= -10 \\ x &= -2 \end{aligned}$

The solution of the system is $(-2, 5)$.

37. $\begin{cases} x + z = 4 & (1) \\ 2x - y = 4 & (2) \\ x + y - z = 0 & (3) \end{cases}$

Adding E2 and E3 gives $3x - z = 4$ (4)
Adding E1 and E4 gives $4x = 8$ or $x = 2$
Replace x with 2 in E1.

$\begin{aligned} 2 + z &= 4 \\ z &= 2 \end{aligned}$

Replace x with 2 and z with 2 in E3.

$\begin{aligned} 2 + y - 2 &= 0 \\ y &= 0 \end{aligned}$

The solution is $(2, 0, 2)$.

38. $\begin{cases} 2x+5y \quad = 4 \quad (1) \\ x-5y+z=-1 \ (2) \\ 4x \quad -z=11 \ (3) \end{cases}$

Add E2 and E3.

$5x-5y=10 \quad (4)$

Add E1 and E4.

$7x=14$

$x=2$

Replace x with 2 in E1.

$2(2)+5y=4$

$4+5y=4$

$5y=0$

$y=0$

Replace x with 2 in E3.

$4(2)-z=11$

$8-z=11$

$z=-3$

The solution is $(2,\,0,\,-3)$.

39. $\begin{cases} 4y+2z=5 \quad (1) \\ 2x+8y \quad =5 \quad (2) \\ 6x+ \quad 4z=1 \quad (3) \end{cases}$

Multiply E1 by -2 and add to E2.

$-8y-4z=-10$

$\underline{2x+8y \qquad =5}$

$2x \qquad -4z=-5 \quad (4)$

Add E3 and E4.

$8x=-4$

$x=-\dfrac{1}{2}$

Replace x with $-\dfrac{1}{2}$ in E2.

$2\left(-\dfrac{1}{2}\right)+8y=5$

$-1+8y=5$

$8y=6$

$y=\dfrac{3}{4}$

Replace x with $-\dfrac{1}{2}$ in E3.

$6\left(-\dfrac{1}{2}\right)+4z=1$

$-3+4z=1$

$4z=4$

$z=1$

The solution is $\left(-\dfrac{1}{2},\,\dfrac{3}{4},\,1\right)$.

40. $\begin{cases} 5x+7y \quad =9 \quad (1) \\ 14y-z=28 \quad (2) \\ 4x+ \quad 2z=-4 \quad (3) \end{cases}$

Dividing E3 by 2 gives $2x+z=-2$.

Adding this equation to E2.

$2x \qquad +z=-2$

$\underline{\quad 14y-z=28}$

$2x+14y=26 \ \text{or} \ x+7y=13 \quad (4)$

Multiply E4 by -1 and add to E1.

$-x-7y=-13$

$\underline{5x+7y=9}$

$4x \qquad =-4$

$x \qquad =-1$

Replace x with -1 in E4.

$$1 + 7y = 13$$
$$7y = 14$$
$$y = 2$$

Replace x with -1 in E3.

$$4(-1) + 2z = -4$$
$$-4 + 2z = -4$$
$$2z = 0$$
$$z = 0$$

The solution is $(-1, 2, 0)$.

41. $\begin{cases} 3x - 2y + 2z = 5 & (1) \\ -x + 6y + z = 4 & (2) \\ 3x + 14y + 7z = 20 & (3) \end{cases}$

Multiply E2 by 3 and add to E1.

$$3x - 2y + 2z = 5$$
$$\underline{-3x + 18y + 3z = 12}$$
$$16y + 5z = 17 \quad (4)$$

Multiply E3 by -1 and add to E1.

$$3x - 2y + 2z = 5$$
$$\underline{-3x - 14y - 7z = -20}$$
$$-16y - 5z = -15 \quad (5)$$

Add E4 and E5.

$$16x + 5z = 17$$
$$\underline{-16x - 5z = -15}$$
$$0 = 2 \quad \text{False}$$

The system is inconsistent.
The solution is $\varnothing$.

42. $\begin{cases} x + 2y + 3z = 11 & (1) \\ y + 2z = 3 & (2) \\ 2x + 7z = 20 & (3) \end{cases}$

Multiply E2 by -2 and add to E1.

$$x + 2y + 3z = 11$$
$$\underline{-2y - 4z = -6}$$
$$x - z = 5 \quad (4)$$

Multiply E4 by 2 and add to E3.

$$2x + 2z = 10$$
$$\underline{2x - 2z = 10}$$
$$4x = 20$$
$$x = 5$$

Replace x with 5 in E3.

$$2(5) + 2z = 10$$
$$10 + 2z = 10$$
$$2z = 0$$
$$z = 0$$

Replace z with 0 in E2.

$$y + 2(0) = 3$$
$$y + 0 = 3$$
$$y = 3$$

The solution is $(5, 3, 0)$.

43. $\begin{cases} 7x - 3y + 2z = 0 & (1) \\ 4x - 4y - z = 2 & (2) \\ 5x + 2y + 3z = 1 & (3) \end{cases}$

Multiply E2 by 2 and add to E1.

$$7x - 3y + 2z = 0$$
$$\underline{8x - 8y - 2z = 4}$$
$$15x - 11y = 4 \quad (4)$$

Multiply E2 by 3 and add to E3.

$12x - 12y - 3z = 6$

$\underline{5x + 2y + 3z = 1}$

$17x - 10y \quad = 7 \quad (5)$

Solve the new system.

$\begin{cases} 15x - 11y = 4 & (4) \\ 17x - 10y = 7 & (5) \end{cases}$

Multiply E4 by -10, multiply E5 by 11, and add.

$-150x + 110y = -40$

$\underline{187x - 110y = 77}$

$37x \quad\quad = 37$

$x \quad\quad = 1$

Replace x with 1 in E4.

$15(1) - 11y = 4$

$15 - 11y = 4$

$-11y = -11$

$y = 1$

Replace x with 1 and y with 1 in E1.

$7(1) - 3(1) + 2z = 0$

$4 + 2z = 0$

$2z = -4$

$z = -2$

The solution is $(1, 1, -2)$.

44. $\begin{cases} x - 3y - 5z = -5 & (1) \\ 4x - 2y + 3z = 13 & (2) \\ 5x + 3y + 4z = 22 & (3) \end{cases}$

Multiply E1 by -4 and add to E2.

$-4x + 12y + 20z = 20$

$\underline{4x - 2y + 3z = 13}$

$10y + 23z = 33 \quad (4)$

Multiply E1 by -5 and add to E3.

$-5x + 15y + 20z = 25$

$\underline{5x + 3y + 4z = 22}$

$18y + 29z = 47 \quad (5)$

Solve the new system.

$\begin{cases} 10y + 23z = 33 & (4) \\ 18y + 29z = 47 & (5) \end{cases}$

Multiply E4 by 9, multiply E5 by -5 and add.

$90y + 207z = 297$

$\underline{-90y - 145z = -235}$

$62z = 62$

$z = 1$

Replace z with 1 in E4.

$10y + 23(1) = 33$

$10y = 10$

$y = 1$

Replace y with 1 and z with 1 in E1.

$x - 3(1) - 5(1) = -5$

$x - 8 = -5$

$x = 3$

The solution is $(3, 1, 1)$.

45. Let x = the larger number and y = the smaller number.

$$\begin{cases} x + y = 16 \\ 3x - y = 72 \end{cases}$$

$$\underline{4x = 88}$$

$$x = 22$$

Let $x = 22$ in the first equation.

$$22 + y = 16$$

$$y = -6$$

The numbers are -6 and 22.

46. Let x = the number of orchestra seats and y = the number of balcony seats.

$$\begin{cases} x + y = 360 \\ 45x + 35y = 15,150 \end{cases}$$

Solve the first equation for x.

$$x = 360 - y$$

Substitute $360 - y$ for x in the second equation.

$$45(360 - y) + 35y = 15,150$$

$$16,200 - 45y + 35y = 15,150$$

$$-10y = -1050$$

$$y = 105$$

Let $y = 105$ in $x = 360 - y$.

$$x = 360 - 105$$

$$x = 255$$

There are 255 people.

47. Let x = the riverboat's speed in still water and y = the rate of the current.

	d	$=$ r	$\cdot$ t
Downriver	340	$x + y$	14
Upriver	340	$x - y$	19

$$\begin{cases} 14(x + y) = 340 \\ 19(x - y) = 340 \end{cases}$$

Multiply the first equation by $\dfrac{1}{14}$ and the second equation by $\dfrac{1}{19}$.

$$\begin{cases} x + y = \dfrac{340}{14} \approx 24.29 \\ x - y = \dfrac{340}{19} \approx 17.89 \end{cases}$$

$$\underline{2x \approx 42.18}$$

$$x \approx 21.09$$

Multiply the second equation of the simplified system by -1.

$$\begin{cases} x + y \approx 24.29 \\ -x + y \approx -17.89 \end{cases}$$

$$\underline{2y \approx 6.4}$$

$$y \approx 3.2$$

The riverboat's speed in still water is 21.1 mph. The rate of the current is 3.2 mph.

48. Let $x =$ amount invested at 6% and

$y =$ amount invested at 10%.

$$\begin{cases} x + y = 9000 \\ 0.06x + 0.10y = 652.80 \end{cases}$$

Multiply the first equation by -6

and the second equation by 100.

$$-6x - 6y = -54,000$$
$$\underline{6x + 10y = 65,280}$$
$$4y = 11,280$$
$$y = 2820$$

Let $y = 2820$ in the first equation.

$$x + 2820 = 9000$$
$$x = 6180$$

$6180 invested at 6% and

$2820 invested at 10%.

49. Let $x =$ liters of 6% solution and

$y =$ liters of 14% solution.

	Concentration Rate	Ounces of Solution	Ounces of Pure Acid
First solution	0.06	x	$0.06x$
Second solution	0.14	y	$0.14y$
Mixture	0.12	50	$0.12(50)$

$$\begin{cases} x + y = 50 \\ 0.06x + 0.14y = 0.12(50) \end{cases}$$

Multiply the first equation by -6

and the second equation by 100.

$$-6x - 6y = -300$$
$$\underline{6x + 14y = 600}$$
$$8y = 300$$
$$y = 37.5$$

Let $y = 37.5$ in the first equation.

$$x + 37.5 = 50$$
$$x = 12.5$$

12.5 cc of 6% solution and

37.5 cc of 14% solution.

50. Let $x =$ the cost of an egg and

$y =$ the cost of a strip of bacon.

$$\begin{cases} 3x + 4y = 3.80 \\ 2x + 3y = 2.75 \end{cases}$$

Multiply the first equation by -2 and

the second equation by 3.

$$-6x - 8y = -7.60$$
$$\underline{6x + 9y = 8.25}$$
$$y = 0.65$$

Let $y = 0.65$ in the first equation.

$$3x + 4(0.65) = 3.80$$
$$3x + 2.60 = 3.80$$
$$3x = 1.20$$
$$x = 0.40$$

An egg costs 40¢ and a strip of

bacon costs 65¢.

51. Let $x =$ the time spent walking and

$y =$ the time spent jogging.

	r	$\cdot$	t	$=$	d
Walking	4		x		$4x$
Jogging	7.5		y		$7.5y$

$$\begin{cases} x + y = 3 \\ 4x + 7.5y = 15 \end{cases}$$

Multiply the first equation by -4.

$$-4x - 4y = -12$$
$$\underline{4x + 7.5y = 15}$$
$$3.5y = 3$$
$$y \approx 0.857$$

Let $y = 0.857$ in the first equation.

$$x + 0.857 = 3$$
$$x = 2.143$$

He spent 2.14 hours walking and 0.86 hours jogging.

52. Let x = length of the equal side and y = length of the third side.

$$\begin{cases} 2x + y = 73 & (1) \\ y = x + 7 & (2) \end{cases}$$

Replace y with $x + 7$ in E1.

$$2x + x + 7 = 73$$
$$3x = 66$$
$$x = 22$$

Replace x with 22 in E2.

$$y = 22 + 7 = 29$$

Two sides of the triangle have length 22 cm and the third side has length 29 cm.

53. Let x = the first number,
y = the second number, and
z = the third number.

$$\begin{cases} x + y + z = 98 & (1) \\ x + y = z + 2 & (2) \\ y = 4x & (3) \end{cases}$$

Replace y with $4x$ in E1 and E2.

$$x + 4x + z = 98$$
$$5x + z = 98$$
$$x + 4x = z + 2$$

Add E4 and E5.

$$5x + z = 98$$
$$\underline{5x - z = 2}$$
$$10x = 100$$
$$x = 10$$

Replace x with 10 in E3.

$$y = 4(10) = 40$$

Replace x with 10 and y with 40 in E2.

$$10 + 40 = z + 2$$
$$50 = z + 2$$
$$48 = z$$

The numbers are 10, 40, and 48.

54. Let x = the number of pennies,
      y = the number of nickels, and
      z = the number of dimes.

$$\begin{cases} x + y + z = 53 & (1) \\ 0.01x + 0.05y + 0.10z = 2.77 & (2) \\ y = z + 4 & (3) \end{cases}$$

Clear the decimals from E2 by multiplying by 100.

$$x + 5y + 10z = 277 \quad (4)$$

Replace y with $z + 4$ in E1.

$$x + z + 4 + z = 53$$
$$x + 2z = 49 \quad (5)$$

Replace y with $z + 4$ in E4.

$$x + 5(z + 4) + 10z = 277$$
$$x + 15 = 257 \quad (6)$$

Solve the new system.

$$\begin{cases} x + 2z = 49 & (5) \\ x + 15z = 257 & (6) \end{cases}$$

Multiply E5 by -1 and add to E6.

$$\begin{aligned} -x - 2z &= -49 \\ \underline{x + 15z} &= \underline{257} \\ 13z &= 208 \\ z &= 16 \end{aligned}$$

Replace z with 16 in E3.

$$x + 2(16) = 49$$
$$x + 32 = 49$$
$$x = 17$$

Replace z with 16 in E3.

$$y = 16 + 4 = 20$$

He has 17 pennies, 20 nickels, and 16 dimes in his jar.

55. Let c = pounds of chocolate used,
      n = pounds of nuts used, and
      r = pounds of raisins used.

$$\begin{cases} r = 2n & (1) \\ c + n + r = 45 & (2) \\ 3.00c + 2.70n + 2.25r = 2.80(45) & (3) \end{cases}$$

Replace r with $2n$ and c with $-3n + 45$ in E3.

$$3.00(-3n + 45) + 2.70n + 2.25(2n) = 126$$
$$-9n + 135 + 2.7n + 4.5n = 126$$
$$-1.8n + 135 = 126$$
$$-1.8n = -9$$
$$n = 5$$

Replace n with 5 in E1.

$$r = 2(5) = 10$$

Replace n with 5 and r with 10 in E2.

$$c + 5 + 10 = 45$$
$$c + 15 = 45$$
$$c = 30$$

She should use 30 pounds of creme-filled chocolates, 5 pounds of chocolate-covered nuts, and 10 pounds of chocolate-covered raisins.

56. Let f = the first number,

s = the second number, and

t = the third number.

$$\begin{cases} f + s + t = 9295 & (1) \\ f = s + 5 & (2) \\ f = 2t & (3) \end{cases}$$

Solve E2 for s and E3 for t.

$s = f - 5$

$t = \dfrac{f}{2}$

Replace s with $f - 5$ and t with $\dfrac{f}{2}$ in E1.

$f + f - 5 + \dfrac{f}{2} = 295$

$\dfrac{5}{2} f = 300$

$f = 120$

Replace f with 300 in the equation

$s = f - 5$.

$s = 120 - 5 = 115$

Replace f with 120 in the equation $\dfrac{f}{2}$.

$t = \dfrac{120}{2} = 60$

The first number is 120, the second number is 115, and the third number is 60.

Chapter 4 Test

1. False

2. False

3. True

4. False

5. Let $x = 1$ and $y = -1$.

$$2x - 3y = 5 \qquad\qquad 6x + y = 1$$

$$2(1) - 3(-1) \overset{?}{=} 5 \qquad 6(1) + (-1) \overset{?}{=} 1$$

$$2 + 3 \overset{?}{=} 5 \qquad\qquad 6 - 1 \overset{?}{=} 1$$

$$5 = 5 \qquad\qquad\qquad 5 = 1$$

$$\text{True} \qquad\qquad\qquad \text{False}$$

$(1, -1)$ is not a solution of the system.

6. Let $x = 3$ and $y = -4$.

$$4x - 3y = 24 \qquad\qquad 4x + 5y = -8$$

$$4(3) - 3(-4) \overset{?}{=} 24 \qquad 4(3) + 5(-4) \overset{?}{=} -8$$

$$12 + 12 \overset{?}{=} 24 \qquad\qquad 12 - 20 \overset{?}{=} -8$$

$$24 = 24 \qquad\qquad\qquad -8 = -8$$

$$\text{True} \qquad\qquad\qquad \text{True}$$

$(3, -4)$ is a solution of the system.

7. $\begin{cases} y - x = 6 \\ y + 2x = -6 \end{cases}$

$(-4, 2)$ is the solution of the system.

8. $\begin{cases} 3x - 2y = -14 \\ x + 3y = -1 \end{cases}$

Solve the second equation for x.

$x = -3y - 1$

Substitute $-3y - 1$ for x in the first equation.

$3(-3y - 1) - 2y = -14$

$-9y - 3 - 2y = -14$

$-11y = -11$

$y = 1$

Let $y = 1$ in $x = -3y - 1$.

$x = -3(1) - 1$

$x = -4$

The solution is $(-4, 1)$.

9. $\begin{cases} \dfrac{1}{2}x + 2y = -\dfrac{15}{4} \\ 4x = -y \end{cases}$

Multiply the first equation by 4 and the second equation by -1.

$\begin{cases} 2x + 8y = -15 \\ -4x = y \end{cases}$

Substitute $-4x$ for y in the first equation.

$2x + 8(-4x) = -15$

$2x - 32x = -15$

$-30x = -15$

$x = \dfrac{1}{2}$

Let $x = \dfrac{1}{2}$ in the second equation.

$-4\left(\dfrac{1}{2}\right) = y$

$y = -2$

The solution is $\left(\dfrac{1}{2}, -2\right)$.

10. $\begin{cases} 3x + 5y = 2 \\ 2x - 3y = 14 \end{cases}$

Multiply the first equation by 2 and the second equation by -3.

$6x + 10y = 4$

$\underline{-6x + 9y = -42}$

$19y = -38$

$y = -2$

Let $y = -2$ in the first equation.

$3x + 5(-2) = 2$

$3x - 10 = 2$

$3x = 12$

$x = 4$

The solution of the system is $(4, -2)$.

11. $\begin{cases} 4x - 6y = 7 \\ -2x + 3y = 0 \end{cases}$

Multiply the second equation by 2.

$4x - 6y = 7$

$\underline{-4x + 6y = 0}$

$0 = 7$

The system is inconsistent.

There is no solution.

12. $\begin{cases} 3x + y = 7 \\ 4x + 3y = 1 \end{cases}$

Solve the first equation for y.

$y = 7 - 3x$

Substitute $7 - 3x$ for y in the second equation.

$4x + 3(7 - 3x) = 1$

$4x + 21 - 9x = 1$

$-5x = -20$

$x = 4$

Let $x = 4$ in $y = 7 - 3x$.

$y = 7 - 3(4)$

$y = -5$

The solution is $(4, -5)$.

13. $\begin{cases} 3(2x + y) = 4x + 20 \\ 6x + 3y = 4x + 20 \\ 2x + 3y = 20 \\ \\ x - 2y = 3 \end{cases}$

Multiply the second equation by -2.

$\quad 2x + 3y = 20$

$\underline{-2x + 4y = -6}$

$\quad\quad 7y = 14$

$\quad\quad y = 2$

Let $y = 2$ in the second equation.

$x - 2(2) = 3$

$x - 4 = 3$

$x = 7$

The solution of the system is $(7, 2)$.

14. $\begin{cases} \dfrac{x - 3}{2} = \dfrac{2 - y}{4} \\ \dfrac{7 - 2x}{3} = \dfrac{y}{2} \end{cases}$

Multiply the first equation by 4 and the second equation by 6

$\begin{cases} 2(x - 3) = 2 - y \\ 2x - 6 = 2 - y \\ 2x + y = 8 \\ \\ 2(7 - 2x) = 3y \\ 14 - 4x = 3y \\ 4x + 3y = 14 \end{cases}$

Multiply the first equation by -3.

$\quad -6x - 3y = -24$

$\underline{\quad 4x + 3y = \;\; 14}$

$\quad -2x \quad\quad = -10$

$\quad\quad\quad x = 5$

Let $x = 5$ in the first equation.

$2(5) + y = 8$

$10 + y = 8$

$y = -2$

The solution of the system is $(5, -2)$.

15. Let $x =$ cc's of 12% solution and $y =$ cc's of 16% solution.

	Concentration Rate	cc's of Solution	cc's of Salt
First solution	12%	x	$0.12x$
Second solution	22%	80	$0.22(80)$
Mixture	16%	y	$0.16y$

$$\begin{cases} x + 80 = y \\ 0.12x + 0.22(80) = 0.16y \end{cases}$$

Multiply the first equation by -16
and the second equation by 100.

$$-16x - 1280 = -16y$$
$$\underline{12x + 1760 = 16y}$$
$$-4x + 480 = 0$$
$$-4x = -480$$
$$x = 120$$

Should add 120 cc of 12% solution.

16. Let x = amount invested at 5% and
y = amount invested at 9%.

$$\begin{cases} x + y = 4000 \\ 0.05x + 0.09y = 311 \end{cases}$$

Multiply the first equation by -5
and the second equation by 100.

$$-5x - 5y = -20,000$$
$$\underline{5x + 9y = 31,100}$$
$$4y = 11,100$$
$$y = 2775$$

Let y = 2775 in the first equation.

$$x + 2775 = 4000$$
$$x = 1225$$

$1225 invested at 5% and
$2775 invested at 9%.

17. Let x = the number of thousands of farms
in Texas and y = the number thousands
of farms in Missouri.

$$\begin{cases} x + y = 336 \\ x - y = 116 \end{cases}$$
$$2x = 452$$
$$x = 226$$

Let x = 226 in the first equation.

$$226 + y = 336$$
$$y = 110$$

There are 226,000 farms in Texas and
110,000 farms in Missouri.

18. $$\begin{cases} 2x - 3y = 4 & (1) \\ 3y + 2z = 2 & (2) \\ x - z = -5 & (3) \end{cases}$$

Add E1 and E2.

$$2x + 2z = 6 \text{ or } x + z = 3 \quad (4)$$

Add E3 and E4.

$$x + z = 3$$
$$\underline{x - z = -5}$$
$$2x = -2$$
$$x = -1$$

Replace x with -1 in E3.

$$-1 - z = -5$$
$$-z = -4 \text{ so } z = 4$$

Replace x with -1 in E1.

$$2(-1) - 3y = 4$$
$$-2 - 3y = 4$$
$$-3y = 6$$
$$y = -2$$

The solution is $(-1, -2, 4)$.

19. $\begin{cases} 3x - 2y - z = -1 & (1) \\ 2x - 2y = 4 & (2) \\ 2x - 2z = -12 & (3) \end{cases}$

Multiply E2 by -1 and add to E1.

$2x + 2z = 6$ or $x + z = 3$ (4)

Add E3 and E4.

$x + z = 3$

$\underline{x - z = -5}$

$2x = -2$

$x = -1$

Replace x with -1 in E3.

$3x - 2y - z = -1$

$\underline{-2x + 2y = -4}$

$x - z = -5$ (4)

Multiply E4 by -2 and add to E3.

$2x - 2z = -12$

$\underline{-2x + 2z = 10}$

$0 = -2$ False

The system is inconsistent.

The solution is $\varnothing$.

20. Let $x =$ measure of the smallest angle.

Then $5x - 3 =$ measure of the largest angle

and $2x - 1 =$ measure of the remaining angle.

$x + (5x - 3) + (2x - 1) = 180$

$8x - 4 = 180$

$8x = 184$

$x = 23$

$5x - 3 = 5(23) - 3 = 112$

$2x - 1 = 2(23) - 1 = 45$

The angles measure $23°$, $45°$, $112°$.

Chapter 4 Cumulative Review

1. a. $-1 < 0$ **b.** $7 = \dfrac{14}{2}$ **c.** $-5 > -6$

2. a. $5^2 = 5 \cdot 5 = 25$ **b.** $2^5 = 2 \cdot 2 \cdot 2 \cdot 2 \cdot 2 = 32$

3. a. commutative property of multiplication

   b. associative property of addition

   c. identity element for addition

   d. commutative property of multiplication

   e. multiplicative inverse property

   f. additive inverse property

   g. commutative and associative properties
       of multiplication

4. Let $x = 8$, $y = 5$

   $y^2 - 3x = 5^2 - 3(8) = 25 - 24 = 1$

5. $(2x - 3) - (4x - 2) = 2x - 3 - 4x + 2$

   $\qquad\qquad\qquad\qquad = -2x - 1$

6. $7 - 12 + (-5) - 2 + (-2)$

   $= 7 + (-12) + (-5) + (-2) + (-2)$

   $= 7 + (-21)$

   $= -14$

7. $\quad 7 = -5(2a - 1) - (-11a + 6)$

   $\quad 7 = -10a + 5 + 11a - 6$

   $\quad 7 = a - 1$

   $7 + 1 = a - 1 + 1$

   $\quad 8 = a$

8. Let $x = -7$, $y = -3$

$$2y^2 - x^2 = 2(-3)^2 - (-7)^2$$
$$= 2(9) - 49$$
$$= 18 - 49$$
$$= -31$$

9. $\dfrac{5}{2}x = 15$

$$5x = 30$$
$$x = 6$$

10. $0.4y - 6.7 + y - 0.3 - 2.6y$

$$= 0.4y + y + (-2.6y) + (-6.7) + (-0.3)$$
$$= -1.2y - 7.0$$

11. $\dfrac{x}{2} - 1 = \dfrac{2}{3}x - 3$

$$6\left(\dfrac{x}{2} - 1\right) = 6\left(\dfrac{2}{3}x - 3\right)$$
$$3x - 6 = 4x - 18$$
$$-x - 6 = -18$$
$$-x = -12$$
$$x = 12$$

12. $7(x - 2) - 6(x + 1) = 20$

$$7x - 14 - 6x - 6 = 20$$
$$x - 20 = 20$$
$$x = 40$$

13.   Let $x =$ the number

$$2(x + 4) = 4x - 12$$
$$2x + 8 = 4x - 12$$
$$-2x + 8 = -12$$
$$-2x = -20$$
$$x = 10$$

The number is 10.

14. $5(y - 5) = 5y + 10$

$$5y - 25 = 5y + 10$$
$$-25 = 10$$

False statement. There is no solution.

15. $y = mx + b$

$$y - b = mx + b - b$$
$$y - b = mx$$
$$\dfrac{y - b}{m} = \dfrac{mx}{m}$$
$$\dfrac{y - b}{m} = x$$

16.   Let $x =$ the number

$$5(x - 1) = 6x$$
$$5x - 5 = 6x$$
$$-x - 5 =$$
$$-x = 5$$
$$x = -5$$

The number is -5.

17. $-2x \le -4$

$$\dfrac{-2x}{-2} \ge \dfrac{-4}{-2}$$
$$x \ge 2, \ [2, \infty)$$

18. $P = a + b + c$

$$P - a - c = a + b + c - a - c$$
$$P - a - c = b$$

19. $x = -2y$

x	y
0	0
-4	2

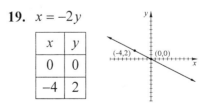

20. $3x + 7 \geq x - 9$

$\qquad 2x + 7 \geq -9$

$\qquad\qquad 2x \geq -16$

$\qquad\qquad x \geq -8, \left[-8, \infty\right)$

21. $(-1, 5)$ and $(2, -3)$

$$m = \frac{y_2 - y_1}{x_2 - x_1} = \frac{-3 - 5}{2 - (-1)} = \frac{-8}{3} = -\frac{8}{3}$$

22. $x - 3y = 3$

x	y
0	−1
3	0
9	2

23. $y = \dfrac{3}{4}x + 6$

$\qquad y = mx + b$

$\qquad m = \dfrac{3}{4}$

24. $(-1, 3)$ and $(2, -8)$

$$m = \frac{y_2 - y_1}{x_2 - x_1} = \frac{-8 - 3}{2 - (-1)} = -\frac{11}{3}$$

Parallel line has the same slope.

Slope is $-\dfrac{11}{3}$

25. $3x - 4y = 4$

$\qquad -4y = -3x + 4$

$\qquad y = \dfrac{-3x}{-4} + \dfrac{4}{-4}$

$\qquad y = \dfrac{3}{4}x - 1$

$y = mx + b$

$m = \dfrac{3}{4}, \, b = -1$

Slope is $\dfrac{3}{4}$, y-intercept is $(0, -1)$

26. $y = 7x + 0$

$\qquad y = mx + b$

$\qquad m = 7, \; b = 0$

Slope is 7. y-intercept is $(0, 0)$.

27. $m = -2$, with point $(-1, 5)$

$\qquad y - y_1 = m(x - x_1)$

$\qquad y - 5 = -2\left[x - (-1)\right]$

$\qquad y - 5 = -2x - 2$

$\qquad 2x + y = 3$

28. Line 1: $y = 4x - 5 \Rightarrow m_1 = 4$

Line 2: $-4x + y = 7 \Rightarrow y = 4x + 7$

$\qquad\qquad\qquad \Rightarrow m_2 = 4$

$m_2 = m_1$. The lines are parallel.

29. A vertical line has an equation $x = c$.

Point, $(-1, 5)$

$x = -1$

30. $m = -5$, with point $(-2, 3)$

$\qquad y - y_1 = m(x - x_1)$

$\qquad y - 3 = -5\left[x - (-2)\right]$

$\qquad y - 3 = -5x - 10$

$\qquad 5x + y = -7$

31. Domain is $\{-1, 0, 3\}$

Range is $\{-2, 0, 2, 3\}$

32. $f(x) = 5x^2 - 6$

$f(0) = 5(0)^2 - 6 = -6$

$f(-2) = 5(-2)^2 - 6 = 5(4) - 6 = 14$

33. **a.** function

b. not a function

34. **a.** not a function

b. function

c. not a function

35. $\begin{cases} 3x - y = 4 \\ y = 3x - 4, \ m = 3 \\ \\ x + 2y = 8 \\ y = -\dfrac{1}{2}x + 4, \ m = -\dfrac{1}{2} \end{cases}$

Because they have different slopes, there is only one solution.

36. **a.** Let $x = 1$ and $y = -4$.

$2x - y = 6 \qquad\qquad 3x + 2y = -5$

$2(1) - (-4) \overset{?}{=} 6 \quad 3(1) + 2(-4) \overset{?}{=} -5$

$2 + 4 \overset{?}{=} 6 \qquad\qquad 3 - 8 \overset{?}{=} -5$

$6 = 6 \qquad\qquad\quad -5 = -5$

True$\qquad\qquad\qquad$ True

$(1, -4)$ is a solution of the system.

b. Let $x = 0$ and $y = 6$.

$2x - y = 6 \qquad\qquad 3x + 2y = -5$

$2(0) - (6) \overset{?}{=} 6 \qquad$ Test not needed

$-6 \overset{?}{=} 6$

$-6 = 6$

False

$(0, 6)$ is not a solution of the system

c. Let $x = 3$ and $y = 0$.

$2x - y = 6 \qquad\qquad 3x + 2y = -5$

$2(3) - (0) \overset{?}{=} 6 \qquad 3(3) + 2(0) \overset{?}{=} -5$

$6 - 0 \overset{?}{=} 6 \qquad\qquad 9 + 0 \overset{?}{=} -5$

$6 = 6 \qquad\qquad\qquad 9 = -5$

True$\qquad\qquad\qquad$ False

$(3, 0)$ is not a solution of the system

37. $\begin{cases} x + 2y = 7 \\ 2x + 2y = 13 \end{cases}$

Solve the first equation for x.

$x = 7 - 2y$

Substitute $7 - 2y$ for x

in the second equation.

$2(7 - 2y) + 2y = 13$

$14 - 4y + 2y = 13$

$-2y = -1$

$y = \dfrac{1}{2}$

Let $y = \dfrac{1}{2}$ in $x = 7 - 2y$.

$x = 7 - 2\left(\dfrac{1}{2}\right)$

$x = 6$

The solution is $\left(6, \dfrac{1}{2}\right)$.

38. $\begin{cases} 3x - 4y = 10 \\ y = 2x \end{cases}$

Substitute $2x$ for y in the first equation.

$3x - 4(2x) = 10$

$3x - 8x = 10$

$-5x = 10$

$x = -2$

Let $x = -2$ in the second equation.

$y = 2(-2) = -4$

The solution is $(-2, -4)$.

39. $\begin{cases} x + y = 7 \\ x - y = 5 \end{cases}$

$2x = 12$

$x = 6$

Let $x = 6$ in the first equation.

$6 + y = 7$

$y = 1$

The solution to the system is $(6, 1)$.

40. $\begin{cases} x = 5y - 3 \\ x = 8y + 4 \end{cases}$

Substitute $8y + 4$ for x
in the first equation.

$8y + 4 = 5y - 3$

$3y + 4 = -3$

$3y = -7$

$y = -\dfrac{7}{3}$

Let $y = -\dfrac{7}{3}$ in the second equation.

$x = 8\left(-\dfrac{7}{3}\right) + 4$

$x = -\dfrac{56}{3} + \dfrac{12}{3}$

$x = -\dfrac{44}{3}$

The solution is $\left(-\dfrac{44}{3}, -\dfrac{7}{3}\right)$.

41. $\begin{cases} 3x - y + z = -15 & (1) \\ x + 2y - z = 1 & (2) \\ 2x + 3y - 2z = 0 & (3) \end{cases}$

Add E1 and E2.

$4x + y = -14 \quad (4)$

Multiply E1 by 2 and add to E3.

$6x - 2y + 2z = -30$

$\underline{2x + 3y - 2z = 0}$

$8x + y = -30 \quad (5)$

Solve the new system:

$$\begin{cases} 4x + y = -14 & (4) \\ 8x + y = -30 & (5) \end{cases}$$

Multiply E4 by -1 and add to E5.

$$-4x - y = 14$$
$$\underline{8x + y = -30}$$
$$4x \qquad = -16$$
$$x = -4$$

Replace x with -4 in E4.

$$4(-4) + y = -14$$
$$-16 + y = -14$$
$$y = 2$$

Replace x with y with 2 in E1.

$$3(-4) - (2) + z = -15$$
$$-12 - 2 + z = -15$$
$$-14 + z = -15$$
$$z = -1$$

The solution is $(-4, 2, -1)$.

42. $\begin{cases} x - 2y + z = 0 & (1) \\ 3x - y - 2z = -15 & (2) \\ 2x - 3y + 3z = 7 & (3) \end{cases}$

Multiply E1 by 2 and add to E2.

$$2x - 4y + 2z = 0$$
$$\underline{3x - y - 2z = -15}$$
$$5x - 5y \qquad = -15 \text{ or } x - y = -3 \ (4)$$

Multiply E1 by -3 and add to E3.

$$-3x + 6y - 3z = 0$$
$$\underline{2x - 3y + 3z = 7}$$
$$-x + 3y \qquad = 7 \quad (5)$$

Add E4 and E5.

$$2y = 4$$
$$y = 2$$

Replace y with 2 in E4.

$$x - 2 = -3$$
$$x = -1$$

Replace x with -1 and y with 2 in E1.

$$-1 - 2(2) + z = 0$$
$$-5 + z = 0$$
$$z = 5$$

The solution is $(-1, 2, 5)$.

43. Let x = the first number and
  y = the second number.

$$x = y - 4$$
$$4x = 2x + 6$$

Substitute $y - 4$ for x in the second equation .

$$4(y - 4) = 2y + 6$$
$$4y - 16 = 2y + 6$$
$$2y - 16 = 6$$
$$2y = 22$$
$$y = 11$$

Substitute 11 for y in the first equation

$$x = 11 - 4 \quad x = 7$$

The numbers are 7 and 11.

44. Let x = one number and
  y = the other number.

$$x + y = 37$$
$$\underline{x - y = 21}$$
$$2x \quad = 58 \quad x = 29$$
$$29 + y = 37 \quad y = 8$$

The numbers are 29 and 8.

Chapter 5

Section 5.1

Mental Math

1. 3^2

base: 3

exponent: 2

2. 5^4

base: 5

exponent: 4

3. $(-3)^6$

base: -3

exponent: 6

4. -3^7

base: 3

exponent: 7

5. -4^2

base: 4

exponent: 2

6. $(-4)^3$

base: -4

exponent: 3

7. $5 \cdot 3^4$

base: 5; exponent: 1

base: 3; exponent: 4

8. $9 \cdot 7^6$

base: 9; exponent: 1

base: 7; exponent: 6

9. $5x^2$

base: 5; exponent: 1

base: x; exponent: 2

10. $(5x)^2$

base: $5x$

exponent: 2

Exercise Set 5.1

1. $7^2 = 7 \cdot 7 = 49$

3. $(-5)^1 = -5$

5. $-2^4 = -2 \cdot 2 \cdot 2 \cdot 2 = -16$

7. $(-2)^4 = (-2)(-2)(-2)(-2) = 16$

9. $(0.1)^5 = (0.1)(0.1)(0.1)(0.1)(0.1)$
$= 0.00001$

11. $\left(\frac{1}{3}\right)^4 = \left(\frac{1}{3}\right)\left(\frac{1}{3}\right)\left(\frac{1}{3}\right)\left(\frac{1}{3}\right) = \frac{1}{81}$

13. $7 \cdot 2^5 = 7 \cdot 2 \cdot 2 \cdot 2 \cdot 2 \cdot 2 = 224$

15. $-2 \cdot 5^3 = -2 \cdot 5 \cdot 5 \cdot 5 = -250$

17. Answers may vary.

19. $x^2 = (-2)^2 = (-2)(-2) = 4$

21. $5x^3 = 5(3)^3 = 5 \cdot 3 \cdot 3 \cdot 3 = 135$

23. $2xy^2 = 2(3)(5)^2 = 2(3)(5)(5) = 150$

25. $\dfrac{2z^4}{5} = \dfrac{2(-2)^4}{5} = \dfrac{2(-2)(-2)(-2)(-2)}{5} = \dfrac{32}{5}$

27. $V = x^3 = 7^3 = 7 \cdot 7 \cdot 7 = 343$

The volume is 343 cubic meters.

29. We use the volume formula.

31. $x^2 \cdot x^8 = x^{2+8} = x^{10}$

33. $(-3)^3 \cdot (-3)^9 = (-3)^{3+9} = (-3)^{12}$

35. $(5y^4)(3y) = 5(3)y^{4+1} = 15y^5$

37. $(4z^{10})(-6z^7)(z^3) = 4(-6)z^{10+7+3} = -24z^{20}$

39. $(pq)^7 = p^7 q^7$

41. $\left(\dfrac{m}{n}\right)^9 = \dfrac{m^9}{n^9}$

43. $(x^2 y^3)^5 = x^{2\cdot 5} y^{3\cdot 5} = x^{10} y^{15}$

45. $\left(\dfrac{-2xz}{y^5}\right)^2 = \dfrac{(-2)^2 x^2 z^2}{y^{5\cdot 2}} = \dfrac{4x^2 z^2}{y^{10}}$

47. $\dfrac{x^3}{x} = \dfrac{x^3}{x^1} = x^{3-1} = x^2$

49. $\dfrac{(-2)^5}{(-2)^3} = (-2)^{5-3} = (-2)^2 = 4$

51. $\dfrac{p^7 q^{20}}{pq^{15}} = p^{7-1} q^{20-15} = p^6 q^5$

53. $\dfrac{7x^2 y^6}{14x^2 y^3} = \dfrac{7}{14} x^{2-2} y^{6-3} = \dfrac{1}{2} x^0 y^3 = \dfrac{y^3}{2}$

55. $(2x)^0 = 1$

57. $-2x^0 = -2(1) = -2$

59. $5^0 + y^0 = 1 + 1 = 2$

61. $\left(\dfrac{-3a^2}{b^3}\right)^3 = \dfrac{(-3)^3 a^{2\cdot 3}}{b^{3\cdot 3}} = -\dfrac{27a^6}{b^9}$

63. $\dfrac{\left(x^5\right)^7 \cdot x^8}{x^4} = \dfrac{x^{5\cdot 7} \cdot x^8}{x^4}$

$= \dfrac{x^{35} x^8}{x^4}$

$= x^{35+8-4}$

$= x^{39}$

65. $\dfrac{\left(z^3\right)^6}{(5z)^4} = \dfrac{z^{3\cdot 6}}{5^4 z^4} = \dfrac{z^{18}}{625z^4} = \dfrac{z^{18-4}}{625} = \dfrac{z^{14}}{625}$

67. $\dfrac{(6mn)^5}{mn^2} = \dfrac{6^5 \cdot m^5 \cdot n^5}{mn^2}$

$= 7776m^{5-1}n^{5-2}$

$= 7776m^4 n^3$

69. $-5^2 = -5 \cdot 5 = -25$

71. $\left(\dfrac{1}{4}\right)^3 = \dfrac{1^3}{4^3} = \dfrac{1}{64}$

73. $(9xy)^2 = 9^2 x^2 y^2 = 81x^2 y^2$

75. $(6b)^0 = 1$

77. $2^3 + 2^5 = 8 + 32 = 40$

79. $b^4 b^2 = b^{4+2} = b^6$

81. $a^2 a^3 a^4 = a^{2+3+4} = a^9$

83. $(2x^3)(-8x^4) = 2(-8)x^{3+4} = -16x^7$

85. $(4a)^3 = 4^3 a^3 = 64a^3$

87. $(-6xyz^3)^2 = (-6)^2 x^2 y^2 z^{3\cdot2} = 36x^2 y^2 z^6$

89. $\left(\dfrac{3y^5}{6x^4}\right)^3 = \dfrac{3^3 y^{5\cdot3}}{6^3 x^{4\cdot3}} = \dfrac{27y^{15}}{216x^{12}} = \dfrac{y^{15}}{8x^{12}}$

91. $\dfrac{x^5}{x^4} = x^{5-4} = x$

93. $\dfrac{2x^3 y^2 z}{xyz} = 2x^{3-1} y^{2-1} z^{1-1} = 2x^2 y$

95. $\dfrac{(3x^2 y^5)^5}{x^3 y} = \dfrac{3^5 x^{2\cdot5} y^{5\cdot5}}{x^3 y}$

$\qquad = \dfrac{243x^{10} y^{25}}{x^3 y}$

$\qquad = 243x^{10-3} y^{25-1}$

$\qquad = 243x^7 y^{24}$

97. Answers may vary.

99. $y - 10 + y = 2y - 10$

101. $7x + 2 - 8x - 6 = -x - 4$

103. $2(x - 5) + 3(5 - x) = 2x - 10 + 15 - 3x$

$\qquad\qquad\qquad\qquad\quad = -x + 5$

105. $(4x^2)(5x^3) = 4(5)x^{2+3} = 20x^5$ sq ft

107. $\pi(5y)^2 = \pi(5)^2 y^2 = 25y^2 \pi$ sq. cm

109. $(3y^4)^3 = 3^3 y^{4\cdot3} = 27y^{12}$

The volume is $27y^{12}$ cubic feet.

111. $x^{5a} x^{4a} = x^{5a+4a} = x^{9a}$

113. $\dfrac{x^{9a}}{x^{4a}} = x^{9a-4a} = x^{5a}$

115. $\left(x^a y^b z^c\right)^{5a} = x^{5a\cdot a} y^{5a\cdot b} z^{5a\cdot c} = x^{5a^2} y^{5ab} z^{5ac}$

117. $A = P\left(1 + \dfrac{r}{12}\right)^6$

$\qquad A = 1000\left(1 + \dfrac{0.09}{12}\right)^6$

$\qquad A = 1000(1.0075)^6$

$\qquad A = 1045.85$

You need $1045.85 to pay off the loan.

Section 5.2

Graphing Calculator Explorations

1. $(2x^2 + 7x + 6) + (x^3 - 6x^2 - 14)$
   $= x^3 - 4x^2 + 7x - 8$

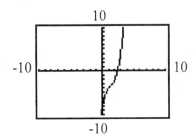

3. $(1.8x^2 - 6.8x - 1.7) - (3.9x^2 - 3.6x)$
   $- -2.1x^2 - 3.2x - 1.7$

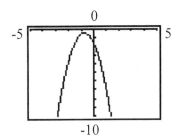

5. $(1.29x - 5.68) + (7.69x^2 - 2.55x + 10.98)$
   $= 7.69x^2 - 1.26x + 5.3$

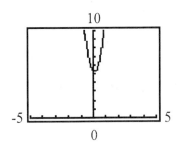

Exercise Set 5.2

1. 4 has degree 0.

3. $5x^2$ has degree 2.

5. $-3xy^2$ has degree $1 + 2 = 3$.

7. $6x + 3$ has degree 1 and is a binomial.

9. $3x^2 - 2x + 5$ has degree 2 and is a trinomial.

11. $-xyz$ has degree $1 + 1 + 1 = 3$ and is a monomial.

13. $x^2y - 4xy^2 + 5x + y$ has degree $2 + 1 = 3$ and is none of these.

15. Answers may vary.

17. $P(x) = x^2 + x + 1$
   $P(7) = 7^2 + 7 + 1$
   $\quad\ = 49 + 7 + 1$
   $\quad\ = 57$

19. $Q(x) = 5x^2 - 1$
   $Q(-10) = 5(-10)^2 - 1$
   $\quad\quad\ = 5(100) - 1$
   $\quad\quad\ = 500 - 1$
   $\quad\quad\ = 499$

21. $P(x) = x^2 + x + 1$
   $P(0) = 0^2 + 0 + 1$
   $\quad\ = 0 + 0 + 1$
   $\quad\ = 1$

23. $Q(x) = 5x^2 - 1$

$$Q\left(\frac{1}{4}\right) = 5\left(\frac{1}{4}\right)^2 - 1$$

$$= 5\left(\frac{1}{16}\right) - 1$$

$$= \frac{5}{16} - \frac{16}{16}$$

$$= -\frac{11}{16}$$

25. $P(t) = -16t^2 + 1053$

$$P(2) = -16(2)^2 + 1053$$

$$= -16(4) + 1053$$

$$= -64 + 1053$$

$$= 989 \text{ ft}$$

27. $P(t) = -16t^2 + 1053$

$$P(6) = -16(6)^2 + 1053$$

$$= -16(36) + 1053$$

$$= -576 + 1053$$

$$= 477 \text{ ft}$$

29. $5y + y = 6y$

31. $4x + 7x - 3 = 11x - 3$

33. $4xy + 2x - 3xy - 1 = xy + 2x - 1$

35. $7x^2 - 2xy + 5y^2 - x^2 + xy + 11y^2$
$= 6x^2 - xy + 16y^2$

37. $\left(9y^2 - 8\right) + \left(9y^2 - 9\right) = 18y^2 - 17$

39. $\begin{array}{r} x^2 + xy - y^2 \\ 2x^2 - 4xy + 7y^2 \\ \hline 3x^2 - 3xy + 6y^2 \end{array}$

41. $\begin{array}{r} x^2 - 6x + 3 \\ +(2x + 5) \\ \hline x^2 - 4x + 8 \end{array}$

43. $(9y^2 - 7y + 5) - (8y^2 - 7y + 2)$
$= 9y^2 - 7y + 5 - 8y^2 + 7y - 2$
$= y^2 + 3$

45. $(4x^2 + 2x) - (6x^2 - 3x)$
$= 4x^2 + 2x - 6x^2 + 3x$
$= -2x^2 + 5x$

47. $\begin{array}{r} 3x^2 - 4x + 8 \\ -5x^2 + 7 \\ \hline -2x^2 - 4x + 15 \end{array}$

49. $(5x - 11) + (-x - 2) = 5x - 11 - x - 2$
$ = 4x - 13$

51. $(7x^2 + x + 1) - (6x^2 + x - 1)$
$= 7x^2 + x + 1 - 6x^2 - x + 1$
$= x^2 + 2$

53. $(7x^3 - 4x + 8) + (5x^3 + 4x + 8x)$
$= 7x^3 - 4x + 8 + 5x^3 + 4x + 8x$
$= 12x^3 + 8x + 8$

55. $\begin{array}{r} 9x^3 - 2x^2 + 4x - 7 \\ -2x^3 + 6x^2 + 4x - 3 \\ \hline 7x^3 + 4x^2 + 8x - 10 \end{array}$

57. $(y^2 + 4yx + 7) + (-19y^2 + 7yx + 7)$
$= y^2 + 4yx + 7 - 19y^2 + 7yx + 7$
$= -18y^2 + 11yx + 14$

59. $(3x^3 - b + 2a - 6) + (-4x^3 + b + 6a - 6)$
$= 3x^3 - b + 2a - 6 - 4x^3 + b + 6a - 6$
$= -x^3 + 8a - 12$

61. $(4x^2 - 6x + 2) - (-x^2 + 3x + 5)$
$= 4x^2 - 6x + 2 + x^2 - 3x - 5$
$= 5x^2 - 9x - 3$

63. $(-3x + 8) + (-3x^2 + 3x - 5)$
$= -3x + 8 - 3x^2 + 3x - 5$
$= -3x^2 + 3$

65. $(-3 + 4x^2 + 7xy^2) + (2x^3 - x^2 + xy^2)$
$= -3 + 4x^2 + 7xy^2 + 2x^3 - x^2 + xy^2$
$= 2x^3 + 3x^2 + 8xy^2 - 3$

67. $\begin{array}{r} 6y^2 - 6y + 4 \\ y^2 + 6y - 7 \\ \hline 7y^2 \qquad - 3 \end{array}$

69. $\begin{array}{r} 3x^2 + 15x + 8 \\ 2x^2 + 7x^2 + 8 \\ \hline 5x^2 + 22x + 16 \end{array}$

71. $\begin{array}{l} \dfrac{1}{2}x^2 - \dfrac{1}{3}x^2y \qquad\qquad + 2y^3 \\ \dfrac{1}{4}x^2 \qquad\qquad - \dfrac{8}{3}x^2y^2 - \dfrac{1}{2}y^3 \\ \hline \dfrac{3}{4}x^2 - \dfrac{1}{3}x^2y - \dfrac{8}{3}x^2y^2 + \dfrac{3}{2}y^3 \end{array}$

73. $(5q^4 - 2q^2 - 3q) + (-6q^4 + 3q^2 + 5)$
$= 5q^4 - 2q^2 - 3q - 6q^4 + 3q^2 + 5$
$= -q^4 + q^2 - 3q + 5$

75. $\begin{array}{r} 7x^2 + 4x + 9 \\ + \; 8x^2 + 7x - 8 \\ \hline 15x^2 + 11x + 1 \\ - \qquad\quad 3x + 7 \\ \hline 15x^2 + 8x - 6 \end{array}$

77. $(4x^4 - 7x^2 + 3) + (2 - 3x^4)$
$= 4x^4 - 7x^2 + 3 + 2 - 3x^4$
$= x^4 - 7x^2 + 5$

79. $\left(\dfrac{2}{3}x^2 - \dfrac{1}{6}x + \dfrac{5}{6}\right) - \left(\dfrac{1}{3}x^2 + \dfrac{5}{6}x - \right)$
$= \dfrac{2}{3}x^2 - \dfrac{1}{6}x + \dfrac{5}{6} - \dfrac{1}{3}x^2 - \dfrac{5}{6}x + \dfrac{1}{6}$
$= \dfrac{1}{3}x^3 - x + 1$

81. If $L = 5$, $W = 4$, and $H = 9$, then

$2HL + 2LW + 2HW$
$= 2(9)(5) + 2(5)(4) + 2(9)(4)$
$= 90 + 40 + 72$
$= 202$

The surface area is 202 square inches.

83. $P(t) = -16t^2 + 300t$

    a. $P(1) = -16(1)^2 + 300(1) = 284$ feet

    b. $P(2) = -16(2)^2 + 300(2) = 536$ feet

    c. $P(1) = -16(3)^2 + 300(3) = 756$ feet

    d. $P(4) = -16(4)^2 + 300(4) = 944$ feet

    e. Answers may vary.

    f.
$$0 = -16t^2 + 300t$$
$$0 = -4t(4t - 75)$$
$$4t - 75 = 0$$
$$4t = 75$$
$$t = \frac{75}{4} = 18.75$$

    19 sec

85. $P(x) = 45x - 100,000$
$$P(4000) = 45(4000) - 100,000$$
$$= 80,000$$

The profit is \$80,000.

87. $R(x) = 2x$
$$R(20,000) = 2(20,000)$$
$$= 40,000$$

The revenue is \$40,000.

89. $x \cdot x = x^2$

91. $x^2 \cdot x^2 = x^4$

93. $5(3x - 2) = 15x - 10$

95. $-2(x^2 - 5x + 6) = -2x^2 + 10x - 12$

97. $(4x^{2a} - 3x^a + 0.5) - (x^{2a} - 5x^a - 0.2)$
$$= 4x^{2a} - 3x^a + 0.5 - x^{2a} + 5x^a + 0.2$$
$$= 3x^{2a} + 2x^a + 0.7$$

99. $(8y^{2y} - 7x^y + 3) + (-4x^{2y} + 9x^y - 14)$
$$= 8y^{2y} - 7x^y + 3 - 4x^{2y} + 9x^y - 14$$
$$= 4y^{2y} + 2x^y - 11$$

101. $P = 2l + 2w$
$$= 2(3x^2 - x + 2y) + 2(x + 5y)$$
$$= 6x^2 - 2x + 4y + 2x + 10y$$
$$= 6x^2 + 14y$$

    The perimeter is $P = (6x^2 + 14y)$ units.

103. $P(x) + Q(x) = (3x + 3) + (4x^2 - 6x + 3)$
$$= 3x + 3 + 4x^2 - 6x + 3$$
$$= 4x^2 - 3x + 6$$

105. $Q(x) - R(x) = (4x^2 - 6x + 3) - (5x^2 - 7)$
$$= 4x^2 - 6x + 3 - 5x^2 + 7$$
$$= -x^2 - 6x + 10$$

107. $2[Q(x)] - R(x)$
$$= 2(4x^2 - 6x + 3) - (5x^2 - 7)$$
$$= 8x^2 - 12x + 6 - 5x^2 + 7$$
$$= 3x^2 - 12x + 13$$

109. $3[R(x)] + 4[P(x)]$
$$= 3(5x^2 - 7) + 4(3x + 3)$$
$$= 15x^2 - 21 + 12x + 12$$
$$= 15x^2 + 12x - 9$$

111. $P(x) = 2x - 3$

    a. $P(a) = 2a - 3$

    b. $P(-x) = 2(-x) - 3 = -2x - 3$

    c. $P(x + h) = 2(2x + h) - 3 = 4x + 2h - 3$

113. $P(x) = 4x$

    a. $P(a) = 4a$

    b. $P(-x) = 4(-x) = -4x$

    c. $P(x + h) = 4(x + h) = 4x + 4h$

115. $P(x) = 4x - 1$

    a. $P(a) = 4a - 1$

    b. $P(-x) = 4(-x) - 1 = -4x - 1$

    c. $P(x + h) = 4(x + h) - 1 = 4x + 4h - 1$

117. $f(x) = -246.7x^2 + 1887.9x + 1016.9$

    a. 1998 means that $x = 0$:

$$f(0) = -246.7(0)^2 + 1887.9(0) + 1016.9$$
$$= 1016.9$$
$$\approx 1017 \text{ stations}$$

    b. 2000 means that $x = 2$:

$$f(2) = -246.7(2)^2 + 1887.9(2) + 1016.9$$
$$= 3805.9$$
$$\approx 3806 \text{ stations}$$

    c. 2006 means $x = 8$:

$$f(8) = -246.7(8)^2 + 1887.9(8) + 1016.9$$
$$= 331.3$$
$$\approx 331 \text{ stations}$$

d. Answers may vary.

119. $f(x) = 0.014x^2 + 0.12x + 0.85$

    a. 1999 means that $x = 9$:

$$f(9) = 0.014(9)^2 + 0.12(9) + 0.85$$
$$= 3.064$$
$$\approx 3.1 \text{ million SUV's}$$

    b. 2005 means that $x = 15$:

$$f(15) = 0.014(15)^2 + 0.12(15) + 0.85$$
$$= 5.8 \text{ million SUV's}$$

121. $f(x) = 1.4x^2 + 129.6x + 939$

    a. 1985 means that $x = 5$:

$$f(5) = 1.4(5)^2 + 129.6(5) + 939$$
$$= \$1622$$

    b. 1995 means that $x = 15$:

$$f(15) = 1.4(15)^2 + 129.6(15) + 939$$
$$= \$3198$$

    c. 2010 means $x = 30$:

$$f(30) = 1.4(30)^2 + 129.6(30) + 939$$
$$= \$6087$$

    d. No; $f(x)$ is not linear.

Section 5.3

Mental Math

1. $10xy$

2. $28ab$

3. x^7

4. z^5

5. $18x^3$

6. $15a^4$

7. $-27x^5$

8. $32x^8$

9. a^7

10. a^{10}

11. Cannot simplify

12. a^3

Exercise Set 5.3

1. $2a(2a-4) = 2a(2a) - 2a(4) = 4a^2 - 8a$

3. $7x(x^2 + 2x - 1)$
$$= 7x(x^2) + 7x(2x) + 7x(-1)$$
$$= 7x^3 + 14x^2 - 7x$$

5. $3x^2(2x^2 - x) = 3x^2(2x^2) + 3x^2(-x)$
$$= 6x^4 - 3x^3$$

7. $x(x+3) = x^2 + 3x$

9. $(a+7)(a-2) = a(a)a(-2) + 7(a) + 7(-2)$
$$= a^2 - 2a + 7a - 14$$
$$= a^2 + 5a - 14$$

11. $(2y-4)^2 = (2y-4)(2y-4)$
$$= 2y(2y) + 2y(-4) - 4(2y) - 4(-4)$$
$$= 4y^2 - 8y - 8y + 16$$
$$= 4y^2 - 16y + 16$$

13. $(5x-9y)(6x-5y)$
$$= 5x(6x) + 5x(-5y) - 9y(6x) - 9y(-5y)$$
$$= 30x^2 - 25xy - 54xy + 45y^2$$
$$= 30x^2 - 79xy + 45y^2$$

15. $(2x^2 - 5)^2 = (2x^2 - 5)(2x^2 - 5)$
$$= 2x^2(2x^2) + 2x^2(-5) - 5(2x^2) - 5(-5)$$
$$= 4x^4 - 10x^2 - 10x^2 + 25$$
$$= 4x^4 - 20x^2 + 25$$

17. $x \cdot x + 3 \cdot x + 2 \cdot x + 2 \cdot 3 = x^2 + 5x + 6$

19. $(x-2)(x^2 - 3x + 7)$
$$= x(x^2) + x(-3x) + x(7)$$
$$\qquad -2(x^2) - 2(-3x) - 2(7)$$
$$= x^3 - 3x^2 + 7x - 2x^2 + 6x - 14$$
$$= x^3 - 5x^2 + 13x - 14$$

21. $(x+5)(x^3-3x+4) = x(x^3)+x(-3x)$

         $+x(4)+5(x^3)+5(-3x)+5(4)$

    $= x^4-3x^2+4x+5x^3-15x+20$

    $= x^4+5x^3-3x^2-11x+20$

23. $(2a-3)(5a^2-6a+4) = 2a(5a^2)+2a(-6a)$

        $+2a(4)-3(5a^2)-3(-6a)-3(4)$

    $= 10a^3-12a^2+8a-15a^2+18a-12$

    $= 10a^3-27a^2+26a-12$

25. $(x+2)^3 = (x+2)(x+2)(x+2)$

        $= (x^2+2x+2x+4)(x+2)$

        $= (x^2+4x+4)(x+2)$

        $= (x^2+4x+4)x+(x^2+4x+4)2$

        $= x^3+4x^2+4x+2x^2+8x+8$

        $= x^3+6x^2+12x+8$

27. $(2y-3)^3 = (2y-3)(2y-3)(2y-3)$

    $= (4y^2-6y-6y+9)(2y-3)$

    $= (4y^2-12y+9)(2y-3)$

    $= (4y^2-12y+9)2y+(4y^2-12y+9)(-3)$

    $= 8y^3-24y^2+18y-12y^2+36y-27$

    $= 8y^3-36y^2+54y-27$

29.

$$
\begin{array}{r}
2x^2+\ 4x-1 \\
x+3 \\
\hline
6x^2+12x-3 \\
2x^3+\ 4x^2-\ \ x \\
\hline
2x^3+10x^2+11x-3
\end{array}
$$

31.

$$
\begin{array}{r}
x^3+5x-7 \\
\times \qquad\quad x^2-9 \\
\hline
-9x^3\qquad\ -45x+63 \\
x^5+5x^3-7x^2 \\
\hline
x^5-4x^3-7x^2-45x+63
\end{array}
$$

33. a. $(2+3)^2 = 5^2 = 25$

       $2^2+3^2 = 4+9 = 13$

   b. $(8+10)^2 = (18)^2 = 324$

       $8^2+10^2 = 64+100 = 164$

   c. No; Answers may vary.

35. $2a(a+4) = 2a(a)+2a(4) = 2a^2+8a$

37. $3x(2x^2-3x+4)$

    $= 3x(2x^2)+3x(-3x)+3x(4)$

    $= 6x^3-9x^2+12x$

39. $(5x+9y)(3x+2y)$

    $= 5x(3x)+5x(2y)+9y(3x)+9y(2y)$

    $= 15x^2+10xy+27xy+18y^2$

    $= 15x^2+37xy+18y^2$

41. $(x+2)(x^2+5x+6)$

    $= x(x^2)+x(5x)+x(6)+2(x^2)+2(5x)+2(6)$

    $= x^3+5x^2+6x+2x^2+10x+12$

    $= x^3+7x^2+16x+12$

43. $(7x+4)^2 = (7x+4)(7x+4)$
$$= 7x(7x)+7x(4)+4(7x)+4(4)$$
$$= 49x^2 + 28x + 28x + 16$$
$$= 49x^2 + 56x + 16$$

45. $-2a^2(3a^2 - 2a + 3)$
$$= -2a^2(3a^2) - 2a^2(-2a) - 2a^2(3)$$
$$= -6a^4 + 4a^3 - 6a^2$$

47. $(x+3)(x^2 + 7x + 12)$
$$= x(x^2) + x(7x) + x(12) + 3(x^2) + 3(7x) + 3(12)$$
$$= x^3 + 7x^2 + 12x + 3x^2 + 21x + 36$$
$$= x^3 + 10^2 + 33x + 36$$

49. $(a+1)^3 = (a+1)(a+1)(a+1)$
$$= (a^2 + a + a + 1)(a+1)$$
$$= (a^2 + 2a + 1)(a+1)$$
$$= (a^2 + 2a + 1)a + (a^2 + 2a + 1)1$$
$$= a^3 + 2a^2 + a + a^2 + 2a + 1$$
$$= a^3 + 3a^2 + 3a + 1$$

51. $(x+y)(x+y) = x(x) + x(y) + y(x) + y(y)$
$$= x^2 + xy + xy + y^2$$
$$= x^2 + 2xy + y^2$$

53. $(x-7)(x-6) = x(x) + x(-6) - 7(x) - 7(-6)$
$$= x^2 - 6x - 7x + 42$$
$$= x^2 - 13x + 42$$

55. $3a(a^2 + 2) = 3a(a^2) + 3a(2) = 3a^3 + 6a$

57. $-4y(y^2 + 3y - 11)$
$$= -4y(y^2) - 4y(3y) - 4y(-11)$$
$$= -4y^3 - 12y^2 + 44y$$

59. $(5x+1)(5x-1)$
$$= 5x(5x) + 5x(-1) + 1(5x) + 1(-1)$$
$$= 25x^2 - 5x + 5x - 1$$
$$= 25x^2 - 1$$

61. $(5x+4)(x^2 - x + 4)$
$$= 5x(x^2) + 5x(-x) + 5x(4) + 4(x^2) + 4(-x) + 4(4)$$
$$= 5x^3 - 5x^2 + 20x + 4x^2 - 4x + 16$$
$$= 5x^3 - x^2 + 16x + 16$$

63. $(2x-5)^3 = (2x-5)(2x-5)(2x-5)$
$$= (4x^2 - 10x + 25)(2x-5)$$
$$= (4x^2 - 20x + 25)(2x-5)$$
$$= (4x^2 - 20x + 25)2x + (4x^2 - 20x + 25)(-5)$$
$$= 8x^3 - 40x^2 + 50x - 20x^2 + 100x - 125$$
$$= 8x^3 - 60x^2 + 150x - 125$$

65. $(4x+5)(8x^2 + 2x - 4)$
$$= 4x(8x^2) + 4x(2x) + 4x(-4) + 5(8x^2) + 5(2x) + 5(-4)$$
$$= 32x^3 + 8x^2 - 16x + 40x^2 + 10x - 20$$
$$= 32x^3 + 48x^2 - 6x - 20$$

67. $(7xy - y)^2 = (7xy - y)(7xy - y)$
$$= 7xy(7xy) + 7xy(-y) - y(7xy) - y(-y)$$
$$= 49x^2y^2 - 7xy^2 - 7xy^2 + y^2$$
$$= 49x^2y^2 - 14xy^2 + y^2$$

69.
$$
\begin{array}{r}
5y^2 - y + 3 \\
\times \quad y^2 - 3y - 2 \\
\hline
-10y^2 + 2y - 6 \\
-15y^3 + 3y^2 - 9y \\
5y^4 - \quad y^3 + 3y^2 \\
\hline
5y^4 - 16y^3 - 4y^2 - 7y - 6
\end{array}
$$

71.
$$
\begin{array}{r}
3x^2 + 2x - 4 \\
\times \quad 2x^2 - 4x + 3 \\
\hline
9x^2 + 6x - 12 \\
-12x^3 - 8x^2 + 16x \\
6x^4 + 4x^3 - 8x^2 \\
\hline
6x^4 - 8x^3 - 7x^2 + 22 - 12
\end{array}
$$

73. $(5x)^2 = 5^2 x^2 = 25x^2$

75. $(-3y^3)^2 = (-3)^2 y^{3\cdot2} = 9y^6$

77. At $t = 0$, value = \$7000

79. At $t = 0$, value $= \$7000$
At $t = 1$, value $= \$6500$
\$7000 − \$6500 = \$500

81. Answers may vary.

83. $(2x - 5)(2x + 5)$
$= 2x(2x) + 2x(5) - 5(2x) - 5(5)$
$= 4x^2 + 10x - 10x - 25 = 4x^2 - 25$
$(4x^2 - 25)$ square yards

85. $\dfrac{1}{2}(3x - 2)(4x) = 2x(3x - 2)$
$= 2x(3x) + 2x(-2) = 6x^2 - 4x$
$(6x^2 - 4x)$ square inches

87. $(x + 3)(x + 3) - 2 \cdot 2$
$= x^2 + 3x + 3x + 9 - 4$
$= (x^2 + 6x + 5)$ square units

89. a. $(a + b)(a - b) = a^2 - ab + ab - b^2$
$= a^2 - b^2$

b. $(2x + 3y)(2x - 3y)$
$= (2x)^2 - 6xy + 6xy - (3y)^2$
$= 4x^2 - 9y^2$

c. $(4x + 7)(4x - 7)$
$= (4x)^2 - 28x + 28x - 7^2$
$= 16x^2 - 49$

d. Answers may vary.

Section 5.4

Mental Math

1. False

2. True

3. False

4. False

Exercise Set 5.4

1. $(x+3)(x+4) = x^2 + 4x + 3x + 12$
$$= x^2 + 7x + 12$$

3. $(x-5)(x+10) = x^2 + 10x - 5x - 50$
$$= x^2 + 5x - 50$$

5. $(5x-6)(x+2) = 5x^2 + 10x - 6x - 12$
$$= 5x^2 + 4x - 12$$

7. $(y-6)(4y-1) = 4y^2 - 1y - 24y + 6$
$$= 4y^2 - 25y + 6$$

9. $(2x+5)(3x-1) = 6x^2 - 2x + 15x - 5$
$$= 6x^2 + 13x - 5$$

11. $(x-2)^2 = x^2 - 2(x)(2) + 2^2$
$$= x^2 - 4x + 4$$

13. $(2x-1)^2 = (2x)^2 - 2(2x)(1) + (1)^2$
$$= 4x^2 - 4x + 1$$

15. $(3a-5)^2 = (3a)^2 - 2(3a)(5) + 5^2$
$$= 9a^2 - 30a + 25$$

17. $(5x+9)^2 = (5x)^2 + 2(5x)(9) + 9^2$
$$= 25x^2 + 90x + 81$$

19. Answers may vary.

21. $(a-7)(a+7) = a^2 - 7^2 = a^2 - 49$

23. $(3x-1)(3x-1) = (3x)^2 - 1^2 = 9x^2 - 1$

25. $\left(3x - \dfrac{1}{2}\right)\left(3x + \dfrac{1}{2}\right) = (3x)^2 - \left(\dfrac{1}{2}\right)^2$
$$= 9x^2 - \dfrac{1}{4}$$

27. $(9x+y)(9x-y) = (9x)^2 - y^2 = 81x^2 - y^2$

29. $(2x+0.1)(2x-0.1) = (2x)^2 - (0.1)^2$
$$= 4x^2 - 0.01$$

31. $(a+5)(a+4) = a^2 + 4a + 5a + 20$
$$= a^2 + 9a + 20$$

33. $(a+7)^2 = a^2 + 2(a)(7) + 7^2$
$$= a^2 + 14a + 49$$

35. $(4a+1)(3a-1) = 12a^2 - 4a + 3a - 1$
$$= 12a^2 - a - 1$$

37. $(x+2)(x-2) = x^2 - 2^2 = x^2 - 4$

39. $(3a+1)^2 = (3a)^2 + 2(3a)(1) + 1^2$
$$= 9a^2 + 6a + 1$$

41. $\left(x^2 + y\right)\left(4x - y^4\right)$
$$= x^2(4x) - x^2 y^4 + y(4x) - y \cdot y^4$$
$$= 4x^3 - x^2 y^4 + 4xy - y^5$$

43. $(x+3)\left(x^2 - 6x + 1\right)$
$$= x\left(x^2\right) + x(-6x) + x(1) + 3\left(x^2\right) + 3(-6x) + 3(1)$$
$$= x^3 - 6x^2 + x + 3x^2 - 18x + 3$$
$$= x^3 - 3x^2 - 17x + 3$$

45. $(2a-3)^2 = (2a)^2 - 2(2a)(3) + (3)^2$

$\qquad = 4a^2 - 12a + 9$

47. $(5x-6z)(5x+6z) = (5x)^2 - (6z)^2$

$\qquad = 25x^2 - 36z^2$

49. $(x^5-3)(x^5-5) = x^{10} - 5x^5 - 3x^5 + 15$

$\qquad = x^{10} - 8x^5 + 15$

51. $\left(x - \dfrac{1}{3}\right)\left(x + \dfrac{1}{3}\right) = (x)^2 - \left(\dfrac{1}{3}\right)^2 = x^2 - \dfrac{1}{9}$

53. $(a^3+11)(a^4-3) = a^7 - 3a^3 + 11a^4 - 33$

55. $3(x-2)^2 = \left[(x)^2 - 2(x)(2) + (2)^2\right]$

$\qquad = 3(x^2 - 4x + 4)$

$\qquad = 3x^2 - 12x + 12$

57. $(3b+7)(2b-5) = 6b^2 - 15b + 14b - 35$

$\qquad = 6b^2 - b - 35$

59. $(7p-8)(7p+8) = (7p)^2 - (8)^2$

$\qquad = 49p^2 - 64$

61. $\left(\dfrac{1}{3}a^2 - 7\right)\left(\dfrac{1}{3}a^2 + 7\right) = \left(\dfrac{1}{3}a^2\right) - (7)^2$

$\qquad = \dfrac{1}{9}a^4 - 49$

63. $5x^3(3x^2 - x + 2)$

$\qquad = 5x^2(3x^2) + 5x^2(-x) + 5x^2(2)$

$\qquad = 15x^4 - 5x^3 + 10x^2$

65. $(2r-3s)(2r+3s) = (2r)^2 - (3s)^2$

$\qquad = 4r^2 - 9s^2$

67. $(3x-7y)^2 = (3x)^2 - 2(3x)(7y) + (7y)^2$

$\qquad = 9x^2 - 42xy + 49y^2$

69. $(4x+5)(4x-5) = (4x)^2 - 5^2$

$\qquad = 16x^2 - 25$

71. $(8x+4)^2 = (8x)^2 + 2(8x)(4) + (4)^2$

$\qquad = 64x^2 + 64x + 16$

73. $\left(a - \dfrac{1}{2}y\right)\left(a + \dfrac{1}{2}y\right) = a^2 - \left(\dfrac{1}{2}y\right)^2$

$\qquad = a^2 - \dfrac{1}{4}y^2$

75. $\left(\dfrac{1}{5}x - y\right)\left(\dfrac{1}{5}x + y\right) = \left(\dfrac{1}{5}x\right)^2 - (y)^2$

$\qquad = \dfrac{1}{25}x^2 - y^2$

77. $(a+1)(3a^2 - a + 1)$

$\qquad = a(3a^2) + a(-a) + a(1) + 1(3a^2) + 1(-a) + 1(1)$

$\qquad = 3a^3 - a^2 + a + 3a^2 - a + 1$

$\qquad = 3a^3 + 2a^2 + 1$

79. $\dfrac{50b^{10}}{70b^5} = \dfrac{5 \cdot 10b^{10-5}}{7 \cdot 10} = \dfrac{5b^5}{7}$

81. $\dfrac{8a^{17}b^{15}}{-4a^7b^{10}} = \dfrac{4 \cdot 2a^{17-7}b^{15-10}}{-4}$

$\qquad = -\dfrac{2a^{10}b^{+5}}{1}$

$\qquad = -2a^{10}b^5$

83. $\dfrac{2x^4 y^{12}}{3x^4 y^4} = \dfrac{2x^{4-4} y^{12-4}}{3} = \dfrac{2x^0 y^8}{3} = \dfrac{2y^8}{3}$

85. $(-1,1)$ and $(2,2)$

$m = \dfrac{y_2 - y_1}{x_2 - x_1} = \dfrac{2-1}{2-(-1)} = \dfrac{1}{3}$

87. $(-1,-2)$ and $(1,0)$

$m = \dfrac{y_2 - y_1}{x_2 - x_1} = \dfrac{0-(-2)}{1-(-1)} = \dfrac{2}{2} = 1$

89. $(2x+1)^2 = (2x)^2 + 2(2x)(1) + 1^2$

$= 4x^2 + 4x + 1$

$(4x^2 + 4x + 1)$ square feet

91. $\dfrac{1}{2}(5a+b)(5a-b)$

$= \dfrac{1}{2}(25a^2 - b^2)$

$= \left(\dfrac{25a^2}{2} - \dfrac{b^2}{2}\right)$ square units

93. $(5x-3)^2 - (x+1)^2$

$= \left[(5x)^2 - 2(5x)(3) + 3^2\right]$

$\qquad\qquad - \left[x^2 + 2(x)(1) + 1^2\right]$

$= (25x^2 - 30x + 9) - (x^2 + 2x + 1)$

$= 25x^2 - 30x + 9 - x^2 - 2x - 1$

$= (24x^2 - 32x + 8)$ square meters

95. $x \cdot x + x(5) + x(5) + 5 \cdot 5$

$= x^2 + 5x + 5x + 25$

$= (x^2 + 10x + 25)$ square units

97. $(x+2)(x+2) = x^2 + 2(x)(2) + (2)^2$

$= x^2 + 4x + 4$

$(x+2)(x-2) = (x)^2 - (2)^2 = x^2 - 4$

Answers may vary.

99. $\left[(a+c)-5\right]\left[(a+c)+5\right]$

$= (a+c)^2 - 5^2$

$= a^2 + 2ac + c^2 - 25$

101. $\left[(x-2)+y\right]\left[(x-2)-y\right]$

$= (x-2)^2 - y^2$

$= x^2 - 4x + 4 - y^2$

Integrated Review

1. $(5x^2)(7x^3) = (5 \cdot 7)(x^2 \cdot x^3)$

$= 35x^5$

2. $(4y^2)(8y^7) = (4 \cdot 8)(y^2 \cdot y^7)$

$= 32y^9$

3. $-4^2 = -(4 \cdot 4) = -16$

4. $(-4)^2 = (-4)(-4) = 16$

5. $(x-5)(2x+1) = 2x^2 + x - 10x - 5$

$= 2x^2 - 9x - 5$

6. $(3x-2)(x+5) = 3x^2 + 15x - 2x - 10$

$= 3x^2 + 13x - 10$

7. $(x-5) + (2x+1) = x - 5 + 2x + 1$

$= 3x - 4$

8. $(3x-2)+(x+5)=3x-2+x+5$
$$=4x+3$$

9. $\dfrac{7x^9y^{12}}{x^3y^{10}}=7x^{9-3}y^{12-10}$
$$=7x^6y^2$$

10. $\dfrac{20a^2b^8}{14a^2b^2}=\dfrac{10a^{2-2}b^{8-2}}{7}$
$$=\dfrac{10b^6}{7}$$

11. $\left(12m^7n^6\right)^2=12^2m^{7\cdot2}n^{6\cdot2}$
$$=144m^{14}n^{12}$$

12. $\left(4y^9z^{10}\right)^3=4^3y^{9\cdot3}z^{10\cdot3}$
$$=64y^{27}z^{30}$$

13. $3(4y-3)(4y+3)=3\left[(4y)^2-3^2\right]$
$$=3(16y^2-9)$$
$$=48y^2-27$$

14. $2(7x-1)(7x+1)=2\left[(7x)^2-1^2\right]$
$$=2(49x^2-1)$$
$$=98x^2-2$$

15. $\left(x^7y^5\right)^9=x^{63}y^{45}$

16. $\left(3^1x^9\right)^3=3^3x^{27}$
$$=27x^{27}$$

17. $\left(7x^2-2x+3\right)-\left(5x^2+9\right)$
$$=7x^2-2x+3-5x^2-9$$
$$=2x^2-2x-6$$

18. $\left(10x^2+7x-9\right)-\left(4x^2-6x+2\right)$
$$=10x^2+7x-9-4x^2+6x-2$$
$$=6x^2+13x-11$$

19. $0.7y^2-1.2+1.8y^2-6y+1$
$$=2.5y^2-6y-0.2$$

20. $7.8x^2-6.8x+3.3+0.6x^2-9$
$$=8.4x^2-6.8x-5.7$$

21. $(x+4y)^2=(x+4y)(x+4y)$
$$=x^2+2(x)(4y)+(4y)^2$$
$$=x^2+8xy+16y^2$$

22. $(y-9z)^2=(y-9z)(y-9z)$
$$=y^2-2(y)(9z)+(9z)^2$$
$$=y^2-18yz+81z^2$$

23. $(x+4y)+(x+4y)=x+4y+x+4y$
$$=2x+8y$$

24. $(y-9z)+(y-9z)=y-9z+y-9z$
$$=2y-18z$$

25. $7x^2-6xy+4\left(y^2-xy\right)$
$$=7x^2-6xy+4y^2-4xy$$
$$=7x^2-10xy+4y^2$$

26. $5a^2 - 3ab + 6\left(b^2 - a^2\right)$

$= 5a^2 - 3ab + 6b^2 - 6a^2$

$= -a^2 - 3ab + 6b^2$

27. $(x-3)\left(x^2 + 5x - 1\right)$

$= x\left(x^2\right) + x(5x) + x(-1) - 3\left(x^2\right)$

$\qquad -3(5x) - 3(-1)$

$= x^3 + 5x^2 - x - 3x^2 - 15x + 3$

$= x^3 + 2x^2 - 16x + 3$

28. $(x+1)\left(x^2 - 3x - 2\right)$

$= x\left(x^2\right) + x(-3x) + x(-2) + 1\left(x^2\right)$

$\qquad + 1(-3x) + 1(-2)$

$= x^3 - 3x^2 - 2x + x^2 - 3x - 2$

$= x^3 - 2x^2 - 5x - 2$

29. $\left(2x^3 - 7\right)\left(3x^2 + 10\right)$

$= 2x^3\left(3x^2\right) + 2x^3(10) - 7\left(3x^2\right) - 7(10)$

$= 6x^5 + 20x^3 - 21x^2 - 70$

30. $\left(5x^3 - 1\right)\left(4x^4 + 5\right)$

$= 5x^3\left(4x^4\right) + 5x^3(5) - 1\left(4x^4\right) - 1(5)$

$= 20x^7 + 25x^3 - 4x^4 - 5$

31. $(2x-7)\left(x^2 - 6x + 1\right)$

$= 2x\left(x^2\right) - 2x(6x) + 2x(1) - 7\left(x^2\right)$

$\qquad -7(-6x) - 7(1)$

$= 2x^3 - 12x^2 + 2x - 7x^2 + 42x - 7$

$= 2x^3 - 19x^2 + 44x - 7$

32. $(5x-1)\left(x^2 + 2x - 3\right)$

$= 5x\left(x^2\right) + 5x(2x) + 5x(-3) - 1\left(x^2\right)$

$\qquad -1(2x) - 1(-3)$

$= 5x^3 + 10x^2 - 15x - x^2 - 2x + 3$

$= 5x^3 + 9x^2 - 17x + 3$

33. Cannot simplify

34. $\left(5x^3\right)\left(5y^3\right) = 25x^3 y^3$

35. $\left(5x^3\right)^3 = 5^3 x^{3\cdot3} = 125x^9$

36. $\dfrac{5x^3}{5y^3} = \dfrac{x^3}{y^3}$

37. $x + x = 2x$

38. $x \cdot x = x^2$

Section 5.5

Calculator Explorations

1. $5.31 \times 10^3 = 5.31$ EE 3

3. $6.6 \times 10^{-9} = 6.6$ EE -9

5. $3,000,000 \times 5,000,000 = 1.5 \times 10^{13}$

7. $\left(3.26 \times 10^6\right)\left(2.5 \times 10^{13}\right) = 8.15 \times 10^{19}$

Mental Math

1. $5x^{-2} = \dfrac{5}{x^2}$

2. $3x^{-3} = \dfrac{3}{x^3}$

3. $\dfrac{1}{y^{-6}} = y^6$

4. $\dfrac{1}{x^{-3}} = x^3$

5. $\dfrac{4}{y^{-3}} = 4y^3$

6. $\dfrac{16}{y^{-7}} = 16y^7$

Exercise Set 5.5

1. $4^{-3} = \dfrac{1}{4^3} = \dfrac{1}{64}$

3. $(-2)^{-4} = \dfrac{1}{(-2)^4} = \dfrac{1}{16}$

5. $7x^{-3} = 7 \cdot \dfrac{1}{x^3} = \dfrac{7}{x^3}$

7. $\left(\dfrac{1}{2}\right)^{-5} = \dfrac{1^{-5}}{2^{-5}} = \dfrac{2^5}{1^5} = 32$

9. $\left(-\dfrac{1}{4}\right)^{-3} = \dfrac{(-1)^{-3}}{(4)^{-3}} = \dfrac{4^3}{(-1)^3} = \dfrac{64}{-1} = -64$

11. $3^{-1} + 2^{-1} = \dfrac{1}{3} + \dfrac{1}{2} = \dfrac{2}{6} + \dfrac{3}{6} = \dfrac{5}{6}$

13. $\dfrac{1}{p^{-3}} = p^3$

15. $\dfrac{p^{-5}}{q^{-4}} = \dfrac{q^4}{p^5}$

17. $\dfrac{x^{-2}}{x} = x^{-2-1} = x^{-3} = \dfrac{1}{x^3}$

19. $2^0 + 3^{-1} = 1 + \dfrac{1}{3} = \dfrac{3}{3} + \dfrac{1}{3} = \dfrac{4}{3}$

21. $\dfrac{-1}{p^{-4}} = 1\left(p^4\right) = -p^4$

23. $-2^0 - 3^0 = -1(1) - 1 = -2$

25. $\dfrac{x^2 x^5}{x^3} = x^{2+5-3} = x^4$

27. $\dfrac{p^2 p}{p^{-1}} = p^{2+1-(-1)} = p^{2+1+1} = p^4$

29. $\dfrac{\left(m^5\right)^4 m}{m^{10}} = m^{5(4)+1-10} = m^{20+1-10} = m^{11}$

31. $\dfrac{r}{r^{-3} r^{-2}} = r^{1-(-3)-(-2)} = r^{1+3+2} = r^6$

33. $\left(x^5 y^3\right)^{-3} = x^{5(-3)} y^{3(-3)} = x^{-15} y^{-9} = \dfrac{1}{x^{15} y^9}$

35. $\dfrac{\left(x^2\right)^3}{x^{10}} = \dfrac{x^6}{x^{10}} = x^{6-10} = x^{-4} = \dfrac{1}{x^4}$

37. $\dfrac{\left(a^5\right)^2}{\left(a^3\right)^4} = \dfrac{a^{10}}{a^{12}} = a^{10-12} = a^{-2} = \dfrac{1}{a^2}$

39. $\dfrac{8k^4}{2k} = \dfrac{8}{2} \cdot k^{4-1} = 4k^3$

41. $\dfrac{-6m^4}{-2m^3} = \dfrac{-6}{-2} \cdot m^{4-3} = 3m$

43. $\dfrac{-24a^6b}{6ab^2} = \dfrac{-24}{6} \cdot a^{6-1}b^{1-2} = -4a^5b^{-1}$

$= -\dfrac{4a^5}{b}$

45. $\left(-2x^3y^{-4}\right)\left(3x^{-1}y\right) = -2(3)x^3x^{-1}y^{-4}y$

$= -6x^2y^{-3}$

$= -\dfrac{6x^2}{y^3}$

47. $\left(a^{-5}b^2\right)^{-6} = a^{-5(-6)}b^{2(-6)} = a^{30}b^{-12} = \dfrac{a^{30}}{b^{12}}$

49. $\left(\dfrac{x^{-2}y^4}{x^3y^7}\right)^2 = \dfrac{x^{-2(2)}y^{4(2)}}{x^{3(2)}y^{7(2)}} = \dfrac{x^{-4}y^8}{x^6y^{14}}$

$= x^{-4-6}y^{8-14} = x^{-10}y^{-6} = \dfrac{1}{x^{10}y^6}$

51. $\dfrac{4^2z^{-3}}{4^3z^{-5}} = 4^{2-3}z^{-3-(-5)} = 4^{-1}z^2 = \dfrac{z^2}{4}$

53. $\dfrac{2^{-3}x^{-4}}{2^2x} = 2^{-3-2}x^{-4-1} = 2^{-5}x^{-5} = \dfrac{1}{2^5x^5} = \dfrac{1}{32x^5}$

55. $\dfrac{7ab^{-4}}{7^{-1}a^{-3}b^2} = 7^{1-(-1)}a^{1-(-3)}b^{-4-2}$

$= 7^2a^4b^{-6}$

$= \dfrac{49a^4}{b^6}$

57. $\left(\dfrac{a^{-5}b}{ab^3}\right)^{-4} = \dfrac{a^{-5(-4)}b^{-4}}{a^{-4}b^{3(-4)}} = \dfrac{a^{20}b^{-4}}{a^{-4}b^{-12}}$

$= a^{20-(-4)}b^{-4-(-12)}$

$= a^{24}b^8$

59. $\dfrac{\left(xy^3\right)^5}{\left(xy\right)^{-4}} = \dfrac{x^5y^{3(5)}}{x^{-4}y^{-4}} = \dfrac{x^5y^{15}}{x^{-4}y^{-4}}$

$= x^{5-(-4)}y^{15-(-4)}$

$= x^9y^{19}$

61. $\dfrac{\left(-2xy^{-3}\right)^{-3}}{\left(xy^{-1}\right)^{-1}} = \dfrac{(-2)^{-3}x^{-3}y^9}{x^{-1}y^1}$

$= (-2)^{-3}x^{-3-(-1)}y^{9-1}$

$= -\dfrac{y^8}{8x^2}$

63. $\dfrac{6x^2y^3}{-7xy^5} = -\dfrac{6}{7}x^{2-1}y^{3-5} = -\dfrac{6}{7}x^1y^{-2}$

$= -\dfrac{6x}{7y^2}$

65. $78,000 = 7.8 \times 10^4$

67. $0.00000167 = 1.67 \times 10^{-6}$

69. $0.00635 = 6.35 \times 10^{-3}$

71. $1,160,000 = 1.16 \times 10^6$

73. $20,000,000 = 2.0 \times 10^7$

75. $15,600,000 = 1.56 \times 10^7$

77. $13,600 = 1.36 \times 10^4$

79. $292,000,000 = 2.92 \times 10^8$

81. $8.673 \times 10^{-10} = 0.0000000008673$

83. $3.3 \times 10^{-2} = 0.033$

85. $2.032 \times 10^4 = 20,320$

87. $6.25 \times 10^{18} = 6,250,000,000,000,000,000$

89. $9.460 \times 10^{12} = 9,460,000,000,000$

91. $\left(1.2 \times 10^{-3}\right)\left(3 \times 10^{-2}\right) = 1.2 \cdot 3 \cdot 10^{-3} \cdot 10^{-2}$
$$= 3.6 \times 10^{-5}$$
$$= 0.000036$$

93. $\left(4 \times 10^{-10}\right)\left(7 \times 10^{-9}\right) = 4 \cdot 7 \cdot 10^{-10} \cdot 10^{-9}$
$$= 28 \times 10^{-19}$$
$$= 0.0000000000000000028$$

95. $\dfrac{8 \times 10^{-1}}{16 \times 10^5} = \dfrac{8}{16} \times 10^{-1-5}$
$$= 0.5 \times 10^{-6}$$
$$= 0.0000005$$

97. $\dfrac{1.4 \times 10^{-2}}{7 \times 10^{-8}} = \dfrac{1.4}{7} \times 10^{-2-(-8)}$
$$= 0.2 \times 10^6$$
$$= 200,000$$

99. $\dfrac{5x^7}{3x^4} = \dfrac{5x^{7-4}}{3} = \dfrac{5x^3}{3}$

101. $\dfrac{15z^4 y^3}{21zy} = \dfrac{3 \cdot 5}{3 \cdot 7} z^{4-1} y^{3-1} = \dfrac{5z^3 y^2}{7}$

103. $\dfrac{1}{y}\left(5y^2 - 6y + 5\right)$
$$= \dfrac{1}{y}\left(5y^2\right) + \dfrac{1}{y}\left(-6y\right) + \dfrac{1}{y}\left(5\right)$$
$$= 5y - 6 + \dfrac{5}{y}$$

105. $\left(\dfrac{3x^{-2}}{z}\right)^3 = \dfrac{3^3 x^{-6}}{z^3} = \dfrac{27}{x^6 z^3}$

The volume is $\dfrac{27}{x^6 z^3}$ cubic inches.

107. $\left(2.63 \times 10^{12}\right)\left(-1.5 \times 10^{-10}\right)$
$$= 2.63 \cdot (-1.5) \cdot 10^{12} \cdot 10^{-10}$$
$$= -3.945 \times 10^2 = -394.5$$

109. $d = r \cdot t$
$$238,857 = \left(1.86 \times 10^5\right)t$$
$$t = \dfrac{238,857}{1.86 \times 10^5}$$
$$t = \dfrac{2.38857}{1.86} \times 10^{5-5}$$
$$t = 1.3 \text{ seconds}$$

111. Answers may vary.

113. $7.5 \times 10^5 \dfrac{\text{gallons}}{\text{second}}\left(\dfrac{3600 \text{ seconds}}{1 \text{ hour}}\right)$
$$= 27,000 \times 10^5 = 2.7 \times 10^4 \times 10^5$$
$$= 2.7 \times 10^9$$

2.7×10^9 gallons flows over Niagra Falls in one hour.

115. Answers may vary.

117. $a^{-4m} \cdot a^{5m} = a^{-4m+5m} = a^{m}$

119. $\dfrac{y^{4a}}{y^{-a}} = y^{4a-(-a)} = y^{5a}$

121. $\left(z^{3a+2}\right)^{-2} = z^{-2(3a+2)} = \dfrac{1}{z^{6a+4}}$

Section 5.6

Mental Math

1. $\dfrac{a^6}{a^4} = a^2$

2. $\dfrac{y^2}{y} = y$

3. $\dfrac{a^3}{a} = a^2$

4. $\dfrac{p^8}{p^3} = p^5$

5. $\dfrac{k^5}{k^2} = k^3$

6. $\dfrac{k^7}{k^5} = k^2$

7. $\dfrac{p^8}{p^3} = p^5$

8. $\dfrac{k^5}{k^2} = k^3$

9. $\dfrac{k^7}{k^5} = k^2$

Exercise Set 5.6

1. $\dfrac{15p^3 + 18p^2}{3p} = \dfrac{15p^3}{3p} + \dfrac{18p^2}{3p} = 5p^2 + 6p$

3. $\dfrac{-9x^4 + 18x^5}{6x^5} = \dfrac{-9x^4}{6x^5} + \dfrac{18x^5}{6x^5} = -\dfrac{3}{2x} + 3$

5. $\dfrac{-9x^5 + 3x^4 - 12}{3x^3} = \dfrac{-9x^5}{3x^3} + \dfrac{3x^4}{3x^3} - \dfrac{12}{3x^3}$

$= -3x^2 + x - \dfrac{4}{x^3}$

7. $\dfrac{4x^4 - 6x^3 + 7}{-4x^4} = \dfrac{4x^4}{-4x^4} - \dfrac{6x^3}{-4x^4} + \dfrac{7}{-4x^4}$

$= -1 + \dfrac{3}{2x} - \dfrac{7}{4x^4}$

9. $\dfrac{25x^5 - 15x^3 + 5}{5x^2} = \dfrac{25x^5}{5x^2} - \dfrac{15x^3}{5x^2} + \dfrac{5}{5x^2}$

$= 5x^3 - 3x + \dfrac{1}{x^2}$

11. $\dfrac{12x^3 + 4x - 16}{4} = \dfrac{12x^3}{4} + \dfrac{4x}{x} - \dfrac{16}{4}$

$= 3x^3 + x - 4$

Each side is $\left(3x^3 + x - 4\right)$ feet.

13.
$$x+3 \overline{)x^2 + 4x + 3} \quad \frac{x+1}{}$$

$$\underline{x^2 + 3x}$$
$$x + 3$$
$$\underline{x + 3}$$
$$0$$

$$\frac{x^2 + 4x + 3}{x + 3} = x + 1$$

15.
$$x+5 \overline{)2x^2 + 13x + 15} \quad \frac{2x+3}{}$$

$$\underline{2x^2 + 10x}$$
$$3x + 15$$
$$\underline{3x + 15}$$
$$0$$

$$\frac{2x^2 + 13x + 15}{x + 5} = 2x + 3$$

17.
$$x-4 \overline{)2x^2 - 7x + 3} \quad \frac{2x+1}{}$$

$$\underline{2x^2 - 8x}$$
$$x + 3$$
$$\underline{x - 4}$$
$$7$$

$$\frac{2x^2 - 7x + 3}{x - 4} = 2x + 1 + \frac{7}{x - 4}$$

19.
$$2x-3 \overline{)8x^2 + 6x - 27} \quad \frac{4x+9}{}$$

$$\underline{8x^2 - 12x}$$
$$18x - 27$$
$$\underline{18x - 27}$$
$$0$$

$$\frac{8x^2 + 6x - 27}{2x - 3} = 4x + 9$$

21.
$$3a+2 \overline{)9a^3 - 3a^2 - 3a + 4} \quad \frac{3a^2 - 3a + 1}{}$$

$$\underline{9a^3 + 6a^2}$$
$$-9a^2 - 3a$$
$$\underline{-9a^2 - 6a}$$
$$3a + 4$$
$$\underline{3a + 2}$$
$$2$$

$$\frac{9a^3 - 3a^2 - 3a + 4}{3a + 2} = 3a^2 - 3a + 1 + \frac{2}{3a + 2}$$

23.
$$b+4 \overline{)2b^3 + 9b^2 + 6b - 4} \quad \frac{2b^2 + b + 2}{}$$

$$\underline{2b^3 + 8b^2}$$
$$b^2 + 6b$$
$$\underline{b^2 + 4b}$$
$$2b - 4$$
$$\underline{2b + 8}$$
$$-12$$

$$\frac{2b^3 + 9b^2 + 6b - 4}{b + 4} = 2b^2 + b + 2 - \frac{12}{b + 4}$$

25. Answers may vary.

27.
$$5x+3 \overline{)10x^2 + 31x + 15} \quad \frac{2x+5}{}$$

$$\underline{10x^2 + 6x}$$
$$25 + 15$$
$$\underline{25 + 15}$$
$$0$$

The height is $(2x + 5)$ meters.

29. $\dfrac{20x^2 + 5x + 9}{5x^3} = \dfrac{20x^2}{5x^3} + \dfrac{5x}{5x^3} + \dfrac{9}{5x^3}$

$$= \dfrac{4}{x} + \dfrac{1}{x^2} + \dfrac{9}{5x^3}$$

31.

$$
\require{enclose}
\begin{array}{r}
5x - 2 \\
x + 6 \enclose{longdiv}{5x^2 + 28x - 10} \\
\underline{5x^2 + 30x} \\
-2x - 10 \\
\underline{-2x - 12} \\
2
\end{array}
$$

$$5x - 2 + \dfrac{2}{x + 6}$$

33. $\dfrac{10x^3 - 24x^2 - 10x}{10x} = \dfrac{10x^3}{10x} - \dfrac{24x^2}{10x} - \dfrac{10x}{10x}$

$$= x^2 - \dfrac{12x}{5} - 1$$

35.

$$
\begin{array}{r}
6x - 1 \\
x + 3 \enclose{longdiv}{6x^2 + 17x - 4} \\
\underline{6x^2 + 18x} \\
-\ x - 4 \\
\underline{-\ x - 3} \\
-1
\end{array}
$$

$$6x - 1 - \dfrac{1}{x + 3}$$

37. $\dfrac{12x^4 + 3x^2}{3x^2} = \dfrac{12x^4}{3x^2} + \dfrac{3x^2}{3x^2} = 4x^2 + 1$

39.

$$
\begin{array}{r}
2x^2 + 6x - 5 \\
x - 2 \enclose{longdiv}{2x^3 + 2x^2 - 17x + 8} \\
\underline{2x^3 - 4x^2} \\
6x^2 - 17x \\
\underline{6x^2 - 12x} \\
-5x + 8 \\
\underline{-5x + 10} \\
-2
\end{array}
$$

$$2x^2 + 6x - 5 - \dfrac{2}{x - 2}$$

41.

$$
\begin{array}{r}
6x - 1 \\
5x - 2 \enclose{longdiv}{30x^2 - 17x + 2} \\
\underline{30x^2 - 12x} \\
-\ 5x + 2 \\
\underline{-\ 5x + 2} \\
0
\end{array}
$$

$$6x - 1$$

43. $\dfrac{3x^4 - 9x^3 + 12}{-3x} = \dfrac{3x^4}{-3x} + \dfrac{-9x^3}{-3x} + \dfrac{12}{-3x}$

$$= -x^3 + 3x^2 - \dfrac{4}{x}$$

45.

$$
\begin{array}{r}
4x + 3 \\
2x + 1 \enclose{longdiv}{8x^2 + 10x + 1} \\
\underline{8x^2 + 4x} \\
6x + 1 \\
\underline{6x + 3} \\
-2
\end{array}
$$

$$\dfrac{8x^2 + 10x + 1}{2x + 1} = 4x + 3 - \dfrac{2}{2x + 1}$$

47.
$$2x - 9 \overline{)4x^2 + 0x - 81} \quad \frac{2x+9}{}$$
$$\underline{4x^2 - 18x}$$
$$18x - 81$$
$$\underline{18x - 81}$$
$$0$$
$$2x + 9$$

49.
$$2x + 3 \overline{)4x^3 + 12x^2 + x - 12} \quad \frac{2x^2 + 3x - 4}{}$$
$$\underline{4x^3 + 6x^2}$$
$$6x^2 + x$$
$$\underline{6x^2 + 9x}$$
$$-8x - 12$$
$$\underline{-8x - 12}$$
$$0$$
$$2x^2 + 3x - 4$$

51.
$$x - 3 \overline{)x^3 + 0x^2 + 0x - 27} \quad \frac{x^2 + 3x + 9}{}$$
$$\underline{x^3 - 3x^2}$$
$$3x^2 + 0x$$
$$\underline{3x^2 - 9x}$$
$$9x - 27$$
$$\underline{9x - 27}$$
$$0$$
$$\frac{x^3 - 27}{x - 3} = x^2 + 3x + 9$$

53.
$$x + 1 \overline{)x^3 + 0x^2 + 0x + 1} \quad \frac{x^2 - x + 1}{}$$
$$\underline{x^3 + x^2}$$
$$-x^2 + 0x$$
$$\underline{-x^2 - x}$$
$$x + 1$$
$$\underline{x + 1}$$
$$0$$
$$\frac{x^3 + 1}{x + 1} = x^2 - x + 1$$

55.
$$x + 2 \overline{)-3x^2 + 0x + 1} \quad \frac{-3x + 6}{}$$
$$\underline{-3x^2 - 6x}$$
$$6x + 1$$
$$\underline{6x + 12}$$
$$-11$$
$$\frac{1 - 3x^2}{x + 2} = -3x + 6 - \frac{11}{x + 2}$$

57.
$$2b - 1 \overline{)4b^2 - 4b - 5} \quad \frac{2b - 1}{}$$
$$\underline{4b^2 - 2b}$$
$$-2b - 5$$
$$\underline{-2b + 1}$$
$$-6$$
$$\frac{-4b + 4b^2 - 5}{2b - 1} = 2b - 1 - \frac{6}{2b - 1}$$

59. $2a(a^2 + 1) = 2a(a^2) + 2a(1)$
$$= 2a^3 + 2a$$

61. $2x\left(x^2 + 7x - 5\right)$

$= 2x\left(x^2\right) + 2x\left(7x\right) + 2x\left(-5\right)$

$= 2x^3 + 14x^2 - 10x$

63. $-3xy\left(xy^2 + 7x^2y + 8\right)$

$= -3xy\left(xy^2\right) - 3xy\left(7x^2y\right) - 3xy\left(8\right)$

$= -3x^2y^3 - 21x^3y^2 - 24xy$

65. $9ab\left(ab^2c + 4bc - 8\right)$

$= 9ab\left(ab^2c\right) + 9ab\left(4bc\right) + 9ab\left(-8\right)$

$= 9a^2b^3c + 36ab^2c - 72ab$

67. The Rolling Stones (1994)

69. $110 million

71.
$$
\begin{array}{r}
x^3 - x^2 + x \\
x^2 + x \overline{\smash{\big)}\ x^5 + 0x^4 + 0x^3 + x^2} \\
\underline{x^5 + \ x^4} \\
-x^4 + 0x^3 \\
\underline{-x^4 - \ x^3} \\
x^3 + x^2 \\
\underline{x^3 + x^2} \\
0
\end{array}
$$

$\dfrac{x^5 + x^2}{x^2 + x} = x^3 - x^2 + x$

73. $\dfrac{18x^{10a} - 12x^{8a} + 14x^{5a} - 2x^{3a}}{2x^{3a}}$

$= \dfrac{18x^{10a}}{2x^{3a}} - \dfrac{12x^{8a}}{2x^{3a}} + \dfrac{14x^{5a}}{2x^{3a}} - \dfrac{2x^{3a}}{2x^{3a}}$

$= 9x^{7a} - 6x^{5a} + 7x^{2a} - 1$

Exercise Set 5.7

1.
$$
\begin{array}{r}
5\underline{|}\ \ 1 \quad\ \ 3 \quad -40 \\
 5 \quad\ \ 40 \\
\hline
1 \quad\ \ 8 \quad\ \ \ 0
\end{array}
$$

$x + 8$

3.
$$
\begin{array}{r}
-6\underline{|}\ \ 1 \quad\ \ 5 \quad -6 \\
 -6 \quad\ \ 6 \\
\hline
1 \quad -1 \quad\ \ 0
\end{array}
$$

$x - 1$

5.
$$
\begin{array}{r}
2\underline{|}\ \ 1 \quad -7 \quad -13 \quad\ \ 5 \\
 2 \quad -10 \quad -46 \\
\hline
1 \quad -5 \quad -23 \quad -41
\end{array}
$$

$x^2 - 5x - 23 - \dfrac{41}{x - 2}$

7.
$$
\begin{array}{r}
2\underline{|}\ \ 4 \quad\ \ 0 \quad -9 \\
 8 \quad\ \ 16 \\
\hline
4 \quad\ \ 8 \quad\ \ 7
\end{array}
$$

$4x + 8 + \dfrac{7}{x - 2}$

9. a. $P(2) = 3(2)^2 - 4(2) - 1 = 12 - 8 - 1 = 3$

b.
$$
\begin{array}{r}
2\underline{|}\ \ 3 \quad -4 \quad -1 \\
 6 \quad\ \ 4 \\
\hline
3 \quad\ \ 2 \quad\ \ 3
\end{array}
$$

$P(2) = 3$

11. a. $P(-2) = 4(-2)^4 + 7(-2)^2 + 9(-2) - 1$
$$= 64 + 28 - 18 - 1$$
$$= 73$$

b.
$$\underline{-2}\begin{array}{rrrrr} 4 & 0 & 7 & 9 & -1 \\ & -8 & 16 & -46 & 74 \\ \hline 4 & -8 & 23 & -37 & 73 \end{array}$$

$$P(-2) = 73$$

13. a. $P(-1) = (-1)^5 + 3(-1)^4 + 3(-1) - 7$
$$= -1 + 3 - 3 - 7$$
$$= -8$$

b.
$$\underline{-1}\begin{array}{rrrrrr} 1 & 3 & 0 & 0 & 3 & -7 \\ & -1 & -2 & 2 & -2 & -1 \\ \hline 1 & 2 & -2 & 2 & 1 & -8 \end{array}$$

$$P(-1) = -8$$

15.
$$\underline{3}\begin{array}{rrrr} 1 & -3 & 0 & 2 \\ & 3 & 0 & 0 \\ \hline 1 & 0 & 0 & 2 \end{array}$$

$$x^2 + \frac{2}{x-3}$$

17.
$$\underline{-1}\begin{array}{rrr} 6 & 13 & 8 \\ & -6 & -7 \\ \hline 6 & 7 & 1 \end{array}$$

$$6x + 7 + \frac{1}{x+1}$$

19.
$$\underline{5}\begin{array}{rrrrr} 2 & -13 & 16 & -9 & 20 \\ & 10 & -15 & 5 & -20 \\ \hline 2 & -3 & 1 & -4 & 0 \end{array}$$

$$2x^3 - 3x^2 + x - 4$$

21.
$$\underline{-3}\begin{array}{rrr} 3 & 0 & -15 \\ & -9 & 27 \\ \hline 3 & -9 & 12 \end{array}$$

$$3x - 9 + \frac{12}{x+3}$$

23.
$$\underline{\tfrac{1}{2}}\begin{array}{rrrr} 3 & -6 & 4 & 5 \\ & \tfrac{3}{2} & -\tfrac{9}{4} & \tfrac{7}{8} \\ \hline 3 & -\tfrac{9}{2} & \tfrac{7}{4} & \tfrac{47}{8} \end{array}$$

$$3x^2 - \frac{9}{2}x + \frac{7}{4} + \frac{47}{8\left(x - \frac{1}{2}\right)}$$

25.
$$\underline{\tfrac{1}{3}}\begin{array}{rrrr} 3 & 2 & -4 & 1 \\ & 1 & 1 & -1 \\ \hline 3 & 3 & -3 & 0 \end{array}$$

$$3x^2 + 3x - 3$$

27.
$$\underline{-1}\begin{array}{rrrr} 3 & 7 & -4 & 12 \\ & -3 & -4 & 8 \\ \hline 3 & 4 & -8 & 20 \end{array}$$

$$3x^2 + 4x - 8 + \frac{20}{x+1}$$

29.

$$
\begin{array}{r}
1\,\rvert\ \ 1 \quad\ \ 0 \quad\ \ 0 \quad -1 \\
\ \ 1 \quad\ \ 1 \quad\ \ 1 \\
\hline
1 \quad\ \ 1 \quad\ \ 1 \quad\ \ 0
\end{array}
$$

$$x^2 + x + 1$$

31.

$$
\begin{array}{r}
-6\,\rvert\ \ 1 \quad\ \ 0 \quad -36 \\
\ -6 \quad\ \ 36 \\
\hline
1 \quad -6 \quad\ \ 0
\end{array}
$$

$$x - 6$$

33.

$$
\begin{array}{r}
1\,\rvert\ \ 1 \quad\ \ 3 \quad -7 \quad\ \ 4 \\
\ \ 1 \quad\ \ 4 \quad -3 \\
\hline
1 \quad\ \ 4 \quad -3 \quad\ \ 1
\end{array}
$$

Thus, $P(1) = 1$.

35.

$$
\begin{array}{r}
-3\,\rvert\ \ 3 \quad -7 \quad -2 \quad\ \ 5 \\
\ -9 \quad\ \ 48 \quad -138 \\
\hline
3 \quad -16 \quad\ \ 46 \quad -133
\end{array}
$$

Thus, $P(-3) = -133$.

37.

$$
\begin{array}{r}
-1\,\rvert\ \ 4 \quad\ \ 0 \quad\ \ 1 \quad\ \ 0 \quad -2 \\
\ -4 \quad\ \ 4 \quad -5 \quad\ \ 5 \\
\hline
4 \quad -4 \quad\ \ 5 \quad -5 \quad\ \ 3
\end{array}
$$

Thus, $P(-1) = 3$.

39.

$$
\begin{array}{r}
\frac{1}{3}\,\rvert\ \ 2 \quad\ \ 0 \quad -3 \quad\ \ 0 \quad -2 \\
\ \frac{2}{3} \quad\ \ \frac{2}{9} \quad -\frac{25}{27} \quad -\frac{25}{81} \\
\hline
2 \quad\ \ \frac{2}{3} \quad -\frac{25}{9} \quad -\frac{25}{27} \quad -\frac{187}{81}
\end{array}
$$

Thus, $P\!\left(\dfrac{1}{3}\right) = -\dfrac{187}{81}$.

41.

$$
\begin{array}{r}
\frac{1}{2}\,\rvert\ \ 1 \quad\ \ 1 \quad -1 \quad\ \ 0 \quad\ \ 0 \quad\ \ 3 \\
\ \frac{1}{2} \quad\ \ \frac{3}{4} \quad -\frac{1}{8} \quad -\frac{1}{16} \quad -\frac{1}{32} \\
\hline
1 \quad\ \ \frac{3}{2} \quad -\frac{1}{4} \quad -\frac{1}{8} \quad -\frac{1}{16} \quad\ \ \frac{95}{32}
\end{array}
$$

Thus, $P\!\left(\dfrac{1}{2}\right) = \dfrac{95}{32}$.

43. Answers may vary.

45.
$$
\begin{aligned}
7x + 2 &= x - 3 \\
7x - x &= -3 - 2 \\
6x &= -5 \\
x &= -\frac{5}{6}
\end{aligned}
$$

The solution is $-\dfrac{5}{6}$.

47.
$$
\begin{aligned}
\frac{x}{3} - 5 &= 13 \\
3\!\left(\frac{x}{3} - 5\right) &= (13)\cdot 3 \\
x - 15 &= 39 \\
x &= 54
\end{aligned}
$$

The solution is 54.

49. $2^3 = 8$

51. $(-2)^5 = -32$

53. $3 \cdot 4^2 = 48$

55. If $x = -5$ then $x^2 = (-5)^2 = 25$

57. If $x = -1$ then $2x^3 = 2(-1)^3 = 2(-1) = -2$

59.

$$
\begin{array}{r|rrrr}
-3 & 1 & 3 & 4 & 12 \\
  &   & -3 & 0 & -12 \\
\hline
  & 1 & 0 & 4 & 0
\end{array}
$$

Remainder = 0 and

$$(x+3)(x^2+4) = x^3 + 3x^2 + 4x + 12$$

61. $P(x)$ is equal to the remainder when

$P(x)$ is divided by $x-c$. Therefore,

$P(c) = 0$.

63. Multiply $(x^2 - x + 10)$ by $(x+3)$ and

add the remainder, –2.

$$
\begin{aligned}
&(x^2 - x + 10)(x+3) - 2 \\
&= (x^3 + 3x^2 - x^2 - 3x + 10x + 30) - 2 \\
&= x^3 + 2x^2 + 7x + 28
\end{aligned}
$$

65. $V = lwh$ so $w = \dfrac{V}{lh}$

$$
\begin{aligned}
&= \frac{x^4 + 6x^3 - 7x^2}{x^2(x+7)} \\
&= \frac{x^4 + 6x^3 - 7x^2}{x^3 + 7x^2}
\end{aligned}
$$

$$
\begin{array}{r}
x - 1 \\
x^3 + 7x^2 \overline{\smash{\big)}\, x^4 + 6x^3 - 7x^2} \\
\underline{x^4 + 7x^3} \\
-x^3 - 7x^2 \\
\underline{-x^3 - 7x^2} \\
0
\end{array}
$$

The width is $(x-1)$ meters.

Chapter 5 Review

1. 3^2

base: 3

exponent: 2

2. $(-5)^4$

base: -5

exponent: 4

3. -5^4

base: 5

exponent: 4

4. $8^3 = 8 \cdot 8 \cdot 8 = 512$

5. $(-6)^2 = (-6)(-6) = 36$

6. $-6^2 = -6 \cdot 6 = -36$

7. $-4^3 - 4^0 = -4 \cdot 4 \cdot 4 - 1 = -65$

8. $(3b)^0 = 1$

9. $\dfrac{8b}{8b} = 1$

10. $5b^3 b^5 a^6 = 5a^6 b^8$

11. $2^3 \cdot x^0 = 8 \cdot 1 = 8$

12. $\left[(-3)^2\right]^3 = (9)^3 = 9 \cdot 9 \cdot 9 = 729$

13. $\left(2x^3\right)\left(-5x^2\right) = (2)(-5)\left(x^3\right)\left(x^2\right) = -10x^5$

14. $\left(\dfrac{mn}{q}\right)^2 \cdot \left(\dfrac{mn}{q}\right) = \dfrac{m^2 n^2}{q^2} \cdot \dfrac{mn}{q} = \dfrac{m^3 n^3}{q^3}$

15. $\left(\dfrac{3ab^2}{6ab}\right)^4 = \left(\dfrac{b}{2}\right)^4 = \dfrac{b^4}{2^4} = \dfrac{b^4}{16}$

16. $\dfrac{x^9}{x^4} = x^{9-4} = x^5$

17. $\dfrac{2x^7 y^8}{8xy^2} = \dfrac{2}{8} \cdot x^{7-1} y^{8-2}$
$\qquad\qquad = \dfrac{x^6 y^6}{4}$

18. $\dfrac{3x^4 y^{10}}{12xy^6} = \dfrac{3}{12} \cdot x^{4-1} \cdot y^{10-6} = \dfrac{x^3 y^4}{4}$

19. $5a^7 \left(2a^4\right)^3$
$= 5a^7 \left(2^3 a^{4\cdot3}\right)$
$= 5a^7 \left(8a^{12}\right)$
$= 5 \cdot 8a^{7+12}$
$= 40a^{19}$

20. $(2x)^2 (9x)$
$= \left(2^2 \cdot x^2\right)(9x)$
$= 4x^2 \cdot 9x$
$= 4 \cdot 9 \cdot x^{2+1}$
$= 36x^3$

21. $\dfrac{(-4)^2 \left(3^3\right)}{(4)^5 \left(3^2\right)} = 4^{2-5} 3^{3-2} = 4^{-3} \cdot 3 = \dfrac{3}{4^3} = \dfrac{3}{64}$

22. $\dfrac{(-7)^2 \left(3^5\right)}{(-7)^3 \left(3^4\right)} = \dfrac{3}{-7} = -\dfrac{3}{7}$

23. $\dfrac{(2x)^0 (-4)^2}{16x} = \dfrac{1 \cdot 16}{16x} = \dfrac{1}{x}$

24. $\dfrac{(8xy)(3xy)}{18x^2 y^2} = \dfrac{24x^2 y^2}{18x^2 y^2} = \dfrac{4}{3}$

25. $m^0 + p^0 + 3q^0 = 1 + 1 + 3(1)$
$\qquad\qquad\qquad = 1 + 1 + 3$
$\qquad\qquad\qquad = 5$

26. $(-5a)^0 + 7^0 + 8^0 = 1 + 1 + 1 = 3$

27. $\left(3xy^2 + 8x + 9\right)^0 = 1$

28. $8x^0 + 9^0 = 8(1) + 1 = 9$

29. $6\left(a^2 b^3\right)^3 = 6\left(a^6 b^9\right) = 6a^6 b^9$

30. $\dfrac{\left(x^3 z\right)^a}{x^2 z^2} = \dfrac{x^{3a} z^a}{x^2 z^2} = x^{3a-2} z^{a-2}$

31. $-5x^4 y^3$
The degree is $4 + 3 = 7$.

32. $95xyz$
The degree is $1 + 1 + 1 = 3$.

33. $-14x^2 y - 28x^2 y^3 - 42x^2 y^2$
The degree is $2 + 3 = 5$.

32. $95xyz$
The degree is $1 + 1 + 1 = 3$.

35. The degree is 5 because y^5 is the term with the highest degree.

36. The degree is 2 because $9y^2$ is the term with the highest degree.

37. The degree is 5 because $-28x^2y^3$ is the term with the highest degree.

38. The degree is 6 because $6x^2y^2z^2$ is the term with the highest degree.

39. **a.** $3a^2b - 2a^2 + ab - b^2 - 6$

Term	Numerical Coefficient	Degree of Term
$3a^2b$	3	3
$-2a^2$	-2	2
ab	1	2
$-b^2$	-1	2
-6	-6	0

b. Degree 3

40. **a.** $x^2y^2 + 5x^2 - 7y^2 + 11xy - 1$

Term	Numerical Coefficient	Degree of Term
x^2y^2	1	4
$5x^2$	5	2
$-7y^2$	-7	2
$11xy$	11	2
-1	-1	0

b. Degree 4

41. $2x^2 + 20x$:

$x = 1: \ 2(1)^2 + 20(1) = 22$

$x = 3: \ 2(3)^2 + 20(3) = 78$

$x = 5.1: \ 2(5.1)^2 + 20(5.1) = 154.02$

$x = 10: \ 2(10)^2 + 20(10) = 400$

42. $6a^2b^2 + 4ab + 9a^2b^2 = (6+9)a^2b^2 + 4ab$
$$= 15a^2b^2 + 4ab$$

43. $21x^2y^3 + 3xy + x^2y^3 + 6$
$$= (21+1)x^2y^3 + 3xy + 6$$
$$= 22x^2y^3 + 3xy + 6$$

44. $4a^2b - 3b^2 - 8q^2 - 10a^2b + 7q^2$
$$= (4a^2b - 10a^2b) - 3b^2 + (-8q^2 + 7q^2)$$
$$= -6a^2b - 3b^2 - q^2$$

45. $2s^{14} + 3s^{13} + 12s^{12} - s^{10}$
Cannot be combined.

46. $(3k^2 + 2k + 6) + (5k^2 + k)$
$$= 3k^2 + 2k + 6 + 5k^2 + k$$
$$= 8k^2 + 3k + 6$$

47. $(2s^5 + 3s^4 + 4s^3 + 5s^2) - (4s^2 + 7s + 6)$
$$= 2s^5 + 3s^4 + 4s^3 + 5s^2 - 4s^2 - 7s - 6$$
$$= 2s^5 + 3s^4 + 4s^3 + s^2 - 7s - 6$$

48. $(2m^7 + 3x^4 + 7m^6) - (8m^7 + 4m^2 + 6x^4)$
$$= 2m^7 + 3x^4 + 7m^6 - 8m^7 - 4m^2 - 6x^4$$
$$= -6m^7 - 3x^4 + 7m^6 - 4m^2$$

49. $(11r^2 + 16rs - 2s^2) - (3r^2 + 5rs - 9s^2)$
$$= 11r^2 + 16rs - 2s^2 - 3r^2 - 5rs + 9s^2$$
$$= 8r^2 + 11rs + 7s^2$$

50. $(3x^2 - 6xy + y^2) - (11x^2 - xy + 5y^2)$
$$= 3x^2 - 6xy + y^2 - 11x^2 + xy - 5y^2$$
$$= -8x^2 - 5xy - 4y^2$$

51. $(7x - 14y) - (3x - y)$
$= 7x - 14y - 3x + y$
$= 4x - 13y$

52. $\left[(x^2 + 7x + 9) + (x^2 + 4)\right] - (4x^2 + 8x - 7)$
$= x^2 + 7x + 9 + x^2 + 4 - 4x^2 - 8x + 7$
$= -2x^2 - x + 20$

53. Let $x = 20$.
$72.5x^2 - 17.5x + 120$
$= 72.5(20)^2 - 17.5(20) + 120 = 28,770$
Expect 28,770,000 trademark
registrations in 2010.

54. $9x(x^2 y) = 9x^3 y$

55. $-7(8xz^2) = -56xz^2$

56. $(6xa^2)(xya^3) = 6x^2 ya^5$

57. $(4xy)(-3xa^2 y^3) = -12x^2 a^2 y^4$

58. $6(x + 5) = 6x + 6(5)$
$= 6x + 30$

59. $9(x - 7) = 9x - 9(7)$
$= 9x - 63$

60. $4(2a + 7) = 4(2a) + 4(7)$
$= 8a + 28$

61. $9(6a - 3) = 9(6a) - 9(3)$
$= 54a - 27$

62. $-7x(x^2 + 5) = -7(x^2) - 7x(5)$
$= -7x^3 - 35x$

63. $-8y(4y^2 - 6) = -8y(4y^2) - 8y(-6)$
$= -32y^3 + 48y$

64. $-2(x^3 - 9x^2 + x) = -2(x^3) - 2(-9x^2) - 2(x)$
$= -2x^3 + 18x^2 - 2x$

65. $-3a(a^2 b + ab + b^2)$
$= -3a(a^2 b) - 3a(ab) - 3a(b^2)$
$= -3a^3 b - 3a^2 b - 3ab^2$

66. $(3a^3 - 4a + 1)(-2a)$
$= 3a^3(-2a) - 4a(-2a) + 1(-2a)$
$= -6a^4 + 8a^2 - 2a$

67. $(6b^3 - 4b + 2)(7b)$
$= 6b^3(7b) - 4b(7b) + 2(7b)$
$= 42b^4 - 28b^2 + 14b$

68. $(2x + 5)(x - 7)$
$= 2x(x) + 2x(-7) + 5(x) + 5(-7)$
$= 2x^2 - 9x - 35$

69. $(2x - 5)(3x + 2)$
$= 2x(3x) + 2x(2) - 5(3x) - 5(2)$
$= 6x^2 + 4x - 15x - 10$
$= 6x^2 - 11x - 10$

70. $(4a - 1)(a + 7) = 4a^2 + 28a - a - 7$
$= 4a^2 + 27a - 7$

71. $(6a-1)(7a+3) = 42a^2 + 18a - 7a - 3$
$$= 42a^2 + 11a - 3$$

72. $(x+7)(x^3 + 4x - 5)$
$$= x^4 + 4x^2 - 5x + 7x^3 + 28x - 35$$
$$= x^4 + 7x^3 + 4x^2 + 23x - 35$$

73. $(x+2)(x^5 + x + 1)$
$$= x^6 + x^2 + x + 2x^5 + 2x + 2$$
$$= x^6 + 2x^5 + x^2 + 3x + 2$$

74. $(x^2 + 2x + 4)(x^2 + 2x - 4)$
$$= x^4 + 2x^3 - 4x^2 + 2x^3 + 4x^2 - 8x$$
$$\quad + 4x^2 + 8x - 16$$
$$= x^4 + 4x^3 + 4x^2 - 16$$

75. $(x^3 + 4x + 4)(x^3 + 4x - 4)$
$$= x^6 + 4x^4 - 4x^3 + 4x^4 + 16x^2 - 16x$$
$$\quad + 4x^3 + 16x - 16$$
$$= x^6 + 8x^4 + 16x^2 - 16$$

76. $(x+7)^3$
$$= (x+7)(x+7)(x+7)$$
$$= (x^2 + 7x + 7x + 49)(x+7)$$
$$= (x^2 + 14x + 49)(x+7)$$
$$= x^3 + 7x^2 + 14x^2 + 98x + 49x + 343$$
$$= x^3 + 21x^2 + 147x + 343$$

77. $(2x-5)^3$
$$= (2x-5)(2x-5)(2x-5)$$
$$= (4x^2 - 10x - 10x + 25)(2x-5)$$
$$= (4x^2 - 20x + 25)(2x-5)$$
$$= 8x^3 - 20x^2 - 40x^2 + 100x + 50x - 125$$
$$= 8x^3 - 60x^2 + 150x - 125$$

78. $2x(3x^2 - 7x + 1)$
$$= 2x(3x^2) - 2x(7x) + 2x(1)$$
$$= 6x^3 - 14x^2 + 2x$$

79. $3y(5y^2 - y + 2)$
$$= 3y(5y^2) - 3y(y) + 3y(2)$$
$$= 15y^3 - 3y^2 + 6y$$

80. $(6x^5 - 1)(4x^2 + 3)$
$$= 6x^5(4x^2) + 6x^5(3) - 1(4x^2) - 1(3)$$
$$= 24x^7 + 18x^5 - 4x^2 - 3$$

81. $(4a^3 - 1)(3a^2 + 7)$
$$= 4a^3(3a^2) + 4a^3(7) - 1(3a^2) - 1(7)$$
$$= 12a^5 + 28a^3 - 3a^2 - 7$$

82. $(x^2 + 7y)^2 = (x^2)^2 + 2(x^2)(7y) + (7y)^2$
$$= x^4 + 14x^2 y + 49y^2$$

83. $(x^3 - 5y)^2 = (x^3)^2 - 2(x^3)(5y) + (5y)^2$
$$= x^6 - 10x^3 y + 25y^2$$

84. $(3x-7)^2 = (3x)^2 - 2(3x)(7) + 7^2$
$$= 9x^2 - 42x + 49$$

85. $(4x+2)^2 = (4x)^2 + 2(4x)(2) + 2^2$
$$= 16x^2 + 16x + 4$$

86. $(y+1)(y^2 - 6y - 5)$
$$= y(y^2) - y(6y) - y(5) + y^2 - 6y - 5$$
$$= y^3 - 6y^2 - 5y + y^2 - 6y - 5$$
$$= y^3 - 5y^2 - 11y - 5$$

87. $(x-2)(x^2 - x - 2)$
$$= x(x^2) - x(x) - 2x - 2x^2 + 2x - 2(-2)$$
$$= x^3 - x^2 - 2x - 2x^2 + 2x + 4$$
$$= x^3 - 3x^2 + 4$$

88. $(5x-9)^2 = (5x)^2 - 2(5x)(9) + 9^2$
$$= 25x^2 - 90x + 81$$

89. $(5x+1)(5x-1) = (5x)^2 - 1^2$
$$= 25x^2 - 1$$

90. $(7x+4)(7x-4) = (7x)^2 - 4^2$
$$= 49x^2 - 16$$

91. $(a+2b)(a-2b) = a^2 - (2b)^2$
$$= a^2 - 4b^2$$

92. $(2x-6)(2x+6) = (2x)^2 - 6^2$
$$= 4x^2 - 36$$

93. $(4a^2 - 2b)(4a^2 + 2b) = (4a^2)^2 - (2b)^2$
$$= 16a^4 - 4b^2$$

94. $7^{-2} = \dfrac{1}{7^2} = \dfrac{1}{49}$

95. $-7^{-2} = -\dfrac{1}{7^2} = -\dfrac{1}{49}$

96. $2x^{-4} = \dfrac{2}{x^4}$

97. $(2x)^{-4} = \dfrac{1}{(2x)^4} = \dfrac{1}{16x^4}$

98. $\left(\dfrac{1}{5}\right)^{-3} = \dfrac{1^{-3}}{5^{-3}} = \dfrac{5^3}{1^3} = 125$

99. $\left(\dfrac{-2}{3}\right)^{-2} = \dfrac{(-2)^{-2}}{3^{-2}} = \dfrac{3}{(-2)^2} = \dfrac{9}{4}$

100. $2^0 + 2^{-4} = 1 + \dfrac{1}{2^4} = \dfrac{16}{16} + \dfrac{1}{16} = \dfrac{17}{16}$

101. $6^{-1} - 7^{-1} = \dfrac{1}{6} - \dfrac{1}{7} = \dfrac{7}{42} - \dfrac{6}{42} = \dfrac{1}{42}$

102. $\dfrac{1}{(2q)^{-3}} = 1 \cdot (2q)^3 = 1 \cdot 2^3 \cdot q^3 = 1 \cdot 8 \cdot q^3 = 8q^3$

103. $\dfrac{-1}{(qr)^{-3}} = -1 \cdot (qr)^3 = -1 \cdot q^3 \cdot r^3 = -q^3 r^3$

104. $\dfrac{r^{-3}}{s^{-4}} = \dfrac{s^4}{r^3}$

105. $\dfrac{rs^{-3}}{r^{-4}} = r^{1-(-4)} s^{-3} = \dfrac{r^5}{s^3}$

106. $(-x^{-3} y^5)(5xy^{-2}) = -5x^{-3+1} y^{5-2}$
$$= -5x^{-2} y^3$$
$$= -\dfrac{5y^3}{x^2}$$

107. $\left(-3x^5y^{-2}\right)\left(-4x^{-5}y\right) = 12x^{5-5}y^{-2+1}$

$$= 12x^0y^{-1}$$

$$= \frac{12}{y}$$

108. $\left(2x^{-5}\right)^{-3} = 2^{-3}x^{15} = \frac{x^{15}}{2^3} = \frac{x^{15}}{8}$

109. $\left(3y^{-6}\right)^{-1} = 3^{-1}y^6 = \frac{y^6}{3}$

110. $\left(3a^{-1}b^{-1}c^{-2}\right)^{-2} = 3^{-2}a^2b^2c^4$

$$= \frac{a^2b^2c^4}{3^2}$$

$$= \frac{a^2b^2c^4}{9}$$

111. $\left(4x^{-2}y^{-3}z\right)^{-3} = 4^{-3}x^6y^9z^{-3} = \frac{x^6y^9}{4^3z^3} = \frac{x^6y^9}{64z^3}$

112. $\frac{5^{-2}x^8}{5^{-3}x^{11}} = 5^{-2-(-3)}x^{8-11}$

$$= 5^{-2+3}x^{8-11}$$

$$= 5^1x^{-3}$$

$$= \frac{5}{x^3}$$

113. $\frac{7^5y^{-2}}{7^7y^{-10}} = 7^{5-7}\cdot y^{-2-(-10)} = 7^{-2}\cdot y^8 = \frac{y^8}{7^2} = \frac{y^8}{49}$

114. $\left(\frac{bc^{-2}}{bc^{-3}}\right)^4 = \frac{b^4c^{-8}}{b^4c^{-12}} = b^{4-4}c^{-8-(-12)} = c^4$

115. $\left(\frac{x^{-3}y^{-4}}{x^{-2}y^{-5}}\right)^{-3} = \frac{x^9y^{12}}{x^6y^{15}} = x^{9-6}y^{12-15} = \frac{x^3}{y^3}$

116. $\frac{x^{-4}y^{-6}z^3}{x^2y^7z^3} = x^{-4-2}y^{-6-7}z^{3-3}$

$$= x^{-6}y^{-13}z^0$$

$$= \frac{1}{x^6y^{13}}$$

117. $\frac{a^5b^{-5}c^4}{a^{-5}b^5c^4} = a^{5-(-5)}b^{-5-5}c^{4-4}$

$$= a^{10}b^{-10}c^0$$

$$= \frac{a^{10}}{b^{10}}$$

118. $-2^0 + 2^{-4} = -1\cdot 2^0 + \frac{1}{2^4}$

$$= -1\cdot 1 + \frac{1}{16}$$

$$= -1 + \frac{1}{16}$$

$$= -\frac{16}{16} + \frac{1}{16}$$

$$= -\frac{15}{16}$$

119. $-3^{-2} - 3^{-3} = -\frac{1}{3^2} - \frac{1}{3^3}$

$$= -\frac{1}{9} - \frac{1}{27}$$

$$= -\frac{3}{27} - \frac{1}{27}$$

$$= -\frac{4}{27}$$

120. $a^{6m}a^{5m} = a^{6m+5m} = a^{11m}$

121.
$$\frac{\left(x^{5+h}\right)^3}{x^5} = \frac{x^{3(5+h)}}{x^5}$$
$$= \frac{x^{15+3h}}{x^5}$$
$$= x^{15+3h-5}$$
$$= x^{10+3h}$$

122. $\left(3xy^{2z}\right)^3 = 3^3 x^3 y^{2z(3)} = 27x^3 y^{6z}$

123. $a^{m+2} \cdot a^{m+3} = a^{m+2m+3} = a^{2m+5}$

124. $0.00027 = 2.7 \times 10^{-4}$

125. $0.8868 = 8.868 \times 10^{-1}$

126. $80,800,000 = 8.08 \times 10^7$

127. $868,000 = 8.68 \times 10^5$

128. $109,379,000 = 1.09379 \times 10^8$ kg

129. $150,000 = 1.5 \times 10^5$ light years

130. $8.67 \times 10^5 = 867,000$

131. $3.86 \times 10^{-3} = 0.00386$

132. $8.6 \times 10^{-4} = 0.00086$

133. $8.936 \times 10^5 = 893,600$

134. 1.43128×10^{15}
$$= 1,431,280,000,000,000 \text{ cu km}$$

135. $1 \times 10^{-10} = 0.0000000001$ m

136.
$$\left(8 \times 10^4\right)\left(2 \times 10^{-7}\right)$$
$$= \left(8 \times 2\right) \times \left(10^4 \times 10^{-7}\right)$$
$$= 16 \times 10^{-3}$$
$$= 0.016$$

137.
$$\frac{8 \times 10^4}{2 \times 10^{-7}}$$
$$= \frac{8}{2} \times \left(10^{4-(-7)}\right)$$
$$= 4 \times 10^{11}$$
$$= 400,000,000,000$$

138.
$$\frac{x^2 + 21x + 49}{7x^2} = \frac{x^2}{7x^2} + \frac{21x}{7x^2} + \frac{49}{7x^2}$$
$$= \frac{1}{7} + \frac{3}{x} + \frac{7}{x^2}$$

139.
$$\frac{5a^3b - 15ab^2 + 20ab}{-5ab}$$
$$= \frac{5a^3b}{-5ab} - \frac{15ab^2}{-5ab} + \frac{20ab}{-5ab}$$
$$= -a^2 + 3b - 4$$

140.
$$\begin{array}{r} a+1 \\ a-2\overline{)a^2 - a + 4} \\ \underline{a^2 - 2a} \\ a + 4 \\ \underline{a - 2} \\ 6 \end{array}$$

$$\left(a^2 - a + 4\right) \div \left(a - 2\right) = a + 1 + \frac{6}{a-2}$$

141.

$$x+5 \overline{)\begin{array}{r} 4x \\ 4x^2 + 20x + 7 \end{array}}$$

$$\underline{4x^2 + 20x}$$

$$7$$

$$\left(4x^2 + 20x + 7\right) \div \left(x+5\right) = 4x + \frac{7}{x+5}$$

142.

$$a-2 \overline{)\begin{array}{r} a^2 + 3a + 8 \\ a^3 + a^2 + 2a + 6 \end{array}}$$

$$\underline{a^3 - 2a^2}$$

$$3a^2 + 2a$$

$$\underline{3a^2 - 6a}$$

$$8a + 6$$

$$\underline{8a - 16}$$

$$22$$

$$\frac{a^3 + a^2 + 2a + 6}{a-2} = a^2 + 3a + 8 + \frac{22}{a-2}$$

143.

$$3b-2 \overline{)\begin{array}{r} 3b^2 - 4b \\ 9b^3 - 18b^2 + 8b - 1 \end{array}}$$

$$\underline{9b^3 - 6b^2}$$

$$-12b^2 + 8b$$

$$\underline{-12b^2 + 8b}$$

$$-1$$

$$\frac{9b^3 - 18b^2 + 8b - 1}{3b-2} = 3b^2 - 4b - \frac{1}{3b-2}$$

144.

$$2x-1 \overline{)\begin{array}{r} 2x^3 - x^2 + 2 \\ 4x^4 - 4x^3 + x^2 + 4x - 3 \end{array}}$$

$$\underline{4x^4 - 2x^3}$$

$$-2x^3 + x^2$$

$$\underline{-2x^2 + x^2}$$

$$4x - 3$$

$$\underline{4x - 2}$$

$$-1$$

$$\frac{4x^4 - 4x^3 + x^2 + 4x - 3}{2x-1}$$

$$= 2x^3 - x^2 + 2 - \frac{1}{2x-1}$$

145.

$$x-6 \overline{)\begin{array}{r} -x^2 - 16x - 117 \\ -x^3 - 10x^2 - 21x + 18 \end{array}}$$

$$\underline{-x^3 + 10x^2}$$

$$-16x^2 - 21x$$

$$\underline{-16x^2 + 96x}$$

$$-117x + 18$$

$$\underline{-117x + 702}$$

$$-684$$

$$\frac{-10x^2 - x^3 - 21x + 18}{x-6}$$

$$= -x^2 - 16x - 117 - \frac{684}{x-6}$$

146.

$$\begin{array}{r|rrrr} 2 & 3 & 0 & 12 & -4 \\ & & 6 & 12 & 48 \\ \hline & 3 & 6 & 24 & 44 \end{array}$$

Answer: $3x^2 + 6x + 24 + \dfrac{44}{x-2}$

147.

$$\begin{array}{r|rrrr} -\frac{3}{2} & 3 & 2 & -4 & -1 \\ & & -\frac{9}{2} & \frac{15}{4} & \frac{3}{8} \\ \hline & 3 & -\frac{5}{2} & -\frac{1}{4} & -\frac{5}{8} \end{array}$$

Answer: $3x^2 - \dfrac{5}{2}x - \dfrac{1}{4} - \dfrac{5}{8\left(x + \dfrac{3}{2}\right)}$

148.
$$
\begin{array}{r|rrrrrr}
-1 & 1 & 0 & 0 & 0 & 0 & -1 \\
& & -1 & 1 & -1 & 1 & -1 \\
\hline
& 1 & -1 & 1 & -1 & 1 & -2
\end{array}
$$

Answer: $x^4 - x^3 + x^2 - x + 1 - \dfrac{2}{x+1}$

149.
$$
\begin{array}{r|rrrr}
3 & 1 & 0 & 0 & -81 \\
& & 3 & 9 & 27 \\
\hline
& 1 & 3 & 9 & -54
\end{array}
$$

Answer: $x^2 + 3x + 9 - \dfrac{54}{x-3}$

150.
$$
\begin{array}{r|rrrrr}
4 & 3 & 1 & -1 & 0 & -2 \\
& & 12 & 52 & 204 & 816 \\
\hline
& 3 & 13 & 51 & 204 & 814
\end{array}
$$

Answer: $3x^3 + 13x^2 + 51x + 204 + \dfrac{814}{x-4}$

151.
$$
\begin{array}{r|rrrrr}
-2 & 3 & 0 & -2 & 0 & 10 \\
& & -6 & 12 & -20 & 40 \\
\hline
& 3 & -6 & 10 & -20 & 50
\end{array}
$$

Answer: $3x^3 - 6x^2 + 10x - 20 + \dfrac{50}{x+2}$

152.
$$
\begin{array}{r|rrrrrr}
4 & 3 & 0 & 0 & 0 & -9 & 7 \\
& & 12 & 48 & 192 & 768 & 3036 \\
\hline
& 3 & 12 & 48 & 192 & 759 & 3043
\end{array}
$$

Thus, $P(4) = 3043$.

153.
$$
\begin{array}{r|rrrrrr}
-5 & 3 & 0 & 0 & 0 & -9 & 7 \\
& & -15 & 75 & -375 & 1875 & -9330 \\
\hline
& 3 & -15 & 75 & -375 & 1866 & -9323
\end{array}
$$

Thus, $P(-5) = -9323$.

Chapter 5 Test

1. $2^5 = 2 \cdot 2 \cdot 2 \cdot 2 \cdot 2 = 32$

2. $(-3)^4 = (-3)(-3)(-3)(-3) = 81$

3. $-3^4 = -3 \cdot 3 \cdot 3 \cdot 3 = -81$

4. $4^{-3} = \dfrac{1}{4^3} = \dfrac{1}{64}$

5. $\left(3x^2\right)\left(-5x^9\right) = (3)(-5)\left(x^2 \cdot x^9\right)$
$$= -15x^{11}$$

6. $\dfrac{y^7}{y^2} = y^{7-2} = y^5$

7. $\dfrac{r^{-8}}{r^{-3}} = r^{-8-(-3)} = r^{-5} = \dfrac{1}{r^5}$

8. $\left(\dfrac{x^2 y^3}{x^3 y^{-4}}\right)^2 = \dfrac{x^4 y^6}{x^6 y^{-8}}$
$$= x^{4-6} y^{6-(-8)}$$
$$= x^{-2} y^{14}$$
$$= \dfrac{y^{14}}{x^2}$$

9. $\left(\dfrac{6^2 x^{-4} y^{-1}}{6^3 x^{-3} y^7}\right) = 6^{2-3} x^{-4-(-3)} y^{-1-7}$
$$= 6^{2-3} x^{-4-(-3)} y^{-1-7}$$
$$= 6^{-1} x^{-1} y^{-8}$$
$$= \dfrac{1}{6xy^8}$$

10. $563{,}000 = 5.63 \times 10^5$

11. $0.0000863 = 8.63 \times 10^{-5}$

12. $1.5 \times 10^{-3} - 0.0015$

13. $6.23 \times 10^4 = 62,300$

14. $\left(1.2 \times 10^5\right)\left(3 \times 10^{-7}\right)$
$= (1.2)(3) \times 10^{5-7}$
$= 3.6 \times 10^{-2}$
$= 0.036$

15. **a.** $4xy^2 + 7xyz + x^3y - 2$

Term	Numerical Coefficient	Degree of Term
$4xy^2$	4	3
$7xyz$	7	3
x^3y	1	4
-2	-2	0

b. Degree 4

16. $5x^2 + 4xy - 7x^2 + 11 + 8xy$
$= \left(5x^2 - 7x^2\right) + \left(4xy + 8xy\right) + 11$
$= -2x^2 + 12xy + 11$

17. $\left(8x^3 + 7x^2 + 4x - 7\right) + \left(8x^3 - 7x - 6\right)$
$= 8x^3 + 7x^2 + 4x - 7 + 8x^3 - 7x - 6$
$= 16x^3 + 7x^2 - 3x - 13$

18. $\quad 5x^3 + x^2 + 5x - 2$
$\underline{-\left(8x^3 - 4x^2 + x - 7\right)}$

$\quad 5x^3 + x^2 + 5x - 2$
$\underline{-\ 8x^3 + 4x^2 - x + 7}$
$-3x^3 + 5x^2 + 4x + 5$

19. $\left[\left(8x^2 + 7x + 5\right) + \left(x^3 - 8\right)\right] - \left(4x + 2\right)$
$= 8x^2 + 7x + 5 + x^3 - 8 - 4x - 2$
$= x^3 + 8x^2 + 3x - 5$

20. $(3x + 7)\left(x^2 + 5x + 2\right)$
$= 3x^3 + 15x^2 + 6x + 7x^2 + 35x + 14$
$= 3x^3 + 22x^2 + 41x + 14$

21. $3x^2\left(2x^2 - 3x + 7\right)$
$= 3x^2\left(2x^2\right) - 3x^2\left(3x\right) + 3x^2\left(7\right)$
$= 6x^4 - 9x^3 + 21x^2$

22. $(x + 7)(3x - 5) = 3x^2 - 5x + 21x - 35$
$= 3x^2 + 16x - 35$

23. $(4x - 2)^2 = (4x)^2 - 2(4x)(2) + 2^2$
$= 16x^2 - 16x + 4$

24. $\left(x^2 - 9b\right)\left(x^2 + 9b\right) = \left(x^2\right)^2 - (9b)^2$
$= x^4 - 81b^2$

25. $-16t^2 + 1001$
$t = 0: \ -16(0)^2 + 1001 = 1001$ ft
$t = 1: \ -16(1)^2 + 1001 = 985$ ft
$t = 3: \ -16(3)^2 + 1001 = 857$ ft
$t = 5: \ -16(5)^2 + 1001 = 601$ ft

26. $\dfrac{4x^2 + 24xy - 7x}{8xy} = \dfrac{4x^2}{8xy} + \dfrac{24xy}{8xy} - \dfrac{7x}{8xy}$
$= \dfrac{x}{2y} + 3 - \dfrac{7}{8y}$

27.
$$\begin{array}{r} x+2 \\ x+5{\overline{\smash{\big)}\,x^2+7x+10}} \\ \underline{x^2+5x} \\ 2x+10 \\ \underline{2x+10} \\ 0 \end{array}$$

$$\frac{x^2+7x+10}{x+5}=x+2$$

28.
$$\begin{array}{r} 9x^2-6x+4 \\ 3x+2{\overline{\smash{\big)}\,27x^3+0x^2+0x-8}} \\ \underline{27x^3+18x^2} \\ -18x^2+\ 0x \\ \underline{-18x^2-12x} \\ 12x-8 \\ \underline{12x+8} \\ -16 \end{array}$$

$$\frac{27x^3-8}{3x+2}=9x^2-6x+4-\frac{16}{3x+2}$$

29. $h(t)=-16t^2+96t+880$

a. $-16(1)^2+96(1)+880=-16+96+880$
$$=960 \text{ feet}$$

b. $-16(5.1)^2+96(5.1)+880$
$$=-416.16+489.6+880$$
$$=953.44 \text{ feet}$$

c.
$$0=-16t^2+96t+880$$
$$16t^2-96t-880=0$$
$$16(t^2-6t-55)=0$$
$$(t-11)(t+5)=0$$

$$t-11=0 \quad \text{or} \quad t+5=0$$
$$t=11 \quad \text{or} \quad t=-5$$

Disregard the negative. The pebble will hit the ground in 11 seconds.

30.
$$\begin{array}{r|rrrrr} -3 & 4 & -3 & 0 & -1 & -1 \\ & & -12 & 45 & -135 & 408 \\ \hline & 4 & -15 & 45 & -136 & 407 \end{array}$$

Answer: $4x^3-15x^2+45x-136+\dfrac{407}{x+3}$

31.
$$\begin{array}{r|rrrrr} -2 & 4 & 0 & 7 & -2 & -5 \\ & & -8 & 16 & -46 & 96 \\ \hline & 4 & -8 & 23 & -48 & 91 \end{array}$$

Thus, $P(-2)=91$.

Chapter 5 Cumulative Review

1. a. True
   b. True
   c. False
   d. True

2. a. $\left|-7.2\right|=7.2$

   b. $\left|0\right|=0$

   c. $\left|-\dfrac{1}{2}\right|=\dfrac{1}{2}$

3. a. $\dfrac{4}{5}\div\dfrac{5}{16}=\dfrac{4}{5}\cdot\dfrac{16}{5}=\dfrac{64}{25}$

   b. $\dfrac{7}{10}\div 14=\dfrac{7}{10}\cdot\dfrac{1}{14}=\dfrac{7}{10\cdot 7\cdot 2}=\dfrac{1}{20}$

   c. $\dfrac{3}{8}\div\dfrac{3}{10}=\dfrac{3}{8}\cdot\dfrac{10}{3}=\dfrac{3\cdot 2\cdot 5}{3\cdot 2\cdot 4}=\dfrac{5}{4}$

4. a. $\dfrac{3}{4} \cdot \dfrac{7}{21} = \dfrac{3 \cdot 7}{4 \cdot 3 \cdot 7} = \dfrac{1}{4}$

    b. $\dfrac{1}{2} \cdot 4\dfrac{5}{6} = \dfrac{1}{2} \cdot \dfrac{29}{6} = \dfrac{29}{12} = 2\dfrac{5}{12}$

5. a. $3^2 = 3 \cdot 3 = 9$

    b. $5^3 = 5 \cdot 5 \cdot 5 = 125$

    c. $2^4 = 2 \cdot 2 \cdot 2 \cdot 2 = 16$

    d. $7^1 = 7$

    e. $\left(\dfrac{3}{7}\right)^2 = \left(\dfrac{3}{7}\right)\left(\dfrac{3}{7}\right) = \dfrac{9}{49}$

6. Let $x = 5$ and $y = 1$.

$$\dfrac{|2x| - |7y|}{x^2} = \dfrac{|2(5)| - |7(1)|}{5^2} = \dfrac{|10| - |7|}{25}$$

$$= \dfrac{10 - 7}{25} = \dfrac{3}{25}$$

7. a. $-3 + (-7) = -10$

    b. $-1 + (-20) = -21$

    c. $-2 + (-10) = -12$

8. $8 + 3(2 \cdot 6 - 1) = 8 + 3(12 - 1) = 8 + 3(11)$

     $= 8 + 33 = 41$

9. $-4 - (8) = -4 + (-8) = -12$

10. $x = 1$

     $5x^2 + 2 = x - 8$

     $5(1)^2 + 2 \overset{?}{=} 1 - 8$

        $5 + 2 \overset{?}{=} -7$

           $7 \overset{?}{=} -7$ False

     $x = 1$ is not a solution.

11. a. $\dfrac{1}{22}$ b. $\dfrac{16}{3}$ c. $-\dfrac{1}{10}$ d. $-\dfrac{13}{9}$

12. a. $7 - 40 = 7 + (-40) = -33$

     b. $-5 - (-10) = -5 + 10 = 5$

13. a. $5 + (4 + 6) = (5 + 4) + 6$

     b. $(-1 \cdot 2) \cdot 5 = -1 \cdot (2 \cdot 5)$

14. $\dfrac{4(-3) + (-8)}{5 + (-5)} = \dfrac{-12 + (-8)}{0}$ is undefined

15. a. $2(x + y) = 2x + 2y$

     b. $-5(-3 + 2z) = 15 - 10z$

     c. $5(x + 3y - z) = 5x + 15y - 5z$

     d. $-1(2 - y) = -2 + y$

     e. $-(3 + x - w) = -3 - x + w$

     f. $4(3x + 7) + 10 = 12x + 28 + 10 = 12 + 38$

16. $-2(x + 3y - z)$

     $= -2(x) + (-2)(3y) - (-2)(z)$

     $= -2x - 6y + 2z$

17. a. $5(x + 2) = 5x + 5(2) = 5x + 10$

     b. $-2(y + 0.3z - 1)$

        $= -2(y) + (-2)(0.3z) - (-2)(1)$

        $= -2y - 0.6z + 2$

     c. $-(x + y - 2z + 6) = -1(x + y - 2z + 6)$

        $= -1(x) + (-1)(y) - (-1)(2z) + (-1)(6)$

        $= -x - y + 2z - 6$

18. $2(6x - 1) - (x - 7) = 12x - 2 - x + 7$

                   $= 11x + 5$

19. $x - 7 = 10$

$$x - 7 + 7 = 10 + 7$$

$$x = 17$$

20. Let $x =$ a number.

$$(x + 7) - 2x$$

21. $\dfrac{5}{2}x = 15$

$$\frac{2}{5} \cdot \frac{5}{2}x = \frac{2}{5} \cdot 15$$

$$x = 6$$

22. $2x + \dfrac{1}{8} = x - \dfrac{3}{8}$

$$x + \frac{1}{8} = -\frac{3}{8}$$

$$x = -\frac{4}{8}$$

$$x = -\frac{1}{2}$$

23. Let $x =$ a number

$$7 + 2x = x - 3$$

$$7 + x = -3$$

$$x = -10$$

The number is -10

24. $10 = 5j - 2$

$$12 = 5j$$

$$\frac{12}{5} = j$$

25. Let $x =$ a number

$$2(x + 4) = 4x - 12$$

$$2x + 8 = 4x - 12$$

$$-2x + 8 = -12$$

$$-2x = -20$$

$$x = 10$$

The number is 10.

26. $\dfrac{7x + 5}{3} = x + 3$

$$3\left(\frac{7x + 5}{3}\right) = 3(x + 3)$$

$$7x + 5 = 3x + 9$$

$$4x + 5 = 9$$

$$4x = 4$$

$$x = 1$$

27. Let $x =$ the width and $3x - 2 =$ the length.

$$2L + 2W = P$$

$$2(3x - 2) + 2x = 28$$

$$6x - 4 + 2x = 28$$

$$8x - 4 = 28$$

$$8x = 32$$

$$x = 4$$

$$3x - 2 = 3(4) - 2 = 10$$

The width is 4 feet and the length is 10 feet.

28. $x < 5, \ (-\infty, 5)$

29. $F = \dfrac{9}{5}C + 32$

$F - 32 = \dfrac{9}{5}C$

$\dfrac{5}{9}(F - 32) = C$

$\dfrac{5F - 160}{9} = C$

30. a. $x = -1$ is a vertical line and the slope is undefined.

b. $y = 7$ is a horizontal line and the slope is zero.

31. $2 < x \le 4$

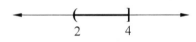

32. $m = \dfrac{y_2 - y_1}{x_2 - x_1} = \dfrac{2}{20} = \dfrac{1}{10} \cdot 100\% = 10\%$

33. $3x + y = 12$

a. $(0,\ ): 3(0) + y = 12$

$y = 12, \qquad (0,12)$

b. $(\ ,6): 3x + (6) = 12$

$3x = 6$

$x = 2, \qquad (2,6)$

c. $(-1,\ ): 3(-1) + y = 12$

$-3 + y = 12$

$y = 15, \qquad (-1,15)$

34. $\begin{cases} 3x + 2y = -8 \\ 2x - 6y = -9 \end{cases}$

Multiply the first equation by 3.

$9x + 6y = -24$

$\underline{2x - 6y = \ -9}$

$11x \qquad = -33$

$x = -3$

Replace x in the first equation with -3

$3(-3) + 2y = -8$

$-9 + 2y = -8$

$2y = 1$

$y = \dfrac{1}{2}$

The solution to the system is $\left(-3, \dfrac{1}{2}\right)$.

35. $2x + y = 5$

x	y
0	5
5/2	0

36. $\begin{cases} x = -3y + 3 \\ 2x + 9y = 5 \end{cases}$

Replace x in the second equation by $-3y + 3$.

$2(-3y + 3) + 9y = 5$

$\quad -6y + 6 + 9y = 5$

$\quad\quad\quad 3y + 6 = 5$

$\quad\quad\quad\quad 3y = -1$

$\quad\quad\quad\quad\; y = -\dfrac{1}{3}$

Replace y in the first equation with $-\dfrac{1}{3}$.

$x = -3\left(-\dfrac{1}{3}\right) + 3$

$x = 4$

The solution to the system is $\left(4, -\dfrac{1}{3}\right)$.

37. $x = 2$ for all values of y

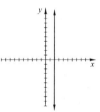

38. a. $(-5)^2 = (-5)(-5) = 25$

b. $-5^2 = -(5)(5) = -25$

c. $2 \cdot 5^2 = 2 \cdot 5 \cdot 5 = 50$

39. $x = 5$ is a vertical line and the slope is undefined.

40. $\dfrac{\left(z^2\right)^3 \cdot z^7}{z^9} = \dfrac{z^6 \cdot z^7}{z^9} = z^{6+7-9} = z^4$

41. $\left(12z^5 - 12z^3 + z\right) - \left(-3z^4 + z^3 + 12z\right)$

$\quad = 12z^5 - 12z^3 + z + 3z^4 - z^3 - 12z$

$\quad = 12z^5 + 3z^4 - 13z^3 - 11z$

42. $\left(5y^2 - 6\right) - \left(y^2 + 2\right) = 5y^2 - 6 - y^2 - 2$

$\quad\quad\quad\quad\quad\quad\quad\quad\quad = 4y^2 - 8$

43. $\left(2x^2\right)\left(-3x^5\right) = -6x^{2+5} = -6x^7$

44. $-x^2$

a. $-(2)^2 = -4$ \qquad b. $-(-2)^2 = -4$

45. $P(x) = 3x^2 - 2x - 5$

a. $P(1) = 3(1)^2 - 2(1) - 5 = -4$

b. $P(-2) = 3(-2)^2 - 2(-2) - 5 = 11$

46. $\left(10x^2 - 3\right)\left(10x^2 + 3\right) = \left(10x^2\right)^2 - 3^2$

$\quad\quad\quad\quad\quad\quad\quad\quad\quad = 100x^4 - 9$

47. $(2x - y)^2 = (2x)^2 - 2(2x)(y) + (y)^2$

$\quad\quad\quad\quad\; = 4x^2 - 4xy + y^2$

48. $\left(10x^2 + 3\right)^2 = \left(10x^2\right)^2 + 2\left(10x^2\right)(3) + 3^2$

$\quad\quad\quad\quad\quad\; = 100x^4 + 60x^2 + 9$

49. $\dfrac{6m^2 + 2m}{2m} = \dfrac{6m^2}{2m} + \dfrac{2m}{2m} = 3m + 1$

50. a. $5^{-1} = \dfrac{1}{5}$ \qquad b. $7^{-2} = \dfrac{1}{7^2} = \dfrac{1}{49}$

Chapter 6

Mental Math

1. $14 = 2 \cdot 7$

2. $15 = 3 \cdot 5$

3. $10 = 2 \cdot 5$

4. $70 = 2 \cdot 5 \cdot 7$

5. $6 = 2 \cdot 3$
   $15 = 3 \cdot 5$
   $GCF = 3$

6. $20 = 2 \cdot 2 \cdot 5$
   $15 = 3 \cdot 5$
   $GCF = 5$

7. $18 = 2 \cdot 3 \cdot 3$
   $3 = 3$
   $GCF = 3$

8. $14 = 2 \cdot 7$
   $35 = 5 \cdot 7$
   $GCF = 7$

Exercise Set 6.1

1. $32 = 2 \cdot 2 \cdot 2 \cdot 2 \cdot 2$
   $36 = 2 \cdot 2 \cdot 3 \cdot 3$
   $GCF = 2 \cdot 2 = 4$

3. $12 = 2 \cdot 2 \cdot 3 = 2^2 \cdot 3$
   $18 = 2 \cdot 3 \cdot 3 = 2 \cdot 3^2$
   $36 = 3 \cdot 2 \cdot 3 \cdot 3 = 2^2 \cdot 3^2$
   $GCF = 2 \cdot 3 = 6$

5. y^2

7. xy^2

9. $8x = 2 \cdot 2 \cdot 2 \cdot x$
   $4 = 2 \cdot 2$
   $GCF = 2 \cdot 2 = 4$

11. $12y^4 = 2 \cdot 2 \cdot 3 \cdot y^4$
    $20y^3 = 2 \cdot 2 \cdot 5 \cdot y^3$
    $GCF = 2 \cdot 2 \cdot y^3 = 4y^3$

13. $12x^3 = 2 \cdot 2 \cdot 3 \cdot x^3$
    $6x^4 = 2 \cdot 3 \cdot x^4$
    $3x^5 = 3 \cdot x^5$
    $GCF = 3 \cdot x^3 = 3x^3$

15. $18x^2y = -2 \cdot 3 \cdot 3 \cdot x^2 \cdot y$
    $9x^3y^3 = 3 \cdot 3 \cdot x^3 \cdot y^3$
    $36x^3y = 2 \cdot 2 \cdot 3 \cdot 3 \cdot x^3 \cdot y$
    $GCF = 3 \cdot 3 \cdot x^2 \cdot y = 9x^2y$

17. $30x - 15 = 15(2x - 1)$

19. $24cd^3 - 18c^2d$; GCF $= 6cd$

$6cd \cdot 4d^2 - 6cd \cdot 3c = 6cd(4d^2 - 3c)$

21. $-24a^4x + 18a^3x$; GCF $= -6a^3x$

$-6a^3x(4a) - 6a^3x(-3) = -6a^3x(4a - 3)$

23. $12x^3 + 16x^2 - 8x = 4x(3x^2 + 4x - 2)$

25. $5x^3y - 15x^2y + 10xy = 5xy(x^2 - 3x + 2)$

27. Answers may vary.

29. $y(x + 2) + 3(x + 2) = (x + 2)(y + 3)$

31. $x(y - 3) - 4(y - 3)$; GCF $= (y - 3)$

$(y - 3)(x - 4)$

33. $2x(x + y) - (x + y)$; GCF $= (x + y)$

$2x(x + y) + (x + y)(-1) = (x + y)(2x - 1)$

35. $5x + 15 + xy + 3y = 5(x + 3) + y(x + 3)$

$= (x + 3)(5 + y)$

37. $2y - 8 + xy - 4x = 2(y - 4) + x(y - 4)$

$= (y - 4)(2 + x)$

39. $3xy - 6x + 8y - 16 = 3x(y - 2) + 8(y - 2)$

$= (y - 2)(3x + 8)$

41. $y^3 + 3y^2 + y + 3 = y^2(y + 3) + 1(y + 3)$

$= (y + 3)(y^2 + 1)$

43. Subtract the area of the innter rectangle from the area of the outer rectangle.

Outer rectangle: $A = l \cdot w$

$$A = 12x \cdot x^2 = 12x^3$$

Inner rectangle: $A = l \cdot w$

$$A = 2 \cdot x = 2x$$

The area of the shaded region is given by the expression $12x^3 - 2x = 2x(6x^2 - 1)$.

45. $20x(10) + \pi \cdot 5^2 = 200x + 25\pi = 25(8x + \pi)$

47. $3x - 6$; GCF $= 3$

$3 \cdot x - 3 \cdot 2 = 3(x - 2)$

49. $32xy - 18x^2$; GCF $= 2x$

$2x(16y) - 2x(9x) = 2x(16y - 9x)$

51. $4x - 8y + 4 = 4(x - 2y + 1)$

53. $8(x + 2) - y(x + 2) = (x + 2)(8 - y)$

55. $-40x^8y^6 - 16x^9y^5$; GCF $= -8x^8y^5$

$-8x^8y^5 \cdot 5y - 8x^8y^5 \cdot 2x = -8x^8y^5(5y + 2x)$

57. $-3x + 12$; GCF $= -3$

$= -3 \cdot x - 3(-4) = -3(x - 4)$

59. $18x^3y^3 - 12x^3y^2 + 6x^5y^2$; GCF $= 6x^3y^2$

$6x^3y^2 \cdot 3y + 6x^3y^2(-2) + 6x^3y^2 \cdot x^2$

$= 6x^3y^2(3y - 2 + x^2)$

61. $y^2(x - 2) + (x - 2)$; GCF $= (x - 2)$

$y^2(x - 2) + 1(x - 2) = (x - 2)(y^2 + 1)$

63. $5xy + 15x + 6y + 18 = 5x(y + 3) + 6(y + 3)$
$$= (y + 3)(5x + 6)$$

65. $4x^2 - 8xy - 3x + 6y = 4x(x - 26) - 3(x = 2y)$
$$= (x - 2y)(4x - 3)$$

67. $126x^3yz + 210y^4z^3$; GCF $= 42yz$
$42yz \cdot 3x^3 + 42yz \cdot 5y^3z^2$
$$= 42yz(3x^3 + 5y^3z^2)$$

69. $3y - 5x + 15 - xy$
$$= 3y + 15 - 5x - xy$$
$$= 3(y + 5) - x(5 + y)$$
$$= 3(y + 5) - x(y + 5)$$
$$= (y + 5)(3 - x) \text{ or } (3 - x)(y + 5)$$

71. $12x^2y - 42x^2 - 4y + 14$; GCF $= 2$
$2[6x^2y - 21x^2 - 2y + 7]$
$$= 2[3x^2(2y - 7) - 1(2y - 7)]$$
$$= 2[(2y - 7)(3x^2 - 1)]$$
$$= 2(2y - 7)(3x^2 - 1)$$

	Two Numbers	Their Product	Their Sum
73.	2, 6	12	8
75.	−1, −8	8	−9
77.	−2, 5	−10	3
79.	−8, 3	−24	−5

81. $(x + 2)(x + 5) = x^2 + 2x + 5x + 10$
$$= x^2 + 7x + 10$$

83. $(a - 7)(a - 8) = a^2 - 8a - 7a + 56$
$$= a^2 - 15a + 56$$

85. Answers may vary.

87. $(a + 6)(b - 2)$ is factored.

89. $5(2y + z) - b(2y + z)$ is not factored.

91. $\dfrac{4n^4 - 24n}{4n} = \dfrac{4n^4}{4n} - \dfrac{25n}{4n} = (n^3 - 6)$ units

93. $x^{2n} + 2x^n + 3x^n + 6$
$$= x^n(x^n + 2) + 3(x^n + 2)$$
$$= (x^n + 2)(x^n + 3)$$

95. $3x^{2n} + 21x^n - 5x^n - 35$
$$= 3x^n(x^n + 7) - 5(x^n + 7)$$
$$= (x^n + 7)(3x^n - 5)$$

97. $8x^2 - 56x + 124$

   a. Let $x = 2$
   $$8(2)^2 - 56(2) + 124$$
   $$= 44 \text{ million}$$

   b. Let $x = 7$
   $$8(7)^2 - 56(7) + 124$$
   $$= 124 \text{ million}$$

   c. $8x^2 - 56x + 124 = 4(2x^2 - 14x + 31)$

Section 6.2

Mental Math

1. $x^2 + 9x + 20 = (x+4)(x+5)$

2. $x^2 + 12x + 35 = (x+5)(x+7)$

3. $x^2 - 7x + 12 = (x-4)(x-3)$

4. $x^2 - 13x + 22 = (x-2)(x-11)$

5. $x^2 + 4x + 4 = (x+2)(x+2)$

6. $x^2 + 10x + 24 = (x+6)(x+4)$

Exercise Set 6.2

1. $x^2 + 7x + 6 = (x+6)(x+1)$

3. $x^2 + x - 20 = (x+5)(x-4)$

5. $x^2 - 8x + 15 = (x-3)(x-5)$

7. $x^2 - 10x + 9 = (x-9)(x-1)$

9. $x^2 - 15x + 5$ is a prime polynomial.

11. $x^2 - 3x - 18 = (x-6)(x+3)$

13. $x^2 + 5x + 2$ is a prime polynomial.

15. $x^2 + 8xy + 15y^2 = (x+5y)(x+3y)$

17. $x^2 - 2xy + y^2 = (x-y)(x-y)$

19. $x^2 - 3xy - 4y^2 = (x-4y)(x+y)$

21. $2z^2 + 20z + 32 = (z^2 + 10z + 16)$
$$= 2(z+8)(z+2)$$

23. $2x^3 - 18x^2 + 40x = 2x(x^2 - 9x + 20)$
$$= 2x(x-5)(x-4)$$

25. $7x^2 + 14xy - 21y^2 = 7(x^2 + 2xy - 3y^2)$
$$= 7(x-y)(x+3y)$$

27. $6 \cdot 7 = 42$
$6 + 7 = 13$
The numbers are 6 and 7.

29. $x^2 + 15x + 36 = (x+12)(x+3)$

31. $x^2 - x - 2 = (x-2)(x+1)$

33. $r^2 - 16r + 48 = (r-12)(r-4)$

35. $x^2 - 4x - 21 = (x-7)(x+3)$

37. $x^2 + 7xy + 10y^2 = (x+5y)(x+2y)$

39. $r^2 - 3r + 6$ is a prime polynomial.

41. $2t^2 + 24t + 64$
$$= 2(t^2 + 12t + 32)$$
$$= 2(t+8)(t+4)$$

43. $x^3 - 2x^2 - 24x = x(x^2 - 2x - 24)$
$$= x(x - 6)(x + 4)$$

45. $x^2 - 16x + 63 = (x - 9)(x - 7)$

47. $x^2 + xy - 2y^2 = (x + 2y)(x - y)$

49. $3x^2 - 60x + 108 = 3(x^2 - 20x + 36)$
$$= 3(x - 18)(x - 2)$$

51. $x^2 - 18x - 144 = (x - 24)(x + 6)$

53. $6x^3 + 54x^2 + 120x = 6x(x^2 + 9x + 20)$
$$= 6x(x + 5)(x + 4)$$

55. $2t^5 - 14t^4 + 24t^3 = 2t^3(t^2 - 7t + 12)$
$$= 2t^3(t - 4)(t - 3)$$

57. $5x^3y - 25x^2y^2 - 120xy^3$
$$= 5xy(x^2 - 5xy - 24y^2)$$
$$= 5xy(x - 8y)(x + 3y)$$

59. $4x^2y + 4xy - 12y = 4y(x^2 + x - 3)$

61. $2a^2b - 20ab^2 + 42b^3$
$$= 2b(a^2 - 10b + 21b^2)$$
$$= 2b(a - 3b)(a - 7b)$$

63. $(2x + 1)(x + 5) = 2x^2 + x + 10x + 5$
$$= 2x^2 + 11x + 5$$

65. $(5y - 4)(3y - 1) = 15y^2 - 12y - 5y + 4$
$$= 15y^2 - 17y + 4$$

67. $(a + 3)(9a - 4) = 9a^2 + 27a - 4a - 12$
$$= 9a^2 + 23a - 12$$

69. $x^2 + bx + 15$ is factorable when b is 8 or 16.

71. $m^2 + bm - 27$ is factorable when b is 6 or 26.

73. $x^2 + 6x + c$ if factorable when c is 5,8 or 9.

75. $y^2 - 4y + c$ if factorable when c is 3 or 4.

77. $P = 2l + 2w$
$l = x^2 + 10x$ and $w = 4x + 33$, so
$$P = 2(x^2 + 10x) + 2(4x + 33)$$
$$= 2x^2 + 20x + 8x + 66$$
$$= 2x^2 + 28x + 66 = (x^2 + 14x + 33)$$
$$= 2(x + 11)(x + 3)$$

The perimeter of the rectangle is given by the polynomial $2x^2 + 28x + 66$ which factors as $2(x + 11)(x + 3)$.

79. Answers may vary.

81. $2x^2y + 30xy + 100y$; GCF $= 2y$
$$2y(x^2 + 15x + 50) = 2y(x + 5)(x + 10)$$

83. $-12x^2y^3 - 24xy^3 - 36y^3$; GCF $= -12y^3$
$$= -12y^3(x^2 + 2x + 3)$$

85. $y^2(x+1) - 2y(x+1) - 15(x+1);$

$\qquad$ GCF $= (x+1)$

$\qquad (x+1)(y^2 - 2y - 15) = (x+1)(y-5)(y+3)$

87. $x^{2n} + 5x^n + 6 = (x^n + 2)(x^n + 3)$

Section 6.3

Mental Math

  1. Yes

  2. Yes

  3. No

  4. No

  5. Yes

  6. Yes

Exercise Set 6.3

  1. $2x^2 + 13x + 15 = (2x+3)(x+5)$

  3. $2x^2 - 9x - 5 = (2x+1)(x-5)$

  5. $2y^2 - y - 6 = (2y+3)(y-2)$

  7. $16a^2 - 24a + 9 = (4a-3)(4a-3)$

$\qquad\qquad\qquad\quad = (4a-3)^2$

  9. $36r^2 - 5r - 24 = (9r-8)(4r+3)$

11. $10x^2 + 17x + 3 = (5x+1)(2x+3)$

13. $21x^2 - 48x - 45 = 3(7x^2 - 16x - 15)$

$\qquad\qquad\qquad\qquad = 3(7x+5)(x-3)$

15. $12x^2 - 14x - 6;$ GCF $= 2$

$\qquad 2(6x^2 - 7x - 3) = 2(2x-3)(3x+1)$

17. $4x^3 - 9x^2 - 9x = x(4x^2 - 9x - 9)$

$\qquad\qquad\qquad\quad = x(4x+3)(x-3)$

19. $x^2 + 22x + 121 = x^2 + 2\cdot x\cdot 11 + 11^2$

$\qquad\qquad\qquad\quad = (x+11)^2$

21. $x^2 - 16x + 64 = x^2 - 2\cdot x\cdot 8 + 8^2$

$\qquad\qquad\qquad\quad = (x-8)^2$

23. $16y^2 - 40y + 25 = (4y)^2 - 2\cdot 4y\cdot 5 + 5^2$

$\qquad\qquad\qquad\quad = (4y-5)^2$

25. $x^2 y^2 - 10xy + 25 = (xy)^2 - 2\cdot xy\cdot 5 + 5^2$

$\qquad\qquad\qquad\quad = (xy-5)^2$

27. Answers may vary.

29. $2x^2 - 7x - 99 = (2x+11)(x-9)$

31. $4x^2 - 8x - 21 = (2x-7)(2x+3)$

33. $30x^2 - 53x + 21 = (6x-7)(5x-3)$

35. $24x^2 - 58x + 9 = (4x-9)(6x-1)$

37. $9x^2 - 24xy + 16y^2$
$= (3x)^2 - 2 \cdot 3x \cdot 4y + (4y)^2$
$= (3x - 4y)^2$

39. $x^2 - 14xy + 49y^2 = x^2 - 2 \cdot x \cdot 7y + (7y)^2$
$= (x - 7y)^2$

41. $2x^2 + 7x + 5 = 2x^2 + 5x + 2x + 5$
$= x(2x + 5) + 1(2x + 5) = (2x + 5)(x + 1)$

43. $3x^2 - 5x + 1$
not factorable, prime

45. $-2y^2 + y + 10 = 10 + y - 2y^2$
$= (5 - 2y)(2 + y)$

47. $16x^2 + 24xy + 9y^2$
$= (4x + 3y)(4x + 3y)$
$= (4x + 3y)^2$

49. $8x^2y + 34xy - 84y = 2y(4x^2 + 17x - 42)$
$= 2y(4x - 7)(x + 6)$

51. $3x^2 + x - 2 = (3x - 2)(x + 1)$

53. $x^2y^2 + 4xy + 4 = (xy + 2)(xy + 2)$
$= (xy + 2)^2$

55. $49y^2 + 42xy + 9x^2 = (7y + 3x)(7y + 3x)$
$= (7y + 3x)^2$

57. $3x^2 - 42x + 63 = 3(x^2 - 14x + 21)$

59. $42a^2 - 43a + 6 = (7a - 6)(6a - 1)$

61. $18x^2 - 9x - 14 = (6x - 7)(3x + 2)$

63. $25p^2 - 70pq + 49q^2 = (5p - 7q)(5p - 7q)$
$= (5p - 7q)^2$

65. $15x^2 - 16x - 15 = (5x + 3)(3x - 5)$

67. $-27t + 7t^2 - 4 = 7t^2 - 27t - 4$
$= (7t + 1)(t - 4)$

69. $a^2 + ab + ab + b^2 = a^2 + 2ab + b^2$

71. $(x - 2)(x + 2) = x^2 - 2x + 2x - 4 = x^2 - 4$

73. $(a + 3)(a^2 - 3a + 9)$
$= a^3 - 3a^2 + 9a + 3a^2 - 9a + 27$
$= a^3 + 27$

75. $(y - 5)(y^2 + 5y + 25)$
$= y^3 + 5y^2 + 25y - 5y^2 - 25y - 125$
$= y^3 - 125$

77. $75,000 and above.

79. Answers may vary

81. $3x^2 + bx - 5$ is factorable when b is 2 or 14.

83. $2z^2 + bz - 7$ is factorable when b is 5 or 13.

85. $5x^2 + 7x + c$ is factorable when c is 2.

87. $3x^2 - 8x + c$ is factorable when c is 4 or 5.

89. $-12x^3y^2 + 3x^2y^2 + 15xy^2$

$= -3xy^2\left(4x^2 - x - 5\right)$

$= -3xy^2\left(4x - 5\right)\left(x + 1\right)$

91. $-30p^3q + 88p^2q^2 + 6pq^3$; GCF $= -2pq$

$-2pq\left(15p^2 - 44pq - 3q^2\right)$

$= -2pq\left(15p + q\right)\left(p - 3q\right)$

93. $4x^2\left(y - 1\right)^2 + 10x\left(y - 1\right)^2 + 25\left(y - 1\right)^2$

$= \left(y - 1\right)^2\left(4x^2 + 10x + 25\right)$

95. $3x^{2n} + 17x^n + 10 = \left(3x^n + 2\right)\left(x^n + 5\right)$

Section 6.4

Graphing Calculator Explorations

x	$x^2 - 2x + 1$	$x^2 - 2x - 1$	$\left(x - 1\right)^2$
5	16	14	16
−3	16	14	16
2.7	2.89	0.89	2.89
−12.1	171.61	169.61	171.61
0	1	−1	1

Mental Math

1. $1 = 1^2$

2. $25 = 5^2$

3. $81 = 9^2$

4. $64 = 8^2$

5. $9 = 3^2$

6. $100 = 10^2$

7. $1 = 1^3$

8. $64 = 4^3$

9. $8 = 2^3$

10. $27 = 3^3$

Exercise Set 6.4

1. $x^2 - 4 = x^2 - 2^2 = \left(x + 2\right)\left(x - 2\right)$

3. $y^2 - 49 = y^2 - 7^2 = \left(y + 7\right)\left(y - 7\right)$

5. $25y^2 - 9 = \left(5y\right)^2 - 3^2 = \left(5y + 3\right)\left(5y - 3\right)$

7. $121 - 100x^2 = 11^2 - \left(10x\right)^2$

$\qquad = \left(11 + 10x\right)\left(11 - 10x\right)$

9. $12x^2 - 27 = 3\left(4x^2 - 9\right) = 3\left(\left(2x\right)^2 - 3^2\right)$

$\qquad = 3\left(2x + 3\right)\left(2x - 3\right)$

11. $169a^2 - 49b^2 = \left(13a\right)^2 - \left(7b\right)^2$

$\qquad = \left(13a + 7b\right)\left(13a - 7b\right)$

13. $x^2y^2 - 1 = \left(xy\right)^2 - 1^2$

$\qquad = \left(xy + 1\right)\left(xy - 1\right)$

15. $x^4 - 9 = \left(x^2\right)^2 - 3^2 = \left(x^2 + 3\right)\left(x^2 - 3\right)$

17. $49a^4 - 16 = \left(7a^2\right)^2 - 4^2 = \left(7a^2 + 4\right)\left(7a^2 - 4\right)$

19. $x^4 - y^{10} = \left(x^2\right)^2 - \left(y^5\right)^2 = \left(x^2 + y^5\right)\left(x^2 - y^5\right)$

21. $x + 6$ since
$$(x + 6)(x - 6) = x^2 + 6x - 6x - 36$$
$$= x^2 - 36 = x^2 - 6^2$$

23. $a^3 + 27 = a^3 + 3^3$
$$= (a + 3)\left(a^2 - 3a + 9\right)$$

25. $8a^3 + 1 = \left(2a\right)^3 + 1^3$
$$= (2a + 1)\left(4a^2 - 2a + 1\right)$$

27. $5k^3 + 40;\ GCF = 5$
$$5\left(k^3 + 8\right) = 5\left(k^3 + 2^3\right)$$
$$= 5(k + 2)\left(k^2 - 2k + 4\right)$$

29. $x^3 y^3 - 64 = \left(xy\right)^3 - 4^3$
$$= (xy - 4)\left(x^2 y^2 + 4xy + 16\right)$$

31. $x^3 + 125 = x^3 + 5^3$
$$= (x + 5)\left(x^2 - 5x + 25\right)$$

33. $24x^4 - 81xy^3;\ GCF = 3x$
$$3x\left(8x^3 - 27y^3\right)$$
$$= 3x\left[\left(2x\right)^3 - \left(3y\right)^3\right]$$
$$= 3x(2x - 3y)\left(4x^2 + 6xy + 9y^2\right)$$

35. $2x + y$, since
$$(2x + y)\left(4x^2 - 2xy + y^2\right)$$
$$= (2x + y)\left(\left(2x\right)^2 - 2x \cdot y + \left(y\right)^2\right)$$
$$= \left(2x\right)^3 + \left(y\right)^3$$

37. $x^2 - 121 = \left(x\right)^2 - 11^2 = (x + 11)(x - 11)$

39. $81 - p^2 = 9^2 - p^2 = (9 + p)(9 - p)$

41. $4r^2 - 1 = \left(2r\right)^2 - 1^2 = (2r + 1)(2r - 1)$

43. $9x^2 - 16 = \left(3x\right)^2 - 4^2 = (3x + 4)(3x - 4)$

45. $16r^2 + 1$ is the sum of two squares, $\left(4r\right)^2 + 1^2$, not the difference of two squares. $16r^2 + 1$ is a prime polynomial.

47. $27 - t^3 = 3^3 - t^3$
$$= (3 - t)\left(9 + 3t + t^2\right)$$

49. $8r^3 - 64;\ GCF = 8$
$$8\left(r^3 - 8\right) = 8\left(r^3 - 2^3\right)$$
$$= 8(r - 2)\left(r^2 + 2r + 4\right)$$

51. $t^3 - 343 = t^3 - 7^3$
$$= (t - 7)\left(t^2 + 7t + 49\right)$$

53. $x^2 - 169y^2 = x^2 - \left(13y\right)^2$
$$= (x + 13y)(x - 13y)$$

55. $x^2y^2 - z^3 = (xy)^2 - z^2$
$$= (xy - z)(xy + z)$$

57. $x^3y^3 + 1 = (xy)^3 + 1^3$
$$= (xy + 1)(x^2y^2 - xy + 1)$$

59. $s^3 - 64t^3 = s^3 - (4t)^3$
$$= (s - 4t)(s^2 + 4st + 16t^2)$$

61. $18r^2 - 8 = 2(9r^2 - 4)$
$$= 2((3r)^2 - 2^2)$$
$$= 2(3r + 2)(3r - 2)$$

63. $9xy^2 - 4x = x(9y^2 - 4)$
$$= x((3y)^2 - 2^2)$$
$$= x(3y + 2)(3y - 2)$$

65. $25y^4 - 100y^2 = 25y^2(y^2 - 4)$
$$= 25y^2(y^2 - 2^2)$$
$$= 25y^2(y + 2)(y - 2)$$

67. $x^3y - 4xy^3 = xy(x^2 - 4y^2)$
$$= xy(x^2 - (2y)^2)$$
$$= xy(x + 2y)(x - 2y)$$

69. $8s^6t^3 + 100s^3t^6$; GCF $= 4s^3t^3$
$$4s^3t^3(2s^3 + 25t^3)$$

71. $27x^2y^3 - xy^2$; GCF $= xy^2$
$$xy^2(27xy - 1)$$

73. $\dfrac{8x^4 + 4x^3 - 2x + 6}{2x}$
$$= \frac{8x^4}{2x} + \frac{4x^3}{2x} - \frac{2x}{2x} + \frac{6}{2x}$$
$$= 4x^3 + 2x^2 - 1 + \frac{3}{x}$$

75.
$$\begin{array}{r}
2x + 1 \\
x - 2 \overline{)2x^2 - 3x - 2} \\
\underline{2x^2 - 4x} \\
x - 2 \\
\underline{x - 2} \\
0
\end{array}$$
$$2x + 1$$

77.
$$\begin{array}{r}
3x + 4 \\
x + 3 \overline{)3x^2 + 13x + 10} \\
\underline{3x^2 + 9x} \\
4x + 10 \\
\underline{4x + 12} \\
-2
\end{array}$$
$$3x + 4 - \frac{2}{x + 3}$$

79. $(x + 2)^2 - y^2 = (x + 2 + y)(x + 2 - y)$

81. $a^2(b - 4) - 16(b - 4) = (b - 4)(a^2 - 16)$
$$= (b - 4)(a^2 - 4^2) = (b - 4)(a + 4)(a - 4)$$

83. $(x^2 + 6x + 9) - 4y^2$
$$= [(x + 3)(x + 3)] - 4y^2 = (x + 3)^2 - (2y)^2$$
$$= (x + 3 + 2y)(x + 3 - 2y)$$

85. $x^{2n} - 100 = \left(x^n\right)^2 - 10^2$

$$= \left(x^n + 10\right)\left(x^n - 10\right)$$

87. a. Let $t = 2$.

$841 - 16t^2 = 841 - 16(2)^2$

$= 841 - 16(4) = 841 - 64 = 777$

After 2 seconds, the height of the object is 777 feet.

b. Let $t = 5$.

$841 - 16t^2 = 841 - 16(5)^2$

$= 841 - 16(25)$

$841 - 400 = 441$

After 5 seconds the height of the object is 441 feet.

c. When the object hits the ground, its height is zero feet. Thus, to find the time, t, when the object's height is zero feet above the ground, we set the expression $841 - 16t^2$ equal to 0 and solve for t.

$841 - 16t^2 = 0$

$841 - 16t^2 + 16t^2 = 0 + 16t^2$

$841 = 16t^2$

$\dfrac{841}{16} = \dfrac{16t^2}{16}$

$52.5625 = t^2$

$\sqrt{52.5625} = \sqrt{t^2}$

$7.25 = t$

Thus, the object will hit the ground after approximately 7 seconds.

d. $841 - 16t^2 = 29^2 - \left(4t\right)^2$

$= \left(29 + 4t\right)\left(29 - 4t\right)$

89. a. Let $t = 3$

$1444 - 16t^2 = 1444 - 16(3)^2 = 1300$

After 3 seconds the height is 1300 feet.

b. Let $t = 7$

$1444 - 16t^2 = 1444 - 16(7)^2$

$= 660$

After 7 seconds the height is 660 feet.

c. When it hits the ground, the height is 0.

Let $0 = 1444 - 16t^2$

$16t^2 = 1444$

$t^2 = 90.25$

$t = \sqrt{90.25}$

$t = 9.5$

Thus, it will hit the ground after about 10 seconds.

d. $1444 - 16t^2 = 4\left(361 - 4t^2\right)$

$= 4\left[\left(19\right)^2 - \left(2t\right)^2\right]$

$= 4\left(19 + 2t\right)\left(19 - 2t\right)$

91. Answers may vary.

Integrated Review

1. $a^2 + 2ab + b^2 = (a+b)(a+b) = (a+b)^2$

2. $a^2 - 2ab + b^2 = (a-b)(a-b) = (a-b)^2$

3. $a^2 + a - 12 = (a-3)(a+4)$

4. $a^2 - 7a + 10 = (a-5)(a-2)$

5. $a^2 - a - 6 = (a-3)(a+2)$

6. $a^2 + 2a + 1 = (a+1)(a+1) = (a+1)^2$

7. $x^2 + 2x + 1 = (x+1)(x+1) = (x+1)^2$

8. $x^2 + x - 2 = (x+2)(x-1)$

9. $x^2 + 4x + 3 = (x+3)(x+1)$

10. $x^2 + x - 6 = (x+3)(x-2)$

11. $x^2 + 7x + 12 = (x+4)(x+3)$

12. $x^2 + x - 12 = (x+4)(x-3)$

13. $x^2 + 3x - 4 = (x+4)(x-1)$

14. $x^2 - 7x + 10 = (x-5)(x-2)$

15. $x^2 + 2x - 15 = (x+5)(x-3)$

16. $x^2 + 11x + 30 = (x+6)(x+5)$

17. $x^2 - x - 30 = (x-6)(x+5)$

18. $x^2 + 11x + 24 = (x+8)(x+3)$

19. $2x^2 - 98 = (x^2 - 49)$
$$= 2(x^2 - 7^2)$$
$$= 2(x+7)(x-7)$$

20. $3x^2 - 75 = 3(x^2 - 25)$
$$= 3(x^2 - 5^2)$$
$$= 3(x+5)(x-5)$$

21. $x^2 + 3x + xy + 3y = x(x+3) + y(x+3)$
$$= (x+3)(x+y)$$

22. $3y - 21 + xy - 7x = 3(y-7) + x(y-7)$
$$= (y-7)(3+x)$$

23. $x^2 + 6x - 16 = (x+8)(x-2)$

24. $x^2 - 3x - 28 = (x-7)(x+4)$

25. $4x^3 + 20x^2 - 56x = 4x(x^2 + 5x - 14)$
$$= 4x(x+7)(x-2)$$

26. $6x^3 - 6x^2 - 120x = 6x(x^2 - x - 20)$
$$= 6x(x-5)(x+4)$$

27. $12x^2 + 34x + 24 = 2(6x^2 + 17x + 12)$
$$= 2(6x^2 + 9x + 8x + 12)$$
$$= 2[3x(2x+3) + 4(2x+3)]$$
$$= 2(2x+3)(3x+4)$$

28. $8a^2 + 6ab - 5b^2 = 8a^2 + 10ab - 4ab - 5b^2$
$$= 2a(4a + 5b) - b(4a + 5b)$$
$$= (4a + 5b)(2a - b)$$

29. $4a^2 - b^2 = (2a)^2 - b^2 = (2a + b)(2a - b)$

30. $28 - 13x - 6x^2 = 28 - 21x + 8x - 6x^2$
$$= 7(4 - 3x) + 2x(4 - 3x) = (4 - 3x)(7 + 2x)$$

31. $20 - 3x - 2x^2 = 20 - 8x + 5x - 2x^2$
$$= 4(5 - 2x) + x(5 - 2x)$$
$$= (5 - 2x)(4 + x)$$

32. $x^2 - 2x + 4$ is a prime polynomial.

33. $a^2 + a - 3$ is a prime polynomial.

34. $6y^2 + y - 15 = 6y^2 + 10y - 9y - 15$
$$= 2y(3y + 5) - 3(3y + 5)$$
$$= (3y + 5)(2y - 3)$$

35. $4x^2 - x - 5 = 4x^2 - 5x + 4x - 5$
$$= x(4x - 5) + 1(4x - 5)$$
$$= (4x - 5)(x + 1)$$

36. $x^2 y - y^3 = y(x^2 - y^2) = y(x - y)(x + y)$

37. $4t^2 + 36;\ \text{GCF} = 4$
$$4(t^2 + 9)$$

38. $x^2 + x + xy + y = x(x + 1) + y(x + 1)$
$$= (x + 1)(x + y)$$

39. $ax + 2x + a + 2 = x(a + 1) + 1(a + 2)$
$$= (a + 2)(x + 1)$$

40. $18x^3 - 63x^2 + 9x = 9x(2x^2 - 7x + 1)$

41. $12a^3 - 24a^2 + 4a = 4a(3a^2 - 6a + 1)$

42. $x^2 + 14x - 32 = (x + 16)(x - 2)$

43. $x^2 - 14x - 48$ is prime

44. $16a^2 - 56ab + 49b^2$
$$= (4a)^2 - 2(4a)(7b) + (7b)^2$$
$$= (4a - 7b)^2$$

45. $25p^2 - 70pq + 49q^2$
$$= (5p)^2 - 2(5p)(7q) + (7q)^2$$
$$= (5p - 7q)^2$$

46. $7x^2 + 24xy + 9y^2 = 7x^2 + 3xy + 21xy + 9y^2$
$$= x(7x + 3y) + 3y(7x + 3y)$$
$$= (7x + 3y)(x + 3y)$$

47. $125 - 8y^3 = 5^3 - (2y)^3$
$$= (5 - 2y)(25 + 10y + 4y^2)$$

48. $64x^3 + 27 = (4x)^3 + 3^3$
$$= (4x + 3)(16x^2 - 12x + 9)$$

49. $-x^2 - x + 30 = -1(x^2 + x - 30)$
$$= -(x + 6)(x - 5)$$

50. $-x^2 + 6x - 8 = -1(x^2 - 6x + 8)$
$$= -(x-2)(x-4)$$

51. $14 + 5x - x^2 = (7-x)(2+x)$

52. $3 - 2x - x^2 = (3+x)(1-x)$

53. $3x^4 y + 6x^3 y - 72x^2 y = 3x^2 y(x^2 + 2x - 24)$
$$= 3x^2 y(x+6)(x-4)$$

54. $2x^3 y + 8x^2 y^2 - 10xy^3 = 2xy(x^2 + 4xy - 5y^2)$
$$= 2xy(x+5y)(x-y)$$

55. $5x^3 y^2 - 40x^2 y^3 + 35xy^4;\ \text{GCF} = 5xy^2$
$$5xy^2(x^2 - 8xy + 7y^2)$$
$$= 5xy^2(x-7y)(x-y)$$

56. $4x^4 y - 8x^3 y - 60x^2 y = 4x^2 y(x^2 - 2x - 15)$
$$= 4x^2 y(x-5)(x+3)$$

57. $12x^3 y + 243xy = 3xy(4x^2 + 81)$

58. $6x^3 y^2 + 8xy^2 = 2xy^2(3x^2 + 4)$

59. $4 - x^2 = 2^2 - x^2 = (2+x)(2-x)$

60. $9 - y^2 = 3^2 - y^2 = (3+y)(3-y)$

61. $3rs - s + 12r - 4 = s(3r-1) + 4(3r-1)$
$$= (3r-1)(s+4)$$

62. $x^3 - 2x^2 + 3x - 6 = x^2(x-2) + 3(x-2)$
$$= (x-2)(x^2 + 3)$$

63. $4x^2 - 8xy - 3x + 6y = 4x(x-2y) - 3(x-2y)$
$$= (x-2y)(4x-3)$$

64. $4x^2 - 2xy - 7yz + 14xz$
$$= 2x(2x-y) + 7z(-y+2x)$$
$$= (2x-y)(2x+7z)$$

65. $6x^2 + 18xy + 12y^2 = 6(x^2 + 3xy + 2y^2)$
$$= 6(x+2)(x+y)$$

66. $12x^2 + 46xy - 8y^2 = 2(6x^2 + 23xy - 4y^2)$
$$= 2(6x^2 + 24xy - xy - 4y^2)$$
$$= 2[6x(x+4y) - y(x+4)]$$
$$= 2(x+4y)(6x-y)$$

67. $xy^2 - 4x + 3y^2 - 12 = x(y^2 - 4) + 3(y^2 - 4)$
$$= (y^2 - 4)(x+3)$$
$$= (y^2 - 2^2)(x+3)$$
$$= (y+2)(y-2)(x+3)$$

68. $x^2 y^2 - 9x^2 + 3y^2 - 27 = x^2(y^2 - 9) + 3(y^2 - 9)$
$$= (y^2 - 9)(x^2 + 3)$$
$$= (y-3)(y+3)(x^2 + 3)$$

69. $5(x+y) + x(x+y) = (x+y)(5+x)$

70. $7(x-y) + y(x-y) = (x-y)(7+y)$

71. $14t^2 - 9t + 1 = 14t^2 - 7t - 2t + 1$
$$= 7t(2t - 1) - 1(2t - 1)$$
$$= (2t - 1)(7t - 1)$$

72. $3t^2 - 5t + 1$ is a prime polynomial

73. $3x^2 + 2x - 5 = 3x^2 + 5x - 3x - 5$
$$= x(3x + 5) - 1(3x + 5)$$
$$= (3x + 5)(x - 1)$$

74. $7x^2 + 19x - 6 = 7x^2 + 21x - 2x - 6$
$$= 7x(x + 3) - 2(x + 3)$$
$$= (x + 3)(7x - 2)$$

75. $x^2 + 9xy - 36y^2 = (x + 12y)(x - 3y)$

76. $3x^2 + 10xy - 8y^2$
$$= 3x^2 - 2xy + 12xy - 8y^2$$
$$= x(3x - 2y) + 4y(3x - 2y)$$
$$= (3x - 2y)(x + 4y)$$

77. $1 - 8ab - 20a^2b^2$
$$= 1 - 10ab + 2ab - 20a^2b^2$$
$$= 1(1 - 10ab) + 2ab(1 - 10ab)$$
$$= (1 - 10ab)(1 + 2ab)$$

78. $1 - 7ab - 60a^2b^2$
$$= 1 - 12ab + 5ab - 60a^2b^2$$
$$= 1(1 - 12ab) + 5ab(1 - 12ab)$$
$$= (1 - 12ab)(1 + 5ab)$$

79. $9 - 10x^2 + x^4 = (9 - x^2)(1 - x^2)$
$$= (3 + x)(3 - x)(1 + x)(1 - x)$$

80. $36 - 13x^2 + x^4 = (9 - x^2)(4 - x^2)$
$$= (3 + x)(3 - x)(2 + x)(2 - x)$$

81. $x^4 - 14x^2 - 32 = (x^2 + 2)(x^2 - 16)$
$$= (x^2 + 2)(x + 4)(x - 4)$$

82. $x^4 - 22x^2 - 75 = (x^2 + 3)(x^2 - 25)$
$$= (x^2 + 3)(x + 5)(x - 5)$$

83. $x^2 - 23x + 120 = (x - 15)(x - 8)$

84. $y^2 + 22y + 96 = (y + 16)(y + 6)$

85. $6x^3 - 28x^2 + 16x$; GCF $= 2x$
$$2x(3x^2 - 14x + 8) = 2x(3x - 2)(x - 4)$$

86. $6y^3 - 8y^2 - 30y = 2y(3y^2 - 4y - 15)$
$$= 2y(3y + 5)(y - 3)$$

87. $27x^3 - 125y^3 = (3x)^3 - (5y)^3$
$$= (3x - 5y)(9x^2 + 15xy + 25y^2)$$

88. $216y^3 - z^3 = (6y)^3 - z^3$
$$= (6y - z)(36y^2 + 6yz + z^2)$$

89. $x^3y^3 + 8z^3 = (xy)^3 + (2z)^3$
$$= (xy + 2z)(x^2y^2 - 2xyz + 4z^2)$$

90. $27a^3b^3 + 8 = (3ab)^3 + 2^3$
$$= (3ab + 2)(9a^2b^2 - 6ab + 4)$$

91. $2xy - 72x^3y = 2xy(1 - 36x^2)$

$$= 2xy\left(1^2 - (6x)^2\right) = 2xy(1 + 6x)(1 - 6x)$$

92. $2x^3 - 18x = 2x(x^2 - 9)$

$$= 2x(x^2 - 3^2)$$

$$= 2x(x + 3)(x - 3)$$

93. $x^3 + 6x^2 - 4x - 24 = x^2(x + 6) - 4(x + 6)$

$$= (x + 6)(x^2 - 4) = (x + 6)(x^2 - 2^2)$$

$$= (x + 6)(x + 2)(x - 2)$$

94. $x^3 - 2x^2 - 36x + 72$

$$= x^2(x - 2) - 36(x - 2)$$

$$= (x - 2)(x^2 - 36) = (x - 2)(x^2 - 6^2)$$

$$= (x - 2)(x + 6)(x - 6)$$

95. $6a^3 + 10a^2 = 2a^2(3a + 5)$

96. $4n^2 - 6n = 2n(2n - 3)$

97. $a^2(a + 2) + 2(a + 2) = (a + 2)(a^2 + 2)$

98. $a - b + x(a - b) = (a - b)(1 + x)$

99. $x^3 - 28 + 7x^2 - 4x = x^3 + 7x^2 - 28 - 4x$

$$= x^2(x + 7) - 4(7 + x)$$

$$= x^2(x + 7) - 4(x + 7)$$

$$= (x + 7)(x^2 - 4)$$

$$= (x + 7)(x + 2)(x - 2)$$

100. $a^3 - 45 - 9a + 5a^2 = a^3 + 5a^2 - 9a - 45$

$$= a^2(a + 5) - 9(a + 5)$$

$$= (a + 5)(a^2 - 9)$$

$$= (a + 5)(a + 3)(a - 3)$$

101. $(x - y)^2 - z^2 = (x - y + z)(x - y - z)$

102. $(x + 2y)^2 - 9 = (x + 2y)^2 - 3^2$

$$= (x + 2y + 3)(x + 2y - 3)$$

103. $81 - (5x + 1)^2 = 9^2 - (5x + 1)^2$

$$= \left[9 + (5x + 1)\right]\left[9 - (5x + 1)\right]$$

$$= (9 + 5x + 1)(9 - 5x - 1)$$

104. $b^2 - (4a + c)^2$

$$= \left[b + (4a + c)\right]\left[b - (4a + c)\right]$$

$$= (b + 4a + c)(b - 4a - c)$$

105. Answers may vary.

106. Yes. $9(x^2 + 9y^2)$

107. A,C

Section 6.5

Graphing Calculator Explorations

1. $-0.9, 2.2$

3. no real solutions

5. $-1.8, 2.8$

Mental Math

1. $(a-3)(a-7) = 0$

$a-3 = 0$ or $a-7 = 0$

     $a = 3$ $a = 7$

The solutions are 3 and 7.

2. $(a-5)(a-2) = 0$

$a-5 = 0$ or $a-2 = 0$

     $a = 5$ $a = 2$

The solutions are 5 and 2.

3. $(x+8)(x+6) = 0$

$x+8 = 0$ or $x+6 = 0$

     $x = -8$ $x = -6$

The solutions are -8 and -6.

4. $(x+2)(x+3) = 0$

$x+2 = 0$ or $x+3 = 0$

     $x = -2$ $x = -3$

The solutions are -2 and -3.

5. $(x+1)(x-3) = 0$

$x+1 = 0$ or $x-3 = 0$

     $x = -1$ $x = 3$

The solutions are -1 and 3.

6. $(x-1)(x+2) = 0$

$x-1 = 0$ or $x+2 = 0$

     $x = 1$ $x = -2$

The solutions are 1 and -2.

Exercise Set 6.5

1. $(x-2)(x+1) = 0$

$x-2 = 0$ or $x+1 = 0$

     $x = 2$ $x = -1$

The solutions are 2 and -1.

3. $x(x+6) = 0$

$x = 0$ or $x+6 = 0$

               $x = -6$

The solutions are 0 and -6.

5. $(2x+3)(4x-5) = 0$

$2x+3 = 0$ or $4x-5 = 0$

    $2x = -3$ $4x = 5$

     $x = -\dfrac{3}{2}$ $x = \dfrac{5}{4}$

The solutions are $-\dfrac{3}{2}$ and $\dfrac{5}{4}$.

7. $(2x-7)(7x+2) = 0$

$2x-7 = 0$ or $7x+2 = 0$

    $2x = 7$ $7x = -2$

     $x = \dfrac{7}{2}$ $x = -\dfrac{2}{7}$

The solutions are $\dfrac{7}{2}$ and $-\dfrac{2}{7}$.

9. If $x = 6$ and $x = -1$ are the solutions, then

$$x = 6 \quad \text{or} \quad x = -1$$
$$x - 6 = 0 \qquad x + 1 = 0$$
$$(x - 6)(x + 1) = 0$$

11. $x^2 - 13x + 36 = 0$
$$(x - 9)(x - 4) = 0$$
$$x - 9 = 0 \quad \text{or} \quad x - 4 = 0$$
$$x = 9 \qquad\qquad x = 4$$
The solutions are 9 and 4.

13. $x^2 + 2x - 8 = 0$
$$(x + 4)(x - 2) = 0$$
$$x + 4 = 0 \quad \text{or} \quad x - 2 = 0$$
$$x = -4 \qquad\qquad x = 2$$
The solutions are -4 and 2.

15. $x^2 - 4x = 32$
$$x^2 - 4x - 32 = 0$$
$$(x - 8)(x + 4) = 0$$
$$x - 8 = 0 \quad \text{or} \quad x + 4 = 0$$
$$x = 8 \qquad\qquad x = -4$$
The solutions are 8 and -4.

17. $x(3x - 1) = 14$
$$3x^2 - x = 14$$
$$3x^2 - x - 14 = 0$$
$$(3x - 7)(x + 2) = 0$$
$$3x - 7 = 0 \quad \text{or} \quad x + 2 = 0$$
$$3x = 7 \qquad\qquad x = -2$$
$$x = \frac{7}{3}$$
The solutions are $\frac{7}{3}$ and -2.

19. $3x^2 + 19x - 72 = 0$
$$(3x - 8)(x + 9) = 0$$
$$3x - 8 = 0 \quad \text{or} \quad x + 9 = 0$$
$$3x = 9 \qquad\qquad x = -9$$
$$x = \frac{8}{3}$$
The solutions are $\frac{8}{3}$ and -9.

21. If the solutions are $x = 5$ and $x = 7$, then, by the zero factor property,

$$x = 5 \quad \text{or} \quad x = 7$$
$$x - 5 = 0 \qquad x - 7 = 0$$
$$(x - 5)(x - 7) = 0$$
$$x^2 - 5x - 7x + 35 = 0$$
$$x^2 - 12x + 35 = 0$$

23. $x^3 - 12x^2 + 32x = 0$
$$x(x^2 - 12x + 32) = 0$$
$$x(x - 8)(x - 4) = 0$$
$$x = 0 \quad \text{or} \quad x - 8 = 0 \quad \text{or} \quad x - 4 = 0$$
$$x = 0 \quad \text{or} \qquad x = 8 \quad \text{or} \qquad x = 4$$

25. $(4x - 3)(16x^2 - 24x + 9) = 0$
$$(4x - 3)(4x - 3)^2 = 0$$
$$(4x - 3)^3 = 0$$
$$4x - 3 = 0$$
$$4x = 3$$
$$x = \frac{3}{4}$$
The solution is $\frac{3}{4}$.

27.
$$4x^3 - x = 0$$
$$x(4x^2 - 1) = 0$$
$$x(2x + 1)(2x - 1) = 0$$
$$x = 0 \quad \text{or} \quad 2x + 1 = 0 \quad \text{or} \quad 2x - 1 = 0$$
$$x = 0 \quad \text{or} \qquad 2x = -1 \quad \text{or} \qquad 2x = 1$$
$$x = 0 \quad \text{or} \qquad x = -\frac{1}{2} \quad \text{or} \qquad x = \frac{1}{2}$$

29.
$$32x^3 - 4x^2 - 6x = 0$$
$$2x(16x^2 - 2x - 3) = 0$$
$$2x(2x - 1)(8x + 3) = 0$$
$$2x = 0 \quad \text{or} \quad 2x - 1 = 0 \quad \text{or} \quad 8x + 3 = 0$$
$$x = \frac{0}{2} \quad \text{or} \qquad 2x = 1 \quad \text{or} \qquad 8x - -3$$
$$x = 0 \quad \text{or} \qquad x = \frac{1}{2} \quad \text{or} \qquad x = -\frac{3}{8}$$

31. $x(x + 7) = 0$
$$x = 0 \quad \text{or} \quad x + 7 = 0$$
$$x = 0 \quad \text{or} \qquad x = -7$$

33. $(x + 5)(x - 4) = 0$
$$x + 5 = 0 \quad \text{or} \quad x - 4 = 0$$
$$x = -5 \quad \text{or} \qquad x = 4$$

35.
$$x^2 - x = 30$$
$$x^2 - x - 30 = 0$$
$$(x - 6)(x + 5) = 0$$
$$x - 6 = 0 \quad \text{or} \quad x + 5 = 0$$
$$x = 6 \quad \text{or} \qquad x = -5$$

37. $6y^2 - 22y - 40 = 0$
$$2(3y^2 - 11y - 20) = 0$$
$$2(3y + 4)(y - 5) = 0$$
$$3y + 4 = 0 \quad \text{or} \quad y - 5 = 0$$
$$3y = -4 \qquad\qquad y = 5$$
$$y = -\frac{4}{3}$$
The solutions are $-\frac{4}{3}$ and 5.

39. $(2x + 3)(2x^2 - 5x - 3) = 0$
$$(2x + 3)(2x + 1)(x - 3) = 0$$
$$2x + 3 = 0 \quad \text{or} \quad 2x + 1 = 0 \quad \text{or} \quad x - 3 = 0$$
$$2x = -3 \qquad\qquad 2x = -1 \qquad\qquad x = 3$$
$$x = -\frac{3}{2} \qquad\qquad x = -\frac{1}{2}$$
The solutions are $-\frac{3}{2}$, $-\frac{1}{2}$, and 3.

41. $x^2 - 15 = -2x$
$$x^2 + 2x - 15 = 0$$
$$(x + 5)(x - 3) = 0$$
$$x + 5 = 0 \quad \text{or} \quad x - 3 = 0$$
$$x = -5 \qquad\qquad x = 3$$
The solutions are -5 and 3.

43. $x^2 - 16x = 0$
$$x(x - 16) = 0$$
$$x = 0 \quad \text{or} \quad x - 16 = 0$$
$$x = 0 \quad \text{or} \qquad x = 16$$

45. $-18y^2 - 33y + 216 = 0$

$-3(6y^2 + 11y - 72) = 0$

$-3(3y - 8)(2y + 9) = 0$

$3y - 8 = 0 \quad$ or $\quad 2y + 9 = 0$

$3y = 8 \quad$ or $\qquad 2y = -9$

$y = \dfrac{8}{3} \quad$ or $\qquad y = -\dfrac{9}{2}$

47. $12x^2 - 59x + 55 = 0$

$(4x - 5)(3x - 11) = 0$

$4x - 5 = 0 \quad$ or $\quad 3x - 11 = 0$

$4x = 5 \quad$ or $\quad 3x = 11$

$x = \dfrac{5}{4} \quad$ or $\quad x = \dfrac{11}{3}$

49. $18x^2 + 9x - 2 = 0$

$(3x + 2)(6x - 1) = 0$

$3x + 2 = 0 \quad$ or $\quad 6x - 1 = 0$

$3x = -2 \quad$ or $\qquad 6x = 1$

$x = -\dfrac{2}{3} \quad$ or $\qquad x = \dfrac{1}{6}$

51. $x(6x + 7) = 5$

$6x^2 + 7x = 5$

$6x^2 + 7x - 5 = 0$

$(3x + 5)(2x - 1) = 0$

$3x + 5 = 0 \quad$ or $\quad 2x - 1 = 0$

$3x = -5 \quad$ or $\qquad 2x = 1$

$x = -\dfrac{5}{3} \quad$ or $\qquad x = \dfrac{1}{2}$

53. $4(x - 7) = 6$

$4x - 28 = 6$

$4x = 34$

$x = \dfrac{34}{4}$

$x = \dfrac{17}{2}$

The solution is $\dfrac{17}{2}$.

55. $5x^2 - 6x - 8 = 0$

$(5x + 4)(x - 2) = 0$

$5x + 4 = 0 \quad$ or $\quad x - 2 = 0$

$5x = -4 \quad$ or $\qquad x = 2$

$x = -\dfrac{4}{5} \quad$ or $\qquad x = 2$

57. $(y - 2)(y + 3) = 6$

$y^2 - 2y + 3y - 6 = 6$

$y^2 + y - 12 = 0$

$(y + 4)(y - 3) = 0$

$y + 4 = 0 \quad$ or $\quad y - 1 = 0$

$y = -4 \qquad\qquad y = 3$

The solutions are -4 and 3.

59. $4y^2 - 1 = 0$

$(2y + 1)(2y - 1) = 0$

$2y + 1 = 0 \quad$ or $\quad 2y - 1 = 0$

$2y = -1 \qquad\qquad 2y = 1$

$y = -\dfrac{1}{2} \qquad\qquad y = \dfrac{1}{2}$

The solutions are $-\dfrac{1}{2}$ and $\dfrac{1}{2}$.

61. $t^2 + 13t + 22 = 0$

$(t+11)(t-+2) = 0$

$t+11 = 0$ or $t+2 = 0$

$t = -11$ or $t = -2$

63. $5t - 3 = 12$

$5t = 12 + 3$

$5t = 15$

$t = \dfrac{15}{5}$

$t = 3$

65. $x^2 + 6x - 17 = -26$

$x^2 + 6x - 17 + 26 = 0$

$x^2 + 6x + 9 = 0$

$(x+3)(x+3) = 0$

$x + 3 = 0$

$x = -3$

67. $12x^2 + 7x - 12 = 0$

$(4x - 3)(3x + 4) = 0$

$4x - 3 = 0$ or $3x + 4 = 0$

$x = \dfrac{3}{4}$ $x = -\dfrac{4}{3}$

The solutions are $-\dfrac{4}{3}$ and $\dfrac{3}{4}$.

69. $10t^3 - 25t - 15t^2 = 0$

$10t^3 - 15t^2 - 25t = 0$

$5t(2t^2 - 3t - 5) = 0$

$5t(2t - 5)(t + 1) = 0$

$5t = 0$ or $2t - 5 = 0$ or $t + 1 = 0$

$t = \dfrac{0}{5}$ or $2t = 5$ or $t = -1$

$t = 0$ or $t = \dfrac{5}{2}$ or $t = -1$

71. Let $y = 0$ and solve for x.

$y = (3x + 4)(x - 1)$

$0 = (3x + 4)(x - 1)$

$3x + 4 = 0$ or $x - 1 = 0$

$3x = -4$ or $x = 1$

$x = -\dfrac{4}{3}$

The x-intercepts are $\left(-\dfrac{4}{3}, 0\right)$ and $(1, 0)$.

73. Let $y = 0$ and solve for x.

$y = x^2 - 3x - 10$

$0 = x^2 - 3x - 10$

$0 = (x - 5)(x + 2)$

$x - 5 = 0$ or $x + 2 = 0$

$x = 5$ or $x = -2$

The x-intercepts are $(5, 0)$ and $(-2, 0)$.

75. Let $y = 0$ and solve for x.

$y = 2x^2 + 11x - 6$

$0 = 2x^2 + 11x - 6$

$0 = (2x - 1)(x + 6)$

$$2x - 1 = 0 \quad \text{or} \quad x + 6 = 0$$
$$2x = 1 \quad \text{or} \quad x = -6$$
$$x = \frac{1}{2}$$

The x-intercepts are $\left(\frac{1}{2}, 0\right)$ and $(-6, 0)$.

77. E; x-intercepts are $(-2, 0), (1, 0)$

79. B; x-intercepts are $(0, 0), (-3, 0)$

81. C; $y = 2x^2 - 8 = 2(x - 2)(x + 2)$
   x-intercepts are $(2, 0), (-2, 0)$

83. $\dfrac{3}{5} + \dfrac{4}{9} = \dfrac{3 \cdot 9}{5 \cdot 9} + \dfrac{4 \cdot 5}{9 \cdot 5}$
$$= \dfrac{27}{45} + \dfrac{20}{45}$$
$$= \dfrac{27 + 20}{45} = \dfrac{47}{45}$$

85. $\dfrac{7}{10} - \dfrac{5}{12} = \dfrac{7 \cdot 6}{10 \cdot 6} - \dfrac{5 \cdot 5}{12 \cdot 5}$
$$= \dfrac{42}{60} - \dfrac{25}{60}$$
$$= \dfrac{42 - 25}{60}$$
$$= \dfrac{17}{60}$$

87. $\dfrac{7}{8} \div \dfrac{7}{15} = \dfrac{7}{8} \cdot \dfrac{15}{7} = \dfrac{15}{8}$

89. $\dfrac{4}{5} \cdot \dfrac{7}{8} = \dfrac{4 \cdot 7}{5 \cdot 8} = \dfrac{4 \cdot 7}{5 \cdot 2 \cdot 4} = \dfrac{7}{10}$

91. $y = -16x^2 + 20x + 300$

a.

time x	0	1	2	3	4	5	6
height y	300	304	276	216	124	0	-156

b. The compass strikes the ground after 5 seconds, when the height, y, is zero feet.

c. The maximum height was approximately 304 feet.

d.

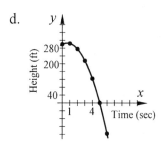

93. $(x - 3)(3x + 4) = (x + 2)(x - 6)$
$$3x^2 - 5x - 12 = x^2 - 4x - 12$$
$$2x^2 - x = 0$$
$$x(2x - 1) = 0$$
$$2x - 1 = 0 \quad \text{or} \quad x = 0$$
$$x = \frac{1}{2}$$
The solutions are $\dfrac{1}{2}$ and 0.

95. $(2x-3)(x+8) = (x-6)(x+4)$

$2x^2 + 13x - 24 = x^2 - 2x - 24$

$x^2 + 15x = 0$

$x(x+15) = 0$

$x + 15 = 0 \quad \text{or} \quad x = 0$

$x = -15$

The solutions are 0 and -15.

Exercise Set 6.6

1. Let x = the width, then $x + 4$ = the length.

3. Let x = the first odd integer, then
$x + 2$ = the next consecutive odd integer.

5. Let x = the base, then $4x + 1$ = the height.

7. Let x = the length of one side.

$A = x^2$

$121 = x^2$

$0 = x^2 - 121$

$0 = x^2 - 11^2$

$0 = (x+11)(x-11)$

$x + 11 = 0 \quad \text{or} \quad x - 11 = 0$

$x = -11 \qquad\qquad x = 11$

Since the length cannot be negative, the sides are 11 units long.

9. The perimeter is the sum of the lengths of the sides.

$120 = (x+5) + (x^2 - 3x) + (3x - 8)(x+3)$

$120 = x + 5 + x^2 - 3x + 3x - 8 + x + 3$

$120 = x^2 + 2x$

$0 = x^2 + 2x - 120$

$x^2 + 2x - 120 = 0$

$(x+12)(x-10) = 0$

$x + 12 = 0 \quad \text{or} \quad x - 10 = 0$

$x = -12 \qquad\qquad x = 10$

Since the dimensions cannot be negative, the lengths of the sides are:

$10 + 5 = 15$ cm, $10^2 - 3(10) = 70$ cm,

$3(10) - 8 = 22$ cm, and $10 + 3 = 13$ cm.

11. $x + 5$ = the base and $x - 5$ = the height.

$A = bh$

$96 = (x+5)(x-5)$

$96 = x^2 + 5x - 5x - 25$

$96 = x^2 - 25$

$0 = x^2 - 121$

$x^2 - 121 = 0$

$(x+11)(x-11) = 0$

$x + 11 = 0 \quad \text{or} \quad x - 11 = 0$

$x = -11 \qquad\qquad x = 11$

Since the dimensions cannot be negative, $x = 11$. The base is $11 + 5 = 16$ miles, and the height is $11 - 5 = 6$ miles.

13. Find t when $h = 0$.

$$h = -16t^2 + 64t + 80$$
$$0 = -16t^2 + 64t + 80$$
$$0 = -16(t^2 - 4t - 5)$$
$$0 = -16(t - 5)(t + 1)$$
$$t - 5 = 0 \quad \text{or} \quad t + 1 = 0$$
$$t = 5 \qquad\qquad t = -1$$

Since the time t cannot be negative, the object hits the ground after 5 seconds.

15. Let $x =$ the width then $2x - 7 =$ the length.

$$A = lw$$
$$30 = (2x - 7)(x)$$
$$30 = 2x^2 - 7x$$
$$0 = 2x^2 - 7x - 30$$
$$0 = (2x + 5)(x - 6)$$
$$2x + 5 = 0 \quad \text{or} \quad x - 6 = 0$$

$$x = -\frac{5}{2} \qquad\qquad x = 6$$

Since the dimensions cannot be negative, the width is 6 cm and the length is $2(6) - 7 = 5$ cm.

17. Let n $= 12$.

$$D = \frac{1}{2}n(n - 3)$$
$$D = \frac{1}{2} \cdot 12(12 - 3) = 6(9) = 54$$

A polygon with 12 sides has 54 diagonals.

19. Let D $= 35$ and solve for n.

$$D = \frac{1}{2}n(n - 3)$$
$$35 = \frac{1}{2}n(n - 3)$$
$$35 = \frac{1}{2}n^2 - \frac{3}{2}n$$
$$0 = \frac{1}{2}n^2 - \frac{3}{2}n - 35$$
$$0 = \frac{1}{2}(n^2 - 3n - 70)$$
$$0 = \frac{1}{2}(n - 10)(n + 7)$$
$$n - 10 = 0 \quad \text{or} \quad n + 7 = 0$$
$$n = 10 \qquad\qquad n = -7$$

The polygon has 10 sides.

21. Let $x =$ the unknown number.

$$x + x^2 = 132$$
$$x^2 + x - 132 = 0$$
$$(x + 12)(x - 11) = 0$$

$$x + 12 = 0 \quad \text{or} \quad x - 11 = 0$$
$$x = -12 \qquad\qquad x = 11$$

The two numbers are -12 and 11.

23. Let $x =$ the rate (in mph) of the slower boat, then $x + 7 =$ the rate (in mph) of the faster boat. After one hour, the slower boat has traveled x miles and the faster boat has traveled $x + 7$ miles. By the Pythagorean theorem,

$$x^2 + (x + 7)^2 = 17^2$$
$$x^2 + x^2 + 14x + 49 = 289$$
$$2x^2 + 14x + 49 = 289$$

$$2x^2 + 14x - 240 = 0$$
$$2\left(x^2 + 7x - 120\right) = 0$$
$$2\left(x + 15\right)\left(x - 8\right) = 0$$
$$x + 15 = 0 \quad \text{or} \quad x - 8 = 0$$
$$x = -15 \qquad\qquad x = 8$$

Since the rate cannot be negative, the slower boat travels at 8 mph. The faster boat travels at $8 + 7 = 15$ mph.

25. Let $x =$ the first number, then $20 - x =$ the other number.

$$x^2 + \left(20 - x\right)^2 = 218$$
$$x^2 + 400 - 40x + x^2 = 218$$
$$2x^2 - 40x + 400 = 218$$
$$2x^2 - 40x + 182 = 0$$
$$2\left(x^2 - 20x + 91\right) = 0$$
$$2\left(x - 13\right)\left(x - 7\right) = 0$$

$$x - 13 = 0 \quad \text{or} \quad x - 7 = 0$$
$$x = 13 \qquad\qquad x = 7$$

The numbers are 13 and 7.

27. Let $x =$ the length of a side of the original square. Then $x + 3 =$ the length of a side of the larger square.

$$64 = \left(x + 3\right)^2$$
$$64 = x^2 + 6x + 9$$
$$0 = x^2 + 6x - 55$$
$$0 = \left(x + 11\right)\left(x - 5\right)$$
$$x + 11 = 0 \quad \text{or} \quad x - 5 = 0$$
$$x = -11 \qquad\qquad x = -5$$

Since the length cannot be negative, the sides of the original square are 5 inches long.

29. Let $x =$ the length of the shorter leg. Then $x + 4 =$ the length of the longer leg and $x + 8 =$ the length of the hypotenuse. By the Pythagorean theorem,

$$x^2 + \left(x + 4\right)^2 = \left(x + 8\right)^2$$
$$x^2 + x^2 + 8x + 16 = x^2 + 16x + 64$$
$$x^2 - 8x - 48 = 0$$
$$\left(x - 12\right)\left(x + 4\right) = 0$$
$$x - 12 = 0 \quad \text{or} \quad x + 4 = 0$$
$$x = 12 \qquad\qquad x = -4$$

Since the length cannot be negative, the sides of the triangle are 12 mm, $12 + 4 = 16$ mm, and $12 + 8 = 20$ mm.

31. Let $x =$ the height of the triangle, then $2x =$ the base.

$$A = \frac{1}{2}bh$$
$$100 = \frac{1}{2}\left(2x\right)\left(x\right)$$
$$100 = x^2$$
$$0 = x^2 - 100$$
$$0 = \left(x + 10\right)\left(x - 10\right)$$
$$x + 10 = 0 \quad \text{or} \quad x - 10 = 0$$
$$x = -10 \qquad\qquad x = 10$$

Since the altitude cannot be negative, the height of the triangle is 10 km.

33. Let x = the length of the shorter leg,
then $x + 12$ = the length of the longer leg
and $2x - 12$ = the length of the hypotenuse.
By the Pythagorean theorem,

$$x^2 + (x + 12)^2 = (2x - 12)^2$$
$$x^2 + x^2 + 24x + 144 = 4x^2 - 48x + 144$$
$$0 = 2x^2 - 72x$$
$$0 = 2x(x - 36)$$
$$2x = 0 \quad \text{or} \quad x - 36 = 0$$
$$x = 0 \qquad\qquad x = 36$$

Since the length cannot be zero feet,
the shorter leg is 36 feet long.

35. Find t when $h = 0$.

$$h = -16t^2 + 625$$
$$0 = -16t^2 + 625$$
$$0 = -(4t + 25)(4t - 25)$$

$$4t + 25 = 0 \quad \text{or} \quad 4t - 25 = 0$$
$$4t = -25 \qquad\qquad 4t = 25$$
$$t = -6.25 \qquad\qquad t = 6.25$$

Since the time cannot be negative, the
solution is 6.25 seconds.

37. Let $P = 100$ and $A = 144$

$$A = P(1 + r)^2$$
$$144 = 100(1 + r)^2$$
$$1.2 = 1 + r$$
$$0.2 = r$$

The interest rate is 20%.

39. Let x = the length and $x - 7$ = the width.

$$A = lw$$
$$120 = (x - 7)(x)$$
$$120 = x^2 - 7x$$
$$0 = x^2 - 7x - 120$$
$$0 = (x + 8)(x - 15)$$
$$x + 8 = 0 \quad \text{or} \quad x - 15 = 0$$
$$x = -8 \qquad\qquad x = 15$$

Since the length cannot be negative,
the length is 15 miles. The width is
$15 - 7 = 8$ miles.

41. Let $C = 9500$

$$C = x^2 - 15x + 50$$
$$9500 = x^2 - 15x + 50$$
$$0 = x^2 - 15x - 9450$$
$$0 = (x + 90)(x - 105)$$

$$x + 90 = 0 \quad \text{or} \quad x - 105 = 0$$
$$x = -90 \qquad\qquad x = 105$$

Since the number of units cannot
be negative the solution is 105 units.

43. 175 acres

45. 6.25 million

47. 1966

49. Answers may vary

51. $\dfrac{24}{32} = \dfrac{2 \cdot 2 \cdot 2 \cdot 3}{2 \cdot 2 \cdot 2 \cdot 2 \cdot 2} = \dfrac{3}{4}$

53. $\dfrac{15}{27} = \dfrac{3 \cdot 5}{3 \cdot 3 \cdot 3} = \dfrac{5}{9}$

55. $\dfrac{45}{50} = \dfrac{3 \cdot 3 \cdot 5}{2 \cdot 5 \cdot 5} = \dfrac{9}{10}$

57. Answers may vary.

59. Let $x =$ the width of the walk, then
$24 - 2x =$ the length of the garden,
and $16 - 2x =$ the width of the garden.

$A = lw$

$180 = (24 - 2x)(16 - 2x)$

$180 = 384 - 80x + 4x^2$

$0 = 204 - 80x + 4x^2$

$0 = 4(51 - 20x + x^2)$

$0 = 4(17 - x)(3 - x)$

$17 - x = 0 \qquad \text{or} \qquad 3 - x = 0$

$\qquad 17 = x \qquad\qquad\qquad x = 3$

Since the walk cannot be wider than
8 yards, the width of the walk is 3 yards.

Chapter 6 Review

1. $6x^2 - 15x = 3x(2x - 5)$

2. $2x^3 y - 6x^2 y^2 - 8xy^3 = 2xy\left(x^2 - 3xy - 4y^2\right)$
$\qquad\qquad\qquad = 2xy(x - 4y)(x + y)$

3. $20x^2 + 12x;\ \text{GCF} = 4x$
$\qquad 4x(5x + 3)$

4. $6x^2 y^2 - 3xy^3 = 3xy^2(2x - y)$

5. $-8x^3 y + 6x^2 y^2 = -2x^2 y(4x - 3y)$

6. $3x(2x + 3) - 5(2x + 3) = (2x + 3)(3x - 5)$

7. $5x(x + 1) - (x + 1) = (x + 1)(5x - 1)$

8. $3x^2 - 3x + 2x - 2 = 3x(x - 1) + 2(x \quad 1)$
$\qquad\qquad\qquad\qquad = (x - 1)(3x + 2)$

9. $6x^2 + 10x - 3x - 5 = 2x(3x + 5) - 1(3x + 5)$
$\qquad\qquad\qquad\qquad = (3x + 5)(2x - 1)$

10. $3a^2 + 9ab + 3b^2 + ab$
$\qquad = 3a(a + 3b) + b(3b + a)$
$\qquad = 3a(a + 3b) + b(a + 3b)$
$\qquad = (a + 3b)(3a + b)$

11. $x^2 + 6x + 8 = (x + 4)(x + 2)$

12. $x^2 - 11x + 24 = (x - 8)(x - 3)$

13. $x^2 + x + 2$ is a prime polynomial.

14. $x^2 - 5x - 6 = (x - 6)(x + 1)$

15. $x^2 + 2x - 8 = (x + 4)(x - 2)$

16. $x^2 + 4xy - 12y^2 = (x + 6y)(x - 2y)$

17. $x^2 + 8xy + 15y^2 = (x + 5y)(x + 3y)$

18. $3x^2 y + 6xy^2 + 3y^3 = 3y\left(x^2 + 2xy + y^2\right)$
$\qquad\qquad\qquad\qquad = 3y(x + y)(x + y)$
$\qquad\qquad\qquad\qquad = 3y(x + y)^2$

19. $72 - 18x - 2x^2 = 2\left(36 - 9x - x^2\right)$
$\qquad\qquad\qquad = 2(3 - x)(12 + x)$

20. $32 + 12x - 4x^2 = 4\left(8 + 3x - x^2\right)$

21. $2x^2 + 11x - 6 = (2x - 1)(x + 6)$

22. $4x^2 - 7x + 4$

not factorable, prime

23. $4x^2 + 4x - 3 = 4x^2 + 6x - 2x - 3$
$$= 2x(2x + 3) - 1(2x + 3)$$
$$= (2x + 3)(2x - 1)$$

24. $6x^2 + 5xy - 4y^2 = 6x^2 + 8xy - 3xy - 4y^2$
$$= 2x(3x + 4y) - y(3x + 4y)$$
$$= (3x + 4y)(2x - y)$$

25. $6x^2 - 25xy + 4y^2 = (6x - y)(x - 4y)$

26. $18x^2 - 60x + 50 = 2\left(9x^2 - 30x + 25\right)$
$$= 2(3x - 5)(3x - 5)$$
$$= 2(3x - 5)^2$$

27. $2x^2 - 23xy - 39y^2$
$$= 2x^2 - 26xy + 3xy - 39y^2$$
$$= 2x(x - 13y) + 3y(x - 13y)$$
$$= (x - 13y)(2x + 3y)$$

28. $4x^2 - 28xy + 49y^2 = (2x - 7y)(2x - 7y)$
$$= (2x - 7y)^2$$

29. $18x^2 - 9xy - 20y^2$
$$= 18x^2 - 24xy + 15xy - 20y^2$$
$$= 6x(3x - 4y) + 5y(3x - 4y)$$
$$= (3x - 4y)(6x + 5y)$$

30. $36x^3y + 24x^2y^2 - 45xy^3$
$$= 3xy\left(12x^2 + 8xy - 15y^2\right)$$
$$= 3xy\left(12x^2 + 18xy - 10y^2 - 15y^2\right)$$
$$= 3xy\left[6x(2x + 3y) - 5y(2x + 3y)\right]$$
$$= 3xy(2x + 3y)(6x - 5y)$$

31. $4x^2 - 9 = (2x)^2 - 3^2 = (2x + 3)(2x - 3)$

32. $9t^2 - 25s^2 = (3t)^2 - (5s)^2$
$$= (3t + 5s)(3t - 5s)$$

33. $16x^2 + y^2$ is a prime polynomial.

34. $x^3 - 8y^3 = x^3 - (2y)^3$
$$= (x - 2y)\left(x^2 + 2xy + 4y^2\right)$$

35. $8x^3 + 27 = (2x)^3 + 3^3$
$$= (2x + 3)\left(4x^2 - 6x + 9\right)$$

36. $2x^3 + 8x = 2x\left(x^2 + 4\right)$

37. $54 - 2x^3y^3$; GCF = 2
$$2\left(27 - x^3y^3\right) = 2\left[3^3 - (xy)^3\right]$$
$$= 2(3 - xy)\left(9 + 3xy + x^2y^2\right)$$

38. $9x^2 - 4y^2$
$$= (3x)^2 - (2y)^2$$
$$= (3x - 2y)(3x + 2y)$$

39. $16x^4 - 1 = \left(4x^2\right)^2 - 1^2$

$\qquad = \left(4x^2 + 1\right)\left(4x^2 - 1\right)$

$\qquad = \left(4x^2 + 1\right)\left(\left(2x\right)^2 - 1^2\right)$

$\qquad = \left(4x^2 + 1\right)\left(2x + 1\right)\left(2x - 1\right)$

40. $x^4 + 16$

not factorable, prime

41. $2x^2 + 5x - 12 = \left(2x - 3\right)\left(x + 4\right)$

42. $3x^2 - 12 = 3\left(x^2 - 4\right) = 3\left(x - 2\right)\left(x + 2\right)$

43. $x\left(x - 1\right) + 3\left(x - 1\right);\ \text{GCF} = \left(x - 1\right)$

$\left(x - 1\right)\left(x + 3\right)$

44. $x^2 + xy - 3x - 3y = x\left(x + y\right) - 3\left(x + y\right)$

$\qquad\qquad\qquad = \left(x + y\right)\left(x - 3\right)$

45. $4x^2 y - 6xy^2;\ \text{GCF} = 2xy$

$2xy\left(2x - 3y\right)$

46. $8x^2 - 15x - x^3 = -x\left(-8x + 15 + x^2\right)$

$\qquad\qquad = -x\left(x^2 - 8x + 15\right)$

$\qquad\qquad = -x\left(x - 5\right)\left(x - 3\right)$

47. $125x^3 + 27 = \left(5x\right)^3 + 3^3$

$\qquad\qquad = \left(5x + 3\right)\left(25x^2 - 15x + 9\right)$

48. $24x^2 - 3x - 18 = 3\left(8x^2 - x - 6\right)$

49. $\left(x + 7\right)^2 - y^2 = \left[\left(x + 7\right) + y\right]\left[\left(x + 7\right) - y\right]$

$\qquad\qquad = \left(x + 7 + y\right)\left(x + 7 - y\right)$

50. $x^2\left(x + 3\right) - 4\left(x + 3\right) = \left(x + 3\right)\left(x^2 - 4\right)$

$\qquad\qquad\qquad = \left(x + 3\right)\left(x - 2\right)\left(x + 2\right)$

51. $\left(x + 6\right)\left(x - 2\right) = 0$

$x + 6 = 0 \quad$ or $\quad x - 2 = 0$

$\quad x = -6 \qquad\qquad x = 2$

The solutions are -6 and 2.

52. $3x\left(x + 1\right)\left(7x - 2\right) = 0$

$3x = 0 \quad$ or $\quad x + 1 = 0 \quad$ or $\quad 7x - 2 = 0$

$x = 0 \qquad\qquad x = -1 \qquad\qquad 7x = 2$

$\qquad\qquad\qquad\qquad\qquad\qquad x = \dfrac{2}{7}$

The solutions are 0, -1, and $\dfrac{2}{7}$.

53. $4\left(5x + 1\right)\left(x + 3\right) = 0$

$5x + 1 = 0 \quad$ or $\quad x + 3 = 0$

$\quad 5x = -1 \qquad\qquad x = -3$

$\quad x = -\dfrac{1}{5}$

The solutions are $-\dfrac{1}{5}$ and -3.

54. $x^2 + 8x + 7 = 0$

$\left(x + 7\right)\left(x + 1\right) = 0$

$x + 7 = 0 \quad$ or $\quad x + 1 = 0$

$x = -7 \qquad\qquad x = -1$

The solutions are -7 and -1.

55. $x^2 - 2x - 24 = 0$

$\left(x - 6\right)\left(x + 4\right) = 0$

$x - 6 = 0 \quad$ or $\quad x + 4 = 0$

$x = 6 \qquad\qquad x = -4$

The solutions are 6 and -4.

56. $x^2 + 10x = -25$

$x^2 + 10x + 25 = 0$

$(x + 5)(x + 5) = 0$

$x + 5 = 0$ or $x + 5 = 0$

$x = -5$ $x = -5$

The solution is -5.

57. $x(x - 10) = -16$

$x^2 - 10x = -16$

$x^2 - 10x + 16 = 0$

$(x - 8)(x - 2) = 0$

$x - 8 = 0$ or $x - 2 = 0$

$x = 8$ $x = 2$

The solutions are 8 and 2.

58. $(3x - 1)(9x^2 + 3x + 1) = 0$

$3x - 1 = 0$ or $9x^2 + 3x + 1 = 0$

$9x^2 + 3x + 1$ is a prime polynomial.

$3x - 1 = 0$

$3x = 1$

$x = \dfrac{1}{3}$

59. $56x^2 - 5x - 6 = 0$

$56x^2 + 16x - 21x - 6 = 0$

$8x(7x + 2)(8x - 3) = 0$

$(7x + 2)(8x - 3) = 0$

$7x + 2 = 0$ or $8x - 3 = 0$

$7x = -2$ $8x = 3$

$x = -\dfrac{2}{7}$ $x = \dfrac{3}{8}$

The solutions are $-\dfrac{2}{7}$ and $\dfrac{3}{8}$.

60. $20x^2 - 7x - 6 = 0$

$(4x - 3)(5x + 2) = 0$

$4x - 3 = 0$ or $5x + 2 = 0$

$4x = 3$ or $5x = -2$

$x = \dfrac{3}{4}$ or $x = -\dfrac{2}{5}$

61. $5(3x + 2) = 4$

$15x + 10 = 4$

$15x = 4 - 10$

$15x = -6$

$x = -\dfrac{6}{15} = -\dfrac{2}{5}$

62. $6x^2 - 3x + 8 = 0$

no real solution

63. $12 - 5t = -3$

$-5t = -3 - 12$

$-5t = -15$

$t = \dfrac{-15}{-5}$

$t = 3$

64. $5x^3 + 20x^2 + 20x = 0$

$5x(x^2 + 4x + 4) = 0$

$5x(x + 2)(x + 2) = 0$

$x + 2 = 0$ or $5x = 0$

$x = -2$ or $x = 0$

65. $4t^3 - 5t^2 - 21t = 0$

$t\left(4t^2 - 5t - 21\right) = 0$

$t\left(4t + 7\right)\left(t - 3\right) = 0$

$t = 0$ or $4t + 7 = 0$ or $t - 3 = 0$

$t = 0$ $4t = -7$ $t = 3$

$t = 0$ $t = -\dfrac{7}{4}$ $t = 3$

66. Let x = the width of the flag. Then

$2x - 15$ = the length of the flag.

$A = lw$

$500 = \left(2x - 15\right)\left(x\right)$

$500 = 2x^2 - 15x$

$0 = 2x^2 - 15x - 500$

$0 = \left(2x + 25\right)\left(x - 20\right)$

$2x + 25 = 0$ or $x - 20 = 0$

$2x = -25$ $x = 20$

$x = -\dfrac{25}{2}$

Since the dimensions cannot be negative,
the width is 20 inches and the length is
$2\left(20\right) - 15 = 25$ inches.

67. Let x = the height of the sail, then

$4x$ = the base of the sail.

$A = \dfrac{1}{2}bh$

$162 = \dfrac{1}{2}\left(4x\right)\left(x\right)$

$162 = 2x^2$

$0 = 2x^2 - 162$

$0 = 2\left(x^2 - 81\right)$

$0 = 2\left(x + 9\right)\left(x - 9\right)$

$x + 9 = 0$ or $x - 9 = 0$

$x = -9$ $x = 9$

Since the dimensions cannot be negative,
the height is 9 yards and the base is
$4 \cdot 9 = 36$ yards.

68. Let x = the first integer. Then

$x + 1$ = the next consecutive
integer.

$x\left(x + 1\right)380$

$x^2 + x = 380$

$x^2 + x - 380 = 0$

$\left(x + 20\right)\left(x - 19\right) = 0$

$x + 20 = 0$ or $x - 19 = 0$

$x = -21$ $x = 19$

The integers are 19 and 20.

69. A. Let h = 2800 and solve for t.

$h = -16t^2 + 440t$

$2800 = -16t^2 + 440t$

$0 = -16t^2 + 440t - 2800$

$0 = -8\left(2t^2 - 55t + 350\right)$

$0 = -8\left(2t - 35\right)\left(t - 10\right)$

$2t - 35 = 0$ or $t - 10 = 0$

$2t = 35$ $t = 10$

$t = \dfrac{35}{2}$

$t = 17.5$

The solutions are 17.5 sec and 10 sec

Answers may vary.

b. Find t when $h = 0$.

$$h = -16t^2 + 440t$$
$$0 = 16t^2 + 440t$$
$$0 = -8t(2t - 55)$$
$$-8t = 0 \quad \text{or} \quad 2t - 55 = 0$$
$$t = 0 \qquad\qquad 2t = 55$$
$$t = \frac{55}{2}$$
$$t = 27.5$$

27.5 seconds after being fired, the rocket will reach the ground again.

70. Let x = the length of the longer leg, then $x - 8$ = the length of the shorter leg and $x + 8$ = the length of the hypotenuse. By the Pythagorean theorem,

$$x^2 + (x-8)^2 = (x+8)^2$$
$$x^2 + x^2 - 16x + 64 = x^2 + 16x + 64$$

$$x^2 - 32x = 0$$
$$x(x - 32) = 0$$
$$x = 0 \quad \text{or} \quad x = 32$$

Since the length cannot be zero cm, the length of the longer leg is 32 cm.

Chapter 6 Test

1. $y^2 - 8y - 48 = (y - 12)(y + 4)$

2. $x^2 + x - 10$
not factorable, prime

3. $9x^3 + 39x^2 + 12x; \text{GCF} = 3x$
$$3x(3x^2 + 13x + 4) = 3x(3x + 1)(x + 4)$$

4. $3a^2 + 3ab - 7a - 7b = 3a(a + b) - 7(a + b)$
$$= (a + b)(3a - 7)$$

5. $3x^2 - 5x + 2 = (3x - 2)(x - 1)$

6. $x^2 + 14xy + 24y^2 = (x + 12y)(x + 2y)$

7. $180 - 5x^2 = 5(36 - x^2)$
$$= 5(6^2 - x^2)$$
$$= 5(6 + x)(6 - x)$$

8. $6t^2 - t - 5 = (6t + 5)(t - 1)$

9. $xy^2 - 7y^2 - 4x + 28$
$$= y^2(x - 7) - 4(x - 7)$$
$$= (x - 7)(y^2 - 4)$$
$$= (x - 7)(y^2 - 2^2)$$
$$= (x - 7)(y + 2)(y - 2)$$

10. $x - x^5 = x\left(1 - x^4\right)$

$\qquad = x\left(1 - \left(x^2\right)^2\right)$

$\qquad = x\left(1 + x^2\right)\left(1 - x^2\right)$

$\qquad = x\left(1 + x^2\right)\left(1 + x\right)\left(1 - x\right)$

11. $-xy^3 - x^3 y; \ \mathrm{GCF} = -xy$

$\qquad -xy\left(y^2 + x^2\right)$

12. $64x^3 - 1 = \left(4x\right)^3 - 1^3 = \left(4x - 1\right)\left(16x^2 + 4x + 1\right)$

13. $8y^3 - 64 = 8\left(y^3 - 8\right) = 8\left(y^3 - 2^3\right)$

$\qquad = 8\left(y - 2\right)\left(y^2 + 2y + 4\right)$

14. $\quad x^2 + 5x = 14$

$\qquad x^2 + 5x - 14 = 0$

$\qquad \left(x + 7\right)\left(x - 2\right) = 0$

$\qquad x + 7 = 0 \quad \text{or} \quad x - 2 = 0$

$\qquad\quad x = -7 \qquad\qquad x = 2$

The solutions are -7 and 2.

15. $x\left(x + 6\right) = 7$

$\qquad x^2 + 6x = 7$

$\qquad x^2 + 6x - 7 = 0$

$\qquad \left(x + 7\right)\left(x - 1\right) = 0$

$\qquad x + 7 = 0 \quad \text{or} \quad x - 1 = 0$

$\qquad\quad x = -7 \qquad\qquad x = 1$

The solutions are -7 and 1.

16. $3x\left(2x - 3\right)\left(3x + 4\right) = 0$

$\quad 3x = 0 \quad \text{or} \quad 2x - 3 = 0 \quad \text{or} \quad 3x + 4 = 0$

$\quad\; x = 0 \qquad\qquad 2x = 3 \qquad\qquad 3x = -4$

$\qquad\qquad\qquad\qquad x = \dfrac{3}{2} \qquad\qquad x = -\dfrac{4}{3}$

The solutions are 0, $\dfrac{3}{2}$, and $-\dfrac{4}{3}$.

17. $5t^3 - 45t = 0$

$\quad 5t\left(t^2 - 0\right) = 0$

$\quad 5t\left(t + 3\right)\left(t - 3\right) = 0$

$\quad 5t = 0 \quad \text{or} \quad t + 3 = 0 \quad \text{or} \quad t - 3 = 0$

$\quad\; t = 0 \qquad\qquad t = -3 \qquad\qquad t = 3$

The solutions are 0, -3, and 3.

18. $t^2 - 2t - 15 = 0$

$\quad \left(t - 5\right)\left(t + 3\right) = 0$

$\quad x - 5 = 0 \quad \text{or} \quad t + 3 = 0$

$\qquad t = 5 \qquad\qquad t = -3$

The solutions are 5 and -3.

19. $\quad 6x^2 = 15x$

$\qquad 6x^2 - 15x = 0$

$\qquad 3x\left(2x - 5\right) = 0$

$\qquad 3x = 0 \quad \text{or} \quad 2x - 5 = 0$

$\qquad\; x = 0 \qquad\qquad x = \dfrac{5}{2}$

The solutions are 0 and $\dfrac{5}{2}$.

20. Let x = the height of the triangle, then $x + 9$ = the base.

$$A = \frac{1}{2}bh$$

$$68 = \frac{1}{2}(x+9)(x)$$

$$68 = \frac{1}{2}x^2 + \frac{9}{2}x$$

$$0 = \frac{1}{2}x^2 + \frac{9}{2}x - 68$$

$$0 = \frac{1}{2}\left(x^2 + 9x - 136\right)$$

$$0 = \frac{1}{2}(x+17)(x-8)$$

$$x + 17 = 0 \quad \text{or} \quad x - 8 = 0$$
$$x = -17 \qquad\qquad x = 8$$

Since the length of the base cannot be negative, the base is $8 + 9 = 17$ feet.

21. Let x = the first number, then $17 - x$ = the other number.

$$x^2 + (17 - x)^2 = 145$$

$$x^2 + 289 - 34x + x^2 = 145$$

$$2x^2 - 34x + 144 = 0$$

$$2\left(x^2 - 17x + 72\right) = 0$$

$$2(x-9)(x-8) = 0$$

$$x - 9 = 0 \quad \text{or} \quad x - 8 = 0$$
$$x = 9 \qquad\qquad x = 8$$

The numbers are 8 and 9.

22. Find t when $h = 0$.

$$h = -16t^2 + 784$$

$$0 = -16t^2 + 784$$

$$16t^2 = 784$$

$$t^2 = 49$$

$$t = 7$$

It reaches the ground after 7 seconds.

Chapter 6 Cumulative Review

1. a. $9 \le 11$

   b. $8 > 1$

   c. $3 \ne 4$

2. a. $|-5| > |-3|$

   b. $|0| < |-2|$

3. a. $\dfrac{42}{49} = \dfrac{6 \cdot 7}{7 \cdot 7} = \dfrac{6}{7}$

   b. $\dfrac{11}{27} = \dfrac{11}{3 \cdot 3 \cdot 3} = \dfrac{11}{27}$

   c. $\dfrac{88}{20} = \dfrac{4 \cdot 22}{4 \cdot 5} = \dfrac{22}{5}$

4. Let $x = 20$ and $y = 10$.

$$\frac{x}{y} + 5x = \frac{20}{10} + 5(20) = 2 + 100 = 102$$

5. $\dfrac{8 + 2 \cdot 3}{2^2 - 1} = \dfrac{8 + 6}{4 - 1} = \dfrac{14}{3}$

6. Let $x = -20$ and $y = 10$.

$$\frac{x}{y} + 5x = \frac{-20}{10} + 5(-20) = -2 - 100 = -102$$

7. a. $3 + (-7) + (-8) = 3 + (-15) = -12$

    b. $\left[7 + (-10) \right] + \left[-2 + |-4| \right]$

        $= -3 + (-2 + 4)$

        $= -3 + 2$

        $= -1$

8. Let $x = -20$ and $y = -10$.

$$\frac{x}{y} + 5x = \frac{-20}{-10} + 5(-20) = 2 - 100 = -98$$

9. a. $(-6)(4) = -24$

    b. $(2)(-1) = -2$

    c. $(-5)(-10) = 50$

10. $5 - 2(3x - 7) = 5 - 6x + 14 = -6x + 19$

11. a. $7x - 3x = (7 - 3)x = 4x$

    b. $10y^2 + y^2 = (10 + 1)y^2 = 11y^2$

    c. $8x^2 + 2x - 3x = 8x^2 + (2 - 3)x$

              $= 8x^2 - x$

12. $0.8y + 0.2(y - 1) = 1.8$

    $0.8y + 0.2y - 0.2 = 1.8$

        $1.0y - 0.2 = 1.8$

             $y = 2.0$

13. $\dfrac{y}{7} = 20$

    $7\left(\dfrac{y}{7}\right) = 7(20)$

        $y = 140$

14. $\dfrac{x}{-7} = -4$

    $-7\left(\dfrac{x}{-7}\right) = -7(-4)$

        $x = 28$

15. $-3x = 33$

    $\dfrac{-3x}{-3} = \dfrac{33}{-3}$

        $x = -11$

16. $-\dfrac{2}{3}x = -22$

    $\left(-\dfrac{3}{2}\right)\left(-\dfrac{2}{3}\right)x = \left(-\dfrac{3}{2}\right)(-22)$

        $x = 33$

17. $8(2 - t) = -5t$

    $16 - 8t = -5t$

    $16 - 8t + 5t = -5t + 5t$

    $16 - 3t = 0$

    $16 - 16 - 3t = -16$

    $-3t = -16$

    $\dfrac{-3t}{-3} = \dfrac{-16}{-3}$

    $t = \dfrac{16}{3}$

18. $-z = \dfrac{7z + 3}{5}$

    $5(-z) = 5\left(\dfrac{7z + 3}{5}\right)$

    $-5z = 7z + 3$

    $-5z - 7z = 7z - 7z + 3$

    $-12z = 3$

$$\frac{-12z}{-12} = \frac{3}{-12}$$

$$z = -\frac{1}{4}$$

19. Let x = the length of the shorter piece
and $4x$ = the length of the longer piece.

$$x + 4x = 10$$

$$5x = 10$$

$$x = 2$$

$$4x = 4(2) = 8$$

The pieces are 2 feet and 8 feet in length.

20. $3x + 9 \le 5(x - 1)$

$$3x + 9 \le 5x - 5$$

$$-2x + 9 \le -5$$

$$-2x \le -14$$

$$\frac{-2x}{-2} \ge \frac{-14}{-2}$$

$$x \ge 7, \ [7, \infty)$$

21. $y = -\frac{1}{3}x$

x	y
0	0
6	-2

22. $-7x - 8y = -9$

$$(-1, 2): -7(-1) - 8(2) \stackrel{?}{=} -9$$

$$7 - 16 \stackrel{?}{=} -9$$

$$-9 = -9 \quad \text{True}$$

$(-1, 2)$ is a solution to the equation.

23. $(-6, 0)$ and $(-2, 3)$

$$m = \frac{y_2 - y_1}{x_2 - x_1} = \frac{3 - 0}{-2 - (-6)} = \frac{3}{4}$$

$(5, 4)$ and $(9, 7)$

$$m = \frac{y_2 - y_1}{x_2 - x_1} = \frac{7 - 4}{9 - 5} = \frac{3}{4}$$

Yes, they are parallel.

24. $(5, -6)$ and $(5, 2)$

$$m = \frac{y_2 - y_1}{x_2 - x_1} = \frac{2 - (-6)}{5 - 5} = \frac{8}{0}$$

The slope is undefined.

25. a. If $x = 5$, $2x^3 = 2(5)^3 = 2(125) = 250$

   b. If $x = -3$, $\dfrac{9}{x^2} = \dfrac{9}{(-3)^2} = \dfrac{9}{9} = 1$

26. $7x - 3y = 2$

$$-3y = -7x + 2$$

$$y = \frac{-7x}{-3} + \frac{2}{-3}$$

$$y = \frac{7}{3}x - \frac{2}{3}$$

$$y = mx + b$$

$$m = \frac{7}{3}, b = -\frac{2}{3}, \left(0, -\frac{2}{3}\right)$$

27. a. $-3x^2$: Degree 2

   b. $5x^3yz$: Degree 5

   c. y : Degree 1

   d. $12x^2yz^3$: Degree 6

   e. 5: Degree 0

28. Vertical line has equation $x = c$

Point $(0, 7)$

$x = 0$

29. $\left(11x^3 - 12x^2 + x - 3\right) + \left(x^3 - 10x + 5\right)$

$= 11x^3 - 12x^2 + x - 3 + x^3 - 10x + 5$

$= 12x^3 - 12x^2 - 9x + 2$

30. $m = 4,\ b = \dfrac{1}{2}$

$y = mx + b$

$y = 4x + \dfrac{1}{2}$

$2y = 8x + 1$

$8x - 2y = -1$

31. $(3x + 2)(2x - 5)$

$= 3x(2x) + 3x(-5) + 2(2x) + 2(-5)$

$= 6x^2 - 15x + 4x - 10$

$= 6x^2 - 11x - 10$

32. $(-4, 0)$ and $(6, -1)$

$m = \dfrac{y_2 - y_1}{x_2 - x_1} = \dfrac{-1 - 0}{6 - (-4)} = -\dfrac{1}{10}$

$m = -\dfrac{1}{10}$, point $(-4, 0)$

$y - y_1 = m(x - x_1)$

$y - 0 = -\dfrac{1}{10}\left[x - (-4)\right]$

$y = -\dfrac{1}{10}x - \dfrac{4}{10}$

$10y = -x - 4$

$x + 10y = -4$

33. $(3y + 1)^2 = (3y)^2 + 2(3y)(1) + 1^2$

$\qquad\qquad = 9y^2 + 6y + 1$

34. $\begin{cases} -x + 3y = 18 \\ -3x + 2y = 19 \end{cases}$

Multiply the first equation by -3

$\quad 3x - 9y = -54$

$\underline{-3x + 2y = 19}$

$\qquad\quad -7y = -35$

$\qquad\qquad y = 5$

Substitute 5 for y in the first equation.

$-x + 3(5) = 18$

$\qquad -x = 3$

$\qquad\ x = -3$

The solution to the system is $(-3, 5)$

35.

a. $3^{-2} = \dfrac{1}{3^2} = \dfrac{1}{9}$

b. $2x^{-3} = \dfrac{2}{x^3}$

c. $2^{-1} + 4^{-1} = \dfrac{1}{2} + \dfrac{1}{4} = \dfrac{1 \cdot 2}{2 \cdot 2} + \dfrac{1}{4}$

$\qquad = \dfrac{2}{4} + \dfrac{1}{4} = \dfrac{2 + 1}{4} = \dfrac{3}{4}$

d. $(-2)^{-4} = \dfrac{1}{(-2)^4} = \dfrac{1}{16}$

36. $\dfrac{\left(5a^7\right)^2}{a^5} = \dfrac{25a^{14}}{a^5} = 25a^9$

37.

a. $367{,}000{,}000 = 3.67 \times 10^8$

b. $0.000003 = 3.0 \times 10^{-6}$

c. $20{,}520{,}000{,}000 = 2.052 \times 10^{10}$

d. $0.00085 = 8.5 \times 10^{-4}$

38. $(3x-7y)^2 = (3x)^2 - 2(3x)(7y) + (7y)^2$
$$= 9x^2 - 42xy + 49y^2$$

39.
$$
\begin{array}{r}
x+4 \\
x+3{\overline{\smash{\big)}\,x^2 + 7x + 12}} \\
\underline{x^2 + 3x} \\
4x + 12 \\
\underline{4x + 12} \\
0
\end{array}
$$

$x+4$

40. $\dfrac{(xy)^{-3}}{(x^5 y^6)^3} = \dfrac{x^{-3} y^{-3}}{x^{15} y^{18}} = x^{-3-15} y^{-3-18} = x^{-18} y^{-21}$

$$= \dfrac{1}{x^{18} y^{21}}$$

41. a. $x^3, x^7, x^5 : \text{GCF} = x^3$
   b. $y, y^4, y^7 : \text{GCF} = y$

42. $z^3 + 7z + z^2 + 7 = z(z^2 + 7) + 1(z^2 + 7)$
$$= (z^2 + 7)(z + 1)$$

43. $x^2 + 7x + 12 = (x+4)(x+3)$

44. $2x^3 + 2x^2 - 84x = 2x(x^2 + x - 42)$
$$= 2x(x+7)(x-6)$$

45. $8x^2 - 22x + 5 = 8x^2 - 20x - 2x + 5$
$$= 4x(2x-5) - 1(2x-5)$$
$$= (2x-5)(4x-1)$$

46. $-4x^2 - 23x + 6 = -1(4x^2 + 23x - 6)$
$$= -(4x^2 - x + 24x - 6)$$
$$= -[x(4x-1) + 6(4x-1)]$$
$$= -(4x-1)(x+6)$$

47. $25a^2 - 9b^2 = (5a)^2 - (3b)^2$
$$= (5a + 3b)(5a - 3b)$$

48. $9xy^2 - 16x = x(9y^2 - 16)$
$$= x[(3y)^2 - 4^2]$$
$$= x(3y + 4)(3y - 4)$$

49. $(x-3)(x+1) = 0$
$$x - 3 = 0 \text{ or } x + 1 = 0$$
$$x = 3 \text{ or } \qquad x = -1$$

50.
$$x^2 - 13x = -36$$
$$x^2 - 13x + 36 = 0$$
$$(x-9)(x-4) = 0$$
$$x - 9 = 0 \text{ or } x - 4 = 0$$
$$x = 9 \text{ or } \qquad x = 4$$

Section 7.1

Graphing Calculator Explorations

1. $f(x) = \dfrac{x+1}{x^2-4} = \dfrac{x+1}{(x+2)(x-2)}$

Domain: $\{x \mid x$ is a real number and

$x \neq -2, x \neq 2\}$

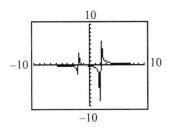

3. $h(x) = \dfrac{x^2}{2x^2+7x-4} = \dfrac{x^2}{(2x-1)(x+4)}$

Domain: $\{x \mid x$ is a real number and

$x \neq -4, x \neq \dfrac{1}{2}\}$

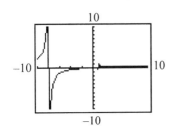

Exercise Set 7.1

1. $f(x) = \dfrac{x+8}{2x-1}$

$f(2) = \dfrac{2+8}{2(2)-1} = \dfrac{10}{4-1} = \dfrac{10}{3}$

$f(0) = \dfrac{0+8}{2(0)-1} = \dfrac{8}{-1} = -8$

$f(-1) = \dfrac{-1+8}{2(-1)-1} = \dfrac{7}{-3} = -\dfrac{7}{3}$

3. $g(x) = \dfrac{x^2+8}{x^3-25x}$

$g(3) = \dfrac{(3)^2+8}{(3)^3-25(3)} = \dfrac{9+8}{27-75} = \dfrac{17}{-48}$

$\qquad = -\dfrac{17}{48}$

$g(-2) = \dfrac{(-2)^2+8}{(-2)^3-25(-2)} = \dfrac{4+8}{-8+50}$

$\qquad = \dfrac{12}{42} = \dfrac{2}{7}$

$g(1) = \dfrac{(1)^2+8}{(1)^3-25(1)} = \dfrac{1+8}{1-25} = \dfrac{9}{-24}$

$\qquad = -\dfrac{3}{8}$

5. $f(x) = \dfrac{5x-7}{4}$

Domain: $\{x \mid x$ is a real number$\}$

7. $s(t) = \dfrac{t^2 + 1}{2t}$

Undefined values when $2t = 0$

$$t = \dfrac{0}{2} = 0$$

Domain: $\{t \mid t \text{ is a real number and } t \neq 0\}$

9. $f(x) = \dfrac{3x}{7 - x}$

Undefined values when $7 - x = 0$

$$7 = x$$

Domain: $\{x \mid x \text{ is a real number and } x \neq 7\}$

11. $C(x) = \dfrac{x + 3}{x^2 - 4x}$

Undefined values when

$$x^2 - 4x = 0$$
$$(x + 2)(x - 2) = 0$$
$$x + 2 = 0 \quad \text{or} \quad x - 2 = 0$$
$$x = -2 \qquad\qquad x = 2$$

Domain:

$\{x \mid x \text{ is a real number and } x \neq -2, \ x \neq 2\}$

13. $g(x) = \dfrac{5x}{3x^2 - 14x - 5}$

Undefined values when

$$3x^2 - 14x - 5 = 0$$
$$(3x + 1)(x - 5) = 0$$
$$3x + 1 = 0 \quad \text{or} \quad x - 5 = 0$$
$$x = -\dfrac{1}{3} \qquad\qquad x = 5$$

Domain:

$\left\{x \mid x \text{ is a real number and } x \neq -\dfrac{1}{3}, \ x \neq 5\right\}$

15. $R(x) = \dfrac{3 + 2x}{x^3 + x^2 - 2x}$

Undefined values when

$$x^3 + x^2 - 2x = 0$$
$$x(x^2 + x - 2) = 0$$
$$x(x + 2)(x - 1) = 0$$
$$x = 0 \quad \text{or} \quad x + 2 = 0 \quad \text{or} \quad x - 1 = 0$$
$$x = -2 \qquad\qquad x = 1$$

Domain:

$\{x \mid x \text{ is a real number and } x \neq -2, x \neq 0, x \neq 1\}$

17. $\dfrac{10x^3}{18x} = \dfrac{2 \cdot 5x^3}{2 \cdot 9 \cdot x} = \dfrac{5x^2}{9}$

19. $\dfrac{9x^6 y^3}{18x^2 y^5} = \dfrac{x^4}{2y^2}$

21. $\dfrac{x + 5}{5 + x} = \dfrac{x + 5}{x + 5} = 1$

23. $\dfrac{x - 1}{1 - x^2} = \dfrac{x - 1}{-(x^2 - 1)} = -\dfrac{x - 1}{(x + 1)(x - 1)} = \dfrac{-1}{x + 1}$

25. $\dfrac{8x - 4}{6x - 3} = \dfrac{4(x - 2)}{3(x - 2)} = \dfrac{4}{3}$

27. $\dfrac{2x - 14}{7 - x} = \dfrac{2(x - 7)}{-1(x - 7)} = \dfrac{2}{-1} = -2$

29. $\dfrac{x^2 - 2x - 3}{x^2 - 6x + 9} = \dfrac{(x - 3)(x + 1)}{(x - 3)^2} = \dfrac{x + 1}{x - 3}$

31. $\dfrac{2x^2+12x+18}{5x^2+17x+6} = \dfrac{2\left(x^2+6x+9\right)}{5x^2+17x+6}$

$\qquad = \dfrac{2(x+3)(x+3)}{(5x+2)(x+3)}$

$\qquad = \dfrac{2(x+3)}{5x+2}$

33. $\dfrac{30x+6}{5x^2+x} = \dfrac{6(5x+1)}{x(5x+1)} = \dfrac{6}{x}$

35. $\dfrac{2x^2-x-3}{2x^3-3x^2+2x-3} = \dfrac{(2x-3)(x+1)}{x^2(2x-3)+1(2x-3)}$

$\qquad = \dfrac{(2x-3)(x+1)}{(2x-3)(x^2+1)}$

$\qquad = \dfrac{x+1}{x^2+1}$

37. $\dfrac{8q^2}{16q^3-16q^2} = \dfrac{8q^2}{16q^2(q-1)} = \dfrac{1}{2(q-1)}$

39. $\dfrac{x^2+6x-40}{10+x} = \dfrac{(x+10)(x-4)}{x+10} = x-4$

41. $\dfrac{x^3-125}{5-x} = \dfrac{(x-5)\left(x^2+5x+25\right)}{-1(x-5)}$

$\qquad = \dfrac{x^2+5x+25}{-1}$

$\qquad = -x^2-5x-25$

43. $\dfrac{8x^3-27}{4x-6} = \dfrac{(2x)^3-3^3}{2(2x-3)}$

$\qquad = \dfrac{(2x-3)\left(4x^2+6x+9\right)}{2(2x-3)}$

$\qquad = \dfrac{4x^2+6x+9}{2}$

45. Let $D=1000$ and $A=8$

$\quad C = \dfrac{DA}{A+12} = \dfrac{1000(8)}{8+12} = \dfrac{8000}{20} = 400$

The child should receive 400 mg.

47. $B = \dfrac{705w}{h^2};\ w=148,\ h=66$

$\quad B = \dfrac{705\cdot148}{(66)^2} = \dfrac{104,340}{4356} \approx 24.0$

No

49. a. $R = \dfrac{150x^2}{x^2+3}$

$\qquad = \dfrac{150(1)^2}{1^2+3}$

$\qquad = \dfrac{150}{4}$

$\qquad = \$37.5$ million

   b. $R = \dfrac{150x^2}{x^2+3}$

$\qquad = \dfrac{150(2)^2}{2^2+3}$

$\qquad = \dfrac{600}{7}$

$\qquad \approx \$85.7$ million

   c. $85.7 = 37.5 = \$48.2$ million

51. $\dfrac{1}{3} \cdot \dfrac{9}{11} = \dfrac{1 \cdot 9}{3 \cdot 11} = \dfrac{3 \cdot 3}{3 \cdot 11} = \dfrac{3}{11}$

53. $\dfrac{1}{3} \div \dfrac{1}{4} = \dfrac{1}{3} \cdot \dfrac{4}{1} = \dfrac{4}{3}$

55. $\dfrac{5}{6} \cdot \dfrac{10}{11} \cdot \dfrac{2}{3} = \dfrac{5 \cdot 10 \cdot 2}{6 \cdot 11 \cdot 3}$

$\qquad = \dfrac{5 \cdot 2 \cdot 5 \cdot 2}{3 \cdot 2 \cdot 11 \cdot 3}$

$\qquad = \dfrac{5 \cdot 5 \cdot 2}{3 \cdot 11 \cdot 3} = \dfrac{50}{99}$

57. $\dfrac{13}{20} \div \dfrac{2}{9} = \dfrac{13}{20} \cdot \dfrac{9}{2} = \dfrac{13 \cdot 9}{20 \cdot 2} = \dfrac{117}{40}$

59-61. Answers may vary

63. $f(x) = \dfrac{1}{x}$

x	$\frac{1}{4}$	$\frac{1}{2}$	1	2	4
y	4	2	1	$\frac{1}{2}$	$\frac{1}{4}$

x	-4	-2	-1	$-\frac{1}{2}$	$-\frac{1}{4}$
y	$-\frac{1}{4}$	$-\frac{1}{2}$	-1	-2	-4

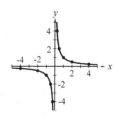

65. $R(x) = \dfrac{1000x^2}{x^2 + 4}$

a. $R(1) = \dfrac{1000(1)^2}{(1)^2 + 4} = \dfrac{1000}{5} = \$200$ million

b. $R(2) = \dfrac{1000(2)^2}{(2)^2 + 4} = \dfrac{4000}{8} = \$500$ million

c. $R(2) - R(1) = \$300$ million

67. $y = \dfrac{x^2 - 16}{x - 4} = \dfrac{(x + 4)(x - 4)}{x - 4}$

$\qquad = x + 4, \ x \neq 4$

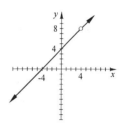

69. $y = \dfrac{x^2 - 6x + 8}{x - 2}$

$\qquad = \dfrac{(x - 2)(x - 4)}{x - 2}$

$\qquad = x - 4, \ x \neq 2$

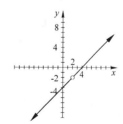

Section 7.2

Mental Math

1. $\dfrac{2}{y} \cdot \dfrac{x}{3} = \dfrac{2x}{3y}$

2. $\dfrac{3x}{4} \cdot \dfrac{1}{y} = \dfrac{3x}{4y}$

3. $\dfrac{5}{7} \cdot \dfrac{y^2}{x^2} = \dfrac{5y^2}{7x^2}$

4. $\dfrac{x^5}{11} \cdot \dfrac{4}{z^3} = \dfrac{4x^5}{11z^3}$

5. $\dfrac{9}{x} \cdot \dfrac{x}{5} = \dfrac{9x}{5x} = \dfrac{9}{5}$

6. $\dfrac{y}{7} \cdot \dfrac{3}{y} = \dfrac{3y}{7y} = \dfrac{3}{7}$

Exercise Set 7.2

1. $\dfrac{8x}{2} \cdot \dfrac{x^5}{4x^2} = \dfrac{8x \cdot x^5}{2 \cdot 4x^2} = \dfrac{2 \cdot 4 \cdot x \cdot x \cdot x^4}{2 \cdot 4 \cdot x \cdot x} = x^4$

3. $-\dfrac{5a^2 b}{30a^2 b^2} \cdot b^3 = \dfrac{5a^2 b \cdot b^3}{30a^2 b^2}$

$= -\dfrac{5 \cdot a^2 \cdot b \cdot b \cdot b^2}{5 \cdot 6 \cdot a^2 \cdot b^2} = -\dfrac{b \cdot b}{6} = -\dfrac{b^2}{6}$

5. $\dfrac{4}{x+2} \cdot \dfrac{x}{7} = \dfrac{4x}{7(x+2)}$

7. $\dfrac{x}{2x-14} \cdot \dfrac{x^2 - 7x}{5} = \dfrac{x \cdot (x^2 - 7x)}{(2x-14) \cdot 5}$

$= \dfrac{x \cdot x(x-7)}{2(x-7) \cdot 5} = \dfrac{x \cdot x}{2 \cdot 5} = \dfrac{x^2}{10}$

9. $\dfrac{6x+6}{5} \cdot \dfrac{10}{36x+36} = \dfrac{(6x+6) \cdot 10}{5 \cdot (36x+36)}$

$= \dfrac{6(x+1) \cdot 2 \cdot 5}{5 \cdot 36(x+1)} = \dfrac{6 \cdot 5 \cdot 2 \cdot (x+1)}{6 \cdot 5 \cdot 2 \cdot 3 \cdot (x+1)}$

$= \dfrac{1}{3}$

11. $\dfrac{m^2 - n^2}{m+n} \cdot \dfrac{m}{m^2 - mn} = \dfrac{(m^2 - n^2) \cdot m}{(m+n) \cdot (m^2 - mn)}$

$= \dfrac{(m-n)(m+n) \cdot m}{(m+n) \cdot m \cdot (m-n)} = 1$

13. $\dfrac{x^2 - 25}{x^2 - 3x - 10} \cdot \dfrac{x+2}{x} = \dfrac{(x^2 - 25) \cdot (x+2)}{(x^2 - 3x - 10) \cdot x}$

$= \dfrac{(x-5)(x+5) \cdot (x+2)}{(x-5)(x+2) \cdot x} = \dfrac{x+5}{x}$

15. $A = \dfrac{2x}{x^2 - 25} \cdot \dfrac{x+5}{9x^3} = \dfrac{2x \cdot (x+5)}{(x^2 - 25) \cdot 9x^3}$

$= \dfrac{2 \cdot x \cdot (x+5)}{9 \cdot x^3 \cdot (x+5)(x-5)} = \dfrac{2}{9x^2 (x-5)}$ sq.ft.

17. $\dfrac{5x^7}{2x^5} \div \dfrac{10x}{4x^3} = \dfrac{5x^7}{2x^5} \cdot \dfrac{4x^3}{10x}$

$= \dfrac{5 \cdot x^2 \cdot x^5 \cdot 2 \cdot 2x \cdot x^2}{2x^5 \cdot 2 \cdot 5 \cdot x}$

$= x^4$

19. $\dfrac{8x^2}{y^3} \div \dfrac{4x^2 y^3}{6} = \dfrac{8x^2}{y^3} \cdot \dfrac{6}{4x^2 y^3}$

$\qquad = \dfrac{2 \cdot 4 \cdot x^2 \cdot 6}{y^3 \cdot 4x^2 y^3}$

$\qquad = \dfrac{12}{y^6}$

21. $\dfrac{(x-6)(x+4)}{4x} \div \dfrac{2x-12}{8x^2}$

$\qquad = \dfrac{(x-6)(x+4)}{4x} \cdot \dfrac{8x^2}{2x-12}$

$\qquad = \dfrac{(x-6)(x+4) \cdot 2 \cdot 4 \cdot x \cdot x}{4x \cdot 2(x-6)}$

$\qquad = x(x+4)$

23. $\dfrac{3x^2}{x^2-1} \div \dfrac{x^5}{(x+1)^2} = \dfrac{3x^2}{x^2-1} \cdot \dfrac{(x+1)^2}{x^5}$

$\qquad = \dfrac{3x^2 \cdot (x+1)(x+1)}{(x-1)(x+1) \cdot x^2 \cdot x^3}$

$\qquad = \dfrac{3(x+1)}{x^3(x-1)}$

25. $\dfrac{m^2-n^2}{m+n} \div \dfrac{m}{m^2+nm}$

$\qquad = \dfrac{m^2-n^2}{m+n} \cdot \dfrac{m^2+nm}{m}$

$\qquad = \dfrac{(m-n)(m+n) \cdot m(m+n)}{(m+n) \cdot m}$

$\qquad = (m-n)(m+n) = m^2-n^2$

27. $\dfrac{x+2}{7-x} \div \dfrac{x^2-5x+6}{x^2-9x+14} = \dfrac{x+2}{7-x} \cdot \dfrac{x^2-9x+14}{x^2-5x+6}$

$\qquad = \dfrac{(x+2) \cdot (x-7)(x-2)}{-1(x-7) \cdot (x-3)(x-2)}$

$\qquad = -\dfrac{x+2}{x-3}$

29. $\dfrac{x^2+7x+10}{1-x} \div \dfrac{x^2+2x-15}{x-1}$

$\qquad = \dfrac{x^2+7x+10}{1-x} \cdot \dfrac{x-1}{x^2+2x-15}$

$\qquad = -\dfrac{(x+5)(x+2) \cdot (x-1)}{(x-1) \cdot (x+5)(x-3)} = -\dfrac{x+2}{x-3}$

31. Answers may vary.

33. $\dfrac{5a^2 b}{30a^2 b^2} \cdot \dfrac{1}{b^3} = \dfrac{5a^2 b}{30a^2 b^2 \cdot b^3} = \dfrac{5 \cdot a^2 \cdot b}{5 \cdot 6 \cdot a^2 \cdot b^2}$

$\qquad = \dfrac{1}{6b^4}$

35. $\dfrac{12x^3 y}{8xy^7} \div \dfrac{7x^5 y}{6x} = \dfrac{12x^3 y}{8xy^7} \cdot \dfrac{6x}{7x^5 y}$

$\qquad = \dfrac{72x^4 y}{56x^6 y^8}$

$\qquad = \dfrac{9}{7x^2 y^7}$

37. $\dfrac{5x-10}{12} \div \dfrac{4x-8}{8} = \dfrac{5x-10}{12} \cdot \dfrac{8}{4x-8}$

$\qquad = \dfrac{5(x-2) \cdot 2 \cdot 4}{6 \cdot 2 \cdot 4(x-2)} = \dfrac{5}{6}$

39. $\dfrac{x^2+5x}{8} \cdot \dfrac{9}{3x+15} = \dfrac{x(x+5) \cdot 3 \cdot 3}{8 \cdot 3(x+5)} \cdot \dfrac{3x}{8}$

41. $\dfrac{7}{6p^2+q} \div \dfrac{14}{18p^2+3q}$

$= \dfrac{7}{6p^2+q} \cdot \dfrac{18p^2+3q}{14}$

$= \dfrac{7\cdot3\left(6p^2+q\right)}{\left(6p^2+q\right)\cdot7\cdot2} = \dfrac{3}{2}$

43. $\dfrac{3x+4y}{x^2+4xy+4y^2} \cdot \dfrac{x+2y}{2}$

$= \dfrac{\left(3x+4y\right)\cdot\left(x+2y\right)}{\left(x+2y\right)\left(x+2y\right)\cdot2}$

$= \dfrac{3x+4y}{2\left(x+2y\right)}$

45. $\dfrac{x^2-9}{x^2+8} \div \dfrac{3-x}{2x^2+16}$

$= \dfrac{x^2-9}{x^2+8} \cdot \dfrac{2x^2+16}{3-x}$

$= -\dfrac{\left(x+3\right)\left(x-3\right)}{x^2+8} \cdot \dfrac{2\left(x^2+8\right)}{-1\left(x-3\right)}$

$= \dfrac{2\cdot\left(x+3\right)\cdot\left(x-3\right)\cdot\left(x^2+8\right)}{-1\cdot\left(x^2+8\right)\cdot\left(x-3\right)}$

$= -2\left(x+3\right)$

47. $\dfrac{\left(x+2\right)^2}{x-2} \div \dfrac{x^2-4}{2x-4} = \dfrac{\left(x+2\right)^2}{x-2} \cdot \dfrac{2x-4}{x^2-4}$

$= \dfrac{\left(x+2\right)\left(x+2\right)\cdot2\left(x-2\right)}{\left(x-2\right)\cdot\left(x+2\right)\left(x-2\right)} = \dfrac{2\left(x+2\right)}{x-2}$

49. $\dfrac{a^2+7a+12}{a^2+5a+6} \cdot \dfrac{a^2+8a+15}{a^2+5a+4}$

$= \dfrac{\left(a+3\right)\left(a+4\right)\cdot\left(a+5\right)\left(a+3\right)}{\left(a+3\right)\left(a+2\right)\cdot\left(a+4\right)\left(a+1\right)}$

$= \dfrac{\left(a+5\right)\left(a+3\right)}{\left(a+2\right)\left(a+1\right)}$

51. $\dfrac{1}{-x-4} \div \dfrac{x^2-7x}{x^2-3x-28}$

$= \dfrac{1}{-x-4} \cdot \dfrac{x^2-3x-28}{x^2-7x}$

$= \dfrac{1}{-1\left(x+4\right)} \cdot \dfrac{\left(x-7\right)\left(x+4\right)}{x\left(x-7\right)}$

$= \dfrac{\left(x-7\right)\cdot\left(x+4\right)}{-1\cdot x\cdot\left(x+4\right)\cdot\left(x-7\right)}$

$= -\dfrac{1}{x}$

53. $\dfrac{x^2-5x-24}{2x^2-2x-24} \cdot \dfrac{4x^2+4x-24}{x^2-10x+16}$

$= \dfrac{\left(x-8\right)\left(x+3\right)}{2\left(x^2-x-12\right)} \cdot \dfrac{4\left(x^2+x-6\right)}{\left(x-8\right)\left(x-2\right)}$

$= \dfrac{\left(x-8\right)\left(x+3\right)}{2\left(x-4\right)\left(x+3\right)} \cdot \dfrac{4\left(x+3\right)\left(x-2\right)}{\left(x-8\right)\left(x-2\right)}$

$= \dfrac{2\cdot2\cdot\left(x-8\right)\cdot\left(x+3\right)\cdot\left(x+3\right)\cdot\left(x-2\right)}{2\cdot\left(x-4\right)\cdot\left(x+3\right)\cdot\left(x-8\right)\cdot\left(x-2\right)}$

$= \dfrac{2\left(x+3\right)}{x-4}$

55. $(x-5) \div \dfrac{5-x}{x^2+2} = (x-5) \cdot \dfrac{x^2+2}{5-x}$

$$= (x-5) \cdot \dfrac{x^2+2}{-1(x-5)}$$

$$= \dfrac{(x-5) \cdot (x^2+2)}{-1 \cdot (x-5)}$$

$$= -(x^2+2)$$

57. $\dfrac{x^2-y^2}{x^2-2xy+y^2} \cdot \dfrac{y-x}{x+y}$

$$= \dfrac{(x+y)(x-y)}{(x-y)(x-y)} \cdot \dfrac{-1(x-y)}{x+y}$$

$$= \dfrac{-1 \cdot (x+y) \cdot (x-y) \cdot (x-y)}{(x-y) \cdot (x-y) \cdot (x+y)}$$

$$= -1$$

59. $\dfrac{x^2-9}{2x} \div \dfrac{x+3}{8x^4} = \dfrac{x^2-9}{2x} \cdot \dfrac{8x^4}{x+3}$

$$= \dfrac{(x+3)(x-3)}{2x} \cdot \dfrac{8x^4}{x+3}$$

$$= \dfrac{2x \cdot 4x^3 \cdot (x+3)(x-3)}{2x \cdot (x+3)}$$

$$= 4x^3(x-3)$$

61. 1 square foot is 12 inches by 12 inches or 144 square inches.

$10 \text{ sq ft} \cdot \dfrac{144 \text{ sq in}}{1 \text{ sq ft}} = 1440 \text{ sq in.}$

63. $3{,}705{,}745$ sq ft

$= 3{,}705{,}745 \text{ sq ft} \left(\dfrac{1 \text{ sq yd}}{9 \text{ sq ft}} \right)$

$\approx 411{,}972$ sq yd

65. $\dfrac{50 \text{ miles}}{1 \text{ hour}} \cdot \dfrac{1 \text{ hour}}{3600 \text{ seconds}} = \dfrac{5280 \text{ feet}}{1 \text{ mile}}$

$= \dfrac{50 \cdot 5280}{3600} \text{ feet/sec} \approx 73 \text{ feet/sec}$

67. $5023 \text{ ft per sec} \cdot \left(\dfrac{3600 \text{ sec}}{1 \text{ hour}} \right) \left(\dfrac{1 \text{ mile}}{5280 \text{ ft}} \right)$

≈ 3424.8 miles per hour

69. $\dfrac{1}{5} + \dfrac{4}{5} = \dfrac{5}{5} = 1$

71. $\dfrac{9}{9} - \dfrac{19}{9} = -\dfrac{10}{9}$

73. $\dfrac{6}{5} + \left(\dfrac{1}{5} - \dfrac{8}{5} \right) = \dfrac{6}{5} + \left(-\dfrac{7}{5} \right) = -\dfrac{1}{5}$

75. $x - 2y = 6$

x	y
0	-3
6	0

77. $\dfrac{a^2+ac+ba+bc}{a-b} \div \dfrac{a+c}{a+b}$

$$= \dfrac{a(a+c)+b(a+c)}{a-b} \cdot \dfrac{a+b}{a+c}$$

$$= \dfrac{(a+c)(a+b)}{a-b} \cdot \dfrac{a+b}{a+c}$$

$$= \dfrac{(a+c) \cdot (a+b) \cdot (a+b)}{(a-b) \cdot (a+c)}$$

$$= \dfrac{(a+b)^2}{a-b}$$

79. $\dfrac{3x^2+8x+5}{x^2+8x+7}\cdot\dfrac{x+7}{x^2+4}$

$=\dfrac{(3x+5)(x+1)}{(x+7)(x+1)}\cdot\dfrac{x+7}{x^2+4}$

$=\dfrac{(3x+5)\cdot(x+1)\cdot(x+7)}{(x+7)\cdot(x+1)\cdot(x^2+4)}$

$=\dfrac{3x+5}{x^2+4}$

81. $\dfrac{x^3+8}{x^2-2x+4}\cdot\dfrac{4}{x^2-4}$

$=\dfrac{(x+2)(x^2-2x+4)}{x^2-2x+4}\cdot\dfrac{4}{(x+2)(x-2)}$

$=\dfrac{4\cdot(x+2)\cdot(x^2-2x+4)}{(x+2)(x-2)(x^2-2x+4)}$

$=\dfrac{4}{x-2}$

83. $\dfrac{a^2-ab}{6a^2+6ab}\div\dfrac{a^3-b^3}{a^2-b^2}$

$=\dfrac{a^2-ab}{6a^2+6ab}\cdot\dfrac{a^2-b^2}{a^3-b^3}$

$\dfrac{a(a-b)}{6a(a+b)}\cdot\dfrac{(a-b)(a+b)}{(a-b)(a^2+ab+b^2)}$

$=\dfrac{a\cdot(a-b)\cdot(a-b)\cdot(a+b)}{6\cdot a\cdot(a+b)\cdot(a-b)\cdot(a^2+ab+b^2)}$

$=\dfrac{a-b}{6(a^2+ab+b^2)}$

85. $\left(\dfrac{x^2-y^2}{x^2+y^2}\div\dfrac{x^2-y^2}{3x}\right)\cdot\dfrac{x^2+y^2}{6}$

$=\dfrac{x^2-y^2}{x^2+y^2}\cdot\dfrac{3x}{x^2-y^2}\cdot\dfrac{x^2+y^2}{6}$

$=\dfrac{(x^2-y^2)\cdot3x\cdot(x^2+y^2)}{(x^2+y^2)\cdot(x^2-y^2)\cdot2\cdot3}=\dfrac{x}{2}$

87. $\left(\dfrac{2a+b}{b^2}\cdot\dfrac{3a^2-2ab}{ab+2b^2}\right)\div\dfrac{a^2-3ab+2b^2}{5ab-10b^2}$

$=\dfrac{2a+b}{b^2}\cdot\dfrac{3a^2-2ab}{ab+2b^2}\cdot\dfrac{5ab-10b^2}{a^2-3ab+2b^2}$

$=\dfrac{(2a+b)\cdot(3a^2-2ab)\cdot(5ab-10b^2)}{b^2\cdot(ab+2b^2)\cdot(a^2-3ab+2b^2)}$

$=\dfrac{(2a+b)\cdot a(3a-2b)\cdot5b(a-2b)}{b^2\cdot b(a+2b)\cdot(a-2b)(a-b)}$

$=\dfrac{5a(2a+b)(3a-2b)}{b^2(a+2b)(a-b)}$

89. $\$2000=\$2000\left(\dfrac{1\text{ euro}}{1.09\text{ dollars}}\right)$

≈1834.86 euros

Section 7.3

Mental Math

1. $\dfrac{2}{3}+\dfrac{1}{3}=\dfrac{3}{3}=1$

2. $\dfrac{5}{11}+\dfrac{1}{11}=\dfrac{6}{11}$

3. $\dfrac{3x}{9}+\dfrac{4x}{9}=\dfrac{7x}{9}$

4. $\dfrac{3y}{8} + \dfrac{2y}{8} = \dfrac{5y}{8}$

5. $\dfrac{8}{9} - \dfrac{7}{9} = \dfrac{1}{9}$

6. $-\dfrac{4}{12} - \dfrac{3}{12} = -\dfrac{7}{12}$

7. $\dfrac{7}{5} - \dfrac{10y}{5} = \dfrac{7-10y}{5}$

8. $\dfrac{12x}{7} - \dfrac{4x}{7} = \dfrac{8x}{7}$

Exercise Set 7.3

1. $\dfrac{a}{13} + \dfrac{9}{13} = \dfrac{a+9}{13}$

3. $\dfrac{9}{3+y} + \dfrac{y+1}{3+y} = \dfrac{9+y+1}{3+y} = \dfrac{y+10}{3+y}$

5. $\dfrac{4m}{3n} + \dfrac{5m}{3n} = \dfrac{4m+5m}{3n} = \dfrac{9m}{3n} = \dfrac{3m}{n}$

7. $\dfrac{2x+1}{x-3} + \dfrac{3x+6}{x-3} = \dfrac{2x+1+3x+6}{x-3} = \dfrac{5x+7}{x-3}$

9. $\dfrac{7}{8} - \dfrac{3}{8} = \dfrac{4}{8} = \dfrac{1}{2}$

11. $\dfrac{4m}{m-6} - \dfrac{24}{m-6} = \dfrac{4m-24}{m-6} = \dfrac{4(m-6)}{m-6} = 4$

13. $\dfrac{2x^2}{x-5} - \dfrac{25+x^2}{x-5} = \dfrac{2x^2 - \left(25+x^2\right)}{x-5}$

$$= \dfrac{2x^2 - 25 - x^2}{x-5}$$

$$= \dfrac{x^2 - 25}{x-5}$$

$$= \dfrac{(x+5)(x-5)}{x-5}$$

$$= x+5$$

15. $\dfrac{-3x^2-4}{x-4} - \dfrac{12-4x^2}{x-4}$

$$= \dfrac{-3x^2 - 4 - \left(12-4x^2\right)}{x-4}$$

$$= \dfrac{-3x^2 - 4 - 12 + 4x^2}{x-4}$$

$$= \dfrac{x^2 - 16}{x-4}$$

$$= \dfrac{(x+4)(x-4)}{x-4}$$

$$= x+4$$

17. $\dfrac{2x+3}{x+1} - \dfrac{x+2}{x+1} = \dfrac{2x+3 - \left(x+2\right)}{x+1}$

$$= \dfrac{2x+3 - x - 2}{x+1}$$

$$= \dfrac{x+1}{x+1}$$

$$= 1$$

19. $\dfrac{3}{x^3} + \dfrac{9}{x^3} = \dfrac{3+9}{x^3} = \dfrac{12}{x^3}$

21. $\dfrac{5}{x+4} - \dfrac{10}{x+4} = \dfrac{5-10}{x+4} = -\dfrac{5}{x+4}$

23. $\dfrac{x}{x+y} - \dfrac{2}{x+y} = \dfrac{x-2}{x+y}$

25. $\dfrac{8x}{2x+5} + \dfrac{20}{2x+5} = \dfrac{8x+20}{2x+5} = \dfrac{4(2x+5)}{2x+5} = 4$

27. $\dfrac{5x+4}{x-1} - \dfrac{2x+7}{x-1} = \dfrac{5x+4-(2x+7)}{x-1}$

$\qquad = \dfrac{5x+4-2x-7}{x-1}$

$\qquad = \dfrac{3x-3}{x-1}$

$\qquad = \dfrac{3(x-1)}{x-1}$

$\qquad = 3$

29. $\dfrac{a}{a^2+2a-15} - \dfrac{3}{a^2+2a-15} = \dfrac{a-3}{a^2+2a-15}$

$\qquad = \dfrac{a-3}{(a+5)(a-3)} = \dfrac{1}{a+5}$

31. $\dfrac{2x+3}{x^2-x-30} - \dfrac{x-2}{x^2-x-30}$

$\qquad = \dfrac{2x+3-(x-2)}{x^2-x-30}$

$\qquad = \dfrac{2x+3-x+2}{x^2-x-30} = \dfrac{x+5}{x^2-x-30}$

$\qquad = \dfrac{x+5}{(x-6)(x+5)} = \dfrac{1}{x-6}$

33. $P = \dfrac{5}{x-2} + \dfrac{5}{x-2} + \dfrac{5}{x-2} + \dfrac{5}{x-2}$

$\qquad = \dfrac{5+5+5+5}{x-2} = \dfrac{20}{x-2}$

The perimeter is $\dfrac{20}{x-2}$ meters.

35. Answers may vary.

37. $3 = 3$

$\quad 33 = 3 \cdot 11$

$\quad$ LCD $= 3 \cdot 11 = 33$

39. $\quad 2x = 2 \cdot x$

$\quad 4x^3 = 2^2 \cdot x^3$

$\quad$ LCD $- 2^2 \cdot x^3 = 4x^3$

41. $\quad 8x = 2^3 \cdot x$

$\quad 2x+4 = 2(x+2)$

$\quad$ LCD $= 2^3 \cdot x \cdot (x+2) = 8x(x+2)$

43. $\quad 3x+3 = 3 \cdot (x+1)$

$\quad 2x^2+4x+2 = 2(x^2+2x+1) = 2 \cdot (x+1)^2$

$\qquad$ LCD $= 2 \cdot 3(x+1)^2$

$\qquad\qquad = 6(x+1)^2$

45. $x-8 = x-8$

$\quad 8-x = -(x-8)$

$\quad$ LCD $= x-8$ or $8-x$

47. $8x^2(x-1)^2 = 2^3 \cdot x^2 \cdot (x-1)^2$

$\quad 10x^3(x-1) = 2 \cdot 5 \cdot x^3 \cdot (x-1)$

$\quad$ LCD $= 2^3 \cdot 5 \cdot x^3 \cdot (x-1)^2 = 40x^3(x-1)^2$

49. $2x + 1 = (2x + 1)$

$\quad\ 2x - 1 = (2x - 1)$

$\quad\ \text{LCD} = (2x + 1)(2x - 1)$

51. $2x^2 + 7x - 4 = (x + 4)(2x - 1)$

$\quad\ 2x^2 + 5x - 3 = (x + 3)(2x - 1)$

$\quad\ \quad\ \text{LCD} = (x + 4)(x + 3)(2x - 1)$

53. Answers may vary

55. $\dfrac{3}{2x} = \dfrac{3(2x)}{2x(2x)} = \dfrac{6x}{4x^2}$

57. $\dfrac{6}{3a} = \dfrac{6(4b^2)}{3a(4b^2)} = \dfrac{24b^2}{12ab^2}$

59. $\dfrac{9}{x + 3} = \dfrac{9(2)}{(x + 3)(2)} = \dfrac{18}{2(x + 3)}$

61. $\dfrac{9a + 2}{5a + 10} = \dfrac{9a + 2}{5(a + 2)}$

$\quad\quad\quad\quad = \dfrac{(9a + 2)(b)}{5(a + 2)(b)}$

$\quad\quad\quad\quad = \dfrac{9ab + 2b}{5b(a + 2)}$

63. $\dfrac{x}{x^2 + 6x + 8} = \dfrac{x}{(x + 4)(x + 2)}$

$\quad\quad\quad\quad = \dfrac{x(x + 1)}{(x + 4)(x + 2)(x + 1)}$

$\quad\quad\quad\quad = \dfrac{x^2 + x}{(x + 4)(x + 2)(x + 1)}$

65. $\dfrac{9y - 1}{15x^2 - 30} = \dfrac{(9y - 1)(2)}{(15x^2 - 30)2} = \dfrac{18y - 2}{30x^2 - 60}$

67. $\dfrac{5}{2x^2 - 9x - 5} = \dfrac{5}{(2x + 1)(x - 5)} \cdot \dfrac{3x(x - 7)}{3x(x - 7)}$

$\quad\quad\quad\quad = \dfrac{15x(x - 7)}{3x(2x + 1)(x - 7)(x - 5)}$

69. $\text{LCD} = 21$

$\quad\ \dfrac{2}{3} + \dfrac{5}{7} = \dfrac{2(7)}{3(7)} + \dfrac{5(3)}{7(3)} = \dfrac{14}{21} + \dfrac{15}{21} = \dfrac{29}{21}$

71. Since $6 = 2 \cdot 3$ and $4 = 2^2$,

$\quad\ \text{LCD} = 2^2 \cdot 3 = 12.$

$\quad\ \dfrac{2}{6} - \dfrac{3}{4} = \dfrac{2(2)}{6(2)} - \dfrac{3(3)}{4(3)} = \dfrac{4}{12} - \dfrac{9}{12} = \dfrac{4 - 9}{12}$

$\quad\quad = -\dfrac{5}{12}$

73. $x(x - 3) = 0$

$\quad\ x = 0 \quad \text{or} \quad x - 3 = 0$

$\quad\quad\quad\quad\quad\quad\quad\quad x = 3$

75. $\quad x^2 + 6x + 5 = 0$

$\quad\ (x + 5)(x + 1) = 0$

$\quad\ x + 5 = 0 \quad \text{or} \quad x + 1 = 0$

$\quad\quad x = -5 \quad \text{or} \quad\quad x = -1$

77. C

79. B

81. $\dfrac{5}{2 - x} = \dfrac{5(-1)}{(2 - x)(-1)} = -\dfrac{5}{x - 2}$

83. $-\dfrac{7+x}{2-x} = \dfrac{7+x}{(-1)(2-x)} = \dfrac{7+x}{x-2}$

85. Since $88 = 2^3 \cdot 11$ and $4332 = 2^2 \cdot 3 \cdot 19^2$ the LCM of 88 and 4332 is $2^3 \cdot 3 \cdot 11 \cdot 19^2 = 95,304$. They will align again in 95,304 Earth days.

87. Answers may vary.

Section 7.4

Mental Math

1. **D**

2. **C**

3. **A**

4. **B**

Exercise Set 7.4

1. $\text{LCD} = 2 \cdot 3 \cdot x = 6x$

$\dfrac{4}{2x} + \dfrac{9}{3x} = \dfrac{4(3)}{2x(3)} + \dfrac{9(2)}{3x(2)} = \dfrac{12}{6x} + \dfrac{18}{6x}$

$= \dfrac{30}{6x} = \dfrac{5(6)}{6x} = \dfrac{5}{x}$

3. $\text{LCD} = 5b$

$\dfrac{15a}{b} + \dfrac{6b}{5} = \dfrac{15a(5)}{b(5)} + \dfrac{6b(b)}{5(b)} = \dfrac{75a}{5b} + \dfrac{6b^2}{5b}$

$= \dfrac{75a + 6b^2}{5b}$

5. $\text{LCD} = 2x^2$

$\dfrac{3}{x} + \dfrac{5}{2x^2} = \dfrac{3(2x)}{x(2x)} + \dfrac{5}{2x^2} = \dfrac{6x}{2x^2} + \dfrac{5}{2x^2}$

$= \dfrac{6x+5}{2x^2}$

7. $2x + 2 = 2(x+1)$

$\text{LCD} = 2(x+1)$

$\dfrac{6}{x+1} + \dfrac{9}{2x+2} = \dfrac{6}{x+1} + \dfrac{9}{2(x+1)}$

$= \dfrac{6(2)}{(x+1)2} + \dfrac{9}{2(x+1)} = \dfrac{12}{2(x+1)} + \dfrac{9}{2(x+1)}$

$= \dfrac{12+9}{2(x+1)} = \dfrac{21}{2(x+1)}$

9. $2x - 4 = 2(x-2)$

$x^2 - 4 = (x-2)(x+2)$

$\text{LCD} = 2(x-2)(x+2)$

$\dfrac{15}{2x-4} + \dfrac{x}{x^2-4} = \dfrac{15}{2(x-2)} + \dfrac{x}{(x-2)(x+2)}$

$= \dfrac{15(x+2)}{2(x-2)(x+2)} + \dfrac{x(2)}{(x-2)(x+2)(x)}$

$= \dfrac{15x+30}{2(x-2)(x+2)} + \dfrac{2x}{2(x-2)(x+2)}$

$= \dfrac{15x+30+2x}{2(x-2)(x+2)} = \dfrac{17x+30}{2(x-2)(x+2)}$

11. $\text{LCD} = 4x(x-2)$

$$\frac{3}{4x} + \frac{8}{x-2} = \frac{3(x-2)}{4x(x-2)} + \frac{8(4x)}{(x-2)(4x)}$$

$$= \frac{3x-6}{4x(x-2)} + \frac{32x}{4x(x-2)} = \frac{3x-6+32x}{4x(x-2)}$$

$$= \frac{35x-6}{4x(x-2)}$$

13. $\text{LCD} = y^2(2y+1)$

$$\frac{5}{y^2} - \frac{y}{2y+1} = \frac{5(2y+1)}{y^2(2y+1)} - \frac{yy^2}{(2y+1)y^2}$$

$$= \frac{10y+5}{y^2(2y+1)} - \frac{y^3}{y^2(2y+1)} = \frac{10y+5-y^3}{y^2(2y+1)}$$

15. Answers may vary.

17. $\dfrac{6}{x-3} + \dfrac{8}{3-x} = \dfrac{6}{x-3} + \dfrac{8}{-(x-3)}$

$$= \frac{6}{x-3} + \frac{-8}{x-3} = \frac{6+(-8)}{x-3} = -\frac{2}{x-3}$$

19. $\dfrac{-8}{x^2-1} - \dfrac{7}{1-x^2} = \dfrac{8}{-(x^2-1)} - \dfrac{7}{1-x^2}$

$$= \frac{8}{1-x^2} - \frac{7}{1-x^2} = \frac{8-7}{1-x^2}$$

$$= \frac{1}{1-x^2} \quad \text{or} \quad -\frac{1}{x^2-1}$$

21. $\dfrac{x}{x^2-4} - \dfrac{2}{4-x^2} = \dfrac{x}{x^2-4} - \dfrac{2}{-(x^2-4)}$

$$= \frac{x}{x^2-4} + \frac{2}{x^2-4} = \frac{x+2}{x^2-4}$$

$$= \frac{x+2}{(x+2)(x-2)} = \frac{1}{x-2}$$

23. $\dfrac{5}{x} + 2 = \dfrac{5}{x} + \dfrac{2}{1} = \dfrac{5}{x} + \dfrac{2(x)}{1(x)} = \dfrac{5+2x}{x}$

25. $\dfrac{5}{x-2} + 6 = \dfrac{5}{x-2} + \dfrac{6}{1} = \dfrac{5}{x-2} + \dfrac{6(x-2)}{1(x-2)}$

$$= \frac{5}{x-2} + \frac{6x-12}{x-2} = \frac{5+6x-12}{x-2} = \frac{6x-7}{x-2}$$

27. $\dfrac{y+2}{y+3} - 2 = \dfrac{y+2}{y+3} - \dfrac{2}{1} = \dfrac{y+2}{y+3} - \dfrac{2(y+3)}{y+3}$

$$= \frac{y+2}{y+3} - \frac{2y+6}{y+3} = \frac{y+2-(2y+6)}{y+3}$$

$$= \frac{y+2-2y-6}{y+3} = \frac{-y-4}{y+3} = \frac{-(y+4)}{y+3}$$

$$= -\frac{y+4}{y+3}$$

29. $90° - \left(\dfrac{40}{x}\right)° = \left(90 - \dfrac{40}{x}\right)°$

$\text{LCD} = x$

$$\left(90 \cdot \frac{x}{x} - \frac{40}{x}\right)° = \left(\frac{90x}{x} - \frac{40}{x}\right)°$$

$$= \left(\frac{90x-40}{x}\right)°$$

31. $\dfrac{5x}{x+2} - \dfrac{3x-4}{x+2} = \dfrac{5x-(3x-4)}{x+2}$

$$= \frac{5x-3x+4}{x+2} = \frac{2x+4}{x+2} = \frac{2(x+2)}{x+2} = 2$$

33. $\dfrac{3x^4}{x} - \dfrac{4x^2}{x^2} = \dfrac{3x^4(x)}{x(x)} - \dfrac{4x^2}{x^2} = \dfrac{3x^5}{x^2} - \dfrac{4x^2}{x^2}$

$$= \frac{3x^5-4x^2}{x^2} = \frac{x^2(3x^3-4)}{x^2} = 3x^3 - 4$$

35. $\dfrac{1}{x+3} - \dfrac{1}{(x+3)^2} = \dfrac{1(x+3)}{(x+3)(x+3)} - \dfrac{1}{(x+3)^2}$

$= \dfrac{x+3}{(x+3)^2} - \dfrac{1}{(x+3)^2} = \dfrac{x+3-1}{(x+3)^2} = \dfrac{x+2}{(x+3)^2}$

37. $\dfrac{4}{5b} + \dfrac{1}{b-1} = \dfrac{4(b-1)}{5b(b-1)} + \dfrac{1(5b)}{(b-1)(5b)}$

$= \dfrac{4b-4}{5b(b-1)} + \dfrac{5b}{5b(b-1)} = \dfrac{4b-4+5b}{5b(b-1)}$

$= \dfrac{9b-4}{5b(b-1)}$

39. $\dfrac{2}{m} + 1 = \dfrac{2}{m} + \dfrac{1}{1} = \dfrac{2}{m} + \dfrac{1(m)}{1(m)} = \dfrac{2+m}{m}$

41. $\dfrac{6}{1-2x} - \dfrac{4}{2x-1} = \dfrac{6}{1-2x} - \dfrac{4}{-(1-2x)}$

$= \dfrac{6}{1-2x} - \dfrac{-4}{1-2x} = \dfrac{6-(-4)}{1-2x} = \dfrac{10}{1-2x}$

43. $\dfrac{7}{(x+1)(x-1)} + \dfrac{8}{(x+1)^2}$

$= \dfrac{7(x+1)}{(x+1)(x-1)(x+1)} + \dfrac{8(x-1)}{(x+1)^2(x-1)}$

$= \dfrac{7x+7}{(x+1)^2(x-1)} + \dfrac{8x-8}{(x+1)^2(x-1)}$

$= \dfrac{7x+7+8x-8}{(x+1)^2(x-1)} = \dfrac{15x-1}{(x+1)^2(x-1)}$

45. $\dfrac{x}{x^2-1} - \dfrac{2}{x^2-2x+1}$

$= \dfrac{x}{(x-1)(x+1)} - \dfrac{2}{(x-1)^2}$

$= \dfrac{x(x-1)}{(x-1)(x+1)(x-1)} - \dfrac{2(x+1)}{(x-1)^2(x+1)}$

$= \dfrac{x^2-x}{(x-1)^2(x+1)} - \dfrac{2x+2}{(x-1)^2(x+1)}$

$= \dfrac{x^2-x-(2x+2)}{(x-1)^2(x+1)} = \dfrac{x^2-x-2x-2}{(x-1)^2(x+1)}$

$= \dfrac{x^2-3x-2}{(x-1)^2(x+1)}$

47. $\dfrac{3a}{2a+6} - \dfrac{a-1}{a+3} = \dfrac{3a}{2(a+3)} - \dfrac{a-1}{a+3}$

$= \dfrac{3a}{2(a+3)} - \dfrac{(a-1)(2)}{(a+3)(2)}$

$= \dfrac{3a}{2(a+3)} - \dfrac{2a-2}{2(a+3)}$

$= \dfrac{3a-(2a-2)}{2(a+3)} = \dfrac{3a-2a+2}{2(a+3)} = \dfrac{a+2}{2(a+3)}$

49. $\dfrac{5}{2-x} + \dfrac{x}{2x-4} = \dfrac{5}{-(x-2)} + \dfrac{x}{2(x-2)}$

$= \dfrac{-5}{x-2} + \dfrac{x}{2(x-2)}$

$= \dfrac{-5(2)}{(x-2)(2)} + \dfrac{x}{2(x-2)}$

$= \dfrac{-10}{2(x-2)} + \dfrac{x}{2(x-2)} = \dfrac{x-10}{2(x-2)}$

51. $\dfrac{-7}{y^2-3y+2}-\dfrac{2}{y-1}=\dfrac{-7}{(y-1)(y-2)}-\dfrac{2}{y-1}$

$\qquad =\dfrac{-7}{(y-1)(y-2)}-\dfrac{2(y-2)}{(y-1)(y-2)}$

$\qquad =\dfrac{-7-(2y-4)}{(y-1)(y-2)}=\dfrac{-7-2y+4}{(y-1)(y-2)}$

$\qquad =\dfrac{-3-2y}{(y-2)(y-1)}$

53. $\dfrac{13}{x^2-5x+6}-\dfrac{5}{x-3}=\dfrac{13}{(x-3)(x-2)}-\dfrac{5}{x-3}$

$\qquad =\dfrac{13}{(x-3)(x-2)}-\dfrac{5(x-2)}{(x-3)(x-2)}$

$\qquad =\dfrac{13-(5x-10)}{(x-3)(x-2)}=\dfrac{213-5x+10}{(x-3)(x-2)}$

$\qquad =\dfrac{-5x+23}{(x-3)(x-2)}$

55. $\dfrac{8}{(x+2)(x-2)}+\dfrac{4}{(x+2)(x-3)}$

$\qquad \text{LCD}=(x+2)(x-2)(x-3)$

$\qquad \dfrac{8}{(x+2)(x-2)}\cdot\dfrac{(x-3)}{(x-3)}+\dfrac{4}{(x+2)(x-3)}\cdot\dfrac{(x-2)}{(x-2)}$

$\qquad \dfrac{8(x-3)}{(x+2)(x-2)(x-3)}+\dfrac{4(x-2)}{(x+2)(x-3)(x-2)}$

$\qquad =\dfrac{8x-24+4x-8}{(x+2)(x-2)(x-3)}$

$\qquad =\dfrac{12x-32}{(x+2)(x-2)(x-3)}$

57. $\dfrac{5}{9x^2-4}+\dfrac{2}{3x-2}=\dfrac{5}{(3x+2)(3x-2)}+\dfrac{2}{3x-2}$

$\qquad \text{LCD}=(3x+2)(3x-2)$

$\qquad =\dfrac{5}{(3x+2)(3x-2)}+\dfrac{2}{(3x-2)}\cdot\dfrac{(3x+2)}{(3x+2)}$

$\qquad =\dfrac{5}{(3x+2)(3x-2)}+\dfrac{2(3x+2)}{(3x+-2)(3x+2)}$

$\qquad =\dfrac{5+6x+4}{(3x+2)(3x-2)}$

$\qquad =\dfrac{6x+9}{(3x+2)(3x-2)}$

59. $\dfrac{x+8}{x^2-5x-6}+\dfrac{x+1}{x^2-4x-5}$

$\qquad =\dfrac{x+8}{(x-6)(x+1)}+\dfrac{x+1}{(x-5)(x+1)}$

$\qquad =\dfrac{(x+8)(x-5)}{(x-6)(x+1)(x-5)}+\dfrac{(x+1)(x-6)}{(x-5)(x+1)(x-6)}$

$\qquad =\dfrac{x^2+3x-40+x^2-5x-6}{(x-6)(x+1)(x-5)}$

$\qquad =\dfrac{2x^2-2x-46}{(x-6)(x+1)(x-5)}$

61. $\qquad 3x+5=7$

$\qquad 3x+5-5=7-5$

$\qquad\qquad\quad 3x=2$

$\qquad\qquad \dfrac{3x}{3}=\dfrac{2}{3}$

$\qquad\qquad\quad x=\dfrac{2}{3}$

63. $2x^2-x-1=0$

$\qquad (2x+1)(x-1)=0$

$\qquad 2x+1=0 \quad\text{or}\quad x-1=0$

$$2x = -1 \qquad\qquad x = 1$$

$$x = -\frac{1}{2}$$

The solutions are $x = -\dfrac{1}{2}$ and $x = 1$.

65. $\dfrac{2+x}{x+2} = \dfrac{x+2}{x+2} = 1$

67. $\dfrac{2-x}{x-2} = \dfrac{-x+2}{x-2} = \dfrac{-(x-2)}{x-2} = -1$

69. $P = 2\left(\dfrac{3}{y-5}\right) + 2\left(\dfrac{2}{y}\right)$

$= \dfrac{6(y)}{(y-5)(y)} + \dfrac{4(y-5)}{y(y-5)}$

$= \dfrac{6y + 4(y-5)}{y(y-5)} = \dfrac{6y + 4y - 20}{y(y-5)}$

$= \dfrac{10y - 20}{y(y-5)}$

The perimeter is $\dfrac{10y-20}{y(y-5)}$ ft

$A = \dfrac{3}{y-5} \cdot \dfrac{2}{y} = \dfrac{6}{y(y-5)}$

The area is $\dfrac{6}{y(y-5)}$ sq ft.

71. $\dfrac{15x}{x+8} \cdot \dfrac{2x+16}{3x} = \dfrac{15x}{x+8} \cdot \dfrac{2(x+8)}{3x}$

$= \dfrac{2 \cdot 5 \cdot 3x \cdot (x+8)}{3x \cdot (x+8)}$

$= 10$

73. $\dfrac{8x+7}{3x+5} - \dfrac{2x-3}{3x+5} = \dfrac{8x + 7(2x-3)}{3x+5}$

$= \dfrac{8x + 7 - 2x + 3}{3x+5}$

$= \dfrac{6x + 10}{3x+5}$

$= \dfrac{2(3x+5)}{3x+5}$

$= 2$

75. $\dfrac{5a+10}{18} \div \dfrac{a^2 - 4}{10a} = \dfrac{5(a+2)}{2 \cdot 9} \cdot \dfrac{2 \cdot 5a}{(a-2)(a+2)}$

$= \dfrac{25a}{9(a-2)}$

77. $\dfrac{5}{x^2 - 3x + 2} + \dfrac{1}{x-2} = \dfrac{5}{(x-2)(x-1)} + \dfrac{1}{x-2}$

LCD $= (x-2)(x-1)$

$\dfrac{5}{(x-2)(x-1)} + \dfrac{1}{(x-2)} \cdot \dfrac{(x-1)}{(x-1)}$

$= \dfrac{5}{(x-2)(x-1)} + \dfrac{x-1}{(x-2)(x-1)}$

$= \dfrac{5 + x - 1}{(x-2)(x-1)}$

$= \dfrac{x+4}{(x-2)(x-1)}$

79. Answers may vary.

81. $\dfrac{5}{x^2-4}+\dfrac{2}{x^2-4x+4}-\dfrac{3}{x^2-x-6}$

$=\dfrac{5}{(x-2)(x+2)}+\dfrac{2}{(x-2)^2}$

$\qquad -\dfrac{3}{(x-3)(x+2)}$

$=\dfrac{5(x-2)(x-3)}{(x-2)(x+2)(x-2)(x-3)}$

$\qquad +\dfrac{2(x+2)(x-3)}{(x-2)^2(x+2)(x-3)}$

$\qquad -\dfrac{3(x-2)^2}{(x-3)(x+2)(x-2)^2}$

$=\dfrac{5(x^2-5x+6)}{(x-2)^2(x+2)(x-3)}$

$\qquad +\dfrac{2(x^2-x-6)}{(x-2)^2(x+2)(x-3)}$

$\qquad -\dfrac{3(x^2-4x+4)}{(x-2)^2(x+2)(x-3)}$

$=\dfrac{5x^2-25x+30}{(x-2)^2(x+2)(x-3)}$

$\qquad +\dfrac{2x^2-2x-12}{(x-2)^2(x+2)(x-3)}$

$\qquad -\dfrac{3x^2-12x+12}{(x-2)^2(x+2)(x-3)}$

$=\dfrac{4x^2-15x+6}{(x-2)^2(x+2)(x-3)}$

83. $\dfrac{5+x}{x^3-27}+\dfrac{x}{x^3+3x^2+9x}$

$=\dfrac{5+x}{(x-3)(x^2+3x+9)}+\dfrac{x}{x(x^2+3x+9)}$

$\text{LCD}=x(x-3)(x^2+3x+9)$

$\dfrac{(5+x)}{(x-3)(x^2+3x+9)}\cdot\dfrac{x}{x}+\dfrac{x}{x(x^2+3x+9)}\cdot\dfrac{(x-3)}{(x-3)}$

$=\dfrac{(5+x)(x)}{x(x-3)(x^2+3x+9)}+\dfrac{x(x-3)}{x(x-3)(x^2+3x+9)}$

$=\dfrac{5x+x^2+x^2-3x}{x(x-3)(x^2+3x+9)}$

$=\dfrac{2x^2+2x}{x(x-3)(x^2+3x+9)}$

$=\dfrac{2x(x+1)}{x(x-3)(x^2+3x+9)}$

$=\dfrac{2(x+1)}{(x-3)(x^2+3x+9)}$

85. $\dfrac{DA}{A+12}-\dfrac{D(A+1)}{24}$

$=\dfrac{24DA}{24(A+12)}-\dfrac{D(A+1)(A+12)}{24(A+12)}$

$=\dfrac{24DA-DA^2-13DA-12D}{24(A+12)}$

$=\dfrac{11DA-DA^2-12D}{24(A+12)}$

Section 7.5

Graphing Calculator Explorations

1. $y_1 = \dfrac{x-4}{2} - \dfrac{x-3}{9}$, $y_2 = \dfrac{5}{18}$.

Use INTERSECT

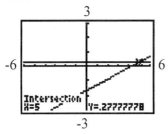

The solution of the equation is 5

3. $y_1 = 3 - \dfrac{6}{x}$, $y_2 = x + 8$.

Use INTERSECT

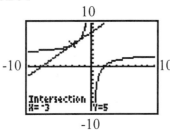

One solution is -3

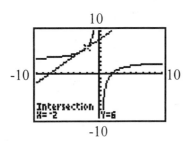

The other solution is -2.

5. $y_1 = x + \dfrac{14}{x-2}$, $y_2 = \dfrac{7x}{x-2} + 1$.

Use INTERSECT

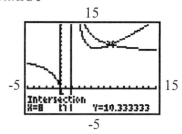

The solution is 8.

Mental Math

1. $\dfrac{x}{5} = 2$

$x = 10$

2. $\dfrac{x}{8} = 4$

$x = 32$

3. $\dfrac{z}{6} = 6$

$z = 36$

4. $\dfrac{y}{7} = 8$

$y = 56$

Exercise Set 7.5

1. $\dfrac{x}{5} + 3 = 9$

$5\left(\dfrac{x}{5} + 3\right) = 5(9)$

$5\left(\dfrac{x}{5}\right) + 5(3) = 5(9)$

$x + 15 = 45$

$x = 30$

Check:

$\dfrac{x}{5} + 3 = 9$

$\dfrac{30}{5} + 3 \overset{?}{=} 9$

$6 + 3 \overset{?}{=} 9$

$9 = 9$ True

The solution is 30.

3. $\dfrac{x}{2} + \dfrac{5x}{4} = \dfrac{x}{12}$

$12\left(\dfrac{x}{2} + \dfrac{5x}{4}\right) = 12\left(\dfrac{x}{12}\right)$

$12\left(\dfrac{x}{2}\right) + 12\left(\dfrac{5x}{4}\right) = 12\left(\dfrac{x}{12}\right)$

$6x + 15x = x$

$21x = x$

$20x = 0$

$x = 0$

Check:

$\dfrac{x}{2} + \dfrac{5x}{4} = \dfrac{x}{12}$

$\dfrac{0}{2} + \dfrac{5 \cdot 0}{4} \overset{?}{=} \dfrac{0}{12}$

$0 + \dfrac{0}{4} \overset{?}{=} 0$

$0 = 0$ True

The solution is 0.

5. $\dfrac{y}{7} - \dfrac{3y}{2} = 1$

$14\left(\dfrac{y}{7} - \dfrac{3y}{2}\right) = 14(1)$

$14\left(\dfrac{y}{7}\right) - 14\left(\dfrac{3y}{2}\right) = 14$

$2y - 21y = 14$

$-19y = 14$

$y = -\dfrac{14}{19}$

Check:

$\dfrac{y}{7} - \dfrac{3y}{2} = 1$

$\dfrac{-(14/19)}{7} - \dfrac{3(-14/19)}{2} \overset{?}{=} 1$

$-\dfrac{2}{19} + \dfrac{21}{19} \overset{?}{=} 1$

$\dfrac{19}{19} \overset{?}{=} 1$

$1 = 1$ True

The solution is $-\dfrac{14}{19}$.

7. $2 + \dfrac{10}{x} = x + 5$

$$x\left(2 + \dfrac{10}{x}\right) = x(x+5)$$

$$x(2) + x\left(\dfrac{10}{x}\right) = x(x+5)$$

$$2x + 10 = x^2 + 5x$$

$$0 = x^2 + 3x - 10$$

$$0 = (x+5)(x-2)$$

$$x + 5 = 0 \quad \text{or} \quad x - 2 = 0$$

$$x = -5 \qquad\qquad x = 2$$

Check:

$x = -5 :$

$$2 + \dfrac{10}{x} = x + 5$$

$$2 + \dfrac{10}{-5} \overset{?}{=} -5 + 5$$

$$2 + (-2) \overset{?}{=} -5 + 5$$

$$0 = 0 \quad \text{True}$$

$x = 2 :$

$$2 + \dfrac{10}{x} = x + 5$$

$$2 + \dfrac{10}{2} \overset{?}{=} 2 + 5$$

$$2 + 5 \overset{?}{=} 2 + 5$$

$$7 = 7 \quad \text{True}$$

Both -5 and 2 are solutions.

9. $\dfrac{a}{5} = \dfrac{a-3}{2}$

$$10\left(\dfrac{a}{5}\right) = 10\left(\dfrac{a-3}{2}\right)$$

$$2a = 5(a-3)$$

$$2a = 5a - 15$$

$$-3a = -15$$

$$a = 5$$

Check:

$$\dfrac{a}{5} = \dfrac{a-3}{2}$$

$$\dfrac{5}{5} \overset{?}{=} \dfrac{5-3}{2}$$

$$\dfrac{5}{5} \overset{?}{=} \dfrac{2}{2}$$

$$1 = 1 \quad \text{True}$$

The solution is 5.

11. $\dfrac{x-3}{5} + \dfrac{x-2}{2} = \dfrac{1}{2}$

$$10\left(\dfrac{x-3}{5} + \dfrac{x-2}{2}\right) = 10\left(\dfrac{1}{2}\right)$$

$$10\left(\dfrac{x-3}{5}\right) + 10\left(\dfrac{x-2}{2}\right) = 10\left(\dfrac{1}{2}\right)$$

$$2(x-3) + 5(x-2) = 5$$

$$2x - 6 + 5x - 10 = 5$$

$$7x - 16 = 5$$

$$7x = 21$$

$$x = 3$$

Check:

$$\dfrac{x-3}{5} + \dfrac{x-2}{2} = \dfrac{1}{2}$$

$$\dfrac{3-3}{5} + \dfrac{3-2}{2} \overset{?}{=} \dfrac{1}{2}$$

$$\dfrac{0}{5} + \dfrac{1}{2} \overset{?}{=} \dfrac{1}{2}$$

$$0 + \dfrac{1}{2} \overset{?}{=} \dfrac{1}{2}$$

$$\dfrac{1}{2} = \dfrac{1}{2} \quad \text{True}$$

The solution is 3.

13. $\dfrac{x+1}{3} - \dfrac{x-2}{4} = \dfrac{5}{6}$

$$12\left(\dfrac{x+1}{3} - \dfrac{x-2}{4}\right) = 12\left(\dfrac{5}{6}\right)$$

$$12\left(\dfrac{x+1}{3}\right) - 12\left(\dfrac{x-2}{4}\right) = 10$$

$$4(x+1) - 3(x-2) = 10$$

$$4x + 4 - 3x + 6 = 10$$

$$x + 10 = 10$$

$$x = 0$$

Check:

$$\dfrac{x+1}{3} - \dfrac{x-2}{4} = \dfrac{5}{6}$$

$$\dfrac{0+1}{3} - \dfrac{0-2}{4} \overset{?}{=} \dfrac{5}{6}$$

$$\dfrac{1}{3} + \dfrac{1}{2} \overset{?}{=} \dfrac{5}{6}$$

$$\dfrac{5}{6} = \dfrac{5}{6} \quad \text{True}$$

The solution is 0.

15. $\dfrac{9}{2a-5} = -2$

$$(2a-5)\left(\dfrac{9}{2a-5}\right) = (2a-5)(-2)$$

$$9 = -4a + 10$$

$$9 - 10 = -4a$$

$$-1 = -4a$$

$$\dfrac{1}{4} = a$$

17. $\dfrac{y}{y+4} + \dfrac{4}{y+4} = 3$

$$(y+4)\left(\dfrac{y}{y+4} + \dfrac{4}{y+4}\right) = (y+4)(3)$$

$$(y+4)\left(\dfrac{y}{y+4}\right) + (y+4)\left(\dfrac{4}{y+4}\right) = (y+4)(3)$$

$$y + 4 = 3y + 12$$

$$y - 3y = 12 - 4$$

$$-2y = 8$$

$$y = \dfrac{8}{-2} = -4$$

−4 makes a denominator zero. There is no solution.

19. $\dfrac{4y}{y-3} - 3 = \dfrac{3y-1}{y+3}$

$(y-3)(y+3)\left(\dfrac{4y}{y-3}\right)$

$\qquad = (y-3)(y+3)\left(\dfrac{3y-1}{y+3}\right)$

$(y-3)(y+3)\left(\dfrac{4y}{y-3}\right) - (y-3)(y+3)(3)$

$\qquad = (y-3)(y+3)\left(\dfrac{3y-1}{y+3}\right)$

$4y(y+3) - 3(y^2-9) = (y-3)(3y-1)$

$4y^2 + 12y - 3y^2 + 27 = 3y^2 - 10y + 3$

$y^2 + 12y + 27 = 3y^2 - 10y + 3$

$0 = 2y^2 - 22y - 24$

$0 = (2y+2)(y-12)$

$2y+2 = 0 \quad$ or $\quad y-12 = 0$

$\qquad 2y = -2$

$\qquad\quad y = -1 \qquad\qquad\quad y = 12$

The solutions are -1 and 12.

21. $\dfrac{4y}{y-4} + 5 = \dfrac{5y}{y-4}$

$(y-4)\left(\dfrac{4y}{y-4} + 5\right) = (y-4)\left(\dfrac{5y}{y-4}\right)$

$(y-4)\left(\dfrac{4y}{y-4}\right) + (y-4)(5) = (y-4)\left(\dfrac{5y}{y-4}\right)$

$4y + 5y - 20 = 5y$

$9y - 20 = 5y$

$4y - 20 = 0$

$4y = 20$

$y = 5$

The solution is 5.

23. $\dfrac{7}{x-2} + 1 = \dfrac{x}{x+2}$

$(x-2)(x+2)\left(\dfrac{7}{x-2} + 1\right)$

$\qquad = (x-2)(x+2)\left(\dfrac{x}{x+2}\right)$

$(x-2)(x+2)\left(\dfrac{7}{x-2}\right) + (x-2)(x+2)$

$\qquad = (x-2)(x+2)\left(\dfrac{x}{x+2}\right)$

$7(x+2) + (x-2)(x+2) = x(x-2)$

$7x + 14 + x^2 - 4 = x^2 - 2x$

$7x + x^2 + 10 = x^2 - 2x$

$9x = -10$

$x = -\dfrac{10}{9}$

The solution is $-\dfrac{10}{9}$.

25. $\dfrac{x+1}{x+3} = \dfrac{2x^2 - 15x}{x^2 + x - 6} - \dfrac{x-3}{x-2}$

$\dfrac{x+1}{x+3} = \dfrac{2x^2 - 15x}{(x+3)(x-2)} - \dfrac{x-3}{x-2}$

$(x+3)(x-2)\left(\dfrac{x+1}{x+3}\right)$

$\qquad = (x+3)(x-2)\left(\dfrac{2x^2 - 15x}{(x+3)(x-2)} - \dfrac{x-3}{x-2}\right)$

$(x+3)(x-2)\left(\dfrac{x+1}{x+3}\right)$

$\qquad = (x+3)(x-2)\left(\dfrac{2x^2 - 15x}{(x+3)(x-2)}\right)$

$\qquad\quad - (x+3)(x-2)\left(\dfrac{x-3}{x-2}\right)$

$$(x-2)(x+1) = 2x^2 - 15x - (x+3)(x-3)$$
$$x^2 - x - 2 = 2x^2 - 15x - (x^2 - 9)$$
$$x^2 - x - 2 = 2x^2 - 15x - x^2 + 9$$
$$x^2 - x - 2 = x^2 - 15x + 9$$
$$14x = 11$$
$$x = \frac{11}{14}$$

The solution is $\frac{11}{14}$.

27. $\dfrac{y}{2y+2} + \dfrac{2y-16}{4y+4} = \dfrac{2y-3}{y+1}$

$$\frac{y}{2(y+1)} + \frac{2y-16}{4(y+1)} = \frac{2y-3}{y+1}$$

$$4(y+1)\left(\frac{y}{2(y+1)} + \frac{2y-16}{4(y+1)}\right)$$
$$= 4(y+1)\left(\frac{2y-3}{y+1}\right)$$

$$4(y+1)\left(\frac{y}{2(y+1)}\right) + 4(y+1)\left(\frac{2y-16}{4(y+1)}\right)$$
$$= 4(y+1)\left(\frac{2y-3}{y+1}\right)$$

$$2y + 2y - 16 = 4(2y-3)$$
$$4y - 16 = 8y - 12$$
$$-4y = 4$$
$$y = -1$$

In the original equation, -1 makes a denominator 0.

This equation has no solution.

29. $\dfrac{2x}{7} - 5x = 9$

$$7\left(\frac{2x}{7} - 5x\right) = 7(9)$$
$$7\left(\frac{2x}{7}\right) + 7(-5x) = 7(9)$$
$$2x - 35x = 63$$
$$-33x = 63$$
$$x = \frac{63}{33} \quad x = -\frac{21}{11}$$

31. $\dfrac{2}{y} + \dfrac{1}{2} = \dfrac{5}{2y}$

$$2y\left(\frac{2}{y} + \frac{1}{2}\right) = 2y\left(\frac{5}{2y}\right)$$
$$2y\left(\frac{2}{y}\right) + 2y\left(\frac{1}{2}\right) = 2y\left(\frac{5}{2y}\right)$$
$$4 + y = 5$$
$$y = 1$$

33. $\dfrac{4x+10}{7} = \dfrac{8}{2}$

$$14\left(\frac{4x+10}{7}\right) = 14\left(\frac{8}{2}\right)$$
$$2(4x+10) = 7(8)$$
$$8x = 56 - 20$$
$$8x = 36$$
$$x = \frac{36}{8} = \frac{9}{2}$$

35.

$$2 + \frac{3}{a-3} = \frac{a}{a-3}$$

$$(a-3)\left(2 + \frac{3}{a-3}\right) = (a-3)\left(\frac{a}{a-3}\right)$$

$$(a-3)(2) + (a-3)\left(\frac{3}{a-3}\right) = a$$

$$2a - 6 + 3 = a$$

$$2a - 3 = a$$

$$-3 = a - 2a$$

$$-3 = -a$$

$$\frac{-3}{-1} = a$$

$$3 = a$$

3 is an extraneous solution. If $a = 3$, the denominator would equal zero. The equation has no solution.

37.

$$\frac{5}{x} + \frac{2}{3} = \frac{7}{2x}$$

$$6x\left(\frac{5}{x} + \frac{2}{3}\right) = 6x\left(\frac{7}{2x}\right)$$

$$6x\left(\frac{5}{x}\right) + 6x\left(\frac{2}{3}\right) = 21$$

$$30 + 4x = 21$$

$$4x = 21 - 30$$

$$4x = -9$$

$$x = -\frac{9}{4}$$

39.

$$\frac{2a}{a+4} = \frac{3}{a-1}$$

$$(a+4)(a-1)\left(\frac{2a}{a+4}\right) = (a+4)(a-1)\left(\frac{3}{a-1}\right)$$

$$(a-1)(2a) = (a+4)(3)$$

$$2a^2 - 2a = 3a + 12$$

$$2a^2 - 2a - 3a - 12 = 0$$

$$2a^2 - 5a - 12 = 0$$

$$(2a+3)(a-4) = 0$$

$$2a + 3 = 0 \quad \text{or} \quad a - 4 = 0$$

$$2a = -3 \qquad\qquad a = 4$$

$$a = -\frac{3}{2}$$

41.

$$\frac{x+1}{3} - \frac{x-1}{6} = \frac{1}{6}$$

$$6\left(\frac{x+1}{3} - \frac{x-1}{6}\right) = 6\left(\frac{1}{6}\right)$$

$$6\left(\frac{x+1}{3}\right) - 6\left(\frac{x-1}{6}\right) = 6\left(\frac{1}{6}\right)$$

$$2(x+1) - (x-1) = 1$$

$$2x + 2 - x + 1 = 1$$

$$x + 3 = 1$$

$$x = -2$$

The solution is -2.

43.

$$\frac{4r-1}{r^2+5r-14} + \frac{2}{r+7} = \frac{1}{r-2}$$

$$\frac{4r-1}{(r+7)(r-2)} + \frac{2}{r+7} = \frac{1}{r-2}$$

$$(r+7)(r-2)\left(\frac{4r-1}{(r+7)(r-2)} + \frac{2}{r+7}\right)$$

$$= (r+7)(r-2)\left(\frac{1}{r-2}\right)$$

$$(r+7)(r-2)\left(\frac{4r-1}{(r+7)(r-2)}\right)$$
$$+(r+7)(r-2)\left(\frac{2}{r+7}\right)$$
$$=(r+7)(r-2)\left(\frac{1}{r-2}\right)$$
$$4r-1+2(r-2)=(r+7)(1)$$
$$4r-1+2r-4=r+7$$
$$6r-5=r+7$$
$$5r=12$$
$$r=\frac{12}{5}$$

$$12(x+3)\left(\frac{x}{2(x+3)}\right)$$
$$+12(x+3)\left(\frac{x+1}{3(x+3)}\right)=3(2)$$
$$6x+4(x+1)=6$$
$$6x+4x+4=6$$
$$10x+4=6$$
$$10x=6-4$$
$$10x=2$$
$$x=\frac{2}{10}=\frac{1}{5}$$

45.
$$\frac{t}{t-4}=\frac{t+4}{6}$$
$$6(t-4)\left(\frac{t}{t-4}\right)=6(t-4)\left(\frac{t+4}{6}\right)$$
$$6t=(t-4)(t+4)$$
$$6t=t^2-16$$
$$0=t^2-16$$
$$0=(t-8)(t+2)$$
$$t+2=0 \quad\text{or}\quad t-8=0$$
$$t=-2 \qquad\quad t=8$$

47.
$$\frac{x}{2x+6}+\frac{x+1}{3x+9}=\frac{2}{4x+12}$$
$$\frac{x}{2(x+3)}+\frac{x+1}{3(x+3)}=\frac{2}{4(x+3)}$$
$$12(x+3)\left(\frac{x}{2(x+3)}+\frac{x+1}{3(x+3)}\right)$$
$$=12(x+3)\left(\frac{2}{4(x+3)}\right)$$

49.
$$\frac{D}{R}=T$$
$$R\left(\frac{D}{R}\right)=R(T)$$
$$D=RT$$
$$\frac{D}{T}=R$$

51.
$$\frac{3}{x}=\frac{5y}{x+2}$$
$$x(x+2)\left(\frac{3}{x}\right)=x(x+2)\left(\frac{5y}{x+2}\right)$$
$$(x+2)(3)=x(5y)$$
$$3x+6=5xy$$
$$\frac{3x+6}{5x}=y$$

53.
$$\frac{3a+2}{3b-2} = -\frac{4}{2a}$$

$$2a(3b-2)\left(\frac{3a+2}{3b-2}\right) = 2a(3b-2)\left(-\frac{4}{2a}\right)$$

$$2a(3a+2) = (3b-2)(-4)$$

$$6a^2 + 4a = -12b + 8$$

$$6a^2 + 4a - 8 = -12b$$

$$\frac{6a^2 + 4 - 8}{2 \cdot 6} = b$$

$$-\frac{2(3a^2 + 2a - 4)}{2 \cdot 6} = b$$

$$-\frac{3a^2 + 2a - 4}{6} = b$$

55.
$$\frac{A}{BH} = \frac{1}{2}$$

$$2BH\left(\frac{A}{BH}\right) = 2BH\left(\frac{1}{2}\right)$$

$$2A = BH$$

$$\frac{2A}{H} = B$$

57.
$$\frac{C}{\pi r} = 2$$

$$\pi r\left(\frac{C}{\pi r}\right) = \pi r(2)$$

$$C = 2\pi r$$

$$\frac{C}{2\pi} = \frac{2\pi r}{2\pi}$$

$$\frac{C}{2\pi} = r$$

59.
$$\frac{1}{a} = \frac{1}{b} + \frac{1}{c}$$

$$abc\left(\frac{1}{a}\right) = abc\left(\frac{1}{b} + \frac{1}{c}\right)$$

$$bc = abc\left(\frac{1}{b}\right) + abc\left(\frac{1}{c}\right)$$

$$bc = ac + ab$$

$$bc = a(c+b)$$

$$\frac{bc}{c+b} = a$$

61.
$$\frac{m^2}{6} - \frac{n}{3} = \frac{p}{2}$$

$$6\left(\frac{m^2}{6} - \frac{n}{3}\right) = 6\left(\frac{p}{2}\right)$$

$$6\left(\frac{m^2}{6}\right) - 6\left(\frac{n}{3}\right) = 3p$$

$$m^2 - 2n = 3p$$

$$m^2 - 3p = 2n$$

$$\frac{m^2 - 3p}{2} = n$$

63. The graph crosses the x-axis at $x = 2$. It crosses the y-axis at $y = -2$. The x-intercept is $(2, 0)$ and the y-intercept is $(0, -2)$.

65. The graph crosses the x-axis at $x = -4$, $x = -2$ and $x = 3$. It crosses the y-axis at $y = 4$. The x-intercept are $(-4, 0)$, $(-2, 0)$ and $(3, 0)$, and the y-intercept is $(0, 4)$.

67. Answers may vary.

69. expression

$$\frac{1}{x} + \frac{5}{9} = \frac{1(9)}{x(9)} + \frac{5x}{9x} = \frac{5x+9}{9x}$$

71. equation

$$\frac{5}{x-1} - \frac{2}{x} = \frac{5}{x(x-1)}$$

$$x(x-1)\left(\frac{5}{x-1}\right) - x(x-1)\left(\frac{2}{x}\right)$$

$$= x(x-1)\left(\frac{5}{x(x-1)}\right)$$

$$5x - 2(x-1) = 5$$

$$5x - 2x + 2 = 5$$

$$3x = 3$$

$$x = 1$$

1 makes a denominator zero.

There is no solution

73. $\dfrac{20x}{3} + \dfrac{32x}{6} = 180$

$$6\left(\frac{180}{3} + \frac{32x}{6}\right) = 6(180)$$

$$6\left(\frac{20x}{3}\right) + 6\left(\frac{32x}{6}\right) = 6(180)$$

$$40x + 32x = 1080$$

$$72x = 1080$$

$$\frac{72x}{72} = \frac{1080}{72}$$

$$x = 15$$

$$\frac{20x}{3} = \frac{20(15)}{3} = 100$$

$$\frac{32x}{6} = \frac{32(15)}{6} = 80$$

The angles are $100°$ and $80°$.

75. $\dfrac{150}{x} + \dfrac{450}{x} = 90$

$$x\left(\frac{150}{x} + \frac{450}{x}\right) = x(90)$$

$$x\left(\frac{150}{x}\right) + x\left(\frac{450}{x}\right) = x(90)$$

$$150 + 450 = 90x$$

$$600 = 90x$$

$$\frac{600}{90} = x$$

$$\frac{20}{3} = x$$

$$\frac{150}{x} = \frac{150}{\frac{20}{3}} = 150\left(\frac{3}{20}\right) = \frac{45}{2} = 22.5$$

$$\frac{450}{x} = \frac{450}{\frac{20}{3}} = 450\left(\frac{3}{20}\right) = \frac{135}{2} = 67.5$$

The angles are $22.5°$ and $67.5°$.

77. $\dfrac{5}{a^2+4a+3}+\dfrac{2}{a^2+a-6}-\dfrac{3}{a^2-a-2}=0$

$$\dfrac{5}{(a+3)(a+1)}+\dfrac{2}{(a+3)(a-2)}$$
$$-\dfrac{3}{(a-2)(a+1)}=0$$

$$(a+3)(a+1)(a-2)\left(\dfrac{\dfrac{5}{(a+3)(a+1)}}{\dfrac{}{}}+\dfrac{2}{(a+3)(a-2)}-\dfrac{3}{(a-2)(a+1)}\right)$$
$$=(a+3)(a+1)(a-2)(0)$$

$$(a+3)(a+1)(a-2)\left(\dfrac{5}{(a+3)(a+1)}\right)$$
$$+(a+3)(a+1)(a-2)\left(\dfrac{2}{(a+3)(a-2)}\right)$$
$$-(a+3)(a+1)(a-2)\left(\dfrac{3}{(a-2)(a+1)}\right)$$
$$=0$$
$$5(a-2)+2(a+1)-3(a+3)=0$$
$$5a-10+2a+2-3a-9=0$$
$$4a-17=0$$
$$4a=17$$
$$a=\dfrac{17}{4}$$

Integrated Review

1. expression

$$\dfrac{1}{x}+\dfrac{2}{3}=\dfrac{1(3)}{x(3)}+\dfrac{2(x)}{3(x)}=\dfrac{3}{3x}+\dfrac{2x}{3x}=\dfrac{3+2x}{3x}$$

2. expression

$$\dfrac{3}{a}+\dfrac{5}{6}=\dfrac{3(6)}{a(6)}+\dfrac{5(a)}{6(a)}=\dfrac{18}{6a}+\dfrac{5a}{6a}=\dfrac{18+5a}{6a}$$

3. equation

$$\dfrac{1}{x}+\dfrac{2}{3}=\dfrac{3}{x}$$
$$3x\left(\dfrac{1}{x}+\dfrac{2}{3}\right)=3x\left(\dfrac{3}{x}\right)$$
$$3x\left(\dfrac{1}{x}\right)+3x\left(\dfrac{2}{3}\right)=3x\left(\dfrac{3}{x}\right)$$
$$3+2x=9$$
$$2x=6$$
$$x=3$$

The solution is 3.

4. equation

$$\dfrac{3}{a}+\dfrac{5}{6}=1$$
$$6a\left(\dfrac{3}{a}+\dfrac{5}{6}\right)=6a(1)$$
$$6a\left(\dfrac{3}{a}\right)+6a\left(\dfrac{5}{6}\right)=6a$$
$$18+5a=6a$$
$$18=a$$

The solution is 18.

5. expression

$$\frac{2}{x-1} - \frac{1}{x} = \frac{2(x)}{(x-1)(x)} - \frac{1(x-1)}{x(x-1)}$$

$$= \frac{2x - (x-1)}{x(x-1)} = \frac{x+1}{x(x-1)}$$

6. expression

$$\frac{4}{x-3} - \frac{1}{x} = \frac{4(x)}{(x-3)(x)} - \frac{1(x-3)}{x(x-3)}$$

$$= \frac{4x - (x-3)}{x(x-3)} = \frac{4x - x + 3}{x(x-3)} = \frac{3x+3}{x(x-3)}$$

$$= \frac{3(x+1)}{x(x-3)}$$

7. equation

$$\frac{2}{x+1} - \frac{1}{x} = 1$$

$$x(x+1)\left(\frac{2}{x+1} - \frac{1}{x}\right) = x(x+1)(1)$$

$$x(x+1)\left(\frac{2}{x+1}\right) - x(x+1)\left(\frac{1}{x}\right) = x(x+1)$$

$$2x - (x+1) = x(x+1)$$

$$2x - x - 1 = x^2 + x$$

$$x - 1 = x^2 + x$$

$$-1 = x^2$$

There is no real number solution.

8. equation

$$\frac{4}{x-3} - \frac{1}{x} = \frac{6}{x(x-3)}$$

$$x(x-3)\left(\frac{4}{x-3} - \frac{1}{x}\right) = x(x-3)\left(\frac{6}{x(x-3)}\right)$$

$$x(x-3)\left(\frac{4}{x-3}\right) - x(x-3)\left(\frac{1}{x}\right) = 6$$

$$4x - (x-3) = 6$$

$$4x - x + 3 = 6$$

$$3x + 3 = 6$$

$$3x = 3$$

$$x = 1$$

The solution is 1.

9. expression

$$\frac{15x}{x+8} \cdot \frac{2x+16}{3x} = \frac{15x \cdot (2x+16)}{(x+8) \cdot 3x}$$

$$= \frac{3 \cdot 5 \cdot x \cdot 2 \cdot (x+8)}{(x+8) \cdot 3 \cdot x} = 5 \cdot 2 = 10$$

10. expression

$$\frac{9z+5}{15} \cdot \frac{5z}{81z^2 - 25} = \frac{(9z+5) \cdot 5z}{15 \cdot (81z^2 - 25)}$$

$$= \frac{(9z+5) \cdot 5 \cdot z}{5 \cdot 3 \cdot (9z+5)(9z-5)} = \frac{z}{3(9z-5)}$$

11. expression

$$\frac{2x+1}{x-3} + \frac{3x+6}{x-3} = \frac{2x+1+3x+6}{x-3}$$

$$= \frac{5x+7}{x-3}$$

12. expression

$$\frac{4p-3}{2p+7}+\frac{3p+8}{2p+7}=\frac{4p-3+3p+8}{2p+7}$$

$$=\frac{7p+5}{2p+7}$$

13. equation

$$\frac{x+5}{7}=\frac{8}{2}$$

$$14\left(\frac{x+5}{7}\right)=14\left(\frac{8}{2}\right)$$

$$2(x+5)=56$$

$$2x+10=56$$

$$2x=46$$

$$x=23$$

The solution is 23.

14. equation

$$\frac{1}{2}=\frac{x-1}{8}$$

$$8\left(\frac{1}{2}\right)=8\left(\frac{x-1}{8}\right)$$

$$4=x-1$$

$$5=x$$

The solution is 5.

15. expression

$$\frac{5a+10}{18}\div\frac{a^2-4}{10a}=\frac{5a+10}{18}\cdot\frac{10a}{a^2-4}$$

$$=\frac{5(a+2)\cdot2\cdot5\cdot a}{2\cdot9(a+2)(a-2)}$$

$$=\frac{5\cdot5\cdot a}{9(a-2)}=\frac{25a}{9(a-2)}$$

16. expression

$$\frac{9}{x^2-1}+\frac{12}{3x+3}$$

$$=\frac{9(3)}{(x+1)(x-1)(3)}+\frac{12(x-1)}{3(x+1)(x-1)}$$

$$=\frac{27+12x-12}{3(x-1)(x+1)}$$

$$=\frac{15+12x}{3(x+1)(x-1)}=\frac{3(5+4x)}{3(x+1)(x-1)}$$

$$=\frac{4x+5}{(x+1)(x-1)}$$

17. expression

$$\frac{x+2}{3x-1}+\frac{5}{(3x-1)^2}$$

$$=\frac{(x+2)(3x-1)}{(3x-1)(3x-1)}+\frac{5}{(3x-1)^2}$$

$$=\frac{3x^2+5x-2+5}{(3x-1)^2}$$

$$=\frac{3x^2+5x+3}{(3x-1)^2}$$

18. expression

$$\frac{4}{(2x-5)^2}+\frac{x+1}{2x-5}$$

$$=\frac{4}{(2x-5)^2}+\frac{(x+1)(2x-5)}{(2x-5)(2x-5)}$$

$$=\frac{4+2x^2-3x-5}{(2x-5)^2}$$

$$=\frac{2x^2-3x-1}{(2x-5)^2}$$

19. expression

$$\frac{x-7}{x} - \frac{x+2}{5x} = \frac{(x-7)(5)}{x(5)} - \frac{x+2}{5x}$$

$$= \frac{5x-25-x-2}{5x}$$

$$= \frac{4x-37}{5x}$$

20. equation

$$\frac{9}{x^2-4} + \frac{2}{x+2} = \frac{-1}{x-2}$$

$$\left(x^2-4\right)\left(\frac{9}{x^2-4}\right) + \left(x^2-4\right)\left(\frac{2}{x+2}\right)$$

$$= \left(x^2-4\right)\left(\frac{-1}{x-2}\right)$$

$$9 + (x-2)(2) = (x+2)(-1)$$

$$9 + 2x - 4 = -x - 2$$

$$2x + 5 = -x - 2$$

$$3x + 5 = -2$$

$$3x = -7$$

$$x = -\frac{7}{3}$$

The solution is $-\frac{7}{3}$.

21. equation

$$\frac{3}{x+3} = \frac{5}{x^2-9} - \frac{2}{x-3}$$

$$\left(x^2-9\right)\left(\frac{3}{x+3}\right)$$

$$= \left(x^2-9\right)\left(\frac{5}{x^2-9}\right) - \left(x^2-9\right)\left(\frac{2}{x-3}\right)$$

$$(x-3)(3) = 5 - (x+3)(2)$$

$$3x - 9 = 5 - 2x - 6$$

$$3x - 9 = -2x - 1$$

$$5x - 9 = -1$$

$$5x = 8$$

$$x = \frac{8}{5}$$

The solution is $\frac{8}{5}$.

22. expression

$$\frac{10x-9}{x} - \frac{x-4}{3x} = \frac{(10x-9)(3)}{x(3)} - \frac{x-4}{3x}$$

$$= \frac{30x-27-x+4}{3x}$$

$$= \frac{29x-23}{3x}$$

Exercise Set 7.6

1. $\dfrac{2}{3} = \dfrac{x}{6}$

$12 = 3x$

$4 = x$

3. $\dfrac{x}{10} = \dfrac{5}{9}$

$9x = 50$

$x = \dfrac{50}{9}$

5. $\dfrac{x+1}{2x+3} = \dfrac{2}{3}$

$3(x+1) = 2(2x+3)$

$3x+3 = 4x+6$

$3 = x+6$

$-3 = x$

7. $\dfrac{9}{5} = \dfrac{12}{3x+2}$

$9(3x+2) = 5(12)$

$27x+18 = 60$

$27x = 42$

$x = \dfrac{42}{27} = \dfrac{14}{9}$

9. a

11. Let x = the elephant's weight on Pluto.

$\dfrac{100}{3} = \dfrac{4100}{x}$

$100x = 3(4100)$

$100x = 12,300$

$x = 123$

The elephant's weight is 123 pounds.

13. Let x = the number of calories in 42.6 grams.

$\dfrac{110}{28.4} = \dfrac{x}{42.6}$

$110(42.6) = 28.4x$

$4686 = 28.4x$

$165 = x.$

There are 165 calories in 42.6 grams.

15. $\dfrac{16}{10} = \dfrac{34}{y}$

$16y = 340$

$y = 21.25$

17. $\dfrac{y}{20} = \dfrac{8}{28}$

$y = \dfrac{20 \cdot 8}{28}$

$y = \dfrac{40}{7}$

$y = 5\dfrac{5}{7}$ ft

19. $3 \cdot \dfrac{1}{x} = 9 \cdot \dfrac{1}{6}$

$\dfrac{3}{x} = \dfrac{9}{6}$

$6x\left(\dfrac{3}{x}\right) = 6x\left(\dfrac{9}{6}\right)$

$18 = 9x$

$x = 2$

The unknown number is 2.

21. $\dfrac{3+2x}{x+1} = \dfrac{3}{2}$

$2(x+1)\left(\dfrac{3+2x}{x+1}\right) = 2(x+1)\left(\dfrac{3}{2}\right)$

$2(3+2x) = 3(x+1)$

$6 + 4x = 3x + 3$

$x = -3$

The unknown number is -3.

23.

	Hours to Complete Total Job	Part of Job Completed in 1 Hour
Experienced	4	1/4
Apprentice	5	1/5
Together	x	1/x

$\dfrac{1}{4} + \dfrac{1}{5} = \dfrac{1}{x}$

$20x\left(\dfrac{1}{4}\right) + 20x\left(\dfrac{1}{5}\right) = 20x\left(\dfrac{1}{x}\right)$

$5x + 4x = 20$

$9x = 20$

$x = \dfrac{20}{9}$ or $2\dfrac{2}{9}$

The experienced surveyor and apprentice surveyor, working together, can survey the road in $2\dfrac{2}{9}$ hours.

25.

	Minutes to Complete Total Job	Part of Job Completed in 1 Minute
Larger belt	2	1/2
Smaller belt	6	1/6
Both belts	x	1/x

$\dfrac{1}{2} + \dfrac{1}{6} = \dfrac{1}{x}$

$6x\left(\dfrac{1}{2}\right) + 6x\left(\dfrac{1}{6}\right) = 6x\left(\dfrac{1}{x}\right)$

$3x + x = 6$

$4x = 6$

$x = \dfrac{6}{4} = \dfrac{3}{2} = 1\dfrac{1}{2}$

Both belts together can move the cans to the storage area in $1\dfrac{1}{2}$ minute.

27.

	Distance =	rate ·	time
Trip to Park	12	12 / x	x
Return Trip	18	$18/(x+1)$	$x+1$

$\dfrac{12}{x} = \dfrac{18}{x+1}$

$12(x+1) = 18x$

$12x + 12 = 18x$

$12 = 6x$

$2 = x$

The jogger spends 2 hours on her trip to the park, so her rate is $\dfrac{12}{2} = 6$ miles per hour.

29.

	Distance =	rate ·	time
1st portion	20	r	20/r
Cooldown portion	16	$r-2$	$16/r-2$

$\dfrac{20}{r} = \dfrac{16}{r-2}$

$20(r-2) = 16r$

$20r - 40 = 16r$

$-40 = -4r$

$r = 10$ and $r - 2 = 10 - 2 = 8$

His speed was 10 miles per hour during the first portion and 8 miles per hour during the cooldown portion.

31. Let x = the minimum floor space needed by 40 students.

$$\frac{1}{9} = \frac{40}{x}$$

$$1x = 9(40)$$

$$x = 360$$

40 students need 360 square feet.

33. $\dfrac{1}{4} = \dfrac{x}{8}$

$$8\left(\frac{1}{4}\right) = 8\left(\frac{x}{8}\right)$$

$$2 = x$$

The unknown number is 2.

35.

	Hours to Complete Total Job	Part of Job Completed in 1 Hour
Marcus	6	1/6
Tony	4	1/4
Together	x	$1/x$

$$\frac{1}{6} + \frac{1}{4} = \frac{1}{x}$$

$$12x\left(\frac{1}{6}\right) + 12x\left(\frac{1}{4}\right) = 12x\left(\frac{1}{x}\right)$$

$$2x + 3x = 12$$

$$5x = 12$$

$$x = \frac{12}{5} = 2\frac{2}{5}$$

$$45\left(\frac{12}{5}\right) = 108$$

Together Marcus and Tony work for $2\dfrac{2}{5}$ hours at \$45 per hour. The labor estimate should be \$108.

37. Let r = the speed of the car in still air.

Distance	=	rate	·	time

	Distance =	rate ·	time
Into the wind	10	$r-3$	$10/r-3$
With the wind	11	$r+3$	$11/r+3$

$$\frac{10}{r-3} = \frac{11}{r+3}$$

$$10(r+3) = 11(r-3)$$

$$10r + 30 = 11r - 33$$

$$63 = r$$

The speed of the car in still air is 63 miles per hour.

39. $\dfrac{y}{25 \text{ ft}} = \dfrac{3 \text{ ft}}{2 \text{ ft}}$

$$y \cdot 2 \text{ ft} = 25 \text{ ft} \cdot 3 \text{ ft}$$

$$y \cdot 2 \text{ ft} = 75 \text{ sq. ft}$$

$$y = \frac{75 \text{ sq. ft}}{2 \text{ ft}}$$

$$y = 37\frac{1}{2} \text{ ft}$$

41. Let x = the number of rushing yards in one game.

$$\frac{x}{1} = \frac{4045}{12}$$

$$12x = 1(4045)$$

$$12x = 4045$$

$$x \approx 337$$

Ken averaged 337 yards per game.

43. $\dfrac{2}{x-3} - \dfrac{4}{x+3} = 8 \cdot \dfrac{1}{x^2-9}$

$$(x-3)(x+3)\left(\frac{2}{x-3} - \frac{4}{x+3}\right)$$

$$= (x-3)(x+3)\left(\frac{8}{x^2-9}\right)$$

$$(x-3)(x+3)\left(\frac{2}{x-3}\right)$$

$$-(x-3)(x+3)\left(\frac{4}{x+3}\right)=8$$

$$2(x+3)-4(x-3)=8$$

$$2x+6-4x+12=8$$

$$-2x=-10$$

$$x=5$$

The unknown number is 5.

45.

	Distance	=	rate	·	time
With wind	630		$r+35$		$\frac{630}{r+35}$
Against wind	455		$r-35$		$\frac{455}{r-35}$

$$\frac{630}{r+35}=\frac{455}{r-35}$$

$$630(r-35)=455(r+35)$$

$$630r-22,050=455r+15,925$$

$$175r=37,975$$

$$r=217$$

The speed in still air is 217 mph.

47. Let x = the number of gallons of water needed.

$$\frac{8}{2}=\frac{36}{x}$$

$$8x=2(36)$$

$$8x=72$$

$$x=9$$

Need to mix 9 gallons of water with the entire box.

49. Let w = the rate of the wind.

	Distance	=	rate	·	time
With the wind	48		$16+w$		$48/16+w$
Into the wind	16		$16-w$		$16/16-w$

$$\frac{48}{16+w}=\frac{16}{16-w}$$

$$48(16-w)=16(16+w)$$

$$768-48w=256+16w$$

$$512=64w$$

$$w=8$$

The rate of the wind is 8 miles per hour.

51.

	Hours to Complete Total Job	Part of Job Completed in 1 Hour
Custodian	3	$1/3$
2nd Worker	x	$1/x$
Together	$1\frac{1}{2}$ or $\frac{3}{2}$	$2/3$

$$\frac{1}{3}+\frac{1}{x}=\frac{2}{3}$$

$$3x\left(\frac{1}{3}\right)+3x\left(\frac{1}{x}\right)=3x\left(\frac{2}{3}\right)$$

$$x+3=2x$$

$$3=x$$

It takes the second worker 3 hours to do the job alone.

53. $\dfrac{x}{8}=\dfrac{20}{6}$

$$x=\frac{160}{6}$$

$$x=\frac{80}{3}=26\frac{2}{3}$$

The side is $26\frac{2}{3}$ feet long.

55. $\dfrac{3}{2} = \dfrac{324}{x}$

$3 \cdot x = 2 \cdot 324$

$3x = 648$

$x = \dfrac{648}{3} = 216$ nuts

57.

	Hours to Complete Total Job	Part of Job Completed in 1 Hour
1st Pipe	20	$1/20$
2nd Pipe	15	$1/15$
3rd Pipe	x	$1/x$
3 Pipes Together	6	$1/6$

$\dfrac{1}{20} + \dfrac{1}{15} + \dfrac{1}{x} = \dfrac{1}{6}$

$60x\left(\dfrac{1}{20}\right) + 60x\left(\dfrac{1}{15}\right) + 60x\left(\dfrac{1}{x}\right)$

$\qquad\qquad\qquad = 60x\left(\dfrac{1}{6}\right)$

$3x + 4x + 60 = 10x$

$7x + 60 = 10x$

$60 = 3x$

$20 = x$

It takes the third pipe 20 hours to fill the pond.

59.

	Time	In one hour
Andew	2	$\dfrac{1}{2}$
Timothy	3	$\dfrac{1}{3}$
Together	x	$\dfrac{1}{x}$

$\dfrac{1}{2} + \dfrac{1}{3} = \dfrac{1}{x}$

$6x\left(\dfrac{1}{2} + \dfrac{1}{3}\right) = 6x\left(\dfrac{1}{x}\right)$

$6x\left(\dfrac{1}{2}\right) + 6x\left(\dfrac{1}{3}\right) = 6$

$3x + 2x = 6$

$5x = 6$

$\dfrac{5x}{5} = \dfrac{6}{5}$

$x = \dfrac{6}{5} = 1\dfrac{1}{5}$

Together it will take them $1\dfrac{1}{5}$ hours.

61.

	Time	In one hour
First cook	6	$\dfrac{1}{6}$
Second cook	7	$\dfrac{1}{7}$
Third cook	x	$\dfrac{1}{x}$
Together	2	$\dfrac{1}{2}$

$\dfrac{1}{6} + \dfrac{1}{7} + \dfrac{1}{x} = \dfrac{1}{2}$

$42x\left(\dfrac{1}{6} + \dfrac{1}{7} + \dfrac{1}{x}\right) = 42x\left(\dfrac{1}{2}\right)$

$42x\left(\dfrac{1}{6}\right) + 42x\left(\dfrac{1}{7}\right) + 42x\left(\dfrac{1}{x}\right) = 21x$

$7x + 6x + 42 = 21x$

$13x + 42 = 21x$

$42 = 21x - 13x$

$42 = 8x$

$$\frac{42}{8} = x$$

$$\frac{21}{4} = x$$

$$5\frac{1}{4} = x$$

The third cook can prepare the pies in

$5\frac{1}{4}$ hours.

63.

	Minutes to Complete Total Job	Part of Job Completed in 1 Minute
1st Pump	$3x$	$1/3x$
2nd Pump	x	$1/x$
Together	21	$1/21$

$$\frac{1}{3x} + \frac{1}{x} = \frac{1}{21}$$

$$21x\left(\frac{1}{3x}\right) + 21x\left(\frac{1}{x}\right) = 21x\left(\frac{1}{21}\right)$$

$$7 + 21 = x$$

$$28 = x, \quad 3x = 3(28) = 84$$

The 1st pump takes 28 minutes and the 2nd takes 84 minutes.

65. $(0, 4), (2, 10)$

$$m = \frac{10 - 4}{2 - 0} = \frac{6}{2} = 3$$

Since the slope is positive, the lines moves upward.

67. $(-2, 7), (3, -2)$

$$m = \frac{-2 - 7}{3 - (-2)} = \frac{-9}{5} = -\frac{9}{5}$$

Since the slope is negative, the lines moves downward.

69. $(0, -4), (2, -4)$

$$m = \frac{-4 - (-4)}{2 - 0} = \frac{0}{2} = 0$$

The slope is zero.
Since the slope is zero, the line is horizontal.

71. The capacity in 2000 is approximately 2650 megawatts. The capacity in 2002 is approximately 4685 megawatts. The increase is approximately $4685 - 2650 = 2035$ megawatts.

73. Answers may vary.

75. $\quad \dfrac{1}{6}x + \dfrac{1}{12}x + \dfrac{1}{7}x + 5 + \dfrac{1}{2}x + 4 = x$

$$\frac{1}{6}x + \frac{1}{12}x + \frac{1}{7}x + \frac{1}{2}x + 9 = x$$

$$84\left(\frac{1}{6}x + \frac{1}{12}x + \frac{1}{7}x + \frac{1}{2}x + 9\right) = 84x$$

$$14x + 7x + 12x + 42x + 756 = 84x$$

$$75x + 756 = 84x$$

$$756 = 9x$$

$$\frac{756}{9} = \frac{9x}{9}$$

$$84 = x$$

He died when he was 84 years old.

77.
$$4+\frac{1}{2}x+\frac{1}{6}x+3+\frac{1}{10}x=x$$

$$30\left(7+\frac{1}{2}x+\frac{1}{6}x+\frac{1}{10}x\right)=(30)(x)$$

$$30\cdot7+30\left(\frac{1}{2}x\right)+30\left(\frac{1}{6}x\right)+30\left(\frac{1}{10}x\right)=30x$$

$$210+15x+5x+3x=30x$$

$$210+23x=30x$$

$$210=30x-23x$$

$$210=7x$$

$$30=x$$

You are 30 years old.

79.

Distance	=	rate	·	time
H	$d+0.5$	40		$\frac{d+0.5}{40}$
G	d	32		$\frac{d}{32}$

$$\frac{d+0.5}{40}=\frac{d}{32}$$

$$32(d+0.5)=40d$$

$$32d+16=40d$$

$$16=8d$$

$$2=d, \quad \frac{d}{32}=\frac{2}{32}=\frac{1}{16}$$

It will take the hyena $\dfrac{1}{16}$ hour or

3.75 minutes to overtake the giraffe.

Exercise Set 7.7

1. $\dfrac{\dfrac{10}{3x}}{\dfrac{5}{6x}}=\dfrac{10}{3x}\cdot\dfrac{6x}{5}=\dfrac{2}{1}\cdot\dfrac{2}{1}=4$

3. $\dfrac{1+\dfrac{2}{5}}{2+\dfrac{3}{5}}=\dfrac{\left(1+\dfrac{2}{5}\right)5}{\left(2+\dfrac{3}{5}\right)5}=\dfrac{5+2}{10+3}=\dfrac{7}{13}$

5. $\dfrac{\dfrac{4}{x-1}}{\dfrac{x}{x-1}}=\dfrac{4}{x-1}\cdot\dfrac{x-1}{x}=\dfrac{4}{x}$

7. $\dfrac{1-\dfrac{2}{x}}{x+\dfrac{4}{9x}}=\dfrac{\left(1-\dfrac{2}{x}\right)9x}{\left(x+\dfrac{4}{9x}\right)9x}=\dfrac{9x-18}{9x^2+4}=\dfrac{9(x-2)}{9x^2+4}$

9. $\dfrac{\dfrac{4x^2-y^2}{xy}}{\dfrac{2}{y}-\dfrac{1}{x}}=\dfrac{\left(\dfrac{4x^2-y^2}{xy}\right)xy}{\left(\dfrac{2}{y}-\dfrac{1}{x}\right)xy}$

$$=\dfrac{4x^2-y^2}{2x-y}$$

$$=\dfrac{(2x+y)(2x-y)}{2x-y}$$

$$=2x+y$$

11. $\dfrac{\dfrac{x+1}{3}}{\dfrac{2x-1}{6}}=\dfrac{x+1}{3}\cdot\dfrac{6}{2x-1}$

$$=\dfrac{x+1}{1}\cdot\dfrac{2}{2x-1}$$

$$=\dfrac{2(x+1)}{2x-1}$$

13. $\dfrac{\dfrac{2}{x}+\dfrac{3}{x^2}}{\dfrac{4}{x^2}-\dfrac{9}{x}} = \dfrac{\left(\dfrac{2}{x}+\dfrac{3}{x^2}\right)x^2}{\left(\dfrac{4}{x^2}-\dfrac{9}{x}\right)x^2} = \dfrac{2x+3}{4-9x}$

$= \dfrac{(x+2)(x-1)-2x}{(x+1)(x-1)+x(x+1)}$

$= \dfrac{x^2+x-2-2x}{x^2-1+x^2+x}$

$= \dfrac{x^2-x-2}{2x^2+x-1}$

$= \dfrac{(x-2)(x+1)}{(2x-1)(x+1)} = \dfrac{x-2}{2x-1}$

15. $\dfrac{\dfrac{1}{x}+\dfrac{2}{x^2}}{x+\dfrac{8}{x^2}} = \dfrac{\left(\dfrac{1}{x}+\dfrac{2}{x^2}\right)x^2}{\left(x+\dfrac{8}{x^2}\right)x^2}$

$= \dfrac{x+2}{x^3+8}$

$= \dfrac{x+2}{(x+2)(x^2-2x+4)}$

$= \dfrac{1}{x^2-2x+4}$

21. $\dfrac{\dfrac{2}{x}+3}{\dfrac{4}{x^2}-9} = \dfrac{\left(\dfrac{2}{x}+3\right)x^2}{\left(\dfrac{4}{x^2}-9\right)x^2}$

$= \dfrac{2x+3x^2}{4-9x^2}$

$= \dfrac{x(2+3x)}{(2+3x)(2-3x)}$

$= \dfrac{x}{2-3x}$

17. $\dfrac{\dfrac{4}{5-x}+\dfrac{5}{x-5}}{\dfrac{2}{x}+\dfrac{3}{x-5}} = \dfrac{-\dfrac{4}{x-5}+\dfrac{5}{x-5}}{\dfrac{2(x-5)+3x}{x(x-5)}}$

$= \dfrac{\dfrac{1}{x-5}}{\dfrac{2x-10+3x}{x(x-5)}}$

$= \dfrac{1}{x-5}\cdot\dfrac{x(x-5)}{5x-10}$

$= \dfrac{x}{5x-10}$

23. $\dfrac{1-\dfrac{x}{y}}{\dfrac{x^2}{y^2}-1} = \dfrac{\left(1-\dfrac{x}{y}\right)y^2}{\left(\dfrac{x^2}{y^2}-1\right)y^2}$

$= \dfrac{y^2-xy}{x^2-y^2}$

$= \dfrac{y(y-x)}{(x+y)(x-y)}$

$= \dfrac{-y(x-y)}{(x+y)(x-y)} = -\dfrac{y}{x+y}$

19. $\dfrac{\dfrac{x+2}{x}-\dfrac{2}{x-1}}{\dfrac{x+1}{x}+\dfrac{x+1}{x-1}} = \dfrac{\left(\dfrac{x+2}{x}-\dfrac{2}{x-1}\right)x(x-1)}{\left(\dfrac{x+1}{x}+\dfrac{x+1}{x-1}\right)x(x-1)}$

25. $\dfrac{\frac{-2x}{x-y}}{\frac{y}{x^2}} = \dfrac{-2x}{x-y} \cdot \dfrac{x^2}{y} = -\dfrac{2x^3}{y(x-y)}$

27. $\dfrac{\frac{2}{x}+\frac{1}{x^2}}{\frac{y}{x^2}} = \dfrac{\left(\frac{2}{x}+\frac{1}{x^2}\right)x^2}{\left(\frac{y}{x^2}\right)x^2} = \dfrac{2x+1}{y}$

29. $\dfrac{\frac{x}{9}-\frac{1}{x}}{1+\frac{3}{x}} = \dfrac{\left(\frac{x}{9}-\frac{1}{x}\right)9x}{\left(1+\frac{3}{x}\right)9x}$

$= \dfrac{x^2-9}{9x+27}$

$= \dfrac{(x+3)(x-3)}{9(x+3)}$

$= \dfrac{x-3}{9}$

31. $\dfrac{\frac{x-1}{x^2-4}}{1+\frac{1}{x-2}} = \dfrac{\frac{x-1}{x^2-4}}{\frac{(x-2)+1}{x-2}}$

$= \dfrac{x-1}{(x+2)(x-2)} \cdot \dfrac{x-2}{x-1}$

$= \dfrac{1}{x+2}$

33. $\dfrac{x^{-1}}{x^{-2}+y^{-2}} = \dfrac{\left(\frac{1}{x}\right)x^2y^2}{\left(\frac{1}{x^2}+\frac{1}{y^2}\right)x^2y^2} = \dfrac{xy^2}{y^2+x^2}$

35. $\dfrac{2a^{-1}+3b^{-2}}{a^{-1}-b^{-1}} = \dfrac{\left(\frac{2}{a}+\frac{3}{b^2}\right)ab^2}{\left(\frac{1}{a}-\frac{1}{b}\right)ab^2}$

$= \dfrac{2b^2+3a}{b^2-ab}$

$= \dfrac{2b^2+3a}{b(b-a)}$

37. $\dfrac{1}{x-x^{-1}} = \dfrac{1 \cdot x}{\left(x-\frac{1}{x}\right)x}$

$= \dfrac{x}{x^2-1}$

$= \dfrac{x}{(x+1)(x-1)}$

39. $\dfrac{a^{-1}+1}{a^{-1}-1} = \dfrac{\left(\frac{1}{a}+1\right)a}{\left(\frac{1}{a}-1\right)a} = \dfrac{1+a}{1-a}$

41. $\dfrac{3x^{-1}+(2y)^{-1}}{x^{-2}} = \dfrac{\left(\frac{3}{x}+\frac{1}{2y}\right)2x^2y}{\left(\frac{1}{x^2}\right)2x^2y}$

$= \dfrac{6xy+x^2}{2y}$

$= \dfrac{x(6y+x)}{2y}$

43. $\dfrac{2a^{-1}+(2a)^{-1}}{a^{-1}+2a^{-2}} = \dfrac{\left(\frac{2}{a}+\frac{1}{2a}\right)2a^2}{\left(\frac{1}{a}+\frac{2}{a^2}\right)2a^2}$

$= \dfrac{4a+a}{2a+4}$

$= \dfrac{5a}{2a+4}$

45. $\dfrac{5x^{-1}+2y^{-1}}{x^{-2}y^{-2}} = \dfrac{\frac{5}{x}+\frac{2}{y}}{\frac{1}{x^2y^2}}$

$= \dfrac{\frac{5y+2x}{xy}}{\frac{1}{x^2y^2}}$

$= \dfrac{5y+2x}{xy} \cdot \dfrac{x^2y^2}{1}$

$= xy(5y+2x) \text{ or } 5xy^2+2x^2y$

47. $\dfrac{5x^{-1}-2y^{-1}}{25x^{-2}-4y^{-2}} = \dfrac{\left(\dfrac{5}{x}-\dfrac{2}{y}\right)x^2y^2}{\left(\dfrac{25}{x^2}-\dfrac{4}{y^2}\right)x^2y^2}$

$= \dfrac{5xy^2-2x^2y}{25y^2-4x^2}$

$= \dfrac{xy(5y-2x)}{(5y+2x)(5y-2x)}$

$= \dfrac{xy}{2x+5y}$

49. $\dfrac{3x^3y^2}{12x} = \dfrac{x^{3-1}y^2}{4} = \dfrac{x^2y^2}{4}$

51. $\dfrac{144x^5y^5}{-16x^2y} = -9x^{5-2}y^{5-1} = -9x^3y^4$

53. $P(x) = -x^2$

$P(-3) = -(-3)^2 = -9$

55. $\dfrac{a}{1-\dfrac{s}{770}} = \dfrac{a}{\dfrac{770}{770}-\dfrac{s}{770}}$

$= \dfrac{a}{\dfrac{770-s}{770}}$

$= a \cdot \dfrac{770}{770-s}$

$= \dfrac{770a}{770-s}$

57. $\dfrac{\dfrac{1}{x}}{\dfrac{3}{y}}$; a and b

59. $\dfrac{1}{1+(1+x)^{-1}} = \dfrac{1(1+x)}{\left(1+\dfrac{1}{1+x}\right)(1+x)}$

$= \dfrac{1+x}{(1+x)+1}$

$= \dfrac{1+x}{2+x}$

61. $\dfrac{x}{1-\dfrac{1}{1+\dfrac{1}{x}}} = \dfrac{x}{1-\dfrac{1}{\left(\dfrac{x+1}{x}\right)}} = \dfrac{x}{1-\dfrac{x}{x+1}}$

$= \dfrac{x(x+1)}{\left(1-\dfrac{x}{x+1}\right)(x+1)}$

$= \dfrac{x(x+1)}{x+1-x}$

$= x(x+1) = x^2+x$

63. $\dfrac{\dfrac{2}{y^2}-\dfrac{5}{xy}-\dfrac{3}{x^2}}{\dfrac{2}{y^2}+\dfrac{7}{xy}+\dfrac{3}{x^2}} = \dfrac{\left(\dfrac{2}{y^2}-\dfrac{5}{xy}-\dfrac{3}{x^2}\right)x^2y^2}{\left(\dfrac{2}{y^2}+\dfrac{7}{xy}+\dfrac{3}{x^2}\right)x^2y^2}$

$= \dfrac{2x^2-5xy-3y^2}{2x^2+7xy+3y^2}$

$= \dfrac{(2x+y)(x-3y)}{(2x+y)(x+3y)}$

$= \dfrac{x-3y}{x+3y}$

65. $\dfrac{3(a+1)^{-1}+4a^{-2}}{(a^3+a^2)^{-1}} = \dfrac{\dfrac{3}{a+1}+\dfrac{4}{a^2}}{\dfrac{1}{a^3+a^2}}$

$= \dfrac{\dfrac{3a^2+4(a+1)}{a^2(a+1)}}{\dfrac{1}{a^2(a+1)}}$

$= \dfrac{3a^2+4a+4}{1}$

$= 3a^2+4a+4$

67. $f(x) = \dfrac{1}{x}$

    a. $f(a+h) = \dfrac{1}{a+h}$

    b. $f(a) = \dfrac{1}{a}$

    c. $\dfrac{f(a+h)-f(a)}{h} = \dfrac{\dfrac{1}{a+h} - \dfrac{1}{a}}{h}$

    d. $\dfrac{f(a+h)-f(a)}{h} = \dfrac{\left(\dfrac{1}{a+h} - \dfrac{1}{a}\right)a(a+h)}{h \cdot a(a+h)}$

          $= \dfrac{a-(a+h)}{ha(a+h)}$

          $= \dfrac{-h}{ha(a+h)}$

          $= \dfrac{-1}{a(a+h)}$

69. $f(x) = \dfrac{3}{x+1}$

    a. $f(a+h) = \dfrac{3}{a+h+1}$

    b. $f(a) = \dfrac{3}{a+1}$

    c. $\dfrac{f(a+h)-f(a)}{h} = \dfrac{\dfrac{3}{a+h+1} - \dfrac{3}{a+1}}{h}$

    d. $\dfrac{f(a+h)-f(a)}{h}$

$= \dfrac{\left(\dfrac{3}{a+h+1} - \dfrac{3}{a+1}\right)(a+h+1)(a+1)}{h \cdot (a+h+1)(a+1)}$

$= \dfrac{3(a+1)-3(a+h+1)}{h(a+h+1)(a+1)}$

$= \dfrac{3a+3-3a-3h-3}{h(a+h+1)(a+1)}$

$= \dfrac{-3h}{h(a+h+1)(a+1)}$

$= \dfrac{-3}{(a+h+1)(a+1)}$

Chapter 7 Review

1. $F(x) = \dfrac{-3x^2}{x-5}$
Undefined values when
$x - 5 = 0$
$x = 5$
Domain $\{x \mid x$ is a real number and $x \neq 5\}$

2. $h(x) = \dfrac{4x}{3x-12}$
Undefined values when
$3x - 12 = 0$
$3x = 12$
$x = 4$
Domain $\{x \mid x$ is a real number and $x \neq 4\}$

3. $f(x) = \dfrac{x^3+2}{x^2+8x}$
Undefined values when
$x^2 + 8x = 0$
$x(x+8) = 0$
$x = 0$ or $x + 8 = 0$
             $x = -8$
Domain
$\{x \mid x$ is a real number and $x \neq 0, x \neq -8\}$

4. $G(x) = \dfrac{20}{3x^2 - 48}$

Undefined values when

$3x^2 - 48 = 0$

$3(x^2 - 16) = 0$

$3(x + 4)(x - 4) = 0$

$x + 4 = 0$ or $x - 4 = 0$

$x = -4$ or $x = 4$

Domain

$\{x \mid x \text{ is a real number and } x \neq -4, x \neq 4\}$

5. $\dfrac{15x^4}{45x^2} = \dfrac{15x^{4-2}}{45} = \dfrac{x^2}{3}$

6. $\dfrac{x+2}{2+x} = \dfrac{x+2}{x+2} = 1$

7. $\dfrac{18m^6 p^2}{10m^4 p} = \dfrac{2 \cdot 9m^{6-4} p^{2-1}}{2 \cdot 5} = \dfrac{9m^2 p}{5}$

8. $\dfrac{x-12}{12-x} = \dfrac{x-12}{-1(x-12)} = \dfrac{1}{-1} = -1$

9. $\dfrac{5x-15}{25x-75} = \dfrac{5(x-3)}{5 \cdot 5(x-3)} = \dfrac{1}{5}$

10. $\dfrac{22x+8}{11x+4} = \dfrac{2(11x+4)}{11x+4} = 2$

11. $\dfrac{2x}{2x^2 - 2x} = \dfrac{2x}{2x(x-1)} = \dfrac{1}{x-1}$

12. $\dfrac{x+7}{x^2-49} = \dfrac{x+7}{(x+7)(x-7)} = \dfrac{1}{x-7}$

13. $\dfrac{2x^2 + 4x - 30}{x^2 + x - 20} = \dfrac{2(x^2 + 2x - 15)}{(x+5)(x-4)}$

$= \dfrac{2(x+5)(x-3)}{(x+5)(x-4)}$

$= \dfrac{2(x-3)}{x-4}$

14. $\dfrac{xy - 3x + 2y - 6}{x^2 + 4x + 4} = \dfrac{x(y-3) + 2(y-3)}{(x+2)^2}$

$= \dfrac{(y-3)(x+2)}{(x+2)^2} = \dfrac{y-3}{x+2}$

15. $C(x) = \dfrac{35x + 4200}{x}$

   a. $C(50) = \dfrac{35(50) + 4200}{50} = \$119$

   b. $C(100) = \dfrac{35(100) + 4200}{100} = \$77$

   c. decrease

16. $\dfrac{15x^3 y^2}{z} \cdot \dfrac{z}{5xy^3} = \dfrac{15x^3 y^2 \cdot z}{z \cdot 5xy^3}$

$= \dfrac{3 \cdot 5 \cdot x^2 \cdot x \cdot y^2 \cdot z}{z \cdot 5 \cdot x \cdot y^2 \cdot y} = \dfrac{3x^2}{y}$

17. $\dfrac{-y^3}{8} \cdot \dfrac{9x^2}{y^3} = \dfrac{y^3 \cdot 9x^2}{8 \cdot y^3} = -\dfrac{9x^2}{8}$

18. $\dfrac{x^2 - 9}{x^2 - 4} \cdot \dfrac{x-2}{x+3} = \dfrac{(x^2-9) \cdot (x-2)}{(x^2-4) \cdot (x+3)}$

$= \dfrac{(x-3)(x+3)(x-2)}{(x+2)(x-2)(x+3)} = \dfrac{x-3}{x+2}$

19. $\dfrac{2x+5}{x-6} \cdot \dfrac{2x}{-x+6} = \dfrac{2x+5}{x-6} \cdot \dfrac{2x}{-(x-6)}$

$= \dfrac{2x+5}{x-6} \cdot \dfrac{-2x}{x-6} = \dfrac{(2x+5) \cdot (-2x)}{(x-6) \cdot (x-6)}$

$= \dfrac{-2x(2x+5)}{(x-6)^2}$

20. $\dfrac{x^2-5x-24}{x^2-x-12} \div \dfrac{x^2-10x+16}{x^2+x-6}$

$= \dfrac{x^2-5x-24}{x^2-x-12} \cdot \dfrac{x^2+x-6}{x^2-10x+16}$

$= \dfrac{(x-8)(x+3)\cdot(x+3)(x-2)}{(x-4)(x+3)\cdot(x-8)(x-2)}$

$= \dfrac{x+3}{x-4}$

21. $\dfrac{4x+4y}{xy^2} \div \dfrac{3x+3y}{x^2y} = \dfrac{4x+4y}{xy^2} \cdot \dfrac{x^2y}{3x+3y}$

$= \dfrac{4(x+y)\cdot x\cdot x\cdot y}{x\cdot y\cdot y\cdot 3(x+y)} = \dfrac{4x}{3y}$

22. $\dfrac{x^2+x-42}{x-3} \cdot \dfrac{(x-3)^2}{x+7}$

$= \dfrac{(x+7)(x-6)\cdot(x-3)(x-3)}{(x-3)\cdot(x+7)}$

$= (x-6)(x-3)$

23. $\dfrac{2a+2b}{3} \cdot \dfrac{a-b}{a^2-b^2} = \dfrac{2(a+b)\cdot(a-b)}{3\cdot(a+b)(a-b)} = \dfrac{2}{3}$

24. $\dfrac{x^2-9x+14}{x^2-5x+6} \cdot \dfrac{x+2}{x^2-5x-14}$

$= \dfrac{(x-7)(x-2)\cdot(x+2)}{(x-3)(x-2)\cdot(x-7)(x+2)} = \dfrac{1}{x-3}$

25. $(x-3) \cdot \dfrac{x}{x^2+3x-18}$

$= \dfrac{(x-3)\cdot x}{(x-3)(x+6)} = \dfrac{x}{x+6}$

26. $\dfrac{2x^2-9x+9}{8x-12} \div \dfrac{x^2-3x}{2x}$

$= \dfrac{2x^2-9x+9}{9x-12} \cdot \dfrac{2x}{x^2-3x}$

$= \dfrac{(2x-3)(x-3)\cdot 2x}{4(2x-3)\cdot x(x-3)}$

$= \dfrac{2}{4} = \dfrac{1}{2}$

27. $\dfrac{x^2-y^2}{x^2+xy} \div \dfrac{3x^2-2xy-y^2}{3x^2+6x}$

$= \dfrac{x^2-y^2}{x^2+xy} \cdot \dfrac{3x^2+6x}{3x^2-2xy-y^2}$

$= \dfrac{(x-y)(x+y)\cdot 3x(x+2)}{x(x+y)\cdot(3x+y)(x-y)}$

$= \dfrac{3(x+2)}{3x+y}$

28. $\dfrac{x^2-y^2}{8x^2-16xy+8y^2} \div \dfrac{x+y}{4x-y}$

$= \dfrac{(x-y)(x+y)}{8(x-y)(x-y)} \cdot \dfrac{4x-y}{x+y}$

$= \dfrac{(x-y)(x+y)(4x-y)}{8(x-y)(x-y)(x+y)}$

$= \dfrac{4x-y}{8(x-y)}$

29. $\dfrac{x-y}{4} \div \dfrac{y^2-2y-xy+2x}{16x+24}$

$= \dfrac{x-y}{4} \cdot \dfrac{16x+24}{y^2-2y-xy+2x}$

$= \dfrac{x-y}{4} \cdot \dfrac{8(2x+3)}{y(y-2)-x(y-2)}$

$= \dfrac{x-y}{4} \cdot \dfrac{8(2x+3)}{(y-2)(y-x)}$

$= -\dfrac{y-x}{4} \cdot \dfrac{8(2x+3)}{(y-2)(y-x)}$

$= -\dfrac{2\cdot 4(y-x)(2x+3)}{4(y-2)(y-x)}$

$= -\dfrac{2(2x+3)}{y-2}$

30. $\dfrac{y-3}{4x+3} \div \dfrac{9-y^2}{4x^2-x-3} = \dfrac{y-3}{4x+3} \cdot \dfrac{4x^2-x-3}{9-y^2}$

$= \dfrac{y-3}{4x+3} \cdot \dfrac{(4x+3)(x-1)}{-1(y-3)(y+3)}$

$= \dfrac{(y-3)(4x+3)(x-1)}{-(4x+3)(y-3)(y+3)}$

$= -\dfrac{x-1}{y+3}$

31. $\dfrac{5x-4}{3x-1} + \dfrac{6}{3x-1} = \dfrac{5x-4+6}{3x-1} = \dfrac{5x+2}{3x-1}$

32. $\dfrac{4x-5}{3x^2} - \dfrac{2x+5}{3x^2} = \dfrac{4x-5-(2x+5)}{3x^2}$

$= \dfrac{4x-5-2x-5}{3x^2} = \dfrac{2x-10}{3x^2}$

33. $\dfrac{9x+7}{6x^2} - \dfrac{3x+4}{6x^2} = \dfrac{9x+7-(3x+4)}{6x^2}$

$= \dfrac{9x+7-3x-4}{6x^2} = \dfrac{6x+3}{6x^2} = \dfrac{3(2x+1)}{3\cdot 2x^2}$

$= \dfrac{2x+1}{2x^2}$

34. $2x = 2\cdot x$

$7x = 7\cdot x$

$\text{LCD} = 2\cdot 7\cdot x = 14x$

35. $x^2-5x-24 = (x-8)(x+3)$

$x^2+11x+24 = (x+8)(x+3)$

$\text{LCD} = (x-8)(x+3)(x+8)$

36. $\dfrac{x+2}{x^2+11x+18} = \dfrac{x+2}{(x+9)(x+2)}$

$= \dfrac{(x+2)(x-5)}{(x+9)(x+2)(x-5)}$

$= \dfrac{x^2-3x-10}{(x+2)(x-5)(x+9)}$

37. $\dfrac{3x-5}{x^2+4x+4} = \dfrac{3x-5}{(x+2)^2}$

$= \dfrac{(3x-5)(x+3)}{(x+2)^2(x+3)} = \dfrac{3x^2+4x-15}{(x+2)^2(x+3)}$

$= \dfrac{x^2-3x-10}{(x+2)(x-5)(x+9)}$

38. $\dfrac{4}{5x^2} - \dfrac{6}{y} = \dfrac{4(y)}{5x^2(y)} - \dfrac{6(5x^2)}{y(5x^2)} = \dfrac{4y-30x^2}{5x^2y}$

39. $\dfrac{2}{x-3} - \dfrac{4}{x-1}$

$= \dfrac{2(x-1)}{(x-3)(x-1)} - \dfrac{4(x-3)}{(x-1)(x-3)}$

$= \dfrac{2(x-1)-4(x-3)}{(x-3)(x-1)} = \dfrac{2x-2-4x+12}{(x-3)(x-1)}$

$= \dfrac{-2x+10}{(x-3)(x-1)}$

40. $\dfrac{x+7}{x+3} - \dfrac{x-3}{x+7}$

$= \dfrac{(x+7)(x+7)}{(x+3)(x+7)} - \dfrac{(x-3)(x+3)}{(x+7)(x+3)}$

$= \dfrac{x^2+14x+49-(x^2-9)}{(x+3)(x+7)}$

$= \dfrac{x^2+14x+49-x^2+9}{(x+3)(x+7)} = \dfrac{14x+58}{(x+3)(x+7)}$

41. $\dfrac{4}{x+3} - 2 = \dfrac{4}{x+3} - \dfrac{2(x+3)}{x+3}$

$= \dfrac{4-2(x+3)}{x+3} = \dfrac{4-2x-6}{x+3} = \dfrac{-2x-2}{x+3}$

42. $\dfrac{3}{x^2+2x-8} + \dfrac{2}{x^2-3x+2}$

$= \dfrac{3}{(x+4)(x-2)} + \dfrac{2}{(x-1)(x-2)}$

$= \dfrac{3(x-1)}{(x+4)(x-2)(x-1)}$

$\qquad + \dfrac{2(x+4)}{(x-1)(x-2)(x+4)}$

$= \dfrac{3(x-1)+2(x+4)}{(x+4)(x-2)(x-1)}$

$= \dfrac{3x-3+2x+8}{(x+4)(x-2)(x-1)}$

$= \dfrac{5x+5}{(x+4)(x-2)(x-1)}$

43. $\dfrac{2x-5}{6x+9} - \dfrac{4}{2x^2+3x}$

$= \dfrac{2x-5}{3(2x+3)} - \dfrac{4}{x(2x+3)}$

$= \dfrac{(2x-5)(x)}{3(2x+3)(x)} - \dfrac{4(3)}{x(2x+3)(3)}$

$= \dfrac{2x^2-5x-12}{3x(2x+3)} = \dfrac{(2x+3)(x-4)}{3x(2x+3)}$

$= \dfrac{x-4}{3x}$

44. $\dfrac{x-1}{x^2-2x+1} - \dfrac{x+1}{x-1} = \dfrac{x-1}{(x-1)^2} - \dfrac{x+1}{x-1}$

$= \dfrac{1}{x-1} = \dfrac{x+1}{x-1} = \dfrac{1-(x+1)}{x-1}$

$= \dfrac{1-x-1}{x-1} = \dfrac{-x}{x-1} = -\dfrac{x}{x-1}$

45. $\dfrac{x-1}{x^2+4x+4} + \dfrac{x-1}{x+2}$

$= \dfrac{x-1}{(x+2)^2} + \dfrac{(x-1)(x+2)}{(x+2)(x+2)}$

$= \dfrac{x-1+(x-1)(x+2)}{(x+2)^2}$

$$= \frac{x - 1 + x^2 + x - 2}{(x + 2)^2}$$

$$= \frac{x^2 + 2x - 3}{(x + 2)^2}$$

46. $P = 2l + 2w$

$$P = 2\left(\frac{2}{8}\right) + 2\left(\frac{x + 2}{4x}\right)$$

$$= \frac{x}{4} + \frac{2(x + 2)}{4x}$$

$$= \frac{x \cdot x}{4 \cdot x} + \frac{2x + 4}{4x}$$

$$= \frac{x^2 + 2x + 4}{4x}$$

$A = l \cdot w$

$$A = \frac{x}{8} \cdot \frac{x + 2}{4x} = \frac{x \cdot (x + 2)}{8 \cdot 4x} = \frac{x + 2}{32}$$

The perimeter is $\dfrac{x^2 + 2x + 4}{4x}$ units

and the area is $\dfrac{x + 2}{32}$ square units.

47. $P = \dfrac{3x}{4x - 4} + \dfrac{2x}{3x - 3} + \dfrac{x}{x - 1}$

$$= \frac{3x}{4(x - 1)} + \frac{2x}{3(x - 1)} + \frac{x}{x - 1}$$

$$= \frac{3x(3)}{4(x - 1)(3)} + \frac{2x(4)}{3(x - 1)(4)} + \frac{x(12)}{(x - 1)(12)}$$

$$= \frac{9x + 8x + 12x}{12(x - 1)} = \frac{29x}{12(x - 1)}$$

$$A = \frac{1}{2} \cdot b \cdot h$$

$$A = \frac{1}{2} \cdot \frac{x}{x - 1} \cdot \frac{6y}{5}$$

$$= \frac{1 \cdot x \cdot 2 \cdot 3y}{2 \cdot (x - 1) \cdot 5}$$

$$= \frac{3xy}{5(x - 1)}$$

The perimeter is $\dfrac{29x}{12(x - 1)}$ units and the

area is $\dfrac{3xy}{5(x - 1)}$ square units.

48. $\dfrac{x + 4}{9} = \dfrac{5}{9}$

$$9\left(\frac{x + 4}{9}\right) = 9\left(\frac{5}{9}\right)$$

$$x + 4 = 5$$

$$x = 1$$

49. $\dfrac{n}{10} = 9 - \dfrac{n}{5}$

$$10\left(\frac{n}{10}\right) = 10\left(9 - \frac{n}{5}\right)$$

$$10\left(\frac{n}{10}\right) = 10(9) - 10\left(\frac{n}{5}\right)$$

$$n = 90 - 2n$$

$$3n = 90$$

$$n = 30$$

50. $\dfrac{5y-3}{7} = \dfrac{15y-2}{28}$

$28\left(\dfrac{5y-3}{7}\right) = 28\left(\dfrac{15y-2}{28}\right)$

$4(5y-3) = 15y-2$

$20y-12 = 15y-2$

$5y = 10$

$y = 2$

51. $\dfrac{2}{x+1} - \dfrac{1}{x-2} = -\dfrac{1}{2}$

$2(x+1)(x-2)\left(\dfrac{2}{x+1} - \dfrac{1}{x-2}\right)$

$\qquad = 2(x+1)(x-2)\left(-\dfrac{1}{2}\right)$

$2(x+1)(x-2)\left(\dfrac{2}{x+1}\right)$

$\qquad -2(x+1)(x-2)\left(\dfrac{1}{x-2}\right)$

$\qquad = 2(x+1)(x-2)\left(-\dfrac{1}{2}\right)$

$4(x-2) - 2(x+1) = -(x+1)(x-2)$

$4x-8-2x-2 = -(x^2 - x - 2)$

$2x-10 = -x^2 + x + 2$

$x^2 + x - 12 = 0$

$(x+4)(x-3) = 0$

$x+4 = 0 \quad$ or $\quad x-3 = 0$

$\qquad x = -4 \qquad\qquad x = 3$

52. $\dfrac{1}{a+3} + \dfrac{1}{a-3} = -\dfrac{5}{a^2 - 9}$

$(a-3)(a+3)\left(\dfrac{1}{a+3} + \dfrac{1}{a-3}\right)$

$\qquad = (a-3)(a+3)\left(-\dfrac{5}{(a-3)(a+3)}\right)$

$(a-3)(a+3)\left(\dfrac{1}{a+3}\right)$

$\qquad + (a-3)(a+3)\left(\dfrac{1}{a-3}\right) = -5$

$a-3+a+3 = -5$

$2a = -5$

$a = -\dfrac{5}{2}$

53. $\dfrac{y}{2y+2} + \dfrac{2y-16}{4y+4} = \dfrac{y-3}{y+1}$

$\dfrac{y}{2(y+1)} + \dfrac{2y-16}{4(y+1)} = \dfrac{y-3}{y+1}$

$4(y+1)\left(\dfrac{y}{2(y+1)} + \dfrac{2y-16}{4(y+1)}\right)$

$\qquad = 4(y+1)\left(\dfrac{y-3}{y+1}\right)$

$4(y+1)\left(\dfrac{y}{2(y+1)}\right) + 4(y+1)\left(\dfrac{2y-16}{4(y+1)}\right)$

$\qquad = 4(y+1)\left(\dfrac{y-3}{y+1}\right)$

$2y + 2y - 16 = 4(y-3)$

$4y - 16 = 4y - 12$

$-16 = -12 \quad$ False

This equation has no solution.

54. $\dfrac{4}{x+3} + \dfrac{8}{x^2-9} = 0$

$(x-3)(x+3)\left(\dfrac{4}{x+3} + \dfrac{8}{(x-3)(x+3)}\right)$

$\qquad = (x-3)(x+3)(0)$

$(x-3)(x+3)\left(\dfrac{4}{x+3}\right)$

$\quad + (x-3)(x+3)\left(\dfrac{8}{(x-3)(x+3)}\right) = 0$

$4(x-3) + 8 = 0$

$4x - 12 + 8 = 0$

$4x - 4 = 0$

$4x = 4$

$x = 1$

55. $\dfrac{2}{x-3} - \dfrac{4}{x+3} = \dfrac{8}{x^2-9}$

$(x-3)(x+3)\left(\dfrac{2}{x-3} - \dfrac{4}{x+3}\right)$

$\quad = (x-3)(x+3)\left(\dfrac{8}{(x-3)(x+3)}\right)$

$(x-3)(x+3)\left(\dfrac{2}{x-3}\right)$

$\quad - (x-3)(x+3)\left(\dfrac{4}{x+3}\right) = 8$

$2(x+3) - 4(x-3) = 8$

$2x + 6 - 4x + 12 = 8$

$-2x + 18 = 8$

$-2x = -10$

$x = 5$

56. $\dfrac{x-3}{x+1} - \dfrac{x-6}{x+5} = 0$

$(x+1)(x+5)\left(\dfrac{x-3}{x+1} - \dfrac{x-6}{x+5}\right)$

$\qquad = (x+1)(x+5)(0)$

$(x+1)(x+5)\left(\dfrac{x-3}{x+1}\right)$

$\quad - (x+1)(x+5)\left(\dfrac{x-6}{x+5}\right) = 0$

$(x+5)(x-3) - (x+1)(x-6) = 0$

$x^2 + 2x - 15 - (x^2 - 5x - 6) = 0$

$x^2 + 2x - 15 - x^2 + 5x + 6 = 0$

$7x - 9 = 0$

$7x = 9$

$x = \dfrac{9}{7}$

57. $x + 5 = \dfrac{6}{x}$

$x(x+5) = x\left(\dfrac{6}{x}\right)$

$x^2 + 5x = 6$

$x^2 + 5x - 6 = 0$

$(x+6)(x-1) = 0$

$x + 6 = 0 \quad$ or $\quad x - 1 = 0$

$\quad x = -6 \qquad\qquad x = 1$

58. $\dfrac{4A}{5b} = x^2$

$4A = 5bx^2$

$\dfrac{4A}{5x^2} = \dfrac{5bx^2}{5x^2}$

$\dfrac{4A}{5x^2} = b$

59.
$$\frac{x}{7} + \frac{y}{8} = 10$$

$$56\left(\frac{x}{7}\right) + 56\left(\frac{y}{8}\right) = 56(10)$$

$$8x + 7y = 560$$

$$7y = 560 - 8x$$

$$y = \frac{560 - 8x}{7}$$

60.
$$\frac{x}{2} = \frac{12}{4}$$

$$4x = 24$$

$$x = 6$$

61.
$$\frac{20}{1} = \frac{x}{25}$$

$$500 = x$$

62.
$$\frac{2}{x-1} = \frac{3}{x+3}$$

$$2(x+3) = 3(x-1)$$

$$2x + 6 = 3x - 3$$

$$6 = x - 3$$

$$9 = x$$

63.
$$\frac{4}{y-3} = \frac{2}{y-3}$$

$$4(y-3) = 2(y-3)$$

$$4y - 12 = 2y - 6$$

$$2y - 12 = -6$$

$$2y = 6$$

$$y = 3$$

$y = 3$ doesn't check.

No solution

64. Let x = the number of parts processed in 45 minutes.

$$\frac{300}{20} = \frac{x}{45}$$

$$13,500 = 20x$$

$$675 = x$$

675 parts can be processed in 45 minutes.

65. Let x = the charge for 3 hours.

$$\frac{90.00}{8} = \frac{x}{3}$$

$$270.00 = 8x$$

$$33.75 = x$$

He charges $33.75 for 3 hours.

66. Let x = the number of letters addressed in 55 minutes.

$$\frac{100}{35} = \frac{x}{55}$$

$$5500 = 35x$$

$$157 \approx x$$

He can address 157 letters in 55 minutes.

67.
$$5 \cdot \frac{1}{x} = \frac{3}{2} \cdot \frac{1}{x} + \frac{7}{6}$$

$$\frac{5}{x} = \frac{3}{2x} + \frac{7}{6}$$

$$6x\left(\frac{5}{x}\right) = 6x\left(\frac{3}{2x}\right) + 6x\left(\frac{7}{6}\right)$$

$$30 = 9 + 7x$$

$$21 = 7x$$

$$x = 3$$

The unknown number is 3.

68. $\dfrac{1}{x} = \dfrac{1}{4-x}$

$4 - x = x$

$4 = 2x$

$2 = x$

The unknown number is 2.

69.

Distance	= rate	·	time
1st car	90	r	$90/r$
2nd car	60	$r-10$	$60/r-10$

$\dfrac{90}{r} = \dfrac{60}{r-10}$

$90(r-10) = 60r$

$90r - 900 = 60r$

$-900 = -30r$

$30 = r$

$r - 10 = 30 - 10 = 20$

The rate of the first car is 30 miles per hour and the rate of the second car is 20 miles per hour.

70.

Distance	= rate	·	time
Upstream	48	$r-4$	$48/r-4$
Downstream	72	$r+4$	$72/r+4$

$\dfrac{48}{r-4} = \dfrac{72}{r+4}$

$48(r+4) = 72(r-4)$

$48r + 192 = 72r - 288$

$480 = 24r$

$r = 20$

The speed of the boat in still water is 20 miles per hour.

71.

	Hours to Complete Total Job	Part of Job Completed in 1 Hour
Mark	7	$1/7$
Maria	x	$1/x$
Together	5	$1/5$

$\dfrac{1}{7} + \dfrac{1}{x} = \dfrac{1}{5}$

$35x\left(\dfrac{1}{7}\right) + 35x\left(\dfrac{1}{x}\right) = 35x\left(\dfrac{1}{5}\right)$

$5x + 35 = 7x$

$35 = 2x$

$x = \dfrac{35}{2}$ or $17\dfrac{1}{2}$

It takes Maria $17\dfrac{1}{2}$ hours to complete the job alone.

72.

	Days to Complete Total Job	Part of Job Completed in 1 Day
Pipe A	20	$1/20$
Pipe B	15	$1/15$
Together	x	$1/x$

$\dfrac{1}{20} + \dfrac{1}{25} = \dfrac{1}{x}$

$60x\left(\dfrac{1}{20}\right) + 60x\left(\dfrac{1}{15}\right) = 60x\left(\dfrac{1}{x}\right)$

$3x + 4x = 60$

$7x = 60$

$x = \dfrac{60}{7} = 8\dfrac{4}{7}$

Both pipes fill the pond in $8\dfrac{4}{7}$ days.

73. $\dfrac{2}{4} = \dfrac{10}{x}$

$2x = 40$

$x = 20$

The missing length is 20.

74. $\dfrac{12}{4} = \dfrac{18}{x}$

$12x = 72$

$x = 6$

The missing length is 6.

75. $\dfrac{9}{7\frac{1}{5}} = \dfrac{x}{12}$

$108 = 7\dfrac{1}{5}x$

$108 = \dfrac{36}{5}x$

$540 = 36x$

$15 = x$

The missing length is 15.

76. $\dfrac{x}{5} = \dfrac{30}{2.5}$

$2.5x = 150$

$x = 60$

The missing length is 60.

77. $\dfrac{\frac{5x}{27}}{-\frac{10xy}{21}} = \dfrac{5x}{27} \cdot -\dfrac{21}{10xy} = -\dfrac{5x \cdot 3 \cdot 7}{3 \cdot 9 \cdot 5 \cdot 2 \cdot x \cdot y}$

$= -\dfrac{7}{18y}$

78. $\dfrac{\frac{8x}{x^2-9}}{\frac{4}{x+3}} = \dfrac{8x}{x^2-9} \cdot \dfrac{x+3}{4}$

$= \dfrac{2 \cdot 4 \cdot x(x+3)}{(x-3)(x+3) \cdot 4} = \dfrac{2x}{x-3}$

79. $\dfrac{\frac{3}{5}+\frac{2}{7}}{\frac{1}{5}+\frac{5}{6}} = \dfrac{\frac{21}{35}+\frac{10}{35}}{\frac{6}{30}+\frac{25}{30}} = \dfrac{\frac{31}{35}}{\frac{31}{30}} = \dfrac{31}{35} \cdot \dfrac{30}{31}$

$= \dfrac{31 \cdot 5 \cdot 6}{5 \cdot 7 \cdot 31} = \dfrac{6}{7}$

80. $\dfrac{\frac{2}{a}+\frac{1}{2a}}{a+\frac{a}{2}} = \dfrac{2a\left(\frac{2}{a}+\frac{1}{2a}\right)}{2a\left(a+\frac{a}{2}\right)} = \dfrac{4+1}{2a^2+a^2} = \dfrac{5}{3a^2}$

81. $\dfrac{\frac{1}{a}+\frac{1}{b}}{\frac{1}{ab}} = \dfrac{ab\left(\frac{1}{a}+\frac{1}{b}\right)}{ab\left(\frac{1}{ab}\right)} = \dfrac{b+a}{1} = b+a$

82. $\dfrac{\frac{6}{x+2}+4}{\frac{8}{x+2}-4} = \dfrac{(x+2)\left(\frac{6}{x+2}+4\right)}{(x+2)\left(\frac{8}{x+2}-4\right)}$

$= \dfrac{(x+2)\left(\frac{6}{x+2}\right)+(x+2)(4)}{(x+2)\left(\frac{8}{x+2}\right)-(x+2)(4)}$

$= \dfrac{6+4x+8}{8-4x-8} = \dfrac{4x+14}{-4x} = -\dfrac{2(2x+7)}{2 \cdot 2x}$

$= -\dfrac{2x+7}{2x}$

83. $\dfrac{\frac{x-3}{x+3}+\frac{x+3}{x-3}}{\frac{x-3}{x+3}-\frac{x+3}{x-3}} = \dfrac{\left(\frac{x-3}{x+3}+\frac{x+3}{x-3}\right)(x+3)(x-3)}{\left(\frac{x-3}{x+3}-\frac{x+3}{x-3}\right)(x+3)(x-3)}$

$= \dfrac{(x-3)^2+(x+3)^2}{(x-3)^2-(x+3)^2}$

$= \dfrac{x^2-6x+9+x^2+6x+9}{x^2-6x+9-x^2-6x-9}$

$= \dfrac{2x^2+18}{-12x}$

$= -\dfrac{2(x^2+9)}{12x}$

$= -\dfrac{x^2+9}{6x}$

84. $\dfrac{\dfrac{3}{x-1} - \dfrac{2}{1-x}}{\dfrac{2}{x-1} - \dfrac{2}{x}} = \dfrac{\dfrac{3}{x-1} + \dfrac{2}{x-1}}{\dfrac{2}{x-1} - \dfrac{2}{x}}$

$$= \dfrac{\dfrac{5}{x-1}}{\dfrac{2x - 2(x-1)}{x(x-1)}}$$

$$= \dfrac{5}{x-1} \cdot \dfrac{x(x-1)}{2}$$

$$= \dfrac{5x}{2}$$

85. $\dfrac{x + y^{-1}}{\dfrac{x}{y}} = \dfrac{\left(x + \frac{1}{y}\right)(y)}{\left(\frac{x}{y}\right)(y)}$

$$= \dfrac{x+1}{x}$$

86. $\dfrac{x - xy^{-1}}{\dfrac{1+x}{y}} = \dfrac{\left(x - \frac{x}{y}\right)(y)}{\left(\frac{1+x}{y}\right)(y)}$

$$= \dfrac{xy - x}{1+x}$$

87. $\dfrac{y^{-2}}{1 - y^{-2}} = \dfrac{\left(\frac{1}{y^2}\right)\left(y^2\right)}{\left(1 - \frac{1}{y^2}\right)\left(y^2\right)}$

$$= \dfrac{1}{y^2 - 1}$$

88. $\dfrac{4 + x^{-1}}{3 + x^{-1}} = \dfrac{\left(4 + \frac{1}{x}\right)(x)}{\left(3 + \frac{1}{x}\right)(x)}$

$$= \dfrac{4x+1}{3x+1}$$

Chapter 7 Test

1. $f(x) = \dfrac{5x^2}{1-x}$

Undefined values when

$1 - x = 0$

$1 = x$

Domain $\{x | x \text{ is a real number and } x \neq 1\}$

2. $f(x) = \dfrac{9x^2 - 9}{x^2 + 4x + 3}$

Undefined values when

$x^2 + 4x + 3 = 0$

$(x+3)(x+1) = 0$

$x + 3 = 0 \quad \text{or} \quad x + 1 = 0$

$\quad x = -3 \qquad\qquad x = -1$

Domain

$\{x | x \text{ is a real number and } x \neq -3, \ x \neq -1\}$

3. $\dfrac{3x - 6}{5x - 10} = \dfrac{3(x-2)}{5(x-2)} = \dfrac{3}{5}$

4. $\dfrac{x+6}{x^2 + 12x + 36} = \dfrac{x+6}{(x+6)^2} = \dfrac{1}{x+6}$

5. $\dfrac{x+3}{x^3 + 27} = \dfrac{x+3}{(x+3)(x^2 - 3x + 9)} = \dfrac{1}{x^2 - 3x + 9}$

6. $\dfrac{2m^3 - 2m^2 - 12m}{m^2 - 5m + 6} = \dfrac{2m(m^2 - m - 6)}{(m-3)(m-2)}$

$$= \dfrac{2m(m-3)(m+2)}{(m-3)(m-2)} = \dfrac{2m(m+2)}{m-2}$$

7. $\dfrac{ay+3a+2y+6}{ay+3a+5y+15}$

$= \dfrac{a(y+3)+2(y+3)}{a(y+3)+5(y+3)}$

$= \dfrac{(a+2)(y+3)}{(a+5)(y+3)}$

$= \dfrac{a+2}{a+5}$

8. $\dfrac{y-x}{x^2-y^2} = \dfrac{-(x-y)}{(x-y)(x+y)} = -\dfrac{1}{x+y}$

9. $\dfrac{3}{x-1} \cdot (5x-5) = \dfrac{3}{x-1} \cdot 5(x-1)$

$= \dfrac{3 \cdot 5(x-1)}{x-1} = 15$

10. $\dfrac{y^2-5y+6}{2y+4} \cdot \dfrac{y+2}{2y-6}$

$= \dfrac{(y-3)(y-2) \cdot (y+2)}{2(y+2) \cdot 2(y-3)} = \dfrac{y-2}{4}$

11. $\dfrac{5}{2x+5} - \dfrac{6}{2x+5} = \dfrac{5-6}{2x+5} = \dfrac{-1}{2x+5}$

12. $\dfrac{5a}{a^2-a-6} - \dfrac{2}{a-3}$

$= \dfrac{5a}{(a-3)(a+2)} - \dfrac{2(a+2)}{(a-3)(a+2)}$

$= \dfrac{5a-2(a+2)}{(a-3)(a+2)} = \dfrac{5a-2a-4}{(a-3)(a+2)}$

$= \dfrac{3a-4}{(a-3)(a+2)}$

13. $\dfrac{6}{x^2-1} + \dfrac{3}{x+1}$

$= \dfrac{6}{(x+1)(x-1)} + \dfrac{3(x-1)}{(x+1)(x-1)}$

$= \dfrac{6+3x-3}{(x+1)(x-1)} = \dfrac{3x+3}{(x+1)(x-1)}$

$= \dfrac{3(x+1)}{(x+1)(x-1)} = \dfrac{3}{x-1}$

14. $\dfrac{x^2-9}{x^2-3x} \div \dfrac{xy+5x+3y+15}{2x+10}$

$= \dfrac{x^2-9}{x^2-3x} \cdot \dfrac{2x+10}{xy+5x+3y+15}$

$= \dfrac{(x-3)(x+3) \cdot 2(x+5)}{x(x-3) \cdot (x+3)(y+5)}$

$= \dfrac{2(x+3)(x+5)}{x(x+3)(y+5)}$

$= \dfrac{2(x+5)}{x(y+5)}$

15. $\dfrac{x+2}{x^2+11x+18} + \dfrac{5}{x^2-3x-10}$

$= \dfrac{x+2}{(x+9)(x+2)} + \dfrac{5}{(x-5)(x+2)}$

$$= \frac{(x+2)(x-5)}{(x+9)(x+2)(x-5)}$$

$$+ \frac{5(x+9)}{(x-5)(x+2)(x+9)}$$

$$= \frac{(x+2)(x-5) + 5(x+9)}{(x+9)(x+2)(x-5)}$$

$$= \frac{x^2 - 3x - 10 + 5x + 45}{(x+9)(x+2)(x-5)}$$

$$= \frac{x^2 + 2x + 35}{(x+9)(x+2)(x-5)}$$

16. $\dfrac{4}{y} - \dfrac{5}{3} = -\dfrac{1}{5}$

$$15y\left(\frac{4}{y} - \frac{5}{3}\right) = 15y\left(-\frac{1}{5}\right)$$

$$15y\left(\frac{4}{y}\right) - 15y\left(\frac{5}{3}\right) = 15y\left(-\frac{1}{5}\right)$$

$$60 - 25y = -3y$$

$$60 = 22y$$

$$\frac{60}{22} = y$$

$$y = \frac{30}{11}$$

17. $\dfrac{5}{y+1} = \dfrac{4}{y+2}$

$$5(y+2) = 4(y+1)$$

$$5y + 10 = 4y + 4$$

$$y = -6$$

18. $\dfrac{a}{a-3} = \dfrac{3}{a-3} - \dfrac{3}{2}$

$$2(a-3)\left(\frac{a}{a-3}\right) = 2(a-3)\left(\frac{3}{a-3} - \frac{3}{2}\right)$$

$$2a = 2(a-3)\left(\frac{3}{a-3}\right) - 2(a-3)\left(\frac{3}{2}\right)$$

$$2a = 6 - 3(a-3)$$

$$2a = 6 - 3a + 9$$

$$2a = 15 - 3a$$

$$5a = 15$$

$$a = 3$$

In the original equation, 3 makes a denominator 0. This equation has no solution.

19.
$$x - \frac{14}{x-1} = 4 - \frac{2x}{x-1}$$

$$(x-1)\left(x - \frac{14}{x-1}\right) = (x-1)\left(4 - \frac{2x}{x-1}\right)$$

$$x(x-1) - 14 = 4(x-1) - 2x$$

$$x^2 - x - 14 = 4x - 4 - 2x$$

$$x^2 - x - 14 = 2x - 4$$

$$x^2 - 3x - 10 = 0$$

$$(x-5)(x+2) = 0$$

$$x - 5 = 0 \text{ or } x + 2 = 0$$

$$x = 5 \text{ or } \quad x = -2$$

20. $\dfrac{\frac{5x^2}{yz^2}}{\frac{10x}{z^3}} = \dfrac{5x^2}{yz^2} \cdot \dfrac{z^3}{10x} = -\dfrac{5 \cdot x \cdot x \cdot z \cdot z^2}{y \cdot z^2 \cdot 2 \cdot 5 \cdot x}$

$$= \frac{xz}{2y}$$

21. $\dfrac{5 - \frac{1}{y^2}}{\frac{1}{y} + \frac{2}{y^2}} = \dfrac{y^2\left(5 - \frac{1}{y^2}\right)}{y^2\left(\frac{1}{y} + \frac{2}{y^2}\right)} = \dfrac{y^2(5) - y^2\left(\frac{1}{y^2}\right)}{y^2\left(\frac{1}{y}\right) + y^2\left(\frac{2}{y^2}\right)}$

$= \dfrac{5y^2 - 1}{y + 2}$

22. Let x = the number of defective bulbs.

$\dfrac{85}{3} = \dfrac{510}{x}$

$85x = 1530$

$x = 18$

Expect to find 18 defective bulbs.

23. $x + 5 \cdot \dfrac{1}{x} = 6$

$x + \dfrac{5}{x} = 6$

$x\left(x + \dfrac{5}{x}\right) = x(6)$

$x(x) + x\left(\dfrac{5}{x}\right) = x(6)$

$x^2 + 5 = 6x$

$x^2 - 6x + 5 = 0$

$(x - 5)(x - 1) = 0$

$x - 5 = 0 \quad$ or $\quad x - 1 = 0$

$x = 5 \qquad\qquad x = 1$

The unknown number is 5 or 1.

24.

	Distance	=	rate	$\cdot$	time
Upstream	14		$r - 2$		$14/r - 2$
Downstream	16		$r + 2$		$16/r + 2$

$\dfrac{14}{r - 2} = \dfrac{16}{r + 2}$

$14(r + 2) = 16(r - 2)$

$14r + 28 = 16r - 32$

$60 = 2r$

$r = 30$

The speed of the boat in still water is 30 miles per hour.

25.

	Hours to Complete Total Job	Part of Job Completed in 1 Hour
1st pipe	12	$1/12$
2nd pipe	15	$1/15$
Together	x	$1/x$

$\dfrac{1}{12} + \dfrac{1}{15} = \dfrac{1}{x}$

$60x\left(\dfrac{1}{12}\right) + 60x\left(\dfrac{1}{15}\right) = 60x\left(\dfrac{1}{x}\right)$

$5x + 4x = 60$

$9x = 60$

$x = \dfrac{60}{9} = \dfrac{20}{3} = 6\dfrac{2}{3}$

Together, the pipes can fill the tank in $6\dfrac{2}{3}$ hours.

26. $\dfrac{8}{x} = \dfrac{10}{15}$

$8(15) = 10x$

$120 = 10x$

$12 = x$

Chapter 7 Cumulative Review

1. a. $\dfrac{15}{x} = 4$

  b. $12 - 3 = x$

  c. $4x + 17 = 21$

2. a. $12 - x = -45$

  b. $12x = -45$

  c. $x - 10 = 2x$

3. Let x = the amount invested at 9%
for one year.

Principal · Rate = Interest

	x	.09	.09x
9%	x	.09	$.09x$
7%	$20,000 - x$	.07	$.07(20,000-x)$
Total	20,000		1550

$.09x + .07(20,000 - x) = 1550$

$.09x + 1400 - .07x = 1550$

$.02x + 1400 = 1550$

$.02x = 150$

$x = 7500$

$20,000 - x = 20,000 - 7500 = 12,500$

He invested \$7500 @ 9% and
\$12,500 @ 7%

4. Let x = the number of bankruptcies in
1994 and $2x - 80,000$ = the number in
2002.

$x + 2x - 80,000 = 2,290,000$

$3x - 80,000 = 2,290,000$

$3x = 2,370,000$

$x = 790,000$

$2x - 80,000 = 2(790,000) - 80,000$

$= 1,500,000$

There were 790,000 bankruptcies in 1994
and 1,500,000 in 2002.

5. $x - 3y = 6$

x	y
0	-2
6	0

6. $7x + 2y = 9$

$2y = -7x + 9$

$y = -\dfrac{7}{2}x + \dfrac{9}{2}$

$y = mx + b$

$m = -\dfrac{7}{2}$

7. a. $4^2 \cdot 4^5 = 4^{2+5} = 4^7$

  b. $x^4 \cdot x^6 = x^{4+6} = x^{10}$

  c. $y^3 \cdot y = y^{3+1} = y^4$

  d. $y^3 \cdot y^2 \cdot y^7 = y^{3+2+7} = y^{12}$

  e. $(-5)^7 \cdot (-5)^8 = (-5)^{7+8} = (-5)^{15}$

  f. $a^2 \cdot b^2 = a^2 b^2$

8. a. $\dfrac{x^9}{x^7} = x^{9-7} = x^2$

b. $\dfrac{x^{19}y^5}{xy} = x^{19-1} \cdot y^{5-1} = x^{18}y^4$

c. $\left(x^5y^2\right)^3 = x^{5\cdot3}y^{2\cdot3} = x^{15}y^6$

d. $\left(-3a^2b\right)\left(5a^3b\right) = -15a^{2+3}b^{1+1}$

$= -15a^5b^2$

9. $\left(12z^5 - 12x^3 + z\right) - \left(-3z^4 + z^3 + 12z\right)$

$= 12z^5 - 12x^3 + z + 3z^4 - z^3 - 12z$

$= 12z^5 + 3z^4 - 13z^3 - 11z$

10. $\left(x+1\right) - \left(9x^2 - 6x + 2\right)$

$= x + 1 - 9x^2 + 6x - 2$

$= -9x^2 + 7x - 1$

11. $\left(3a+b\right)^3 = \left(3a+b\right)\left(3a+b\right)^2$

$= \left(3a+b\right)\left[\left(3a\right)^2 + 2\left(3a\right)\left(b\right) + \left(b\right)^2\right]$

$= \left(3a+b\right)\left(9a^2 + 6ab + b^2\right)$

$= 27a^3 + 18a^2b + 3ab^2 + 9a^2b + 6ab^2 + b^3$

$= 27a^3 + 27a^2b + 9ab^2 + b^3$

12. $\left(2x+1\right)\left(5x^2 - x + 2\right)$

$= 2x\left(5x^2 - x + 2\right) + 1\left(5x^2 - x + 2\right)$

$= 10x^3 - 2x^2 + 4x + 5x^2 - x + 2$

$= 10x^3 + 3x^2 + 3x + 2$

13. a. $\left(t+2\right)^2 = \left(t\right)^2 + 2\left(t\right)\left(2\right) + \left(2\right)^2$

$= t^2 + 4t + 4$

b. $\left(p-q\right)^2 = \left(p\right)^2 + 2\left(p\right)\left(q\right) + \left(q\right)^2$

$= p^2 + 2pq + q$

c. $\left(2x+5\right)^2 = \left(2x\right)^2 + 2\left(2x\right)\left(5\right) + \left(5\right)^2$

$= 4x^2 + 20x + 25$

d. $\left(x^2 - 7y\right)^2 = \left(x^2\right)^2 - 2\left(x^2\right)\left(7y\right) + \left(7y\right)^2$

$= x^4 - 14x^2y + 49y^2$

14. a. $\left(x+9\right)^2 = \left(x\right)^2 + 2\left(x\right)\left(9\right) + \left(9\right)^2$

$= x^2 + 18x + 81$

b. $\left(2x+1\right)\left(2x-1\right) = \left(2x\right)^2 - \left(1\right)^2$

$= 4x^2 - 1$

c. $8x\left(x^2+1\right)\left(x^2-1\right) = 8x\left[\left(x^2\right)^2 - \left(1\right)^2\right]$

$= 8x\left[x^4 - 1\right]$

$= 8x^5 - 8x$

15. a. $\dfrac{1}{x^{-3}} = x^3$

b. $\dfrac{1}{3^{-4}} = 3^4 = 81$

c. $\dfrac{p^{-4}}{q^{-9}} = \dfrac{q^9}{p^4}$

d. $\dfrac{5^{-3}}{2^{-5}} = \dfrac{2^5}{5^3} = \dfrac{32}{125}$

16. a. $5^{-3} = \dfrac{1}{5^3} = \dfrac{1}{125}$

    b. $\dfrac{9}{x^{-7}} = 9x^7$

    c. $\dfrac{11^{-1}}{7^{-2}} = \dfrac{7^2}{11^1} = \dfrac{49}{11}$

17.
$$2x+3\overline{\smash{\big)}\,8x^3 + 4x^2 + 0x + 7}$$

$$\underline{8x^3 + 12x^2}$$
$$-8x^2 + 0x$$
$$\underline{-8x^2 - 12x}$$
$$12x + 7$$
$$\underline{12x + 18}$$
$$-11$$

quotient: $4x^2 - 4x + 6$

$$\dfrac{8x^3 + 4x^2 + 7}{2x+3} = 4x^2 - 4x + 6 - \dfrac{11}{2x+3}$$

18.
$$x-4\overline{\smash{\big)}\,4x^3 + 0x^2 - 9x + 2}$$

$$\underline{4x^3 - 16x^2}$$
$$16x^2 - 9x$$
$$\underline{16x^2 - 64x}$$
$$55x + 2$$
$$\underline{55x - 220}$$
$$222$$

quotient: $4x^2 + 16x + 55$

$$\dfrac{4x^3 - 9x + 2}{x-4} = 4x^2 + 16x + 55 + \dfrac{222}{x-4}$$

19. a. $28 = 2 \cdot 2 \cdot 7$

    $40 = 2 \cdot 2 \cdot 2 \cdot 5$

    $\text{GCF} = 2^2 = 4$

    b. $55 = 5 \cdot 11$

    $21 = 3 \cdot 7$

    $\text{GCF} = 1$

c. $15 = 3 \cdot 5$

    $18 = 2 \cdot 3 \cdot 3$

    $66 = 2 \cdot 3 \cdot 11$

    $\text{GCF} = 3$

20. $9x^2 = 3 \cdot 3 \cdot x^2$

    $6x^3 = 2 \cdot 3 \cdot x^3$

    $21x^5 = 3 \cdot 7 \cdot x^5$

    $\text{GCF} = 3x^2$

21. $-9a^5 + 18a^2 - 3a = -3a\left(3a^4 - 6a + 1\right)$

22. $7x^6 - 7x^5 + 7x^4 = 7x^4\left(x^2 - x + 1\right)$

23. $3m^2 - 24m - 60$

    $= 3\left(m^2 - 8m - 20\right)$

    $= 3\left(m^2 - 10m + 2m - 20\right)$

    $= 3\left[m(m-10) + 2(m-10)\right]$

    $= 3(m-10)(m+2)$

24. $-2a^2 + 10a + 12 = -2\left(a^2 - 5a - 6\right)$

    $= -2(a+1)(a-6)$

25. $3x^2 + 11x + 6 = 3x^2 + 2x + 9x + 6$

    $= x(3x+2) + 3(3x+2)$

    $= (3x+2)(x+3)$

26. $10m^2 - 7m + 1$

    $= 10m^2 - 2m - 5m + 1$

    $= 2m(5m-1) - 1(5m-1)$

    $= (2m-1)(5m-1)$

27. $x^2 + 12x + 36 = x^2 + 2 \cdot x \cdot 6 + 6^2 = (x+6)^2$

28. $4x^2 + 12x + 9 = (2x)^2 + 2(2x)(3) + (3)^2$
$$= (2x+3)^2$$

29. $x^2 + 4$ is a prime polynomial

30. $x^2 - 4 = (x)^2 - (2)^2 = (x+2)(x-2)$

31. $x^3 + 8 = x^3 + 2^3$
$$= (x+2)(x^2 - x \cdot 2 + 2^2)$$
$$= (x+2)(x^2 - 2x + 4)$$

32. $27y^3 - 1 = (3y)^3 - (1)^3$
$$= (3y-1)\left[(3y)^2 + 3y(1) + (1)^2\right]$$
$$= (3y-1)(9y^2 + 3y + 1)$$

33. $2x^3 + 3x^2 - 2x - 3$
$$= x^2(2x+3) - 1(2x+3)$$
$$= (2x+3)(x^2 - 1)$$
$$= (2x+3)(x^2 - 1^2)$$
$$= (2x+3)(x+1)(x-1)$$

34. $3x^3 + 5x^2 - 12x - 20$
$$= x^2(3x+5) - 4(3x+5)$$
$$= (3x+5)(x^2 - 4)$$
$$= (3x+5)(x^2 - 2^2)$$
$$= (3x+5)(x+2)(x-2)$$

35. $12m^2 - 3n^2 = 3(4m^2 - n^2)$
$$= 3\left[(2m)^2 - (n)^2\right]$$
$$= 3(2m+n)(2m-n)$$

36. $x^5 - x = x(x^4 - 1)$
$$= x\left[(x^2)^2 - 1^2\right]$$
$$= x(x^2 + 1)(x^2 - 1)$$
$$= x(x^2 + 1)(x+1)(x-1)$$

37.
$$x(2x-7) = 4$$
$$2x^2 - 7x = 4$$
$$2x^2 - 7x - 4 = 0$$
$$2x^2 - 8x + x - 4 = 0$$
$$2x(x-4) + 1(x-4) = 0$$
$$(x-4)(2x+1) = 0$$
$$2x+1 = 0 \text{ or } x - 4 = 0$$
$$2x = -1 \text{ or } \quad x = 4$$
$$x = -\frac{1}{2}$$

38.
$$3x^2 + 5x = 2$$
$$3x^2 + 5x - 2 = 0$$
$$3x^2 + 6x - x - 2 = 0$$
$$3x(x+2) - 1(x+2) = 0$$
$$(x+2)(3x-1) = 0$$
$$3x - 1 = 0 \text{ or } x + 2 = 0$$
$$3x = 1 \text{ or } \quad x = -2$$
$$x = \frac{1}{3}$$

39. $y = x^2 - 5x + 4$

$0 = x^2 - 5x + 4$

$0 = (x - 4)(x - 1)$

$x - 1 = 0$ or $x - 4 = 0$

$x = 1$ or $x = 4$

The x-intercepts are $(1, 0)$ and $(4, 0)$.

40. $y = x^2 - x - 6$

$0 = x^2 - x - 6$

$0 = (x - 3)(x + 2)$

$x + 2 = 0$ or $x - 3 = 0$

$x = -2$ or $x = 3$

The x-intercepts are $(-2, 0)$ and $(3, 0)$.

41.   Let x = the base and $2x - 2$ = the height.

$$A = \frac{1}{2}bh$$

$$30 = \frac{1}{2}x(2x - 2)$$

$$30 = \frac{1}{2}(2x)(x - 1)$$

$$30 = x(x - 1)$$

$$30 = x^2 - x$$

$$0 = x^2 - x - 30$$

$$0 = (x + 5)(x - 6)$$

$x - 6 = 0$ or $x + 5 = 0$

$x = 6$ or $x = -5$

Length cannot be negative, so $x = 6$

$2x - 2 = 2(6) - 2 = 10$

The base is 6 m and the height is 10 m.

42.   Let x = the base and $3x + 5$ = the height.

$$A = bh$$

$$182 = x(3x + 5)$$

$$182 = 3x^2 + 5x$$

$$0 = 3x^2 + 5x - 182$$

$$0 = 3x^2 + 26x - 21x - 182$$

$$0 = x(3x + 26) - 7(3x + 26)$$

$$0 = (x - 7)(3x + 26)$$

$x - 7 = 0$ or $3x + 26 = 0$

$x = 7$ or $x = -\dfrac{26}{3}$

Length cannot be negative, so $x = 7$

$3x + 5 = 3(7) + 5 = 26$

The base is 7 ft and the height is 26 ft.

43. $\dfrac{18 - 2x^2}{x^2 - 2x - 3} = \dfrac{-2x^2 + 18}{x^2 - 2x - 3}$

$$= \frac{-2(x^2 - 9)}{(x - 3)(x + 1)}$$

$$= \frac{-2(x + 3)(x - 3)}{(x - 3)(x + 1)}$$

$$= \frac{-2(x + 3)}{x + 1}$$

44. $\dfrac{2x^2 - 50}{4x^4 - 20x^3} = \dfrac{2(x^2 - 25)}{4x^3(x - 5)}$

$$= \frac{2(x + 5)(x - 5)}{4x^3(x - 5)}$$

$$= \frac{x + 5}{2x^3}$$

45. $\dfrac{6x+2}{x^2-1} \div \dfrac{3x^2+x}{x-1}$

$= \dfrac{6x+2}{x^2-1} \cdot \dfrac{x-1}{3x^2+x}$

$= \dfrac{2(3x+1)}{(x+1)(x-1)} \cdot \dfrac{x-1}{x(3x+1)}$

$= \dfrac{2}{x(x+1)}$

46. $\dfrac{6x^2-18x}{3x^2-2x} \cdot \dfrac{15x-10}{x^2-9}$

$= \dfrac{6x(x-3)}{x(3x-2)} \cdot \dfrac{5(3x-2)}{(x+3)(x-3)}$

$= \dfrac{30}{x+3}$

47. $\dfrac{(2x)^{-1}+1}{2x^{-1}-1} = \dfrac{\left(\frac{1}{2x}+1\right)(2x)}{\left(\frac{2}{x}-1\right)(2x)}$

$\qquad\qquad = \dfrac{1+2x}{4-2x}$

48. $\dfrac{\dfrac{m}{3}+\dfrac{n}{6}}{\dfrac{m+n}{12}} = \dfrac{12}{12} \cdot \dfrac{\dfrac{m}{3}+\dfrac{n}{6}}{\dfrac{m+n}{12}}$

$\qquad = \dfrac{12\left(\dfrac{m}{3}\right)+12\left(\dfrac{n}{6}\right)}{12\left(\dfrac{m+n}{12}\right)}$

$\qquad = \dfrac{4m+2n}{m+n} \ \text{or} \ \dfrac{2(2m+n)}{m+n}$

Chapter 8

Section 8.1

Mental Math

1. $m = -4$, $b = 12$

2. $m = \frac{2}{3}$, $b = -\frac{7}{2}$

3. $m = 5$, $b = 0$

4. $m = -1$, $b = 0$

5. $m = \frac{1}{2}$, $b = 6$

6. $m = -\frac{2}{3}$, $b = 5$

7. Parallel

8. Parallel

9. Neither

10. Neither

Exercise Set 8.1

1. $f(x) = -2x$

x	0	-1	1
y	0	2	-2

Plot the points to obtain the graph.

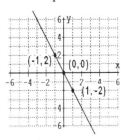

3. $f(x) = -2x + 3$

x	0	1	-1
y	3	1	5

Plot the points to obtain the graph.

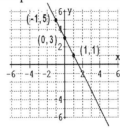

5. $f(x) = \frac{1}{2}x$

x	0	2	-2
y	0	1	-1

Plot the points to obtain the graph.

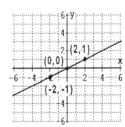

7. $f(x) = \frac{1}{2}x - 4$

x	0	2	4
y	-4	-3	-2

Plot the points to obtain the graph.

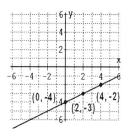

9. C

11. D

13. $x - y = 3$

Let $x = 0$ Let $y = 0$ Let $x = 2$
$0 - y = 3$ $x - 0 = 3$ $2 - y = 3$
$\quad y = -3$ $\quad x = 3$ $\quad y = -1$

x	0	3	2
y	-3	0	-1

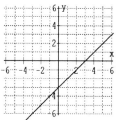

$f(x) = x - 3$

15. $x = 5y$

Let $x = 0$ Let $x = 5$ Let $x = -5$
$0 = 5y$ $5 = 5y$ $-5 = 5y$
$y = 0$ $y = 1$ $y = -1$

x	0	5	-5
y	0	1	-1

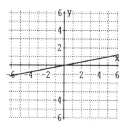

$f(x) = \dfrac{x}{5}$

17. $-x + 2y = 6$

Let $x = 0$ Let $y = 0$ Let $x = 2$
$-0 + 2y = 6$ $-x + 2(0) = 6$ $-2 + 2y = 6$
$\quad y = 3$ $\quad x = -6$ $\quad y = 2$

x	0	-6	2
y	3	0	2

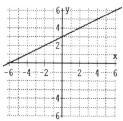

$f(x) = \dfrac{1}{2}x + 3$

19. $2x - 4y = 8$

Let $x = 0$ Let $y = 0$
$2(0) - 4y = 8$ $2x - 4(0) = 8$
$\quad y = -2$ $\quad x = 4$

Let $x = 2$
$2(2) - 4y = 8$
$\quad y = -1$

x	0	4	2
y	-2	0	-1

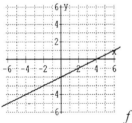

$$f(x) = \frac{1}{2}x - 2$$

21. Answers may vary.

23. $m = \dfrac{6-0}{4-2} = \dfrac{6}{2} = 3$

$y - 0 = 3(x-2)$
$y = 3x - 6$
$f(x) = 3x - 6$

25. $m = \dfrac{13-5}{-6-(-2)} = \dfrac{8}{-4} = -2$

$y - 5 = -2[x - (-2)]$
$y - 5 = -2(x+2)$
$y - 5 = -2x - 4$
$y = -2x + 1$
$f(x) = -2x + 1$

27. $m = \dfrac{-9-(-8)}{-6-(-3)} = \dfrac{-1}{-3} = \dfrac{1}{3}$

$y - (-8) = \dfrac{1}{3}[x - (-3)]$

$y + 8 = \dfrac{1}{3}(x+3)$

$3y + 24 = x + 3$
$3y = x - 21$
$y = \dfrac{1}{3}x - 7$

$f(x) = \dfrac{1}{3}x - 7$

29. $y = 4x - 2$ so $m = 4$

$y - 8 = 4(x-3)$
$y - 8 = 4x - 12$
$y = 4x - 4$
$f(x) = 4x - 4$

31. $3y = x - 6$ or $y = \dfrac{1}{3}x - 2$

so $m = \dfrac{1}{3}$ and $m_\perp = -3$

$y - (-5) = -3(x-2)$
$y + 5 = -3x + 6$
$y = -3x + 1$
$f(x) = -3x + 1$

33. $3x + 2y = 5$

$2y = -3x + 5$

$y = -\dfrac{3}{2}x + \dfrac{5}{2}$ so $m = -\dfrac{3}{2}$

$y - (-3) = -\dfrac{3}{2}[x - (-2)]$

$2(y+3) = -3(x+2)$
$2y + 6 = -3(x+2)$
$2y + 6 = -3x - 6$

$y = -\dfrac{3}{2}x - 6$

$f(x) = -\dfrac{3}{2}x - 6$

35. $x + 2y = 8$

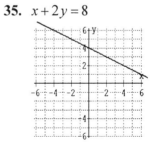

37. $3x + 5y = 7$

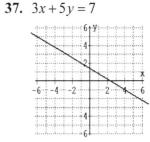

39. $5 = 6x - y$

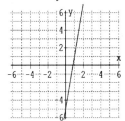

41. $-x + 10y = 11$

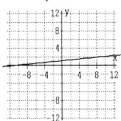

43. $x + 3 = 0,$ or $x = -3$

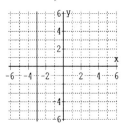

45. $f(x) = \dfrac{3}{4}x + 2$

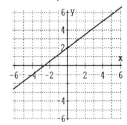

47. $f(x) = x$

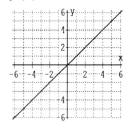

49. $f(x) = \dfrac{1}{2}x$

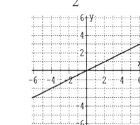

51. $f(x) = 4x - \dfrac{1}{3}$

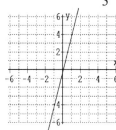

53.
$$y - 3 = 2[x - (-2)]$$
$$y - 3 = 2(x + 2)$$
$$y - 3 = 2x + 4$$
$$2x - y = -7$$

55. $m = \dfrac{-6 - (-4)}{0 - (-7)} = \dfrac{-2}{7} = -\dfrac{2}{7}$

$$y = -\dfrac{2}{7}x - 6$$
$$7y = -2x - 42$$
$$2x + 7y = -42$$

57. $2x + 4y = 8$

$\qquad 4y = -2x + 8$

$\qquad\qquad y = -\dfrac{1}{2}x + 2$ so $m = -\dfrac{1}{2}$

$y - (-2) = -\dfrac{1}{2}(x - 6)$

$2(y + 2) = -(x - 6)$

$\quad 2y + 4 = -x + 6$

$\quad x + 2y = 2$

59. $8x - y = 9$

$\qquad\qquad y = 8x - 9$ so $m = 8$

$\qquad y - 1 = 8(x - 6)$

$\qquad y - 1 = 8x - 48$

$\quad 8x - y = 47$

61. $f(x) = x^2$

$\qquad f(3) = (3)^2 = 9$

63. $g(x) = |x|$

$\qquad g(-2) = |-2| = 2$

65. $h(x) = 3x - 4$

$\qquad h(1) = 3(1) - 4 = -1$

67. $2x + 3y = 1500$

    a. $2(0) + 3y = 1500$

$\qquad\qquad 3y = 1500$

$\qquad\qquad\; y = 500$

    (0, 500); If no tables are produced, 500 chairs can be produced.

    b. $2x + 3(0) = 1500$

$\qquad\qquad 2x = 1500$

$\qquad\qquad\; x = 750$

    (750, 0); If no chairs are produced, 750 tables can be produced.

    c. $2(50) + 3y = 1500$

$\qquad\quad 100 + 3y = 1500$

$\qquad\qquad\;\; 3y = 1400$

$\qquad\qquad\quad y = 466.7$

    466 chairs

69. $C(x) = 0.2x + 24$

    a. $C(200) = 0.2(200) + 24$

$\qquad\qquad\quad = 40 + 24$

$\qquad\qquad\quad = 64$

    $64

    b.

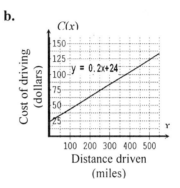

    c. The line moves upward from left to right.

71. $f(x) = 53.6x + 849.88$

    a. $f(20) = 53.6(20) + 849.88$

$\qquad\qquad\; = 1072 + 849.88$

$\qquad\qquad\; = 1921.88$

    $1921.88

    b. $\quad 2000 = 53.6x + 849.88$

$\qquad 1150.12 = 53.6x$

$\qquad\; 21.46 = x$

$\qquad 1990 + 22 = 2012$

    c. Answers may vary.

73. $D(x) = \dfrac{136}{25}x$

$\qquad D(30) = \dfrac{136}{25}(30) = 163.2$ milligrams

75. $C(x) = 1.69x + 87.54$

    a. $C(5) = 1.69(5) + 87.54 = 95.99$

    The per capita consumption of poultry was about 95.99 lb in 2000.

b. 2002 gives $x = 7$
$$C(7) = 1.69(7) + 87.54$$
$$= 99.37 \text{ pounds}$$

77. a. $(0, 133{,}300), (3, 147{,}802)$
$$m = \frac{147{,}802 - 133{,}300}{3 - 0}$$
$$= \frac{14502}{3}$$
$$= 4834$$
$$y - 133{,}300 = 4834(x - 0)$$
$$y = 4834x + 133{,}300$$

b. $x = 2008 - 1999 = 9$
$$y = 4834(9) + 133{,}300$$
$$= \$176{,}806$$

c. The median price of a home is rising $4834 every year.

79. a. $(0, 757), (10, 1052)$
$$m = \frac{1052 - 757}{10 - 0} = \frac{295}{10} = 29.5$$
$$y - 757 = 29.5(x - 0)$$
$$y = 29.5x + 757$$

b. $x = 2004 - 2000 = 4$
$$y = 29.5(4) + 757$$
$$= 875 \text{ thousand people}$$

81. $m = \dfrac{1 - (-1)}{-5 - 3} = \dfrac{2}{-8} = -\dfrac{1}{4}$ so $m_\perp = 4$

$$M((3, -1), (5, 1)) = \left(\frac{3 - 5}{2}, \frac{-1 + 1}{2} \right) = (1, 0)$$

$$y - 0 = 4[x - (-1)]$$
$$y = 4(x + 1)$$
$$y = 4x + 4$$
$$-4x + y = 4$$

83. $m = \dfrac{-4 - 6}{-22 - (-2)} = \dfrac{-10}{-20} = \dfrac{1}{2}$ so $m_\perp = -2$

$$M((-2, 6), (-22, -4)) = \left(\frac{-2 - 22}{2}, \frac{6 - 4}{2} \right)$$
$$= (-12, 1)$$

$$y - 1 = -2[x - (-12)]$$
$$y - 1 = -2(x + 12)$$
$$y - 1 = -2x - 24$$
$$2x + y = -23$$

85. $m = \dfrac{7 - 3}{-4 - 2} = \dfrac{4}{-6} = -\dfrac{2}{3}$ so $m_\perp = \dfrac{3}{2}$

$$M((2, 3), (-4, 7)) = \left(\frac{2 - 4}{2}, \frac{3 + 7}{2} \right)$$
$$= (-1, 5)$$

$$y - 5 = \frac{3}{2}[x - (-1)]$$
$$2(y - 5) = 3(x + 1)$$
$$2y - 10 = 3x + 3$$
$$3x - 2y = -13$$

87. They agree.

89. They agree.

91. a. $y = -4x + 2$ is a line parallel to $y = -4x$ but with y-intercept $(0, 2)$.

b. $y = -4x - 5$ is a line parallel to $y = -4x$ but with y-intercept $(0, -5)$.

Section 8.2

Graphing Calculator Explorations

1.

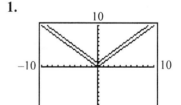

3.

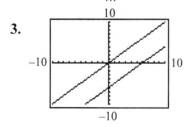

5.

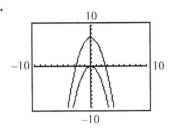

Exercise Set 8.2

1. $f(1) = 0$

3. $f(-1) = -4$

5. $f(x) = 4$, $x = 3$

7. $(1, -10)$

9. $(4, 56)$

11. $f(-1) = -2$

13. $g(2) = 0$

15. $-4, 0$

17. 3

19. $f(x) = x^2 + 3$

x	y
-2	7
-1	4
0	3
1	4
2	7

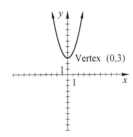

21. $h(x) = |x| - 2$

x	y
-2	0
-1	-1
0	-2
1	-1
2	0

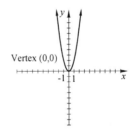

23. $g(x) = 2x^2$

x	y
-2	8
-1	2
0	0
1	2
2	8

25. $f(x) = 5x - 1$

x	y
-1	-6
0	-1
1	4
2	9

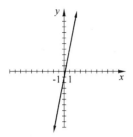

27. $h(x) = |x + 1| + 3$

x	y
−3	5
−2	4
−1	3
0	4
1	5

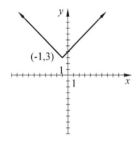

(-1,3)

29. $g(x) = -|x|$

x	y
−2	−2
−1	−1
0	0
1	−1
2	−2

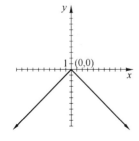

(0,0)

31. $h(x) = x^3$

x	y
−2	−8
−1	−1
0	0
1	1
2	8

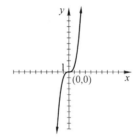

(0,0)

33. d

35. c

37. A function may have an infinite number of x-intercepts. Answers may vary.

39. $A(r) = \pi r^2$
$A(5) = \pi(5)^2 = 25\pi$ square centimeters

41. $V(x) = x^3$
$V(14) = (14)^3 = 2744$ cubic inches

43. $(5x - 7) - (9x - 1) = 5x - 7 - 9x + 1$
$= -4x - 6$

45. $(5x - 7) + (9x - 1) = 5x - 7 + 9x - 1$
$= 14x - 8$

47. B

49. C

51. 1991: In April it rose to $4.25.

53. Answers may vary.

55. $y = x^2 - 4x + 7$

x	y

0	7
1	4
2	3
3	4
4	7

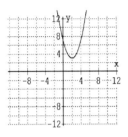

57. $f(x) = [x]$

x	y
-2	-2
-1	-1
0	0
1	1
2	2

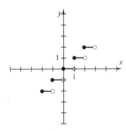

Integrated Review-Summary on Functions

1. $f(x) = 3x - 5$
$m = 3,\ (0, -5)$

2. $f(x) = \dfrac{5}{2}x - \dfrac{7}{2}$
$m = \dfrac{5}{2},\ \left(0, -\dfrac{7}{2}\right)$

3. $y = 8x - 6$ $y = 8x + 6$
$m = 8$ $m = 8$
Parallel, since their slopes are equal.

4. $y = \dfrac{2}{3}x + 1$ $2y + 3x = 1$
$m = \dfrac{2}{3}$ $2y = -3x + 1$
                    $y = -\dfrac{3}{2}x + \dfrac{1}{2}$
                 $m = -\dfrac{3}{2}$

Perpendicular, since the product of their slopes is -1.

5. $m = \dfrac{2 - 6}{5 - 1} = \dfrac{-4}{4} = -1$
$y - 1 = -1(x - 6)$
$y - 1 = -x + 6$
$y = -x + 7$

6. $m = \dfrac{-5 - (-8)}{-6 - 2} = \dfrac{3}{-8} = -\dfrac{3}{8}$
$y - (-8) = -\dfrac{3}{8}(x - 2)$
$8(y + 8) = -3(x - 2)$
$8y + 64 = -3x + 6$
$8y = -3x - 58$
$y = -\dfrac{3}{8}x - \dfrac{29}{4}$

7. $3x - y = 5$
$y = 3x - 5$
$m = 3$
$y - (-5) = 3[x - (-1)]$
$y + 5 = 3(x + 1)$
$y + 5 = 3x + 3$
$y = 3x - 2$

8. $4x - 5y = 10$
$-5y = -4x + 10$
$y = \dfrac{4}{5}x - 2;\ m = \dfrac{4}{5}$ so $m_\perp = -\dfrac{5}{4}$
Therefore, $y = -\dfrac{5}{4}x + 4$.

9. $4x + y = \dfrac{2}{3}$
$y = -4x + \dfrac{2}{3};\ m = -4$ so $m_\perp = \dfrac{1}{4}$

$$y - (-3) = \frac{1}{4}(x - 2)$$
$$4(y + 3) = x - 2$$
$$4y + 12 = x - 2$$
$$4y = x - 14$$
$$y = \frac{1}{4}x - \frac{7}{2}$$

10. $5x + 2y = 2$
$$2y = -5x + 2$$
$$y = -\frac{5}{2}x + 1$$
$$m = -\frac{5}{2}$$
$$y - 0 = -\frac{5}{2}[x - (-1)]$$
$$y = -\frac{5}{2}(x + 1)$$
$$2y = -5(x + 1)$$
$$2y = -5x - 5$$
$$y = -\frac{5}{2}x - \frac{5}{2}$$

11. Linear

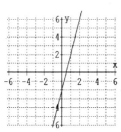

12. Linear

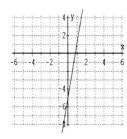

13. Not linear

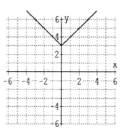

14. Not linear

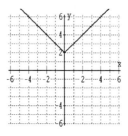

15. Not linear

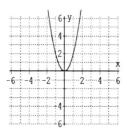

16. Not linear

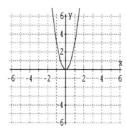

17. Not linear

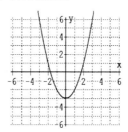

18. Not linear

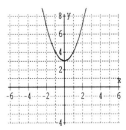

19. Linear

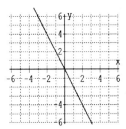

20. Linear

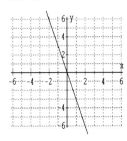

21. Not linear

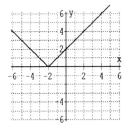

22. Not linear

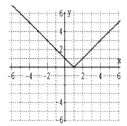

23. Linear

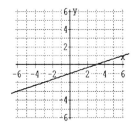

24. Linear

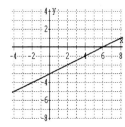

25. Linear

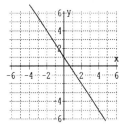

26. Linear

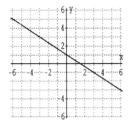

Section 8.3

Mental Math

1. C

2. E

3. F

4. A

5. D

6. B

Exercise Set 8.3

1. **a.** $(f+g)(x) = x-7+2x+1 = 3x-6$

    b. $(f-g)(x) = x-7-(2x+1)$
$$= x-7-2x-1$$
$$= -x-8$$

    c. $(f \cdot g)(x) = (x-7)(2x+1)$
$$= 2x^2 -13x-7$$

    d. $\left(\dfrac{f}{g}\right)(x) = \dfrac{x-7}{2x+1}$, where $x \neq \dfrac{1}{2}$

3. **a.** $(f+g)(x) = x^2 +5x+1$

    b. $(f-g)(x) = x^2 -5x+1$

    c. $(f \cdot g)(x) = (x^2 +1)(5x) = 5x^3 +5x$

    d. $\left(\dfrac{f}{g}\right)(x) = \dfrac{x^2 +1}{5x}$, where $x \neq 0$

5. $f(x) = |x|,\ g(x) = x+5$

    a. $(f+g)(x) = |x| +5$

    b. $(f-g)(x) = |x| -5$

    c. $(f \cdot g)(x) = |x|(x+5)$

    d. $(f/g)(x) = \dfrac{|x|}{x+5}$

7. $f(x) = -3x,\ g(x) = 5x^2$

    a. $(f+g)(x) = -3x+5x^2$

    b. $(f-g)(x) = -3x-5x^2$

    c. $(f \cdot g)(x) = -3x(5x^2) = -15x^3$

    d. $(f/g)(x) = \dfrac{-3x}{5x^2} = -\dfrac{3}{5x}$

9. $(f \circ g)(2) = f(g(2))$
$$= f(-4)$$
$$= (-4)^2 -6(-4)+2$$
$$= 16+24+2$$
$$= 42$$

11. $(g \circ f)(-1) = g(f(-1))$
$$= g(9)$$
$$= -2(9)$$
$$= -18$$

13. $(g \circ h)(0) = g(h(0))$
$$= g(0)$$
$$= -2(0)$$
$$= 0$$

15. $(f \circ g)(x) = f(g(x))$
$$= f(5x)$$
$$= (5x)^2 +1$$
$$= 25x^2 +1$$

17. $(f \circ g)(x) = f(g(x))$
$$= f(x+7)$$
$$= 2(x+7)-3$$
$$= 2x+14-3$$
$$= 2x+11$$

19. $(f \circ g)(x) = f(g(x))$
$\qquad\qquad = f(-2x)$
$\qquad\qquad = (-2x)^3 + (-2x) - 2$
$\qquad\qquad = -8x^3 - 2x - 2$

21. $(f \circ g)(x) = f(g(x))$
$\qquad\qquad = f(-5x + 2)$
$\qquad\qquad = |-5x + 2|$

23. $H(x) = (g \circ h)(x)$
$\qquad\quad = g(h(x))$
$\qquad\quad = g(x^2 + 2)$
$\qquad\quad = |x^2 + 2|$

25. $F(x) = (h \circ f)(x)$
$\qquad\quad = h(f(x))$
$\qquad\quad = h(3x)$
$\qquad\quad = (3x)^2 + 2$
$\qquad\quad = 9x^2 + 2$

27. $G(x) = (f \circ g)(x)$
$\qquad\quad = f(g(x))$
$\qquad\quad = f(|x|)$
$\qquad\quad = 3|x|$

29. Answers may vary. For example, $g(x) = x + 2$ and $f(x) = x^2$.

31. Answers may vary. For example, $g(x) = x + 5$ and $f(x) = |x| + 2$.

33. Answers may vary. For example, $g(x) = 2x - 3$ and $f(x) = \dfrac{1}{x}$.

35. $y = x - 2$

37. $y = \dfrac{x}{3}$

39. $y = -\dfrac{x + 7}{2}$

41. 6

43. 4

45. 48

47. -1

49. Answers may vary.

51. $P(x) = R(x) - C(x)$

Section 8.4

Mental Math

1. $y = 5x$ represents direct variation

2. $y = \dfrac{300}{x}$ represents inverse variation

3. $y = 5xz$ represents joint variation

4. $y = \dfrac{1}{2}abc$ represents joint variation

5. $y = \dfrac{9.1}{x}$ represents inverse variation

6. $y = 2.3x$ represents direct variation

7. $y = \dfrac{2}{3}x$ represents direct variation

8. $y = 3.1st$ represents joint variation

Exercise Set 8.4

1. $y = kx$

3. $a = \dfrac{k}{b}$

5. $y = kxz$

7. $y = \dfrac{k}{x^3}$

9. $y = \dfrac{kx}{p^2}$

11. $y = kx$
$4 = k(20)$
$k = \dfrac{1}{5}$
$y = \dfrac{1}{5}x$

13. $y = kx$
$6 = k(4)$
$k = \dfrac{3}{2}$
$y = \dfrac{3}{2}x$

15. $y = kx$
$7 = k\left(\dfrac{1}{2}\right)$
$k = 14$
$y = 14x$

17. $y = kx$
$0.2 = k(0.8)$
$k = 0.25$
$y = 0.25x$

19. $W = kr^3$
$1.2 = k(2)^3$
$1.2 = 8k$
$k = \dfrac{1.2}{8} = 0.15$
$W = 0.15r^3$
$\quad = 0.15(3)^3$
$\quad = 0.15(27)$
$\quad = 4.05 \text{ lb}$

21. $\qquad P = kN$
$260,000 = k(450,000)$
$\qquad k = \dfrac{260,000}{450,000} = \dfrac{26}{45}$

$P = \dfrac{26}{45}N = \dfrac{26}{45}(980,000)$
$\qquad\qquad = 566,222 \text{ tons}$

23. $y = \dfrac{k}{x}$
$6 = \dfrac{k}{5}$
$k = 30$
$y = \dfrac{30}{x}$

25. $y = \dfrac{k}{x}$
$100 = \dfrac{k}{7}$
$k = 700$
$y = \dfrac{700}{x}$

27. $y = \dfrac{k}{x}$
$\dfrac{1}{8} = \dfrac{k}{16}$
$k = 2$
$y = \dfrac{2}{x}$

29. $y = \dfrac{k}{x}$
$0.2 = \dfrac{k}{0.7}$
$k = 0.14$
$y = \dfrac{0.14}{x}$

31. $R = \dfrac{k}{T}$
$45 = \dfrac{k}{6}$
$k = 45(6) = 270$
$R = \dfrac{270}{T} = \dfrac{270}{5} = 54 \text{ mph}$

33. $I = \dfrac{k}{R}$
$40 = \dfrac{k}{270}$
$k = 40(270) = 10,800$
$I = \dfrac{10,800}{R} = \dfrac{10,800}{150} = 72 \text{ amps}$

35. $I_1 = \dfrac{k}{d^2}$

Replace d with $2d$.

$I_2 = \dfrac{k}{(2d)^2} = \dfrac{k}{4d^2} = \dfrac{1}{4} \cdot \dfrac{k}{d^2} = \dfrac{1}{4} I_1$

Thus, the intensity is divided by 4.

37. $x = kyz$

39. $r = kst^3$

41. $y = kx^3$

$9 = k(3)^3$

$9 = 27k$

$k = \dfrac{9}{27} = \dfrac{1}{3}$

$y = \dfrac{1}{3} x^3$

43. $y = k\sqrt{x}$

$0.4 = k\sqrt{4}$

$0.4 = 2k$

$k = \dfrac{0.4}{2} = 0.2$

$y = 0.2\sqrt{x}$

45. $y = \dfrac{k}{x^2}$

$0.052 = \dfrac{k}{(5)^2}$

$0.052 = \dfrac{k}{25}$

$k = 0.052(25) = 1.3$

$y = \dfrac{1.3}{x^2}$

47. $y = kxz^3$

$120 = k(5)(2)^3$

$120 = 40k$

$k = \dfrac{120}{40} = 3$

$y = 3xz^3$

49. $W = \dfrac{kwh^2}{l}$

$12 = \dfrac{k\left(\dfrac{1}{2}\right)\left(\dfrac{1}{3}\right)^2}{10}$

$120 = \dfrac{1}{18} k$

$k = 120(18) = 2160$

$W = \dfrac{2160 wh^2}{l}$

$= \dfrac{2160\left(\dfrac{2}{3}\right)\left(\dfrac{1}{2}\right)^2}{16}$

$= \dfrac{360}{16}$

$= \dfrac{45}{2}$ tons or 22.5 tons

51. $V = kr^2 h$

$32\pi = k(4)^2(6)$

$32\pi = 96k$

$k = \dfrac{32\pi}{96} = \dfrac{\pi}{3}$

$V = \dfrac{\pi}{3} r^2 h = \dfrac{\pi}{3}(3)^2(5)$

$= \dfrac{45\pi}{3}$

$= 15\pi$ cu. in.

53. $H = ksd^3$

$40 = k(120)(3)^3$

$40 = 1080k$

$k = \dfrac{40}{1080} = \dfrac{1}{24}$

$H = \dfrac{1}{24} sd^3 = \dfrac{1}{24}(80)(3)^3 = 90$ hp

55. $y = \dfrac{k}{x}$

$400 = \dfrac{k}{8}$

$k = 400(8) = 3200$

$y = \dfrac{3200}{x} = \dfrac{3200}{4} = 800$ millibars

57. $r = 6$ cm
$C = 2\pi r = 2\pi(6) = 12\pi$ cm
$A = \pi r^2 = \pi(6)^2 = 36\pi$ sq. cm

59. $r = 7$ m
$C = 2\pi r = 2\pi(7) = 14\pi$ m
$A = \pi r^2 = \pi(7)^2 = 49\pi$ sq. m

61. $|-3| = 3$

63. $|0| = 0$

65. $-\left|\dfrac{1}{5}\right| = -\dfrac{1}{5}$

67. $\left(\dfrac{5}{11}\right)^2 = \dfrac{25}{121}$

69. $V_1 = khr^2$
$V_2 = k\left(\dfrac{1}{2}h\right)(2r)^2$
$\quad = k\left(\dfrac{1}{2}h\right)(4r^2) = 2(khr^2) = 2V_1$
It is multiplied by 2.

71. $y_1 = kx^2$
$y_2 = k(2x)^2 = k(4x^2) = 4(kx^2) = 4y_1$
It is multiplied by 4.

73.

x	$\dfrac{1}{4}$	$\dfrac{1}{2}$	1	2	4
$y = \dfrac{3}{x}$	12	6	3	$\dfrac{3}{2}$	$\dfrac{3}{4}$

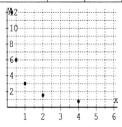

75.

x	$\dfrac{1}{4}$	$\dfrac{1}{2}$	1	2	4
$y = \dfrac{1}{2x}$	2	1	$\dfrac{1}{2}$	$\dfrac{1}{4}$	$\dfrac{1}{8}$

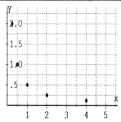

Chapter 8 Review Exercises

1. $f(x) = x$ or $y = x$
$m = 1, \; b = 0$

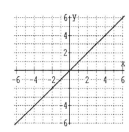

2. $f(x) = -\dfrac{1}{3}x$ or $y = -\dfrac{1}{3}x$
$m = -\dfrac{1}{3}, \; b = 0$

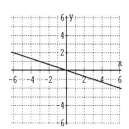

3. $g(x) = 4x - 1$ or $y = 4x - 1$
    $m = 4, \ b = -1$

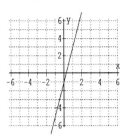

4. $f(x) = -\dfrac{2}{3}x + 2$ or $y = -\dfrac{2}{3}x + 2$

$m = -\dfrac{2}{3}, \ b = 2$

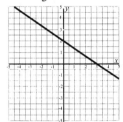

5. C

6. A

7. B

8. D

9. $f(x) = \dfrac{2}{5}x - \dfrac{4}{3}$

$m = \dfrac{2}{5}, \ b = -\dfrac{4}{3}$

10. $f(x) = -\dfrac{2}{7}x + \dfrac{3}{2}$

$m = -\dfrac{2}{7}, \ b = \dfrac{3}{2}$

11. $y = -\dfrac{2}{3}x + 4$

$f(x) = -\dfrac{2}{3}x + 4$

12. $y = -x - 2$

$f(x) = -x - 2$

13. $6x + 3y = 5$

$3y = -6x + 5$

$y = -2x + \dfrac{5}{3}$ so $m = -2$

$y - (-6) = -2(x - 2)$

$y + 6 = -2x + 4$

$y = -2x - 2$

$f(x) = -2x - 2$

14. $3x + 2y = 8$

$2y = -3x + 8$

$y = -\dfrac{3}{2}x + 4$ so $m = -\dfrac{3}{2}$

$y - (-2) = -\dfrac{3}{2}[x - (-4)]$

$2(y + 2) = -3(x + 4)$

$2y + 4 = -3x - 12$

$2y = -3x - 16$

$y = -\dfrac{3}{2}x - 8$

$f(x) = -\dfrac{3}{2}x - 8$

15. $4x + 3y = 5$

$3y = -4x + 5$

$y = -\dfrac{4}{3}x + \dfrac{5}{3}$

so $m = -\dfrac{4}{3}$ and $m_\perp = \dfrac{3}{4}$

$y - (-1) = \dfrac{3}{4}[x - (-6)]$

$4(y + 1) = 3(x + 6)$

$4y + 4 = 3x + 18$

$4y = 3x + 14$

$y = \dfrac{3}{4}x + \dfrac{7}{2}$

$f(x) = \dfrac{3}{4}x + \dfrac{7}{2}$

16. $2x - 3y = 6$

$-3y = -2x + 6$

$y = \dfrac{2}{3}x - 2$

so $m = \dfrac{2}{3}$ and $m_\perp = -\dfrac{3}{2}$

$$y - 5 = -\frac{3}{2}[x - (-4)]$$
$$2(y - 5) = -3(x + 4)$$
$$2y - 10 = -3x - 12$$
$$2y = -3x - 2$$
$$y = -\frac{3}{2}x - 1$$
$$f(x) = -\frac{3}{2}x - 1$$

17. a. Use ordered pairs $(0, 65)$ and $(12, 81)$

$$m = \frac{81 - 65}{12 - 0} = \frac{16}{12} = \frac{4}{3} \text{ and } b = 65$$
$$y = \frac{4}{3}x + 65$$

b. $x = 2009 - 1990 = 19$

$$y = \frac{4}{3}(19) + 65 \approx 90.3$$

About 90% of US drivers will be wearing seat belts.

18. a. Use ordered pairs $(0, 43)$ and $(22, 60)$

$$m = \frac{60 - 43}{22 - 0} = \frac{17}{22} \text{ and } b = 43$$
$$y = \frac{17}{22}x + 43$$

b. $x = 2010 - 1998 = 12$

$$y = \frac{17}{22}(12) + 43 \approx 52.3$$

There will be about 52 million people reporting arthritis.

19. $f(x) = -2x + 6 \qquad g(x) = 2x - 1$
$m = -2 \qquad\qquad\quad m = 2$
Neither; The slopes are not the same and their product is not -1.

20. $-x + 3y = 2 \qquad\qquad 6x - 18y = 3$

$$y = \frac{1}{3}x + \frac{2}{3} \qquad\qquad y = \frac{1}{3}x - \frac{1}{6}$$

$$m = \frac{1}{3} \qquad\qquad\qquad m = \frac{1}{3}$$

Parallel, since their slopes are equal.

21. $C(x) = 0.3x + 42$

$$C(150) = 0.3(150) + 42$$
$$= 45 + 42$$
$$= 87$$
$$\$87$$

22. $m = 0.3,\ b = 42$

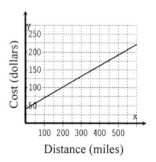

23. $f(-1) = 0$

24. $f(1) = -2$

25. $f(x) = 1$
$f(-2) = f(4) = 1$
$x = -2, 4$

26. $f(x) = -1$
$f(0) = f(2) = -1$
$x = 0, 2$

27. $y = 3x$; Linear

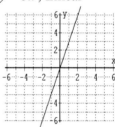

28. $y = 5x$; Linear

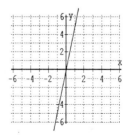

29. $y = |x| + 4$; Nonlinear

x	–3	–2	–1	0	1	2	3
y	7	6	5	4	5	6	7

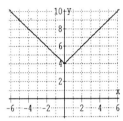

30. $y = x^2 + 4$; Nonlinear

x	–3	–2	–1	0	1	2	3
y	13	8	5	4	5	8	13

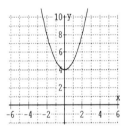

31. $y = -\dfrac{1}{2}x + 2$; Linear

Find three ordered pair solutions, or find x- and y-intercepts, or find m and b.

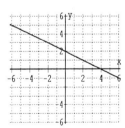

32. $y = -x + 5$; Linear

Find three ordered pair solutions, or find x- and y-intercepts, or find m and b.

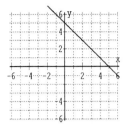

33. $y = -1.36x$; Linear

Find three ordered pair solutions, or find x- and y-intercepts, or find m and b.

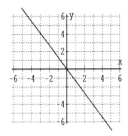

34. $y = 2.1x + 5.9$

Find three ordered pair solutions, or find x- and y-intercepts, or find m and b.

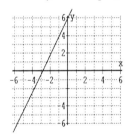

35. $y = (x - 2)^2$; Nonlinear

x	–1	0	1	2	3	4	5
y	9	4	1	0	1	4	9

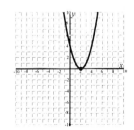

36. $y = -|x-3|$; Nonlinear

x	0	1	2	3	4	5	6
y	-3	-2	-1	0	-1	-2	-3

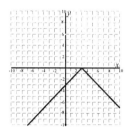

37. $(f+g)(x) = f(x) + g(x)$
$$= (x-5) + (2x+1)$$
$$= x - 5 + 2x + 1$$
$$= 3x - 4$$

38. $(f-g)(x) = f(x) - g(x)$
$$= (x-5) + (2x+1)$$
$$= x - 5 - 2x - 1$$
$$= -x - 6$$

39. $(f \cdot g)(x) = f(x) \cdot g(x)$
$$= (x-5)(2x+1)$$
$$= 2x^2 + x - 10x - 5$$
$$= 2x^2 - 9x - 5$$

40. $\left(\dfrac{g}{f}\right)(x) = \dfrac{g(x)}{f(x)} = \dfrac{2x+1}{x-5}, x \neq 5$

41. $(f \circ g)(x) = f(g(x))$
$$= f(x+1)$$
$$= (x+1)^2 - 2$$
$$= x^2 + 2x - 1$$

42. $(g \circ f)(x) = g(f(x))$
$$= g(x^2 - 2)$$
$$= x^2 - 2 + 1$$
$$= x^2 - 1$$

43. $(h \circ g)(2) = h(g(2))$
$$= h(3)$$
$$= 3^3 - 3^2$$
$$= 18$$

44. $(f \circ f)(x) = f(f(x))$
$$= f(x^2 - 2)$$
$$= (x^2 - 2)^2 - 2$$
$$= x^4 - 4x^2 + 4 - 2$$
$$= x^4 - 4x^2 + 2$$

45. $(f \circ g)(-1) = f(g(-1))$
$$= f(0)$$
$$= 0^2 - 2$$
$$= -2$$

46. $(h \circ h)(2) = h(h(2))$
$$= h(4)$$
$$= 4^3 - 4^2$$
$$= 48$$

47. $A = kB$
$$6 = k(14)$$
$$k = \frac{6}{14} = \frac{3}{7}$$
$$A = \frac{3}{7}B = \frac{3}{7}(21) = 3(3) = 9$$

48. $C = \dfrac{k}{D}$
$$12 = \frac{k}{8}$$
$$96 = k$$
$$C = \frac{96}{D}$$
$$C = \frac{96}{24} = 4$$

Chapter 9

Exercise Set 9.1

1. $C \cup D = \{2, 3, 4, 5, 6, 7\}$

3. $A \cap D = \{4, 6\}$

5. $A \cup D = \{..., -2, -1, 0, 1, ...\}$

7. $B \cap D = \{5, 7\}$

9. $B \cup C$
$= \{x \mid x \text{ is an odd integer or } x = 2 \text{ or } x = 4\}$

11. $A \cap C = \{2, 4\}$

13. $x < 5 \text{ and } x > -2$
$-2 < x < 5$
$(-2, 5)$

15. $x + 1 \geq 7 \text{ and } 3x - 1 \geq 5$
$\quad x \geq 6 \text{ and } \quad 3x \geq 6$
$\quad\quad\quad\quad\quad\quad\quad x \geq 2$
$x \geq 6$
$[6, \infty)$

17. $4x + 2 \leq -10 \text{ and } 2x \leq 0$
$\quad 4x \leq -12 \text{ and } \quad x \leq 0$
$\quad\quad x \leq -3$
$x \leq -3$
$(-\infty, -3]$

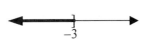

19. $5 < x - 6 < 11$
$11 < x < 17$
$(11, 17)$

21. $-2 \leq 3x - 5 \leq 7$
$\quad 3 \leq 3x \leq 12$
$\quad 1 \leq x \leq 4$
$[1, 4]$

23. $1 \le \dfrac{2}{3}x + 3 \le 4$

$-2 \le \dfrac{2}{3}x \le 1$

$-3 \le x \le \dfrac{3}{2}$

$\left[-3, \dfrac{3}{2} \right]$

25. $-5 \le \dfrac{x+1}{4} \le -2$

$-20 \le x + 1 \le -8$

$-21 \le x \le -9$

$[-21, 9]$

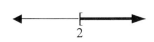

27. $x < -1$ or $x > 0$

$(-\infty, -1) \cup (0, \infty)$

29. $-2x \le -4$ or $5x - 20 \ge 5$

$x \ge 2$ or $5x \ge 25$

$x \ge 5$

$x \ge 2$

$[2, \infty)$

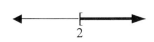

31. $3(x-1) < 12$ or $x + 7 > 10$

$3x - 3 < 12$ or $x > 3$

$3x < 15$

$x < 5$

All real numbers.

$(-\infty, \infty)$

33. Answers may vary.

35. $x < 2$ and $x > -1$

$-1 < x < 2$

$(-1, 2)$

37. $x < 2$ or $x > -1$

All real numbers.

$(-\infty, \infty)$

39. $x \ge -5$ and $x \ge -1$

$x \ge -1$

$[-1, \infty)$

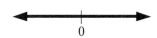

41. $x \geq -5$ or $x \geq -1$

$x \geq -5$

$[-5, \infty)$

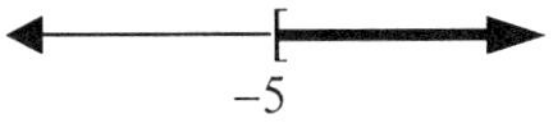

43. $0 \leq 2x - 3 \leq 9$

$3 \leq 2x \leq 12$

$\frac{3}{2} \leq x \leq 6$

$\left[\frac{3}{2}, 6\right]$

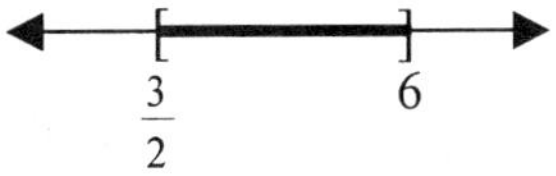

45. $\frac{1}{2} < x - \frac{3}{4} < 2$

$4\left(\frac{1}{2}\right) < 4\left(x - \frac{3}{4}\right) < 4(2)$

$2 < 4x - 3 < 8$

$5 < 4x < 11$

$\frac{5}{4} < x < \frac{11}{4}$

$\left(\frac{5}{4}, \frac{11}{4}\right)$

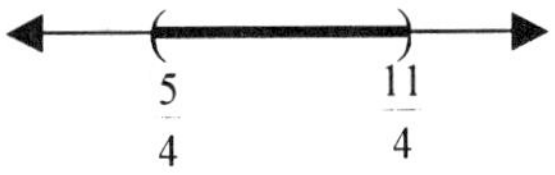

47. $x + 3 \geq 3$ and $x + 3 \leq 2$

$x \geq 0$ and $x \leq -1$

No solution exist.

$\varnothing$

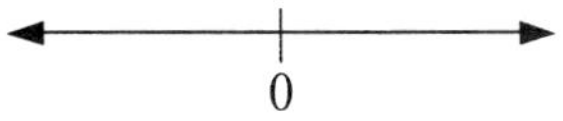

49. $3x \geq 5$ or $-x - 6 < 1$

$x \geq \frac{5}{3}$ or $-x < 7$

$x > -7$

$x > -7$

$(-7, \infty)$

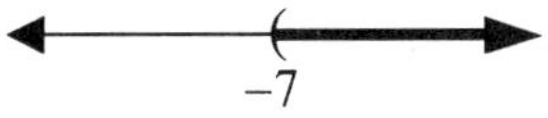

51. $0 < \frac{5 - 2x}{3} < 5$

$0 < 5 - 2x < 15$

$\frac{-5}{-2} > \frac{-2x}{-2} > \frac{10}{-2}$

$\frac{5}{2} > x > -5$

$-5 < x < \frac{5}{2}$

$\left(-5, \frac{5}{2}\right)$

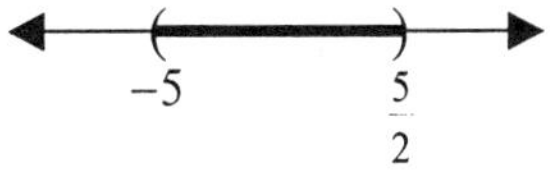

53. $-6 < 3(x-2) \le 8$

$\quad -6 < 3x - 6 \le 8$

$\quad\quad 0 < 3x \le 14$

$\quad\quad 0 < x < \dfrac{14}{3}$

$\quad\quad \left(0, \dfrac{14}{3}\right]$

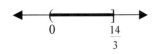

55. $-x + 5 > 6 \quad \text{and} \quad 1 + 2x \le -5$

$\quad\quad -x > 1 \quad\quad\quad\quad 2x \le -6$

$\quad\quad x < -1 \quad\quad\quad\quad x \le -3$

$\quad\quad x \le -3$

$\quad\quad (-\infty, -3]$

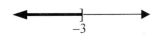

57. $3x + 2 \le 5 \quad \text{or} \quad 7x < 29$

$\quad\quad 3x \le 3 \quad\quad\quad x > \dfrac{29}{7}$

$\quad\quad x \le 1 \quad\quad\quad x > \dfrac{29}{7}$

$\quad (-\infty, 1] \cup \left(\dfrac{29}{7}, \infty\right)$

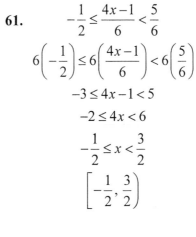

59. $5 - x > 7 \quad \text{and} \quad 2x + 3 \ge 13$

$\quad\quad -x > 2 \quad\quad\quad\quad 2x \ge 10$

$\quad\quad x < -2 \quad\quad\quad\quad x \ge 5$

No solution exist.

$\varnothing$

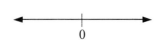

61. $\qquad -\dfrac{1}{2} \le \dfrac{4x-1}{6} < \dfrac{5}{6}$

$\quad 6\left(-\dfrac{1}{2}\right) \le 6\left(\dfrac{4x-1}{6}\right) < 6\left(\dfrac{5}{6}\right)$

$\quad\quad -3 \le 4x - 1 < 5$

$\quad\quad -2 \le 4x < 6$

$\quad\quad -\dfrac{1}{2} \le x < \dfrac{3}{2}$

$\quad\quad \left[-\dfrac{1}{2}, \dfrac{3}{2}\right)$

63.

$$\frac{1}{15} < \frac{8-3x}{15} < \frac{4}{5}$$

$$15\left(\frac{1}{15}\right) < 15\left(\frac{8-3x}{15}\right) < 15\left(\frac{4}{5}\right)$$

$$1 < 8-3x < 12$$

$$-7 < -3x < 4$$

$$-\frac{4}{3} < x < \frac{7}{3}$$

$$\left(-\frac{4}{3}, \frac{7}{3}\right)$$

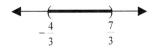

65. $0.3 < 0.2x - 0.9 < 1.5$

$$1.2 < 0.2x < 2.4$$

$$6 < x < 12$$

$$(6, 12)$$

67. $|-7| - |19| = 7 - 19 = -12$

69. $-(-6) - |-10| = 6 - 10 = -4$

71. $|x| = 7$

$x = -7, 7$

73. $|x| = 0$

$x = 0$

75. $-29 \le C \le 35$

$$-29 \le \frac{5}{9}(F - 32) \le 35$$

$$-52.5 \le F - 32 \le 63$$

$$-20.2 \le F \le 95$$

$$-20.2° \le F \le 95°$$

77. $70 \le \dfrac{68 + 65 + 75 + 78 + 2x}{6} \le 79$

$$420 \le 286 + 2x \le 474$$

$$134 \le 2x \le 188$$

$$67 \le x \le 94$$

If Christian scores between 67 and 94 inclusive on his final exam, he will receive a C in the course.

79. The years that the consumption of pork was greater than 48 pounds per person were 1994, 1995, 1998, and 1999. The years that the consumption of chicken was greater than 48 pounds per person were 1994, 1995, 1996, 1997, 1998, 1999, 2000, and 2001. The years in common are 1994, 1995, 1998, and 1999.

81. $2x - 3 < 3x + 1 < 4x - 5$

$2x - 3 < 3x + 1$ and $3x + 1 < 4x - 5$

$\quad -x < 4 \qquad\qquad -x < -6$

$\quad\quad x > -4 \qquad\qquad x > 6$

$\quad\quad x > -6$

$(6, \infty)$

83. $-3(x - 2) \le 3 - 2x \le 10 - 3x$

$-3x + 6 \le 3 - 2x$ and $3 - 2x \le 10 - 3x$

$\quad -x \le -3 \qquad\qquad x \le 7$

$\quad\quad x \ge 3$

$3 \le x \le 7$

$[3, 7]$

85. $5x - 8 < 2(2 + x) < -2(1 + 2x)$

$5x - 8 < 4 + 2x$ and $4 + 2x < -2 - 4x$

$\quad 3x < 12 \qquad\qquad 6x < -6$

$\quad\quad x < -4 \qquad\qquad x < -1$

$x < -1$

$(-\infty, -1)$

Section 9.2

Mental Math

1. $|-7| = 7$

2. $|-8| = 8$

3. $-|5| = -5$

4. $-|10| = -10$

5. $-|-6| = -6$

6. $-|-3| = -3$

7. $|-3| + |-2| + |-7| = 3 + 2 + 7 = 12$

8. $|-1| + |-6| + |-8| = 1 + 6 + 8 = 15$

Exercise Set 9.2

1. $|x| = 7$

$x = 7$ or $x = -7$

The solution set is $\{-7, 7\}$.

3. $|3x| = 12.6$

$3x = 12.6$ or $3x = -12.6$

$x = 4.2$ or $x = -4.2$

The solution set is $\{-4.2, 4.2\}$.

5. $|2x-5| = 9$

$2x - 5 = 9$ or $2x - 5 = -9$

$2x = 14$ or $2x = -4$

$x = 7$ or $x = -2$

The solution set is $\{-2, 7\}$.

7. $\left|\dfrac{x}{2} - 3\right| = 1$

$\dfrac{x}{2} - 3 = 1$ or $\dfrac{x}{2} - 3 = -1$

$2\left(\dfrac{x}{2} - 3\right) = 2(1)$ or $2\left(\dfrac{x}{2} - 3\right) = 2(-1)$

$x - 6 = 2$ or $x - 6 = -2$

$x = 8$ or $x = 4$

The solution set is $\{4, 8\}$.

9. $|z| + 4 = 9$

$|z| = 5$

$z = -5$ or $z = -5$

The solution set is $\{-5, 5\}$.

11. $|3x| + 5 = 14$

$|3x| = 9$

$3x = 9$ or $3x = -9$

$x = 3$ or $x = -3$

The solution set is $\{-3, 3\}$.

13. $|2x| = 0$

$2x = 0$

$x = 0$

The solution set is $\{0\}$.

15. $|4n+1| + 10 = 4$

$|4n+1| = -6$ which is impossible.

The solution set is $\varnothing$.

17. $|5x - 1| = 0$

$5x - 1 = 0$

$5x = 1$

$x = \dfrac{1}{5}$

The solution set is $\left\{\dfrac{1}{5}\right\}$.

19. $|x| = 5$

21. $|5x - 7| = |3x + 11|$

$5x - 7 = 3x + 11$ or $5x - 7 = -(3x + 11)$

$2x = 18$ or $5x - 7 = -3x - 11$

$x = 9$ or $8x = -4$

$x = -\dfrac{1}{2}$

The solution set is $\left\{-\dfrac{1}{2}, 9\right\}$.

23. $|z + 8| = |z - 3|$

$z + 8 = z - 3$ or $z + 8 = -(z - 3)$

$8 = -3$ or $z + 8 = -z + 3$

$2z = -5$

$z = -\dfrac{5}{2}$

The solution set is $\left\{-\dfrac{5}{2}\right\}$.

25. Answers may vary.

27. $|x| = 4$

$x = 4$ or $x = -4$

The solution set is $\{-4, 4\}$.

29. $|y| = 0;\ y = 0$

The solution set is $\{0\}$.

31. $|z| = -2$ is impossible.

The solution set is $\varnothing$.

33. $|7 - 3x| = 7$

$7 - 3x = 7$ or $7 - 3x = -7$

$-3x = 0$ or $-3x = -14$

$x = 0$ or $x = \dfrac{14}{3}$

The solution set is $\left\{0, \dfrac{14}{3}\right\}$.

35. $|6x| - 1 = 11$

$|6x| = 12$

$6x = 12$ or $6x = -12$

$x = 2$ or $x = -2$

The solution set is $\{-2, 2\}$.

37. $|4p| = -8$ is impossible.

The solution set is $\varnothing$.

39. $|x - 3| + 3 = 7$

$|x - 3| = 4$

$x - 3 = 4$ or $x - 3 = -4$

$x = 7$ or $x = -1$

The solution set is $\{-1, 7\}$.

41. $\left|\dfrac{z}{4} + 5\right| = -7$ is impossible.

The solution set is $\varnothing$.

43. $|9v - 3| = -8$ is impossible.

The solution set is $\varnothing$.

45. $|8n + 1| = 0$

$8n + 1 = 0$

$8n = -1$ so $n = -\dfrac{1}{8}$

The solution set is $\left\{-\dfrac{1}{8}\right\}$.

47. $|1 + 6c| - 7 = -3$

$|1 + 6c| = 4$

$1 + 6c = 4$ or $1 + 6c = -4$

$6c = 3$ $6c = -5$

$c = \dfrac{1}{2}$ $c = -\dfrac{5}{6}$

The solution set is $\left\{-\dfrac{5}{6}, \dfrac{1}{2}\right\}$.

49. $|5x + 1| = 11$

$5x + 1 = 11$ or $5x + 1 = -11$

$5x = 10$ or $5x = -12$

$x = 2$ or $x = -\dfrac{12}{5}$

The solution set is $\left\{-\dfrac{12}{5}, 2\right\}$.

51. $|4x - 2| = |-10|$

$|4x - 2| = 10$

$4x - 2 = 10$ or $4x - 2 = -10$

$4x = 12$ or $\quad 4x = -8$

$x = 3$ or $\quad\quad x = -2$

The solution set is $\{-2, 3\}$.

53. $|5x + 1| = |4x - 7|$

$5x + 1 = 4x - 7$ or $5x + 1 = -(4x - 7)$

$x = -8\quad$ or $5x + 1 = -4x + 7$

$9x = 6$

$x = \dfrac{2}{3}$

The solution set is $\left\{-8, \dfrac{2}{3}\right\}$.

55. $|6 + 2x| = -|-7|$

$|6 + 2x| = -7$ which is impossible.

The solution set is $\varnothing$.

57. $|2x - 6| = |10 - 2x|$

$2x - 6 = 10 - 2x$ or $2x - 6 = -(10 - 2x)$

$4x = 16\quad\quad$ or $2x - 6 = -10 + 2x$

$x = 4\quad\quad$ or $\quad -6 = -10$

$-6 = -10$ is impossible.

The solution set is $\{4\}$.

59. $\left|\dfrac{2x - 5}{3}\right| = 7$

$\dfrac{2x - 5}{3} = 7\quad$ or $\quad\dfrac{2x - 5}{3} = -7$

$2x - 5 = 21$ or $2x - 5 = -21$

$2x = 26$ or $\quad\quad 2x = -16$

$x = 13$ or $\quad\quad x = -8$

The solution set is $\{-8, 13\}$.

61. $2 + |5n| = 17$

$|5n| = 15$

$5n = 15$ or $5n = -15$

$n = 3$ or $\quad n = -3$

The solution set is $\{-3, 3\}$.

63. $\left|\dfrac{2x - 1}{3}\right| = |-5|$

$\left|\dfrac{2x - 1}{3}\right| = 5$

$\dfrac{2x - 1}{3} = 5\quad$ or $\quad\dfrac{2x - 1}{3} = -5$

$2x - 1 = 15$ or $\quad 2x - 1 = -15$

$2x = 16\quad$ or $\quad\quad 2x = -14$

$x = 8\quad$ or $\quad\quad x = -7$

The solution set is $\{-7, 8\}$.

65. $|2y - 3| = |9 - 4y|$

$2y - 3 = 9 - 4y$ or $2y - 3 = -(9 - 4y)$

$6y = 12\quad\quad$ or $2y - 3 = -9 + 4y$

$y = 2\quad\quad$ or $\quad -2y = -6$

$y = 3$

The solution set is $\{2, 3\}$.

67. $\left|\dfrac{3n+2}{8}\right| = |-1|$

$\left|\dfrac{3n+2}{8}\right| = 1$

$\dfrac{3n+2}{8} = 1$ or $\dfrac{3n+2}{8} = -1$

$3n+2 = 8$ $3n+2 = -8$

$\quad 3n = 6$ $\quad 3n = -10$

$\quad\quad n = 2$ $\quad\quad n = -\dfrac{10}{3}$

69. $|x+4| = |7-x|$

$x+4 = 7-x$ or $x+4 = -(7-x)$

$2x = 3$ or $x+4 = -7+x$

$x = \dfrac{3}{2}$ or $4 = -7$

$4 = -7$ is impossible.

The solution set is $\left\{\dfrac{3}{2}\right\}$.

71. $\left|\dfrac{8c-7}{3}\right| = -|-5|$

$\left|\dfrac{8c-7}{3}\right| = -5$ which is impossible.

The solution set is $\varnothing$.

73. Answers may vary.

75. 33% of cheese consumption came from cheddar cheese.

77. $32\% \cdot (120 \text{ pounds}) = 0.32(120 \text{ pounds})$
$= 38.4 \text{ pounds}$

We might expect they consumed 38.4 pounds.

79. Answers may vary.

81. $|y| < 0$
No solution.

83. $|x| = 2$

85. $|2x-1| = 4$

Section 9.3

1. D

2. E

3. C

4. B

5. A

Exercise Set 9.3

1. $|x| \le 4$

$-4 \le x \le 4$

$[-4, 4]$

3. $|x-3| < 2$

$-2 < x-3 < 2$

$1 < x < 5$

$(1, 5)$

5. $|x+3| < 2$

$-2 < x+3 < 2$

$-5 < x < -1$

$(-5, -1)$

7. $|2x+7| \le 13$

$-13 \le 2x+7 \le 13$

$-20 \le 2x \le 6$

$-10 \le x \le 3$

$[-10, 3]$

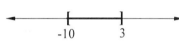

9. $|x|+7 \le 12$

$|x| \le 5$

$-5 \le x \le 5$

$[-5, 5]$

11. $|3x-1| < -5$

No real solutions; $\varnothing$

13. $|x-6|-7 \le -1$

$|x-6| \le 6$

$-6 \le x-6 \le 6$

$0 \le x \le 12$

$[0, 12]$

15. $|x| > 3$

$x < -3$ or $x > 3$

$(-\infty, -3) \cup (3, \infty)$

17. $|x+10| \ge 14$

$x+10 \le -14$ or $x+10 \ge 14$

$x \le -24$ $x \ge 4$

$(-\infty, -24] \cup [4, \infty)$

19. $|x|+2 > 6$

$|x| > 4$

$x < -4$ or $x > 4$

$(-\infty, -4) \cup (4, \infty)$

21. $|5x| > -4$

All real numbers.

$(-\infty, \infty)$

23. $|6x - 8| + 3 > 7$

$|6x - 8| > 4$

$6x - 8 < -4$ or $6x - 8 > 4$

$6x < 4$ $\qquad\qquad$ $6x > 12$

$x < \dfrac{2}{3}$ $\qquad\qquad$ $x > 2$

$\left(-\infty, \dfrac{2}{3}\right) \cup (2, \infty)$

25. $|x| \leq 0$

$|x| = 0$

$x = 0$

27. $|8x + 3| > 0$ only excludes $|8x + 3| = 0$

$8x + 3 = 0$

$8x = -3$

$x = -\dfrac{3}{8}$

All real numbers except $-\dfrac{3}{8}$.

$\left(-\infty, -\dfrac{3}{8}\right) \cup \left(-\dfrac{3}{8}, \infty\right)$

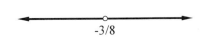

29. $|x| \leq 2$

$-2 \leq x \leq 2$

$[-2, 2]$

31. $|y| > 1$

$y < -1$ or $y > 1$

$(-\infty, -1) \cup (1, \infty)$

33. $|x - 3| < 8$

$-8 < x - 3 < 8$

$-5 < x < 11$

$(5, 11)$

35. $|0.6x - 3| > 0.6$

$0.6x - 3 < -0.6$ or $0.6x - 3 > 0.6$

$0.6x < -2.4$ $\qquad$ $0.6x > 3.6$

$x < -4$ $\qquad\qquad$ $x > 6$

$(-\infty, -4) \cup (6, \infty)$

37. $5 + |x| \leq 2$

$|x| \leq -3$

No real solution.

$\varnothing$

39. $|x| > -4$

All real numbers.

$(-\infty, \infty)$

41. $|2x - 7| \le 11$

$-11 \le 2x - 7 \le 11$

$-4 \le 2x \le 18$

$-2 \le x \le 9$

$[-2, 9]$

43. $|x + 5| + 2 \ge 8$

$|x + 5| \ge 6$

$x + 5 \le -6 \quad \text{or} \quad x + 5 \ge 6$

$\quad x \le -11 \qquad\qquad x \ge 5$

$(-\infty, -11] \cup [1, \infty)$

45. $|x| > 0$ only excludes $|x| = 0$, or $x = 0$

All real numbers except $x = 0$.

$(-\infty, 0) \cup (0, \infty)$

47. $9 + |x| > 7$

All real numbers.

$(-\infty, \infty)$

49. $6 + |4x - 1| \le 9$

$|4x - 1| \le 3$

$-3 \le 4x - 1 \le 3$

$-2 \le 4x \le 4$

$-\dfrac{1}{2} \le x \le 1$

$\left[-\dfrac{1}{2}, 1\right]$

51. $\left|\dfrac{2}{3}x + 1\right| > 1$

$\dfrac{2}{3}x + 1 < -1 \quad \text{or} \quad \dfrac{2}{3}x + 1 > 1$

$\quad \dfrac{2}{3}x < -2 \qquad\qquad \dfrac{2}{3}x > 0$

$\quad\quad x < -3 \qquad\qquad\quad x > 0$

$(-\infty, -3) \cup (0, \infty)$

53. $|5x + 3| < -6$

No real solution.

$\varnothing$

55. $|8x + 3| \ge 0$

All real numbers.

$(-\infty, \infty)$

57. $|1+3x|+4<5$

$$|1+3x|<1$$

$$-1<1+3x<1$$

$$-2<3x<0$$

$$-\frac{2}{3}<x<0$$

$$\left(-\frac{2}{3},\,0\right)$$

59. $\left|\dfrac{x+6}{3}\right|>2$

$$\frac{x+6}{3}<-2 \quad\text{or}\quad \frac{x+6}{3}>2$$

$$x+6<-6 \qquad\quad x+6>6$$

$$x<-12 \qquad\qquad x>0$$

$$(-\infty,\,-12)\cup(0,\,\infty)$$

61. $-15+|2x-7|\le-6$

$$|2x-7|\le 9$$

$$-9\le 2x-7\le 9$$

$$-2\le 2x\le 16$$

$$-1\le x\le 8$$

$$[-1,\,8]$$

63. $\left|2x+\dfrac{3}{4}-7\right|\le-2$

$$\left|2x+\frac{3}{4}\right|\le 5$$

$$-5\le 2x+\frac{3}{4}\le 5$$

$$-20\le 8x+3\le 20$$

$$-23\le 8x\le 17$$

$$-\frac{23}{8}\le x\le\frac{17}{8}$$

$$\left[-\frac{23}{8},\,\frac{17}{8}\right]$$

65. $|2x-3|<7$

$$-7<2x-3<7$$

$$-4<2x<10$$

$$-2<x<5$$

The solution set is $(-2,\,5)$.

67. $|2x-3|=7$

$$2x-3=7 \quad\text{or}\quad 2x-3=-7$$

$$2x=10 \qquad\qquad 2x=-4$$

$$x=5 \qquad\qquad\quad x=-2$$

The solution set is $\{-2,\,5\}$.

69. $|x-5|\ge 12$

$$x-5\le-12 \quad\text{or}\quad x-5\ge 12$$

$$x\le-7 \qquad\qquad x\ge 17$$

The solution set is $(-\infty,\,-7]\cup[17,\,\infty)$.

71. $|9 + 4x| = 0$

$9 + 4x = 0$

$4x = -9$

$x = -\dfrac{9}{4}$

The solution set is $\left\{-\dfrac{9}{4}\right\}$.

73. $|2x + 1| + 4 < 7$

$|2x + 1| < 3$

$-3 < 2x + 1 < 3$

$-4 < 2x < 2$

$-2 < x < 1$

The solution set is $(-2, 1)$.

75. $|3x - 5| + 4 = 5$

$|3x - 5| = 1$

$3x - 5 = 1$ or $3x - 5 = -1$

$3x = 6$ $3x = 4$

$x = 2$ $x = \dfrac{4}{3}$

The solution set is $\left\{\dfrac{4}{3}, 2\right\}$.

77. $|x + 11| = -1$ is impossible.

The solution set is $\varnothing$.

79. $\left|\dfrac{2x - 1}{3}\right| = 6$

$\left|\dfrac{2x - 1}{3}\right| = 6$ or $\dfrac{2x - 1}{3} = -6$

$2x - 1 = 18$ or $2x - 1 = -18$

$2x = 19$ $2x = -17$

$x = \dfrac{19}{2}$ $x = -\dfrac{17}{2}$

The solution set is $\left\{-\dfrac{17}{2}, \dfrac{19}{2}\right\}$.

81. $\left|\dfrac{3x - 5}{6}\right| > 5$

$\dfrac{3x - 5}{6} < -5$ or $\dfrac{3x - 5}{6} > 5$

$3x - 5 < -30$ $3x - 5 > 30$

$x < -\dfrac{25}{3}$ $x > \dfrac{35}{3}$

$\left(-\infty, -\dfrac{25}{3}\right) \cup \left(\dfrac{35}{3}, \infty\right)$

83. $P(\text{rolling a 2}) = \dfrac{1}{6}$

85. $P(\text{rolling a 7}) = 0$

87. $P(\text{rolling a 1 or 3}) = \dfrac{1}{3}$

89.
$$3x - 4y = 12$$
$$3(2) - 4y = 12$$
$$6 - 4y = 12$$
$$-4y = 6$$
$$y = -\frac{3}{2} = -1.5$$

91.
$$3x - 4y = 12$$
$$3x - 4(-3) = 12$$
$$3x + 12 = 12$$
$$3x = 0$$
$$x = 0$$

93. $|x| < 7$

95. $|x| \le 5$

97. Answers may vary

99.
$$|3.5 - x| < 0.05$$
$$-0.05 < 3.5 - x < 0.05$$
$$-3.55 < -x < -3.45$$
$$3.55 > x > 3.45$$
$$3.45 < x < 3.55$$

Integrated Review

1. $x < 7$ and $x > 5$
$$(5, 7)$$

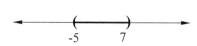

2. $x < 7$ or $x > -5$
$$(-\infty, \infty)$$

3. $|4x - 3| = 1$

$4x - 3 = 1$	$4x - 3 = -1$
$4x = 4$	$4x = 2$
$x = 1$	$x = \frac{1}{2}$

4. $|2x + 1| < 5$
$$-5 < 2x + 1 < 5$$
$$-6 < 2x < 4$$
$$-3 < x < 2$$
$$(-3, 2)$$

5. $|6x| - 9 \geq -3$

$|6x| \geq 6$

$6x \geq 6$ or $6x \leq -6$

$x \geq 1$ $x \leq -1$

$(-\infty, -1] \cup (1, \infty]$

6. $|x - 7| = |2x + 11|$

$x - 7 = 2x + 11$ $x - 7 = -(2x + 11)$

$-x - 7 = 11$ $x - 7 = -2x - 11$

$-x = 18$ $3x - 7 = -11$

$x = -18$ $3x = -4$

$x = -\dfrac{4}{3}$

7. $-5 \leq \dfrac{3x - 8}{2} \leq 2$

$-10 \leq 3x - 8 \leq 4$

$-2 \leq 3x \leq 12$

$-\dfrac{2}{3} \leq x \leq 4$

$\left[-\dfrac{2}{3}, 4\right]$

8. $|9x - 1| = -3$

The absolute value cannot be negative.

9. $3x + 2 \leq 5$ or $-3x \geq 0$

$3x \leq 3$ or $x \leq 0$

$x \leq 1$ or $x \leq 0$

$(-\infty, 1]$

10. $3x + 2 \leq 5$ and $-3x \geq 0$

$3x \leq 3$ $x \leq 0$

$x \leq 1$ $x \leq 0$

$(-\infty, 0]$

11. $|3 - x| - 5 \leq -2$

$|3 - x| \leq 3$

$-3 \leq 3 - x \leq 3$

$-6 \leq -x \leq 0$

$6 \geq x \geq 0$

$0 \leq x \leq 6$

$[0, 6]$

12. $\left|\dfrac{4x + 1}{5}\right| = |-1|$

$\left|\dfrac{4x + 1}{5}\right| = 1$

$\dfrac{4x + 1}{5} = 1$ $\dfrac{4x + 1}{5} = -1$

$4x + 1 = 5$ $4x + 1 = -5$

$4x = 4$ $4x = -6$

$x = 1$ $x = -\dfrac{3}{2}$

13. B

14. E

15. A

16. C

17. D

Section 9.4

Mental Math

1. Yes

2. No

3. Yes

4. No

5. $x + y > -5, \quad (0,0)$

$$0 + 0 \overset{?}{>} -5$$

$$0 \overset{?}{>} -5$$

Yes

6. $2x + 3y < 10, \quad (0,0)$

$$2(0) + 3(0) \overset{?}{<} 10$$

$$0 \overset{?}{<} 10$$

Yes

7. $x - y \leq -1, \quad (0,0)$

$$0 - 0 \overset{?}{\leq} -1$$

$$0 \overset{?}{\leq} -1$$

No

8. $\dfrac{2}{3}x + \dfrac{5}{6}y > 4, \quad (0,0)$

$$\dfrac{2}{3}(0) + \dfrac{5}{6}(0) \overset{?}{>} 4$$

$$0 \overset{?}{>} 4$$

No

Exercise Set 9.4

1. $x - y > 3$

$$(2,-1), \ 2 - (-1) \overset{?}{>} 3$$

$$2 + 1 \overset{?}{>} 3$$

$$3 > 3, \text{ False}$$

$(2,-1)$ is not a solution

$$(5,1), \ 5 - 1 \overset{?}{>} 3$$

$$4 > 3, \text{ True}$$

$(5,1)$ is a solution

3. $3x - 5y \leq -4$

$$(-1,-1), \ 3(-1) - 5(-1) \overset{?}{\leq} -4$$

$$-3 + 5 \overset{?}{\leq} -4$$

$$2 \overset{?}{\leq} -4, \text{ False}$$

$(-1,-1)$ is not a solution

$$(4,0), \ 3(4) - 5(0) \overset{?}{\leq} -4$$

$$12 - 0 \overset{?}{\leq} -4$$

$$12 \leq -4, \text{ False}$$

$(4,0)$ is not a solution

5. $x < -y$

$$(0,2), \ 0 \overset{?}{<} -2, \text{ False}$$

$(0,2)$ is not a solution

$$(-5,1), \ -5 \overset{?}{<} -1, \text{ True}$$

$(-5,1)$ is a solution

7. $x + y \leq 1$

Test $(0,0)$

$0 + 0 \overset{?}{\leq} 1$, True

Shade below.

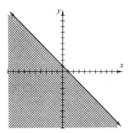

9. $2x + y > -4$

Test $(0,0)$

$2(0) + 0 \overset{?}{>} -4$

True

Shade above.

11. $x + 6y \leq -6$

Test $(0,0)$

$0 + 6(0) \overset{?}{\leq} -6$

False

Shade below.

13. $2x + 5y > -10$

Test $(0,0)$

$2(0) + 5(0) \overset{?}{>} -10$

True

Shade above.

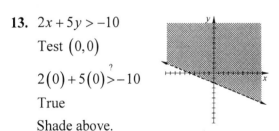

15. $x + 2y \leq 3$

Test $(0,0)$

$0 + 2(0) \overset{?}{\leq} 3$

True

Shade below.

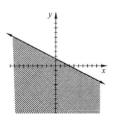

17. $2x + 7y > 5$

Test $(0,0)$

$2(0) + 7(0) \overset{?}{>} 5$

False

Shade above.

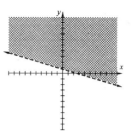

19. $x - 2y \geq 3$

Test $(0,0)$

$(0) - 2(0) \overset{?}{\geq} 3$

False

Shade below.

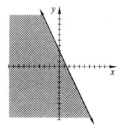

21. $5x + y < 3$

Test $(0,0)$

$5(0) + 0 \overset{?}{<} 3$

True

Shade below.

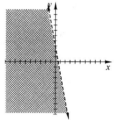

23. $4x + y < 8$

Test $(0,0)$

$4(0) + 0 \overset{?}{<} 8$

True

Shade below.

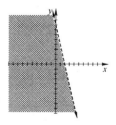

25. $y \geq 2x$

Test $(1,0)$

$0 \overset{?}{\geq} 2(1)$

False

Shade above.

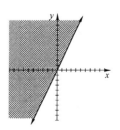

27. $x \geq 0$

Test $(1, 0)$

$1 \overset{?}{\geq} 0$

True

Shade right.

29. $y \leq -3$

Shade below.

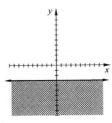

31. $2x - 7y > 0$

Test $(1, 0)$

$2(1) - 7(0) \overset{?}{>} 0$

True

Shade below.

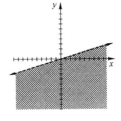

33. $3x - 7y \geq 0$

Test $(1, 0)$

$3(1) - 7(0) \overset{?}{\geq} 0$

True

Shade below.

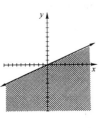

35. $x > y$

Test $(0, 1)$

$0 \overset{?}{>} 1$

False

Shade below.

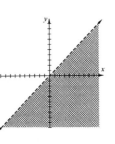

37. $x - y \leq 6$

Test $(0, 0)$

$0 - 0 \overset{?}{\leq} 6$

True

Shade above.

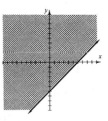

39. $-\dfrac{1}{4}y + \dfrac{1}{3}x > 1$

Test $(0, 0)$

$-\dfrac{1}{4}(0) + \dfrac{1}{3}(0) \overset{?}{>} 1$

False

Shade below.

41. $-x < 0.4y$

Test $(1, 0)$

$-(1) \overset{?}{<} 0$

True

Shade above.

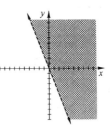

43. E

45. C

47. F

49. $y \geq x + 1$ $y \geq 3 - x$

   Test $(0,0)$ Test $(0,0)$

   $0 \overset{?}{\geq} 0 + 1$ $0 \overset{?}{\geq} 3 - 0$

   False   False

   Shade above.   Shade above.

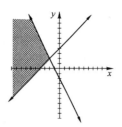

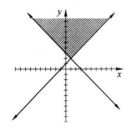

55. $y \geq -x + 2$ $y \leq 2x + 5$

   Test $(0,0)$ Test $(0,0)$

   $0 \overset{?}{\geq} -0 + 2$ $0 \overset{?}{\leq} 2(0) + 5$

   False   True

   Shade above.   Shade below.

51. $y < 3x - 4$ $y \leq x + 2$

   Test $(0,0)$ Test $(0,0)$

   $0 \overset{?}{<} 3(0) - 4$ $0 \overset{?}{\leq} 0 + 2$

   False   True

   Shade below.   Shade below.

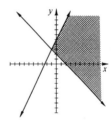

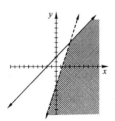

57. $x \geq 3y$ $x + 3y \leq 6$

   Test $(1,0)$ Test $(0,0)$

   $1 \overset{?}{\geq} 3(0)$ $0 + 3(0) \overset{?}{\leq} 6$

   True   True

   Shade below.   Shade below.

53. $y \leq -2x - 2$ $y \geq x + 4$

   Test $(0,0)$ Test $(0,0)$

   $0 \overset{?}{\leq} -2(0) - 2$ $0 \overset{?}{\geq} 0 + 4$

   False   False

   Shade below.   Shade above.

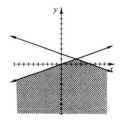

59. $y + 2x \geq 0$ $5x - 3y \leq 12$

Test $(1, 0)$ Test $(0, 0)$

$0 + 2(1) \overset{?}{\geq} 0$ $5(0) - 3(0) \overset{?}{\leq} 12$

True   True

Shade above.   Shade above.

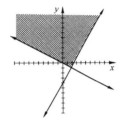

61. $3x - 4y \geq -6$ $2x + y \leq 7$

Test $(0, 0)$ Test $(0, 0)$

$3(0) - 4(0) \overset{?}{\geq} -6$ $2(0) + (0) \overset{?}{\leq} 7$

True   True

Shade below.   Shade below.

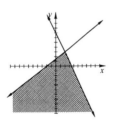

63. $x \leq 2$ $y \geq -3$

Shade left.   Shade above.

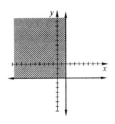

65. $y \geq 1$ $x < -3$

Shade above.   Shade left.

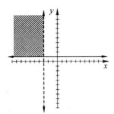

67. $2x + 3y \leq -8$ $x \geq -4$

Test $(0, 0)$ Shade right.

$2(0) + 3(0) \overset{?}{\leq} -8$

False

Shade below.

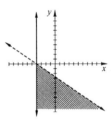

69. $2x - 5y \leq 9$ $y \leq -3$

Test $(0, 0)$ Shade below.

$2(0) - 5(0) \overset{?}{\leq} 9$

True

Shade above.

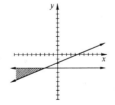

71. $y \geq \dfrac{1}{2}x + 2$ $y \leq \dfrac{1}{2}x - 3$

Test $(0,0)$ Test $(0,0)$

$0 \overset{?}{\geq} \dfrac{1}{2}(0) + 2$ $0 \overset{?}{\leq} \dfrac{1}{2}(0) - 3$

False   False

Shade above.   Shade below.

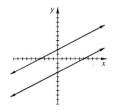

73. $(4)^2 = (4)(4) = 16$

75. $(6x)^2 = (6x)(6x) = 36x^2$

77. $\left(10y^3\right)^2 = \left(10y^3\right)\left(10y^3\right) = 100y^6$

79. C

81. D

83. Answers may vary.

85. $x + y \geq 13$

Test $(0,0)$

$0 + 0 \overset{?}{\geq} 13$

False

Shade above

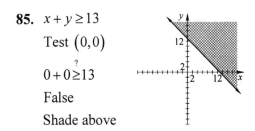

87. Answers may vary

89. $\begin{cases} 2.5x + 0.25y \leq 20 \\ x \geq 0 \\ y \geq 0 \end{cases}$

Test $(0,0)$

$2.5(0) + 0.25(0) \overset{?}{\leq} 20$

True

Shade below

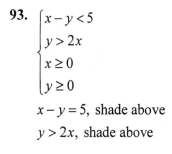

91. Answers may vary

93. $\begin{cases} x - y < 5 \\ y > 2x \\ x \geq 0 \\ y \geq 0 \end{cases}$

$x - y = 5$, shade above

$y > 2x$, shade above

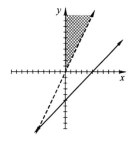

Chapter 9 Review

1. $1 \le 4x - 7 \le 3$

$8 \le 4x \le 10$

$2 \le x \le \dfrac{5}{2}$

$\left[2, \dfrac{5}{2} \right]$

2. $-2 \le 8 + 5x < -1$

$-10 \le 5x \le -9$

$-2 \le x \le -\dfrac{9}{5}$

$\left[-2, \dfrac{9}{5} \right)$

3. $-3 < 4(2x - 1) < 12$

$-3 < 8x - 4 < 12$

$1 < 8x < 16$

$\dfrac{1}{8} < x < 2$

$\left(\dfrac{1}{8}, 2 \right)$

4. $-6 < x - (3 - 4x) < -3$

$-6 < x - 3 + 4x < -3$

$-6 < 5x - 3 < -3$

$-3 < 5x > 0$

$-\dfrac{3}{5} < x < 0$

$\left(-\dfrac{3}{5}, 0 \right)$

5. $\dfrac{1}{6} < \dfrac{4x - 3}{3} \le \dfrac{4}{5}$

$30 \left(\dfrac{1}{6} \right) < 30 \left(\dfrac{4x - 3}{3} \right) < 30 \left(\dfrac{4}{5} \right)$

$5 < 10(4x - 3) \le 24$

$5 < 40x - 30 \le 24$

$35 < 40x < 54$

$\dfrac{7}{8} < x \le \dfrac{27}{20}$

$\left(\dfrac{7}{8}, \dfrac{27}{20} \right)$

6. $0 \le \dfrac{2(3x + 4)}{5} \le 3$

$5(0) \le 5 \left[\dfrac{2(3x + 4)}{5} \right] \le 5(3)$

$0 \le 2(3x + 4) \le 15$

$0 \le 6x + 8 \le 15$

$-8 \le 6x \le 7$

$-\dfrac{4}{3} \le x \le \dfrac{7}{6}$

$\left[-\dfrac{4}{3}, \dfrac{7}{6} \right]$

7. $x \le 2$ and $x > -5$

$-5 < x \le 2$

$(-5, 2]$

8. $x \le 2$ or $x > -5$

$(-\infty, \infty)$

9. $3x - 5 > 6$ or $-x < -5$

$\qquad 3x > 11$ or $\quad x < 5$

$\qquad x > \dfrac{11}{3}$ or $\quad x < 5$

$\qquad x > \dfrac{11}{3}$

$\qquad \left(\dfrac{11}{3}, \infty \right)$

10. $-2x \le 6$ and $-2x + 3 < -7$

$\qquad x \ge -3$ and $\qquad -2x < -10$

$\qquad x \ge -3$ and $\qquad x > 5$

$\qquad x > 5$

$\qquad (5, \infty)$

11. $|x - 7| = 9$

$\qquad x - 7 = 9$ or $x - 7 = -9$

$\qquad x = 16 \qquad\qquad x = -2$

The solution set is $\{-2, 16\}$.

12. $|8 - x| = 3$

$\qquad 8 - x = 3$ or $8 - x = -3$

$\qquad -x = -5 \qquad\qquad -x = -11$

$\qquad x = 5 \qquad\qquad x = 11$

The solution set is $\{5, 11\}$.

13. $|2x + 9| = 9$

$\qquad 2x + 9 = 9$ or $2x + 9 = -9$

$\qquad 2x = 0 \qquad\qquad 2x = -18$

$\qquad x = 0 \qquad\qquad x = -9$

The solution set is $\{-9, 0\}$.

14. $|-3x + 4| = 7$

$\qquad -3x + 4 = 7$ or $-3x + 4 = -7$

$\qquad -3x = 3 \qquad\qquad -3x = -11$

$\qquad x = -1 \qquad\qquad x = \dfrac{11}{3}$

The solution set is $\left\{-1, \dfrac{11}{3}\right\}$.

15. $|3x - 2| + 6 = 10$

$\qquad |3x - 2| = 4$

$\qquad 3x - 2 = 4$ or $3x - 2 = -4$

$\qquad 3x = 6 \qquad\qquad 3x = -2$

$\qquad x = 2 \qquad\qquad x = -\dfrac{2}{3}$

The solution set is $\left\{-\dfrac{2}{3}, 2\right\}$.

16. $5 + |6x + 1| = 5$

$\qquad |6x + 1| = 0$

$\qquad 6x + 1 = 0$

$\qquad 6x = -1$

$\qquad x = -\dfrac{1}{6}$

The solution set is $\left\{-\dfrac{1}{6}\right\}$.

17. $-5 = |4x - 3|$

The solution set is $\varnothing$.

18. $|5 - 6x| + 8 = 3$

$\qquad |5 - 6x| = -5$

The solution set is $\varnothing$.

19. $|7x| - 26 = -5$

$|7x| = 21$

$7x = 21$ or $7x = -21$

$x = 3$ $x = -3$

The solution set is $\{-3, 3\}$.

20. $-8 = |x - 3| - 10$

$2 = |x - 3|$

$x - 3 = 2$ or $x - 3 = -2$

$x = 5$ $x = 1$

The solution set is $\{1, 5\}$.

21. $\left| \dfrac{3x - 7}{4} \right| = 2$

$\dfrac{3x - 7}{4} = 2$ or $\dfrac{3x - 7}{4} = -2$

$3x - 7 = 8$ or $3x - 7 = -8$

$3x = 15$ $3x = -1$

$x = 5$ $x = -\dfrac{1}{3}$

The solution set is $\left\{ -\dfrac{1}{3}, 5 \right\}$.

22. $\left| \dfrac{9 - 2x}{5} \right| = -3$

The solution set is $\varnothing$.

23. $|6x + 1| = |15 + 4x|$

$6x + 1 = 15$ or $6x + 1 = -(15 + 4x)$

$2x = 14$ $6x + 1 = -15 - 4x$

$x = 7$ $10x = -16$

$x = -\dfrac{8}{5}$

The solution set is $\left\{ -\dfrac{8}{5}, 7 \right\}$.

24. $|x - 3| = |7 + 2x|$

$x - 3 = 7$ or $x - 3 = -(7 + 2x)$

$-10 = x$ $x - 3 = -7 - 2x$

$3x = -4$

$x = -\dfrac{4}{3}$

The solution set is $\left\{ -10, -\dfrac{4}{3} \right\}$.

25. $|5x - 1| < 9$

$-9 < 5x - 1 < 9$

$-8 < 5x < 10$

$-\dfrac{8}{5} < x < 2$

$\left(-\dfrac{8}{5}, 2 \right)$

26. $|6 + 4x| \geq 10$

$6 + 4x \leq -10$ or $6 + 4x \geq 10$

$4x \leq -16$ $4x \geq 4$

$x \leq -4$ $x \geq 1$

$(-\infty, -4] \cup [1, \infty)$

27. $|3x| - 8 > 1$

$|3x| > 9$

$3x < -9 \quad \text{or} \quad 3x > 9$

$x < -3 \qquad x > 3$

$(-\infty, -3) \cup (3, \infty)$

30. $|6x - 5| \geq -1$

Since $|6x - 5|$ is nonnegative for all

numbers x the solution set is $(-\infty, \infty)$.

31. $\left|3x + \dfrac{2}{5}\right| \geq 4$

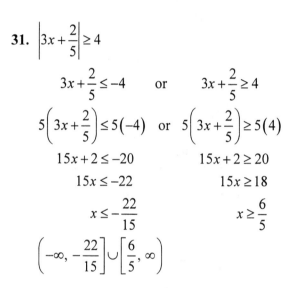

$3x + \dfrac{2}{5} \leq -4 \qquad \text{or} \qquad 3x + \dfrac{2}{5} \geq 4$

$5\left(3x + \dfrac{2}{5}\right) \leq 5(-4) \quad \text{or} \quad 5\left(3x + \dfrac{2}{5}\right) \geq 5(4)$

$15x + 2 \leq -20 \qquad\qquad 15x + 2 \geq 20$

$15x \leq -22 \qquad\qquad 15x \geq 18$

$x \leq -\dfrac{22}{15} \qquad\qquad x \geq \dfrac{6}{5}$

$\left(-\infty, -\dfrac{22}{15}\right] \cup \left[\dfrac{6}{5}, \infty\right)$

28. $9 + |5x| < 24$

$|5x| < 15$

$-15 < 5x < 15$

$-3 < x < 3$

$(-3, 3)$

29. $|6x - 5| \leq -1$

The solution set is $\varnothing$.

32. $\left|\dfrac{4x-3}{5}\right| < 1$

$-1 < \dfrac{4x-3}{5} < 1$

$-5 < 4x - 3 < 5$

$-2 < 4x < 8$

$-\dfrac{1}{2} < x < 2$

$\left(-\dfrac{1}{2},\, 2\right)$

33. $\left|\dfrac{x}{3} + 6\right| - 8 > -5$

$\left|\dfrac{x}{3} + 6\right| > 3$

$\dfrac{x}{3} + 6 < -3$ or $\dfrac{x}{3} + 6 > 3$

$\dfrac{x}{3} < -9$ $\qquad \dfrac{x}{3} > -3$

$x < -27$ $\qquad\quad x > -9$

$(-\infty,\, -27) \cup (-9,\, \infty)$

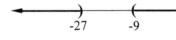

34. $\left|\dfrac{4(x-1)}{7}\right| + 10 < 2$

$\left|\dfrac{4(x-1)}{7}\right| < -8$

The solution set is $\varnothing$.

35. $3x - 4y \le 0$

Test $(1, 0)$

$3(1) - 4(0) \overset{?}{\le} 0$

False

Shade above.

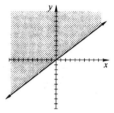

36. $3x - 4y \ge 0$

Test $(1, 0)$

$3(1) - 4(0) \overset{?}{\ge} 0$

True

Shade below.

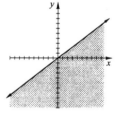

37. $x + 6y < 6$

Test $(0, 0)$

$0 + 6(0) \overset{?}{<} 6$

True

Shade below.

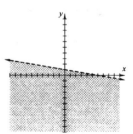

38. $x + y > -2$

Test $(0, 0)$

$0 + 0 \overset{?}{>} -2$

True

Shade above.

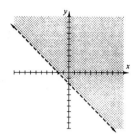

39. $y \geq -7$

Test $(0, 0)$

$0 \overset{?}{\geq} -7$

True

Shade above.

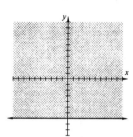

40. $y \leq -4$

Test $(0, 0)$

$0 \overset{?}{\leq} -4$

False

Shade below.

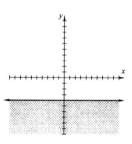

41. $-x \leq y$

Test $(1, 0)$

$-1 \overset{?}{\leq} 0$

True

Shade above.

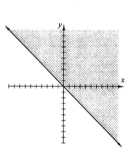

42. $x \geq -y$

Test $(1, 0)$

$1 \overset{?}{\geq} 0$

True

Shade above.

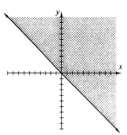

43. $y \geq 2x - 3$ $y \leq -2x + 1$

Test $(0, 0)$ Test $(0, 0)$

$0 \overset{?}{\leq} 2(0) - 3$ $0 \overset{?}{\leq} -2(0) + 1$

True   True

Shade above.   Shade below.

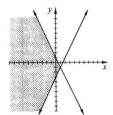

44. $y \leq -3x - 3$ $y \leq 2x + 7$

Test $(0, 0)$ Test $(0, 0)$

$0 \overset{?}{\leq} -3(0) - 3$ $0 \overset{?}{\leq} 2(0) + 7$

False   True

Shade below.   Shade below.

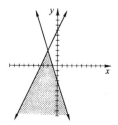

Martin-Gay, Beginning and Intermediate Algebra, 3e   **377**

45. $x + 2y > 0$ $x - y \le 6$

Test $(1,0)$ Test $(0,0)$

$1 + 2(0) \overset{?}{>} 0$ $(0) - (0) \overset{?}{\le} 6$

True   True

Shade above.   Shade above.

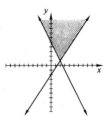

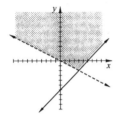

46. $x - 2y \ge 7$ $x + y \le -5$

Test $(0,0)$ Test $(0,0)$

$0 - 2(0) \overset{?}{\ge} 7$ $(0) + (0) \overset{?}{\le} -5$

False   False

Shade below.   Shade below.

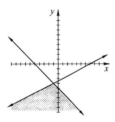

47. $3x - 2y \le 4$ $2x + y \ge 5$

Test $(0,0)$ Test $(0,0)$

$3(0) - 2(0) \overset{?}{\le} 4$ $2(0) + (0) \overset{?}{\ge} 5$

True   False

Shade above.   Shade above.

48. $4x - y \le 0$ $3x - 2y \ge -5$

Test $(1,0)$ Test $(0,0)$

$4(1) - (0) \overset{?}{\le} 4$ $3(0) - 2(0) \overset{?}{\ge} -5$

True   True

Shade above.   Shade below.

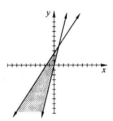

49. $-3x + 2y > -1$ $y < -2$

Test $(0,0)$ Shade below.

$-3(0) + 2(0) \overset{?}{>} -1$

True

Shade above.

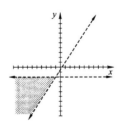

50. $-2x + 3y > -7$ $x \geq -2$

Test $(0,0)$ Shade right.

$$-2(0) + 3(0) \overset{?}{>} -7$$

True

Shade above.

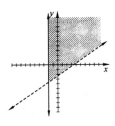

Chapter 9 Test

1. $|6x - 5| - 3 = -2$

$|6x - 5| = 1$

$6x - 5 = 1$ or $6x - 5 = -1$

$6x = 6$ $6x = 4$

$x = 1$ $x = \dfrac{2}{3}$

The solution set is $\left\{ \dfrac{2}{3}, 1 \right\}$.

2. $|8 - 2t| = -6$

No solution.

3. $|x - 5| = |x + 2|$

$x - 5 = x + 2$ or $x - 5 = -(x + 2)$

$-5 = 2$ $x - 5 = -x - 2$

                  $2x = 3$

                  $x = \dfrac{3}{2}$

Since $-5 = 2$ is not possible, the

solution set is $\left\{ \dfrac{3}{2} \right\}$.

4. $-3 < 2(x - 3) \leq 4$

$-3 < 2x - 6 \leq 4$

$3 < 2x \leq 10$

$\dfrac{3}{2} < x \leq 5$

$\left(\dfrac{3}{2}, 5 \right]$

5. $|3x + 1| > 5$

$3x + 1 < -5$ or $3x + 1 > 5$

$3x < -6$ $3x > 4$

$x < -2$ $x > \dfrac{4}{3}$

$(-\infty, -2) \cup \left(\dfrac{4}{3}, \infty \right)$

6. $|x - 5| - 4 < -2$

$|x - 5| < 2$

$-2 < x - 5 < 2$

$3 < x < 7$

$(-3, 7)$

7. $x \le -2$ and $x \le -5$, $(-\infty, -5]$

8. $x \le -2$ or $x \le -5$, $(-\infty, -2]$

9. $-x > 1$ and $3x + 3 \ge x - 3$
$\quad\quad x < -1 \quad\quad\quad 2x \ge -6$
$\quad\quad\quad\quad\quad\quad\quad\quad x \ge -3$
$\quad -3 \le x < -1$
$\quad [-3, -1)$

10. $6x + 1 > 5x + 4$ or $1 - x > -4$
$\quad\quad x > 3 \quad\quad\quad\quad 5 > x$
$\quad (-\infty, \infty)$

11. $\left| \dfrac{5x - 7}{2} \right| = 4$

$\dfrac{5x - 7}{4} = 4 \quad$ or $\quad \dfrac{5x - 7}{4} = -4$

$5x - 7 = 8 \quad\quad\quad 5x - 7 = -8$

$\quad 5x = 15 \quad\quad\quad\quad 5x = -1$

$\quad\quad x = 3 \quad\quad\quad\quad\quad x = -\dfrac{1}{5}$

The solution set is $\left\{ -\dfrac{1}{5}, 3 \right\}$.

12. $\left| 17x - \dfrac{1}{5} \right| > -2$

The solution set is $\{-\infty, \infty\}$.

13. $y \ge -4x$
Test $(1, 0)$
$0 \overset{?}{\ge} -4(1)$
True
Shade above.

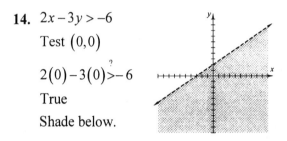

14. $2x - 3y > -6$
Test $(0, 0)$
$2(0) - 3(0) \overset{?}{>} -6$
True
Shade below.

15. $y + 2x \le 4 \quad\quad\quad y \ge 2$
Test $(0, 0)$ Shade above.
$0 + 2(0) \overset{?}{\le} 4$
True
Shade below.

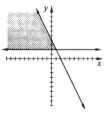

16. $2y - x \ge 1 \quad\quad\quad x + y \ge -4$
Test $(0, 0)$ Test $(0, 0)$
$2(0) - 0 \overset{?}{\ge} 1 \quad\quad (0) + (0) \overset{?}{\ge} -4$
False   True
Shade above.   Shade above.

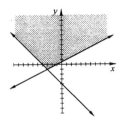

Chapter 9 Cumulative Review

1. $x = 2$ and $y = -5$

a. $\dfrac{x-y}{12+x} = \dfrac{2-(-5)}{12+2} = \dfrac{7}{14} = \dfrac{1}{2}$

b. $x^2 - y = (2)^2 - (-5) = 9$

2. $x = -4$ and $y = 7$

a. $\dfrac{x-y}{7-x} = \dfrac{-4-7}{7-(-4)} = \dfrac{-11}{11} = -1$

b. $x^2 + 2y = (-4)^2 + 2(7) = 30$

3. a. $\dfrac{(-12)(-3)+3}{-7-(-2)} = \dfrac{36+3}{-5} = -\dfrac{39}{5}$

b. $\dfrac{2(-3)^2 - 20}{-5+4} = \dfrac{2(9)-20}{-1} = \dfrac{-2}{-1} = 2$

4. a. $\dfrac{4(-3)-(-6)}{-8+4} = \dfrac{-12+6}{-4} = \dfrac{-6}{-4} = \dfrac{3}{2}$

b. $\dfrac{3+(-3)(-2)^3}{-1-(-4)} = \dfrac{3+(-3)(-8)}{3}$

$= \dfrac{3+24}{3} = \dfrac{27}{3} = 9$

5. a. $2x + 3x + 5 + 2 = 5x + 7$

b. $-5x - 3 + z + 2 = -4x - 1$

c. $4y - 3y^2 = 4y - 3y^2$

d. $2.3x + 5x - 6 = 7.3x - 6$

e. $-\dfrac{1}{2}b + b = \dfrac{1}{2}b$

6. a. $4x - 3 + 7 - 5x = -x + 4$

b. $-6y + 3y - 8 + 8y = 5y - 8$

c. $2 + 8.1a + a - 6 = 9.1a - 4$

d. $2x^2 - 2x = 2x^2 - 2x$

7. $2x + 3x - 5 + 7 = 10x + 3 - 6x - 4$

$5x + 2 = 4x - 1$

$x + 2 = -1$

$x = -3$

8. $6y - 11 + 4 + 2y = 8 + 15y - 8y$

$8y - 7 = 8 + 7y$

$y - 7 = 8$

$y = 15$

9. $y = 3x$

$x = -1, \ y = 3(-1) = -3$

$y = 0, \ 0 = 3x \Rightarrow x = 0$

$y = -9, \ -9 = 3x \Rightarrow x = -3$

x	y
-1	-3
0	0
-3	-9

10. $2x + y = 6$

$\quad x = 0, \; 2(0) + y = 6 \Rightarrow y = 6$

$\quad y = -2, \; 2x - 2 = 6 \Rightarrow x = 4$

$\quad x = 3, \; 2(3) + y = 6 \Rightarrow y = 0$

x	y
0	6
4	-2
3	0

11. a. x-intercept: $(-3, 0)$

$\quad\;$ y-intercept: $(0, 2)$

$\quad$ b. x-intercept: $(-4, 0), (-1, 0)$

$\quad\;$ y-intercept: $(0, 1)$

$\quad$ c. x-intercept: $(0, 0)$

$\quad\;$ y-intercept: $(0, 0)$

$\quad$ d. x-intercept: $(2, 0)$

$\quad\;$ y-intercept: None

$\quad$ e. x-intercept: $(-1, 0), (3, 0)$

$\quad\;$ y-intercept: $(0, -1), (0, 2)$

12. a. x-intercept: $(4, 0)$

$\quad\;$ y-intercept: $(0, 1)$

$\quad$ b. x-intercept: $(-2, 0), (0, 0), (3, 0)$

$\quad\;$ y-intercept: $(0, 0)$

$\quad$ c. x-intercept: None

$\quad\;$ y-intercept: $(0, -3)$

$\quad$ d. x-intercept: $(-3, 0), (3, 0)$

$\quad\;$ y-intercept: $(0, -3), (0, 3)$

13. $y = \dfrac{1}{5}x + 1, \; m_1 = -\dfrac{1}{5}, \; b_1 = 1$

$\quad 2x + 10y = 30$

$\qquad y = -\dfrac{1}{5}x + 3, \; m_2 = -\dfrac{1}{5}, \; b_2 = 3$

$\quad m_2 = m_1, \; b_2 \neq b.$

$\quad$ They are parallel.

14. $y = 3x + 7, \; m_1 = 3, \; b_1 = 7$

$\quad x + 3y = -15$

$\qquad y = -\dfrac{1}{3}x - 5, \; m_2 = -\dfrac{1}{3}$

$\quad m_2 = \dfrac{1}{m_1}$

$\quad$ They are perpendicular.

15. $m = \dfrac{1}{4}, \; b = -3$

$\quad y = \dfrac{1}{4}x - 3$

16. $m = -2, \; b = 4$

$\quad y = -2x + 4$

17. $y = 5$ is a horizontal line.

$\quad$ A horizontal line through $(-2, -3)$

$\quad$ is $y = -3$.

18. $y = 2x + 4$, $m_1 = 2$

Perpendicular line, $m_2 = -\dfrac{1}{2}$

Point: $(1, 5)$

$y - 5 = -\dfrac{1}{2}(x - 1)$

$2y - 10 = x - 1$

$-9 = x - 2y$

$x - 2y = -9$

19. a. $y = x$, function

b. $y = 2x + 1$, function

c. $y = 5$, function

d. $x = -1$, not a function

20. a. $2x + 3 = y$, function

b. $x + 4 = 0$, not a function

c. $\dfrac{1}{2}y = 2x$, function

d. $y = 0$, function

21. $\begin{cases} 2x - 3y = 6 \\ \quad\ x = 2y \end{cases}$

a. $2x - 3y = 6$ $x = 2y$

$2(12) - 3(6) \overset{?}{=} 6$ $12 \overset{?}{=} 2(6)$

$24 - 18 \overset{?}{=} 6$ $12 = 12$ True

$6 = 6$ True

$(12, 6)$ is a solution.

b. $2x - 3y = 6$ $x = 2y$

$2(0) - 3(-2) \overset{?}{=} 6$ $0 \overset{?}{=} 2(-2)$

$6 = 6$ True $0 \overset{?}{=} -4$ False

$(0, -2)$ is not a solution.

22. $\begin{cases} 2x + y = 4 \\ \ \ x + y = 2 \end{cases}$

a. $2x + y = 4$ $x + y = 2$

$2(1) + (1) \overset{?}{=} 4$ $(1) + (1) \overset{?}{=} 2$

$3 \overset{?}{=} 4$ False $2 = 2$ True

$(1, 1)$ is not a solution.

b. $2x + y = 4$ $x + y = 2$

$2(2) + 0 \overset{?}{=} 4$ $2 + 0 \overset{?}{=} 2$

$4 = 4$ True $2 = 2$ True

$(2, 0)$ is a solution.

23. $\left(11x^3 - 12x^2 + x - 3\right) + \left(x^3 - 10x + 5\right)$

$= 11x^3 - 12x^2 + x - 3 + x^3 - 10x + 5$

$= 12x^3 - 12x^2 - 9x + 2$

24. $4a^2 + 3a - 2a^2 + 7a - 5$

$= 20a^2 + 10a - 5$

25. $x^2 + 5yx + 6y^2$

$= (x + 2y)(x + 3y)$

26. $3x^2 + 15x + 18$

$= 3\left(x^2 + 5x + 6\right)$

$= 3(x + 2)(x + 3)$

27. $\dfrac{3x^3y^7}{40} \div \dfrac{4x^3}{y^2} = \dfrac{3x^3y^7}{40} \cdot \dfrac{y^2}{4x^3}$

$\qquad = \dfrac{3y^9}{160}$

28. $\dfrac{12x^2y^3}{5} \div \dfrac{3y^2}{x} = \dfrac{12x^2y^3}{5} \cdot \dfrac{x}{3y^2}$

$\qquad = \dfrac{4x^3y}{5}$

29. $\dfrac{2y}{2y-7} - \dfrac{7}{2y-7} = \dfrac{2y-7}{2y-7} = 1$

30. $\dfrac{-4x^2}{x+1} - \dfrac{4x}{x+1} = \dfrac{-4x^2-4x}{x+1}$

$\qquad = \dfrac{-4x(x+1)}{x+1}$

$\qquad = -4x$

31. $\dfrac{2x}{x^2+2x+1} + \dfrac{x}{x^2-1}$

$\quad = \dfrac{2x}{(x+1)(x+1)} + \dfrac{x}{(x+1)(x-1)}$

$\quad = \dfrac{2x(x-1)}{(x+1)(x+1)(x-1)} + \dfrac{x(x+1)}{(x+1)(x+1)(x-1)}$

$\quad = \dfrac{2x(x-1)+x(x+1)}{(x+1)(x+1)(x-1)}$

$\quad = \dfrac{2x^2-2x+x^2+x}{(x+1)(x+1)(x-1)}$

$\quad = \dfrac{3x^2-x}{(x+1)(x+1)(x-1)}$

32. $\dfrac{3x}{x^2+5x+6} + \dfrac{1}{x^2+2x-3}$

$\quad = \dfrac{3x}{(x+2)(x+3)} + \dfrac{1}{(x+3)(x-1)}$

$\quad = \dfrac{3x(x-1)}{(x+2)(x+3)(x-1)} + \dfrac{x+2}{(x+3)(x+2)(x-1)}$

$\quad = \dfrac{3x^2-2x+2}{(x+2)(x+3)(x-1)}$

$\quad = \dfrac{3x^2-2x+2}{(x+2)(x+3)(x-1)}$

33. $\dfrac{x}{2} + \dfrac{8}{3} = \dfrac{1}{6}$

$\quad 6\left(\dfrac{x}{2}\right) + 6\left(\dfrac{8}{3}\right) = 6\left(\dfrac{1}{6}\right)$

$\qquad 3x + 16 = 1$

$\qquad 3x = -15$

$\qquad x = -5$

34. $\dfrac{1}{21} + \dfrac{x}{7} = \dfrac{5}{3}$

$\quad 21\left(\dfrac{1}{21}\right) + 21\left(\dfrac{x}{7}\right) = 21\left(\dfrac{5}{3}\right)$

$\qquad 1 + 3x = 35$

$\qquad 3x = 34$

$\qquad x = \dfrac{34}{3}$

35. $2x + y = 7$ $2y = -4x$

x	y
0	7
$\frac{7}{2}$	0

x	y
0	0
-3	6

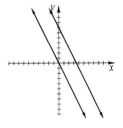

36. $y = x + 2$ $2x + y = 5$

x	y
0	2
2	4

x	y
0	5
$\frac{5}{2}$	0

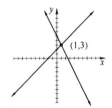

37. $\begin{cases} 7x - 3y = -14 \\ -3x + y = 6 \end{cases}$

Multiply the second equation
by 3 and then add.

$$7x - 3y = -14$$
$$-9x + 3y = 18$$
$$\overline{-2x = 4}$$
$$x = -2$$

Let $x = -2$ in the second equation.

$$-3(-2) + y = 6$$
$$6 + y = 6$$
$$y = 0$$

The solution is $(-2, 0)$.

38. $5x + y = 3$
         $y = -5x$

Let $y = -5x$ in the first equation

$$5x - 5x = 3$$
$$0 = 3$$

$\varnothing$

39. $3x - 2y = 2$
    $-9x + 6y = -6$

Multiply the first equation by
3 and then add.

$$9x - 6y = 6$$
$$-9x + 6y = -6$$
$$\overline{ 0 = 0}$$

There are an infinite number of
solutions.

40. $-2x + y = 7$

$6x - 3y = -21$

Multiply the first equation by 3 and then add.

$-6x + 3y = 21$

$\underline{6x - 3y = -21}$

$0 = 0$

There are an infinite number of solutions.

41. $-3x + 4y < 12$ $x \geq 2$

x	y
0	3
-4	0

Shade right

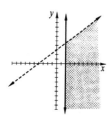

42. $2x - y \leq 6$ $x \geq 2$

x	y
0	-6
3	0

Shade above

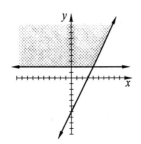

43. a. $\left(\dfrac{-5x^2}{y^3}\right)^2 = \dfrac{25x^4}{y^6}$

  b. $\dfrac{\left(x^3\right)^4 x}{x^7} = \dfrac{x^{12}x}{x^7}$

  c. $\dfrac{\left(2x\right)^5}{x^3} = \dfrac{32x^5}{x^3} = 32x^2$

  d. $\dfrac{\left(a^2b\right)^3}{a^3b^2} = \dfrac{a^6b^3}{a^3b^2} = a^3b$

44. a. $\left(\dfrac{-6x}{y^3}\right)^3 = \dfrac{216x^3}{y^9}$

  b. $\dfrac{a^2b^7}{\left(2b^2\right)^5} = \dfrac{a^2b^7}{32b^{10}} = \dfrac{a^2}{32b^3}$

  c. $\dfrac{\left(3y\right)^2}{y^2} = \dfrac{9y^2}{y^2} = 9$

  d. $\dfrac{\left(x^2y^4\right)^2}{xy^3} = \dfrac{x^4y^8}{xy^3} = x^3y^5$

45. $\left(5x - 1\right)\left(2x^2 + 15x + 18\right) = 0$

$\left(5x - 1\right)\left(2x + 3\right)\left(x + 6\right) = 0$

$5x - 1 = 0$ or $2x + 3 = 0$ or $x + 6 = 0$

    $5x = 1$ $2x = -3$ $x = -6$

    $x = \dfrac{1}{5}$ $x = -\dfrac{3}{2}$ $x = -6$

46. $\left(x + 1\right)\left(2x^2 - 3x - 5\right) = 0$

$\left(x + 1\right)\left(2x - 5\right)\left(x + 1\right) = 0$

$x + 1 = 0$ or $2x - 5 = 0$ or $x + 1 = 0$

    $x = -1$ $2x = 5$ $x = -1$

    $x = -1$ $x = \dfrac{5}{2}$ $x = -1$

47. $\dfrac{45}{x} = \dfrac{5}{7}$

$5x = 7(45)$

$x = 7(9)$

$x = 63$

48. $\dfrac{2x+7}{3} = \dfrac{x-6}{2}$

$6\left(\dfrac{2x+7}{3}\right) = 6\left(\dfrac{x-6}{2}\right)$

$2(2x+7) = 3(x-6)$

$4x+14 = 3x-18$

$x+14 = -18$

$x = -32$

Chapter 10

Section 10.1

Mental Math

1. D

2. A, C

3. D

4. C

Exercise Set 10.1

1. $\sqrt{100} = 10$ because $10^2 = 100$.

3. $\sqrt{\dfrac{1}{4}} = \dfrac{1}{2}$ because $\left(\dfrac{1}{2}\right)^2 = \dfrac{1}{4}$.

5. $\sqrt{0.0001} = 0.01$ because $(0.01)^2 = 0.0001$.

7. $-\sqrt{36} = -1 \cdot \sqrt{36} = -1 \cdot 6 = -6$ because $6^2 = 36$.

9. $\sqrt{x^{10}} = x^5$ because $(x^5)^2 = x^{10}$.

11. $\sqrt{16y^6} = 4y^3$ because $(4y^3)^2 = 16y^6$.

13. $\sqrt{7} \approx 2.646$
    Since $4 < 7 < 9$, then $\sqrt{4} < \sqrt{7} < \sqrt{9}$, or $2 < \sqrt{7} < 3$. The approximation is between 2 and 3 and thus is reasonable.

15. $\sqrt{38} \approx 6.164$
    Since $36 < 38 < 49$, then $\sqrt{36} < \sqrt{38} < \sqrt{49}$, or $6 < \sqrt{38} < 7$. The approximation is between 6 and 7 and thus is reasonable.

17. $\sqrt{200} \approx 14.142$
    Since $196 < 200 < 225$, then $\sqrt{196} < \sqrt{200} < \sqrt{225}$, or $14 < \sqrt{200} < 15$. The approximation is between 14 and 15 and thus is reasonable.

19. $\sqrt[3]{64} = 4$ because $4^3 = 64$.

21. $\sqrt[3]{\dfrac{1}{8}} = \dfrac{1}{2}$ because $\left(\dfrac{1}{2}\right)^3 = \dfrac{1}{8}$.

23. $\sqrt[3]{-1} = -1$ because $(-1)^3 = -1$.

25. $\sqrt[3]{x^{12}} = x^4$ because $(x^4)^3 = x^{12}$.

27. $\sqrt[3]{-27x^9} = -3x^3$ because $(-3x^3)^3 = -27x^9$.

29. $-\sqrt[4]{16} = -2$ because $2^4 = 16$.

31. $\sqrt[4]{-16}$ is not a real number. There is no real number that, when raised to the fourth power, is -16.

33. $\sqrt[5]{-32} = -2$ because $(-2)^5 = -32$.

35. $\sqrt[5]{x^{20}} = x^4$ because $(x^4)^5 = x^{20}$.

37. $\sqrt[6]{64x^{12}} = 2x^2$ because $(2x^2)^6 = 64x^{12}$.

39. $\sqrt{81x^4} = 9x^2$ because $(9x^2)^2 = 81x^4$.

41. $\sqrt[4]{256x^8} = 4x^2$ because $(4x^2)^4 = 256x^8$.

43. $\sqrt{(-8)^2} = |-8| = 8$

45. $\sqrt[3]{(-8)^3} = -8$

47. $\sqrt{4x^2} = 2|x|$

49. $\sqrt[3]{x^3} = x$

51. $\sqrt{(x-5)^2} = |x-5|$

53. $\sqrt{x^2 + 4x + 4} = \sqrt{(x+2)^2} = |x+2|$

55. $-\sqrt{121} = -11$

57. $\sqrt[3]{8x^3} = 2x$

59. $\sqrt{y^{12}} = y^6$

61. $\sqrt{25a^2 b^{20}} = 5ab^{10}$

63. $\sqrt[3]{-27x^{12} y^9} = -3x^4 y^3$

65. $\sqrt[4]{a^{16} b^4} = a^4 b$

67. $\sqrt[5]{-32x^{10} y^5} = -2x^2 y$

69. $\sqrt{\dfrac{25}{49}} = \dfrac{5}{7}$

71. $\sqrt{\dfrac{x^2}{4y^2}} = \dfrac{x}{2y}$

73. $-\sqrt[3]{\dfrac{z^{21}}{27x^3}} = -\dfrac{z^7}{3x}$

75. $\sqrt[4]{\dfrac{x^4}{16}} = \dfrac{x}{2}$

77. $f(x) = \sqrt{2x+3}$
$f(0) = \sqrt{2(0)+3} = \sqrt{3}$

79. $g(x) = \sqrt[3]{x-8}$
$g(7) = \sqrt[3]{7-8} = \sqrt[3]{-1} = -1$

81. $g(x) = \sqrt[3]{x-8}$
$g(-19) = \sqrt[3]{-19-8} = \sqrt[3]{-27} = -3$

83. $f(x) = \sqrt{2x+3}$
$f(2) = \sqrt{2(2)+3} = \sqrt{7}$

85. $f(x) = \sqrt{x} + 2$
Domain: $[0, \infty)$

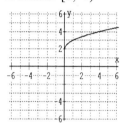

87. $f(x) = \sqrt{x-3}$
Domain: $[3, \infty)$

x	$f(x) = \sqrt{x-3}$
3	$\sqrt{3-3} = \sqrt{0} = 0$
4	$\sqrt{4-3} = \sqrt{1} = 1$
7	$\sqrt{7-3} = \sqrt{4} = 2$
12	$\sqrt{12-3} = \sqrt{9} = 3$

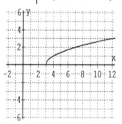

89. $f(x) = \sqrt[3]{x} + 1$
Domain: $(-\infty, \infty)$

91. $g(x) = \sqrt[3]{x} - 1$

Domain: $(-\infty, \infty)$

x	$g(x) = \sqrt[3]{x} - 1$
1	$\sqrt[3]{1} - 1 = \sqrt[3]{0} = 0$
2	$\sqrt[3]{2} - 1 = \sqrt[3]{1} = 1$
0	$\sqrt[3]{0} - 1 = \sqrt[3]{-1} = -1$
9	$\sqrt[3]{9} - 1 = \sqrt[3]{8} = 2$
-7	$\sqrt[3]{-7} - 1 = \sqrt[3]{8} - 8 = -2$

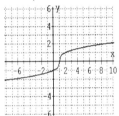

93. $(-2x^3y^2)^5 = (-2)^5 x^{3\cdot5} y^{2\cdot5} = -32x^{15}y^{10}$

95. $(-3x^2y^3z^5)(20x^5y^7) = -3(20)x^{2+5}y^{3+7}z^5$
$$= -60x^7y^{10}z^5$$

97. $\dfrac{7x^{-1}y}{14(x^5y^2)^{-2}} = \dfrac{7x^{-1}y}{14x^{-10}y^{-4}} = \dfrac{x^9y^5}{2}$

99. Answers may vary.

101. $144 < 160 < 169$ so
$\sqrt{144} < \sqrt{160} < \sqrt{169}$, or
$12 < \sqrt{160} < 13$. Thus $\sqrt{160}$ is between
12 and 13. Therefore, the answer is **b.**

103. $\sqrt{30} \approx 5$, $\sqrt{10} \approx 3$, and $\sqrt{90} \approx 10$ so
$P = \sqrt{30} + \sqrt{10} + \sqrt{90} \approx 5 + 3 + 10 = 18$.
Therefore, the answer is **b.**

105. $B = \sqrt{\dfrac{hw}{3131}} = \sqrt{\dfrac{66 \cdot 135}{3131}}$
$$= \sqrt{\dfrac{8910}{3131}} \approx 1.69 \text{ sq meters}$$

107. Answers may vary.

109. $f(x) = \sqrt{x} + 2$

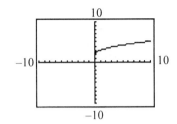

Domain: $[0, \infty)$

111. $f(x) = \sqrt[3]{x} + 1$

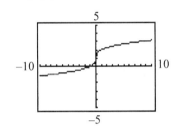

Domain: $(-\infty, \infty)$

Section 10.2

Mental Math

1. A
2. B
3. C
4. A
5. B
6. B
7. B
8. C
9. B
10. B

Exercise Set 10.2

1. $49^{1/2} = \sqrt{49} = 7$

3. $27^{1/3} = \sqrt[3]{27} = 3$

5. $\left(\dfrac{1}{16}\right)^{1/4} = \sqrt[4]{\dfrac{1}{16}} = \dfrac{1}{2}$

7. $169^{1/2} = \sqrt{169} = 13$

9. $2m^{1/3} = 2\sqrt[3]{m}$

11. $(9x^4)^{1/2} = \sqrt{9x^4} = 3x^2$

13. $(-27)^{1/3} = \sqrt[3]{-27} = -3$

15. $-16^{1/4} = -\sqrt[4]{16} = -2$

17. $16^{3/4} = \left(\sqrt[4]{16}\right)^3 = 2^3 = 8$

19. $(-64)^{2/3} = \left(\sqrt[3]{-64}\right)^2 = (-4)^2 = 16$

21. $(-16)^{3/4} = \left(\sqrt[4]{-16}\right)^3$ is not a real number.

23. $(2x)^{3/5} = \sqrt[5]{(2x)^3}$ or $\left(\sqrt[5]{2x}\right)^3$

25. $(7x+2)^{2/3} = \sqrt[3]{(7x+2)^2}$ or $\left(\sqrt[3]{7x+2}\right)^2$

27. $\left(\dfrac{16}{9}\right)^{3/2} = \left(\sqrt{\dfrac{16}{9}}\right)^3 = \left(\dfrac{4}{3}\right)^3 = \dfrac{64}{27}$

29. $8^{-4/3} = \dfrac{1}{8^{4/3}} = \dfrac{1}{\left(\sqrt[3]{8}\right)^4} = \dfrac{1}{2^4} = \dfrac{1}{16}$

31. $(-64)^{-2/3} = \dfrac{1}{(-64)^{2/3}}$
$$= \dfrac{1}{\left(\sqrt[3]{-64}\right)^2} = \dfrac{1}{(-4)^2} = \dfrac{1}{16}$$

33. $(-4)^{-3/2} = \dfrac{1}{(-4)^{3/2}} = \dfrac{1}{\left(\sqrt{-4}\right)^3}$ is not a real number.

35. $x^{-1/4} = \dfrac{1}{x^{1/4}}$

37. $\dfrac{1}{a^{-2/3}} = a^{2/3}$

39. $\dfrac{5}{7x^{-3/4}} = \dfrac{5x^{3/4}}{7}$

41. Answers may vary.

43. $a^{2/3}a^{5/3} = a^{2/3+5/3} = a^{7/3}$

45. $x^{-2/5} \cdot x^{7/5} = x^{-\frac{2}{5}+\frac{7}{5}} = x^{5/5} = x$

47. $3^{1/4} \cdot 3^{3/8} = 3^{\frac{1}{4}+\frac{3}{8}} = 3^{\frac{2}{8}+\frac{3}{8}} = 3^{5/8}$

49. $\dfrac{y^{1/3}}{y^{1/6}} = y^{\frac{1}{3}-\frac{1}{6}} = y^{\frac{2}{6}-\frac{1}{6}} = y^{1/6}$

51. $(4u^2)^{3/2} = 4^{3/2}u^{2(3/2)}$
$$= \left(\sqrt{4}\right)^3 u^3$$
$$= 2^3 u^3$$
$$= 8u^3$$

53. $\dfrac{b^{1/2}b^{3/4}}{-b^{1/4}} = -b^{\frac{1}{2}+\frac{3}{4}-\frac{1}{4}} = -b^{\frac{2}{4}+\frac{3}{4}-\frac{1}{4}} = -b^1 = -b$

55. $\dfrac{(3x^{1/4})^3}{x^{1/12}} = \dfrac{3^3 x^{3/4}}{x^{1/12}}$
$$= 27x^{\frac{3}{4}-\frac{1}{12}}$$
$$= 27x^{\frac{9}{12}-\frac{1}{12}}$$
$$= 27x^{8/12}$$
$$= 27x^{2/3}$$

57. $y^{1/2}(y^{1/2} - y^{2/3}) = y^{1/2}y^{1/2} - y^{1/2}y^{2/3}$
$= y^{1/2+1/2} - y^{1/2+2/3}$
$= y^1 - y^{7/6}$
$= y - y^{7/6}$

59. $x^{2/3}(2x - 2) = 2xx^{2/3} - 2x^{2/3}$
$= 2x^{1+2/3} - 2x^{2/3}$
$= 2x^{5/3} - 2x^{2/3}$

61. $(2x^{1/3} + 3)(2x^{1/3} - 3) = (2x^{1/3})^2 - 3^2$
$= 2^2(x^{1/3})^2 - 9$
$= 4x^{2/3} - 9$

63. $x^{8/3} + x^{10/3} = x^{8/3}(1) + x^{8/3}(x^{2/3})$
$= x^{8/3}(1 + x^{2/3})$

65. $x^{2/5} - 3x^{1/5} = x^{1/5}(x^{1/5}) - x^{1/5}(3)$
$= x^{1/5}(x^{1/5} - 3)$

67. $5x^{-1/3} + x^{2/3} = x^{-1/3}(5) + x^{-1/3}(x^{3/3})$
$= x^{-1/3}(5 + x)$

69. $\sqrt[6]{x^3} = x^{3/6} = x^{1/2} = \sqrt{x}$

71. $\sqrt[6]{4} = 4^{1/6} = (2^2)^{1/6} = 2^{1/3} = \sqrt[3]{2}$

73. $\sqrt[4]{16x^2} = (16x^2)^{1/4}$
$= 16^{1/4}x^{2/4} = 2x^{1/2} = 2\sqrt{x}$

75. $\sqrt[8]{x^4y^4} = (x^4y^4)^{1/8}$
$= x^{4/8}y^{4/8}$
$= x^{1/2}y^{1/2}$
$= (xy)^{1/2}$
$= \sqrt{xy}$

77. $\sqrt[3]{y} \cdot \sqrt[5]{y^2} = y^{1/3} \cdot y^{2/5}$
$= y^{\frac{5}{15}+\frac{6}{15}}$
$= y^{11/15}$
$= \sqrt[15]{y^{11}}$

79. $\dfrac{\sqrt[3]{b^2}}{\sqrt[4]{b}} = \dfrac{b^{2/3}}{b^{1/4}} = b^{\frac{2}{3}-\frac{1}{4}} = b^{\frac{8}{12}-\frac{3}{12}} = b^{5/12} = \sqrt[12]{b^5}$

81. $\dfrac{\sqrt[3]{a^2}}{\sqrt[6]{a}} = \dfrac{a^{2/3}}{a^{1/6}}$
$= a^{\frac{2}{3}-\frac{1}{6}} = a^{\frac{4}{6}-\frac{1}{6}} = a^{3/6} = a^{1/2} = \sqrt{a}$

83. $\sqrt{3} \cdot \sqrt[3]{4} = 3^{1/2} \cdot 4^{1/3}$
$= 3^{3/6} \cdot 4^{2/6}$
$= (3^3 \cdot 4^2)^{1/6}$
$= (432)^{1/6}$
$= \sqrt[6]{432}$

85. $\sqrt[5]{7} \cdot \sqrt[3]{y} = 7^{1/5} \cdot y^{1/3}$
$= 7^{3/15} \cdot y^{5/15}$
$= (7^3 \cdot y^5)^{1/15}$
$= (343y^5)^{1/15}$
$= \sqrt[15]{343y^5}$

87. $75 = 25 \cdot 3$

89. $48 = 4 \cdot 12$ or $16 \cdot 3$

91. $16 = 8 \cdot 2$

93. $54 = 27 \cdot 2$

95. $B(w) = 70w^{3/4}$
$B(60) = 70(60)^{3/4}$
≈ 1509 calories

97. $f(x) = 1.54x^{9/5}$
$f(10) = 1.54(10)^{9/5}$
≈ 97.2 million subscriptions

99. $\square \cdot a^{2/3} = a^{3/3}$
$\square = \dfrac{a^{3/3}}{a^{2/3}}$
$\square = a^{3/3-2/3}$
$\square = a^{1/3}$

101.
$$\frac{\square}{x^{-2/5}} = x^{3/5}$$
$$x^{-2/5}\left(\frac{\square}{x^{-2/5}}\right) = x^{3/5} \cdot x^{-2/5}$$
$$\square = x^{3/5-2/5}$$
$$\square = x^{1/5}$$

103. $8^{1/4} \approx 1.6818$

105. $18^{3/5} \approx 5.6645$

107. $\dfrac{\sqrt{t}}{\sqrt{u}} = \dfrac{t^{1/2}}{u^{1/2}}$

Exercise Set 10.3

1. $\sqrt{7} \cdot \sqrt{2} = \sqrt{7 \cdot 2} = \sqrt{14}$

3. $\sqrt[4]{8} \cdot \sqrt[4]{2} = \sqrt[4]{8 \cdot 2} = \sqrt[4]{16} = 2$

5. $\sqrt[3]{4} \cdot \sqrt[3]{9} = \sqrt[3]{4 \cdot 9} = \sqrt[3]{36}$

7. $\sqrt{2} \cdot \sqrt{3x} = \sqrt{2 \cdot 3x} = \sqrt{6x}$

9. $\sqrt{\dfrac{7}{x}} \cdot \sqrt{\dfrac{2}{y}} = \sqrt{\dfrac{7}{x} \cdot \dfrac{2}{y}} = \sqrt{\dfrac{14}{xy}}$

11. $\sqrt[4]{4x^3} \cdot \sqrt[4]{5} = \sqrt[4]{4x^3 \cdot 5} = \sqrt[4]{20x^3}$

13. $\sqrt{\dfrac{6}{49}} = \dfrac{\sqrt{6}}{\sqrt{49}} = \dfrac{\sqrt{6}}{7}$

15. $\sqrt{\dfrac{2}{49}} = \dfrac{\sqrt{2}}{\sqrt{49}} = \dfrac{\sqrt{2}}{7}$

17. $\sqrt[4]{\dfrac{x^3}{16}} = \dfrac{\sqrt[4]{x^3}}{\sqrt[4]{16}} = \dfrac{\sqrt[4]{x^3}}{2}$

19. $\sqrt[3]{\dfrac{4}{27}} = \dfrac{\sqrt[3]{4}}{\sqrt[3]{27}} = \dfrac{\sqrt[3]{4}}{3}$

21. $\sqrt[4]{\dfrac{8}{x^8}} = \dfrac{\sqrt[4]{8}}{\sqrt[4]{x^8}} = \dfrac{\sqrt[4]{8}}{x^2}$

23. $\sqrt[3]{\dfrac{2x}{81y^{12}}} = \dfrac{\sqrt[3]{2x}}{\sqrt[3]{81y^{12}}}$
$$= \dfrac{\sqrt[3]{2x}}{\sqrt[3]{27y^{12}} \cdot \sqrt[3]{3}}$$
$$= \dfrac{\sqrt[3]{2x}}{3y^4 \sqrt[3]{3}}$$

25. $\sqrt{\dfrac{x^2 y}{100}} = \dfrac{\sqrt{x^2 y}}{\sqrt{100}} = \dfrac{\sqrt{x^2}\sqrt{y}}{10} = \dfrac{x\sqrt{y}}{10}$

27. $\sqrt{\dfrac{5x^2}{4y^2}} = \dfrac{\sqrt{5x^2}}{\sqrt{4y^2}} = \dfrac{\sqrt{5}\sqrt{x^2}}{2y} = \dfrac{\sqrt{5}x}{2y}$

29. $-\sqrt[3]{\dfrac{z^7}{27x^3}} = -\dfrac{\sqrt[3]{z^7}}{\sqrt[3]{27x^3}} = -\dfrac{\sqrt[3]{z^6 z}}{3x} = -\dfrac{z^2\sqrt[3]{z}}{3x}$

31. $\sqrt{32} = \sqrt{16 \cdot 2} = \sqrt{16} \cdot \sqrt{2} = 4\sqrt{2}$

33. $\sqrt[3]{192} = \sqrt[3]{64 \cdot 3} = \sqrt[3]{64} \cdot \sqrt[3]{3} = 4\sqrt[3]{3}$

35. $5\sqrt{75} = 5\sqrt{25 \cdot 3}$
$$= 5\sqrt{25} \cdot \sqrt{3} = 5(5)\sqrt{3} = 25\sqrt{3}$$

37. $\sqrt{24} = \sqrt{4 \cdot 6} = \sqrt{4} \cdot \sqrt{6} = 2\sqrt{6}$

39. $\sqrt{100x^5} = \sqrt{100x^4 \cdot x}$
$$= \sqrt{100x^4} \cdot \sqrt{x} = 10x^2\sqrt{x}$$

41. $\sqrt[3]{16y^7} = \sqrt[3]{8y^6 \cdot 2y}$
$$= \sqrt[3]{8y^6} \cdot \sqrt[3]{2y} = 2y^2\sqrt[3]{2y}$$

43. $\sqrt[4]{a^8 b^7} = \sqrt[4]{a^8 b^4 \cdot b^3}$
$$= \sqrt[4]{a^8 b^4} \cdot \sqrt[4]{b^3} = a^2 b\sqrt[4]{b^3}$$

45. $\sqrt{y^5} = \sqrt{y^4 \cdot y} = \sqrt{y^4} \cdot \sqrt{y} = y^2\sqrt{y}$

47. $\sqrt{25a^2 b^3} = \sqrt{25a^2 b^2 \cdot b}$
$$= \sqrt{25a^2 b^2} \cdot \sqrt{b} = 5ab\sqrt{b}$$

49. $\sqrt[5]{-32x^{10}y} = \sqrt[5]{-32x^{10} \cdot y}$
$= \sqrt[5]{-32x^{10}} \cdot \sqrt[5]{y} = -2x^2\sqrt[5]{y}$

51. $\sqrt[3]{50x^{14}} = \sqrt[3]{x^{12} \cdot 50x^2}$
$= \sqrt[3]{x^{12}} \cdot \sqrt[3]{50x^2} = x^4\sqrt[3]{50x^2}$

53. $-\sqrt{32a^8b^7} = -\sqrt{16a^8b^6 \cdot 2b}$
$= -\sqrt{16a^8b^6} \cdot \sqrt{2b}$
$= -4a^4b^3\sqrt{2b}$

55. $\sqrt{9x^7y^9} = \sqrt{9x^6y^8 \cdot xy}$
$= \sqrt{9x^6y^8} \cdot \sqrt{xy} = 3x^3y^4\sqrt{xy}$

57. $\sqrt[3]{125r^9s^{12}} = 5r^3s^4$

59. $\dfrac{\sqrt{14}}{\sqrt{7}} = \sqrt{\dfrac{14}{7}} = \sqrt{2}$

61. $\dfrac{\sqrt[3]{24}}{\sqrt[3]{3}} = \sqrt[3]{\dfrac{24}{3}} = \sqrt[3]{8} = 2$

63. $\dfrac{5\sqrt[4]{48}}{\sqrt[4]{3}} = 5\sqrt[4]{\dfrac{48}{3}} = 5\sqrt[4]{16} = 5(2) = 10$

65. $\dfrac{\sqrt{x^5y^3}}{\sqrt{xy}} = \sqrt{\dfrac{x^5y^3}{xy}} = \sqrt{x^4y^2} = x^2y$

67. $\dfrac{8\sqrt[3]{54m^7}}{\sqrt[3]{2m}} = 8\sqrt[3]{\dfrac{54m^7}{2m}}$
$= 8\sqrt[3]{27m^6} = 8(3m^2) = 24m^2$

69. $\dfrac{3\sqrt{100x^2}}{2\sqrt{2x^{-1}}} = \dfrac{3}{2}\sqrt{\dfrac{100x^2}{2x^{-1}}}$
$= \dfrac{3}{2}\sqrt{50x^3}$
$= \dfrac{3}{2}\sqrt{25x^2 \cdot 2}$
$= \dfrac{3}{2}(5x)\sqrt{2}$
$= \dfrac{15x}{2}\sqrt{2}$

71. $\dfrac{\sqrt[4]{96a^{10}b^3}}{\sqrt[4]{3a^2b^3}} = \sqrt[4]{\dfrac{96a^{10}b^3}{3a^2b^3}}$
$= \sqrt[4]{32a^8}$
$= \sqrt[4]{16a^8 \cdot 2}$
$= 2a^2\sqrt[4]{2}$

73. $6x + 8x = 14x$

75. $(2x+3)(x-5) = 2x^2 - 10x + 3x - 15$
$= 2x^2 - 7x - 15$

77. $9y^2 - 8y^2 = 1y^2 = y^2$

79. $-3(x+5) = -3x - 3(5) = -3x - 15$

81. $(x-4)^2 = x^2 - 2(x)(4) + 4^2$
$= x^2 - 8x + 16$

83. $A = \pi r\sqrt{r^2 + h^2}$

  a. $A = \pi(4)\sqrt{4^2 + 3^2}$
  $= 4\pi\sqrt{16 + 9}$
  $= 4\pi\sqrt{25}$
  $= 4\pi(5)$
  $= 20\pi$ sq. centimeters

  b. $A = \pi(6.8)\sqrt{(6.8)^2 + (7.2)^2}$
  $= 6.8\pi\sqrt{46.24 + 51.84}$
  $= 6.8\pi\sqrt{98.08}$
  ≈ 211.57 sq. feet

85. $F(x) = 0.6\sqrt{49 - x^2}$

  a. $F(3) = 0.6\sqrt{49 - 3^2}$
  $= 0.6\sqrt{49 - 9}$
  $= 0.6\sqrt{40} \approx 3.8$ times

  b. $F(5) = 0.6\sqrt{49 - 5^2}$
  $= 0.6\sqrt{49 - 25}$
  $= 0.6\sqrt{24} \approx 2.9$ times

  c. Answers may vary.

Section 10.4

Mental Math

1. $2\sqrt{3} + 4\sqrt{3} = 6\sqrt{3}$

3. $8\sqrt{x} - 5\sqrt{x} = 3\sqrt{x}$

5. $7\sqrt[3]{x} + 5\sqrt[3]{x} = 12\sqrt[3]{x}$

7. $\sqrt{11} + \sqrt{11} = 2\sqrt{11}$

9. $9\sqrt{13} - \sqrt{13} = 8\sqrt{13}$

11. $8\sqrt[3]{2x} + 3\sqrt[3]{2x} - \sqrt[3]{2x} = 10\sqrt[3]{2x}$

Exercise Set 10.4

1. $\begin{aligned}[t] \sqrt{8} - \sqrt{32} &= \sqrt{4 \cdot 2} - \sqrt{16 \cdot 2} \\ &= \sqrt{4} \cdot \sqrt{2} - \sqrt{16} \cdot \sqrt{2} \\ &= 2\sqrt{2} - 4\sqrt{2} \\ &= -2\sqrt{2} \end{aligned}$

3. $\begin{aligned}[t] 2\sqrt{2x^3} + 4x\sqrt{8x} &= 2\sqrt{x^2 \cdot 2x} + 4x\sqrt{4 \cdot 2x} \\ &= 2\sqrt{x^2} \cdot \sqrt{2x} + 4x\sqrt{4} \cdot \sqrt{2x} \\ &= 2x\sqrt{2x} + 4x(2)\sqrt{2x} \\ &= 2x\sqrt{2x} + 8x\sqrt{2x} \\ &= 10x\sqrt{2x} \end{aligned}$

5. $\begin{aligned}[t] 2\sqrt{50} - 3\sqrt{125} + \sqrt{98} &= 2\sqrt{25 \cdot 2} - 3\sqrt{25 \cdot 5} + \sqrt{49 \cdot 2} \\ &= 2\sqrt{25} \cdot \sqrt{2} - 3\sqrt{25} \cdot \sqrt{5} + \sqrt{49} \cdot \sqrt{2} \\ &= 2(5)\sqrt{2} - 3(5)\sqrt{5} + 7\sqrt{2} \\ &= 10\sqrt{2} - 15\sqrt{5} + 7\sqrt{2} \\ &= 17\sqrt{2} - 15\sqrt{5} \end{aligned}$

7. $\begin{aligned}[t] \sqrt[3]{16x} - \sqrt[3]{54x} &= \sqrt[3]{8 \cdot 2x} - \sqrt[3]{27 \cdot 2x} \\ &= \sqrt[3]{8} \cdot \sqrt[3]{2x} - \sqrt[3]{27} \cdot \sqrt[3]{2x} \\ &= 2\sqrt[3]{2x} - 3\sqrt[3]{2x} \\ &= -\sqrt[3]{2x} \end{aligned}$

9. $\begin{aligned}[t] \sqrt{9b^3} - \sqrt{25b^3} + \sqrt{49b^3} &= \sqrt{9b^2 \cdot b} - \sqrt{25b^2 \cdot b} + \sqrt{49b^2 \cdot b} \\ &= \sqrt{9b^2} \cdot \sqrt{b} - \sqrt{25b^2} \cdot \sqrt{b} + \sqrt{49b^2} \cdot \sqrt{b} \\ &= 3b\sqrt{b} - 5b\sqrt{b} + 7b\sqrt{b} \\ &= 5b\sqrt{b} \end{aligned}$

11. $\begin{aligned}[t] \frac{5\sqrt{2}}{3} + \frac{2\sqrt{2}}{5} &= \frac{5\left(5\sqrt{2}\right) + 3\left(2\sqrt{2}\right)}{3(5)} \\ &= \frac{25\sqrt{2} + 6\sqrt{2}}{15} \\ &= \frac{31\sqrt{2}}{15} \end{aligned}$

13. $\begin{aligned}[t] \sqrt[3]{\frac{11}{8}} - \frac{\sqrt[3]{11}}{6} &= \frac{\sqrt[3]{11}}{\sqrt[3]{8}} - \frac{\sqrt[3]{11}}{6} \\ &= \frac{\sqrt[3]{11}}{2} - \frac{\sqrt[3]{11}}{6} \\ &= \frac{3\sqrt[3]{11} - \sqrt[3]{11}}{6} \\ &= \frac{2\sqrt[3]{11}}{6} \\ &= \frac{\sqrt[3]{11}}{3} \end{aligned}$

15. $\begin{aligned}[t] \frac{\sqrt{20x}}{9} + \sqrt{\frac{5x}{9}} &= \frac{\sqrt{4 \cdot 5x}}{9} + \frac{\sqrt{5x}}{\sqrt{9}} \\ &= \frac{2\sqrt{5x}}{9} + \frac{\sqrt{5x}}{3} \\ &= \frac{2\sqrt{5x} + 3\sqrt{5x}}{9} \\ &= \frac{5\sqrt{5x}}{9} \end{aligned}$

17. $\begin{aligned}[t] 7\sqrt{9} - 7 + \sqrt{3} &= 7(3) - 7 + \sqrt{3} \\ &= 21 - 7 + \sqrt{3} \\ &= 14 + \sqrt{3} \end{aligned}$

19. $\begin{aligned}[t] 2 + 3\sqrt{y^2} - 6\sqrt{y^2} + 5 &= 2 + 3y - 6y + 5 \\ &= 7 - 3y \end{aligned}$

Martin-Gay, Beginning and Intermediate Algebra, 3e   395

21. $3\sqrt{108} - 2\sqrt{18} - 3\sqrt{48}$
$= 3\sqrt{36 \cdot 3} + 2\sqrt{9 \cdot 2} - 3\sqrt{16 \cdot 3}$
$= 3\sqrt{36} \cdot \sqrt{3} + 2\sqrt{9} \cdot \sqrt{2} - 3\sqrt{16} \cdot \sqrt{3}$
$= 3(6)\sqrt{3} + 2(3)\sqrt{2} - 3(4)\sqrt{3}$
$= 18\sqrt{3} + 6\sqrt{2} - 12\sqrt{3}$
$= 6\sqrt{3} + 6\sqrt{2}$

23. $-5\sqrt[3]{625} + \sqrt[3]{40} = -5\sqrt[3]{125 \cdot 5} + \sqrt[3]{8 \cdot 5}$
$= -5(5)\sqrt[3]{5} + 2\sqrt[3]{5}$
$= -25\sqrt[3]{5} + 2\sqrt[3]{5}$
$= -23\sqrt[3]{5}$

25. $\sqrt{9b^3} - \sqrt{25b^3} + \sqrt{16b^3}$
$= \sqrt{9b^2 \cdot b} - \sqrt{25b^2 \cdot b} + \sqrt{16b^2 \cdot b}$
$= 3b\sqrt{b} - 5b\sqrt{b} + 4b\sqrt{b}$
$= 2b\sqrt{b}$

27. $5y\sqrt{8y} + 2\sqrt{50y^3}$
$= 5y\sqrt{4 \cdot 2y} + 2\sqrt{25y^2 \cdot 2y}$
$= 5y(2)\sqrt{2y} + 2(5y)\sqrt{2y}$
$= 10y\sqrt{2y} + 10y\sqrt{2y}$
$= 20y\sqrt{2y}$

29. $\sqrt[3]{54xy^3} - 5\sqrt[3]{2xy^3} + y\sqrt[3]{128x}$
$= \sqrt[3]{27y^3 \cdot 2x} - 5\sqrt[3]{y^3 \cdot 2x} + y\sqrt[3]{64 \cdot 2x}$
$= 3y\sqrt[3]{2x} - 5y\sqrt[3]{2x} + 4y\sqrt[3]{2x}$
$= 2y\sqrt[3]{2x}$

31. $6\sqrt[3]{11} + 8\sqrt{11} - 12\sqrt{11} = 6\sqrt[3]{11} - 4\sqrt{11}$

33. $-2\sqrt[4]{x^7} + 3\sqrt[4]{16x^7}$
$= -2\sqrt[4]{x^4 \cdot x^3} + 3\sqrt[4]{8x^4 \cdot x^3}$
$= -2x\sqrt[4]{x^3} + 3(2x)\sqrt[4]{x^3}$
$= -2x\sqrt[4]{x^3} + 6x\sqrt[4]{x^3}$
$= 4x\sqrt[4]{x^3}$

35. $\dfrac{4\sqrt{3}}{3} - \dfrac{\sqrt{12}}{3} = \dfrac{4\sqrt{3}}{3} - \dfrac{\sqrt{4 \cdot 3}}{3}$
$= \dfrac{4\sqrt{3} - 2\sqrt{3}}{3} = \dfrac{2\sqrt{3}}{3}$

37. $\dfrac{\sqrt[3]{8x^4}}{7} + \dfrac{3x\sqrt[3]{x}}{7} = \dfrac{\sqrt[3]{8x^3 \cdot x}}{7} + \dfrac{3x\sqrt[3]{x}}{7}$
$= \dfrac{2x\sqrt[3]{x} + 3x\sqrt[3]{x}}{7}$
$= \dfrac{5x\sqrt[3]{x}}{7}$

39. $\sqrt{\dfrac{28}{x^2}} + \sqrt{\dfrac{7}{4x^2}} = \dfrac{\sqrt{4 \cdot 7}}{x} + \dfrac{\sqrt{7}}{2x}$
$= \dfrac{2\sqrt{7}}{x} + \dfrac{\sqrt{7}}{2x}$
$= \dfrac{2\left(2\sqrt{7}\right) + \sqrt{7}}{2x}$
$= \dfrac{4\sqrt{7} + \sqrt{7}}{2x} = \dfrac{5\sqrt{7}}{2x}$

41. $\sqrt[3]{\dfrac{16}{27}} - \dfrac{\sqrt[3]{54}}{6} = \dfrac{\sqrt[3]{8 \cdot 2}}{\sqrt[3]{27}} - \dfrac{\sqrt[3]{27 \cdot 2}}{6}$
$= \dfrac{2\sqrt[3]{2}}{3} - \dfrac{3\sqrt[3]{2}}{6}$
$= \dfrac{2\left(2\sqrt[3]{2}\right) - 3\sqrt[3]{2}}{6}$
$= \dfrac{4\sqrt[3]{2} - 3\sqrt[3]{2}}{6} = \dfrac{\sqrt[3]{2}}{6}$

43. $-\dfrac{\sqrt[3]{2x^4}}{9} + \sqrt[3]{\dfrac{250x^4}{27}}$
$= -\dfrac{\sqrt[3]{x^3 \cdot 2x}}{9} + \dfrac{\sqrt[3]{125x^3 \cdot 2x}}{\sqrt[3]{27}}$
$= \dfrac{-x\sqrt[3]{2x}}{9} + \dfrac{5x\sqrt[3]{2x}}{3}$
$= \dfrac{-x\sqrt[3]{2x} + 3\left(5x\sqrt[3]{2x}\right)}{9}$
$= \dfrac{-x\sqrt[3]{2x} + 15x\sqrt[3]{2x}}{9} = \dfrac{14x\sqrt[3]{2x}}{9}$

45.
$$P = 2\sqrt{12} + \sqrt{12} + 2\sqrt{27} + 3\sqrt{3}$$
$$= 2\sqrt{4 \cdot 3} + \sqrt{4 \cdot 3} + 2\sqrt{9 \cdot 3} + 3\sqrt{3}$$
$$= 2(2)\sqrt{3} + 2\sqrt{3} + 2(3)\sqrt{3} + 3\sqrt{3}$$
$$= 4\sqrt{3} + 2\sqrt{3} + 6\sqrt{3} + 3\sqrt{3}$$
$$= 15\sqrt{3} \text{ inches}$$

47.
$$\sqrt{7}\left(\sqrt{5} + \sqrt{3}\right) = \sqrt{7}\sqrt{5} + \sqrt{7}\sqrt{3}$$
$$= \sqrt{35} + \sqrt{21}$$

49.
$$\left(\sqrt{5} - \sqrt{2}\right)^2 = \left(\sqrt{5}\right)^2 - 2\sqrt{5}\sqrt{2} + \left(\sqrt{2}\right)^2$$
$$= 5 - 2\sqrt{10} + 2$$
$$= 7 - 2\sqrt{10}$$

51.
$$\sqrt{3x}\left(\sqrt{3} - \sqrt{x}\right) = \sqrt{3x}\sqrt{3} - \sqrt{3x}\sqrt{x}$$
$$= \sqrt{9x} - \sqrt{3x^2}$$
$$= 3\sqrt{x} - x\sqrt{3}$$

53.
$$\left(2\sqrt{x} - 5\right)\left(3\sqrt{x} + 1\right)$$
$$= 2\sqrt{x}\left(3\sqrt{x}\right) + 2\sqrt{x} \cdot 1 - 5\left(3\sqrt{x}\right) - 5(1)$$
$$= 6x + 2\sqrt{x} - 15\sqrt{x} - 5$$
$$= 6x - 13\sqrt{x} - 5$$

55.
$$\left(\sqrt[3]{a} - 4\right)\left(\sqrt[3]{a} + 5\right)$$
$$= \sqrt[3]{a}\left(\sqrt[3]{a}\right) + \sqrt[3]{a} \cdot 5 - 4\sqrt[3]{a} - 4(5)$$
$$= \sqrt[3]{a^2} + 5\sqrt[3]{a} - 4\sqrt[3]{a} - 20$$
$$= \sqrt[3]{a^2} + \sqrt[3]{a} - 20$$

57. $6\left(\sqrt{2} - 2\right) = 6\sqrt{2} - 6(2) = 6\sqrt{2} - 12$

59.
$$\sqrt{2}\left(\sqrt{2} + x\sqrt{6}\right) = \sqrt{2}\sqrt{2} + \sqrt{2}\left(x\sqrt{6}\right)$$
$$= 2 + x\sqrt{12}$$
$$= 2 + x\sqrt{4 \cdot 3}$$
$$= 2 + 2x\sqrt{3}$$

61.
$$\left(2\sqrt{7} + 3\sqrt{5}\right)\left(\sqrt{7} - 2\sqrt{5}\right)$$
$$= 2\sqrt{7}\sqrt{7} + 2\sqrt{7}\left(-2\sqrt{5}\right) + 3\sqrt{5}\sqrt{7}$$
$$\qquad\qquad + 3\sqrt{5}\left(-2\sqrt{5}\right)$$
$$= 2(7) - 4\sqrt{35} + 3\sqrt{35} - 6(5)$$
$$= 14 - \sqrt{35} - 30$$
$$= -16 - \sqrt{35}$$

63. $\left(\sqrt{x} - y\right)\left(\sqrt{x} + y\right) = \left(\sqrt{x}\right)^2 - y^2 = x - y^2$

65.
$$\left(\sqrt{3} + x\right)^2 = \left(\sqrt{3}\right)^2 + 2\sqrt{3} \cdot x + x^2$$
$$= 3 + 2x\sqrt{3} + x^2$$

67.
$$\left(\sqrt{5x} - 3\sqrt{2}\right)\left(\sqrt{5x} - 3\sqrt{3}\right)$$
$$= \left(\sqrt{5x}\right)^2 + \sqrt{5x}\left(-3\sqrt{3}\right) - 3\sqrt{2}\left(\sqrt{5x}\right)$$
$$\qquad\qquad - 3\sqrt{2}\left(-3\sqrt{3}\right)$$
$$= 5x - 3\sqrt{15x} - 3\sqrt{10x} + 9\sqrt{6}$$

69.
$$\left(\sqrt[3]{4} + 2\right)\left(\sqrt[3]{2} - 1\right)$$
$$= \sqrt[3]{4}\left(\sqrt[3]{2}\right) + \sqrt[3]{4} \cdot (-1) + 2\sqrt[3]{2} + 2(-1)$$
$$= \sqrt[3]{8} - \sqrt[3]{4} + 2\sqrt[3]{2} - 2$$
$$= 2 - \sqrt[3]{4} + 2\sqrt[3]{2} - 2$$
$$= 2\sqrt[3]{2} - \sqrt[3]{4}$$

71.
$$\left(\sqrt[3]{x} + 1\right)\left(\sqrt[3]{x} - 4\sqrt{x} + 7\right)$$
$$= \left(\sqrt[3]{x}\right)^2 + \sqrt[3]{x}\left(-4\sqrt{x}\right) + \sqrt[3]{x} \cdot 7$$
$$\qquad\qquad + 1\left(\sqrt[3]{x}\right) + 1\left(-4\sqrt{x}\right) + 1(7)$$
$$= \sqrt[3]{x^2} - 4x^{1/3}x^{1/2} + 8\sqrt[3]{x} - 4\sqrt{x} + 7$$
$$= \sqrt[3]{x^2} - 4x^{5/6} + 8\sqrt[3]{x} - 4\sqrt{x} + 7$$
$$= \sqrt[3]{x^2} - 4\sqrt[6]{x^5} + 8\sqrt[3]{x} - 4\sqrt{x} + 7$$

73. $\left(\sqrt{x-1}+5\right)^2$

$= \sqrt{x-1}^2 + 2\sqrt{x-1}\cdot 5 + 5^2$
$= (x-1) + 10\sqrt{x-1} + 25$
$= x + 10\sqrt{x-1} + 24$

75. $\left(\sqrt{2x+5}-1\right)^2$

$= \sqrt{2x+5}^2 - 2\sqrt{2x+5}\cdot 1 + 1^2$
$= (2x+5) - 2\sqrt{2x+5} + 1$
$= 2x - 2\sqrt{2x+5} + 6$

77. $\dfrac{2x-14}{2} = \dfrac{2(x-7)}{2} = x-7$

79. $\dfrac{7x-7y}{x^2-y^2} = \dfrac{7(x-y)}{(x+y)(x-y)} = \dfrac{7}{x+y}$

81. $\dfrac{6a^2b-9ab}{3ab} = \dfrac{3ab(2a-3)}{3ab} = 2a-3$

83. $\dfrac{-4+2\sqrt{3}}{6} = \dfrac{2\left(-2+\sqrt{3}\right)}{6} = \dfrac{-2+\sqrt{3}}{3}$

85. $P = 2l + 2w$
$= 2\left(3\sqrt{20}\right) + 2\left(\sqrt{125}\right)$
$= 6\sqrt{4\cdot 5} + 2\sqrt{25\cdot 5}$
$= 6(2)\sqrt{5} + 2(5)\sqrt{5}$
$= 12\sqrt{5} + 10\sqrt{5}$
$= 22\sqrt{5}$ feet

$A = lw$
$= \left(3\sqrt{20}\right)\left(\sqrt{125}\right)$
$= 3\sqrt{4\cdot 5}\sqrt{25\cdot 5}$
$= 3(2)\sqrt{5}\cdot 5\sqrt{5}$
$= 30\cdot 5$
$= 150$ square feet

87. a. $\sqrt{3} + \sqrt{3} = 2\sqrt{3}$

b. $\sqrt{3}\cdot\sqrt{3} = \sqrt{9} = 3$

c. Answers may vary.

89. Answer may vary.

Section 10.5
Mental Math

1. The conjugate of $\sqrt{2}+x$ is $\sqrt{2}-x$.

2. The conjugate of $\sqrt{3}+y$ is $\sqrt{3}-y$.

3. The conjugate of $5-\sqrt{a}$ is $5+\sqrt{a}$.

4. The conjugate of $6-\sqrt{b}$ is $6+\sqrt{b}$.

5. The conjugate of $7\sqrt{5}+8\sqrt{x}$ is $7\sqrt{5}-8\sqrt{x}$.

6. The conjugate of $9\sqrt{2}-6\sqrt{y}$ is $9\sqrt{2}+6\sqrt{y}$.

Exercise Set 10.5

1. $\dfrac{\sqrt{2}}{\sqrt{7}} = \dfrac{\sqrt{2}\cdot\sqrt{7}}{\sqrt{7}\cdot\sqrt{7}} = \dfrac{\sqrt{14}}{\sqrt{49}} = \dfrac{\sqrt{14}}{7}$

3. $\sqrt{\dfrac{1}{5}} = \dfrac{\sqrt{1}}{\sqrt{5}} = \dfrac{1\cdot\sqrt{5}}{\sqrt{5}\cdot\sqrt{5}} = \dfrac{\sqrt{5}}{5}$

5. $\sqrt[3]{\dfrac{3}{4}} = \dfrac{\sqrt[3]{3}}{\sqrt[3]{4}} = \dfrac{\sqrt[3]{3}\cdot\sqrt[3]{2}}{\sqrt[3]{4}\cdot\sqrt[3]{2}} = \dfrac{\sqrt[3]{6}}{\sqrt[3]{8}} = \dfrac{\sqrt[3]{6}}{2}$

7. $\dfrac{4}{\sqrt[3]{3}} = \dfrac{4\cdot\sqrt[3]{9}}{\sqrt[3]{3}\cdot\sqrt[3]{9}} = \dfrac{4\sqrt[3]{9}}{\sqrt[3]{27}} = \dfrac{4\sqrt[3]{9}}{3}$

9. $\dfrac{3}{\sqrt{8x}} = \dfrac{3\cdot\sqrt{2x}}{\sqrt{8x}\cdot\sqrt{2x}} = \dfrac{3\sqrt{2x}}{\sqrt{16x^2}} = \dfrac{3\sqrt{2x}}{4x}$

11. $\dfrac{3}{\sqrt[3]{4x^2}} = \dfrac{3\cdot\sqrt[3]{2x}}{\sqrt[3]{4x^2}\cdot\sqrt[3]{2x}} = \dfrac{3\sqrt[3]{2x}}{\sqrt[3]{8x^3}} = \dfrac{3\sqrt[3]{2x}}{2x}$

13. $\sqrt{\dfrac{4}{x}} = \dfrac{\sqrt{4}}{\sqrt{x}} = \dfrac{2 \cdot \sqrt{x}}{\sqrt{x} \cdot \sqrt{x}} = \dfrac{2\sqrt{x}}{\sqrt{x^2}} = \dfrac{2\sqrt{x}}{x}$

15. $\dfrac{9}{\sqrt{3a}} = \dfrac{9 \cdot \sqrt{3a}}{\sqrt{3a} \cdot \sqrt{3a}} = \dfrac{9\sqrt{3a}}{3a} = \dfrac{3\sqrt{3a}}{a}$

17. $\dfrac{3}{\sqrt[3]{2}} = \dfrac{3 \cdot \sqrt[3]{4}}{\sqrt[3]{2} \cdot \sqrt[3]{4}} = \dfrac{3\sqrt[3]{4}}{\sqrt[3]{8}} = \dfrac{3\sqrt[3]{4}}{2}$

19. $\dfrac{2\sqrt{3}}{\sqrt{7}} = \dfrac{2\sqrt{3} \cdot \sqrt{7}}{\sqrt{7} \cdot \sqrt{7}} = \dfrac{2\sqrt{21}}{\sqrt{49}} = \dfrac{2\sqrt{21}}{7}$

21. $\sqrt{\dfrac{2x}{5y}} = \dfrac{\sqrt{2x}}{\sqrt{5y}} = \dfrac{\sqrt{2x} \cdot \sqrt{5y}}{\sqrt{5y} \cdot \sqrt{5y}} = \dfrac{\sqrt{10xy}}{5y}$

23. $\sqrt[4]{\dfrac{81}{8}} = \dfrac{\sqrt[4]{81}}{\sqrt[4]{8}} = \dfrac{3 \cdot \sqrt[4]{2}}{\sqrt[4]{8} \cdot \sqrt[4]{2}} = \dfrac{3\sqrt[4]{2}}{\sqrt[4]{16}} = \dfrac{3\sqrt[4]{2}}{2}$

25. $\sqrt[4]{\dfrac{16}{9x^7}} = \dfrac{\sqrt[4]{16}}{\sqrt[4]{9x^7}}$

$\qquad = \dfrac{2 \cdot \sqrt[4]{9x}}{\sqrt[4]{9x^7} \cdot \sqrt[4]{9x}} = \dfrac{2\sqrt[4]{9x}}{\sqrt[4]{81x^8}} = \dfrac{2\sqrt[4]{9x}}{3x^2}$

27. $\dfrac{5a}{\sqrt[5]{8a^9 b^{11}}} = \dfrac{5a \cdot \sqrt[5]{4ab^4}}{\sqrt[5]{8a^9 b^{11}} \cdot \sqrt[5]{4ab^4}}$

$\qquad = \dfrac{5a\sqrt[5]{4ab^4}}{\sqrt[5]{32a^{10}b^{15}}} = \dfrac{5a\sqrt[5]{4ab^4}}{2a^2 b^3}$

29. $\dfrac{6}{2 - \sqrt{7}} = \dfrac{6\left(2 + \sqrt{7}\right)}{\left(2 - \sqrt{7}\right)\left(2 + \sqrt{7}\right)}$

$\qquad = \dfrac{6\left(2 + \sqrt{7}\right)}{2^2 - \left(\sqrt{7}\right)^2}$

$\qquad = \dfrac{6\left(2 + \sqrt{7}\right)}{4 - 7}$

$\qquad = \dfrac{6\left(2 + \sqrt{7}\right)}{-3} = -2\left(2 + \sqrt{7}\right)$

31. $\dfrac{-7}{\sqrt{x} - 3} = \dfrac{-7\left(\sqrt{x} + 3\right)}{\left(\sqrt{x} - 3\right)\left(\sqrt{x} + 3\right)}$

$\qquad = \dfrac{-7\left(\sqrt{x} + 3\right)}{\left(\sqrt{x}\right)^2 - \left(3\right)^2}$

$\qquad = \dfrac{-7\left(\sqrt{x} + 3\right)}{x - 9}$

33. $\dfrac{\sqrt{2} - \sqrt{3}}{\sqrt{2} + \sqrt{3}} = \dfrac{\left(\sqrt{2} - \sqrt{3}\right)\left(\sqrt{2} - \sqrt{3}\right)}{\left(\sqrt{2} + \sqrt{3}\right)\left(\sqrt{2} - \sqrt{3}\right)}$

$\qquad = \dfrac{\left(\sqrt{2}\right)^2 - 2\sqrt{2}\sqrt{3} + \left(\sqrt{3}\right)^2}{\left(\sqrt{2}\right)^2 - \left(\sqrt{3}\right)^2}$

$\qquad = \dfrac{2 - 2\sqrt{6} + 3}{2 - 3}$

$\qquad = \dfrac{5 - 2\sqrt{6}}{-1} = -5 + 2\sqrt{6}$

35. $\dfrac{\sqrt{a} + 1}{2\sqrt{a} - \sqrt{b}}$

$\qquad = \dfrac{\left(\sqrt{a} + 1\right)\left(2\sqrt{a} + \sqrt{b}\right)}{\left(2\sqrt{a} - \sqrt{b}\right)\left(2\sqrt{a} + \sqrt{b}\right)}$

$\qquad = \dfrac{\sqrt{a} \cdot 2\sqrt{a} + \sqrt{a}\sqrt{b} + 1 \cdot 2\sqrt{a} + 1 \cdot \sqrt{b}}{\left(2\sqrt{a}\right)^2 - \left(\sqrt{b}\right)^2}$

$\qquad = \dfrac{2a + \sqrt{ab} + 2\sqrt{a} + \sqrt{b}}{4a - b}$

37. $\dfrac{8}{1 + \sqrt{10}} = \dfrac{8\left(1 - \sqrt{10}\right)}{\left(1 + \sqrt{10}\right)\left(1 - \sqrt{10}\right)}$

$\qquad = \dfrac{8\left(1 - \sqrt{10}\right)}{1^2 - \left(\sqrt{10}\right)^2}$

$\qquad = \dfrac{8\left(1 - \sqrt{10}\right)}{1 - 10} = -\dfrac{8\left(1 - \sqrt{10}\right)}{9}$

39. $\dfrac{\sqrt{x}}{\sqrt{x}+\sqrt{y}} = \dfrac{\sqrt{x}\left(\sqrt{x}-\sqrt{y}\right)}{\left(\sqrt{x}+\sqrt{y}\right)\left(\sqrt{x}-\sqrt{y}\right)}$

$\qquad = \dfrac{\sqrt{x}\left(\sqrt{x}-\sqrt{y}\right)}{\left(\sqrt{x}\right)^2-\left(\sqrt{y}\right)^2}$

$\qquad = \dfrac{\sqrt{x}\left(\sqrt{x}-\sqrt{y}\right)}{x-y}$

41. $\dfrac{2\sqrt{3}+\sqrt{6}}{4\sqrt{3}-\sqrt{6}} = \dfrac{\left(2\sqrt{3}+\sqrt{6}\right)\left(4\sqrt{3}+\sqrt{6}\right)}{\left(4\sqrt{3}-\sqrt{6}\right)\left(4\sqrt{3}+\sqrt{6}\right)}$

$\qquad = \dfrac{8\cdot3+2\sqrt{18}+4\sqrt{18}+6}{\left(4\sqrt{3}\right)^2-\left(\sqrt{6}\right)^2}$

$\qquad = \dfrac{30+6\sqrt{18}}{16\cdot3-6}$

$\qquad = \dfrac{30+6(3)\sqrt{2}}{42}$

$\qquad = \dfrac{30+18\sqrt{2}}{42}$

$\qquad = \dfrac{6\left(5+3\sqrt{2}\right)}{42}$

$\qquad = \dfrac{5+3\sqrt{2}}{7}$

43. $\sqrt{\dfrac{5}{3}} = \dfrac{\sqrt{5}}{\sqrt{3}} = \dfrac{\sqrt{5}\cdot\sqrt{5}}{\sqrt{3}\cdot\sqrt{5}} = \dfrac{\sqrt{25}}{\sqrt{15}} = \dfrac{5}{\sqrt{15}}$

45. $\sqrt{\dfrac{18}{5}} = \dfrac{\sqrt{18}}{\sqrt{5}}$

$\qquad = \dfrac{\sqrt{9}\cdot\sqrt{2}}{\sqrt{5}}$

$\qquad = \dfrac{3\sqrt{2}}{\sqrt{5}}$

$\qquad = \dfrac{3\sqrt{2}\cdot\sqrt{2}}{\sqrt{5}\cdot\sqrt{2}}$

$\qquad = \dfrac{3\cdot2}{\sqrt{10}} = \dfrac{6}{\sqrt{10}}$

47. $\dfrac{\sqrt{4x}}{7} = \dfrac{2\sqrt{x}}{7} = \dfrac{2\sqrt{x}\cdot\sqrt{x}}{7\cdot\sqrt{x}} = \dfrac{2\sqrt{x^2}}{7\sqrt{x}} = \dfrac{2x}{7\sqrt{x}}$

49. $\dfrac{\sqrt[3]{5y^2}}{\sqrt[3]{4x}} = \dfrac{\sqrt[3]{5y^2}\cdot\sqrt[3]{5^2y}}{\sqrt[3]{4x}\cdot\sqrt[3]{5^2y}} = \dfrac{\sqrt[3]{5^3y^3}}{\sqrt[3]{100xy}} = \dfrac{5y}{\sqrt[3]{100xy}}$

51. $\sqrt{\dfrac{2}{5}} = \dfrac{\sqrt{2}}{\sqrt{5}} = \dfrac{\sqrt{2}\cdot\sqrt{2}}{\sqrt{5}\cdot\sqrt{2}} = \dfrac{\sqrt{4}}{\sqrt{10}} = \dfrac{2}{\sqrt{10}}$

53. $\dfrac{\sqrt{2x}}{11} = \dfrac{\sqrt{2x}\cdot\sqrt{2x}}{11\cdot\sqrt{2x}} = \dfrac{\sqrt{4x^2}}{11\sqrt{2x}} = \dfrac{2x}{11\sqrt{2x}}$

55. $\sqrt[3]{\dfrac{7}{8}} = \dfrac{\sqrt[3]{7}}{\sqrt[3]{8}} = \dfrac{\sqrt[3]{7}}{2}$

$\qquad = \dfrac{\sqrt[3]{7}\cdot\sqrt[3]{7^2}}{2\cdot\sqrt[3]{7^2}} = \dfrac{\sqrt[3]{7^3}}{2\sqrt[3]{49}} = \dfrac{7}{2\sqrt[3]{49}}$

57. $\dfrac{\sqrt[3]{3x^5}}{10} = \dfrac{\sqrt[3]{x^3\cdot3x^2}}{10}$

$\qquad = \dfrac{x\sqrt[3]{3x^2}\cdot\sqrt[3]{3^2x}}{10\cdot\sqrt[3]{3^2x}}$

$\qquad = \dfrac{x\sqrt[3]{3^3x^3}}{10\sqrt[3]{9x}}$

$\qquad = \dfrac{x\cdot3x}{10\sqrt[3]{9x}}$

$\qquad = \dfrac{3x^2}{10\sqrt[3]{9x}}$

59. $\sqrt{\dfrac{18x^4y^6}{3z}} = \dfrac{\sqrt{18x^4y^6}}{\sqrt{3z}}$

$\qquad = \dfrac{\sqrt{9x^4y^6\cdot2}}{\sqrt{3z}}$

$\qquad = \dfrac{3x^2y^3\sqrt{2}}{\sqrt{3z}}$

$\qquad = \dfrac{3x^2y^3\sqrt{2}\cdot\sqrt{2}}{\sqrt{3z}\cdot\sqrt{2}}$

$\qquad = \dfrac{3x^2y^3\cdot2}{\sqrt{6z}}$

61. Answers may vary.

63.
$$\frac{2-\sqrt{11}}{6} = \frac{\left(2-\sqrt{11}\right)\left(2+\sqrt{11}\right)}{6\left(2+\sqrt{11}\right)}$$
$$= \frac{4-11}{12+6\sqrt{11}}$$
$$= \frac{-7}{12+6\sqrt{11}}$$

65.
$$\frac{2-\sqrt{7}}{-5} = \frac{\left(2-\sqrt{7}\right)\left(2+\sqrt{7}\right)}{-5\left(2+\sqrt{7}\right)}$$
$$= \frac{4-7}{-5\left(2+\sqrt{7}\right)}$$
$$= \frac{-3}{-10-5\sqrt{7}} = \frac{3}{10+5\sqrt{7}}$$

67.
$$\frac{\sqrt{x}+3}{\sqrt{x}} = \frac{\left(\sqrt{x}+3\right)\left(\sqrt{x}-3\right)}{\sqrt{x}\left(\sqrt{x}-3\right)}$$
$$= \frac{\sqrt{x^2}-9}{\sqrt{x^2}-3\sqrt{x}}$$
$$= \frac{x-9}{x-3\sqrt{x}}$$

69.
$$\frac{\sqrt{2}-1}{\sqrt{2}+1} = \frac{\left(\sqrt{2}-1\right)\left(\sqrt{2}+1\right)}{\left(\sqrt{2}+1\right)\left(\sqrt{2}+1\right)}$$
$$= \frac{\sqrt{4}-1}{\sqrt{4}+2\sqrt{2}+1}$$
$$= \frac{2-1}{2+2\sqrt{2}+1} = \frac{1}{3+2\sqrt{2}}$$

71.
$$\frac{\sqrt{x}+1}{\sqrt{x}-1} = \frac{\left(\sqrt{x}+1\right)\left(\sqrt{x}-1\right)}{\left(\sqrt{x}-1\right)\left(\sqrt{x}-1\right)}$$
$$= \frac{\sqrt{x^2}-1}{\sqrt{x^2}-2\sqrt{x}+1}$$
$$= \frac{x-1}{x-2\sqrt{x}+1}$$

73.
$$2x-7 = 3(x-4)$$
$$2x-7 = 3x-12$$
$$-x-7 = -12$$
$$-x = -5$$
$$x = 5$$
The solution is 5.

75.
$$(x-6)(2x+1) = 0$$
$$x-6 = 0 \ \text{ or } \ 2x+1 = 0$$
$$x = 6 \ \text{ or } \quad\quad 2x = -1$$
$$x = -\frac{1}{2}$$
The solutions are $-\frac{1}{2}, 6$.

77.
$$x^2 - 8x = -12$$
$$x^2 - 8x + 12 = 0$$
$$(x-6)(x-2) = 0$$
$$x-6 = 0 \ \text{ or } \ x-2 = 0$$
$$x = 6 \ \text{ or } \quad\quad x = 2$$
The solutions are 2, 6.

79.
$$r = \sqrt{\frac{A}{4\pi}}$$
$$= \frac{\sqrt{A}}{\sqrt{4\pi}}$$
$$= \frac{\sqrt{A}}{2\sqrt{\pi}}$$
$$= \frac{\sqrt{A} \cdot \sqrt{\pi}}{2\sqrt{\pi} \cdot \sqrt{\pi}} = \frac{\sqrt{A\pi}}{2\pi}$$

81. Answers may vary.

Integrated Review

1. $\sqrt{81} = 9$

2. $\sqrt[3]{-8} = -2$

3. $\sqrt[4]{\frac{1}{16}} = \frac{1}{2}$

4. $\sqrt{x^6} = x^3$

5. $\sqrt[3]{y^9} = y^3$

6. $\sqrt{4y^{10}} = 2y^5$

7. $\sqrt[5]{-32y^5} = -2y$

8. $\sqrt[4]{81b^{12}} = 3b^3$

9. $36^{1/2} = \sqrt{36} = 6$

10. $(3y)^{1/4} = \sqrt[4]{3y}$

11. $64^{-2/3} = \dfrac{1}{64^{2/3}} = \dfrac{1}{\left(\sqrt[3]{64}\right)^2} = \dfrac{1}{(4)^2} = \dfrac{1}{16}$

12. $(x+1)^{3/5} = \sqrt[5]{(x+1)^3}$

13. $y^{-1/6} \cdot y^{7/6} = y^{-1/6+7/6} = y^{6/6} = y^1 = y$

14. $\dfrac{(2x^{1/3})^4}{x^{5/6}} = \dfrac{2^4 x^{4/3}}{x^{5/6}}$
$= 16x^{\frac{4}{3}-\frac{5}{6}}$
$= 16x^{\frac{8}{6}-\frac{5}{6}}$
$= 16x^{3/6}$
$= 16x^{1/2}$

15. $\dfrac{x^{1/4} x^{3/4}}{x^{-1/4}} = x^{\frac{1}{4}+\frac{3}{4}-\left(-\frac{1}{4}\right)} = x^{\frac{1+3+1}{4}} = x^{5/4}$

16. $4^{1/3} \cdot 4^{2/5} = 4^{\frac{1}{3}+\frac{2}{5}} = 4^{\frac{5+6}{15}} = 4^{11/15}$

17. $\sqrt[3]{8x^6} = (8x^6)^{1/3}$
$= (2^3 x^6)^{1/3} = 2^{3(1/3)} x^{6(1/3)} = 2x^2$

18. $\sqrt[12]{a^9 b^6} = (a^9 b^6)^{1/12}$
$= a^{9(1/12)} b^{6(1/12)}$
$= a^{3/4} b^{1/2}$
$= a^{3/4} b^{2/4}$
$= (a^3 b^2)^{1/4}$
$= \sqrt[4]{a^3 b^2}$

19. $\sqrt[4]{x} \cdot \sqrt{x} = x^{1/4} \cdot x^{1/2}$
$= x^{\frac{1}{4}+\frac{1}{2}} = x^{\frac{1+2}{4}} = x^{3/4} = \sqrt[4]{x^3}$

20. $\sqrt{5} \cdot \sqrt[3]{2} = 5^{1/2} \cdot 2^{1/3}$
$= 5^{3/6} \cdot 2^{2/6}$
$= (5^3 \cdot 2^2)^{1/6}$
$= \sqrt[6]{125 \cdot 4}$
$= \sqrt[6]{500}$

21. $\sqrt{40} = \sqrt{4 \cdot 10} = \sqrt{4} \cdot \sqrt{10} = 2\sqrt{10}$

22. $\sqrt[4]{16x^7 y^{10}} = \sqrt[4]{16x^4 y^8 \cdot x^3 y^2}$
$= \sqrt[4]{16x^4 y^8} \cdot \sqrt[4]{x^3 y^2}$
$= 2xy^2 \sqrt[4]{x^3 y^2}$

23. $\sqrt[3]{54x^4} = \sqrt[3]{27x^3 \cdot 2x}$
$= \sqrt[3]{27x^3} \cdot \sqrt[3]{2x}$
$= 3x\sqrt[3]{2x}$

24. $\sqrt[5]{-64b^{10}} = \sqrt[5]{-32b^{10} \cdot 2} = -2b^2 \sqrt[5]{2}$

25. $\sqrt{5} \cdot \sqrt{x} = \sqrt{5x}$

26. $\sqrt[3]{8x} \cdot \sqrt[3]{8x^2} = \sqrt[3]{64x^3} = 4x$

27. $\dfrac{\sqrt{98y^6}}{\sqrt{2y}} = \sqrt{\dfrac{98y^6}{2y}}$
$= \sqrt{49y^5}$
$= \sqrt{49y^4 \cdot y}$
$= 7y^2 \sqrt{y}$

28. $\dfrac{\sqrt[4]{48a^9 b^3}}{\sqrt[4]{ab^3}} = \sqrt[4]{\dfrac{48a^9 b^3}{ab^3}}$
$= \sqrt[4]{48a^8}$
$= \sqrt[4]{16a^8 \cdot 4}$
$= 2a^2 \sqrt[4]{4}$

29. $\sqrt{20} - \sqrt{75} + 5\sqrt{7} = \sqrt{4 \cdot 5} - \sqrt{25 \cdot 3} + 5\sqrt{7}$
$$= 2\sqrt{5} - 5\sqrt{3} + 5\sqrt{7}$$

30. $\sqrt[3]{54y^4} - y\sqrt[3]{16y} = \sqrt[3]{27y^3 \cdot 2y} - y\sqrt[3]{8 \cdot 2y}$
$$= 3y\sqrt[3]{2y} - 2y\sqrt[3]{2y}$$
$$= y\sqrt[3]{2y}$$

31. $\sqrt{3}\left(\sqrt{5} - \sqrt{2}\right) = \sqrt{3}\sqrt{5} - \sqrt{3}\sqrt{2}$
$$= \sqrt{15} - \sqrt{6}$$

32. $\left(\sqrt{7} + \sqrt{3}\right)^2 = \left(\sqrt{7}\right)^2 + 2\sqrt{7}\sqrt{3} + \left(\sqrt{3}\right)^2$
$$= 7 + 2\sqrt{21} + 3$$
$$= 10 + 2\sqrt{21}$$

33. $\left(2x - \sqrt{5}\right)\left(2x + \sqrt{5}\right) = (2x)^2 - \left(\sqrt{5}\right)^2$
$$= 4x^2 - 5$$

34. $\left(\sqrt{x+1} - 1\right)^2 = \sqrt{x+1}^2 - 2\sqrt{x+1} \cdot 1 + 1^2$
$$= x + 1 - 2\sqrt{x+1} + 1$$
$$= x + 2 - 2\sqrt{x+1}$$

35. $\sqrt{\dfrac{7}{3}} = \dfrac{\sqrt{7}}{\sqrt{3}} = \dfrac{\sqrt{7} \cdot \sqrt{3}}{\sqrt{3} \cdot \sqrt{3}} = \dfrac{\sqrt{21}}{\sqrt{9}} = \dfrac{\sqrt{21}}{3}$

36. $\dfrac{5}{\sqrt[3]{2x^2}} = \dfrac{5 \cdot \sqrt[3]{2^2 x}}{\sqrt[3]{2x^2} \cdot \sqrt[3]{2^2 x}} = \dfrac{5\sqrt[3]{4x}}{\sqrt[3]{2^3 x^3}} = \dfrac{5\sqrt[3]{4x}}{2x}$

37. $\dfrac{\sqrt{3} - \sqrt{7}}{2\sqrt{3} + \sqrt{7}}$
$$= \dfrac{\left(\sqrt{3} - \sqrt{7}\right)\left(2\sqrt{3} - \sqrt{7}\right)}{\left(2\sqrt{3} + \sqrt{7}\right)\left(2\sqrt{3} - \sqrt{7}\right)}$$
$$= \dfrac{2\sqrt{9} - \sqrt{3}\sqrt{7} - \sqrt{7} \cdot 2\sqrt{3} + \sqrt{49}}{\left(2\sqrt{3}\right)^2 - \left(\sqrt{7}\right)^2}$$
$$= \dfrac{2(3) - \sqrt{21} - 2\sqrt{21} + 7}{4 \cdot 3 - 7}$$
$$= \dfrac{6 - 3\sqrt{21} + 7}{12 - 7} = \dfrac{13 - 3\sqrt{21}}{5}$$

38. $\sqrt{\dfrac{7}{3}} = \dfrac{\sqrt{7}}{\sqrt{3}} = \dfrac{\sqrt{7} \cdot \sqrt{7}}{\sqrt{3} \cdot \sqrt{7}} = \dfrac{\sqrt{49}}{\sqrt{21}} = \dfrac{7}{\sqrt{21}}$

39. $\sqrt[3]{\dfrac{9y}{11}} = \dfrac{\sqrt[3]{9y}}{\sqrt[3]{11}}$
$$= \dfrac{\sqrt[3]{9y} \cdot \sqrt[3]{3y^2}}{\sqrt[3]{11} \cdot \sqrt[3]{3y^2}} = \dfrac{\sqrt[3]{27y^3}}{\sqrt[3]{33y^2}} = \dfrac{3y}{\sqrt[3]{33y^2}}$$

40. $\dfrac{\sqrt{x} - 2}{\sqrt{x}} = \dfrac{\left(\sqrt{x} - 2\right)\left(\sqrt{x} + 2\right)}{\sqrt{x}\left(\sqrt{x} + 2\right)}$
$$= \dfrac{\left(\sqrt{x}\right)^2 - 2^2}{\sqrt{x^2} + 2\sqrt{x}} = \dfrac{x - 4}{x + 2\sqrt{x}}$$

Section 10.6

Graphing Calculator Explorations

1.

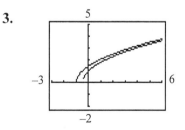

The solution is 3.19.

3.

There is no solution.

5.

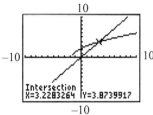

The solution is 3.23.

Exercise Set 10.6

1. $\sqrt{2x} = 4$
$\left(\sqrt{2x}\right)^2 = 4^2$
$2x = 16$
$x = 8$
The solution is 8.

3. $\sqrt{x-3} = 2$
$\left(\sqrt{x-3}\right)^2 = 2^2$
$x - 3 = 4$
$x = 7$
The solution is 7.

5. $\sqrt{2x} = -4$
No solution since a principle square root does not yield a negative number.

7. $\sqrt{4x-3} - 5 = 0$
$\sqrt{4x-3} = 5$
$\left(\sqrt{4x-3}\right)^2 = 5^2$
$4x - 3 = 25$
$4x = 28$
$x = 7$
The solution is 7.

9. $\sqrt{2x-3} - 2 = 1$
$\sqrt{2x-3} = 3$
$\left(\sqrt{2x-3}\right)^2 = 3^2$
$2x - 3 = 9$
$2x = 12$
$x = 6$
The solution is 6.

11. $\sqrt[3]{6x} = -3$
$\left(\sqrt[3]{6x}\right)^3 = (-3)^3$
$6x = -27$
$x = -\dfrac{27}{6} = -\dfrac{9}{2}$
The solution is $-\dfrac{9}{2}$.

13. $\sqrt[3]{x-2} - 3 = 0$
$\sqrt[3]{x-2} = 3$
$\left(\sqrt[3]{x-2}\right)^3 = 3^2$
$x - 2 = 27$
$x = 29$
The solution is 29.

15. $\sqrt{13-x} = x - 1$
$\left(\sqrt{13-x}\right)^2 = (x-1)^2$
$13 - x = x^2 - 2x + 1$
$0 = x^2 - x - 12$
$0 = (x-4)(x+3)$
$x - 4 = 0$ or $x + 3 = 0$
$x = 4$ or $x = -3$
We discard –3 as extraneous.
The solution is 4.

17. $x - \sqrt{4 - 3x} = -8$
$$x + 8 = \sqrt{4 - 3x}$$
$$(x + 8)^2 = \left(\sqrt{4 - 3x}\right)^2$$
$$x^2 + 16x + 64 = 4 - 3x$$
$$x^2 + 19x + 60 = 0$$
$$(x + 4)(x + 15) = 0$$
$$x + 4 = 0 \quad \text{or} \ \ x + 15 = 0$$
$$x = -4 \ \ \text{or} \qquad x = -15$$

We discard –15 as extraneous.
The solution is –4.

19. $\sqrt{y + 5} = 2 - \sqrt{y - 4}$
$$\left(\sqrt{y + 5}\right)^2 = \left(2 - \sqrt{y - 4}\right)^2$$
$$y + 5 = 4 - 4\sqrt{y - 4} + (y - 4)$$
$$y + 5 = y - 4\sqrt{y - 4}$$
$$5 = -4\sqrt{y - 4}$$
$$5^2 = \left(-4\sqrt{y - 4}\right)^2$$
$$25 = 16(y - 4)$$
$$25 = 16y - 64$$
$$89 = 16y$$
$$\frac{89}{16} = y$$

which we discard as extraneous.
There is no solution.

21. $\sqrt{x - 3} + \sqrt{x + 2} = 5$
$$\sqrt{x - 3} = 5 - \sqrt{x + 2}$$
$$\left(\sqrt{x - 3}\right)^2 = \left(5 - \sqrt{x + 2}\right)^2$$
$$x - 3 = 25 - 10\sqrt{x + 2} + (x + 2)$$
$$x - 3 = 27 - 10\sqrt{x + 2} + x$$
$$-30 = -10\sqrt{x + 2}$$
$$3 = \sqrt{x + 2}$$
$$3^2 = \left(\sqrt{x + 2}\right)^2$$
$$9 = x + 2$$
$$7 = x$$

The solution is 7.

23. $\sqrt{3x - 2} = 5$
$$\left(\sqrt{3x - 2}\right)^2 = 5^2$$
$$3x - 2 = 25$$
$$3x = 27$$
$$x = 9$$
The solution is 9.

25. $-\sqrt{2x} + 4 = -6$
$$10 = \sqrt{2x}$$
$$10^2 = \left(\sqrt{2x}\right)^2$$
$$100 = 2x$$
$$50 = x$$
The solution is 50.

27. $\sqrt{3x + 1} + 2 = 0$
$$\sqrt{3x + 1} = -2$$
No solution since a principle square root
does not yield a negative number.

29. $\sqrt[4]{4x + 1} - 2 = 0$
$$\sqrt[4]{4x + 1} = 2$$
$$\left(\sqrt[4]{4x + 1}\right)^4 = 2^4$$
$$4x + 1 = 16$$
$$4x = 15$$
$$x = \frac{15}{4}$$

The solution is $\frac{15}{4}$.

31. $\sqrt{4x - 3} = 7$
$$\left(\sqrt{4x - 3}\right)^2 = 7^2$$
$$4x - 3 = 49$$
$$4x = 52$$
$$x = 13$$
The solution is 13.

33. $\sqrt[3]{6x-3} - 3 = 0$

$\quad\quad \sqrt[3]{6x-3} = 3$

$\quad\quad \left(\sqrt[3]{6x-3}\right)^3 = 3^3$

$\quad\quad\quad 6x-3 = 27$

$\quad\quad\quad\quad 6x = 30$

$\quad\quad\quad\quad\quad x = 5$

The solution is 5.

35. $\sqrt[3]{2x-3} - 2 = -5$

$\quad\quad \sqrt[3]{2x-3} = -3$

$\quad\quad \left(\sqrt[3]{2x-3}\right)^3 = (-3)^3$

$\quad\quad\quad 2x-3 = -27$

$\quad\quad\quad\quad 2x = -24$

$\quad\quad\quad\quad\quad x = -12$

The solution is -12

37. $\quad \sqrt{x+4} = \sqrt{2x-5}$

$\quad \left(\sqrt{x+4}\right)^2 = \left(\sqrt{2x-5}\right)^2$

$\quad\quad\quad x+4 = 2x-5$

$\quad\quad\quad\quad -x = -9$

$\quad\quad\quad\quad\quad x = 9$

The solution is 9.

39. $\quad x - \sqrt{1-x} = -5$

$\quad\quad\quad x+5 = \sqrt{1-x}$

$\quad\quad (x+5)^2 = \left(\sqrt{1-x}\right)^2$

$\quad x^2 + 10x + 25 = 1 - x$

$\quad x^2 + 11x + 24 = 0$

$\quad (x+8)(x+3) = 0$

$\quad x+8 = 0 \quad$ or $\quad x+3 = 0$

$\quad\quad x = -8 \quad$ or $\quad\quad x = -3$

We discard -8 as extraneous.

The solution is -3.

41. $\quad \sqrt[3]{-6x-1} = \sqrt[3]{-2x-5}$

$\quad \left(\sqrt[3]{-6x-1}\right)^3 = \left(\sqrt[3]{-2x-5}\right)^3$

$\quad\quad\quad -6x-1 = -2x-5$

$\quad\quad\quad\quad -4x = -4$

$\quad\quad\quad\quad\quad x = 1$

The solution is 1.

43. $\sqrt{5x-1} - \sqrt{x+2} = 3$

$\quad\quad \sqrt{5x-1} = \sqrt{x}+1$

$\quad\quad \left(\sqrt{5x-1}\right)^2 = \left(\sqrt{x}+1\right)^2$

$\quad\quad\quad 5x-1 = x + 2\sqrt{x} + 1$

$\quad\quad\quad 4x-2 = 2\sqrt{x}$

$\quad\quad\quad 2x-1 = \sqrt{x}$

$\quad\quad (2x-1)^2 = \left(\sqrt{x}\right)^2$

$\quad\quad 4x^2 - 4x + 1 = x$

$\quad\quad 4x^2 - 5x + 1 = 0$

$\quad\quad (4x-1)(x-1) = 0$

$\quad 4x-1 = 0 \quad$ or $\quad x-1 = 0$

$\quad\quad 4x = 1 \quad$ or $\quad\quad x = 1$

$\quad\quad x = \dfrac{1}{4}$

We discard $\dfrac{1}{4}$ as extraneous.

The solution is 1.

45. $\quad \sqrt{2x-1} = \sqrt{1-2x}$

$\quad \left(\sqrt{2x-1}\right)^2 = \left(\sqrt{1-2x}\right)^2$

$\quad\quad\quad 2x-1 = 1-2x$

$\quad\quad\quad\quad 4x = 2$

$\quad\quad\quad\quad x = \dfrac{2}{4} = \dfrac{1}{2}$

The solution is $\dfrac{1}{2}$.

47.
$$\sqrt{3x+4}-1=\sqrt{2x+1}$$
$$\sqrt{3x+4}=\sqrt{2x+1}+1$$
$$\left(\sqrt{3x+4}\right)^2=\left(\sqrt{2x+1}+1\right)^2$$
$$3x+4=(2x+1)+2\sqrt{2x+1}+1$$
$$3x+4=2x+2+2\sqrt{2x+1}$$
$$x+2=2\sqrt{2x+1}$$
$$(x+2)^2=\left(2\sqrt{2x+1}\right)^2$$
$$x^2+4x+4=4(2x+1)$$
$$x^2+4x+4=8x+4$$
$$x^2-4x=0$$
$$x(x-4)=0$$
$$x=0 \ \text{ or } \ x-4=0$$
$$x=4$$
The solutions are 0 and 4.

49.
$$\sqrt{y+3}-\sqrt{y-3}=1$$
$$\sqrt{y+3}=1+\sqrt{y-3}$$
$$\left(\sqrt{y+3}\right)^2=\left(1+\sqrt{y-3}\right)^2$$
$$y+3=1+2\sqrt{y-3}+(y-3)$$
$$y+3=-2+2\sqrt{y-3}+y$$
$$5=2\sqrt{y-3}$$
$$(5)^2=\left(2\sqrt{y-3}\right)^2$$
$$25=4(y-3)$$
$$25=4y-12$$
$$37=4y$$
$$\frac{37}{4}=y$$
The solution is $\frac{37}{4}$.

51. Let c = length of the hypotenuse.
$$6^2+3^2=c^2$$
$$36+9=c^2$$
$$45=c^2$$
$$\sqrt{45}=\sqrt{c^2}$$
$$\sqrt{9\cdot5}=c$$
$$3\sqrt{5}=c \ \text{ so } \ c=3\sqrt{5} \ \text{feet}$$

53. Let b = length of the unknown leg.
$$3^2+b^2=7^2$$
$$9+b^2=49$$
$$b^2=40$$
$$\sqrt{b^2}=\sqrt{40}$$
$$b=\sqrt{4\cdot10}$$
$$b=2\sqrt{10} \ \text{meters}$$

55. Let b = length of the unknown leg.
$$9^2+b^2=\left(11\sqrt{5}\right)^2$$
$$81+b^2=121\cdot5$$
$$81+b^2=605$$
$$b^2=525$$
$$\sqrt{b^2}=\sqrt{524}$$
$$b=\sqrt{4\cdot131}$$
$$b=2\sqrt{131}\approx22.9 \ \text{meters}$$

57. Let c = length of the hypotenuse.
$$\left(7.2\right)^2+7^2=c^2$$
$$51.84+49=c^2$$
$$100.84=c^2$$
$$\sqrt{100.84}=\sqrt{c^2}$$
$$\sqrt{100.84}=c$$
$$c=\sqrt{100.84}\approx10.0 \ \text{mm}$$

59. Let c = amount of cable needed.
$$15^2 + 8^2 = c^2$$
$$225 + 64 = c^2$$
$$289 = c^2$$
$$\sqrt{289} = \sqrt{c^2}$$
$$17 = c$$
Thus, 17 feet of cable is needed.

61. Let c = length of the ladder.
$$12^2 + 5^2 = c^2$$
$$144 + 25 = c^2$$
$$169 = c^2$$
$$\sqrt{169} = \sqrt{c^2}$$
$$13 = c$$
A 13-foot ladder is needed.

63.
$$r = \sqrt{\frac{A}{4\pi}}$$
$$1080 = \sqrt{\frac{A}{4\pi}}$$
$$(1080)^2 = \left(\sqrt{\frac{A}{4\pi}}\right)^2$$
$$1{,}166{,}400 = \frac{A}{4\pi}$$
$$14{,}657{,}415 \approx A$$
The surface area is 14,657,415 sq. miles.

65.
$$v = \sqrt{2gh}$$
$$80 = \sqrt{2(32)h}$$
$$(80)^2 = \left(\sqrt{64h}\right)^2$$
$$6400 = 64h$$
$$100 = h$$
The object fell 100 feet.

67.
$$S = 2\sqrt{I} - 9$$
$$11 = 2\sqrt{I} - 9$$
$$20 = 2\sqrt{I}$$
$$10 = \sqrt{I}$$
$$10^2 = \left(\sqrt{I}\right)^2$$
$$100 = I$$
The estimated IQ is 100.

69.
$$P = 2\pi\sqrt{\frac{l}{32}}$$
$$= 2\pi\sqrt{\frac{2}{32}}$$
$$= 2\pi\sqrt{\frac{1}{16}}$$
$$= 2\pi\left(\frac{1}{4}\right)$$
$$= \frac{\pi}{2} \text{ sec} \approx 1.57 \text{ sec}$$

71.
$$P = 2\pi\sqrt{\frac{l}{32}}$$
$$4 = 2\pi\sqrt{\frac{l}{32}}$$
$$\frac{4}{2\pi} = \sqrt{\frac{l}{32}}$$
$$\left(\frac{2}{\pi}\right)^2 = \left(\sqrt{\frac{l}{32}}\right)^2$$
$$\frac{4}{\pi^2} = \frac{l}{32}$$
$$l = 32\left(\frac{4}{\pi^2}\right) \approx 12.97 \text{ feet}$$

73. Answers may vary.

75. $s = \frac{1}{2}(6+10+14) = \frac{1}{2}(30) = 15$

$A = \sqrt{s(s-a)(s-b)(s-c)}$

$\quad = \sqrt{15(15-6)(15-10)(15-14)}$

$\quad = \sqrt{15(9)(5)(1)}$

$\quad = \sqrt{675}$

$\quad = \sqrt{225 \cdot 3}$

$\quad = 15\sqrt{3}$ sq. mi ≈ 25.98 sq. mi.

77. Answers may vary.

79.
$$D(h) = 111.7\sqrt{h}$$
$$80 = 111.7\sqrt{h}$$
$$\frac{80}{111.7} = \sqrt{h}$$
$$\left(\frac{80}{111.7}\right)^2 = \left(\sqrt{h}\right)^2$$
$$0.5129483389 = h$$
$$h \approx 0.51 \text{ km}$$

81. Function

83. Function

85. Not a function

87. $\dfrac{\dfrac{x}{6}}{\dfrac{2x}{3}+\dfrac{1}{2}} = \dfrac{\left(\dfrac{x}{6}\right)6}{\left(\dfrac{2x}{3}+\dfrac{1}{2}\right)6} = \dfrac{x}{4x+3}$

89. $\dfrac{\dfrac{z}{5}+\dfrac{1}{10}}{\dfrac{z}{20}-\dfrac{z}{5}} = \dfrac{\left(\dfrac{z}{5}+\dfrac{1}{10}\right)20}{\left(\dfrac{z}{20}-\dfrac{z}{5}\right)20}$

$\quad = \dfrac{4z+2}{z-4z}$

$\quad = \dfrac{4z+2}{-3z}$

$\quad = -\dfrac{4z+2}{3z}$

91.
$$\sqrt{\sqrt{x+3}+\sqrt{x}} = \sqrt{3}$$
$$\left(\sqrt{\sqrt{x+3}+\sqrt{x}}\right)^2 = \left(\sqrt{3}\right)^2$$
$$\sqrt{x+3}+\sqrt{x} = 3 \cdot$$
$$\sqrt{x+3} = 3-\sqrt{x}$$
$$\left(\sqrt{x+3}\right)^2 = \left(3-\sqrt{x}\right)^2$$
$$x+3 = 9-6\sqrt{x}+x$$
$$-6 = -6\sqrt{x}$$
$$(-6)^2 = \left(-6\sqrt{x}\right)^2$$
$$36 = 36x$$
$$1 = x$$

93.
$$C(x) = 80\sqrt[3]{x}+500$$
$$1620 = 80\sqrt[3]{x}+500$$
$$1120 = 80\sqrt[3]{x}$$
$$14 = \sqrt[3]{x}$$
$$14^3 = \left(\sqrt[3]{x}\right)^3$$
$$2744 = x$$

Thus, 2743 deliveries will keep overhead below \$1620.

95. $3\sqrt{x^2-8x} = x^2-8x$

Let $t = x^2-8x$. Then
$$3\sqrt{u} = u$$
$$\left(3\sqrt{u}\right)^2 = u^2$$
$$9u = u^2$$
$$0 = u^2-9u$$
$$0 = u(u-9)$$
$$t = 0 \qquad \text{or} \qquad t = 9$$

Replace t with x^2-8x.

$x^2-8x = 0 \qquad$ or $\qquad x^2-8x = 9$

$x(x-8) = 0 \qquad\qquad x^2-8x-9 = 0$

$x = 0$ or $x = 8 \qquad (x-9)(x+1) = 0$

$\qquad\qquad\qquad\qquad x = 9$ or $x = -1$

The solutions are -1, 0, 8 and 9.

97. $7 - (x^2 - 3x) = \sqrt{(x^2 - 3x) + 5}$

Let $t = x^2 - 3x$. Then

$$7 - t = \sqrt{t + 5}$$
$$(7 - t)^2 = \left(\sqrt{t + 5}\right)^2$$
$$49 - 14t + t^2 = t + 5$$
$$t^2 - 15t + 44 = 0$$
$$(t - 11)(t - 4) = 0$$
$$t = 11 \text{ or } t = 4$$

Replace t with $x^2 - 3x$.

$x^2 - 3x = 11$	or	$x^2 - 3x = 4$
$x^2 - 3x - 11 = 0$		$x^2 - 3x - 4 = 0$
Can't factor		$(x - 4)(x + 1) = 0$
		$x = 4 \text{ or } x = -1$

The solutions are -1 and 4.

Section 10.7

Mental Math

1. $\sqrt{-81} = 9i$

2. $\sqrt{-49} = 7i$

3. $\sqrt{-7} = i\sqrt{7}$

4. $\sqrt{-3} = i\sqrt{3}$

5. $-\sqrt{16} = -4$

6. $-\sqrt{4} = -2$

7. $\sqrt{-64} = 8i$

8. $\sqrt{-100} = 10i$

Exercise Set 10.7

1. $\sqrt{-24} = \sqrt{-1 \cdot 24}$
   $= \sqrt{-1}\sqrt{4 \cdot 6} = i \cdot 2\sqrt{6} = 2i\sqrt{6}$

3. $-\sqrt{-36} = -\sqrt{-1 \cdot 36}$
   $= -\sqrt{-1}\sqrt{36} = -i \cdot 6 = -6i$

5. $8\sqrt{-63} = 8\sqrt{-1 \cdot 63}$
   $= 8\sqrt{-1}\sqrt{9 \cdot 7} = 8i \cdot 3\sqrt{7} = 24i\sqrt{7}$

7. $-\sqrt{54} = -\sqrt{9 \cdot 6} = -3\sqrt{6}$

9. $\sqrt{-2} \cdot \sqrt{-7} = i\sqrt{2} \cdot i\sqrt{7}$
   $= i^2\sqrt{14} = (-1)\sqrt{14} = -\sqrt{14}$

11. $\sqrt{-5} \cdot \sqrt{-10} = i\sqrt{5} \cdot i\sqrt{10}$
   $= i^2\sqrt{50} = (-1)\sqrt{25 \cdot 2}$
   $= -5\sqrt{2}$

13. $\sqrt{16} \cdot \sqrt{-1} = 4i$

15. $\dfrac{\sqrt{-9}}{\sqrt{3}} = \dfrac{i\sqrt{9}}{\sqrt{3}}$
   $= \dfrac{3i}{\sqrt{3}} = \dfrac{3i \cdot \sqrt{3}}{\sqrt{3} \cdot \sqrt{3}} = \dfrac{3i\sqrt{3}}{3} = i\sqrt{3}$

17. $\dfrac{\sqrt{-80}}{\sqrt{-10}} = \dfrac{i\sqrt{80}}{i\sqrt{10}}$
   $= \sqrt{\dfrac{80}{10}} = \sqrt{8} = \sqrt{4 \cdot 2} = 2\sqrt{2}$

19. $(4 - 7i) + (2 + 3i) = (4 + 2) + (-7 + 3)i$
   $= 6 + (-4)i$
   $= 6 - 4i$

21. $(6 + 5i) - (8 - i) = 6 + 5i - 8 + i$
   $= (6 - 8) + (5 + 1)i$
   $= -2 + 6i$

23. $6 - (8 + 4i) = 6 - 8 - 4i$
   $= (6 - 8) - 4i$
   $= -2 - 4i$

25. $6i(2 - 3i) = 12i - 18i^2$
   $= 12i - 18(-1)$
   $= 18 + 12i$

27. $\left(\sqrt{3}+2i\right)\left(\sqrt{3}-2i\right)$
$= \sqrt{3}\cdot\sqrt{3}-\sqrt{3}\cdot 2i+\sqrt{3}\cdot 2i-4i^2$
$= 3-4(-1)$
$= 3+4$
$= 7$

29. $\left(4-2i\right)^2 = (4-2i)(4-2i)$
$= 16-4\cdot 2i-4\cdot 2i+4i^2$
$= 16-8i-8i+4(-1)$
$= 16-16i-4$
$= 12-16i$

31. $\dfrac{4}{i} = \dfrac{4(-i)}{i(-i)} = \dfrac{-4i}{-i^2} = \dfrac{-4i}{-(-1)} = -4i$

33. $\dfrac{7}{4+3i} = \dfrac{7(4-3i)}{(4+3i)(4-3i)}$
$= \dfrac{28-21i}{4^2-9i^2}$
$= \dfrac{28-21i}{16+9}$
$= \dfrac{28-21i}{25}$
$= \dfrac{28}{25}-\dfrac{21}{25}i$

35. $\dfrac{3+5i}{1+i} = \dfrac{(3+5i)(1-i)}{(1+i)(1-i)}$
$= \dfrac{3-3i+5i-5i^2}{1^2-i^2}$
$= \dfrac{3+2i+5}{1+1}$
$= \dfrac{8+2i}{2} = 4+i$

37. $\dfrac{5-i}{3-2i} = \dfrac{(5-i)(3+2i)}{(3-2i)(3+2i)}$
$= \dfrac{15+10i-3i-2i^2}{3^2-4i^2}$
$= \dfrac{15+7i+2}{9+4}$
$= \dfrac{17+7i}{13} = \dfrac{17}{13}+\dfrac{7}{13}i$

39. $(7i)(-9i) = -63i^2 = -63(-1) = 63$

41. $(6-3i)-(4-2i) = 6-3i-4+2i = 2-i$

43. $(6-2i)(3+i) = 18+6i-6i-2i^2$
$= 18+2$
$= 20$

45. $(8-3i)+(2+3i) = 8-3i+2+3i = 10$

47. $(1-i)+(1+i) = 1+i-i-i^2 = 1+1 = 2$

49. $\dfrac{16+15i}{-3i} = \dfrac{(16+15i)(3i)}{-3i(3i)}$
$= \dfrac{48i+45i^2}{-9i^2}$
$= \dfrac{-45+48i}{9}$
$= \dfrac{-45}{9}+\dfrac{48}{9}i = -5+\dfrac{16}{3}i$

51. $(9+8i)^2 = 9^2+2(9)(8i)+(8i)^2$
$= 81+144i+64i^2$
$= 81+144i-64$
$= 17+144i$

53. $\dfrac{2}{3+i} = \dfrac{2(3-i)}{(3+i)(3-i)}$
$= \dfrac{6-2i}{3^2-i^2}$
$= \dfrac{6-2i}{9+1}$
$= \dfrac{6-2i}{10}$
$= \dfrac{6}{10}-\dfrac{2}{10}i = \dfrac{3}{5}-\dfrac{1}{5}i$

55. $(5-6i)-4i = 5-6i-4i = 5-10i$

57. $\dfrac{2-3i}{2+i} = \dfrac{(2-3i)(2-i)}{(2+i)(2-i)}$
$= \dfrac{4-2i-6i+3i^2}{2^2-i^2}$
$= \dfrac{4-8i-3}{4+1}$
$= \dfrac{1-8i}{5} = \dfrac{1}{5}-\dfrac{8}{5}i$

59. $(2+4i)+(6-5i)=2+4i+6-5i=8-i$

61. $i^8=(i^4)^2=1^2=1$

63. $i^{21}=i^{20}\cdot i=(i^4)^5\cdot i=1^5\cdot i=i$

65. $i^{11}=i^8\cdot i^3=(i^4)^2\cdot i^3=1^2\cdot(-i)=-i$

67. $i^{-6}=\dfrac{1}{i^6}=\dfrac{1}{i^4\cdot i^2}=\dfrac{1}{1\cdot(-1)}=-1$

69. $(2i)^6=2^6i^6=64i^4\cdot i^2=64(1)(-1)=-64$

71. $(-3i)^5=(-3)^5i^5$
$=-243i^4\cdot i=-243(1)i=-243i$

73. $x+50^\circ+90^\circ=180^\circ$
$x+140^\circ=180^\circ$
$x=40^\circ$

75.
$$\begin{array}{r|rrrr}1 & 1 & -6 & 3 & -4\\ & & 1 & -5 & -2\\\hline & 1 & -5 & -2 & -6\end{array}$$

Answer: $x^2-5x-2-\dfrac{6}{x-1}$

77. 5 people

79. $5+9=14$ people

81. $\dfrac{5\text{ people}}{30\text{ people}}=\dfrac{1}{6}\approx0.1666$

About 16.7% of the people reported an average checking balance of $201 to $300.

83. $i^3+i^4=-i+1=1-i$

85. $i^6+i^8=i^4\cdot i^2+(i^4)^2$
$=1(-1)+1^2=-1+1=0$

87. $2+\sqrt{-9}=2+i\sqrt9=2+3i$

89. $\dfrac{6+\sqrt{-18}}{3}=\dfrac{6+i\sqrt{9\cdot2}}{3}$
$=\dfrac{6+3i\sqrt2}{3}$
$=\dfrac{6}{3}+\dfrac{3\sqrt2}{3}i$
$=2+i\sqrt2$

91. $\dfrac{5-\sqrt{-75}}{10}=\dfrac{5-i\sqrt{25\cdot3}}{10}$
$=\dfrac{5-5i\sqrt3}{10}$
$=\dfrac{5}{10}-\dfrac{5\sqrt3}{10}i$
$=\dfrac{1}{2}-\dfrac{\sqrt3}{2}i$

93. Answers may vary.

95. $\left(8-\sqrt{-4}\right)-\left(2+\sqrt{-16}\right)=8-2i-2-4i$
$=6-6i$

97. $x^2+2x=-2$
$(-1+i)^2+2(-1+i)=-2$
$(1-2i+i^2)-2+2i=-2$
$1-1-2=-2$
$-2=-2$, which is true.

Yes, $-1+i$ is a solution.

Chapter 10 Review

1. $\sqrt{81}=9$ because $9^2=81$.

2. $\sqrt[4]{81}=3$ because $3^4=81$.

3. $\sqrt[3]{-8}=-2$ because $(-2)^4=-8$.

4. $\sqrt[4]{16}=2$ because $2^4=16$.

5. $-\sqrt{\dfrac{1}{49}}=-\dfrac{1}{7}$ because $\left(\dfrac{1}{7}\right)^2=\dfrac{1}{49}$.

6. $\sqrt{x^{64}} = x^{32}$ because $(x^{32})^2 = x^{32\cdot2} = x^{64}$.

7. $-\sqrt{36} = -6$ because $6^2 = 36$.

8. $\sqrt[3]{64} = 4$ because $4^3 = 64$.

9. $\sqrt[3]{-a^6 b^9} = \sqrt[3]{-1}\sqrt[3]{a^6}\sqrt[3]{b^9}$
$$= -1a^2 b^3$$
$$= -a^2 b^3$$

10. $\sqrt{16a^4 b^{12}} = \sqrt{16}\sqrt{a^4}\sqrt{b^{12}} = 4a^2 b^6$

11. $\sqrt[5]{32a^5 b^{10}} = \sqrt[5]{32}\sqrt[5]{a^5}\sqrt[5]{b^{10}} = 2ab^2$

12. $\sqrt[5]{-32x^{15} y^{20}} = \sqrt[5]{-32}\sqrt[5]{x^{15}}\sqrt[5]{y^{20}} = -2x^3 y^4$

13. $\sqrt{\dfrac{x^{12}}{36y^2}} = \dfrac{\sqrt{x^{12}}}{\sqrt{36y^2}} = \dfrac{x^6}{6y}$

14. $\sqrt[3]{\dfrac{27y^3}{z^{12}}} = \dfrac{\sqrt[3]{27y^3}}{\sqrt[3]{z^{12}}} = \dfrac{3y}{z^4}$

15. $\sqrt{(-x)^2} = |-x|$

16. $\sqrt[4]{(x^2-4)^4} = |x^2-4|$

17. $\sqrt[3]{(-27)^3} = -27$

18. $\sqrt[5]{(-5)^5} = -5$

19. $-\sqrt[5]{x^5} = -x$

20. $\sqrt[4]{16(2y+z)^{12}} = \sqrt[4]{16}\sqrt[4]{(2y+z)^{12}}$
$$= 2|2y+z|^3$$

21. $\sqrt{25(x-y)^{10}} = \sqrt{25}\sqrt{(x-y)^{10}}$
$$= 5|(x-y)^5|$$

22. $\sqrt[5]{-y^5} = \sqrt[5]{-1}\sqrt[5]{y^5} = -1y = -y$

23. $\sqrt[9]{-x^9} = \sqrt[9]{-1}\sqrt[9]{x^9} = -1x = -x$

24. $f(x) = \sqrt{x} + 3$

Domain: $[0, \infty)$

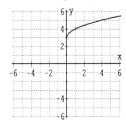

25. $g(x) = \sqrt[3]{x} - 3$

Domain: $(-\infty, \infty)$

x	-5	2	3	4	11
$g(x)$	-2	-1	0	1	2

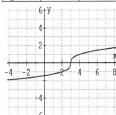

26. $\left(\dfrac{1}{81}\right)^{1/4} = \dfrac{1}{81^{1/4}} = \dfrac{1}{\sqrt[4]{81}} = \dfrac{1}{3}$

27. $\left(-\dfrac{1}{27}\right)^{1/3} = -\dfrac{1}{27^{1/3}} = -\dfrac{1}{\sqrt[3]{27}} = -\dfrac{1}{3}$

28. $(-27)^{-1/3} = \dfrac{1}{(-27)^{1/3}} = \dfrac{1}{\sqrt[3]{-27}} = \dfrac{1}{-3} = -\dfrac{1}{3}$

29. $(-64)^{-1/3} = \dfrac{1}{(-64)^{1/3}} = \dfrac{1}{\sqrt[3]{-64}} = \dfrac{1}{-4} = -\dfrac{1}{4}$

30. $-9^{3/2} = -\left(\sqrt{9}\right)^3 = -3^3 = -27$

31. $64^{-1/3} = \dfrac{1}{64^{1/3}} = \dfrac{1}{\sqrt[3]{64}} = \dfrac{1}{4}$

32. $(-25)^{5/2} = \left(\sqrt{-25}\right)^5$ is not a real number, since there is no real number whose 4th power is negative.

33. $\left(\dfrac{25}{49}\right)^{-3/2} = \dfrac{1}{\left(\dfrac{25}{49}\right)^{3/2}}$

$= \dfrac{1}{\left(\sqrt{\dfrac{25}{49}}\right)^3} = \dfrac{1}{\left(\dfrac{5}{7}\right)^3} = \dfrac{1}{\dfrac{125}{343}} = \dfrac{343}{125}$

34. $\left(\dfrac{8}{27}\right)^{-2/3} = \dfrac{1}{\left(\dfrac{8}{27}\right)^{2/3}}$

$= \dfrac{1}{\left(\sqrt[3]{\dfrac{8}{27}}\right)^2} = \dfrac{1}{\left(\dfrac{2}{3}\right)^2} = \dfrac{1}{\dfrac{8}{9}} = \dfrac{9}{8}$

35. $\left(-\dfrac{1}{36}\right)^{-1/4} = \dfrac{1}{\left(-\dfrac{1}{36}\right)^{1/4}} = \dfrac{1}{\sqrt[4]{-\dfrac{1}{36}}}$ is not a real

number, since there is no real number whose 4th power is negative.

36. $\sqrt[3]{x^2} = x^{2/3}$

37. $\sqrt[5]{5x^2 y^3} = (5x^2 y^3)^{1/5}$
$= 5^{1/5}(x^2)^{1/5}(y^3)^{1/5}$
$= 5^{1/5} x^{2/5} y^{3/5}$

38. $y^{4/5} = \sqrt[5]{y^4}$

39. $5(xy^2 z^5)^{1/3} = 5\sqrt[3]{xy^2 z^5}$

40. $(x+2y)^{-1/2} = \dfrac{1}{(x+2y)^{1/2}} = \dfrac{1}{\sqrt{x+2y}}$

41. $a^{1/3} a^{4/3} a^{1/2} = a^{\frac{1}{3}+\frac{4}{3}+\frac{1}{2}} = a^{\frac{2}{6}+\frac{8}{6}+\frac{3}{6}} = a^{13/6}$

42. $\dfrac{b^{1/3}}{b^{4/3}} = b^{1/3-4/3} = b^{-3/3} = b^{-1} = \dfrac{1}{b}$

43. $(a^{1/2} a^{-2})^3 = (a^{1/2-2})^3$
$= (a^{1/2-4/2})^3$
$= (a^{-3/2})^3$
$= a^{-9/2}$
$= \dfrac{1}{a^{9/2}}$

44. $(x^{-3} y^6)^{1/3} = (x^{-3})^{1/3}(y^6)^{1/3} = x^{-1} y^2 = \dfrac{y^2}{x}$

45. $\left(\dfrac{b^{3/4}}{a^{-1/2}}\right)^8 = (a^{1/2} b^{3/4})^8$
$= (a^{1/2})^8 (b^{3/4})^8$
$= a^4 b^6$

46. $\dfrac{x^{1/4} x^{-1/2}}{x^{2/3}} = x^{1/4+(-1/2)-2/3}$
$= x^{\frac{3}{12}-\frac{6}{12}-\frac{8}{12}}$
$= x^{-11/12}$
$= \dfrac{1}{x^{11/12}}$

47. $\left(\dfrac{49c^{5/3}}{a^{-1/4} b^{5/6}}\right)^{-1} = \dfrac{49^{-1} c^{-5/3}}{a^{1/4} b^{-5/6}}$
$= \dfrac{b^{5/6}}{49 a^{1/4} c^{5/3}}$

48. $a^{-1/4}(a^{5/4} - a^{9/4})$
$= a^{-1/4}(a^{5/4}) - a^{-1/4}(a^{9/4})$
$= a^{-1/4+5/4} - a^{-1/4+9/4}$
$= a^{4/4} - a^{8/4}$
$= a - a^2$

49. $\sqrt{20} \approx 4.472$

50. $\sqrt[3]{-39} \approx -3.391$

51. $\sqrt[4]{726} \approx 5.191$

52. $56^{1/3} \approx 3.826$

53. $-78^{3/4} \approx -26.246$

54. $105^{-2/3} \approx 0.045$

55. $\sqrt[3]{2} \cdot \sqrt{7} = 2^{1/3} \cdot 7^{1/2}$
$= 2^{2/6} \cdot 7^{3/6}$
$= (2^2 \cdot 7^3)^{1/6}$
$= \sqrt[6]{1372}$

56. $\sqrt[3]{3} \cdot \sqrt[4]{x} = 3^{1/3} \cdot x^{1/4}$
$= 3^{4/12} \cdot x^{3/12}$
$= (3^4 \cdot x^3)^{1/12}$
$= \sqrt[12]{81x^3}$

57. $\sqrt{3} \cdot \sqrt{8} = \sqrt{24} = \sqrt{4 \cdot 6} = 2\sqrt{6}$

58. $\sqrt[3]{7y} \cdot \sqrt[3]{x^2 z} = \sqrt[3]{7y \cdot x^2 z} = \sqrt[3]{7x^2 yz}$

59. $\dfrac{\sqrt{44x^3}}{\sqrt{11x}} = \sqrt{\dfrac{44x^3}{11x}} = \sqrt{4x^2} = 2x$

60. $\dfrac{\sqrt[4]{a^6 b^{13}}}{\sqrt[4]{a^2 b}} = \sqrt[4]{\dfrac{a^6 b^{13}}{a^2 b}} = \sqrt[4]{a^4 b^{12}} = ab^3$

61. $\sqrt{60} = \sqrt{4 \cdot 15} = 2\sqrt{15}$

62. $-\sqrt{75} = -\sqrt{25 \cdot 3} = -5\sqrt{3}$

63. $\sqrt[3]{162} = \sqrt[3]{27 \cdot 6} = 3\sqrt[3]{6}$

64. $\sqrt[3]{-32} = \sqrt[3]{-8 \cdot 4} = -2\sqrt[3]{4}$

65. $\sqrt{36x^7} = \sqrt{36x^6 \cdot x} = 6x^3 \sqrt{x}$

66. $\sqrt[3]{24a^5 b^7} = \sqrt[3]{8a^3 b^6 \cdot 3a^2 b^2} = 2ab^2 \sqrt[3]{3a^2 b}$

67. $\sqrt{\dfrac{p^{17}}{121}} = \dfrac{\sqrt{p^{17}}}{\sqrt{121}} = \dfrac{\sqrt{p^{16} \cdot p}}{11} = \dfrac{p^8 \sqrt{p}}{11}$

68. $\sqrt[3]{\dfrac{y^5}{27x^6}} = \dfrac{\sqrt[3]{y^5}}{\sqrt[3]{27x^6}} = \dfrac{\sqrt[3]{y^3 y^2}}{\sqrt[3]{27x^6}} = \dfrac{y\sqrt[3]{y^2}}{3x^2}$

69. $\sqrt[4]{\dfrac{xy^6}{81}} = \dfrac{\sqrt[4]{xy^6}}{\sqrt[4]{81}} = \dfrac{\sqrt[4]{y^4 \cdot xy^2}}{3} = \dfrac{y\sqrt[4]{xy^2}}{3}$

70. $\sqrt{\dfrac{2x^3}{49y^4}} = \dfrac{\sqrt{2x^3}}{\sqrt{49y^4}} = \dfrac{\sqrt{x^2 \cdot 2x}}{7y^2} = \dfrac{x\sqrt{2x}}{7y^2}$

71. $r = \sqrt{\dfrac{A}{\pi}}$

  a. $r = \sqrt{\dfrac{25}{\pi}} = \dfrac{\sqrt{25}}{\sqrt{\pi}} = \dfrac{5}{\sqrt{\pi}}$ meters, or

    $r = \dfrac{5}{\sqrt{\pi}} = \dfrac{5\sqrt{\pi}}{\sqrt{\pi}\sqrt{\pi}} = \dfrac{5\sqrt{\pi}}{\pi}$ meters

  b. $r = \sqrt{\dfrac{104}{\pi}} \approx 5.75$ inches

72. $x\sqrt{75xy} - \sqrt{27x^3 y}$
$= x\sqrt{25 \cdot 3xy} - \sqrt{9x^2 \cdot 3xy}$
$= x \cdot 5\sqrt{3xy} - 3x\sqrt{3xy}$
$= 2x\sqrt{3xy}$

73. $2\sqrt{32x^2 y^3} - xy\sqrt{98y}$
$= 2\sqrt{16x^2 y^2 \cdot 2y} - xy\sqrt{49 \cdot 2y}$
$= 2 \cdot 4xy\sqrt{2y} - xy \cdot 7\sqrt{2y}$
$= 8xy\sqrt{2y} - 7xy\sqrt{2y}$
$= xy\sqrt{2y}$

74. $\sqrt[3]{128} + \sqrt[3]{250} = \sqrt[3]{64 \cdot 2} + \sqrt[3]{125 \cdot 2}$
$= 4\sqrt[3]{2} + 5\sqrt[3]{2}$
$= 9\sqrt[3]{2}$

75. $3\sqrt[4]{32a^5} - a\sqrt[4]{162a}$
$= 3\sqrt[4]{16a^4 \cdot 2a} - a\sqrt[4]{81 \cdot 2a}$
$= 3 \cdot 2a\sqrt[4]{2a} - 3a\sqrt[4]{2a}$
$= 6a\sqrt[4]{2a} - 3a\sqrt[4]{2a}$
$= 3a\sqrt[4]{2a}$

76. $\dfrac{5}{\sqrt{4}} + \dfrac{\sqrt{3}}{3} = \dfrac{5}{2} + \dfrac{\sqrt{3}}{3}$
$= \dfrac{5 \cdot 3 + 2\sqrt{3}}{6} = \dfrac{15 + 2\sqrt{3}}{6}$

77. $\sqrt{\dfrac{8}{x^2}} - \sqrt{\dfrac{50}{16x^2}} = \dfrac{\sqrt{8}}{\sqrt{x^2}} - \dfrac{\sqrt{50}}{\sqrt{16x^2}}$

$$= \dfrac{\sqrt{4 \cdot 2}}{x} - \dfrac{\sqrt{25 \cdot 2}}{4x}$$

$$= \dfrac{2\sqrt{2} \cdot 4}{x \cdot 4} - \dfrac{5\sqrt{2}}{4x}$$

$$= \dfrac{8\sqrt{2} - 5\sqrt{2}}{4x}$$

$$= \dfrac{3\sqrt{2}}{4x}$$

78. $2\sqrt{50} - 3\sqrt{125} + \sqrt{98}$

$$= 2\sqrt{25 \cdot 2} - 3\sqrt{25 \cdot 5} + \sqrt{49 \cdot 2}$$

$$= 2 \cdot 5\sqrt{2} - 3 \cdot 5\sqrt{5} + 7\sqrt{2}$$

$$= 10\sqrt{2} - 15\sqrt{5} + 7\sqrt{2}$$

$$= 17\sqrt{2} - 15\sqrt{5}$$

79. $2a\sqrt[4]{32b^5} - 3b\sqrt[4]{162a^4b} + \sqrt[4]{2a^4b^5}$

$$= 2a\sqrt[4]{16b^4 \cdot 2b} - 3b\sqrt[4]{81a^4 \cdot 2b}$$
$$+ \sqrt[4]{a^4b^4 \cdot 2b}$$

$$= 2a \cdot 2b\sqrt[4]{2b} - 3b \cdot 3a\sqrt[4]{2b} + ab\sqrt[4]{2b}$$

$$= 4ab\sqrt[4]{2b} - 9ab\sqrt[4]{2b} + ab\sqrt[4]{2b}$$

$$= -4ab\sqrt[4]{2b}$$

80. $\sqrt{3}\left(\sqrt{27} - \sqrt{3}\right) = \sqrt{3}\left(\sqrt{9 \cdot 3} - \sqrt{3}\right)$

$$= \sqrt{3}\left(3\sqrt{3} - \sqrt{3}\right)$$

$$= \sqrt{3}\left(2\sqrt{3}\right)$$

$$= 2\sqrt{9}$$

$$= 2(3)$$

$$= 6$$

81. $\left(\sqrt{x} - 3\right)^2 = \left(\sqrt{x}\right)^2 - 2 \cdot \sqrt{x} \cdot 3 + 3^2$

$$= x - 6\sqrt{x} + 9$$

82. $\left(\sqrt{5} - 5\right)\left(2\sqrt{5} + 2\right)$

$$= 2\sqrt{25} + 2\sqrt{5} - 10\sqrt{5} - 10$$

$$= 2(5) - 8\sqrt{5} - 10$$

$$= 10 - 8\sqrt{5} - 10$$

$$= -8\sqrt{5}$$

83. $\left(2\sqrt{x} - 3\sqrt{y}\right)\left(2\sqrt{x} + 3\sqrt{y}\right)$

$$= \left(2\sqrt{x}\right)^2 - \left(3\sqrt{y}\right)^2$$

$$= 2^2\left(\sqrt{x}\right)^2 - 3^2\left(\sqrt{y}\right)^2$$

$$= 4x - 9y$$

84. $\left(\sqrt{a} - 3\right)\left(\sqrt{a} + 3\right) = \left(\sqrt{a}\right)^2 - (3)^2$

$$= a - 9$$

85. $\left(\sqrt[3]{a} + 2\right)^2 = \left(\sqrt[3]{a}\right)^2 + 2 \cdot \sqrt[3]{a} \cdot 2 + 2^2$

$$= \sqrt[3]{a^2} + 4\sqrt[3]{a} + 4$$

86. $\left(\sqrt[3]{5x} + 9\right)\left(\sqrt[3]{5x} - 9\right) = \left(\sqrt[3]{5x}\right)^2 - 9^2$

$$= \sqrt[3]{(5x)^2} - 81$$

$$= \sqrt[3]{25x^2} - 81$$

87. $\left(\sqrt[3]{a} + 4\right)\left(\sqrt[3]{a^2} - 4\sqrt[3]{a} + 16\right)$

$$= \left(\sqrt[3]{a}\right)\left(\sqrt[3]{a^2}\right) - 4 \cdot \left(\sqrt[3]{a}\right)^2 + 16\sqrt[3]{a}$$
$$+ 4\sqrt[3]{a^2} \quad - 16\sqrt[3]{a} + 64$$

$$= \sqrt[3]{a^3} - 4\sqrt[3]{a^2} + 4\sqrt[3]{a^2} + 64$$

$$= a + 64$$

88. $\dfrac{3}{\sqrt{7}} = \dfrac{3 \cdot \sqrt{7}}{\sqrt{7} \cdot \sqrt{7}} = \dfrac{3\sqrt{7}}{7}$

89. $\sqrt{\dfrac{x}{12}} = \dfrac{\sqrt{x}}{\sqrt{12}}$

$= \dfrac{\sqrt{x}}{\sqrt{4\cdot 3}}$

$= \dfrac{\sqrt{x}}{2\sqrt{3}}$

$= \dfrac{\sqrt{x}\cdot\sqrt{3}}{2\sqrt{3}\cdot\sqrt{3}} = \dfrac{\sqrt{3x}}{2\cdot 3} = \dfrac{\sqrt{3x}}{6}$

90. $\dfrac{5}{\sqrt[3]{4}} = \dfrac{5\cdot\sqrt[3]{2}}{\sqrt[3]{4}\cdot\sqrt[3]{2}} = \dfrac{5\sqrt[3]{2}}{\sqrt[3]{8}} = \dfrac{5\sqrt[3]{2}}{2}$

91. $\sqrt{\dfrac{24x^5}{3y^2}} = \sqrt{\dfrac{8x^5}{y^2}}$

$= \dfrac{\sqrt{8x^5}}{\sqrt{y^2}}$

$= \dfrac{\sqrt{4x^4\cdot 2x}}{y} = \dfrac{2x\sqrt{2x}}{y}$

92. $\sqrt[3]{\dfrac{15x^6y^7}{z^2}} = \dfrac{\sqrt[3]{15x^6y^7}}{\sqrt[3]{z^2}}$

$= \dfrac{\sqrt[3]{15x^6y^7}\cdot\sqrt[3]{z}}{\sqrt[3]{z^2}\cdot\sqrt[3]{z}}$

$= \dfrac{\sqrt[3]{15x^6y^7 z}}{\sqrt[3]{z^3}}$

$= \dfrac{\sqrt[3]{15x^6y^6\cdot yz}}{z} = \dfrac{x^2y^2\sqrt[3]{15yz}}{z}$

93. $\dfrac{5}{2-\sqrt{7}} = \dfrac{5\left(2+\sqrt{7}\right)}{\left(2-\sqrt{7}\right)\left(2+\sqrt{7}\right)}$

$= \dfrac{5\left(2+\sqrt{7}\right)}{2^2 - \left(\sqrt{7}\right)^2}$

$= \dfrac{10+5\sqrt{7}}{4-7}$

$= \dfrac{10+5\sqrt{7}}{-3} = -\dfrac{10+5\sqrt{7}}{3}$

94. $\dfrac{3}{\sqrt{y}-2} = \dfrac{3\left(\sqrt{y}+2\right)}{\left(\sqrt{y}-2\right)\left(\sqrt{y}+2\right)}$

$= \dfrac{3\left(\sqrt{y}+2\right)}{\left(\sqrt{y}\right)^2 - 2^2} = \dfrac{3\sqrt{y}+6}{y-4}$

95. $\dfrac{\sqrt{2}-\sqrt{3}}{\sqrt{2}+\sqrt{3}} = \dfrac{\left(\sqrt{2}-\sqrt{3}\right)\left(\sqrt{2}-\sqrt{3}\right)}{\left(\sqrt{2}+\sqrt{3}\right)\left(\sqrt{2}-\sqrt{3}\right)}$

$= \dfrac{2-\sqrt{2}\sqrt{3}-\sqrt{3}\sqrt{2}+3}{\left(\sqrt{2}\right)^2 - \left(\sqrt{3}\right)^2}$

$= \dfrac{5-2\sqrt{6}}{-1}$

$= -5+2\sqrt{6}$

96. $\dfrac{\sqrt{11}}{3} = \dfrac{\sqrt{11}\cdot\sqrt{11}}{3\cdot\sqrt{11}} = \dfrac{11}{3\sqrt{11}}$

97. $\sqrt{\dfrac{18}{y}} = \dfrac{\sqrt{18}}{\sqrt{y}}$

$= \dfrac{3\sqrt{2}}{\sqrt{y}} = \dfrac{3\sqrt{2}\cdot\sqrt{2}}{\sqrt{y}\cdot\sqrt{2}} = \dfrac{3\cdot 2}{\sqrt{2y}} = \dfrac{6}{\sqrt{2y}}$

98. $\dfrac{\sqrt[3]{9}}{7} = \dfrac{\sqrt[3]{9}\cdot\sqrt[3]{3}}{7\cdot\sqrt[3]{3}} = \dfrac{\sqrt[3]{27}}{7\sqrt[3]{3}} = \dfrac{3}{7\sqrt[3]{3}}$

99. $\sqrt{\dfrac{24x^5}{3y^2}} = \sqrt{\dfrac{8x^5}{y^2}}$

$= \dfrac{\sqrt{4x^4\cdot 2x}}{\sqrt{y^2}}$

$= \dfrac{2x^2\sqrt{2x}}{y}$

$= \dfrac{2x^2\sqrt{2x}\cdot\sqrt{2x}}{y\cdot\sqrt{2x}}$

$= \dfrac{2x^2\cdot 2x}{y\sqrt{2x}} = \dfrac{4x^3}{y\sqrt{2x}}$

100. $\sqrt[3]{\dfrac{xy^2}{10z}} = \dfrac{\sqrt[3]{xy^2}}{\sqrt[3]{10z}}$

$= \dfrac{\sqrt[3]{xy^2} \cdot \sqrt[3]{x^2 y}}{\sqrt[3]{10z} \cdot \sqrt[3]{x^2 y}}$

$= \dfrac{\sqrt[3]{x^3 y^3}}{\sqrt[3]{10x^2 yz}}$

$= \dfrac{xy}{\sqrt[3]{10x^2 yz}}$

101. $\dfrac{\sqrt{x}+5}{-3} = \dfrac{\left(\sqrt{x}+5\right)\left(\sqrt{x}-5\right)}{-3\left(\sqrt{x}-5\right)}$

$= \dfrac{\left(\sqrt{x}\right)^2 - 5^2}{-3\sqrt{x}+15}$

$= \dfrac{x-25}{-3\sqrt{x}+15}$

102. $\sqrt{y-7} = 5$

$\left(\sqrt{y-7}\right)^2 = 5^2$

$y - 7 = 25$

$y = 32$

The solution is 32.

103. $\sqrt{2x}+10 = 4$

$\sqrt{2x} = -6$

No solution exist since the principle square root of a number is not negative.

104. $\sqrt[3]{2x-6} = 4$

$\left(\sqrt[3]{2x-6}\right)^3 = 4^3$

$2x - 6 = 64$

$2x = 70$

$x = 35$

The solution is 35.

105. $\sqrt{x+6} = \sqrt{x+2}$

$\left(\sqrt{x+6}\right)^2 = \left(\sqrt{x+2}\right)^2$

$x + 6 = x + 2$

$6 = 2$, which is false.

There is no solution.

106. $2x - 5\sqrt{x} = 3$

$2x - 3 = 5\sqrt{x}$

$(2x-3)^2 = \left(5\sqrt{x}\right)^2$

$4x^2 - 12x + 9 = 25x$

$4x^2 - 37x + 9 = 0$

$(4x-1)(x-9) = 0$

$4x - 1 = 0 \ \text{ or } \ x - 9 = 0$

$4x = 1 \ \text{ or } \qquad x = 9$

$x = \dfrac{1}{4}$

Discard the solution $\dfrac{1}{4}$ as extraneous.

The solution is 9.

107. $\sqrt{x+9} = 2 + \sqrt{x-7}$

$\left(\sqrt{x+9}\right)^2 = \left(2 + \sqrt{x-7}\right)^2$

$x + 9 = 4 + 4\sqrt{x-7} + (x-7)$

$x + 9 = x - 3 + 4\sqrt{x-7}$

$12 = 4\sqrt{x-7}$

$3 = \sqrt{x-7}$

$3^2 = \left(\sqrt{x-7}\right)^2$

$9 = x - 7$

$16 = x$

The solution is 16.

108. Let $c =$ length of the hypotenuse.

$3^2 + 3^2 = c^2$

$18 = c^2$

$\sqrt{18} = \sqrt{c^2}$

$3\sqrt{2} = c$

109. Let c = length of the hypotenuse.
$$7^2 + \left(8\sqrt{3}\right)^2 = c^2$$
$$49 + 64 \cdot 3 = c^2$$
$$241 = c^2$$
$$\sqrt{241} = \sqrt{c^2}$$
$$\sqrt{241} = c$$

110. Let b = width of the lake.
$$a^2 + b^2 = c^2$$
$$40^2 + b^2 = 65^2$$
$$1600 + b^2 = 4225$$
$$b^2 = 2625$$
$$\sqrt{b^2} = \sqrt{2625}$$
$$b = 51.23475$$
The width is about 51.2 feet.

111. Let c = length of the shortest pipe.
$$a^2 + b^2 = c^2$$
$$3^2 + 3^2 = c^2$$
$$18 = c^2$$
$$\sqrt{18} = \sqrt{c^2}$$
$$4.24264 = c$$
The shortest possible pipe is 4.24 feet.

112. $\sqrt{-8} = i\sqrt{4 \cdot 2} = 2i\sqrt{2}$

113. $-\sqrt{-6} = -i\sqrt{6}$

114. $\sqrt{-4} + \sqrt{-16} = 2i + 4i = 6i$

115. $\sqrt{-2} \cdot \sqrt{-5} = i\sqrt{2} \cdot i\sqrt{5}$
$$= i^2\sqrt{10} =$$
$$= -1 \cdot \sqrt{10}$$
$$= -\sqrt{10}$$

116. $(12 - 6i) + (3 + 2i) = (12 + 3) + (-6 + 2)i$
$$= 15 + (-4)i$$
$$= 15 - 4i$$

117. $(-8 - 7i) - (5 - 4i) = -8 - 7i - 5 + 4i$
$$= -13 - 3i$$

118. $\left(\sqrt{3} + \sqrt{2}\right) + \left(3\sqrt{2} - \sqrt{-8}\right)$
$$= \sqrt{3} + \sqrt{2} + 3\sqrt{2} - i\sqrt{4 \cdot 2}$$
$$= \sqrt{3} + 4\sqrt{2} - 2i\sqrt{2}$$

119. $2i(2 - 5i) = 4i - 10i^2$
$$= 4i - 10(-1)$$
$$= 10 + 4i$$

120. $-3i(6 - 4i) = -18i + 12i^2$
$$= -18i + 12(-1)$$
$$= -12 - 18i$$

121. $(3 + 2i)(1 + i) = 3 + 3i + 2i + 2i^2$
$$= 3 + 5i + 2(-1)$$
$$= 1 + 5i$$

122. $(2 - 3i)^2 = 4 - 12i + 9i^2$
$$= 4 - 12i + 9(-1)$$
$$= -5 - 12i$$

123. $\left(\sqrt{6} - 9i\right)\left(\sqrt{6} + 9i\right) = \left(\sqrt{6}\right)^2 - (9i)^2$
$$= 6 - 81i^2$$
$$= 6 + 81$$
$$= 87$$

124. $\dfrac{2 + 3i}{2i} = \dfrac{(2 + 3i) \cdot (-2i)}{2i \cdot (-2i)}$
$$= \dfrac{-4i - 6i^2}{-4i^2}$$
$$= \dfrac{-4i + 6}{4}$$
$$= \dfrac{6}{4} - \dfrac{4}{4}i$$
$$= \dfrac{3}{2} - i$$

125. $\dfrac{1 + i}{-3i} = \dfrac{(1 + i) \cdot (3i)}{-3i \cdot (3i)}$
$$= \dfrac{3i + 3i^2}{-9i^2}$$
$$= \dfrac{3i - 3}{9}$$
$$= \dfrac{-3}{9} - \dfrac{3}{9}i = -\dfrac{1}{3} + \dfrac{1}{3}i$$

Chapter 10 Test

1. $\sqrt{216} = \sqrt{36 \cdot 6} = 6\sqrt{6}$

2. $-\sqrt[4]{x^{64}} = -x^{16}$

3. $\left(\dfrac{1}{125}\right)^{1/3} = \dfrac{1}{125^{1/3}} = \dfrac{1}{\sqrt[3]{125}} = \dfrac{1}{5}$

4. $\left(\dfrac{1}{125}\right)^{-1/3} = \dfrac{1}{\left(\dfrac{1}{125}\right)^{1/3}} = \dfrac{1}{\dfrac{1}{5}} = 5$

5. $\left(\dfrac{8x^3}{27}\right)^{2/3} = \dfrac{(8x^3)^{2/3}}{27^{2/3}}$

$= \dfrac{\left(\sqrt[3]{8x^3}\right)^2}{\left(\sqrt[3]{27}\right)^2}$

$= \dfrac{(2x)^2}{3^3}$

$= \dfrac{4x^2}{9}$

6. $\sqrt[3]{-a^{18}b^9} = \sqrt[3]{-1a^{18}b^9} = (-1)a^6b^3 = -a^6b^3$

7. $\left(\dfrac{64c^{4/3}}{a^{-2/3}b^{5/6}}\right)^{1/2} = \left(\dfrac{64a^{2/3}c^{4/3}}{b^{5/6}}\right)^{1/2}$

$= \dfrac{64^{1/2}(a^{2/3})^{1/2}(c^{4/3})^{1/2}}{(b^{5/6})^{1/2}}$

$= \dfrac{\sqrt{64}\,a^{1/3}c^{2/3}}{b^{5/12}}$

$= \dfrac{8a^{1/3}c^{2/3}}{b^{5/12}}$

8. $a^{-2/3}(a^{5/4} - a^3) = a^{-2/3}a^{5/4} - a^{-2/3}a^3$

$= a^{-\frac{2}{3}+\frac{5}{4}} - a^{-\frac{2}{3}+3}$

$= a^{-\frac{8}{12}+\frac{15}{12}} - a^{-\frac{2}{3}+\frac{9}{3}}$

$= a^{7/12} - a^{7/3}$

9. $\sqrt[4]{(4xy)^4} = |4xy| = 4|xy|$

10. $\sqrt[3]{(-27)^3} = -27$

11. $\sqrt{\dfrac{9}{y}} = \dfrac{\sqrt{9}}{\sqrt{y}} = \dfrac{3}{\sqrt{y}} = \dfrac{3 \cdot \sqrt{y}}{\sqrt{y} \cdot \sqrt{y}} = \dfrac{3\sqrt{y}}{y}$

12. $\dfrac{4 - \sqrt{x}}{4 + 2\sqrt{x}} = \dfrac{4 - \sqrt{x}}{2(2 + \sqrt{x})}$

$= \dfrac{(4 - \sqrt{x})(2 - \sqrt{x})}{2(2 + \sqrt{x})(2 - \sqrt{x})}$

$= \dfrac{8 - 4\sqrt{x} - 2\sqrt{x} + x}{2\left[2^2 - \left(\sqrt{x}\right)^2\right]}$

$= \dfrac{8 - 6\sqrt{x} + x}{2(4 - x)}$

13. $\dfrac{\sqrt[3]{ab}}{\sqrt[3]{ab^2}} = \sqrt[3]{\dfrac{ab}{ab^2}}$

$= \sqrt[3]{\dfrac{1}{b}}$

$= \dfrac{1}{\sqrt[3]{b}}$

$= \dfrac{1 \cdot \sqrt[3]{b^2}}{\sqrt[3]{b} \cdot \sqrt[3]{b^2}} = \dfrac{\sqrt[3]{b^2}}{b}$

14. $\dfrac{\sqrt{6} + x}{8} = \dfrac{\left(\sqrt{6} + x\right)\left(\sqrt{6} - x\right)}{8\left(\sqrt{6} - x\right)}$

$= \dfrac{\left(\sqrt{6}\right)^2 - x^2}{8\left(\sqrt{6} - x\right)}$

$= \dfrac{6 - x^2}{8\left(\sqrt{6} - x\right)}$

15. $\sqrt{125x^3} - 3\sqrt{20x^3}$
$= \sqrt{25x^2 \cdot 5x} - 3\sqrt{4x^2 \cdot 5x}$
$= 5x\sqrt{5x} - 3 \cdot 2x\sqrt{5x}$
$= 5x\sqrt{5x} - 6x\sqrt{5x}$
$= -x\sqrt{5x}$

16. $\sqrt{3}\left(\sqrt{16} - \sqrt{2}\right) = \sqrt{3}\left(4 - \sqrt{2}\right)$
$\qquad = 4\sqrt{3} - \sqrt{3}\sqrt{2}$
$\qquad = 4\sqrt{3} - \sqrt{6}$

17. $\left(\sqrt{x} + 1\right)^2 = \left(\sqrt{x}\right)^2 + 2\sqrt{x} + 1^2$
$\qquad = x + 2\sqrt{x} + 1$

18. $\left(\sqrt{2} - 4\right)\left(\sqrt{3} + 1\right)$
$= \sqrt{2}\sqrt{3} + 1 \cdot \sqrt{2} - 4\sqrt{3} - 4$
$= \sqrt{6} + \sqrt{2} - 4\sqrt{3} - 4$

19. $\left(\sqrt{5} + 5\right)\left(\sqrt{5} - 5\right) = \left(\sqrt{5}\right)^2 - 5^2$
$\qquad = 5 - 25$
$\qquad = -20$

20. $\sqrt{561} \approx 23.685$

21. $386^{-2/3} \approx 0.019$

22.
$$x = \sqrt{x - 2} + 2$$
$$x - 2 = \sqrt{x - 2}$$
$$(x - 2)^2 = \left(\sqrt{x - 2}\right)^2$$
$$x^2 - 4x + 4 = x - 2$$
$$x^2 - 5x + 6 = 0$$
$$(x - 2)(x - 3) = 0$$
$$x = 2 \text{ or } x = 3$$
The solutions are 2 and 3.

23. $\sqrt{x^2 - 7} + 3 = 0$
$\qquad \sqrt{x^2 - 7} = -3$
No solution exists since the principle
square root of a number is not negative.

24.
$$\sqrt[3]{x + 5} = \sqrt[3]{2x - 1}$$
$$\left(\sqrt[3]{x + 5}\right)^3 = \left(\sqrt[3]{2x - 1}\right)^3$$
$$x + 5 = 2x - 1$$
$$-x = -6$$
$$x = 6$$
The solution is 6.

25. $\sqrt{-2} = i\sqrt{2}$

26. $-\sqrt{-8} = -i\sqrt{4 \cdot 2} = 2i\sqrt{2}$

27. $(12 - 6i) - (12 - 3i) = 12 - 6i - 12 + 3i$
$\qquad = -3i$

28. $(6 - 2i)(6 + 2i) = 6^2 - (2i)^2$
$\qquad = 36 - 4i^2$
$\qquad = 36 + 4$
$\qquad = 40$

29. $(4 + 3i)^2 = 4^2 + 2 \cdot 4 \cdot 3i + (3i)^2$
$\qquad = 16 + 24i + 9i^2$
$\qquad = 16 + 24i - 9$
$\qquad = 7 + 24i$

30. $\dfrac{1 + 4i}{1 - i} = \dfrac{(1 + 4i)(1 + i)}{(1 - i)(1 + i)}$
$\qquad = \dfrac{1 + i + 4i + 4i^2}{1^2 - i^2}$
$\qquad = \dfrac{1 + 5i - 4}{1 - (-1)}$
$\qquad = \dfrac{-3 + 5i}{2}$
$\qquad = -\dfrac{3}{2} + \dfrac{5}{2}i$

31. $x^2 + x^2 = 5^2$
$\qquad 2x^2 = 25$
$\qquad x^2 = \dfrac{25}{2}$
$\qquad \sqrt{x^2} = \sqrt{\dfrac{25}{2}}$
$\qquad x = \dfrac{5}{\sqrt{2}} = \dfrac{5 \cdot \sqrt{2}}{\sqrt{2} \cdot \sqrt{2}} = \dfrac{5\sqrt{2}}{2}$

32. $g(x) = \sqrt{x+2}$

Domain: $[-2, \infty)$

x	-2	-1	2	7
$g(x)$	0	1	2	3

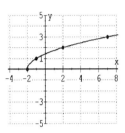

33. $V(r) = \sqrt{2.5r}$

$V(300) = \sqrt{2.5(300)} \approx 27$ mph

34. $V(r) = \sqrt{2.5r}$

$30 = \sqrt{2.5r}$

$30^2 = \left(\sqrt{2.5r}\right)^2$

$900 = 2.5r$

$r = \dfrac{900}{2.5} = 360$ feet

Chapter 10 Cumulative Review

1. a. $\dfrac{(-12)(-3)+3}{-7-(-2)} = \dfrac{36+3}{-7+2}$

$= \dfrac{39}{-5} = -\dfrac{39}{5}$

b. $\dfrac{2(-3)^2 - 20}{-5+4} = \dfrac{2(9)-20}{-1}$

$= \dfrac{18-20}{-1} = \dfrac{-2}{-1} = 2$

c. $(2.1x - 5.6) - (-x - 5.3)$

$= 2.1x - 5.6 + x + 5.3$

$= 3.1x - 0.3$

2. a. $2(x-3) + (5x+3) = 2x - 6 + 5x + 3$

$\qquad = 7x - 3$

b. $4(3x+2) - 3(5x-1)$

$= 12x + 8 - 15x + 3$

$= -3x + 11$

c. $7x + 2(x-7) - 3x$

$= 7x + 2x - 14 - 3x$

$= 6x - 14$

3. $\dfrac{x}{2} - 1 = \dfrac{2}{3}x - 3$

$6\left(\dfrac{x}{2}\right) - 6(1) = 6\left(\dfrac{2}{3}x\right) - 6(3)$

$3x - 6 = 4x - 18$

$-x - 6 = -18$

$-x = -12$

$x = 12$

4. $\dfrac{a-1}{2} + a = 2 - \dfrac{2a+7}{8}$

$8\left(\dfrac{a-1}{2} + a\right) = 8\left(2 - \dfrac{2a+7}{8}\right)$

$4(a-1) + 8a = 16 - (2a+7)$

$4a - 4 + 8a = 16 - 2a - 7$

$12a - 4 = 9 - 2a$

$14a = 13$

$a = \dfrac{13}{14}$

5. Let x = the length of the longer piece and y = the length of the shorter piece.

$x + y = 10$

$x = 4y$

Substitute $4y$ for x in the first equation.

$4y + y = 10$

$5y = 10$

$y = 2$

$x = 4y = 4(2) = 8$

The lengths are 2 ft and 8 ft.

6. Let r = their average speed.

$$t_{going} + t_{returning} = 4.5 \text{ hrs}$$

$$\frac{121.5}{r} + \frac{121.5}{r} = 4.5$$

$$\frac{243}{r} = 4.5$$

$$243 = 4.5r$$

$$r = \frac{243}{4.5} = 54$$

Their average speed was 54 mph.

7. $\begin{cases} 3x - y = 4 \\ x + 2y = 8 \end{cases}$

$$\begin{array}{ll} 3x - y = 4 & x + 2y = 8 \\ -y = -3x + 4 & 2y = -x + 8 \\ y = 3x - 4 & y = -\frac{1}{2}x + 4 \\ \text{Slope} = 3 & \text{Slope} = \frac{1}{2} \end{array}$$

Because the equations have different

Slopes, their graphs will intersect at

One point. Therefore, there is only

One solution to the system.

8. $|3x - 2| + 5 = 5$

$$|3x - 2| = 0$$

$$3x - 2 = 0$$

$$3x = 2$$

$$x = \frac{2}{3}$$

9. $\begin{cases} x + 2y = 7 \\ 2x + 2y = 13 \end{cases}$

Multiply the first equation by -1 and add.

$$-x - 2y = -7$$

$$2x + 27 = 13$$

$$\overline{x \qquad = 6}$$

Let $x = 6$ in the first equation.

$$6 + 2y = 7$$

$$2y = 1$$

$$y = \frac{1}{2}$$

The solution is $\left(6, \frac{1}{2}\right)$.

10. $\left| \frac{x}{2} - 1 \right| \le 0$

$$\frac{x}{2} - 1 = 0$$

$$\frac{x}{2} = 1$$

$$x = 2$$

11. $\begin{cases} 2x - y = 7 \\ 7x - 4y = 1 \end{cases}$

Multiply the first equation by -4 and add.

$$-8x + 4y = -28$$

$$\underline{8x - 4y = 1}$$

$$0 = -27$$

There is no solution.

12. $y = |x - 2|$

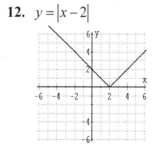

13. Let x = the amount of 5% solution and
   y = the amount of 25% solution

No. of liters x strength = amt of salt.

5%	x	0.05	$0.05x$
25%	y	0.25	$0.25y$
20%	10	0.20	$10(0.2)$

$$\begin{cases} x + y = 10 \\ 0.05x + 0.25y = 2 \end{cases}$$

Solve the first equation for $x = 10 - y$
and substitute into the second equation.

$0.05(10 - y) + 0.25y = 2$
$0.05 - 0.05y + 0.25y = 2$
$0.05 + 0.2y = 2$
$0.2y = 1.5$
$y = 7.5$

Let $y = 7.5$ in $x = 10 - y$.
   $x = 10.75 = 2.5$
2.5L of 5% solution and
7.5L of 25% solution.

14. a. Domain: $(-\infty, 0]$, Range: $(-\infty, \infty)$

  b. Domain: $(-\infty, \infty)$, Range: $(-\infty, \infty)$

  c. Domain: $(-\infty, -2] \cup [2, \infty)$
     Range: $(-\infty, \infty)$

15. $P(x) = 3x^2 - 2x - 5$

  a. $P(1) = 3(1)^2 - 2(1) - 5 = -4$

  b. $P(-2) = 3(-2)^2 - 2(-2) - 5 = 11$

16. $f(x) = -2$
   $y = -2$

17. $\dfrac{10x^3 - 5x^2 + 20x}{5x}$

$$= \frac{10x^3}{5x} - \frac{5x^2}{5x} + \frac{20x}{5x}$$

$$= 2x^2 - x + 4$$

18. $y = -3$ is a horizontal line. Thus, the slope is $m = 0$.

19.
$$\begin{array}{r|rrrr} 3 & 2 & -1 & -13 & 0 \\ & & 6 & 15 & 6 \\ \hline & 2 & 5 & 2 & 6 \end{array}$$

Answer: $2x^2 + 5x + 2 + \dfrac{6}{x - 3}$

20.
$$\begin{cases} \dfrac{x}{6} - \dfrac{y}{2} = 1 \quad (1) \\ \dfrac{x}{3} - \dfrac{y}{4} = 2 \quad (2) \end{cases}$$

Multiply E1 by 6 and E2 by 12 to clear fractions.

$$\begin{cases} x - 3y = 6 & (1) \\ 4x - 3y = 24 & (2) \end{cases}$$

Solve the new E1 for x.

$x = 3y + 6$

Replace x with $3y + 6$ in the new E2.

$4(3y + 6) - 3y = 24$

$12y + 24 - 3y = 24$

$9y = 0$

$y = 0$

Replace y with 0 in the equation

$x = 3y + 6$.

$x = 3(0) + 6 = 6$

The solution is $(6, 0)$.

21. $x^2 + 5yx + 6y^2 = (x + 2y)(x + 3y)$

22. Let x = number of tee-shirts and y = number of shorts.

$$\begin{cases} x + y = 9 & (1) \\ 3.50x + 4.25y = 33.75 & (2) \end{cases}$$

Solve E1 for y: $y = 9 - x$.

Replace y with $9 - x$ in E2

$3.50x + 4.25(9 - x) = 33.75$

$3.50x + 38.25 - 4.25x = 33.75$

$-0.75x = -4.5$

$x = 6$

Replace x with 6 in the equation

$y = 9 - x$.

$y = 9 - 6 = 3$

She bought 6 tee-shirts and 3 shorts.

23. a. $\dfrac{x^3 + 8}{2 + x} = \dfrac{(x + 2)(x^2 - 2x + 4)}{x + 2}$

$\qquad = x^2 - 2x + 4$

24. $\dfrac{0.0000035 \times 4000}{0.28}$

$= \dfrac{(3.5 \times 10^{-6}) \times (4 \times 10^3)}{2.8 \times 10^{-1}}$

$= \dfrac{3.5 \times 4}{2.8} \times 10^{-6+3-(-1)}$

$= 5 \times 10^{-2}$

25. $\dfrac{3x^3 y^7}{40} \div \dfrac{4x^3}{y^2} = \dfrac{3x^3 y^7}{40} \cdot \dfrac{y^2}{4x^3}$

$\qquad = \dfrac{3y^9}{160}$

26. $[(5x^2 - 3x + 6) + (4x^2 + 5x - 3)] - (2x - 5)$

$= (9x^2 + 2x + 3) - (2x - 5)$

$= 9x^2 + 2x + 3 - 2x + 5$

$= 9x^2 + 8$

27. $\dfrac{2y}{2y - 7} - \dfrac{7}{2y - 7} = \dfrac{2y - 7}{2y - 7} = 1$

28. a. $(y - 2)(3y + 4) = 3y^2 + 4y - 6x - 8$

$\qquad = 3y^2 - 6y - 8$

b. $(3y - 1)(2y^2 + 3y - 1)$

$= 6y^3 + 9y^2 - 3y$

$\qquad - 2y^2 - 3y + 1$

$= 6y^3 + 7x^2 - 6y + 1$

29. $\dfrac{2x}{x^2 + 2x + 1} + \dfrac{x}{x^2 - 1}$

$= \dfrac{2x}{(x + 1)(x + 1)} + \dfrac{x}{(x + 1)(x - 1)}$

$= \dfrac{2x(x - 1)}{(x + 1)(x + 1)(x - 1)}$

$\quad + \dfrac{x(x + 1)}{(x + 1)(x + 1)(x - 1)}$

$= \dfrac{2x(x - 1) + x(x + 1)}{(x + 1)(x + 1)(x - 1)}$

$= \dfrac{2x^2 - 2x + x^2 + x}{(x + 1)(x + 1)(x - 1)}$

$= \dfrac{3x^2 - x}{(x + 1)(x + 1)(x - 1)}$

$= \dfrac{x(3x - 1)}{(x + 1)(x + 1)(x - 1)}$

30. $x^3 - x^2 + 4x - 4 = (x^3 - x^2) + (4x - 4)$
$$= x^2(x-1) + 4(x-1)$$
$$= (x-1)(x^2 + 4)$$

31. $\dfrac{\frac{5x}{x+2}}{\frac{10}{x-2}} = \dfrac{\left(\frac{5x}{x+2}\right)(x+2)(x-2)}{\left(\frac{10}{x-2}\right)(x+2)(x-2)}$
$$= \dfrac{5x(x-2)}{10(x+2)}$$
$$= \dfrac{x(x-2)}{2(x+2)}$$

32. a. $\dfrac{a^3 - 8}{2-a} = \dfrac{a^3 - 2^3}{2-a}$
$$= \dfrac{(a-2)(a^2 + 2a + 4)}{2-a}$$
$$= -1(a^2 + 2a + 4)$$
$$= -a^2 - 2a - 4$$

b. $\dfrac{3a^2 - 3}{a^3 + 5a^2 - a - 5} = \dfrac{3(a^2 - 1)}{a^2(a+5) - 1(a+5)}$
$$= \dfrac{2(a^2 - 1)}{(a+5)(a^2 - 1)}$$
$$= \dfrac{2}{a+5}$$

33. $\dfrac{x}{2} + \dfrac{8}{3} = \dfrac{1}{6}$
$$6\left(\dfrac{x}{2}\right) + 6\left(\dfrac{8}{3}\right) = 6\left(\dfrac{1}{6}\right)$$
$$3x + 16 = 1$$
$$3x = -15$$
$$x = -5$$

34. a. $\dfrac{3}{xy^2} - \dfrac{2}{3x^2 y} = \dfrac{3 \cdot 3x}{xy^2 \cdot 3x} - \dfrac{2 \cdot y}{3x^2 y \cdot y}$
$$= \dfrac{9x - 2y}{3x^2 y^2}$$

b. $\dfrac{5x}{x+3} - \dfrac{2x}{x-3} = \dfrac{5x(x-3) - 2x(x+3)}{(x+3)(x-3)}$
$$= \dfrac{5x^2 - 15x - 2x^2 - 6x}{(x+3)(x-3)}$$
$$= \dfrac{3x^2 - 21x}{(x+3)(x-3)}$$
$$\text{or } \dfrac{3x(x-7)}{(x+3)(x-3)}$$

c. $\dfrac{x}{x-2} - \dfrac{5}{2-x} = \dfrac{x}{x-2} + \dfrac{5}{x-2} = \dfrac{x+5}{x-2}$

35. $\dfrac{x}{10} = \dfrac{3}{2}$
$$x = \dfrac{30}{2}$$
$$x = 15 \text{ yards}$$

36. a. $\dfrac{\frac{y-2}{16}}{\frac{2y+3}{12}} = \dfrac{y-2}{16} \cdot \dfrac{12}{2y+3} = \dfrac{3(y-2)}{4(2y+3)}$

b. $\dfrac{\frac{x}{16} - \frac{1}{x}}{1 - \frac{4}{x}} = \dfrac{\left(\frac{x}{16} - \frac{1}{x}\right)16x}{\left(1 - \frac{4}{x}\right)16x}$
$$= \dfrac{x^2 - 16}{16x - 64}$$
$$= \dfrac{(x+4)(x-4)}{16(x-4)}$$
$$= \dfrac{x+4}{16}$$

37. a. $\sqrt[3]{1} = 1$

b. $\sqrt[3]{-64} = -4$

c. $\sqrt[3]{\frac{8}{125}} = \frac{2}{5}$

d. $\sqrt[3]{x^2} = x^{2/3}$

e. $\sqrt[3]{-27x^9} = -3x^3$

38.
$$
\begin{array}{r}
x^2 + 3 \\
x-2 \overline{\smash{\big)}\, x^3 - 2x^2 + 3x - 6} \\
\underline{x^3 - 2x^2} \\
3x - 6 \\
\underline{3x - 6} \\
0
\end{array}
$$

Answer: $x^2 + 3$

39. a. $16^{-3/4} = \dfrac{1}{16^{3/4}} = \dfrac{1}{2^3} = \dfrac{1}{8}$

b. $(-27)^{-2/3} = \dfrac{1}{(-27)^{2/3}}$

$= \dfrac{1}{(-3)^2} = \dfrac{1}{9}$

40.
$$
\begin{array}{r|rrrr}
3 & 4 & -12 & -1 & 12 \\
  &   & 12 & 0 & -3 \\
\hline
  & 4 & 0 & -1 & 9
\end{array}
$$

Answer: $4y^2 - 1 + \dfrac{9}{y-3}$

41. $\dfrac{\sqrt{x}+2}{5} = \dfrac{\sqrt{x}+2}{5} \cdot \dfrac{\sqrt{x}-2}{\sqrt{x}-2}$

$= \dfrac{x-4}{5\sqrt{x}-10}$

42.
$$\frac{28}{9-a^2} = \frac{2a}{a-3} + \frac{6}{a+3}$$

$$\frac{28}{-(a^2-9)} = \frac{2a}{a-3} + \frac{6}{a+3}$$

$$\frac{-28}{(a+3)(a-3)} = \frac{2a}{a-3} + \frac{6}{a+3}$$

$-28 = 2a(a+3) + 6(a-3)$
$-28 = 2a^2 + 6a + 6a - 18$
$0 = 2a^2 + 12a + 10$
$0 = 2(a^2 + 6a + 5)$
$0 = 2(a+5)(a+1)$
$a = -5$ or $a = -1$
The solutions are -5 and -1.

43. $u = \dfrac{k}{w}$, $u = 3$ when $w = 5$

$3 = \dfrac{k}{5}$

$15 = k$

$u = \dfrac{15}{w}$

44. $y = kx$
$0.51 = k(3)$
$k = \dfrac{0.51}{3} = 0.17$

$y = 0.17x$

Chapter 11

Section 11.1

Graphing Calculator Explorations

1. $-1.27, 6.27$

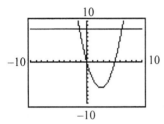

3. $-1.10, 0.90$

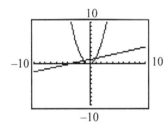

5. No real solutions, or $\varnothing$

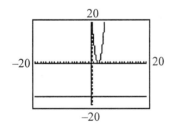

Exercise Set 11.1

1. $x^2 = 16$
$x = \pm\sqrt{16}$
$x = \pm 4$

3. $x^2 - 7 = 0$
$x^2 = 7$
$x = \pm\sqrt{7}$

5. $x^2 = 18$
$x = \pm\sqrt{18}$
$x = \pm\sqrt{9 \cdot 2}$
$x = \pm 3\sqrt{2}$

7. $3z^2 - 30 = 0$
$3z^2 = 30$
$z^2 = 10$
$z = \pm\sqrt{10}$

9. $(x + 5)^2 = 9$
$x + 5 = \pm\sqrt{9}$
$x + 5 = \pm 3$
$x = -5 \pm 3$
$x = -8 \ \text{or} \ x = -2$

11. $(z - 6)^2 = 18$
$z - 6 = \pm\sqrt{18}$
$z - 6 = \pm 3\sqrt{2}$
$z = 6 \pm 3\sqrt{2}$

13. $(2x - 3)^2 = 8$
$2x - 3 = \pm\sqrt{8}$
$2x - 3 = \pm 2\sqrt{2}$
$2x = 3 \pm 2\sqrt{2}$
$x = \dfrac{3 \pm 2\sqrt{2}}{2}$

15. $x^2 + 9 = 0$
$x^2 = -9$
$x = \pm\sqrt{-9}$
$x = \pm 3i$

17. $x^2 - 6 = 0$
$$x^2 = 6$$
$$x = \pm\sqrt{6}$$

19. $2z^2 + 16 = 0$
$$2z^2 = -16$$
$$z^2 = -8$$
$$z = \pm\sqrt{-8}$$
$$z = \pm i\sqrt{8}$$
$$z = \pm 2i\sqrt{2}$$

21. $(x-1)^2 = -16$
$$x - 1 = \pm\sqrt{-16}$$
$$x - 1 = \pm 4i$$
$$x = 1 \pm 4i$$

23. $(z+7)^2 = 5$
$$z + 7 = \pm\sqrt{5}$$
$$z = -7 \pm \sqrt{5}$$

25. $(x+3)^2 = -8$
$$x + 3 = \pm\sqrt{-8}$$
$$x + 3 = \pm i\sqrt{8}$$
$$x + 3 = \pm 2i\sqrt{2}$$
$$x = -3 \pm 2i\sqrt{2}$$

27. $x^2 + 16x + \left(\dfrac{16}{2}\right)^2 = x^2 + 16x + 64$
$$= (x+8)^2$$

29. $z^2 - 12z + \left(\dfrac{12}{2}\right)^2 = z^2 - 12z + 36$
$$= (z-6)^2$$

31. $p^2 + 9p + \left(\dfrac{9}{2}\right)^2 = p^2 + 9p + \dfrac{81}{4}$
$$= \left(p + \dfrac{9}{2}\right)^2$$

33. $x^2 + x + \left(\dfrac{1}{2}\right)^2 = x^2 + 16x + \dfrac{1}{4}$
$$= \left(x + \dfrac{1}{2}\right)^2$$

35. $x^2 + 8x = -15$
$$x^2 + 8x + \left(\dfrac{8}{2}\right)^2 = -15 + 16$$
$$x^2 + 8x + 16 = 1$$
$$(x+4)^2 = 1$$
$$x + 4 = \pm\sqrt{1}$$
$$x = -4 \pm 1$$
$$x = -5 \ \text{ or } \ x = -3$$

37. $x^2 + 6x + 2 = 0$
$$x^2 + 6x = -2$$
$$x^2 + 6x + \left(\dfrac{6}{2}\right)^2 = -2 + 9$$
$$x^2 + 6x + 9 = 7$$
$$(x+3)^2 = 7$$
$$x + 3 = \pm\sqrt{7}$$
$$x = -3 \pm \sqrt{7}$$

39. $x^2 + x - 1 = 0$
$$x^2 + x = 1$$
$$x^2 + x + \left(\dfrac{1}{2}\right)^2 = -2 + \dfrac{1}{4}$$
$$x^2 + x + \dfrac{1}{4} = \dfrac{5}{4}$$
$$\left(x + \dfrac{1}{2}\right)^2 = \dfrac{5}{4}$$
$$x + \dfrac{1}{2} = \pm\sqrt{\dfrac{5}{4}}$$
$$x = -\dfrac{1}{2} \pm \dfrac{\sqrt{5}}{2} = \dfrac{-1 \pm \sqrt{5}}{2}$$

41.
$$x^2 + 2x - 5 = 0$$
$$x^2 + 2x = 5$$
$$x^2 + 2x + \left(\frac{2}{2}\right)^2 = 5 + 1$$
$$x^2 + 2x + 1 = 6$$
$$(x+1)^2 = 6$$
$$x + 1 = \pm\sqrt{6}$$
$$x = -1 \pm \sqrt{6}$$

43.
$$3p^2 - 12p + 2 = 0$$
$$3p^2 - 12p = -2$$
$$p^2 - 4p = -\frac{2}{3}$$
$$p^2 - 4p + \left(\frac{-4}{2}\right)^2 = -\frac{2}{3} + 4$$
$$(p-2)^2 = \frac{10}{3}$$
$$p - 2 = \pm\sqrt{\frac{10}{3}}$$
$$p - 2 = \pm\frac{\sqrt{10} \cdot \sqrt{3}}{\sqrt{3} \cdot \sqrt{3}}$$
$$p - 2 = \pm\frac{\sqrt{30}}{3}$$
$$p = 2 \pm \frac{\sqrt{30}}{3} = \frac{6 \pm \sqrt{30}}{3}$$

45.
$$4y^2 - 12y - 2 = 0$$
$$4y^2 - 12y = 2$$
$$y^2 - 3y = \frac{1}{2}$$
$$y^2 - 3y + \left(\frac{-3}{2}\right)^2 = \frac{1}{2} + \frac{9}{4}$$
$$y^2 - 3y + \frac{9}{4} = \frac{11}{4}$$
$$\left(y - \frac{3}{2}\right)^2 = \frac{11}{4}$$
$$y - \frac{3}{2} = \pm\sqrt{\frac{11}{4}}$$
$$y = \frac{3}{2} \pm \frac{\sqrt{11}}{2} = \frac{3 \pm \sqrt{11}}{2}$$

47.
$$2x^2 + 7x = 4$$
$$x^2 + \frac{7}{2}x = 2$$
$$x^2 + \frac{7}{2}x + \left(\frac{\frac{7}{2}}{2}\right)^2 = 2 + \frac{49}{16}$$
$$x^2 + \frac{7}{2}x + \frac{49}{16} = \frac{81}{16}$$
$$\left(x + \frac{7}{4}\right)^2 = \frac{81}{16}$$
$$x + \frac{7}{4} = \pm\sqrt{\frac{81}{16}}$$
$$x = -\frac{7}{4} \pm \frac{9}{4} = \frac{-7 \pm 9}{4}$$
$$x = -4, \frac{1}{2}$$

49.
$$x^2 - 4x - 5 = 0$$
$$x^2 - 4x = 5$$
$$x^2 - 4x + \left(\frac{-4}{2}\right)^2 = 5 + 4$$
$$x^2 - 4x + 4 = 9$$
$$(x-2)^2 = 9$$
$$x = 2 \pm 3$$
$$x = -1, 5$$

51.
$$x^2 + 8x + 1 = 0$$
$$x^2 + 8x = -1$$
$$x^2 + 8x + \left(\frac{8}{2}\right)^2 = -1 + 16$$
$$x^2 + 8x + 16 = 15$$
$$(x+4)^2 = 15$$
$$x + 4 = \pm\sqrt{15}$$
$$x = -4 \pm \sqrt{15}$$

53.
$$3y^2 + 6y - 4 = 0$$
$$3y^2 + 6y = 4$$
$$y^2 + 2y = \frac{4}{3}$$
$$y^2 + 2y + \left(\frac{2}{2}\right)^2 = \frac{4}{3} + 1$$
$$y^2 + 2y + 1 = \frac{7}{3}$$
$$(y+1)^2 = \frac{7}{3}$$
$$y + 1 = \pm\sqrt{\frac{7}{3}}$$
$$y + 1 = \pm\frac{\sqrt{7}\cdot\sqrt{3}}{\sqrt{3}\cdot\sqrt{3}}$$
$$y + 1 = \pm\frac{\sqrt{21}}{3}$$
$$y = -1 \pm \frac{\sqrt{21}}{3} = \frac{-3 \pm \sqrt{21}}{3}$$

55.
$$2x^2 - 3x - 5 = 0$$
$$2x^2 - 3x = 5$$
$$x^2 - \frac{3}{2}x = \frac{5}{2}$$
$$x^2 - \frac{3}{2}x + \left(\frac{\frac{3}{2}}{2}\right)^2 = \frac{5}{2} + \frac{9}{16}$$
$$x^2 - \frac{3}{2}x + \frac{9}{16} = \frac{49}{16}$$
$$\left(x - \frac{3}{4}\right)^2 = \frac{49}{16}$$
$$x - \frac{3}{4} = \pm\sqrt{\frac{49}{16}}$$
$$x = \frac{3}{4} \pm \frac{7}{4} = \frac{3 \pm 7}{4}$$
$$x = -1, \frac{5}{2}$$

57.
$$y^2 + 2y + 2 = 0$$
$$y^2 + 2y = -2$$
$$y^2 + 2y + \left(\frac{2}{2}\right)^2 = -2 + 1$$
$$y^2 + 2y + 1 = -1$$
$$(y+1)^2 = -1$$
$$y + 1 = \pm\sqrt{-1}$$
$$y = -1 \pm i$$

59.
$$x^2 - 6x + 3 = 0$$
$$x^2 - 6x = -3$$
$$x^2 - 6x + \left(\frac{-6}{2}\right)^2 = -3 + 9$$
$$x^2 - 6x + 9 = 6$$
$$(x-3)^2 = 6$$
$$x - 3 = \pm\sqrt{6}$$
$$x = 3 \pm \sqrt{6}$$

61.
$$2a^2 + 8a = -12$$
$$a^2 + 4a = -6$$
$$a^2 + 4a + \left(\frac{4}{2}\right)^2 = -6 + 4$$
$$a^2 + 4a + 4 = -2$$
$$(a+2)^2 = -2$$
$$a + 2 = \pm\sqrt{-2}$$
$$a + 2 = \pm i\sqrt{2}$$
$$a = -2 \pm i\sqrt{2}$$

63.
$$5x^2 + 15x - 1 = 0$$
$$5x^2 + 15x = 1$$
$$x^2 + 3x = \frac{1}{5}$$
$$x^2 + 3x + \left(\frac{3}{2}\right)^2 = \frac{1}{5} + \frac{9}{4}$$
$$x^2 + 3x + \frac{9}{4} = \frac{49}{20}$$
$$\left(x + \frac{3}{2}\right)^2 = \frac{49}{20}$$
$$x + \frac{3}{2} = \pm\sqrt{\frac{49}{20}}$$
$$x + \frac{3}{2} = \pm\frac{7}{\sqrt{20}}$$
$$x + \frac{3}{2} = \pm\frac{7}{2\sqrt{5}}$$
$$x + \frac{3}{2} = \pm\frac{7 \cdot \sqrt{5}}{2\sqrt{5} \cdot \sqrt{5}}$$
$$x + \frac{3}{2} = \pm\frac{7\sqrt{5}}{10}$$
$$x = -\frac{3}{2} \pm \frac{7\sqrt{5}}{10} = \frac{-15 \pm 7\sqrt{5}}{10}$$

65.
$$2x^2 - x + 6 = 0$$
$$2x^2 - x = -6$$
$$x^2 - \frac{1}{2}x = -3$$
$$x^2 - \frac{1}{2}x + \left(\frac{-\frac{1}{2}}{2}\right)^2 = -3 + \frac{1}{16}$$
$$x^2 - \frac{1}{2}x + \frac{1}{16} = -\frac{47}{16}$$
$$\left(x - \frac{1}{4}\right)^2 = -\frac{47}{16}$$
$$x - \frac{1}{4} = \pm\sqrt{-\frac{47}{16}}$$
$$x - \frac{1}{4} = \pm\frac{\sqrt{47}}{4}i$$
$$x = \frac{1}{4} \pm \frac{\sqrt{47}}{4}i$$

67.
$$x^2 + 10x + 28 = 0$$
$$x^2 + 10x = -28$$
$$x^2 + 10x + \left(\frac{10}{2}\right)^2 = -28 + 25$$
$$(x + 5)^2 = -3$$
$$x + 5 = \pm\sqrt{-3}$$
$$x = -5 \pm i\sqrt{3}$$

69.
$$z^2 + 3z - 4 = 0$$
$$z^2 + 3z = 4$$
$$z^2 + 3z + \left(\frac{3}{2}\right)^2 = 4 + \frac{9}{4}$$
$$z^2 + 3z + \frac{9}{4} = \frac{25}{4}$$
$$\left(z + \frac{3}{2}\right)^2 = \frac{25}{4}$$
$$z + \frac{3}{2} = \pm\sqrt{\frac{25}{4}}$$
$$z = -\frac{3}{2} \pm \frac{5}{2} = \frac{-3 \pm 5}{2}$$
$$z = -4, 1$$

71.
$$2x^2 - 4x + 3 = 0$$
$$2x^2 - 4x = -3$$
$$x^2 - 2x = -\frac{3}{2}$$
$$x^2 - 2x + \left(\frac{-2}{2}\right)^2 = -\frac{3}{2} + 1$$
$$x^2 - 2x + 1 = -\frac{1}{2}$$
$$(x - 1)^2 = -\frac{1}{2}$$
$$x - 1 = \pm\sqrt{-\frac{1}{2}}$$
$$x - 1 = \pm\frac{1}{\sqrt{2}}i$$
$$x - 1 = \pm\frac{1 \cdot \sqrt{2}}{\sqrt{2} \cdot \sqrt{2}}i$$
$$x - 1 = \pm\frac{\sqrt{2}}{2}i$$
$$x = 1 \pm \frac{\sqrt{2}}{2}i$$

73.
$$3x^2 + 3x = 5$$
$$x^2 + x = \frac{5}{3}$$
$$x^2 + x + \left(\frac{1}{2}\right)^2 = \frac{5}{3} + \frac{1}{4}$$
$$x^2 + x + \frac{1}{4} = \frac{23}{12}$$
$$\left(x + \frac{1}{2}\right)^2 = \frac{23}{12}$$
$$x + \frac{1}{2} = \pm\sqrt{\frac{23}{12}}$$
$$x + \frac{1}{2} = \pm\frac{\sqrt{23}}{2\sqrt{3}}$$
$$x + \frac{1}{2} = \pm\frac{\sqrt{23} \cdot \sqrt{3}}{2\sqrt{3} \cdot \sqrt{3}}$$
$$x + \frac{1}{2} = \pm\frac{\sqrt{69}}{6}$$
$$x = -\frac{1}{2} \pm \frac{\sqrt{69}}{6} = \frac{-3 \pm \sqrt{69}}{6}$$

75.
$$A = P(1+r)^t$$
$$4320 = 3000(1+r)^2$$
$$\frac{4320}{3000} = (1+r)^2$$
$$1.44 = (1+r)^2$$
$$\pm\sqrt{1.44} = 1+r$$
$$\pm1.2 = 1+r$$
$$-1 \pm 1.2 = r$$
$$-2.2 = r \quad \text{or} \quad 0.2 = r$$
Rate cannot be negative, so the rate is
$r = 0.2 = 20\%$.

77.
$$A = P(1+r)^t$$
$$1000 = 810(1+r)^2$$
$$\frac{1000}{810} = (1+r)^2$$
$$\frac{100}{81} = (1+r)^2$$
$$\pm\sqrt{\frac{100}{81}} = 1+r$$
$$\pm\frac{10}{9} = 1+r$$
$$-1 \pm \frac{10}{9} = r$$
$$-\frac{19}{9} = r \quad \text{or} \quad \frac{1}{9} = r$$
Rate cannot be negative, so the rate is
$r = \frac{1}{9}$, or $11\frac{1}{9}\%$.

79. Answers may vary.

81. Simple

83. $\frac{3}{5} + \sqrt{\frac{16}{25}} = \frac{3}{5} + \frac{4}{5} = \frac{7}{5}$

85. $\frac{9}{10} - \sqrt{\frac{49}{100}} = \frac{9}{10} - \frac{7}{10} = \frac{2}{10} = \frac{1}{5}$

87. $\frac{10 - 20\sqrt{3}}{2} = \frac{10}{2} - \frac{20\sqrt{3}}{2} = 5 - 10\sqrt{3}$

89. $\frac{12 - 8\sqrt{7}}{16} = \frac{12}{16} - \frac{8\sqrt{7}}{16}$
$$= \frac{3}{4} - \frac{\sqrt{7}}{2}$$
$$= \frac{3}{4} - \frac{2\sqrt{7}}{4} = \frac{3 - 2\sqrt{7}}{4}$$

91. $\sqrt{b^2 - 4ac} = \sqrt{(6)^2 - 4(1)(2)}$
$$= \sqrt{36 - 8}$$
$$= \sqrt{28}$$
$$= \sqrt{4 \cdot 7}$$
$$= 2\sqrt{7}$$

93. $\sqrt{b^2 - 4ac} = \sqrt{(-3)^2 - 4(1)(-1)}$

$\qquad\qquad\quad = \sqrt{9 + 4}$

$\qquad\qquad\quad = \sqrt{13}$

95. $y^2 + \underline{\quad} + 9$

$\left(\dfrac{b}{2}\right)^2 = 9$

$\dfrac{b}{2} = \pm\sqrt{9}$

$\dfrac{b}{2} = \pm 3$

$b = \pm 6$

Answer: $\pm 6y$

97. $x^2 + \underline{\quad} + \dfrac{1}{4}$

$\left(\dfrac{b}{2}\right)^2 = \dfrac{1}{4}$

$\dfrac{b}{2} = \pm\sqrt{\dfrac{1}{4}}$

$\dfrac{b}{2} = \pm\dfrac{1}{2}$

$b = \pm 1$

Answer: $\pm x$

99. $s(t) = 16t^2$

$1053 = 16t^2$

$t^2 = \dfrac{1053}{16}$

$t = \pm\sqrt{\dfrac{1053}{16}}$

$t \approx 8.11$ or -8.11 (disregard)

It would take 8.11 seconds.

101. $s(t) = 16t^2$

$725 = 16t^2$

$t^2 = \dfrac{725}{16}$

$t = \pm\sqrt{\dfrac{725}{16}}$

$t \approx 6.73$ or -6.73 (disregard)

It would take 6.73 seconds.

103. $A = \pi r^2$

$36\pi = \pi r^2$

$r^2 = \dfrac{36\pi}{\pi}$

$r^2 = 36$

$r = \pm\sqrt{36}$

$r = 6$ or -6 (disregard)

The radius is 6 inches.

105. $a^2 + b^2 = c^2$

$(4x)^2 + (3x)^2 = 27^2$

$16x^2 + 9x^2 = 729$

$25x^2 = 729$

$x^2 = \dfrac{729}{25}$

$x = \pm\sqrt{\dfrac{729}{25}} = \pm\dfrac{27}{5}$

$x = 5.4$ or -5.4 (disregard)

$3x = 3(5.4) = 16.2$

$4x = 4(5.4) = 21.6$

The sides are 16.2 in. and 21.6 in.

107. $p = -x^2 + 15$

$7 = -x^2 + 15$

$x^2 = 8$

$x = \pm\sqrt{8}$

$x \approx \pm 2.828$

Demand cannot be negative. Therefore, the demand is approximately 2.828 thousand (or 2828) units.

Section 11.2

Mental Math

1. $x^2 + 3x + 1$

$a = 1, b = 3, c = 1$

2. $2x^2 - 5x - 7$

$a = 2, b = -5, c = -7$

3. $7x^2 - 4 = 0$

$a = 7, b = 0, c = -4$

4. $x^2 + 9 = 0$

$a = 1, b = 0, c = 9$

5. $6x^2 - x = 0$

$a = 6, b = -1, c = 0$

6. $5x^2 + 3x = 0$

$a = 5, b = 3, c = 0$

Exercise Set 11.2

1. $m^2 + 5m - 6 = 0$

$a = 1, b = 5, c = -6$

$m = \dfrac{-5 \pm \sqrt{(5)^2 - 4(1)(-6)}}{2(1)}$

$= \dfrac{-5 \pm \sqrt{25 + 24}}{2}$

$= \dfrac{-5 \pm \sqrt{49}}{2}$

$= \dfrac{-5 \pm 7}{2} = -6 \text{ or } 1$

The solutions are –6 and 1.

3. $2y = 5y^2 - 3$

$5y^2 - 2y - 3 = 0$

$a = 5, b = -2, c = -3$

$y = \dfrac{2 \pm \sqrt{(-2)^2 - 4(5)(-3)}}{2(5)}$

$= \dfrac{2 \pm \sqrt{4 + 60}}{10}$

$= \dfrac{2 \pm \sqrt{64}}{10}$

$= \dfrac{2 \pm 8}{10} = -\dfrac{3}{5} \text{ or } 1$

The solutions are $-\dfrac{3}{5}$ and 1.

5. $x^2 - 6x + 9 = 0$

$a = 1, b = -6, c = 9$

$x = \dfrac{6 \pm \sqrt{(-6)^2 - 4(1)(9)}}{2(1)}$

$= \dfrac{6 \pm \sqrt{36 - 36}}{2}$

$= \dfrac{6 \pm \sqrt{0}}{2} = \dfrac{6}{2} = 3$

The solution is 3.

7. $x^2 + 7x + 4 = 0$

$a = 1, b = 7, c = 4$

$x = \dfrac{-7 \pm \sqrt{(7)^2 - 4(1)(4)}}{2(1)}$

$= \dfrac{-7 \pm \sqrt{49 - 16}}{2}$

$= \dfrac{-7 \pm \sqrt{33}}{2}$

The solutions are $\dfrac{-7 + \sqrt{33}}{2}$ and $\dfrac{-7 - \sqrt{33}}{2}$.

9. $8m^2 - 2m = 7$

$8m^2 - 2m - 7 = 0$

$a = 8, b = -2, c = -7$

$m = \dfrac{2 \pm \sqrt{(-2)^2 - 4(8)(-7)}}{2(8)}$

$= \dfrac{2 \pm \sqrt{4 + 224}}{16}$

$= \dfrac{2 \pm \sqrt{228}}{16}$

$= \dfrac{2 \pm \sqrt{4 \cdot 57}}{16}$

$= \dfrac{2 \pm 2\sqrt{57}}{16} = \dfrac{1 \pm \sqrt{57}}{8}$

The solutions are $\dfrac{1 + \sqrt{57}}{8}$ and $\dfrac{1 - \sqrt{57}}{8}$.

11. $3m^2 - 7m = 3$
$3m^2 - 7m - 3 = 0$
$a = 3, b = -7, c = -3$

$m = \dfrac{7 \pm \sqrt{(-7)^2 - 4(3)(-3)}}{2(3)}$

$= \dfrac{7 \pm \sqrt{49 + 36}}{6}$

$= \dfrac{7 \pm \sqrt{85}}{6}$

The solutions are $\dfrac{7 + \sqrt{85}}{6}$ and $\dfrac{7 - \sqrt{85}}{6}$.

13. $\dfrac{1}{2}x^2 - x - 1 = 0$
$x^2 - 2x - 2 = 0$
$a = 1, b = -2, c = -2$

$x = \dfrac{2 \pm \sqrt{(-2)^2 - 4(1)(-2)}}{2(1)}$

$= \dfrac{2 \pm \sqrt{4 + 8}}{2}$

$= \dfrac{2 \pm \sqrt{12}}{2}$

$= \dfrac{2 \pm 2\sqrt{3}}{2} = 1 \pm \sqrt{3}$

The solutions are $1 + \sqrt{3}$ and $1 - \sqrt{3}$.

15. $\dfrac{2}{5}y^2 + \dfrac{1}{5}y = \dfrac{3}{5}$
$2y^2 + y - 3 = 0$
$a = 2, b = 1, c = -3$

$y = \dfrac{-1 \pm \sqrt{(1)^2 - 4(2)(-3)}}{2(2)}$

$= \dfrac{-1 \pm \sqrt{1 + 24}}{4}$

$= \dfrac{-1 \pm \sqrt{25}}{4}$

$= \dfrac{-1 \pm 5}{4} = -\dfrac{3}{2}$ or 1

The solutions are $-\dfrac{3}{2}$ and 1.

17. $\dfrac{1}{3}y^2 - y - \dfrac{1}{6} = 0$
$2y^2 - 6y - 1 = 0$
$a = 2, b = -6, c = -1$

$x = \dfrac{6 \pm \sqrt{(-6)^2 - 4(2)(-1)}}{2(2)}$

$= \dfrac{6 \pm \sqrt{36 + 8}}{4}$

$= \dfrac{6 \pm \sqrt{44}}{4}$

$= \dfrac{6 \pm 2\sqrt{11}}{4} = \dfrac{3 \pm \sqrt{11}}{2}$

The solutions are $\dfrac{3 + \sqrt{11}}{2}$ and $\dfrac{3 - \sqrt{11}}{2}$.

19. $m^2 + 5m - 6 = 0$
$(m + 6)(m - 1) = 0$
$m + 6 = 0$ or $m - 1 = 0$
  $m = -6$ or $m = 1$

The results are the same. Answers may vary.

21. $6 = -4x^2 + 3x$
$4x^2 - 3x + 6 = 0$
$a = 4, b = -3, c = 6$

$x = \dfrac{3 \pm \sqrt{(-3)^2 - 4(4)(6)}}{2(4)}$

$= \dfrac{3 \pm \sqrt{9 - 96}}{8}$

$= \dfrac{3 \pm \sqrt{-87}}{8}$

$= \dfrac{3 \pm i\sqrt{87}}{8} = \dfrac{3}{8} \pm \dfrac{\sqrt{87}}{8}i$

The solutions are $\dfrac{3 + i\sqrt{87}}{8}$ and $\dfrac{3 - i\sqrt{87}}{8}$.

23. $(x + 5)(x - 1) = 2$
$x^2 + 4x - 5 = 2$
$x^2 + 4x - 7 = 0$
$a = 1, b = 4, c = -7$

$$x = \frac{-4 \pm \sqrt{(4)^2 - 4(1)(-7)}}{2(1)}$$
$$= \frac{-4 \pm \sqrt{16 + 28}}{2}$$
$$= \frac{-4 \pm \sqrt{44}}{2}$$
$$= \frac{-4 \pm 2\sqrt{11}}{2} = -2 \pm \sqrt{11}$$

The solutions are $-2 + \sqrt{11}$ and $-2 - \sqrt{11}$.

25. $10y^2 + 10y + 3 = 0$
$a = 10, b = 10, c = 3$
$$y = \frac{-10 \pm \sqrt{(10)^2 - 4(10)(3)}}{2(10)}$$
$$= \frac{-10 \pm \sqrt{100 - 120}}{20}$$
$$= \frac{-10 \pm \sqrt{-20}}{20}$$
$$= \frac{-10 \pm i\sqrt{4 \cdot 5}}{20}$$
$$= \frac{-10 \pm 2i\sqrt{5}}{20} =$$
$$= \frac{-5 \pm i\sqrt{5}}{10} = -\frac{1}{2} \pm \frac{\sqrt{5}}{10}i$$

The solutions are $\frac{-5 + i\sqrt{5}}{10}$ and $\frac{-5 - i\sqrt{5}}{10}$.

27. $9x - 2x^2 + 5 = 0$
$-2x^2 + 9x + 5 = 0$
$a = -2, b = 9, c = 5$
$b^2 - 4ac = 9^2 - 4(-2)(5)$
$= 81 + 40$
$= 121 > 0$
Therefore, there are two real solutions.

29. $4x^2 + 12x = -9$
$4x^2 + 12x + 9 = 0$
$a = 4, b = 12, c = 9$

$b^2 - 4ac = 12^2 - 4(4)(9)$
$= 144 - 144$
$= 0$
Therefore, there is one real solution.

31. $3x = -2x^2 + 7$
$2x^2 + 3x - 7 = 0$
$a = 2, b = 3, c = -7$
$b^2 - 4ac = 3^2 - 4(2)(-7)$
$= 9 + 56$
$= 65 > 0$
Therefore, there are two real solutions.

33. $6 = 4x - 5x^2$
$5x^2 - 4x + 6 = 0$
$a = 5, b = -4, c = 6$
$b^2 - 4ac = (-4)^2 - 4(5)(6)$
$= 16 - 120$
$= -104 < 0$
Therefore, there are two complex but not real solutions.

35. $x^2 + 5x = -2$
$x^2 + 5x + 2 = 0$
$a = 1, b = 5, c = 2$
$$x = \frac{-5 \pm \sqrt{(5)^2 - 4(1)(2)}}{2(1)}$$
$$= \frac{-5 \pm \sqrt{25 - 8}}{2}$$
$$= \frac{-5 \pm \sqrt{17}}{2}$$
The solutions are $\frac{-5 + \sqrt{17}}{2}$ and $\frac{-5 - \sqrt{17}}{2}$.

37. $(m + 2)(2m - 6) = 5(m - 1) - 12$
$2m^2 - 6m + 4m - 12 = 5m - 5 - 12$
$2m^2 - 7m + 5 = 0$
$a = 2, b = -7, c = 5$

$$m = \frac{7 \pm \sqrt{(-7)^2 - 4(2)(5)}}{2(2)}$$
$$= \frac{7 \pm \sqrt{49 - 40}}{4}$$
$$= \frac{7 \pm \sqrt{9}}{4}$$
$$= \frac{7 \pm 3}{4} = 1 \text{ or } \frac{5}{2}$$

The solutions are 1 and $\frac{5}{2}$.

39. $\frac{x^2}{3} - x = \frac{5}{3}$
$x^2 - 3x = 5$
$x^2 - 3x - 5 = 0$
$a = 1, b = -3, c = -5$
$$x = \frac{3 \pm \sqrt{(-3)^2 - 4(1)(-5)}}{2(1)}$$
$$= \frac{3 \pm \sqrt{9 + 20}}{2}$$
$$= \frac{3 \pm \sqrt{29}}{2}$$

The solutions are $\frac{3 + \sqrt{29}}{2}$ and $\frac{3 - \sqrt{29}}{2}$.

41. $x(6x + 2) - 3 = 0$
$6x^2 + 2x - 3 = 0$
$a = 6, b = 2, c = -3$
$$x = \frac{-2 \pm \sqrt{(2)^2 - 4(6)(-3)}}{2(6)}$$
$$= \frac{-2 \pm \sqrt{4 + 72}}{12}$$
$$= \frac{-2 \pm \sqrt{76}}{12}$$
$$= \frac{-2 \pm \sqrt{4 \cdot 19}}{12}$$
$$= \frac{-2 \pm 2\sqrt{19}}{12} = \frac{-1 \pm \sqrt{19}}{6}$$

The solutions are $\frac{-1 + \sqrt{19}}{6}$ and $\frac{-1 - \sqrt{19}}{6}$.

43. $x^2 + 6x + 13 = 0$
$a = 1, b = 6, c = 13$
$$x = \frac{-6 \pm \sqrt{(6)^2 - 4(1)(13)}}{2(1)}$$
$$= \frac{-6 \pm \sqrt{36 - 52}}{2}$$
$$= \frac{-6 \pm \sqrt{-16}}{2}$$
$$= \frac{-6 \pm 4i}{2} = -3 \pm 2i$$

The solutions are $-3 + 2i$ and $-3 - 2i$.

45. $\frac{2}{5}y^2 + \frac{1}{5}y + \frac{3}{5} = 0$
$2y^2 + y + 3 = 0$
$a = 2, b = 1, c = 3$
$$y = \frac{-1 \pm \sqrt{(1)^2 - 4(2)(3)}}{2(2)}$$
$$= \frac{-1 \pm \sqrt{1 - 24}}{4}$$
$$= \frac{-1 \pm \sqrt{-23}}{4}$$
$$= \frac{-1 \pm i\sqrt{23}}{4} = -\frac{1}{4} \pm \frac{\sqrt{23}}{4}i$$

The solutions are $\frac{-1 + i\sqrt{23}}{4}$ and $\frac{-1 - i\sqrt{23}}{4}$.

47. $\frac{1}{2}y^2 = y - \frac{1}{2}$
$y^2 = 2y - 1$
$y^2 - 2y + 1 = 0$
$a = 1, b = -2, c = 1$
$$y = \frac{2 \pm \sqrt{(-2)^2 - 4(1)(1)}}{2(1)}$$
$$= \frac{2 \pm \sqrt{4 - 4}}{2}$$
$$= \frac{2 \pm \sqrt{0}}{2} = \frac{2}{2} = 1$$

The solution is 1.

49.
$$(n-2)^2 = 15n$$
$$n^2 - 4n + 4 = 15n$$
$$n^2 - 19n + 4 = 0$$
$$a = 1, b = -19, c = 4$$
$$n = \frac{19 \pm \sqrt{(-19)^2 - 4(1)(4)}}{2(1)}$$
$$= \frac{19 \pm \sqrt{361 - 16}}{2}$$
$$= \frac{19 \pm \sqrt{345}}{2}$$

The solutions are $\dfrac{19 + \sqrt{345}}{2}$ and

$\dfrac{19 - \sqrt{345}}{2}$.

51.
$$(x+8)^2 + x^2 = 36^2$$
$$(x^2 + 16x + 64) + x^2 = 1296$$
$$2x^2 + 16x - 1232 = 0$$
$$a = 2, b = 16, c = -1232$$
$$x = \frac{-16 \pm \sqrt{(16)^2 - 4(2)(-1232)}}{2(2)}$$
$$= \frac{-16 \pm \sqrt{10,112}}{4}$$
$$x \approx 21 \text{ or } x \approx -29 \text{ (disregard)}$$
$$x + (x+8) = 21 + 21 + 8 = 50$$
$$50 - 36 = 14$$
They saved about 14 feet of walking distance.

53. Let x = length of leg. Then
$x + 2$ = length of hypotenuse
$$x^2 + x^2 = (x+2)^2$$
$$2x^2 = x^2 + 4x + 4$$
$$x^2 - 4x - 4 = 0$$
$$a = 1, b = -4, c = -4$$

$$x = \frac{4 \pm \sqrt{(-4)^2 - 4(1)(-4)}}{2(1)}$$
$$= \frac{4 \pm \sqrt{32}}{2}$$
$$= \frac{4 \pm 4\sqrt{2}}{2}$$
$$= 2 \pm 2\sqrt{2} \text{ (disregard the negative)}$$
$$= 2 + 2\sqrt{2}$$
The sides measure $2 + 2\sqrt{2}$ cm,
$2 + 2\sqrt{2}$ cm, and $4 + 2\sqrt{2}$ cm.

55. Let x = width; then $x + 10$ = length.
Area = length · width
$$400 = (x+10)x$$
$$0 = x^2 + 10x - 400$$
$$a = 1, b = 10, c = -400$$
$$x = \frac{-10 \pm \sqrt{(10)^2 - 4(1)(-400)}}{2(1)}$$
$$= \frac{-10 \pm \sqrt{1700}}{2}$$
$$= \frac{-10 \pm 10\sqrt{17}}{2}$$
$$= -5 \pm 5\sqrt{17}$$
Disregard the negative length. The width is $-5 + 5\sqrt{17}$ ft and the length is $5 + 5\sqrt{17}$ ft.

57. a. Let x = length.
$$x^2 + x^2 = 100^2$$
$$2x^2 - 10,000 = 0$$
$$a = 2, b = 0, c = -10,000$$
$$x = \frac{0 \pm \sqrt{(0)^2 - 4(2)(-10,000)}}{2(2)}$$
$$= \frac{\pm \sqrt{80,000}}{4}$$
$$= \frac{\pm 200\sqrt{2}}{4} = \pm 50\sqrt{2}$$
Disregard the negative length. The side measures $50\sqrt{2}$ meters.

b. Area $= s^2$

$$= \left(50\sqrt{2}\right)^2$$
$$= 2500(2)$$
$$= 5000$$

The area is 5000 square meters.

59. Let $w =$ width; then $w + 1.1 =$ height.
Area $=$ length $\cdot$ width
$$1439.9 = (w + 1.1)w$$
$$0 = w^2 + 1.1w - 1439.9$$
$$a = 1, b = 1.1, c = -1439.9$$
$$w = \frac{-1.1 \pm \sqrt{(1.1)^2 - 4(1)(-1439.9)}}{2(1)}$$
$$= \frac{-1.1 \pm \sqrt{5760.81}}{2}$$
$$= 37.4 \text{ or } -3.608 \text{ (disregard)}$$

Its width is 37.4 ft and its height is 38.5 ft.

61.
$$\frac{x-1}{1} = \frac{1}{x}$$
$$x(x-1) = 1$$
$$x^2 - x - 1 = 0$$
$$a = 1, b = -1, c = -1$$
$$x = \frac{1 \pm \sqrt{(-1)^2 - 4(1)(-1)}}{2(1)}$$
$$= \frac{1 \pm \sqrt{5}}{2} \text{ (disregard the negative)}$$

The value is $\dfrac{1 + \sqrt{5}}{2}$.

63. $h(t) = -16t^2 + 20t + 1100$
$$0 = -16t^2 - 20t + 1100$$
$$a = -16, b = -20, c = 1100$$
$$t = \frac{-20 \pm \sqrt{(20)^2 - 4(-16)(1100)}}{2(-16)}$$
$$= \frac{-20 \pm \sqrt{70,800}}{-32}$$
$$\approx 8.9 \text{ or } -7.7 \text{ (disregard)}$$

It will take about 8.9 seconds.

65. $h(t) = -16t^2 - 20t + 180$
$$0 = -16t^2 - 20t + 180$$
$$a = -16, b = -20, c = 180$$
$$t = \frac{20 \pm \sqrt{(-20)^2 - 4(-16)(180)}}{2(-16)}$$
$$= \frac{20 \pm \sqrt{11,920}}{-32}$$
$$\approx 2.8 \text{ or } -4.0 \text{ (disregard)}$$

It will take about 2.8 seconds.

67.
$$\sqrt{5x - 2} = 3$$
$$\left(\sqrt{5x - 2}\right)^2 = 3^2$$
$$5x - 2 = 9$$
$$5x = 11$$
$$x = \frac{11}{5}$$

69.
$$\frac{1}{x} + \frac{2}{5} = \frac{7}{x}$$
$$5x\left(\frac{1}{x} + \frac{2}{5}\right) = 5x\left(\frac{7}{x}\right)$$
$$5 + 2x = 35$$
$$2x = 30$$
$$x = 15$$

71. $x^4 + x^2 - 20 = (x^2 + 5)(x^2 - 4)$
$$= (x^2 + 5)(x + 2)(x - 2)$$

73. $z^4 - 13z^2 + 36$
$$= (z^2 - 9)(z^2 - 4)$$
$$= (z + 3)(z - 3)(z + 2)(z - 2)$$

75. $2x^2 - 6x + 3 = 0$
$$a = 2, b = -6, c = 3$$
$$x = \frac{6 \pm \sqrt{(-6)^2 - 4(2)(3)}}{2(2)}$$
$$= \frac{6 \pm \sqrt{12}}{4}$$
$$\approx 0.6 \text{ or } 2.4$$

77. From Sunday to Monday

79. Wednesday

81. $f(x) = 3x^2 - 18x + 56$
$f(4) = 3(4)^2 - 18(4) + 56 = 32$
This answers appears to agree with the graph.

83. $f(x) = 112.5x^2 + 498.7x + 5454$

   a. $x = 2002 - 2000 = 2$
$f(2) = 112.5(2)^2 + 498.7(2) + 5454$
$= 6901.4$

   Their net income was $6901.4 million.

   b. $15,000 = 112.5x^2 + 498.7x + 5454$
$0 = 112.5x^2 + 498.7x - 9546$
$a = 112.5, b = 498.7, c = -9546$

$$x = \frac{-498.7 \pm \sqrt{(498.7)^2 - 4(112.5)(-9546)}}{2(112.5)}$$

$$= \frac{-498.7 \pm \sqrt{4,544,401.69}}{225}$$

≈ 7.26 or -11.69 (disregard)
Their income will be $15,000 million in the year 2007.

85. $\dfrac{-b + \sqrt{b^2 - 4ac}}{2a} + \dfrac{-b - \sqrt{b^2 - 4ac}}{2a}$

$$= \frac{-b + \sqrt{b^2 - 4ac} - b - \sqrt{b^2 - 4ac}}{2a}$$

$$= \frac{-2b}{2a}$$

$$= -\frac{b}{a}$$

87. $3x^2 - \sqrt{12}x + 1 = 0$
$a = 3, b = -\sqrt{12}, c = 1$

$$x = \frac{\sqrt{12} \pm \sqrt{\left(-\sqrt{12}\right)^2 - 4(3)(1)}}{2(3)}$$

$$= \frac{\sqrt{12} \pm \sqrt{12 - 12}}{6}$$

$$= \frac{\sqrt{4 \cdot 3} \pm \sqrt{0}}{6} = \frac{2\sqrt{3}}{6} = \frac{\sqrt{3}}{3}$$

The solution is $\dfrac{\sqrt{3}}{3}$.

89. $x^2 + \sqrt{2}x + 1 = 0$
$a = 1, b = \sqrt{2}, c = 1$

$$x = \frac{-\sqrt{2} \pm \sqrt{\left(\sqrt{2}\right)^2 - 4(1)(1)}}{2(1)}$$

$$= \frac{-\sqrt{2} \pm \sqrt{2 - 4}}{2}$$

$$= \frac{-\sqrt{2} \pm \sqrt{-2}}{2}$$

$$= \frac{-\sqrt{2} \pm i\sqrt{2}}{2} = -\frac{\sqrt{2}}{2} \pm \frac{\sqrt{2}}{2}i$$

The solutions are $\dfrac{-\sqrt{2} + i\sqrt{2}}{2}$ and
$\dfrac{-\sqrt{2} - i\sqrt{2}}{2}$.

91. $2x^2 - \sqrt{3}x - 1 = 0$
$a = 2, b = -\sqrt{3}, c = -1$

$$x = \frac{\sqrt{3} \pm \sqrt{\left(-\sqrt{3}\right)^2 - 4(2)(-1)}}{2(2)}$$

$$= \frac{\sqrt{3} \pm \sqrt{3 + 8}}{4}$$

$$= \frac{\sqrt{3} \pm \sqrt{11}}{4}$$

The solutions are $\dfrac{\sqrt{3} + \sqrt{11}}{4}$ and
$\dfrac{\sqrt{3} - \sqrt{11}}{4}$.

93. Exercise 63:

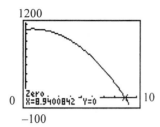

Exercise 65:

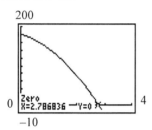

95. $y = 9x - 2x^2 + 5$

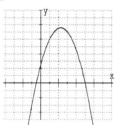

There are two *x*-intercepts.

Exercise Set 11.3

1.
$$2x = \sqrt{10 + 3x}$$
$$4x^2 = 10 + 3x$$
$$4x^2 - 3x - 10 = 0$$
$$(4x + 5)(x - 2) = 0$$
$$4x + 5 = 0 \quad \text{or} \quad x - 2 = 0$$
$$x = -\frac{5}{4} \quad \text{or} \quad x = 2$$

Discard $-\frac{5}{4}$. The solution is 2.

3.
$$x - 2\sqrt{x} = 8$$
$$x - 8 = 2\sqrt{x}$$
$$(x - 8)^2 = \left(2\sqrt{x}\right)^2$$
$$x^2 - 16x + 64 = 4x$$
$$x^2 - 20x + 64 = 0$$
$$(x - 16)(x - 4) = 0$$
$$x - 16 = 0 \quad \text{or} \quad x - 4 = 0$$
$$x = 16 \quad \text{or} \quad x = 4 \text{ (discard)}$$

The solution is 16.

5.
$$\sqrt{9x} = x + 2$$
$$\left(\sqrt{9x}\right)^2 = (x + 2)^2$$
$$9x = x^2 + 4x + 4$$
$$0 = x^2 - 5x + 4$$
$$0 = (x - 4)(x - 1)$$
$$x - 4 = 0 \quad \text{or} \quad x - 1 = 0$$
$$x = 4 \quad \text{or} \quad x = 1$$

The solutions are 1 and 4.

·7. $\dfrac{2}{x} + \dfrac{3}{x - 1} = 1$

Multiply each term by $x(x - 1)$.
$$2(x - 1) + 3x = x(x - 1)$$
$$2x - 2 + 3x = x^2 - x$$
$$0 = x^2 - 6x + 2$$
$$x = \frac{6 \pm \sqrt{(-6)^2 - 4(1)(2)}}{2(1)}$$
$$= \frac{6 \pm \sqrt{28}}{2}$$
$$= \frac{6 \pm 2\sqrt{7}}{2} = 3 \pm \sqrt{7}$$

The solutions are $3 + \sqrt{7}$ and $3 - \sqrt{7}$.

9. $\dfrac{3}{x} + \dfrac{4}{x+2} = 2$

Multiply each term by $x(x+2)$.

$3(x+2) + 4x = 2x(x+2)$

$3x + 6 + 4x = 2x^2 + 4x$

$0 = 2x^2 - 3x - 6$

$x = \dfrac{3 \pm \sqrt{(-3)^2 - 4(2)(-6)}}{2(2)}$

$= \dfrac{3 \pm \sqrt{57}}{4}$

The solutions are $\dfrac{3+\sqrt{57}}{4}$ and $\dfrac{3-\sqrt{57}}{4}$.

11. $\dfrac{7}{x^2 - 5x + 6} = \dfrac{2x}{x-3} - \dfrac{x}{x-2}$

$\dfrac{7}{(x-3)(x-2)} = \dfrac{2x}{x-3} - \dfrac{x}{x-2}$

Multiply each term by $(x-3)(x-2)$.

$7 = 2x(x-2) - x(x-3)$

$7 = 2x^2 - 4x - x^2 + 3x$

$0 = x^2 - x - 7$

$x = \dfrac{1 \pm \sqrt{(-1)^2 - 4(1)(-7)}}{2(1)}$

$= \dfrac{1 \pm \sqrt{29}}{2}$

The solutions are $\dfrac{1+\sqrt{29}}{2}$ and $\dfrac{2-\sqrt{29}}{2}$.

13. $p^4 - 16 = 0$

$(p^2 - 4)(p^2 + 4) = 0$

$(p+2)(p-2)(p^2 + 4) = 0$

$p + 2 = 0 \quad$ or $p - 2 = 0$ or $p^2 + 4 = 0$

$\quad p = -2$ or $\quad\quad p = 2$ or $\quad\quad p^2 = -4$

$\quad\quad\quad\quad\quad\quad\quad\quad\quad\quad\quad p = \pm\sqrt{-4}$

$\quad\quad\quad\quad\quad\quad\quad\quad\quad\quad\quad p = \pm 2i$

The solutions are -2, 2, $-2i$, and $2i$.

15. $4x^4 + 11x^2 = 3$

$4x^2 + 11x^2 - 3 = 0$

$(4x^2 - 1)(x^2 + 3) = 0$

$(2x+1)(2x-1)(x^2 + 3) = 0$

The solutions are $-\dfrac{1}{2}, \dfrac{1}{2}, -i\sqrt{3},$ and $i\sqrt{3}$.

17. $z^4 - 13z^2 + 36 = 0$

$(z^2 - 9)(z^2 - 4) = 0$

$(z+3)(z-3)(z+2)(z-2) = 0$

$z = -3, z = 3, z = -2, z = 2$

The solutions are -3, 3, -2, and 2.

19. $x^{2/3} - 3x^{1/3} - 10 = 0$

Let $y = x^{1/3}$. Then $y^2 = x^{2/3}$ and

$y^2 - 3y - 10 = 0$

$(y-5)(y+2) = 0$

$y - 5 = 0 \quad$ or $\quad y + 2 = 0$

$\quad y = 5 \quad$ or $\quad\quad y = -2$

$\quad x^{1/3} = 5 \quad$ or $\quad x^{1/3} = -2$

$\quad x = 125$ or $\quad\quad x = -8$

The solutions are -8 and 125.

21. $(5n+1)^2 + 2(5n+1) - 3 = 0$

Let $y = 5n + 1$. Then $y^2 = (5n+1)^2$ and

$x^2 + 2x - 3 = 0$

$(x+3)(x-1) = 0$

$y + 3 = 0 \quad$ or $\quad y - 1 = 0$

$\quad y = -3$ or $\quad\quad y = 1$

$5n + 1 = -3$ or $5n + 1 = 1$

$\quad 5n = -4$ or $\quad\quad 5n = 0$

$\quad n = -\dfrac{4}{5} \quad$ or $\quad\quad n = 0$

The solutions are $-\dfrac{4}{5}$ and 0.

23. $2x^{2/3} - 5x^{1/3} = 3$

Let $y = x^{1/3}$. Then $y^2 = x^{2/3}$ and

$$2y^2 - 5y = 3$$
$$2y^2 - 5y - 3 = 0$$
$$(2y+1)(y-3) = 0$$
$$2y+1 = 0 \quad \text{or} \quad y - 3 = 0$$
$$y = -\frac{1}{2} \quad \text{or} \quad y = 3$$
$$x^{1/3} = -\frac{1}{2} \quad \text{or} \quad x^{1/3} = 3$$
$$x = -\frac{1}{8} \quad \text{or} \quad x = 27$$

The solutions are $-\frac{1}{8}$ and 27.

25.
$$1 + \frac{2}{3t-2} = \frac{8}{(3t-2)^2}$$
$$(3t-2)^2 + 2(3t-2) = 8$$
$$(3t-2)^2 + 2(3t-2) - 8 = 0$$

Let $y = 3t - 2$. Then $y^2 = (3t-2)^2$ and

$$y^2 + 2y - 8 = 0$$
$$(y+4)(y-2) = 0$$
$$y + 4 = 0 \quad \text{or} \quad y - 2 = 0$$
$$y = -4 \quad \text{or} \quad y = 2$$
$$3t - 2 = -4 \quad \text{or} \quad 3t - 2 = 2$$
$$3t = -2 \quad \text{or} \quad 3t = 4$$
$$t = -\frac{2}{3} \quad \text{or} \quad t = \frac{4}{3}$$

The solutions are $-\frac{2}{5}$ and $\frac{4}{3}$.

27. $20x^{2/3} - 6x^{1/3} - 2 = 0$

Let $y = x^{1/3}$. Then $y^2 = x^{2/3}$ and

$$20y^2 - 6y - 2 = 0$$
$$2(10y^2 - 3y - 1) = 0$$
$$2(5y+1)(2y-1) = 0$$

$$5y + 1 = 0 \quad \text{or} \quad 2y - 1 = 0$$
$$y = -\frac{1}{5} \quad \text{or} \quad y = \frac{1}{2}$$
$$x^{1/3} = -\frac{1}{5} \quad \text{or} \quad x^{1/3} = \frac{1}{2}$$
$$x = -\frac{1}{125} \quad \text{or} \quad x = \frac{1}{8}$$

The solutions are $\frac{1}{8}$ and $-\frac{1}{125}$.

29.
$$a^4 - 5a^2 + 6 = 0$$
$$(a^2 - 3)(a^2 - 2) = 0$$
$$a^2 - 3 = 0 \quad \text{or} \quad a^2 - 2 = 0$$
$$a^2 = 3 \quad \text{or} \quad a^2 = 2$$
$$a = \pm\sqrt{3} \quad \text{or} \quad a = \pm\sqrt{2}$$

The solutions are $-\sqrt{3}$, $\sqrt{3}$, $-\sqrt{2}$, and $\sqrt{2}$.

31. $\dfrac{2x}{x-2} + \dfrac{x}{x+3} = -\dfrac{5}{x+3}$

Multiply each term by $(x+3)(x-2)$.

$$2x(x+3) + x(x-2) = -5(x-2)$$
$$2x^2 + 6x + x^2 - 2x = -5x + 10$$
$$3x^2 + 9x - 10 = 0$$
$$x = \frac{-9 \pm \sqrt{(9)^2 - 4(3)(-10)}}{2(3)}$$
$$= \frac{-9 \pm \sqrt{201}}{6}$$

The solutions are $\dfrac{-9 + \sqrt{201}}{6}$ and $\dfrac{-9 - \sqrt{201}}{6}$.

33.
$$(p+2)^2 = 9(p+2) - 20$$
$$(p+2)^2 - 9(p+2) + 20 = 0$$

Let $x = p + 2$. Then $x^2 = (p+2)^2$ and

$$x^2 - 9x + 20 = 0$$
$$(x-5)(x-4) = 0$$
$$x = 5 \quad \text{or} \quad x = 4$$
$$p + 2 = 5 \quad \text{or} \quad p + 2 = 4$$
$$p = 3 \quad \text{or} \quad p = 2$$

The solutions are 2 and 3.

35.
$$2x = \sqrt{11x + 3}$$
$$(2x)^2 = \left(\sqrt{11x + 3}\right)^2$$
$$4x^2 = 11x + 3$$
$$4x^2 - 11x - 3 = 0$$
$$(4x + 1)(x - 3) = 0$$
$$x = -\frac{1}{4} \text{ (discard) or } x = 3$$
The solution is 3.

37. $x^{2/3} - 8x^{1/3} + 15 = 0$
Let $y = x^{1/3}$. Then $y^2 = x^{2/3}$ and
$$y^2 - 8y + 15 = 0$$
$$(y - 5)(y - 3) = 0$$
$$y = 5 \quad \text{or} \quad y = 3$$
$$x^{1/3} = 5 \quad \text{or} \quad x^{1/3} = 3$$
$$x = 125 \quad \text{or} \quad x = 27$$
The solutions are 27 and 125.

39.
$$y^3 + 9y - y^2 - 9 = 0$$
$$y(y^2 + 9) - 1(y^2 + 9) = 0$$
$$(y^2 + 9)(y - 1) = 0$$
$$y^2 + 9 = 0 \quad \text{or } y - 1 = 0$$
$$y^2 = -9 \quad \text{or} \quad y = 1$$
$$y = \pm\sqrt{-9}$$
$$y = \pm 3i$$
The solutions are $1, -3i,$ and $3i$.

41. $2x^{2/3} + 3x^{1/3} - 2 = 0$
Let $y = x^{1/3}$. Then $y^2 = x^{2/3}$ and
$$2y^2 + 3y - 2 = 0$$
$$(2y - 1)(y + 2) = 0$$
$$y = \frac{1}{2} \quad \text{or} \quad y = -2$$
$$x^{1/3} = \frac{1}{2} \quad \text{or } x^{1/3} = -2$$
$$x = \frac{1}{8} \quad \text{or} \quad x = -8$$
The solutions are -8 and $\frac{1}{8}$.

43. $x^{-2} - x^{-1} - 6 = 0$
Let $y = x^{-1}$. Then $y^2 = x^{-2}$ and
$$y^2 - y - 6 = 0$$
$$(y - 3)(y + 2) = 0$$
$$y = 3 \quad \text{or} \quad y = -2$$
$$x^{-1} = 3 \quad \text{or} \quad x^{-1} = -2$$
$$\frac{1}{x} = 3 \quad \text{or} \quad \frac{1}{x} = -2$$
$$x = \frac{1}{3} \quad \text{or} \quad x = -\frac{1}{2}$$
The solutions are $-\frac{1}{2}$ and $\frac{1}{3}$.

45.
$$x - \sqrt{x} = 2$$
$$x - 2 = \sqrt{x}$$
$$(x - 2)^2 = x$$
$$x^2 - 4x + 4 = x$$
$$x^2 - 5x + 4 = 0$$
$$(x - 4)(x - 1) = 0$$
$$x = 4 \text{ or } x = 1 \text{ (discard)}$$
The solution is 4.

47.
$$\frac{x}{x - 1} + \frac{1}{x + 1} = \frac{2}{x^2 - 1}$$
$$\frac{x}{x - 1} + \frac{1}{x + 1} = \frac{2}{(x + 1)(x - 1)}$$
$$x(x + 1) + (x - 1) = 2$$
$$x^2 + x + x - 1 = 2$$
$$x^2 + 2x - 3 = 0$$
$$(x + 3)(x - 1) = 0$$
$$x = -3 \text{ or } x = 1 \text{ (discard)}$$
The solution is -3.

49.
$$p^4 - p^2 - 20 = 0$$
$$(p^2 - 5)(p^2 + 4) = 0$$
$$p^2 - 5 = 0 \quad \text{or } p^2 + 4 = 0$$
$$p^2 = 5 \quad \text{or} \quad p^2 = -4$$
$$p = \pm\sqrt{5} \quad \text{or} \quad p = \pm 2i$$
The solutions are $-\sqrt{5}, \sqrt{5}, -2i,$ and $2i$.

51.
$$2x^3 = -54$$
$$x^3 = -27$$
$$x^3 + 27 = 0$$
$$(x+3)(x^2 - 3x + 9) = 0$$
$$x + 3 = 0 \quad \text{or} \quad x^2 - 3x + 9 = 0$$
$$x = -3 \quad \text{or}$$
$$x = \frac{3 \pm \sqrt{(-3)^2 - 4(1)(9)}}{2(1)}$$
$$= \frac{3 \pm \sqrt{-27}}{2}$$
$$= \frac{3 \pm 3i\sqrt{3}}{2} = \frac{3}{2} \pm \frac{3\sqrt{3}}{2}i$$

The solutions are -3, $\dfrac{3 + 3i\sqrt{3}}{2}$, and

$\dfrac{3 - 3i\sqrt{3}}{2}$.

53.
$$1 = \frac{4}{x-7} + \frac{5}{(x-7)^2}$$
$$(x-7)^2 - 4(x-7) - 5 = 0$$
Let $y = x - 7$. Then $y^2 = (x-7)^2$ and
$$y^2 - 4y - 5 = 0$$
$$(y-5)(y+1) = 0$$
$$y = 5 \quad \text{or} \quad y = -1$$
$$x - 7 = 5 \quad \text{or} \quad x - 7 = -1$$
$$x = 12 \quad \text{or} \quad x = 6$$
The solutions are 6 and 12.

55.
$$27y^4 + 15y^2 = 2$$
$$27y^4 + 15y^2 - 2 = 0$$
$$(9y^2 - 1)(3y^2 + 2) = 0$$
$$(3y+1)(3y-1)(3y^2 + 2) = 0$$
$$y = -\frac{1}{3} \quad \text{or} \quad y = \frac{1}{3} \quad \text{or} \quad y^2 = -\frac{2}{3}$$
$$y = \pm\sqrt{-\frac{2}{3}}$$
$$y = \pm\frac{\sqrt{6}}{3}i$$

The solutions are $-\dfrac{1}{3}, \dfrac{1}{3}, -\dfrac{\sqrt{6}}{3}i$, and $\dfrac{\sqrt{6}}{3}i$.

57. Let x = speed on the first part. Then
$x - 1$ = speed on the second part.
$$d = rt \implies t = \frac{d}{r}$$
$$t_{\text{on first part}} + t_{\text{on second part}} = 1\frac{3}{5}$$
$$\frac{3}{x} + \frac{4}{x-1} = \frac{8}{5}$$
$$3 \cdot 5(x-1) + 4 \cdot 5x = 8x(x-1)$$
$$15x - 15 + 20x = 8x^2 - 8x$$
$$0 = 8x^2 - 43x + 15$$
$$0 = (8x - 3)(x - 5)$$
$$8x - 3 = 0 \quad \text{or} \quad x - 5 = 0$$
$$x = \frac{3}{8} \quad \text{or} \quad x = 5$$
$$x - 1 = 4$$

Discard $\dfrac{3}{8}$. Her speeds were 5 mph and

4 mph.

59. Let x = time for hose alone. Then
$x - 1$ = time for the inlet pipe alone.
$$\frac{1}{x} + \frac{1}{x-1} = \frac{1}{8}$$
$$8(x-1) + 8x = x(x-1)$$
$$8x - 8 + 8x = x^2 - x$$
$$0 = x^2 - 17x + 8$$
$$x = \frac{17 \pm \sqrt{(-17)^2 - 4(1)(8)}}{2(1)}$$
$$= \frac{17 \pm \sqrt{257}}{2}$$
$$x \approx 0.5 \text{ (discard)} \quad \text{or} \quad x \approx 16.5$$
$$x - 1 \approx 15.5$$

Hose: 16.5 hrs; Inlet pipe: 15.5 hrs

61. Let x = original speed. Then
$x + 11$ = return speed.

$$d = rt \implies t = \frac{d}{r}$$

$$t_{\text{return}} = t_{\text{original}} - 1$$

$$\frac{330}{x + 11} = \frac{330}{x} - 1$$

$$330x = 330(x + 11) - x(x + 11)$$

$$330x = 330x + 3630 - x^2 - 11x$$

$$x^2 + 11x - 3630 = 0$$

$$x = \frac{-11 \pm \sqrt{(11)^2 - 4(1)(-3630)}}{2(1)}$$

$$= \frac{-11 \pm \sqrt{14,641}}{2}$$

$$= \frac{-11 \pm 121}{2} = 55 \text{ or } -66 \text{ (disregard)}$$

$x + 11 = 55 + 11 = 66$

Original speed: 55 mph
Return speed: 66 mph

63. Let x = time for son alone. Then
$x - 1$ = time for dad alone.

$$\frac{1}{x} + \frac{1}{x - 1} = \frac{1}{4}$$

$$4(x - 1) + 4x = x(x - 1)$$

$$4x - 4 + 4x = x^2 - x$$

$$0 = x^2 - 9x + 4$$

$$x = \frac{9 \pm \sqrt{(-9)^2 - 4(1)(4)}}{2(1)}$$

$$= \frac{9 \pm \sqrt{65}}{2}$$

$$\approx 0.5 \text{ (discard) or } 8.5$$

It takes his son about 8.5 hours.

65. Let x = the number.

$$x(x - 4) = 96$$

$$x^2 - 4x - 96 = 0$$

$$(x - 12)(x + 8) = 0$$

$$x = 12 \text{ or } x = -8$$

The number is 12 or –8.

67. a. length = $x - 3 - 3 = x - 6$

   b. $V = lwh$
   $$300 = (x - 6)(x - 6) \cdot 3$$

   c. $300 = 3(x - 6)^2$
   $$100 = x^2 - 12x + 36$$
   $$0 = x^2 - 12x - 64$$
   $$0 = (x - 16)(x + 4)$$
   $$x = 16 \text{ or } x = -4 \text{ (discard)}$$

   The sheet is 16 cm by 16 cm.
   Check: $V = 3(x - 6)(x - 6)$
   $$= 3(16 - 6)(16 - 6)$$
   $$= 3(10)(10)$$
   $$= 300 \text{ cubic cm}$$

69. Let x = length of the side of the square.

$$\text{Area} = x^2$$

$$920 = x^2$$

$$\sqrt{920} = x$$

Adding another radial line to a different corner would yield a right triangle with legs r and hypotenuse x.

$$r^2 + r^2 = x^2$$

$$2r^2 = \left(\sqrt{920}\right)^2$$

$$2r^2 = 920$$

$$r^2 = 460$$

$$r = \pm\sqrt{460} = \pm 21.4476$$

Disregard the negative. The smallest radius would be 22 feet.

71. $\dfrac{5x}{3} + 2 \leq 7$

$$\frac{5x}{3} \leq 5$$

$$5x \leq 15$$

$$x \leq 3$$

$$(-\infty, 3]$$

73. $\dfrac{y-1}{15} > -\dfrac{2}{5}$

$15\left(\dfrac{y-1}{15}\right) > 15\left(-\dfrac{2}{5}\right)$

$y-1 > -6$

$y > -5$

$(-5, \infty)$

75. Domain: $\{x \mid x \text{ is a real number}\}$ or $(-\infty, \infty)$
Range: $\{y \mid y \text{ is a real number}\}$ or $(-\infty, \infty)$
It is a function.

77. Domain: $\{x \mid x \text{ is a real number}\}$ or $(-\infty, \infty)$
Range: $\{y \mid y \geq -1\}$ or $[-1, \infty)$
It is a function.

79. Answers may vary.

81. a. Let x = Dominguez's fastest lap speed
and $x + 0.88$ = Fernandez's fastest lap speed.

Using $t = \dfrac{d}{r}$, we have

$t_{\text{Dominguez}} = t_{\text{Fernandez}} + 0.38$

$\dfrac{7920}{x} = \dfrac{7920}{x + 0.88} + 0.38$

$7920(x + 0.88) = 7920x + 0.38x(x + 0.88)$

$7920x + 6969.6 = 7920x + 0.38x^2 + 0.3344x$

$0 = 0.38x^2 + 0.3344x - 6969.6$

$x = \dfrac{-0.3344 \pm \sqrt{(0.3344)^2 - 4(0.38)(-6969.6)}}{2(0.38)}$

Using the positive square root, $x \approx 135.0$ feet per second.

b. $x + 0.88 = 135.0 + 0.88$
$= 135.9$ feet per second

c. 5280 ft = 1 mile, and 3600 sec = 1 hr.

Dominguez: $\dfrac{135 \text{ ft}}{\text{sec}} \cdot \dfrac{3600 \text{ sec}}{\text{hr}} \cdot \dfrac{1 \text{ mile}}{5280 \text{ ft}} \approx 92.0$ mph

Fernandez: $\dfrac{135.9 \text{ ft}}{\text{sec}} \cdot \dfrac{3600 \text{ sec}}{\text{hr}} \cdot \dfrac{1 \text{ mile}}{5280 \text{ ft}} \approx 92.7$ mph

Integrated Review

1. $x^2 - 10 = 0$
$$x^2 = 10$$
$$x = \pm\sqrt{10}$$

2. $x^2 - 14 = 0$
$$x^2 = 14$$
$$x = \pm\sqrt{14}$$

3. $(x-1)^2 = 8$
$$x - 1 = \pm\sqrt{8}$$
$$x - 1 = \pm 2\sqrt{2}$$
$$x = 1 \pm 2\sqrt{2}$$

4. $(x+5)^2 = 12$
$$x + 5 = \pm\sqrt{12}$$
$$x + 5 = \pm 2\sqrt{3}$$
$$x = -5 \pm 2\sqrt{3}$$

5. $x^2 + 2x - 12 = 0$
$$x^2 + 2x + \left(\frac{2}{2}\right)^2 = 12 + 1$$
$$x^2 + 2x + 1 = 13$$
$$(x+1)^2 = 13$$
$$x + 1 = \pm\sqrt{13}$$
$$x = -1 \pm \sqrt{13}$$

6. $x^2 - 12x + 11 = 0$
$$x^2 - 12x + \left(\frac{-12}{2}\right)^2 = -11 + 36$$
$$x^2 - 12x + 36 = 25$$
$$(x-6)^2 = \pm\sqrt{25}$$
$$x - 6 = \pm 5$$
$$x = 6 \pm 5$$
$$x = 1 \text{ or } x = 11$$

7. $3x^2 + 3x = 5$
$$x^2 + x = \frac{5}{3}$$
$$x^2 + x + \left(\frac{1}{2}\right)^2 = \frac{5}{3} + \frac{1}{4}$$
$$x^2 + x + \frac{1}{4} = \frac{23}{12}$$
$$\left(x + \frac{1}{2}\right)^2 = \frac{23}{12}$$
$$x + \frac{1}{2} = \pm\sqrt{\frac{23}{12}}$$
$$x + \frac{1}{2} = \pm\frac{\sqrt{23}}{2\sqrt{3}}$$
$$x + \frac{1}{2} = \pm\frac{\sqrt{23} \cdot \sqrt{3}}{2\sqrt{3} \cdot \sqrt{3}}$$
$$x + \frac{1}{2} = \pm\frac{\sqrt{69}}{6}$$
$$x = -\frac{1}{2} \pm \frac{\sqrt{69}}{6} = \frac{-3 \pm \sqrt{69}}{6}$$

8. $16y^2 + 16y = 1$
$$y^2 + y = \frac{1}{16}$$
$$y^2 + y + \left(\frac{1}{2}\right)^2 = \frac{1}{16} + \frac{1}{4}$$
$$y^2 + y + \frac{1}{4} = \frac{5}{16}$$
$$\left(y + \frac{1}{2}\right)^2 = \frac{5}{16}$$
$$y + \frac{1}{2} = \pm\sqrt{\frac{5}{16}}$$
$$y + \frac{1}{2} = \pm\frac{\sqrt{5}}{4}$$
$$y = -\frac{1}{2} \pm \frac{\sqrt{5}}{4} = \frac{-2 \pm \sqrt{5}}{4}$$

9. $2x^2 - 4x + 1 = 0$

$a = 2, b = -4, c = 1$

$x = \dfrac{4 \pm \sqrt{(-4)^2 - 4(2)(1)}}{2(2)}$

$= \dfrac{4 \pm \sqrt{8}}{4}$

$= \dfrac{4 \pm 2\sqrt{2}}{4} = \dfrac{2 \pm \sqrt{2}}{2}$

10. $\dfrac{1}{2}x^2 + 3x + 2 = 0$

$x^2 + 6x + 4 = 0$

$a = 1, b = 6, c = 4$

$x = \dfrac{-6 \pm \sqrt{(6)^2 - 4(1)(4)}}{2(1)}$

$= \dfrac{-6 \pm \sqrt{20}}{2}$

$= \dfrac{-6 \pm 2\sqrt{5}}{2} = -3 \pm \sqrt{5}$

11. $x^2 + 4x = -7$

$x^2 + 4x + 7 = 0$

$a = 1, b = 4, c = 7$

$x = \dfrac{-4 \pm \sqrt{(4)^2 - 4(1)(7)}}{2(1)}$

$= \dfrac{-4 \pm \sqrt{-12}}{2}$

$= \dfrac{-4 \pm i\sqrt{4 \cdot 3}}{2}$

$= \dfrac{-4 \pm 2i\sqrt{3}}{2} = -2 \pm i\sqrt{3}$

12. $x^2 + x = -3$

$x^2 + x + 3 = 0$

$a = 1, b = 1, c = 3$

$x = \dfrac{-1 \pm \sqrt{(1)^2 - 4(1)(3)}}{2(1)}$

$= \dfrac{-1 \pm \sqrt{-11}}{2}$

$= \dfrac{-1 \pm i\sqrt{11}}{2}$ or $-\dfrac{1}{2} \pm \dfrac{\sqrt{11}}{2}i$

13. $x^2 + 3x + 6 = 0$

$a = 1, b = 3, c = 6$

$x = \dfrac{-3 \pm \sqrt{(3)^2 - 4(1)(6)}}{2(1)}$

$= \dfrac{-3 \pm \sqrt{-15}}{2}$

$= \dfrac{-3 \pm i\sqrt{15}}{2}$ or $-\dfrac{3}{2} \pm \dfrac{\sqrt{15}}{2}i$

14. $2x^2 + 18 = 0$

$2x^2 = -18$

$x^2 = -9$

$x = \pm\sqrt{-9}$

$x = \pm 3i$

15. $x^2 + 17x = 0$

$x(x + 17) = 0$

$x = 0$ or $x + 17 = 0$

$x = -17$

$x = 0, -17$

16. $4x^2 - 2x - 3 = 0$

$a = 4, b = -2, c = -3$

$x = \dfrac{2 \pm \sqrt{(-2)^2 - 4(4)(-3)}}{2(4)}$

$= \dfrac{2 \pm \sqrt{52}}{8}$

$= \dfrac{2 \pm 2\sqrt{13}}{8} = \dfrac{1 \pm \sqrt{13}}{4}$

17. $(x - 2)^2 = 27$

$x - 2 = \pm\sqrt{27}$

$x - 2 = \pm 3\sqrt{3}$

$x = 2 \pm 3\sqrt{3}$

18. $\dfrac{1}{2}x^2 - 2x + \dfrac{1}{2} = 0$

$x^2 - 4x + 1 = 0$

$x^2 - 4x + \left(\dfrac{-4}{2}\right)^2 = -1 + 4$

$x^2 - 4x + 4 = 3$

$(x - 2)^2 = 3$

$x - 2 = \pm\sqrt{3}$

$x = 2 \pm \sqrt{3}$

19. $3x^2 + 2x = 8$

$3x^2 + 2x - 8 = 0$

$(3x - 4)(x + 2) = 0$

$3x - 4 = 0 \ \text{ or } \ x + 2 = 0$

$x = \dfrac{4}{3} \ \text{ or } \ \ x = -2$

20. $2x^2 = -5x - 1$

$2x^2 + 5x + 1 = 0$

$a = 2, b = 5, c = 1$

$x = \dfrac{-5 \pm \sqrt{(5)^2 - 4(2)(1)}}{2(2)}$

$= \dfrac{-5 \pm \sqrt{17}}{4}$

21. $x(x - 2) = 5$

$x^2 - 2x = 5$

$x^2 - 2x + \left(\dfrac{-2}{2}\right)^2 = 5 + 1$

$x^2 - 2x + 1 = 6$

$(x - 1)^2 = 6$

$x - 1 = \pm\sqrt{6}$

$x = 1 \pm \sqrt{6}$

22. $x^2 - 31 = 0$

$x^2 = 31$

$x = \pm\sqrt{31}$

23. $5x^2 - 55 = 0$

$5x^2 = 55$

$x^2 = 11$

$x = \pm\sqrt{11}$

24. $5x^2 + 55 = 0$

$5x^2 = -55$

$x^2 = -11$

$x = \pm\sqrt{-11}$

$x = \pm i\sqrt{11}$

25. $x(x + 5) = 66$

$x^2 + 5x = 66$

$x^2 + 5x - 66 = 0$

$(x + 11)(x - 6) = 0$

$x + 11 = 0 \ \ \text{ or } x - 6 = 0$

$x = -11 \ \text{ or } \ \ x = 6$

26. $5x^2 + 6x - 2 = 0$

$a = 5, b = 6, c = -2$

$x = \dfrac{-6 \pm \sqrt{(6)^2 - 4(5)(-2)}}{2(5)}$

$= \dfrac{-6 \pm \sqrt{76}}{10}$

$= \dfrac{-6 \pm \sqrt{4 \cdot 19}}{10}$

$= \dfrac{-6 \pm 2\sqrt{19}}{10} = \dfrac{-3 \pm \sqrt{19}}{5}$

27. $2x^2 + 3x = 1$

$2x^2 + 3x - 1 = 0$

$a = 2, b = 3, c = -1$

$x = \dfrac{-3 \pm \sqrt{(3)^2 - 4(2)(-1)}}{2(2)}$

$= \dfrac{-3 \pm \sqrt{17}}{4}$

28. $a^2 + b^2 = c^2$

$$x^2 + x^2 = 20^2$$
$$2x^2 = 400$$
$$x^2 = 200$$
$$x = \pm\sqrt{200}$$
$$= \pm 10\sqrt{2} \approx 14.1421$$

Disregard the negative. A side of the room is $10\sqrt{2}$ feet ≈ 14.1 feet.

29. Let $x =$ time for Jack alone. Then $x - 2 =$ time for Lucy alone.

$$\frac{1}{x} + \frac{1}{x-2} = \frac{1}{4}$$
$$4(x-2) + 4x = x(x-2)$$
$$4x - 8 + 4x = x^2 - 2x$$
$$0 = x^2 - 10x + 8$$
$$x = \frac{10 \pm \sqrt{(-10)^2 - 4(1)(8)}}{2(1)}$$
$$= \frac{10 \pm \sqrt{68}}{2} \approx 9.1 \text{ or } 0.9 \text{ (disregard)}$$

$x - 2 = 9.1 - 2 = 7.1$

It would take Jack 9.1 hours and Lucy 7.1 hours.

30. Let $x =$ speed on treadmill. Then $x + 1 =$ speed running.

$$t_{\text{treadmill}} + t_{\text{running}} = \frac{4}{3}$$
$$\frac{5}{x} + \frac{2}{x+1} = \frac{4}{3}$$
$$5 \cdot 3(x+1) + 2 \cdot 3x = 4x(x+1)$$
$$15x + 15 + 6x = 4x^2 + 4x$$
$$0 = 4x^2 - 17x - 15$$
$$0 = (4x + 3)(x - 5)$$

$x = -\frac{4}{3}$ (disregard) or $x = 5$

$x + 1 = 5 + 1 = 6$

Treadmill: 5 mph

Running: 6 mph

Exercise Set 11.4

1. $(x+1)(x+5) > 0$

$x + 1 = 0 \quad$ or $\quad x + 5 = 0$

$x = -1 \quad$ or $\quad x = -5$

Region	Test Point	$(x+1)(x+5) > 0$ Result
A: $(-\infty, -5)$	-6	$(-5)(-11) > 0$ True
B: $(-5, -1)$	-2	$(-1)(4) > 0$ False
C: $(-1, \infty)$	0	$(1)(5) > 0$ True

Solution: $(-\infty, -5) \cup (-1, \infty)$

3. $(x-3)(x+4) \le 0$

$x - 3 = 0 \quad$ or $\quad x + 4 = 0$

$x = 3 \quad$ or $\quad x = -4$

Region	Test Point	$(x-3)(x+4) \le 0$ Result
A: $(-\infty, -4)$	-5	$(-8)(-1) \le 0$ False
B: $(-4, 3)$	0	$(-3)(4) \le 0$ True
C: $(3, \infty)$	4	$(1)(9) \le 0$ False

Solution: $[-4, 3]$

5. $x^2 - 7x + 10 \le 0$
$(x-5)(x-2) \le 0$
$x - 5 = 0$ or $x - 2 = 0$
$x = 5$ or $x = 2$

Region	Test Point	$(x-5)(x-2) \le 0$ Result
A: $(-\infty, 2)$	0	$(-5)(-2) \le 0$ False
B: $(2, 5)$	3	$(-2)(1) \le 0$ True
C: $(5, \infty)$	6	$(1)(4) \le 0$ False

$$\underset{2 \qquad\quad 5}{}$$

Solution: $[2, 5]$

7. $3x^2 + 16x < -5$
$3x^2 + 16x + 5 < 0$
$(3x+1)(x+5) < 0$
$3x + 1 = 0$ or $x + 5 = 0$
$x = -\dfrac{1}{3}$ or $x = -5$

Region	Test Point	$(3x+1)(x+5) < 0$ Result
A: $(-\infty, -5)$	-6	$(-17)(-1) < 0$ False
B: $\left(-5, -\dfrac{1}{3}\right)$	-1	$(-2)(4) < 0$ True
C: $\left(-\dfrac{1}{3}, \infty\right)$	0	$(1)(5) < 0$ False

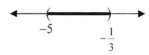

$$\underset{-5 \qquad\qquad -\frac{1}{3}}{}$$

Solution: $\left(-5, -\dfrac{1}{3}\right)$

9. $(x-6)(x-4)(x-2) > 0$
$x - 6 = 0$ or $x - 4 = 0$ or $x - 2 = 0$
$x = 6$ or $x = 4$ or $x = 2$

Region	Test Point	$(x-6)(x-4)(x-2) > 0$ Result
A: $(-\infty, 2)$	0	$(-6)(-4)\,(-2) > 0$ False
B: $(2, 4)$	3	$(-3)(-1)\,(1) > 0$ True
C: $(4, 6)$	5	$(-1)(1)(3) > 0$ False
D: $(6, \infty)$	7	$(1)(3)(5) > 0$ True

$$\underset{2 \quad\; 4 \qquad 6}{}$$

Solution: $(2, 4) \cup (6, \infty)$

11. $x(x-1)(x+4) \le 0$

$\quad x = 0$ or $x - 1 = 0$ or $x + 4 = 0$

$\qquad\qquad\quad x = 1$ or $\quad x = -4$

Region	Test Point	$x(x-1)(x+4) \le 0$ Result
A: $(-\infty, -4)$	-5	$-5(-6)(-1) \le 0$ True
B: $(-4, 0)$	-1	$-1(-2)(3) \le 0$ False
C: $(0, 1)$	$\dfrac{1}{2}$	$\dfrac{1}{2}\left(-\dfrac{1}{2}\right)\left(\dfrac{7}{2}\right) \le 0$ True
D: $(1, \infty)$	2	$2(1)(6) \le 0$ False

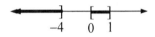

Solution: $(-\infty, -4] \cup [0, 1]$

13. $\qquad\qquad (x^2 - 9)(x^2 - 4) > 0$

$\quad (x+3)(x-3)(x+2)(x-2) > 0$

$\quad x + 3 = 0 \quad$ or $x - 3 = 0$ or $x + 2 = 0 \quad$ or $x - 2 = 0$

$\quad x = -3$ or $\quad x = 3$ or $\quad x = -2$ or $\qquad x = 2$

Region	Test Point	$(x+3)(x-3)(x+2)(x-2) > 0$ Result
A: $(-\infty, -3)$	-4	$(-1)(-7)(-2)(-6) > 0$ True
B: $(-3, -2)$	$-\dfrac{5}{2}$	$\left(\dfrac{1}{2}\right)\left(-\dfrac{11}{2}\right)\left(-\dfrac{1}{2}\right)\left(-\dfrac{9}{2}\right) > 0$ False
C: $(-2, 2)$	0	$(3)(-3)(2)(-2) > 0$ True
D: $(2, 3)$	$\dfrac{5}{2}$	$\left(\dfrac{11}{2}\right)\left(-\dfrac{1}{2}\right)\left(\dfrac{9}{2}\right)\left(\dfrac{1}{2}\right) > 0$ False
E: $(3, \infty)$	4	$(7)(1)(6)(2) > 0$ True

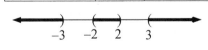

Solution: $(-\infty, -3) \cup (-2, 2) \cup (3, \infty)$

15. $\dfrac{x+7}{x-2} < 0$

$x+7=0$ or $x-2=0$

$x=-7$ or $x=2$

Region	Test Point	$\dfrac{x+7}{x-2} < 0$; Result
A: $(-\infty, -7)$	-8	$\dfrac{-1}{-10} < 0$; False
B: $(-7, 2)$	0	$\dfrac{7}{-2} < 0$; True
C: $(2, \infty)$	3	$\dfrac{10}{1} < 0$; False

Solution: $(-7, 2)$

17. $\dfrac{5}{x+1} > 0$

$x+1=0$

$x=-1$

Region	Test Point	$\dfrac{5}{x+1} > 0$; Result
A: $(-\infty, -1)$	-2	$\dfrac{5}{-1} > 0$; False
B: $(-1, \infty)$	0	$\dfrac{5}{1} > 0$; True

Solution: $(-1, \infty)$

19. $\dfrac{x+1}{x-4} \geq 0$

$x+1=0$ or $x-4=0$

$x=-1$ or $x=4$

Region	Test Point	$\dfrac{x+1}{x-4} \geq 0$; Result
A: $(-\infty, -1)$	-2	$\dfrac{-1}{-6} \geq 0$; True
B: $(-1, 4)$	0	$\dfrac{1}{-4} \geq 0$; False
C: $(4, \infty)$	5	$\dfrac{6}{1} \geq 0$; True

Solution: $(-\infty, -1] \cup (4, \infty)$

21. $\dfrac{3}{x-2} < 4$

The denominator is equal to 0 when

$x-2=0$, or $x=2$.

$\dfrac{3}{x-2}=4$

$3=4x-8$

$11=4x$

$\dfrac{11}{4}=x$

Region	Test Point	$\dfrac{3}{x-2} < 4$; Result
A: $(-\infty, 2)$	0	$\dfrac{3}{-2} < 4$; True
B: $\left(2, \dfrac{11}{4}\right)$	$\dfrac{5}{2}$	$\dfrac{3}{\frac{1}{2}}=6 < 4$; False
C: $\left(\dfrac{11}{4}, \infty\right)$	4	$\dfrac{3}{2} < 4$; True

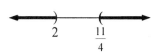

Solution: $(-\infty, 2) \cup \left(\dfrac{11}{4}, \infty\right)$

23. $\dfrac{x^2+6}{5x} \geq 1$

The denominator is equal to 0 when
$5x = 0$, or $x = 0$.

$$\frac{x^2+6}{5x} = 1$$
$$x^2 + 6 = 5x$$
$$x^2 - 5x + 6 = 0$$
$$(x-2)(x-3) = 0$$
$$x - 2 = 0 \text{ or } x - 3 = 0$$
$$x = 2 \text{ or } \quad x = 3$$

Region	Test Point	$\dfrac{x^2+6}{5x} \geq 1$; Result
A: $(-\infty, 0)$	-1	$\dfrac{7}{-5} \geq 1$; F
B: $(0, 2)$	1	$\dfrac{7}{5} \geq 1$; T
C: $(2, 3)$	$\dfrac{5}{2}$	$\dfrac{49/4}{25/2} = \dfrac{49}{50} \geq 1$; F
D: $(3, \infty)$	4	$\dfrac{22}{20} \geq 1$; T

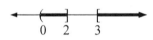

Solution: $(0, 2] \cup [3, \infty)$

25. $(x-8)(x+7) > 0$

$x - 8 = 0 \text{ or } x + 7 = 0$
$\quad x = 8 \text{ or } \quad x = -7$

Region	Test Point	$(x-8)(x+7) > 0$ Result
A: $(-\infty, -7)$	-8	$(-16)(-1) > 0$; T
B: $(-7, 8)$	0	$(-8)(7) > 0$; F
C: $(8, \infty)$	9	$(1)(16) > 0$; T

Solution: $(-\infty, -7) \cup (8, \infty)$

27. $(2x-3)(4x+5) \leq 0$

$2x - 3 = 0 \text{ or } 4x + 5 = 0$
$\quad x = \dfrac{3}{2} \text{ or } \quad x = -\dfrac{5}{4}$

Region	Test Point	$(2x-3)(4x+5) \leq 0$ Result
A: $\left(-\infty, -\dfrac{5}{4}\right)$	-2	$(-7)(-3) \leq 0$ False
B: $\left(-\dfrac{5}{4}, \dfrac{3}{2}\right)$	0	$(-3)(5) \leq 0$ True
C: $\left(\dfrac{3}{2}, \infty\right)$	2	$(1)(13) \leq 0$ False

Solution: $\left[-\dfrac{5}{4}, \dfrac{3}{2}\right]$

29.
$$x^2 > x$$
$$x^2 - x > 0$$
$$x(x-1) > 0$$
$$x = 0 \text{ or } x - 1 = 0$$
$$x = 1$$

Region	Test Point	$x(x-1) > 0$; Result
A: $(-\infty, 0)$	-1	$-1(-2) > 0$; True
B: $(0, 1)$	$\dfrac{1}{2}$	$\dfrac{1}{2}\left(-\dfrac{1}{2}\right) > 0$; False
C: $(1, \infty)$	2	$2(1) > 0$; True

Solution: $(-\infty, 0) \cup (1, \infty)$

31. $(2x-8)(x+4)(x-6) \le 0$

$2x-8=0$ or $x+4=0$ or $x-6=0$

   $x=4$ $x=-4$ or $x=6$

Region	Test Point	$(2x-8)(x+4)(x-6) \le 0$ Result
A: $(-\infty, -4)$	-5	$(-18)(-1)(-11) \le 0$ True
B: $(-4, 4)$	0	$(-8)(4)(-6) \le 0$ False
C: $(4, 6)$	5	$(2)(9)(-1) \le 0$ True
D: $(6, \infty)$	7	$(6)(11)(1) \le 0$ False

       -4 4 6

Solution: $(-\infty, -4] \cup [4, 6]$

33. $6x^2 - 5x \ge 6$

       $6x^2 - 5x - 6 \ge 0$

    $(3x+2)(2x-3) \ge 0$

    $3x+2=0$ or $2x-3=0$

      $x=-\dfrac{2}{3}$ or $x=\dfrac{3}{2}$

Region	Test Point	$(3x+2)(2x-3) \ge 0$ Result
A: $\left(-\infty, -\dfrac{2}{3}\right)$	-1	$(-1)(-5) \ge 0$ True
B: $\left(-\dfrac{2}{3}, \dfrac{3}{2}\right)$	0	$(2)(-3) \ge 0$ False
C: $\left(\dfrac{3}{2}, \infty\right)$	2	$(8)(1) \ge 0$ True

      $-\dfrac{2}{3}$ $\dfrac{3}{2}$

Solution: $\left(-\infty, -\dfrac{2}{3}\right] \cup \left[\dfrac{3}{2}, \infty\right)$

35. $4x^3 + 16x^2 - 9x - 36 > 0$

$4x^2(x+4) - 9(x+4) > 0$

$(x+4)(4x^2 - 9) > 0$

$(x+4)(2x+3)(2x-3) > 0$

$x + 4 = 0$ or $2x + 3 = 0$ or $2x - 3 = 0$

$x = -4$ or $x = -\dfrac{3}{2}$ or $x = \dfrac{3}{2}$

Region	Test Point	$(x+4)(2x+3)(2x-3) > 0$ Result
A: $(-\infty, -4)$	-5	$(-1)(-7)\,(-13) > 0$ False
B: $\left(-4, -\dfrac{3}{2}\right)$	-3	$(1)(-3)\,(-9) > 0$ True
C: $\left(-\dfrac{3}{2}, \dfrac{3}{2}\right)$	0	$(4)(3)(-3) > 0$ False
D: $\left(\dfrac{3}{2}, \infty\right)$	4	$(8)(11)(5) > 0$ True

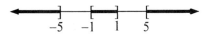

$$-4 \quad -\frac{3}{2} \qquad \frac{3}{2}$$

Solution: $\left(-4, -\dfrac{3}{2}\right) \cup \left(\dfrac{3}{2}, \infty\right)$

37. $x^4 - 26x^2 + 25 \geq 0$

$(x^2 - 25)(x^2 - 1) \geq 0$

$(x+5)(x-5)(x+1)(x-1) \geq 0$

$x = -5$ or $x = 5$ or $x = -1$ or $x = 1$

Region	Test Point	$(x+5)(x-5)(x+1)(x-1) \geq 0$ Result
A: $(-\infty, -5)$	-6	$(-1)(-11)\,(-5)(-7) \geq 0$; True
B: $(-5, -1)$	-2	$(3)(-7)(-1)(-3) \geq 0$; False
C: $(-1, 1)$	0	$(5)(-5)(1)(-1) \geq 0$; True
D: $(1, 5)$	2	$(7)(-3)(3)(1) \geq 0$; False
E: $(5, \infty)$	6	$(11)(1)(6)(2) \geq 0$; True

$$-5 \quad -1 \quad 1 \quad 5$$

Solution: $(-\infty, -5] \cup [-1, 1] \cup [5, \infty)$

39. $(2x-7)(3x+5) > 0$

$2x - 7 = 0$ or $3x + 5 = 0$

$x = \dfrac{7}{2}$ or $x = -\dfrac{5}{3}$

Region	Test Point	$(2x-7)(3x+5) > 0$ Result
A: $\left(-\infty, -\dfrac{5}{3}\right)$	-2	$(-11)(-1) > 0$ True
B: $\left(-\dfrac{5}{3}, \dfrac{7}{2}\right)$	0	$(-7)(5) > 0$ False
C: $\left(\dfrac{7}{2}, \infty\right)$	4	$(1)(17) > 0$ True

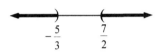

$-\dfrac{5}{3}$ $\dfrac{7}{2}$

Solution: $\left(-\infty, -\dfrac{5}{3}\right) \cup \left(\dfrac{7}{2}, \infty\right)$

41. $\dfrac{x}{x-10} < 0$

$x = 0$ or $x - 10 = 0$

$x = 10$

Region	Test Point	$\dfrac{x}{x-10} < 0$ Result
A: $(-\infty, 0)$	-1	$\dfrac{-1}{-11} < 0$ False
B: $(0, 10)$	5	$\dfrac{5}{-5} < 0$ True
C: $(10, \infty)$	11	$\dfrac{11}{1} < 0$ False

0 10

Solution: $(0, 10)$

43. $\dfrac{x-5}{x+4} \geq 0$

$x - 5 = 0$ or $x + 4 = 0$

$x = 5$ or $x = -4$

Region	Test Point	$\dfrac{x-5}{x+4} \geq 0$ Result
A: $(-\infty, -4)$	-5	$\dfrac{-10}{-1} \geq 0$ True
B: $(-4, 5)$	0	$\dfrac{-5}{4} \geq 0$ False
C: $(5, \infty)$	6	$\dfrac{1}{10} \geq 0$ True

-4 5

Solution: $(-\infty, -4) \cup [5, \infty)$

45. $\dfrac{x(x+6)}{(x-7)(x+1)} \geq 0$

$x = 0$ or $x+6 = 0$ or $x-7 = 0$ or $x+1 = 0$

$x = -6$ or $\quad x = 7$ or $x = -1$

Region	Test Point	$\dfrac{x(x+6)}{(x-7)(x+1)} \geq 0$; Result
A: $(-\infty, -6)$	-7	$\dfrac{-7(-1)}{(-14)(-6)} \geq 0$; True
B: $(-6, -1)$	-3	$\dfrac{-3(3)}{(-10)(-2)} \geq 0$; False
C: $(-1, 0)$	$-\dfrac{1}{2}$	$\dfrac{-\dfrac{1}{2}\left(\dfrac{11}{2}\right)}{\left(-\dfrac{13}{2}\right)\left(\dfrac{1}{2}\right)} \geq 0$; True
D: $(0, 7)$	2	$\dfrac{2(8)}{(-5)(3)} \geq 0$; False
E: $(7, \infty)$	8	$\dfrac{8(14)}{(1)(9)} \geq 0$; True

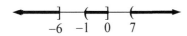

Solution: $(-\infty, -6] \cup (-1, 0] \cup (7, \infty)$

47. $\dfrac{-1}{x-1} > -1$

The denominator is equal to 0 when

$x - 1 = 0$, or $x = 1$.

$\dfrac{-1}{x-1} = -1$

$-1 = -1(x-1)$

$-1 = -x + 1$

$x = 2$

Region	Test Point	$\dfrac{-1}{x-1} > -1$; Result
A: $(-\infty, 1)$	0	$\dfrac{-1}{-1} > -1$; True
B: $(1, 2)$	$\dfrac{3}{2}$	$\dfrac{-1}{\frac{1}{2}} = -2 > -1$; False
C: $(2, \infty)$	3	$\dfrac{-1}{2} > -1$; True

Solution: $(-\infty, 1) \cup (2, \infty)$

49. $\dfrac{x}{x+4} \le 2$

The denominator is equal to 0 when
$x+4=0$, or $x=-4$.

$$\dfrac{x}{x+4} = 2$$
$$x = 2x+8$$
$$-x = 8$$
$$x = -8$$

Region	Test Point	$\dfrac{x}{x+4} \le 2$; Result
A: $(-\infty, -8)$	-9	$\dfrac{-9}{-5} \le 2$; True
B: $(-8, -4)$	-6	$\dfrac{-6}{-2} \le 2$; False
C: $(-4, \infty)$	0	$\dfrac{0}{4} \le 2$; True

Solution: $(-\infty, -8] \cup (-4, \infty)$

51. $\dfrac{z}{z-5} \ge 2z$

The denominator is equal to 0 when
$z-5=0$, or $z=5$.

$$\dfrac{z}{z-5} = 2z$$
$$z = 2z(z-5)$$
$$z = 2z^2 - 10z$$
$$0 = 2z^2 - 11z$$
$$0 = z(2z-11)$$
$$z = 0 \ \text{ or } \ 2z-11 = 0$$
$$z = \dfrac{11}{2}$$

Region	Test Point	$\dfrac{z}{z-5} \ge 2z$; Result
A: $(-\infty, 0)$	-1	$\dfrac{-1}{-6} \ge -2$; True
B: $(0, 5)$	1	$\dfrac{1}{-4} \ge 2$; False
C: $\left(5, \dfrac{11}{2}\right)$	$\dfrac{21}{4}$	$\dfrac{(21/4)}{(1/4)} \ge \dfrac{21}{2}$ $21 \ge \dfrac{21}{2}$; True
D: $\left(\dfrac{11}{2}, \infty\right)$	6	$\dfrac{6}{1} \ge 12$; False

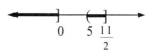

Solution: $(-\infty, 0] \cup \left(5, \dfrac{11}{2}\right]$

53. $\dfrac{(x+1)^2}{5x} > 0$

The denominator is equal to 0 when
$5x=0$, or $x=0$.

$$\dfrac{(x+1)^2}{5x} = 0$$
$$(x+1)^2 = 0$$
$$x+1 = 0$$
$$x = -1$$

Region	Test Point	$\dfrac{(x+1)^2}{5x} > 0$; Result
A: $(-\infty, -1)$	-2	$\dfrac{1}{-10} > 0$; False
B: $(-1, 0)$	$-\dfrac{1}{2}$	$\dfrac{(1/4)}{(-5/2)} > 0$; False
C: $(0, \infty)$	1	$\dfrac{4}{5} > 0$; True

Solution: $(0, \infty)$

55. $g(x) = |x| + 2$

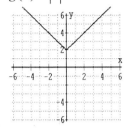

57. $F(x) = |x| - 1$

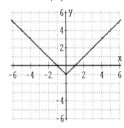

59. $F(x) = x^2 - 3$

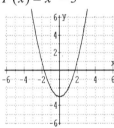

61. $H(x) = x^2 + 1$

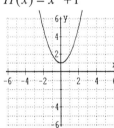

63. Answers may vary.

65. Let x = the number. Then

$\dfrac{1}{x}$ = the reciprocal of the number.

$$x - \frac{1}{x} < 0$$

$$\frac{x^2 - 1}{x} < 0$$

$$\frac{(x+1)(x-1)}{x} < 0$$

$x + 1 = 0$ or $x - 1 = 0$ or $x = 0$
  $x = -1$ or $x = 1$

Region	Test Point	$\dfrac{(x+1)(x-1)}{x} < 0$ Result
A: $(-\infty, -1)$	-2	$\dfrac{(-1)(-3)}{-2} < 0$; True
B: $(-1, 0)$	$-\dfrac{1}{2}$	$\dfrac{\left(\frac{1}{2}\right)\left(-\frac{3}{2}\right)}{\left(-\frac{1}{2}\right)} < 0$; False
C: $(0, 1)$	$\dfrac{1}{2}$	$\dfrac{\left(\frac{3}{2}\right)\left(-\frac{1}{2}\right)}{\left(\frac{1}{2}\right)} < 0$; True
D: $(1, \infty)$	2	$\dfrac{(3)(1)}{2} < 0$; False

The numbers are any number less than -1 or between 0 and 1.

67. $P(x) = -2x^2 + 26x - 44$
  $-2x^2 + 26x - 44 > 0$
  $-2(x^2 + 13x - 22) > 0$
  $-2(x - 11)(x - 2) > 0$
  $x - 11 = 0$ or $x - 2 = 0$
    $x = 11$ or $x = 2$

Region	Test Point	$-2(x-11)(x-2) > 0$ Result
A: $(0, 2)$	1	$-2(-10)(-3) > 0$ False
B: $(2, 11)$	3	$-2(-8)(1) > 0$ True
C: $(11, \infty)$	12	$-2(1)(10) > 0$ False

The company makes a profit when x is between 2 and 11.

69.

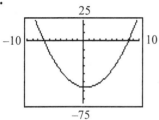

71.

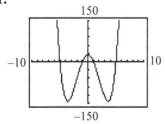

Section 11.5

Graphing Calculator Explorations

1.

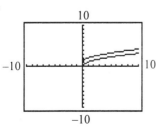

3.

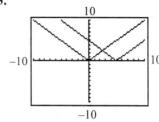

5.

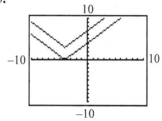

Mental Math

1. $f(x) = x^2$; vertex: $(0, 0)$

2. $f(x) = -5x^2$; vertex: $(0, 0)$

3. $g(x) = (x-2)^2$; vertex: $(2, 0)$

4. $g(x) = (x+5)^2$; vertex: $(-5, 0)$

5. $f(x) = 2x^2 + 3$; vertex: $(0, 3)$

6. $h(x) = x^2 - 1$; vertex: $(0, -1)$

7. $g(x) = (x+1)^2 + 5$; vertex: $(-1, 5)$

8. $h(x) = (x-10)^2 - 7$; vertex: $(10, -7)$

Exercise Set 11.5

1. $f(x) = x^2 - 1$

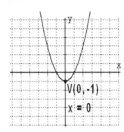

3. $h(x) = x^2 + 5$

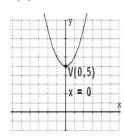

5. $g(x) = x^2 + 7$

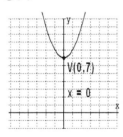

7. $f(x) = (x - 5)^2$

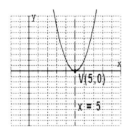

9. $h(x) = (x + 2)^2$

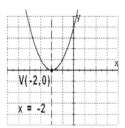

11. $G(x) = (x + 3)^2$

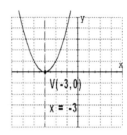

13. $f(x) = (x - 2)^2 + 5$

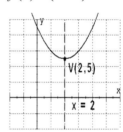

15. $h(x) = (x + 1)^2 + 4$

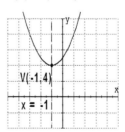

17. $g(x) = (x+2)^2 - 5$

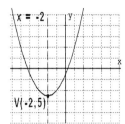

19. $g(x) = -x^2$

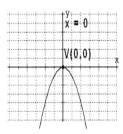

21. $h(x) = \dfrac{1}{3}x^2$

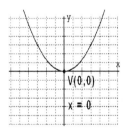

23. $H(x) = 2x^2$

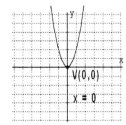

25. $f(x) = 2(x-1)^2 + 3$

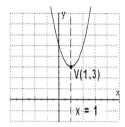

27. $h(x) = -3(x+3)^2 + 1$

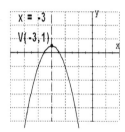

29. $H(x) = \dfrac{1}{2}(x-6)^2 - 3$

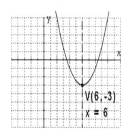

31. $f(x) = -(x-2)^2$

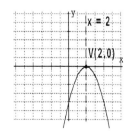

33. $F(x) = -x^2 + 4$

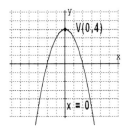

35. $F(x) = 2x^2 - 5$

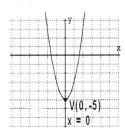

37. $h(x) = (x - 6)^2 + 4$

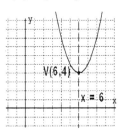

39. $F(x) = \left(x + \dfrac{1}{2}\right)^2 - 2$

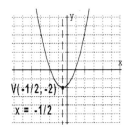

41. $F(x) = \dfrac{3}{2}(x + 7)^2 + 1$

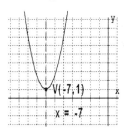

43. $f(x) = \dfrac{1}{4}x^2 - 9$

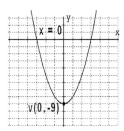

45. $G(x) = 5\left(x + \dfrac{1}{2}\right)^2$

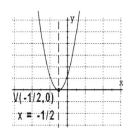

47. $h(x) = -(x - 1)^2 - 1$

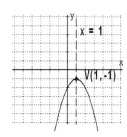

49. $g(x) = \sqrt{3}(x+5)^2 + \dfrac{3}{4}$

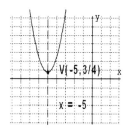

51. $h(x) = 10(x+4)^2 - 6$

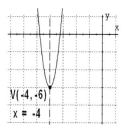

53. $f(x) = -2(x-4)^2 + 5$

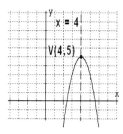

55. $x^2 + 8x$

$$\left[\dfrac{1}{2}(8)\right]^2 = (4)^2 = 16$$

$$x^2 + 8x + 16$$

57. $z^2 - 16z$

$$\left[\dfrac{1}{2}(-16)\right]^2 = (-8)^2 = 64$$

$$z^2 - 16z + 64$$

59. $y^2 + y$

$$\left[\dfrac{1}{2}(1)\right]^2 = \left(\dfrac{1}{2}\right)^2 = \dfrac{1}{4}$$

$$y^2 + y + \dfrac{1}{4}$$

61. $\qquad x^2 + 4x = 12$

$$x^2 + 4x + \left(\dfrac{4}{2}\right)^2 = 12 + 4$$

$$x^2 + 4x + 4 = 16$$

$$(x+2)^2 = 16$$

$$x + 2 = \pm\sqrt{16}$$

$$x + 2 = \pm 4$$

$$x = -2 \pm 4$$

$$x = -6 \ \text{ or } \ 2$$

63. $\qquad z^2 + 10z - 1 = 0$

$$z^2 + 10z = 1$$

$$z^2 + 10z + \left(\dfrac{10}{2}\right)^2 = 1 + 25$$

$$z^2 + 10z + 25 = 26$$

$$(z+5)^2 = 26$$

$$z + 5 = \pm\sqrt{26}$$

$$z = -5 \pm \sqrt{26}$$

65. $\qquad z^2 - 8z = 2$

$$z^2 - 8z + \left(\dfrac{-8}{2}\right)^2 = 2 + 16$$

$$z^2 - 8z + 16 = 18$$

$$(z-4)^2 = 18$$

$$z - 4 = \pm\sqrt{18}$$

$$z - 4 = \pm 3\sqrt{2}$$

$$z = 4 \pm 3\sqrt{2}$$

67. $f(x) = 5(x-2)^2 + 3$

69. $f(x) = 5[x-(-3)]^2 + 6$

$\qquad = 5(x+3)^2 + 6$

71. $y = f(x) + 1$

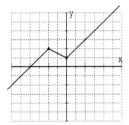

73. $y = f(x - 3)$

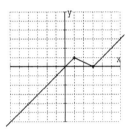

75. $y = f(x + 2) + 2$

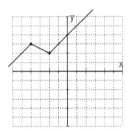

Exercise Set 11.6

1. $f(x) = x^2 + 8x + 7$

$-\dfrac{b}{2a} = \dfrac{-8}{2(1)} = -4$ and

$f(-4) = (-4)^2 + 8(-4) + 7$
$\quad\quad = 16 - 32 + 7$
$\quad\quad = -9$

Thus, the vertex is $(-4, -9)$.

3. $f(x) = -x^2 + 10x + 5$

$-\dfrac{b}{2a} = \dfrac{-10}{2(-1)} = 5$ and

$f(5) = -(5)^2 + 10(5) + 5$
$\quad\quad = -25 + 50 + 5$
$\quad\quad = 30$

Thus, the vertex is $(5, 30)$.

5. $f(x) = 5x^2 - 10x + 3$

$-\dfrac{b}{2a} = \dfrac{-(-10)}{2(5)} = 1$ and

$f(1) = 5(1)^2 - 10(1) + 3$
$\quad\quad = 5 - 10 + 3$
$\quad\quad = -2$

Thus, the vertex is $(1, -2)$.

7. $f(x) = -x^2 + x + 1$

$-\dfrac{b}{2a} = \dfrac{-1}{2(-1)} = \dfrac{1}{2}$ and

$f\left(\dfrac{1}{2}\right) = -\left(\dfrac{1}{2}\right)^2 + \left(\dfrac{1}{2}\right) + 1$

$\quad\quad = -\dfrac{1}{4} + \dfrac{1}{2} + 1$

$\quad\quad = \dfrac{5}{4}$

Thus, the vertex is $\left(\dfrac{1}{2}, \dfrac{5}{4}\right)$.

9. $f(x) = x^2 - 4x + 3$

$-\dfrac{b}{2a} = \dfrac{-(-4)}{2(1)} = 2$ and

$f(2) = (2)^2 - 4(2) + 3 = -1$

The vertex is $(2, -1)$, so the graph is D.

11. $f(x) = x^2 - 2x - 3$

$-\dfrac{b}{2a} = \dfrac{-(-2)}{2(1)} = 1$ and

$f(1) = (1)^2 - 2(1) - 3 = -4$

The vertex is $(, -4)$, so the graph is B.

13. $f(x) = x^2 + 4x - 5$

$-\dfrac{b}{2a} = \dfrac{-4}{2(1)} = -2$ and

$f(-2) = (-2)^2 + 4(-2) - 5 = -9$

Thus, the vertex is $(-2, -9)$.

The graph opens upward ($a = 1 > 0$).

$x^2 + 4x - 5 = 0$

$(x + 5)(x - 1) = 0$

$x + 5 = 0 \quad$ or $\quad x - 1 = 0$

$\quad x = -5 \quad$ or $\qquad x = 1$

x-intercepts: $(-5, 0)$ and $(1, 0)$.

$f(0) = -5$, so the y-intercept is $(0, -5)$.

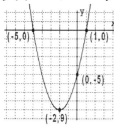

15. $f(x) = -x^2 + 2x - 1$

$-\dfrac{b}{2a} = \dfrac{-2}{2(-1)} = 1$ and

$f(1) = -(1)^2 + 2(1) - 1 = 0$

Thus, the vertex is $(1, 0)$.

The graph opens downward ($a = -1 < 0$).

$-x^2 + 2x - 1 = 0$

$x^2 - 2x + 1 = 0$

$(x - 1)^2 = 0$

$x - 1 = 0$

$x = 1$

x-intercept: $(1, 0)$.

$f(0) = -1$, so the y-intercept is $(0, -1)$.

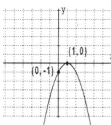

17. $f(x) = x^2 - 4$

$-\dfrac{b}{2a} = \dfrac{-0}{2(1)} = 0$ and

$f(0) = (0)^2 - 4 = -4$

Thus, the vertex is $(0, -4)$.

The graph opens upward ($a = 1 > 0$).

$x^2 - 4 = 0$

$(x + 2)(x - 2) = 0$

$x + 2 = 0 \quad$ or $\quad x - 2 = 0$

$\quad x = -2 \quad$ or $\qquad x = 2$

x-intercepts: $(-2, 0)$ and $(2, 0)$.

$f(0) = -4$, so the y-intercept is $(0, -4)$.

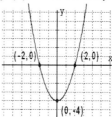

19. $f(x) = 4x^2 + 4x - 3$

$-\dfrac{b}{2a} = \dfrac{-4}{2(4)} = -\dfrac{1}{2}$ and

$f\left(-\dfrac{1}{2}\right) = 4\left(-\dfrac{1}{2}\right)^2 + 4\left(-\dfrac{1}{2}\right) - 3 = -4$

Thus, the vertex is $\left(-\dfrac{1}{2}, -4\right)$.

The graph opens upward ($a = 4 > 0$).

$4x^2 + 4x - 3 = 0$

$(2x + 3)(2x - 1) = 0$

$2x + 3 = 0 \quad$ or $\quad 2x - 1 = 0$

$\quad x = -\dfrac{3}{2} \quad$ or $\qquad x = \dfrac{1}{2}$

x-intercepts: $\left(-\dfrac{3}{2}, 0\right)$ and $\left(\dfrac{1}{2}, 0\right)$.

$f(0) = -3$, so the y-intercept is $(0, -3)$.

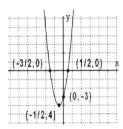

21.
$$f(x) = x^2 + 8x + 15$$
$$y = x^2 + 8x + 15$$
$$y - 15 = x^2 + 8x$$
$$y - 15 + 16 = x^2 + 8x + 16$$
$$y - 1 = (x + 4)^2$$
$$y = (x + 4)^2 + 1$$
$$f(x) = (x + 4)^2 + 1$$

Thus, the vertex is (–4, 1).
The graph opens upward ($a = 1 > 0$).
$$x^2 + 8x + 15 = 0$$
$$(x + 5)(x + 3) = 0$$
$$x + 5 = 0 \quad \text{or} \quad x + 3 = 0$$
$$x = -5 \quad \text{or} \quad x = -3$$

x-intercepts: (–5, 0) and (–3, 0).
$f(0) = 15$, so the y-intercept is (0, 15).

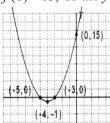

23.
$$f(x) = x^2 - 6x + 5$$
$$y = x^2 - 6x + 5$$
$$y - 5 = x^2 - 6x$$
$$y - 5 + 9 = x^2 - 6x + 9$$
$$y + 4 = (x - 3)^2$$
$$y = (x - 3)^2 - 4$$
$$f(x) = (x - 3)^2 - 4$$

Thus, the vertex is (3, –4).
The graph opens upward ($a = 1 > 0$).

$$x^2 - 6x + 5 = 0$$
$$(x - 5)(x - 1) = 0$$
$$x = -5 \quad \text{or} \quad x = 1$$

x-intercepts: (–5, 0) and (1, 0).
$f(0) = 5$, so the y-intercept is (0, 5).

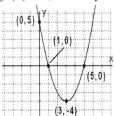

25.
$$f(x) = x^2 - 4x + 5$$
$$y = x^2 - 4x + 5$$
$$y - 5 = x^2 - 4x$$
$$y - 5 + 4 = x^2 - 4x + 4$$
$$y - 1 = (x - 2)^2$$
$$y = (x - 2)^2 + 1$$
$$f(x) = (x - 2)^2 + 1$$

Thus, the vertex is (2, 1).
The graph opens upward ($a = 1 > 0$).
$$x^2 - 4x + 5 = 0$$
$$x = \frac{4 \pm \sqrt{(-4)^2 - 4(1)(5)}}{2(1)} = \frac{4 \pm \sqrt{-4}}{2}$$

which give non-real solutions.
Hence, there are no x-intercepts.
$f(0) = 5$, so the y-intercept is (0, 5).

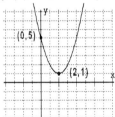

27. $f(x) = 2x^2 + 4x + 5$
$$y = 2x^2 + 4x + 5$$
$$y - 5 = 2(x^2 + 2x)$$
$$y - 5 + 2(1) = 2(x^2 + 2x + 1)$$
$$y - 3 = 2(x+1)^2$$
$$y = 2(x+1)^2 + 3$$
$$f(x) = 2(x+1)^2 + 3$$

Thus, the vertex is (–1, 3).
The graph opens upward ($a = 2 > 0$).
$$2x^2 + 4x + 5 = 0$$
$$x = \frac{-4 \pm \sqrt{(4)^2 - 4(2)(5)}}{2(2)} = \frac{-4 \pm \sqrt{-24}}{4}$$

which give non-real solutions.
Hence, there are no x-intercepts.
$f(0) = 5$, so the y-intercept is (0, 5).

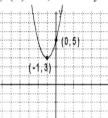

29. $f(x) = -2x^2 + 12x$
$$y = -2(x^2 - 6x)$$
$$y + [-2(9)] = -2(x^2 - 6x + 9)$$
$$y - 18 = -2(x-3)^2$$
$$y = -2(x-3)^2 + 18$$
$$f(x) = -2(x-3)^2 + 18$$

Thus, the vertex is (3, 18).
The graph opens downward ($a = -2 < 0$).
$$-2x^2 + 12x = 0$$
$$-2x(x - 6) = 0$$
$$x = 0 \ \text{ or } \ x - 6 = 0$$
$$x = 6$$

x-intercepts: (0, 0) and (6, 0)

$f(0) = 0$, so the y-intercept is (0, 0).

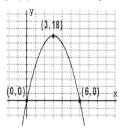

31. $f(x) = x^2 + 1$
$$x = -\frac{b}{2a} = -\frac{0}{2(1)} = 0$$
$$f(0) = (0)^2 + 1 = 1$$

Thus, the vertex is (0, 1).
The graph opens upward ($a = 1 > 0$).
$$x^2 + 1 = 0$$
$$x^2 = -1$$

which give non-real solutions.
Hence, there are no x-intercepts.
$f(0) = 1$, so the y-intercept is (0, 1).

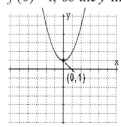

33. $f(x) = x^2 - 2x - 15$
$$y = x^2 - 2x - 15$$
$$y + 15 = x^2 - 2x$$
$$y + 15 + 1 = x^2 - 2x + 1$$
$$y + 16 = (x - 1)^2$$
$$y = (x - 1)^2 - 16$$
$$f(x) = (x - 1)^2 - 16$$

Thus, the vertex is (1, –16).
The graph opens upward ($a = 1 > 0$).
$$x^2 - 2x - 15 = 0$$
$$(x - 5)(x + 3) = 0$$
$$x = 5 \ \text{ or } \ x = -3$$

x-intercepts: (–3, 0) and (5, 0).

$f(0) = -15$ so the y-intercept is $(0, -15)$.

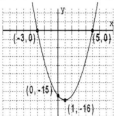

35. $f(x) = -5x^2 + 5x$

$x = -\dfrac{b}{2a} = \dfrac{-5}{2(-5)} = \dfrac{1}{2}$ and

$f\left(\dfrac{1}{2}\right) = -5\left(\dfrac{1}{2}\right)^2 + 5\left(\dfrac{1}{2}\right) = -\dfrac{5}{4} + \dfrac{5}{2} = \dfrac{5}{4}$

Thus, the vertex is $\left(\dfrac{1}{2}, \dfrac{5}{4}\right)$.

The graph opens downward ($a = -5 < 0$).
$-5x^2 + 5x = 0$
$-5x(x - 1) = 0$
$x = 0$ or $x - 1 = 0$
$\phantom{x = 0 \text{ or } x - 1} x = 1$

x-intercepts: $(0, 0)$ and $(1, 0)$
$f(0) = 0$, so the y-intercept is $(0, 0)$.

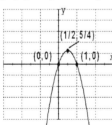

37. $f(x) = -x^2 + 2x - 12$

$x = -\dfrac{b}{2a} = \dfrac{-2}{2(-1)} = 1$ and

$f(1) = -(1)^2 + 2(1) - 12 = -11$

Thus, the vertex is $(1, -11)$.
The graph opens downward ($a = -1 < 0$).
$-x^2 + 2x - 12 = 0$
$x^2 - 2x + 12 = 0$

$x = \dfrac{2 \pm \sqrt{(-2)^2 - 4(1)(12)}}{2(1)} = \dfrac{2 \pm \sqrt{-44}}{2}$

which yields non-real solutions.
Hence, there are no x-intercepts.
$f(0) = -12$ so the y-intercept is $(0, -12)$.

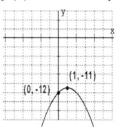

39. $f(x) = 3x^2 - 12 + 15$

$x = -\dfrac{b}{2a} = \dfrac{-(-12)}{2(3)} = \dfrac{12}{6} = 2$ and

$f(2) = 3(2)^2 - 12(2) + 15$
$ = 12 - 24 + 15 = 3$

Thus, the vertex is $(2, 3)$.
The graph opens upward ($a = 3 > 0$).
$3x^2 - 12x + 15 = 0$
$x^2 - 4x + 5 = 0$

$x = \dfrac{4 \pm \sqrt{(-4)^2 - 4(1)(5)}}{2(1)} = \dfrac{4 \pm \sqrt{-4}}{2}$

which yields non-real solutions.
Hence, there are no x-intercepts.
$f(0) = 15$, so the y-intercept is $(0, 15)$.

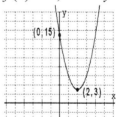

41. $f(x) = x^2 + x - 6$

$$x = -\frac{b}{2a} = \frac{-1}{2(1)} = -\frac{1}{2} \text{ and}$$

$$f\left(-\frac{1}{2}\right) = \left(-\frac{1}{2}\right)^2 + \left(-\frac{1}{2}\right) - 6$$

$$= \frac{1}{4} - \frac{1}{2} - 6 = -\frac{25}{4}$$

Thus, the vertex is $\left(-\frac{1}{2}, -\frac{25}{4}\right)$.

The graph opens upward ($a = 1 > 0$).

$$x^2 + x - 6 = 0$$
$$(x+3)(x-2) = 0$$
$$x = 3 \text{ or } x = -2$$

x-intercepts: $(-3, 0)$ and $(2, 0)$.

$f(0) = -6$ so the y-intercept is $(0, -6)$.

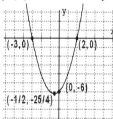

43. $f(x) = -2x^2 - 3x + 35$

$$x = -\frac{b}{2a} = \frac{-(-3)}{2(-2)} = -\frac{3}{4} \text{ and}$$

$$f\left(-\frac{3}{4}\right) = -2\left(-\frac{3}{4}\right)^2 - 3\left(-\frac{3}{4}\right) + 35$$

$$= -\frac{9}{8} + \frac{9}{4} + 35 = -\frac{289}{8}$$

Thus, the vertex is $\left(-\frac{3}{4}, \frac{289}{8}\right)$.

The graph opens downward ($a = -2 < 0$).

$$-2x^2 - 3x + 35 = 0$$
$$2x^2 + 3x - 35 = 0$$
$$(2x-7)(x+5) = 0$$
$$2x - 7 = 0 \text{ or } x + 5 = 0$$
$$x = \frac{7}{2} \text{ or } \quad x = -5$$

x-intercepts: $(-5, 0)$ and $\left(\frac{7}{2}, 0\right)$.

$f(0) = 35$ so the y-intercept is $(0, 35)$.

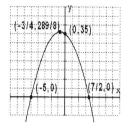

45. $h(t) = -16t^2 + 96t$

$$t = -\frac{b}{2a} = \frac{-96}{2(-16)} = \frac{96}{32} = 3 \text{ and}$$

$$h(3) = -16(3)^2 + 96(3)$$
$$= -144 + 288$$
$$= 144$$

The maximum height is 144 feet.

47. $h(t) = -16t^2 + 32t$

$$t = -\frac{b}{2a} = \frac{-32}{2(-16)} = \frac{32}{32} = 1 \text{ and}$$

$$h(1) = -16(1)^2 + 32(1)$$
$$= -16 + 32$$
$$= 16$$

The maximum height is 16 feet.

49. Let x = one number. Then
$60 - x$ = the other number.
$$f(x) = x(60 - x)$$
$$= 60x - x^2$$
$$= -x^2 + 60x$$
The maximum will occur at the vertex.
$$x = -\frac{b}{2a} = \frac{-60}{2(-1)} = 30$$
$$60 - x = 60 - 30 = 30$$
The numbers are 30 and 30.

51. Let x = one number. Then
$10 + x$ = the other number.
$$f(x) = x(10 + x)$$
$$= 10x + x^2$$
$$= x^2 + 10x$$
The minimum will occur at the vertex.

$$x = -\frac{b}{2a} = \frac{-10}{2(1)} = -5$$

$$10 + x = 10 + (-5) = 5$$

The numbers are –5 and 5.

53. Let x = width. Then
$40 - x$ = the length.
Area = length · width
$$A(x) = (40 - x)x$$
$$= 40x - x^2$$
$$= -x^2 + 40x$$
The maximum will occur at the vertex.

$$x = -\frac{b}{2a} = \frac{-40}{2(-1)} = 20$$

$$40 - x = 40 - 20 = 20$$

The maximum area will occur when the length and width are 20 units each.

55. $f(x) = x^2 + 2$

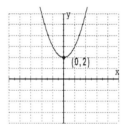

57. $g(x) = x + 2$

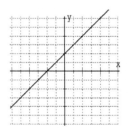

59. $f(x) = (x + 5)^2 + 2$

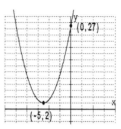

61. $f(x) = 3(x - 4)^2 + 1$

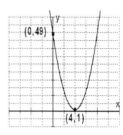

63. $f(x) = -(x - 4)^2 + \dfrac{3}{2}$

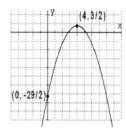

65. $f(x) = x^2 + 10x + 15$

$$x = -\frac{b}{2a} = \frac{-10}{2(1)} = -5 \text{ and}$$

$$f(-5) = (-5)^2 + 10(-5) + 15 = -10$$

Thus, the vertex is (–5, –10).
The graph opens upward ($a = 1 > 0$).
$f(0) = 15$ so the y-intercept is (0, 15).

$$x^2 + 10x + 15 = 0$$

$$x = \frac{-10 \pm \sqrt{(10)^2 - 4(1)(15)}}{2(1)}$$

$$= \frac{-10 \pm \sqrt{40}}{2} \approx -8.2 \text{ or } -1.8$$

The *x*-intercepts are approximately $(-8.2, 0)$ and $(-1.8, 0)$.

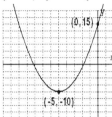

67. $f(x) = 3x^2 - 6x + 7$

$x = -\dfrac{b}{2a} = \dfrac{-(-6)}{2(3)} = 1$ and

$f(1) = 3(1)^2 - 6(1) + 7 = 4$

Thus, the vertex is $(1, 4)$.

The graph opens upward ($a = 1 > 0$).

$f(0) = 7$ so the *y*-intercept is $(0, 7)$.

$3x^2 - 6x + 7 = 0$

$x = \dfrac{6 \pm \sqrt{(-6)^2 - 4(3)(7)}}{2(1)} = \dfrac{6 \pm \sqrt{-48}}{2}$

which yields non-real solutions.

Hence, there are no *x*-intercepts.

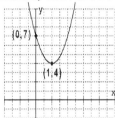

69. $f(x) = 2.3x^2 - 6.1x + 3.2$

minimum ≈ -0.84

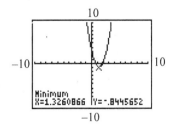

71. $f(x) = -1.9x^2 + 5.6x - 2.7$

maximum ≈ 1.43

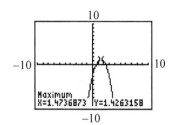

73. $p(x) = -x^2 + 93x + 1128$

a. It will have a maximum; answer may vary (e.g., since $a = -1 < 0$).

b. $x = -\dfrac{b}{2a} = \dfrac{-93}{2(-1)} = 46.5$

$1990 + 46.5 = 2036.5$

In the year 2036.

c. $p(46.5) = -(46.5)^2 + 93(46.5) + 1128$

$= 3290.25$ thousand inmates,

or 32,902,500 inmates

75.

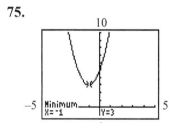

77.

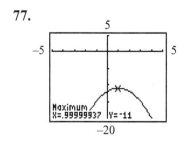

Chapter 11 Review

1. $x^2 - 15x + 14 = 0$
$(x - 14)(x - 1) = 0$
$x - 14 = 0$ or $x - 1 = 0$
$\quad x = 14$ or $\quad x = 1$
The solutions are 1 and 14.

2. $x^2 - x - 30 = 0$
$(x + 5)(x - 6) = 0$
$x + 5 = 0$ or $x - 6 = 0$
$\quad x = -5$ or $\quad x = 6$
The solutions are -5 and 6.

3. $\qquad 10x^2 = 3x + 4$
$10x^2 - 3x - 4 = 0$
$(5x - 4)(2x + 1) = 0$
$5x - 4 = 0$ or $2x + 1 = 0$
$\quad 5x = 4$ or $\quad 2x = -1$
$\quad x = \dfrac{4}{5}$ or $\quad x = -\dfrac{1}{2}$
The solutions are $-\dfrac{1}{2}$ and $\dfrac{4}{5}$.

4. $\qquad 7a^2 = 29a + 30$
$7a^2 - 29a - 30 = 0$
$(7a + 6)(a - 5) = 0$
$7a + 6 = 0$ or $a - 5 = 0$
$\quad 7a = -6$ or $\quad a = 5$
$\quad a = -\dfrac{6}{7}$
The solutions are $-\dfrac{6}{7}$ and 5.

5. $4m^2 = 196$
$\quad m^2 = 49$
$\quad m = \pm\sqrt{49}$
$\quad m = \pm 7$
The solutions are -7 and 7.

6. $9y^2 = 36$
$\quad y^2 = 4$
$\quad y = \pm\sqrt{4}$
$\quad y = \pm 2$
The solutions are -2 and 2.

7. $(9n + 1)^2 = 9$
$\quad 9n + 1 = \pm\sqrt{9}$
$\quad 9n + 1 = \pm 3$
$\quad 9n = -1 \pm 3$
$\quad n = \dfrac{-1 \pm 3}{9} = \dfrac{2}{9}, -\dfrac{4}{9}$
The solutions are $-\dfrac{4}{9}$ and $\dfrac{2}{9}$.

8. $(5x - 2)^2 = 2$
$\quad 5x - 2 = \pm\sqrt{2}$
$\quad 5x = 2 \pm \sqrt{2}$
$\quad x = \dfrac{2 \pm \sqrt{2}}{5}$
The solutions are $\dfrac{2 + \sqrt{2}}{5}$ and $\dfrac{2 - \sqrt{2}}{5}$.

9. $\qquad z^2 + 3z + 1 = 0$
$\qquad z^2 + 3z = -1$
$z^2 + 3z + \left(\dfrac{3}{2}\right)^2 = -1 + \dfrac{9}{4}$
$\qquad \left(z + \dfrac{3}{2}\right)^2 = \dfrac{5}{4}$
$\qquad z + \dfrac{3}{2} = \pm\sqrt{\dfrac{5}{4}}$
$\qquad z + \dfrac{3}{2} = \pm\dfrac{\sqrt{5}}{2}$
$\qquad z = -\dfrac{3}{2} \pm \dfrac{\sqrt{5}}{2}$
The solutions are $-\dfrac{3}{2} + \dfrac{\sqrt{5}}{2}$ and $-\dfrac{3}{2} - \dfrac{\sqrt{5}}{2}$.

10.
$$x^2 + x + 7 = 0$$
$$x^2 + x = -7$$
$$x^2 + x + \left(\frac{1}{2}\right)^2 = -7 + \frac{1}{4}$$
$$\left(x + \frac{1}{2}\right)^2 = -\frac{27}{4}$$
$$x + \frac{1}{2} = \pm\sqrt{-\frac{27}{4}}$$
$$x + \frac{1}{2} = \pm\frac{\sqrt{9 \cdot 3}}{2}i$$
$$x + \frac{1}{2} = \pm\frac{3\sqrt{3}}{2}i$$
$$x = -\frac{1}{2} \pm \frac{3\sqrt{3}}{2}i$$

The solutions are $-\frac{1}{2} + \frac{3\sqrt{3}}{2}i$ and

$-\frac{1}{2} - \frac{3\sqrt{3}}{2}i$.

11.
$$(2x + 1)^2 = x$$
$$4x^2 + 4x + 1 = x$$
$$4x^2 + 3x = -1$$
$$x^2 + \frac{3}{4}x = -\frac{1}{4}$$
$$x^2 + \frac{3}{4}x + \left(\frac{\frac{3}{4}}{2}\right)^2 = -\frac{1}{4} + \frac{9}{64}$$
$$\left(x + \frac{3}{8}\right)^2 = -\frac{7}{64}$$
$$x + \frac{3}{8} = \pm\sqrt{-\frac{7}{64}}$$
$$x + \frac{3}{8} = \pm\frac{\sqrt{7}}{8}i$$
$$x = -\frac{3}{8} \pm \frac{\sqrt{7}}{8}i$$

The solutions are $-\frac{3}{8} + \frac{\sqrt{7}}{8}i$ and

$-\frac{3}{8} - \frac{\sqrt{7}}{8}i$.

12.
$$(3x - 4)^2 = 10x$$
$$9x^2 - 24x + 16 = 10x$$
$$9x^2 - 34x = -16$$
$$x^2 - \frac{34}{9}x = -\frac{16}{9}$$
$$x^2 - \frac{34}{9}x + \left(\frac{-\frac{34}{9}}{2}\right)^2 = -\frac{16}{9} + \frac{289}{81}$$
$$\left(x - \frac{17}{9}\right)^2 = \frac{145}{81}$$
$$x - \frac{17}{9} = \pm\sqrt{\frac{145}{81}}$$
$$x - \frac{17}{9} = \pm\frac{\sqrt{145}}{9}$$
$$x = \frac{17 \pm \sqrt{145}}{9}$$

The solutions are $\frac{17 + \sqrt{145}}{9}$ and

$\frac{17 - \sqrt{145}}{9}$.

13.
$$A = P(1 + r)^2$$
$$2717 = 2500(1 + r)^2$$
$$\frac{2717}{2500} = (1 + r)^2$$
$$(1 + r)^2 = 1.0868$$
$$1 + r = \pm\sqrt{1.0868}$$
$$1 + r = \pm 1.0425$$
$$r = -1 \pm 1.0425$$
$$= 0.0425 \text{ or } -2.0425 \text{ (disregard)}$$
The interest rate is 4.25%.

14. Let x = distance traveled.
$$a^2 + b^2 = c^2$$
$$x^2 + x^2 = (150)^2$$
$$2x^2 = 22,500$$
$$x^2 = 11,250$$
$$x = \pm 75\sqrt{2} \approx \pm 106.1$$
Disregard the negative.
The ship traveled $75\sqrt{2} \approx 106.1$ miles.

15. Two complex but not real solutions exist.

16. Two real solutions exist.

17. Two real solutions exist.

18. One real solution exist.

19. $x^2 - 16x + 64 = 0$
$a = 1, b = -16, c = 64$
$$x = \frac{16 \pm \sqrt{(-16)^2 - 4(1)(64)}}{2(1)}$$
$$= \frac{16 \pm \sqrt{256 - 256}}{2} = \frac{16 \pm \sqrt{0}}{2} = 8$$
The solution is 8.

20. $x^2 + 5x = 0$
$a = 1, b = 5, c = 0$
$$x = \frac{-5 \pm \sqrt{(5)^2 - 4(1)(0)}}{2(1)}$$
$$= \frac{-5 \pm \sqrt{25}}{2}$$
$$= \frac{-5 \pm 5}{2} = 0 \text{ or } -5$$
The solutions are −5 and 0.

21. $x^2 + 11 = 0$
$a = 1, b = 0, c = 11$
$$x = \frac{0 \pm \sqrt{(0)^2 - 4(1)(11)}}{2(1)}$$
$$= \frac{\pm \sqrt{-44}}{2}$$
$$= \frac{\pm 2i\sqrt{11}}{2} = \pm i\sqrt{11}$$
The solutions are $-i\sqrt{11}$ and $i\sqrt{11}$.

22. $2x^2 + 3x = 5$
$2x^2 + 3x - 5 = 0$
$a = 2, b = 3, c = -5$
$$x = \frac{-3 \pm \sqrt{(3)^2 - 4(2)(-5)}}{2(2)}$$
$$= \frac{-3 \pm \sqrt{49}}{4}$$
$$= \frac{-3 \pm 7}{4} = 1 \text{ or } -\frac{5}{2}$$
The solutions are $-\frac{5}{2}$ and 1.

23. $6x^2 + 7 = 5x$
$6x^2 - 5x + 7 = 0$
$a = 6, b = -5, c = 7$
$$x = \frac{5 \pm \sqrt{(-5)^2 - 4(6)(7)}}{2(6)}$$
$$= \frac{5 \pm \sqrt{25 - 168}}{12}$$
$$= \frac{5 \pm \sqrt{-143}}{12} = \frac{5 \pm i\sqrt{143}}{12} = \frac{5}{12} \pm \frac{\sqrt{143}}{12} i$$
The solutions are $\frac{5 + i\sqrt{143}}{12}$ and
$\frac{5 - i\sqrt{143}}{12}$.

24. $9a^2 + 4 = 2a$
$9a^2 - 2a + 4 = 0$
$$a = \frac{2 \pm \sqrt{(-2)^2 - 4(9)(4)}}{2(9)}$$
$$= \frac{2 \pm \sqrt{-140}}{18}$$
$$= \frac{2 \pm i\sqrt{4 \cdot 35}}{18}$$
$$= \frac{2 \pm 2i\sqrt{35}}{18} = \frac{1 \pm i\sqrt{35}}{9} = \frac{1}{9} \pm \frac{\sqrt{35}}{9} i$$
The solutions are $\frac{1 + i\sqrt{35}}{9}$ and
$\frac{1 - i\sqrt{35}}{9}$.

25.
$$(5a-2)^2 - a = 0$$
$$25a^2 - 20a + 4 - a = 0$$
$$25a^2 - 21a + 4 = 0$$
$$a = \frac{21 \pm \sqrt{(-21)^2 - 4(25)(4)}}{2(25)}$$
$$= \frac{21 \pm \sqrt{441 - 400}}{50}$$
$$= \frac{21 \pm \sqrt{41}}{50}$$

The solutions are $\dfrac{21 + \sqrt{41}}{50}$ and $\dfrac{21 - \sqrt{41}}{50}$.

26.
$$(2x-3)^2 = x$$
$$4x^2 - 12x + 9 - x = 0$$
$$4x^2 - 13x + 9 = 0$$
$$a = 4, b = -13, c = 9$$
$$x = \frac{13 \pm \sqrt{(-13)^2 - 4(4)(9)}}{2(4)}$$
$$= \frac{13 \pm \sqrt{169 - 144}}{8}$$
$$= \frac{13 \pm \sqrt{25}}{8}$$
$$= \frac{13 \pm 5}{8} = \frac{9}{4} \text{ or } 1$$

The solutions are 1 and $\dfrac{9}{4}$.

27. $d(t) = -16t^2 + 30t + 6$

   a. $d(1) = -16(1)^2 + 30(1) + 6$
$$= -16 + 30 + 6$$
$$= 20 \text{ feet}$$

   b. $-16t^2 + 30t + 6 = 0$
$$8t^2 - 15t - 3 = 0$$
$$a = 8, b = -15, c = -3$$
$$t = \frac{15 \pm \sqrt{(-15)^2 - 4(8)(-3)}}{2(8)}$$
$$= \frac{15 \pm \sqrt{225 + 96}}{16}$$
$$= \frac{15 \pm \sqrt{321}}{16}$$

Disregarding the negative, we have
$$t = \frac{15 + \sqrt{321}}{16} \text{ seconds} \approx 2.1 \text{ seconds}.$$

28. Let $x =$ length of the legs. Then
$x + 6 =$ length of the hypotenuse.
$$x^2 + x^2 = (x+6)^2$$
$$2x^2 = x^2 + 12x + 36$$
$$x^2 - 12x - 36 = 0$$
$$a = 1, b = -12, c = -36$$
$$x = \frac{12 \pm \sqrt{(-12)^2 - 4(1)(-36)}}{2(1)}$$
$$= \frac{12 \pm \sqrt{144 + 144}}{2}$$
$$= \frac{12 \pm \sqrt{144 \cdot 2}}{2}$$
$$= \frac{12 \pm 12\sqrt{2}}{2} = 6 \pm 6\sqrt{2}$$

Disregard the negative.
The length of each leg is $\left(6 + 6\sqrt{2}\right)$ cm.

29.
$$x^3 = 27$$
$$x^3 - 27 = 0$$
$$(x - 3)(x^2 + 3x + 9) = 0$$
$$x - 3 = 0 \text{ or } x^2 + 3x + 9 = 0$$
$$x = 3 \qquad a = 1, b = 3, c = 9$$
$$x = \frac{-3 \pm \sqrt{(3)^2 - 4(1)(9)}}{2(1)}$$
$$= \frac{-3 \pm \sqrt{9 - 36}}{2}$$
$$= \frac{-3 \pm \sqrt{-27}}{2}$$
$$= \frac{-3 \pm 3i\sqrt{3}}{2} \text{ or } -\frac{3}{2} \pm \frac{3\sqrt{3}}{2} i$$

The solutions are 3, $-\frac{3}{2} + \frac{3\sqrt{3}}{2} i$, and $-\frac{3}{2} - \frac{3\sqrt{3}}{2} i$.

30.
$$y^3 = -64$$
$$y^3 + 64 = 0$$
$$(y + 4)(y^2 - 4y + 16) = 0$$
$$y + 4 = 0 \quad \text{or } y^2 - 4y + 16 = 0$$
$$y = -4 \qquad a = 1, b = -4, c = 16$$
$$y = \frac{4 \pm \sqrt{(-4)^2 - 4(1)(16)}}{2(1)}$$
$$= \frac{4 \pm \sqrt{16 - 64}}{2}$$
$$= \frac{4 \pm \sqrt{-48}}{2} = \frac{4 \pm 4i\sqrt{3}}{2} = 2 \pm 2i\sqrt{3}$$

The solutions are –4, $2 + 2i\sqrt{3}$, and $2 - 2i\sqrt{3}$.

31.
$$\frac{5}{x} + \frac{6}{x - 2} = 3$$
$$x(x - 2)\left(\frac{5}{x} + \frac{6}{x - 2}\right) = 3x(x - 2)$$
$$5(x - 2) + 6x = 3x^2 - 6x$$
$$5x - 10 + 6x = 3x^2 - 6x$$

$$0 = 3x^2 - 17x + 10$$
$$0 = (3x - 2)(x - 5)$$
$$3x - 2 = 0 \text{ or } x - 5 = 0$$
$$x = \frac{2}{3} \text{ or } \qquad x = 5$$

The solutions area $\frac{2}{3}$ and 5.

32.
$$\frac{7}{8} = \frac{8}{x^2}$$
$$7x^2 = 64$$
$$x^2 = \frac{64}{7}$$
$$x = \pm\sqrt{\frac{64}{7}}$$
$$x = \pm\frac{8}{\sqrt{7}} = \pm\frac{8 \cdot \sqrt{7}}{\sqrt{7} \cdot \sqrt{7}} = \pm\frac{8\sqrt{7}}{7}$$

The solutions are $-\frac{8\sqrt{7}}{7}$ and $\frac{8\sqrt{7}}{7}$.

33.
$$x^4 - 21x^2 - 100 = 0$$
$$(x^2 - 25)(x^2 + 4) = 0$$
$$(x + 5)(x - 5)(x^2 + 4) = 0$$
$$x + 5 = 0 \quad \text{or } x - 5 = 0 \text{ or } x^2 + 4 = 0$$
$$x = -5 \text{ or } \qquad x = 5 \text{ or } \qquad x^2 = -4$$
$$x = \pm 2i$$

The solutions are –5, 5 –2i, and 2i.

34.
$$5(x + 3)^2 - 19(x + 3) = 4$$
$$5(x + 3)^2 - 19(x + 3) - 4 = 0$$
Let $y = x + 3$. Then $y^2 = (x + 3)^2$ and
$$5y^2 - 19y - 4 = 0$$
$$(5y + 1)(y - 4) = 0$$
$$5y + 1 = 0 \qquad \text{or } y - 4 = 0$$
$$y = -\frac{1}{5} \text{ or } \qquad y = 4$$
$$x + 3 = -\frac{1}{5} \text{ or } x + 3 = 4$$
$$x = -\frac{16}{5} \text{ or } \qquad x = 1$$

The solutions are $-\frac{16}{5}$ and 1.

35. $x^{2/3} - 6x^{1/3} + 5 = 0$

Let $y = x^{1/3}$. Then $y^2 = x^{2/3}$ and

$y^2 - 6y + 5 = 0$
$(y - 5)(y - 1) = 0$
$y - 5 = 0 \quad \text{or} \quad y - 1 = 0$
$\quad y = 5 \quad \text{or} \quad \quad y = 1$
$\quad x^{1/3} = 5 \quad \text{or} \quad x^{1/3} = 1$
$\quad x = 125 \quad \text{or} \quad \quad x = 1$

The solutions are 1 and 125.

36. $x^{2/3} - 6x^{1/3} = -8$

$x^{2/3} - 6x^{1/3} + 8 = 0$

Let $y = x^{1/3}$. Then $y^2 = x^{2/3}$ and

$y^2 - 6y + 8 = 0$
$(y - 4)(y - 2) = 0$
$y - 4 = 0 \quad \text{or} \quad y - 2 = 0$
$\quad y = 4 \quad \text{or} \quad \quad y = 2$
$\quad x^{1/3} = 4 \quad \text{or} \quad x^{1/3} = 2$
$\quad x = 64 \quad \text{or} \quad \quad x = 8$

The solutions are 8 and 64.

37.
$$a^6 - a^2 = a^4 - 1$$
$$a^6 - a^4 - a^2 + 1 = 0$$
$$a^4(a^2 - 1) - 1(a^2 - 1) = 0$$
$$(a^2 - 1)(a^4 - 1) = 0$$
$$(a + 1)(a - 1)(a^2 + 1)(a^2 - 1) = 0$$
$$(a + 1)(a - 1)(a^2 + 1)(a + 1)(a - 1) = 0$$
$$(a + 1)^2(a - 1)^2(a^2 + 1) = 0$$

$(a + 1)^2 = 0$ or $(a - 1)^2 = 0$ or $a^2 + 1 = 0$
$a + 1 = 0 \quad \text{or} \quad a - 1 = 0 \text{ or} \quad a^2 = -1$
$a = -1 \text{ or} \quad \quad a = 1 \text{ or} \quad \quad a = \pm i$

The solutions are -1, 1, $-i$, and i.

38. $y^{-2} + y^{-1} = 20$

$\dfrac{1}{y^2} + \dfrac{1}{y} = 20$

$1 + y = 20y^2$
$0 = 20y^2 - y - 1$
$0 = (5y + 1)(4y - 1)$

$5y + 1 = 0 \quad \text{or} \quad 4y - 1 = 0$
$\quad y = -\dfrac{1}{5} \text{ or} \quad \quad y = \dfrac{1}{4}$

The solutions are $-\dfrac{1}{5}$ and $\dfrac{1}{4}$.

39. Let x = time for Jerome alone. Then $x - 1$ = time for Tim alone.

$\dfrac{1}{x} + \dfrac{1}{x-1} = \dfrac{1}{5}$

$5(x - 1) + 5x = x(x - 1)$
$5x - 5 + 5x = x^2 - x$
$0 = x^2 - 11x + 5$
$a = 1, b = -11, c = 5$

$x = \dfrac{11 \pm \sqrt{(-11)^2 - 4(1)(5)}}{2(1)}$

$= \dfrac{11 \pm \sqrt{101}}{2}$

≈ 0.475 (disregard) or 10.525

Jerome: 10.5 hours
Tim: 9.5 hours

40. Let x = the number; then

$\dfrac{1}{x}$ = the reciprocal of the number.

$x - \dfrac{1}{x} = -\dfrac{24}{5}$

$5x\left(x - \dfrac{1}{x}\right) = 5x\left(-\dfrac{24}{5}\right)$

$5x^2 - 5 = -24x$
$5x^2 + 24x - 5 = 0$
$(5x - 1)(x + 5) = 0$
$5x - 1 = 0 \quad \text{or} \quad x + 5 = 0$
$\quad x = \dfrac{1}{5} \quad \text{or} \quad \quad x = -5$

Disregard the positive value as extraneous. The number is -5.

41.

$$2x^2 - 50 \le 0$$
$$2(x^2 - 25) \le 0$$
$$2(x + 5)(x - 5) \le 0$$
$$x + 5 = 0 \quad \text{or} \quad x - 5 = 0$$
$$x = -5 \quad \text{or} \quad x = 5$$

Region	Test Point	$2(x+5)(x-5) \le 0$
A: $(-\infty, -5)$	-6	$2(-1)(-11) \le 0$; F
B: $(-5, 5)$	0	$2(5)(-5) \le 0$; T
C: $(5, \infty)$	6	$2(11)(1) \le 0$; F

Solution: $[-5, 5]$

42.

$$\frac{1}{4}x^2 < \frac{1}{16}$$
$$x^2 < \frac{1}{4}$$
$$x^2 - \frac{1}{4} < 0$$
$$\left(x + \frac{1}{2}\right)\left(x - \frac{1}{2}\right) < 0$$
$$x + \frac{1}{2} = 0 \quad \text{or} \quad x - \frac{1}{2} = 0$$
$$x = -\frac{1}{2} \quad \text{or} \quad x = \frac{1}{2}$$

Region	Test Point	$\left(x+\frac{1}{2}\right)\left(x-\frac{1}{2}\right) < 0$
A: $\left(-\infty, -\frac{1}{2}\right)$	-1	$\left(-\frac{1}{2}\right)\left(-\frac{3}{2}\right) < 0$; F
B: $\left(-\frac{1}{2}, \frac{1}{2}\right)$	0	$\left(\frac{1}{2}\right)\left(-\frac{1}{2}\right) < 0$; T
C: $\left(\frac{1}{2}, \infty\right)$	1	$\left(\frac{3}{2}\right)\left(\frac{1}{2}\right) < 0$; F

Solution: $\left(-\frac{1}{2}, \frac{1}{2}\right)$

43.

$$(2x - 3)(4x + 5) \ge 0$$
$$2x - 3 = 0 \quad \text{or} \quad 4x + 5 = 0$$
$$x = \frac{3}{2} \quad \text{or} \quad x = -\frac{5}{4}$$

Region	Test Point	$(2x-3)(4x+5) \ge 0$
A: $\left(-\infty, -\frac{5}{4}\right)$	-2	$(-7)(-3) \ge 0$ True
B: $\left(-\frac{5}{4}, \frac{3}{2}\right)$	0	$(-3)(5) \ge 0$ False
C: $\left(\frac{3}{2}, \infty\right)$	3	$(3)(17) \ge 0$ True

Solution: $\left(-\infty, -\frac{5}{4}\right] \cup \left[\frac{3}{2}, \infty\right)$

44.
$$(x^2 - 16)(x^2 - 1) > 0$$
$$(x + 4)(x - 4)(x + 1)(x - 1) > 0$$
$$x + 4 = 0 \quad \text{or} \quad x - 4 = 0 \quad \text{or} \quad x + 1 = 0 \quad \text{or} \quad x - 1 = 0$$
$$x = -4 \quad \text{or} \quad x = 4 \quad \text{or} \quad x = -1 \quad \text{or} \quad x = 1$$

Region	Test Point	$(x+4)(x-4)(x+1)(x-1) > 0$
A: $(-\infty, -4)$	-5	$(-1)(-9)\,(-4)(-6) > 0$; True
B: $(-4, -1)$	-2	$(2)(-6)(-1)(-3) > 0$; False
C: $(-1, 1)$	0	$(4)(-4)(1)(-1) > 0$; True
D: $(1, 4)$	2	$(6)(-2)(3)(1) > 0$; False
E: $(4, \infty)$	5	$(9)(1)(6)(4) > 0$; True

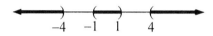

Solution: $(-\infty, -4) \cup (-1, 1) \cup (4, \infty)$

45. $\dfrac{x-5}{x-6} < 0$
$$x - 5 = 0 \quad \text{or} \quad x - 6 = 0$$
$$x = 5 \quad \text{or} \quad x = 6$$

Region	Test Point	$\dfrac{x-5}{x-6} < 0$
A: $(-\infty, 5)$	0	$\dfrac{-5}{-6} < 0$; False
B: $(5, 6)$	$\dfrac{11}{2}$	$\dfrac{\frac{1}{2}}{-\frac{1}{2}} < 0$; True
C: $(6, \infty)$	7	$\dfrac{2}{1} < 0$; False

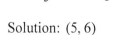

Solution: $(5, 6)$

46. $\dfrac{x(x+5)}{4x-3} \geq 0$
$$x = 0 \quad \text{or} \quad x + 5 = 0 \quad \text{or} \quad 4x - 3 = 0$$
$$x = -5 \quad \text{or} \quad x = \dfrac{3}{4} \cdot$$

Region	Test Point	$\dfrac{x(x+5)}{4x-3} \geq 0$
A: $(-\infty, -5)$	-6	$\dfrac{-6(-1)}{-27} \geq 0$; F
B: $(-5, 0)$	-1	$\dfrac{-1(4)}{-7} \geq 0$; T
C: $\left(0, \dfrac{3}{4}\right)$	$\dfrac{1}{2}$	$\dfrac{\frac{1}{2}\left(\frac{11}{2}\right)}{-1} \geq 0$; F
D: $\left(\dfrac{3}{4}, \infty\right)$	1	$\dfrac{1(6)}{1} \geq 0$; T

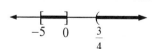

Solution: $[-5, 0] \cup \left(\dfrac{3}{4}, \infty\right)$

47. $\dfrac{(4x+3)(x-5)}{x(x+6)} > 0$

$4x+3=0,\ x-5=0,\ x=0,\ \text{or } x+6=0$

$x=-\dfrac{3}{4},\ x=5,\ x=0,\ \text{or } x=-6$

Region	Test Point	$\dfrac{(4x+3)(x-5)}{x(x+6)} > 0$
A: $(-\infty, -6)$	-7	$\dfrac{(-25)(-12)}{-7(-1)} > 0;\ \text{T}$
B: $\left(-6, -\dfrac{3}{4}\right)$	-3	$\dfrac{(-9)(-8)}{-3(3)} > 0;\ \text{F}$
C: $\left(-\dfrac{3}{4}, 0\right)$	$-\dfrac{1}{2}$	$\dfrac{(1)\left(-\dfrac{11}{2}\right)}{-\dfrac{1}{2}\left(\dfrac{11}{2}\right)} > 0;\ \text{T}$
D: $(0, 5)$	1	$\dfrac{(7)(-4)}{1(7)} > 0;\ \text{F}$
E: $(5, \infty)$	6	$\dfrac{(27)(1)}{6(12)} > 0;\ \text{T}$

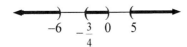

Solution: $(-\infty, -6) \cup \left(-\dfrac{3}{4}, 0\right) \cup (5, \infty)$

48. $(x+5)(x-6)(x+2) \le 0$

$x+5=0 \quad \text{or } x-6=0 \text{ or } x+2=0$

$x=-5 \text{ or} \qquad x=6 \text{ or} \qquad x=-2$

Region	Test Point	$(x+5)(x-6)(x+2) \le 0$
$(-\infty, -5)$	-6	$(-1)(-12)(-4) \le 0;\ \text{T}$
$(-5, -2)$	-3	$(2)(-9)(-1) \le 0;\ \text{F}$
$(-2, 6)$	0	$(5)(-6)(2) \le 0;\ \text{T}$
$(6, \infty)$	7	$(12)(1)(9) \le 0;\ \text{F}$

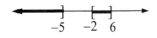

Solution: $(-\infty, -5] \cup [-2, 6]$

49. $x^3 + 3x^2 - 25 - 75 > 0$

$x^2(x+3) - 25(x+3) > 0$

$(x+3)(x^2-25) > 0$

$(x+3)(x+5)(x-5) > 0$

$x+3=0 \quad \text{or } x+5=0 \quad \text{or } x-5=0$

$x=-3 \text{ or} \qquad x=-5 \text{ or} \qquad x=5$

Region	Test Point	$(x+3)(x+5)(x-5) > 0$
$(-\infty, -5)$	-6	$(-3)(-1)(-11) > 0;\ \text{F}$
$(-5, -3)$	-4	$(-1)(1)(-9) > 0;\ \text{T}$
$(-3, 5)$	0	$(3)(5)(-5) > 0;\ \text{F}$
$(5, \infty)$	6	$(9)(11)(1) > 0;\ \text{T}$

Solution: $(-5, -3) \cup (5, \infty)$

50. $\dfrac{x^2+4}{3x} \le 1$

The denominator equals 0 when

$3x=0,\ \text{or } x=0$.

$\dfrac{x^2+4}{3x} = 1$

$x^2+4 = 3x$

$x^2-3x+4 = 0$

$x = \dfrac{3 \pm \sqrt{(-3)^2-4(1)(4)}}{2(1)} = \dfrac{3 \pm \sqrt{-7}}{2(1)}$

which yields non-real solutions.

Region	Test Point	$\dfrac{x^2+4}{3x} \le 1$
A: $(-\infty, 0)$	-1	$\dfrac{5}{-3} \le 1;\ \text{True}$
B: $(0, \infty)$	1	$\dfrac{5}{3} \le 1;\ \text{False}$

Solution: $(\infty, 0)$

51. $\dfrac{(5x+6)(x-3)}{x(6x-5)} < 0$

$x = -\dfrac{6}{5}$ or $x = 3$ or $x = 0$ or $x = \dfrac{5}{6}$

Region	Test Point	$\dfrac{(5x+6)(x-3)}{x(6x-5)} < 0$
A: $\left(-\infty, -\dfrac{6}{5}\right)$	-2	$\dfrac{(-4)(-5)}{-2(-17)} < 0;\ \text{F}$
B: $\left(-\dfrac{6}{5}, 0\right)$	-1	$\dfrac{(1)(-4)}{-1(-11)} < 0;\ \text{T}$
C: $\left(0, \dfrac{5}{6}\right)$	$\dfrac{1}{2}$	$\dfrac{\left(\dfrac{17}{2}\right)\left(-\dfrac{5}{2}\right)}{\dfrac{1}{2}(-2)} < 0;\ \text{F}$
D: $\left(\dfrac{5}{6}, 3\right)$	2	$\dfrac{(16)(-1)}{2(7)} < 0;\ \text{T}$
E: $(3, \infty)$	4	$\dfrac{(26)(1)}{4(19)} < 0;\ \text{F}$

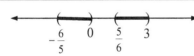

Solution: $\left(-\dfrac{6}{5}, 0\right) \cup \left(\dfrac{5}{6}, 3\right)$

52. $\dfrac{3}{x-2} > 2$

The denominator is equal to 0 when
$x - 2 = 0$, or $x = 2$.

$\dfrac{3}{x-2} = 2$

$3 = 2(x-2)$

$3 = 2x - 4$

$7 = 2x$

$\dfrac{7}{2} = x$

Region	Test Point	$\dfrac{3}{x-2} > 2$
A: $(-\infty, 2)$	0	$\dfrac{3}{-2} > 2;\ \text{False}$
B: $\left(2, \dfrac{7}{2}\right)$	3	$\dfrac{3}{1} > 2;\ \text{True}$
C: $\left(\dfrac{7}{2}, \infty\right)$	5	$\dfrac{3}{3} > 2;\ \text{False}$

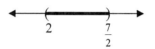

Solution: $\left(2, \dfrac{7}{2}\right)$

53. $f(x) = x^2 - 4$

$x = -\dfrac{b}{2a} = \dfrac{-0}{2(1)} = 0$

$f(0) = (0)^2 - 4 = -4$

Vertex: $(0, -4)$

Axis of symmetry: $x = 0$

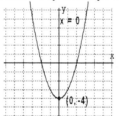

54. $g(x) = x^2 + 7$

$x = -\dfrac{b}{2a} = \dfrac{-0}{2(1)} = 0$

$f(0) = (0)^2 + 7 = 7$

Vertex: $(0, 7)$

Axis of symmetry: $x = 0$

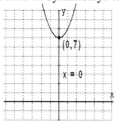

55. $H(x) = 2x^2$

$x = -\dfrac{b}{2a} = \dfrac{-0}{2(2)} = 0$

$f(0) = 2(0)^2 = 0$

Vertex: (0, 0)

Axis of symmetry: $x = 0$

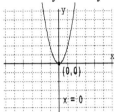

56. $h(x) = -\dfrac{1}{3}x^2$

$x = -\dfrac{b}{2a} = \dfrac{-0}{2(2)} = 0$

$f(0) = -\dfrac{1}{3}(0)^2 = 0$

Vertex: (0, 0)

Axis of symmetry: $x = 0$

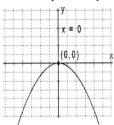

57. $F(x) = (x-1)^2$

Vertex: (1, 0)

Axis of symmetry: $x = 1$

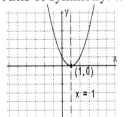

58. $G(x) = (x+5)^2$

Vertex: (−5, 0)

Axis of symmetry: $x = -5$

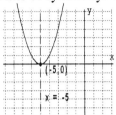

59. $f(x) = (x-4)^2 - 2$

Vertex: (4, −2)

Axis of symmetry: $x = 4$

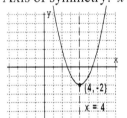

60. $f(x) = -3(x-1)^2 + 1$

Vertex: (1, 1)

Axis of symmetry: $x = 1$

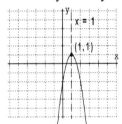

61. $f(x) = x^2 + 10x + 25$

$x = -\dfrac{b}{2a} = \dfrac{-10}{2(1)} = -5$

$f(-5) = (-5)^2 + 10(-5) + 25 = 0$

Vertex: (−5, 0)

$x^2 + 10x + 25 = 0$

$(x+5)^2 = 0$

$x + 5 = 0$

$x = -5$

x-intercept: (–5, 0)

$f(0) = 25$ so the *y*-intercept is (0, 25).

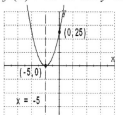

$f(0) = -1$

y-intercept: (0, –1).

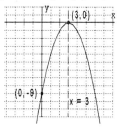

62. $f(x) = -x^2 + 6x - 9$

$x = -\dfrac{b}{2a} = \dfrac{-6}{2(-1)} = 3$

$f(3) = -(3)^2 + 6(3) - 9 = 0$

Vertex: (3, 0)

x-intercept: (3, 0)

$f(0) = -9$

y-intercept: (0, –9).

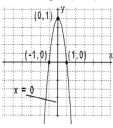

63. $f(x) = 4x^2 - 1$

$x = -\dfrac{b}{2a} = \dfrac{-0}{2(4)} = 0$

$f(0) = 4(0)^2 - 1 = -1$

Vertex: (0, –1)

$4x^2 - 1 = 0$

$(2x + 1)(2x - 1) = 0$

$x = -\dfrac{1}{2}$ or $x = \dfrac{1}{2}$

x-intercepts: $\left(-\dfrac{1}{2}, 0\right), \left(\dfrac{1}{2}, 0\right)$

64. $f(x) = -5x^2 + 5$

$x = -\dfrac{b}{2a} = \dfrac{-0}{2(-5)} = 0$

$f(0) = -5(0)^2 + 5 = 5$

Vertex: (0, 5)

$-5x^2 + 5 = 0$

$-5x^2 = -5$

$x^2 = 1$

$x = \pm 1$

x-intercepts: (–1, 0), (1, 0)

$f(0) = 5$

y-intercept: (0, 5).

65. $f(x) = -3x^2 - 5x + 4$

$x = -\dfrac{b}{2a} = \dfrac{-(-5)}{2(-3)} = -\dfrac{5}{6}$

$f\left(-\dfrac{5}{6}\right) = -3\left(-\dfrac{5}{6}\right)^2 - 5\left(-\dfrac{5}{6}\right) + 4 = \dfrac{73}{12}$

Vertex: $\left(-\dfrac{5}{6}, \dfrac{73}{12}\right)$

The graph opens downward ($a = -3 < 0$).

$f(0) = 4 \Rightarrow$ *y*-intercept: (0, 4)

$$-3x^2 - 5x + 4 = 0$$

$$x = \frac{5 \pm \sqrt{(-5)^2 - 4(-3)(4)}}{2(-3)}$$

$$= \frac{5 \pm \sqrt{73}}{-6} \approx -2.2573 \text{ or } 0.5907$$

x-intercepts: $(-2.3, 0)$, $(0.6, 0)$

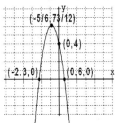

66. $h(t) = -16t^2 + 120t + 300$

a.
$$350 = -16t^2 + 120t + 300$$
$$16t^2 - 120t + 50 = 0$$
$$8t^2 - 60t + 25 = 0$$
$$a = 8, b = -60, c = 25$$
$$t = \frac{60 \pm \sqrt{(-60)^2 - 4(8)(25)}}{2(8)}$$
$$= \frac{60 \pm \sqrt{2800}}{16}$$
$$\approx 0.4 \text{ seconds and } 7.1 \text{ seconds}$$

b. The object will be at 350 feet on the way up and on the way down.

67. Let x = one number; then
$420 - x$ = the other number.
Let $f(x)$ represent their product.
$$f(x) = x(420 - x)$$
$$= 420x - x^2$$
$$= -x^2 + 420x$$
$$x = -\frac{b}{2a} = \frac{-420}{2(-1)} = 210;$$
$$420 - x = 420 - 210 = 210$$
Therefore, the numbers are both 210.

68. $y = a(x - h)^2 + k$
vertex $(-3, 7)$ gives
$$y = a(x + 3)^2 + 7$$
Passing through the origin gives
$$0 = a(0 + 3)^2 + 7$$
$$-7 = 9a$$
$$-\frac{7}{9} = a$$
Thus, $y = -\frac{7}{9}(x + 3)^2 + 7$.

Chapter 11 Test

1.
$$5x^2 - 2x = 7$$
$$5x^2 - 2x - 7 = 0$$
$$(5x - 7)(x + 1) = 0$$
$$5x - 7 = 0 \quad \text{or} \quad x + 1 = 0$$
$$x = \frac{7}{5} \quad \text{or} \quad x = -1$$
The solutions are -1 and $\frac{7}{5}$.

2. $(x + 1)^2 = 10$
$$x + 1 = \pm\sqrt{10}$$
$$x = -1 \pm \sqrt{10}$$
The solutions are $-1 + \sqrt{10}$ and $-1 - \sqrt{10}$.

3. $m^2 - m + 8 = 0$
$$a = 1, b = -1, c = 8$$
$$m = \frac{1 \pm \sqrt{(-1)^2 - 4(1)(8)}}{2(1)}$$
$$= \frac{1 \pm \sqrt{1 - 32}}{2}$$
$$= \frac{1 \pm \sqrt{-31}}{2}$$
$$= \frac{1 \pm i\sqrt{31}}{2} \quad \text{or} \quad \frac{1}{2} \pm \frac{\sqrt{31}}{2}i$$
The solutions are $\frac{1 + i\sqrt{31}}{2}$ and $\frac{1 - i\sqrt{31}}{2}$.

4. $a^2 - 3a = 5$

$a^2 - 3a - 5 = 0$

$a = 1, b = -3, c = -5$

$a = \dfrac{3 \pm \sqrt{(-3)^2 - 4(1)(-5)}}{2(1)}$

$= \dfrac{3 \pm \sqrt{9 + 20}}{2}$

$= \dfrac{3 \pm \sqrt{29}}{2}$

The solutions are $\dfrac{3 + \sqrt{29}}{2}$ and $\dfrac{3 - \sqrt{29}}{2}$.

5. $\dfrac{4}{x+2} + \dfrac{2x}{x-2} = \dfrac{6}{x^2 - 4}$

$\dfrac{4}{x+2} + \dfrac{2x}{x-2} = \dfrac{6}{(x+2)(x-2)}$

$4(x-2) + 2x(x+2) = 6$

$4x - 8 + 2x^2 + 4x = 6$

$2x^2 + 8x - 14 = 0$

$x^2 + 4x - 7 = 0$

$a = 1, b = 4, c = -7$

$x = \dfrac{-4 \pm \sqrt{(4)^2 - 4(1)(-7)}}{2(1)}$

$= \dfrac{-4 \pm \sqrt{16 + 28}}{2}$

$= \dfrac{-4 \pm \sqrt{44}}{2}$

$= \dfrac{-4 \pm 2\sqrt{11}}{2} = -2 \pm \sqrt{11}$

The solutions are $-2 + \sqrt{11}$ and $-2 - \sqrt{11}$.

6. $x^5 + 3x^4 = x + 3$

$x^5 + 3x^4 - x - 3 = 0$

$x^4(x+3) - 1(x+3) = 0$

$(x+3)(x^4 - 1) = 0$

$(x+3)(x^2 + 1)(x^2 - 1) = 0$

$x + 3 = 0$ or $x^2 + 1 = 0$ or $x^2 - 1 = 0$

$x = -3$ or $x^2 = -1$ or $x^2 = 1$

$x = \pm i$ or $x = \pm 1$

The solutions are $-3, -1, 1, -i,$ and i.

7. $(x+1)^2 - 15(x+1) + 56 = 0$

Let $y = x + 1$. Then $y^2 = (x+1)^2$ and

$y^2 - 15x + 56 = 0$

$(y-8)(y-7) = 0$

$y = 8$ or $y = 7$

$x + 1 = 8$ or $x + 1 = 7$

$x = 7$ or $x = 6$

The solutions are 6 and 7.

8. $x^2 - 6x = -2$

$x^2 - 6x + \left(\dfrac{-6}{2}\right)^2 = -2 + 9$

$x^2 - 6x + 9 = 7$

$(x-3)^2 = 7$

$x - 3 = \pm\sqrt{7}$

$x = 3 \pm \sqrt{7}$

The solutions are $3 + \sqrt{7}$ and $3 - \sqrt{7}$.

9. $2a^2 + 5 = 4a$

$2a^2 - 4a = -5$

$a^2 - 2a = -\dfrac{5}{2}$

$a^2 - 2a + \left(\dfrac{-2}{2}\right)^2 = -\dfrac{5}{2} + 1$

$a^2 - 2a + 1 = -\dfrac{3}{2}$

$(a-1)^2 = -\dfrac{3}{2}$

$a - 1 = \pm\sqrt{-\dfrac{3}{2}} = \pm\dfrac{\sqrt{3}}{\sqrt{2}}i$

$a - 1 = \pm\dfrac{\sqrt{6}}{2}i$

$a = 1 \pm \dfrac{\sqrt{6}}{2}i$ or $\dfrac{2 \pm i\sqrt{6}}{2}$

The solutions are $1 + \dfrac{\sqrt{6}}{2}i$ and $1 - \dfrac{\sqrt{6}}{2}i$.

10. $2x^2 - 7x > 15$

$2x^2 - 7x - 15 > 0$

$(2x+3)(x-5) > 0$

$2x+3 = 0 \quad$ or $\quad x - 5 = 0$

$x = -\dfrac{3}{2} \quad$ or $\quad x = 5$

Region	Test Point	$(2x+1)(x-5) > 0$
A: $\left(-\infty, -\dfrac{3}{2}\right)$	−2	$(-3)(-7) > 0$; True
B: $\left(-\dfrac{3}{2}, 5\right)$	0	$(1)(-5) > 0$; False
C: $(5, \infty)$	6	$(13)(1) > 0$; True

$$\xleftarrow{\hspace{2cm}}\overset{\textstyle)}{\underset{-\frac{3}{2}}{}}\hspace{1cm}\overset{\textstyle(}{\underset{5}{}}\xrightarrow{\hspace{2cm}}$$

Solution: $\left(-\infty, -\dfrac{3}{2}\right) \cup (5, \infty)$

11. $(x^2 - 16)(x^2 - 25) \geq 0$

$(x+4)(x-4)(x+5)(x-5) \geq 0$

$x+4 = 0 \quad$ or $\quad x-4 = 0 \quad$ or $\quad x+5 = 0 \quad$ or $\quad x-5 = 0$

$x = -4 \quad$ or $\quad x = 4 \quad$ or $\quad x = -5 \quad$ or $\quad x = 5$

Region	Test Point	$(x+4)(x-4)(x+5)(x-5) \geq 0$
A: $(-\infty, -5)$	−6	$(-2)(-10)(-1)(-11) \geq 0$; True
B: $(-5, -4)$	$-\dfrac{9}{2}$	$\left(-\dfrac{1}{2}\right)\left(-\dfrac{17}{2}\right)\left(\dfrac{1}{2}\right)\left(-\dfrac{19}{2}\right) \geq 0$; False
C: $(-4, 4)$	0	$(4)(-4)(5)(-5) \geq 0$; True
D: $(4, 5)$	$\dfrac{9}{2}$	$\left(\dfrac{17}{2}\right)\left(\dfrac{1}{2}\right)\left(\dfrac{19}{2}\right)\left(-\dfrac{1}{2}\right) \geq 0$; False
E: $(5, \infty)$	6	$(10)(2)(11)(1) \geq 0$; True

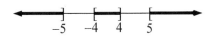

Solution: $(-\infty, -5] \cup [-4, 4] \cup [5, \infty)$

12. $\dfrac{5}{x+3} < 1$

The denominator is equal to 0 when
$x+3=0$, or $x=-3$.

$\dfrac{5}{x+3} = 1$

$5 = x+3$ so $x = 2$

Region	Test Point	$\dfrac{5}{x+3} < 1$
A: $(-\infty, -3)$	-4	$\dfrac{5}{-1} < 1$; True
B: $(-3, 2)$	0	$\dfrac{5}{3} < 1$; False
C: $(2, \infty)$	3	$\dfrac{5}{6} < 1$; True

Solution: $(-\infty, -3) \cup (2, \infty)$

13. $\dfrac{7x-14}{x^2-9} \le 0$

$\dfrac{7(x-2)}{(x+3)(x-3)} \le 0$

$x-2=0$ or $x+3=0$ or $x-3=0$
$x=2$ or $x=-3$ or $x=3$

Region	Test Point	$\dfrac{7(x-2)}{(x+3)(x-3)} \le 0$
A: $(-\infty, -3)$	-4	$\dfrac{7(-6)}{(-1)(-7)} \le 0$; T
B: $(-3, 2)$	0	$\dfrac{7(-2)}{(3)(-3)} \le 0$; F
C: $(2, 3)$	$\dfrac{5}{2}$	$\dfrac{7\left(\frac{1}{2}\right)}{\left(\frac{11}{2}\right)\left(-\frac{1}{2}\right)} \le 0$; T
D: $(3, \infty)$	4	$\dfrac{7(2)}{(7)(1)} \le 0$; F

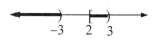

Solution: $(-\infty, -3) \cup [2, 3)$

14. $f(x) = 3x^2$
Vertex: $(0, 0)$

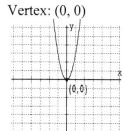

15. $G(x) = -2(x-1)^2 + 5$
Vertex: $(1, 5)$

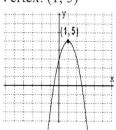

16. $h(x) = x^2 - 4x + 4$

$x = -\dfrac{b}{2a} = \dfrac{-(-4)}{2(1)} = 2$

$h(2) = (2)^2 - 4(2) + 4 = 0$

Vertex: $(2, 0)$

$h(0) = 4 \Rightarrow y\text{-intercept}: (0, 4)$

x-intercept: $(2,0)$

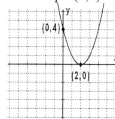

17. $F(x) = 2x^2 - 8x + 9$

$x = -\dfrac{b}{2a} = \dfrac{-(-8)}{2(2)} = 2$

$F(2) = 2(2)^2 - 8(2) + 9 = 1$

Vertex: (2, 1)

$F(0) = 9 \Rightarrow$ y-intercept: (0, 9)

$2x^2 - 8x + 9 = 0$

$a = 2, b = -8, c = 9$

$x = \dfrac{8 \pm \sqrt{(-8)^2 - 4(2)(9)}}{2(2)}$

$= \dfrac{8 \pm \sqrt{-8}}{4}$

which yields non-real solutons.
Therefore, there are no x-intercepts.

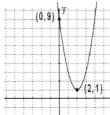

18. Let t = time for Sandy alone. Then
$t - 2$ = time for Dave alone.

$\dfrac{1}{t} + \dfrac{1}{t-2} = \dfrac{1}{4}$

$4(t-2) + 4t = t(t-2)$

$4t - 8 + 4t = t^2 - 2t$

$0 = t^2 - 10t + 8$

$a = 1, b = -10, c = 8$

$t = \dfrac{10 \pm \sqrt{(-10)^2 - 4(1)(8)}}{2(1)}$

$= \dfrac{10 \pm \sqrt{68}}{2}$

$= \dfrac{10 \pm 2\sqrt{17}}{2}$

$= 5 \pm \sqrt{17}$

≈ 9.12 or 0.88 (discard)

It takes her about 9.12 hours.

19. $s(t) = -16t^2 + 32t + 256$

a. $t = -\dfrac{b}{2a} = \dfrac{-32}{2(-16)} = 1$

$s(1) = -16(1)^2 + 32(1) + 256 = 272$

Vertex: (1, 272)
The maximum height is 272 feet.

b. $-16t^2 + 32t + 256 = 0$

$t^2 - 2t - 16 = 0$

$a = 1, b = -2, c = -16$

$t = \dfrac{2 \pm \sqrt{(-2)^2 - 4(1)(-16)}}{2(1)}$

$= \dfrac{2 \pm \sqrt{68}}{2}$

$= \dfrac{2 \pm 2\sqrt{17}}{2}$

$= 1 \pm \sqrt{17} \approx -3.12$ and 5.12

Disregard the negative. Hit will hit
the water in about 5.12 seconds.

20. $a^2 + b^2 = c^2$

$x^2 + (x+8)^2 = (20)^2$

$x^2 + (x^2 + 16x + 64) = 400$

$2x^2 + 16x - 336 = 0$

$x^2 + 8x - 168 = 0$

$a = 1, b = 8, c = -168$

$x = \dfrac{-8 \pm \sqrt{(8)^2 - 4(1)(-168)}}{2(1)}$

$= \dfrac{-8 \pm \sqrt{736}}{2}$

≈ -17.565 or 9.565

Disregard the negative.

$x \approx 9.6$

$x + 8 \approx 9.6 + 8 = 17.6$

$17.6 + 9.6 = 27.2$

$27.2 - 20 = 7.2$

They would save about 7 feet.

Chapter 11 Cumulative Review

1. a. $x = 2$ and $y = -5$

    a. $\dfrac{x - y}{12 + x} = \dfrac{2 - (-5)}{12 + 2} = \dfrac{7}{14} = \dfrac{1}{2}$

    b. $x^2 - y = (2)^2 - (-5) = 4 + 5 = 9$

2. $|3x - 2| = -5$ which is impossible.

    Thus, there is no solution, or $\varnothing$.

3. a. $2x + 3x + 5 + 2 = 5x + 7$

    b. $-5a - 3 + a + 2 = -4a - 1$

    c. $4y - 3y^2 = 4y - 3y^2$

    d. $2.3x + 5x - 6 = 7.3x - b$

    e. $-\dfrac{1}{2}b + b = \dfrac{1}{2}b$

4. $\begin{cases} -6x + \ \ y = 5 \ \ (1) \\ \ \ \ 4x - 2y = 6 \ \ (2) \end{cases}$

    Multiply E1 by 2 and add to E2.

$$\begin{array}{r} -12x + 2y = 10 \\ 4x - 2y = 6 \\ \hline -8x \qquad\quad = 16 \\ x = -2 \end{array}$$

    Replace x with -2 in E1.

$$-6(-2) + y = 5$$
$$12 + y = 5$$
$$y = -7$$

    The solution is $(-2, -7)$.

5. $\begin{cases} 2x + y = 7 \\ \ \ \ 2y = -4x \end{cases}$

No solution.

6. a. $(a^{-2}bc^3)^{-3} = (a^{-2})^{-3}b^{-3}(c^3)^{-3}$
$$= a^6 b^{-3} c^{-9}$$
$$= \dfrac{a^6}{b^3 c^9}$$

    b. $\left(\dfrac{a^{-4}b^2}{c^3}\right)^{-2} = \dfrac{(a^{-4})^{-2}(b^2)^{-2}}{(c^3)^{-2}}$
$$= \dfrac{a^8 b^{-4}}{c^{-6}}$$
$$= \dfrac{a^8 c^6}{b^4}$$

    c. $\left(\dfrac{3a^8 b^2}{12a^5 b^5}\right)^{-2} = \left(\dfrac{a^3}{4b^3}\right)^{-2}$
$$= \dfrac{(a^3)^{-2}}{4^{-2}(b^3)^{-2}}$$
$$= \dfrac{4^2 a^{-6}}{b^{-6}}$$
$$= \dfrac{16 b^6}{a^6}$$

7. $\begin{cases} 7x - 3y = -14 \\ -3x + y = 6 \end{cases}$

    Multiply the second equation by 3 and then add.

$$\begin{array}{r} 7x - 3y = -14 \\ -9x + 3y = 18 \\ \hline -2x \qquad\quad = 4 \\ x = -2 \end{array}$$

    Let $x = -2$ in the second equation.

$$-3(-2) + y = 6$$
$$6 + y = 6$$
$$y = 0$$

    The solution is $(-2, 0)$.

8. a. $(4a-3)(7a-2) = 28a^2 - 8a - 21a + 6$
$$= 28a^2 - 29a + 6$$

b. $(2a+b)(3a-5b)$
$$= 6a^2 - 10ab + 3ab - 5b^2$$
$$= 6a^2 - 7ab - 5b^2$$

9. a. $\left(\dfrac{-5x^2}{y^3}\right)^2 = \dfrac{25x^4}{y^6}$

b. $\dfrac{\left(x^3\right)^4 x}{x^7} = \dfrac{x^{12}x}{x^7} = x^6$

c. $\dfrac{\left(2x\right)^5}{x^3} = \dfrac{32x^5}{x^3} = 32x^2$

d. $\dfrac{\left(a^2b\right)^3}{a^3b^2} = \dfrac{a^6b^3}{a^3b^2} = a^3b$

10. a. $9x^3 + 27x^2 - 15x = 3x(3x^2 + 9x - 5)$

b. $2x(3y-2) - 5(3y-2)$
$$= (3y-2)(2x-5)$$

c. $2xy + 6x - y - 3 = 2x(y+3) - 1(y+3)$
$$= (y+3)(2x-1)$$

11. $P(x) = 2x^3 - 4x^2 + 5$

    a. $P(2) = 2(2)^3 - 4(2)^2 + 5 = 5$

    b.
$$\begin{array}{r|rrrr} 2 & 2 & -4 & 0 & 5 \\ & & 4 & 0 & 0 \\ \hline & 2 & 0 & 0 & 5 \end{array}$$

The remainder is 5.

12. $x^2 - 2x - 48 = (x+6)(x-8)$

13. $(5x-1)\left(2x^2 + 15x + 18\right) = 0$
$$(5x-1)(2x+3)(x+6) = 0$$
$$5x-1=0 \quad \text{or} \quad 2x+3=0 \quad \text{or} \quad x+6=0$$
$$5x=1 \qquad\qquad 2x=-3 \qquad\qquad x=-6$$
$$x=\frac{1}{5} \qquad\qquad x=-\frac{3}{2}$$

The solutions are -6, $-\dfrac{3}{2}$ and $\dfrac{1}{5}$.

14. $2ax^2 - 12axy + 18ay^2 = 2a(x^2 - 6xy + 9y^2)$
$$= 2a(x-3y)(x-3y)$$
$$= 2a(x-3y)^2$$

15. $\dfrac{2x^2}{10x^3 - 2x^2} = \dfrac{2x^2}{2x^2(5x-1)} = \dfrac{1}{5x-1}$

16. $2(a^2+2) - 8 = -2a(a-2) - 5$
$$2a^2 + 4 - 8 = -2a^2 + 4a - 5$$
$$4a^2 - 4a + 1 = 0$$
$$(2a-1)^2 = 0$$
$$2a-1 = 0$$
$$2a = 1$$
$$a = \frac{1}{2}$$

The solution is $\dfrac{1}{2}$.

17. $\dfrac{x^{-1}+2xy^{-1}}{x^{-2}-x^{-2}y^{-1}} = \dfrac{\dfrac{1}{x}+\dfrac{2x}{y}}{\dfrac{1}{x^2}-\dfrac{1}{x^2y}}$

$\qquad = \dfrac{\left(\dfrac{1}{x}+\dfrac{2x}{y}\right)x^2y}{\left(\dfrac{1}{x^2}-\dfrac{1}{x^2y}\right)x^2y}$

$\qquad = \dfrac{xy+2x^3}{y-1}$

18. $f(x) = x^2 + x - 12$

$\quad x = -\dfrac{b}{2a} = \dfrac{-1}{2(1)} = -\dfrac{1}{2}$

$\quad f\left(-\dfrac{1}{2}\right) = \left(-\dfrac{1}{2}\right)^2 + \left(-\dfrac{1}{2}\right) - 12$

$\qquad = \dfrac{1}{4} - \dfrac{1}{2} - 12$

$\qquad = -\dfrac{49}{4}$

$\quad$ Vertex: $\left(-\dfrac{1}{2}, -\dfrac{49}{4}\right)$

19. $4m^2 - 4m + 1$

$\quad = (2m)^2 - 2(2m)(1) + 1^2$

$\quad = (2m-1)^2$

20. $\dfrac{x^2-4x+4}{2-x} = \dfrac{(x-2)^2}{-(x-2)} = \dfrac{x-2}{-1} = 2-x$

21. Let $x = $ a number

$\quad x^2 + 3x = 70$

$\quad x^2 + 3x - 70 = 0$

$\quad (x+10)(x-7) = 0$

$\quad x + 10 = 0 \quad$ or $\quad x - 7 = 0$

$\qquad\quad x = -10 \qquad\qquad x = 7$

$\quad$ The number is -10 or 7.

22. $\dfrac{a+1}{a^2-6a+8} - \dfrac{3}{16-a^2}$

$\quad = \dfrac{a+1}{(a-4)(a-2)} - \dfrac{3}{(4+a)(4-a)}$

$\quad = \dfrac{a+1}{(a-4)(a-2)} + \dfrac{3}{(4+a)(a-4)}$

$\quad = \dfrac{(a+1)(a+4)+3(a-2)}{(a-4)(a-2)(a+4)}$

$\quad = \dfrac{(a^2+4a+a+4)+3a-6}{(a-4)(a-2)(a+4)}$

$\quad = \dfrac{a^2+8a-2}{(a-4)(a-2)(a+4)}$

23. a. $\sqrt{25x^3} = \sqrt{25x^2 \cdot x}$

$\qquad\qquad = \sqrt{25x^2} \cdot \sqrt{x}$

$\qquad\qquad = 5x\sqrt{x}$

$\quad$ b. $\sqrt[3]{54x^6y^8} = \sqrt[3]{27 \cdot 2x^6y^6y^2}$

$\qquad\qquad = \sqrt[3]{27x^6y^6} \cdot \sqrt[3]{2y^2}$

$\qquad\qquad = 3x^2y^2\sqrt[3]{2y^2}$

$\quad$ c. $\sqrt[4]{81z^{11}} = \sqrt[4]{81z^8 \cdot z^3}$

$\qquad\qquad = \sqrt[4]{81z^8} \cdot \sqrt[4]{z^3}$

$\qquad\qquad = 3z^2\sqrt[4]{z^3}$

24. $\dfrac{(2a)^{-1}+b^{-1}}{a^{-1}+(2b)^{-1}} = \dfrac{\dfrac{1}{2a}+\dfrac{1}{b}}{\dfrac{1}{a}+\dfrac{1}{2b}}$

$\qquad = \dfrac{\left(\dfrac{1}{2a}+\dfrac{1}{b}\right)2ab}{\left(\dfrac{1}{a}+\dfrac{1}{2b}\right)2ab}$

$\qquad = \dfrac{b+2a}{2b+a}$

$\qquad = \dfrac{2a+b}{a+2b}$

25. a. $\dfrac{2}{\sqrt{5}} = \dfrac{2 \cdot \sqrt{5}}{\sqrt{5} \cdot \sqrt{5}} = \dfrac{2\sqrt{5}}{5}$

b. $\dfrac{2\sqrt{16}}{\sqrt{9x}} = \dfrac{2 \cdot 4}{3\sqrt{x}} = \dfrac{8 \cdot \sqrt{x}}{3\sqrt{x} \cdot \sqrt{x}} = \dfrac{8\sqrt{x}}{3x}$

c. $\sqrt[3]{\dfrac{1}{2}} = \dfrac{\sqrt[3]{1}}{\sqrt[3]{2}} = \dfrac{1}{\sqrt[3]{2}} = \dfrac{1 \cdot \sqrt[3]{2^2}}{\sqrt[3]{2} \cdot \sqrt[3]{2^2}} = \dfrac{\sqrt[3]{4}}{2}$

26. $\begin{array}{r} x^2 - 6x + 8 \\ x+3\overline{\smash{)}x^3 - 3x^2 - 10x + 24} \\ \underline{x^3 + 3x^2} \\ -6x^2 - 10x \\ \underline{-6x^2 - 18x} \\ 8x + 24 \\ \underline{8x + 24} \\ 0 \end{array}$

Answer: $x^2 - 6x + 8$

27. $\sqrt{2x+5} + \sqrt{2x} = 3$

$\sqrt{2x+5} = 3 - \sqrt{2x}$

$\left(\sqrt{2x+5}\right)^2 = \left(3 - \sqrt{3x}\right)^2$

$2x + 5 = 9 - 6\sqrt{2x} + 2x$

$5 = 9 - 6\sqrt{2x}$

$-4 = -6\sqrt{2x}$

$\dfrac{2}{3} = \sqrt{2x}$

$\left(\dfrac{2}{3}\right)^2 = \left(\sqrt{2x}\right)^2$

$\dfrac{4}{9} = 2x$

$\dfrac{2}{9} = x$

28. $P(x) = 4x^3 - 2x^2 + 3$

a. $P(-2) = 4(-2)^3 - 2(-2)^2 + 3$

$= 4(-8) - 2(4) + 3$

$= -32 - 8 + 3$

$= -37$

b. $\begin{array}{r|rrrr} -2 & 4 & -2 & 0 & 3 \\ & & -8 & 20 & -40 \\ \hline & 4 & -10 & 20 & -37 \end{array}$

Thus, $P(-2) = -37$.

29. $\dfrac{x}{2} + \dfrac{8}{3} = \dfrac{1}{6}$

$6\left(\dfrac{x}{2}\right) + 6\left(\dfrac{8}{3}\right) = 6\left(\dfrac{1}{6}\right)$

$3x + 16 = 1$

$3x = -15$

$x = -5$

30. $\dfrac{x+3}{x^2 + 5x + 6} = \dfrac{3}{2x+4} - \dfrac{1}{x+3}$

$\dfrac{x+3}{(x+3)(x+2)} = \dfrac{3}{2(x+2)} - \dfrac{1}{x+3}$

$2(x+3) = 3(x+3) - 2(x+2)$

$2x + 6 = 3x + 9 - 2x - 4$

$2x + 6 = x + 5$

$x = -1$

31. Let $x = $ a number

$\dfrac{x}{6} - \dfrac{5}{3} = \dfrac{x}{2}$

$6\left(\dfrac{x}{6}\right) - 6\left(\dfrac{5}{3}\right) = 6\left(\dfrac{x}{2}\right)$

$x - 10 = 3x$

$-10 = 2x$

$-5 = x$

32. Let t = time to roof the house together.

$$\frac{1}{24} + \frac{1}{40} = \frac{1}{t}$$

$$120t\left(\frac{1}{24} + \frac{1}{40}\right) = 120t\left(\frac{1}{t}\right)$$

$$5t + 3t = 120$$

$$8t = 120$$

$$t = \frac{120}{8} = 15$$

It would take them 15 hours to roof the house working together.

33. $y = kx$

$5 = k(30)$

$k = \frac{5}{30} = \frac{1}{6}$ and $y = \frac{1}{6}x$

34. $y = \frac{k}{x}$

$8 = \frac{k}{24}$

$k = 8(24) = 192$ and $y = \frac{192}{x}$

35. a. $\sqrt{(-3)^2} = |-3| = 3$

b. $\sqrt{x^2} = |x|$

c. $\sqrt[4]{(x-2)^4} = |x-2|$

d. $\sqrt[3]{(-5)^3} = -5$

e. $\sqrt[5]{(2x-7)^5} = 2x-7$

f. $\sqrt{25x^2} = \sqrt{25} \cdot \sqrt{x^2} = 5|x|$

g. $\sqrt{x^2 + 2x + 1} = \sqrt{(x+1)^2} = |x+1|$

36. a. $\sqrt{(-2)^2} = |-2| = 2$

b. $\sqrt{y^2} = |y|$

c. $\sqrt[4]{(a-3)^4} = |a-3|$

d. $\sqrt[3]{(-6)^3} = -6$

e. $\sqrt[5]{(3x-1)^5} = 3x-1$

37. a. $\sqrt[8]{x^4} = x^{4/8} = x^{1/2} = \sqrt{x}$

b. $\sqrt[6]{25} = (25)^{1/6}$
$= (5^2)^{1/6} = 5^{2/6} = 5^{1/3} = \sqrt[3]{5}$

c. $\sqrt[4]{r^2 s^6} = (r^2 s^6)^{1/4}$
$= r^{2/4} s^{6/4}$
$= r^{1/2} s^{3/2}$
$= (rs^3)^{1/2} = \sqrt{rs^3}$

38. a. $\sqrt[4]{5^2} = 5^{2/4} = 5^{1/2} = \sqrt{5}$

b. $\sqrt[12]{x^3} = x^{3/12} = x^{1/4} = \sqrt[4]{x}$

c. $\sqrt[6]{x^2 y^4} = (x^2 y^4)^{1/6}$
$= x^{2/6} y^{4/6}$
$= x^{1/3} y^{2/3}$
$= (xy^2)^{1/3} = \sqrt[3]{xy^2}$

39. a. $\dfrac{2+i}{1-i} = \dfrac{(2+i)\cdot(1+i)}{(1-i)\cdot(1+i)}$

$= \dfrac{2+2i+1i+i^2}{1^2-i^2}$

$= \dfrac{2+3i-1}{1+1}$

$= \dfrac{1+3i}{2}$ or $\dfrac{1}{2}+\dfrac{3}{2}i$

b. $\dfrac{7}{3i} = \dfrac{7\cdot(-3i)}{3i\cdot(-3i)} = \dfrac{-21i}{-9i^2} = \dfrac{-21i}{9} = -\dfrac{7}{3}i$

40. a. $3i(5-2i) = 15i - 6i^2$

$= 15i + 6$

$= 6 + 15i$

b. $(6-5i)^2 = 6^2 - 2(6)(5i) + (5i)^2$

$= 36 - 60i + 25i^2$

$= 36 - 60i - 25$

$= 11 - 60i$

c. $\left(\sqrt{3}+2i\right)\left(\sqrt{3}-2i\right) = \left(\sqrt{3}\right)^2 - (2i)^2$

$= 3 - 4i^2$

$= 3 + 4$

$= 7$

41. $(x+1)^2 = 12$

$x+1 = \pm\sqrt{12}$

$x+1 = \pm 2\sqrt{3}$

$x = -1 \pm 2\sqrt{3}$

The solutions are $-1+2\sqrt{3}$ and $-1-2\sqrt{3}$.

42. $(y-1)^2 = 24$

$y-1 = \pm\sqrt{24}$

$y-1 = \pm 2\sqrt{6}$

$y = 1 \pm 2\sqrt{6}$

The solutions are $1+2\sqrt{6}$ and $1-2\sqrt{6}$.

43. $x - \sqrt{x} - 6 = 0$

Let $y = \sqrt{x}$. Then $y^2 = x$ and

$y^2 - y - 6 = 0$

$(y-3)(y+2) = 0$

$y-3 = 0$ or $y+2 = 0$

$y = 3$ or $\quad y = -2$

$\sqrt{x} = 3$ or $\sqrt{x} = -2$ (can't happen)

$x = 9$

The solution is 9.

44. $m^2 = 4m + 8$

$m^2 - 4m - 8 = 0$

$a = 1, b = -4, c = -8$

$x = \dfrac{4 \pm \sqrt{(-4)^2 - 4(1)(-8)}}{2(1)}$

$= \dfrac{4 \pm \sqrt{16+32}}{2}$

$= \dfrac{4 \pm \sqrt{48}}{2}$

$= \dfrac{4 \pm 4\sqrt{3}}{2}$

$= 2 \pm 2\sqrt{3}$

The solutions are $2+2\sqrt{3}$ and $2-2\sqrt{3}$.

Chapter 12

Exercise Set 12.1

1. $f = \{(-1,-1),(1,1),(0,2),(2,0)\}$
is a one-to-one function.
$f^{-1} = \{(-1,-1),(1,1),(2,0),(0,2)\}$

3. $h = \{(10,10)\}$
is a one-to-one function.
$h^{-1} = \{(10,10)\}$

5. $f = \{(11,12),(4,3),(3,4),(6,6)\}$
is a one-to-one function.
$f^{-1} = \{(12,11),(3,4),(4,3),(6,6)\}$

7. This function is not one-to-one because there are two months with the same output: (January, 282) and (May, 282).

9. This function is one-to-one.

Rank in population (Input)	1	49	12	2	45
State (Output)	California	Vermont	Virginia	Texas	South Dakota

11. $f(x) = x^3 + 2$

   a. $f(1) = 1^3 + 2 = 3$

   b. $f^{-1}(3) = 1$

13. $f(x) = x^3 + 2$

   a. $f(-1) = (-1)^3 + 2 = 1$

   b. $f^{-1}(1) = -1$

15. The graph represents a one-to-one function because it passes the horizontal line test.

17. The graph does not represent a one-to-one function because it does not pass the horizontal line test.

19. The graph represents a one-to-one function because it passes the horizontal line test.

21. The graph does not represent a one-to-one function because it does not pass the horizontal line test.

23. $f(x) = x + 4$
$$y = x + 4$$
$$x = y + 4$$
$$y = x - 4$$
$$f^{-1}(x) = x - 4$$

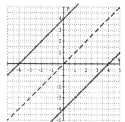

25. $f(x) = 2x - 3$

$$y = 2x - 3$$
$$x = 2y - 3$$
$$2y = x + 3$$
$$y = \frac{x + 3}{2}$$
$$f^{-1}(x) = \frac{x + 3}{2}$$

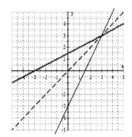

27. $f(x) = \dfrac{1}{2}x - 1$

$$y = \frac{1}{2}x - 1$$
$$x = \frac{1}{2}y - 1$$
$$\frac{1}{2}y = x + 1$$
$$y = 2x + 2$$
$$f^{-1}(x) = 2x + 2$$

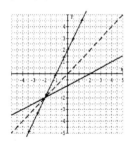

29. $f(x) = x^3$

$$y = x^3$$
$$x = y^3$$
$$y = \sqrt[3]{x}$$
$$f^{-1}(x) = \sqrt[3]{x}$$

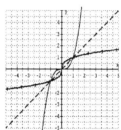

31. $f(x) = 5x + 2$

$$y = 5x + 2$$
$$x = 5y + 2$$
$$5y = x - 2$$
$$y = \frac{x - 2}{5}$$
$$f^{-1}(x) = \frac{x - 2}{5}$$

33. $f(x) = \dfrac{x - 2}{5}$

$$y = \frac{x - 2}{5}$$
$$x = \frac{y - 2}{5}$$
$$5x = y - 2$$
$$y = 5x + 2$$
$$f^{-1}(x) = x^3$$

35. $f(x) = \sqrt[3]{x}$

$$y = \sqrt[3]{x}$$
$$x = \sqrt[3]{y}$$
$$x^3 = y$$
$$f^{-1}(x) = x^3$$

37. $f(x) = \dfrac{5}{3x+1}$

$y = \dfrac{5}{3x+1}$

$x = \dfrac{5}{3y+1}$

$3y+1 = \dfrac{5}{x}$

$3y = \dfrac{5}{x} - 1$

$3y = \dfrac{5-x}{x}$

$y = \dfrac{5-x}{3x}$

$f^{-1}(x) = \dfrac{5-x}{3x}$

39. $f(x) = (x+2)^3$

$y = (x+2)^3$

$x = (y+2)^3$

$\sqrt[3]{x} = y+2$

$\sqrt[3]{x} - 2 = y$

$f^{-1}(x) = \sqrt[3]{x} - 2$

41.

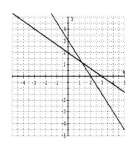

43.

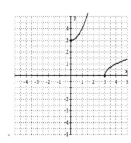

45.

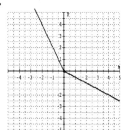

47. $(f \circ f^{-1})(x) = f(f^{-1}(x))$

$\quad = f\left(\dfrac{x-1}{2}\right)$

$\quad = 2\left(\dfrac{x-1}{2}\right) + 1$

$\quad = x - 1 + 1$

$\quad = x$

$(f^{-1} \circ f)(x) = f^{-1}(f(x))$

$\quad = f^{-1}(2x+1)$

$\quad = \dfrac{(2x+1)-1}{2}$

$\quad = \dfrac{2x}{2}$

$\quad = x$

49. $(f \circ f^{-1})(x) = f(f^{-1}(x))$

$\quad = f\left(\sqrt[3]{x-6}\right)$

$\quad = \left(\sqrt[3]{x-6}\right)^3 + 6$

$\quad = x - 6 + 6$

$\quad = x$

$(f^{-1} \circ f)(x) = f^{-1}(f(x))$

$\quad = f^{-1}(x^3 + 6)$

$\quad = \sqrt[3]{(x^3+6)-6}$

$\quad = \sqrt[3]{x^3}$

$\quad = x$

51. 5

53. 8

55. $\dfrac{1}{27}$

57. 9

59. $3^{1/2} \approx 1.73$

61. a. $\left(-2, \dfrac{1}{4}\right), \left(-1, \dfrac{1}{2}\right), (0,1), (1,2),$
$(2,5)$

    b. $\left(\dfrac{1}{4}, -2\right), \left(\dfrac{1}{2}, -1\right), (1,0), (2,1),$
$(5,2)$

    c., d.

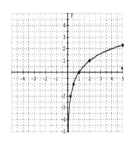

63. Answers may vary.

65. $f(x) = 3x + 1$
$y = 3x + 1$
$x = 3y + 1$
$x - 1 = 3y$
$y = \dfrac{x-1}{3}$
$f^{-1}(x) = \dfrac{x-1}{3}$

67. $f(x) = \sqrt[3]{x+1}$
$y = \sqrt[3]{x+1}$
$x = \sqrt[3]{y+1}$
$x^3 = y + 1$
$y = x^3 - 1$
$f^{-1}(x) = x^3 - 1$

Section 12.2

Graphing Calculator Explorations

1.

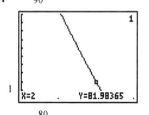

81.98%

3.

30

13   X=15   Y=22.540023   17

20

22.54%

Exercise Set 12.2

1. $y = 4^x$

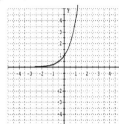

3. $y = 2^x + 1$

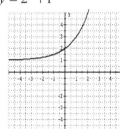

5. $y = \left(\dfrac{1}{4}\right)^x$

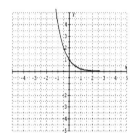

7. $y = \left(\dfrac{1}{2}\right)^x - 2$

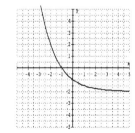

9. $y = -2^x$

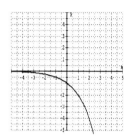

11. $y = -\left(\dfrac{1}{4}\right)^x$

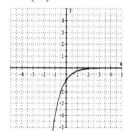

13. $f(x) = 2^{x+1}$

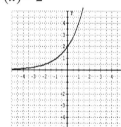

15. $f(x) = 4^{x-2}$

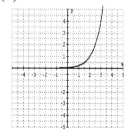

17. C

19. D

21. $3^x = 27$

$3^x = 3^3$

$x = 3$

The solution is 3.

23. $16^x = 8$

$(2^4)^x = 2^3$

$2^{4x} = 2^3$

$4x = 3$

$x = \dfrac{3}{4}$

The solution is $\dfrac{3}{4}$.

25. $32^{2x\ 3} = 2$

$(2^5)^{2x-3} = 2^1$

$10x - 15 = 1$

$10x = 16$

$x = \dfrac{8}{5}$

The solution is $\dfrac{8}{5}$.

27. $\dfrac{1}{4} = 2^{3x}$

$2^{-2} = 2^{3x}$

$3x = -2$

$x = -\dfrac{2}{3}$

The solution is $-\dfrac{2}{3}$.

29. $5^x = 625$

$5^x = 5^4$

$x = 4$

The solution is 4.

31. $4^x = 8$

$(2^2)^x = 2^3$

$2^{2x} = 2^3$

$2x = 3$

$x = \dfrac{3}{2}$

The solution is $\dfrac{3}{2}$.

33. $27^{x+1} = 9$

$(3^3)^{x+1} = 3^2$

$3^{3x+3} = 3^2$

$3x + 3 = 2$

$3x = -1$

$x = -\dfrac{1}{3}$

The solution is $-\dfrac{1}{3}$.

35. $81^{x-1} = 27^{2x}$

$(3^4)^{x-1} = (3^3)^{2x}$

$3^{4x-4} = 3^{6x}$

$4x - 4 = 6x$

$-4 = 2x$

$x = -2$

The solution is -2.

37. $y = 30(2.7)^{-0.004t}, t = 50$

$= 30(2.7)^{-(0.004)(50)}$

$= 30(2.7)^{-0.2}$

≈ 24.6

Therefore, approximately 24.6 pounds of uranium will remain after 50 days.

39. $y = 260(2.7)^{0.025t}$

$y = 260(2.7)^{0.025(10)}$

$y = 260(2.7)^{0.25}$

$y = 333$

There should be about 333 bison in the park in 10 years.

41. $y = 5(2.7)^{-0.15t}, t = 10$

$= 5(2.7)^{-0.15(10)}$

$= 5(2.7)^{-1.5}$

≈ 1.1 grams

43. a. $y = 42.1(1.56)^t$

$= 42.1(1.56)^1$

≈ 65.7

$\$65.7$ billion

b. $y = 42.1(1.56)^t$

$= 42.1(1.56)^9$

≈ 2303.6

$\$2303.6$ billion

45. a. $y = 120.882(1.012)^x$

$= 120.882(1.012)^{40}$

≈ 194.8 million people

b. $y = 120.882(1.012)^x$

$= 120.882(1.012)^{90}$

≈ 353.7 million people

47. $A = P\left(1 + \dfrac{r}{n}\right)^{nt}$

$t = 3, P = 6000, r = 0.08,$ and $n = 12.$

$A = 6000\left(1 + \dfrac{0.08}{12}\right)^{12(3)}$

$= 6000(1.006)^{36}$

≈ 7621.42

Erica would owe $\$7621.42$ after 3 years.

49. $A = P\left(1 + \dfrac{r}{n}\right)^{nt}$, where $P = 2000$

$r = 0.06, n = 2,$ and $t = 12.$

$A = 2000\left(1 + \dfrac{0.06}{2}\right)^{2(12)}$

$= 2000(1.03)^{24}$

$= 4065.59$

Janina has approximately $\$4065.59$ in her savings account.

51. $y = 34(1.254)^x$

$= 34(1.254)^{13}$

≈ 645

There will be approximately 645 million cellular phone users in 2005.

53. 4

55. $\varnothing$

57. 2, 3

59. 3

61. -1

63. Answers may vary

65. $y = \left|3^x\right|$

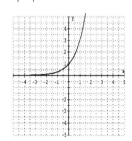

67. $y = 3^{|x|}$

The graphs are the same

since $\left(\dfrac{1}{2}\right)^{-x} = 2^x$.

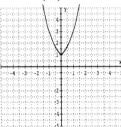

69.

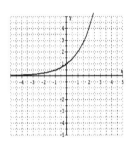

71.

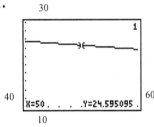

24.60 pounds

73.

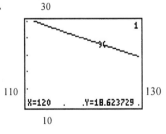

18.62 lbs

75.

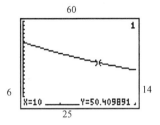

50.41 g

Exercise Set 12.3

1. $\log_6 36 = 2$

$6^2 = 36$

3. $\log_3 \dfrac{1}{27} = -3$

$3^{-3} = \dfrac{1}{27}$

5. $\log_{10} 1000 = 3$

$10^3 = 1000$

7. $\log_e x = 4$

$e^4 = x$

9. $\log_e \dfrac{1}{e^2} = -2$

$e^{-2} = \dfrac{1}{e^2}$

11. $\log_7 \sqrt{7} = \dfrac{1}{2}$

$7^{1/2} = \sqrt{7}$

13. $2^4 = 16$

$\log_2 16 = 4$

15. $10^2 = 100$

$\log_{10} 100 = 2$

17. $e^3 = x$

$\log_e x = 3$

19. $10^{-1} = \dfrac{1}{10}$

$\log_{10} \dfrac{1}{10} = -1$

21. $4^{-2} = \dfrac{1}{16}$

$\log_4 \dfrac{1}{16} = -2$

23. $5^{1/2} = \sqrt{5}$

$\log_5 \sqrt{5} = \dfrac{1}{2}$

25. $\log_2 8 = 3$ since $2^3 = 8$.

27. $\log_3 \dfrac{1}{9} = -2$ since $3^{-2} = \dfrac{1}{9}$.

29. $\log_{25} 5 = \dfrac{1}{2}$ since $25^{1/2} = 5$.

31. $\log_{1/2} 2 = -1$ since $\left(\dfrac{1}{2}\right)^{-1} = 2$.

33. $\log_7 1 = 0$

35. $\log_2 2^4 = 4$

37. $\log_{10} 100 = 2$

39. $3^{\log_3 5} = 5$

41. $\log_3 81 = 4$

43. $\log_4 \left(\dfrac{1}{64}\right) = -3$

45. Answers may vary.

47. $\log_3 9 = x$

49. $\log_3 x = 4$

$x = 3^4 = 81$

51. $\log_x 49 = 2$

$x^2 = 49$

$x = \pm 7$

We discard the negative base.

$x = 7$

53. $\log_2 \dfrac{1}{8} = x$

$2^x = \dfrac{1}{8}$

$2^x = 2^{-3}$

$x = -3$

55. $\log_3 \left(\dfrac{1}{27}\right) = x$

$\dfrac{1}{27} = 3^x$

$3^{-3} = 3^x$

$-3 = x$

57. $\log_8 x = \dfrac{1}{3}$

$x = 8^{1/3} = 2$

59. $\log_4 16 = x$

$4^x = 4^2$

$x = 2$

61. $\log_{3/4} x = 3$

$\left(\dfrac{3}{4}\right)^3 = x$

63. $\log_x 100 = 2$ or $x^2 = 100$
$x = \pm 10$ and we discard the negative base.
$x = 10$

65. $\log_5 5^3 = 3$

67. $2^{\log_2 3} = 3$

69. $\log_9 9 = 1$

71. $y = \log_3 x$
$y = 0$:
$\log_3 x = 0$
$x = 3^0 = 1$ is the only
x-intercept. No y-intercept exists.

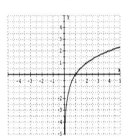

73. $f(x) = \log_{1/4} x$
$0 = \log_{1/4} x$

$x = \left(\dfrac{1}{4}\right)^0 = 1$ is the x-intercept.

No y-intercept exist.

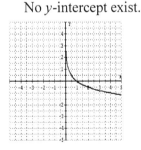

75. $f(x) = \log_5 x$
$x = 0$:
$y = \log_5 0$ is undefined so there is
no y-intercept.
$y = 0$:
$0 = \log_5 x$
$x = 5^0 = 1$ is the x-intercept.

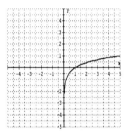

77. $f(x) = \log_{1/16} x$
$x = 0$:
$y = \log_{1/6} 0$ is not defined so there
is no y-intercept.
$y = 0$:
$0 = \log_{1/6} x$

$x = \left(\dfrac{1}{6}\right)^0 = 1$ is the x-intercept.

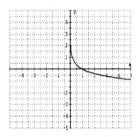

79. 1

81. $\dfrac{x - 4}{2}$

83. $\dfrac{2x + 3}{x^2}$

85. $m - 1$

87. $\log_7(5x-2)=1$

$5x-2=7$

$5x=9$

$x=\dfrac{9}{5}$

89. $\log_3\left(\log_5 125\right)=\log_3(3)=1$

91.

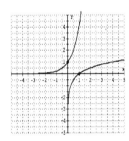

93.

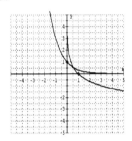

95. $\log_{10}(1-k)=\dfrac{-0.3}{H}$, $H=8$

$\log_{10}(1-k)=\dfrac{-0.3}{8}=-0.0375$

$1-k=10^{-0.0375}$

$1-10^{-0.0375}=k$

$k\approx 0.0827$

97. $\log_3 10$ is between 2 and 3 because

$3^2=9$ and $3^3=27$

Section 12.4

Mental Math

   1. A

   2. C

   3. B

   4. C

   5. A

   6. B

Exercise Set 12.4

   1. $\log_5 2+\log_5 7=\log_5(2\cdot 7)=\log_5 14$

   3. $\log_4 9+\log_4 x=\log_4 9x$

   5. $\log_{10} 5+\log_{10} 2+\log_{10}(x^2+2)$

$=\log_{10}\left[5\cdot 2\left(x^2+2\right)\right]$

$=\log_{10}\left(10x^2+20\right)$

   7. $\log_5 12-\log_5 4=\log_5\left(\dfrac{12}{4}\right)=\log_5 3$

   9. $\log_2 x-\log_2 y=\log_2\dfrac{x}{y}$

   11. $\log_4 2+\log_4 10-\log_4 5$

$=\log_4 2\cdot 10-\log_4 5$

$=\log_4\left(\dfrac{20}{5}\right)$

$=\log_4 4=1$

   13. $\log_3 x^2=2\log_3 x$

   15. $\log_4 5^{-1}=-\log_4 5$

17. $\log_5 \sqrt{y} - \log_5 y^{1/2} = \frac{1}{2}\log_5 y$

19. $2\log_2 5 = \log_2 5^2 = \log_2 25$

21. $3\log_5 x + 6\log_5 z = \log_5 x^3 + \log_5 z^6$
$$= \log_5 x^3 z^6$$

23. $\log_{10} x - \log_{10}(x+1) + \log_{10}(x^2-2)$
$$= \log_{10}\frac{x}{x+1} + \log_{10}(x^2-2)$$
$$= \log_{10}\frac{x(x^2-2)}{x+1}$$
$$= \log_{10}\frac{x^3-2x}{x+1}$$

25. $\log_4 5 + \log_4 7 = \log_4(5\cdot 7) = \log_4 35$

27. $\log_3 8 - \log_3 2 = \log_3\left(\frac{8}{2}\right) = \log_3 4$

29. $\log_7 6 + \log_7 3 - \log_7 4$
$$= \log_7(6\cdot 3) - \log_7 4$$
$$= \log_7\left(\frac{18}{4}\right)$$
$$= \log_7\frac{9}{2}$$

31. $3\log_4 2 + \log_4 6 = \log_4 2^3 + \log_4 6$
$$= \log_4 8 + \log_4 6$$
$$= \log_4(8\cdot 6)$$
$$= \log_4 48$$

33. $3\log_2 x + \frac{1}{2}\log_2 x - 2\log_2(x+1)$
$$= \log_2 x^3 + \log_2 x^{1/2} - \log_2(x+1)^2$$
$$= \log_2(x^3 \cdot x^{1/2}) - \log_2(x+1)^2$$
$$= \log_2 x^{7/2} - \log_2(x+1)^2$$
$$= \log_2\frac{x^{7/2}}{(x+1)^2}$$

35. $2\log_8 x - \frac{2}{3}\log_8 x + \log_8 x$
$$= \left(2 - \frac{2}{3} + 4\right)\log_8 x$$
$$= \frac{16}{3}\log_8 x$$
$$= \log_8 x^{16/3}$$

37. $\log_2\frac{7\cdot 11}{3} = \log_2(7\cdot 11) - \log_2 3$
$$= \log_2 7 + \log_2 11 - \log_2 3$$

39. $\log_3\left(\frac{4y}{5}\right) = \log_3 4y - \log_3 5$
$$= \log_3 4 + \log_3 y - \log_3 5$$

41. $\log_2\left(\frac{x^3}{y}\right) = \log_2 x^3 - \log_2 y$
$$= 3\log_2 x - \log_2 y$$

43. $\log_b \sqrt{7x} = \log_b(7x)^{1/2}$
$$= \frac{1}{2}\log_b(7x)$$
$$= \frac{1}{2}\left[\log_b 7 + \log_b x\right]$$
$$= \frac{1}{2}\log_b 7 + \log_b x$$

45. $\log_7\left(\frac{5x}{4}\right) = \log_7 5x - \log_7 4$
$$= \log_7 5 + \log_7 x - \log_7 4$$

47. $\log_5 x^3(x+1) = \log_5 x^3 + \log_5(x+1)$
$$= 3\log_5 x + \log_5(x+1)$$

49. $\log_6\frac{x^2}{x+3} = \log_6 x^2 - \log_6(x+3)$
$$= 2\log_6 x - \log_6(x+3)$$

51. $\log_b\left(\dfrac{5}{3}\right) = \log_b 5 - \log_b 3$

$= 0.7 - 0.5 = 0.2$

53. $\log_b 15 = \log_b(5 \cdot 3)$

$= \log_b 5 + \log_b 3$

$= 0.7 + 0.5 = 1.2$

55. $\log_b \sqrt[3]{5} = \log_b 5^{1/3}$

$= \dfrac{1}{3}\log_b 5$

$= \dfrac{1}{3}(0.7)$

≈ 0.233

57. $\log_b 8 = \log_b 2^3$

$= 3\log_b 2$

$= 3(0.43)$

$= 1.29$

59. $\log_b\left(\dfrac{3}{9}\right) = \log_b\left(\dfrac{1}{3}\right)$

$= \log_b 3^{-1}$

$= (-1)\log_b 3$

$= -(0.68)$

$= -0.68$

61. $\log_b \sqrt{\dfrac{2}{3}} = \log_b\left(\dfrac{2}{3}\right)^{1/2}$

$= \dfrac{1}{2}\log_b \dfrac{2}{3}$

$= \dfrac{1}{2}\left(\log_b 2 - \log_b 3\right)$

$= \dfrac{1}{2}(0.43 - 0.68)$

$= \dfrac{1}{2}(-0.25)$

$= -0.125$

63.

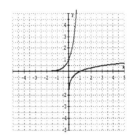

65. $\log_{10} \dfrac{1}{10} = x$

$10^x = \dfrac{1}{10}$

$10^x = 10^{-1}$

$x = -1$

67. $\log_7 \sqrt{7} = x$

$7^x = \sqrt{7}$

$7^x = 7^{1/2}$

$x = \dfrac{1}{2}$

69. False

71. True

73. False

Integrated Review

1. one-to-one;

    $\{(6,-2),(8,4),(-6,2),(3,3)\}$

2. not one-to-one

3. not one-to-one

4. one-to-one

5. not one-to-one

6. $f^{-1}(x) = \dfrac{x}{3}$

7. $f^{-1}(x) = x - 4$

8. $f^{-1}(x) = \dfrac{x+1}{5}$

9. $f^{-1}(x) = \dfrac{x-2}{3}$

10. $y = \left(\dfrac{1}{2}\right)^x$

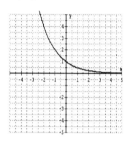

11. $y = 2^x + 1$

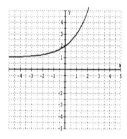

12. $y = \log_3 x$

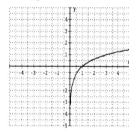

13. $y = \log_{1/3} x$

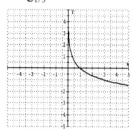

14. $2^x = 8$
$\quad 2^x = 2^3$
$\quad\quad x = 3$
The solution is 3.

15. $9 = 3^{x-5}$
$\quad 3^2 = 3^{x-5}$
$\quad\quad 2 = x - 5$
$\quad\quad 7 = x$
The solution is 7.

16. $\quad 4^{x-1} = 8^{x+2}$
$\quad (2^2)^{x-1} = (2^3)^{x+2}$
$\quad\quad 2^{2x-2} = 2^{3x+6}$
$\quad\quad 2x-2 = 3x+6$
$\quad\quad\quad -8 = x$
The solution is -8.

17. $\quad 25^x = 125^{x-1}$
$\quad (5^2)^x = (5^3)^{x-1}$
$\quad\quad 5^{2x} = 5^{3x-3}$
$\quad\quad 2x = 3x - 3$
$\quad\quad\; 3 = x$
The solution is 3.

18. $\log_4 16 = x$
$\quad\quad 4^x = 16$
$\quad\quad 4^x = 4^2$
$\quad\quad\; x = 2$
The solution is 2.

19. $\log_{49} 7 = x$
$49^x = 7$
$(7^2)^x = 7$
$7^{2x} = 7$
$2x = 1$
$x = \dfrac{1}{2}$

The solution is $\dfrac{1}{2}$.

20. $\log_2 x = 5$
$2^5 = x$
$32 = x$
The solution is 32.

21. $\log_x 64 = 3$
$x^3 = 64$
$x = \sqrt[3]{64} = 4$
The solution is 4.

22. $\log_x \dfrac{1}{125} = -3$
$x^{-3} = \dfrac{1}{125}$
$\dfrac{1}{x^3} = \dfrac{1}{125}$
$x^3 = 125$
$x = \sqrt[3]{125} = 5$
The solution is 5.

23. $\log_3 x = -2$
$3^{-2} = x$
$x = \dfrac{1}{3^2} = \dfrac{1}{9}$
The solution is $\dfrac{1}{9}$.

24. $\log_2 x^5$

25. $\log_2 5^x$

26. $\log_5 \dfrac{x^3}{y^5}$

27. $\log_5 x^9 y^3$

28. $\log_2 \dfrac{x^2 - 3x}{x^2 + 4}$

29. $\log_3 \dfrac{y\left(y^3 + 11\right)}{y + 2}$

30. $\log_7 9 + 2\log_7 x - \log_7 y$

31. $\log_6 5 + \log_6 y - 2\log_6 z$

Exercise Set 12.5

1. $\log 8 \approx 0.9031$

3. $\log 2.31 \approx 0.3636$

5. $\ln 2 \approx 0.6931$

7. $\ln 0.0716 \approx -2.6367$

9. $\log 12.6 \approx 1.1004$

11. $\ln 5 \approx 1.6094$

13. $\log 41.5 \approx 1.6180$

15. Answers may vary.

17. $\log 100 = \log 10^2 = 2$

19. $\log \dfrac{1}{1000} = \log 10^{-3} = -3$

21. $\ln e^2 = 2$

23. $\ln \sqrt[4]{e} = \ln e^{1/4} = \dfrac{1}{4}$

25. $\log 10^3 = 3$

27. $\ln e^2 = 2$

29. $\log 0.0001 = \log 10^{-4} = -4$

31. $\ln \sqrt{e} = \ln e^{1/2} = \dfrac{1}{2}$

33. $\ln 50$ is larger
Answers may vary.

35. $\log x = 1.3$
$\qquad x = 10^{1.3} \approx 19.9526$

37. $\log 2x = 1.1$
$\qquad 2x = 10^{1.1}$
$\qquad x = \dfrac{10^{1.1}}{2} \approx 6.2946$

39. $\ln x = 1.4$
$\qquad x = e^{1.4} \approx 4.0552$

41. $\ln(3x - 4) = 2.3$
$\qquad 3x - 4 = e^{2.3}$
$\qquad 3x = 4 + e^{2.3}$
$\qquad x = \dfrac{4 + e^{2.3}}{3} \approx 4.6581$

43. $\log x = 2.3$
$\qquad x = 10^{2.3} \approx 199.5262$

45. $\ln x = -2.3$
$\qquad x = e^{-2.3} \approx 0.1003$

47. $\log(2x + 1) = -0.5$
$\qquad 2x + 1 = 10^{-0.5}$
$\qquad 2x = 10^{-0.5} - 1$
$\qquad x = \dfrac{10^{-0.5} - 1}{2} \approx 0.3419$

49. $\ln 4x = 0.18$
$\qquad 4x = e^{0.18}$
$\qquad x = \dfrac{e^{0.18}}{4} \approx 0.2993$

51. $\log_2 3 = \dfrac{\ln 3}{\ln 2} \approx 1.5850$

53. $\log_{1/2} 5 = \dfrac{\ln 5}{\ln\left(\dfrac{1}{2}\right)} \approx -2.3219$

55. $\log_4 9 = \dfrac{\ln 9}{\ln 4} \approx 1.5850$

57. $\log_3\left(\dfrac{1}{6}\right) = \log_3 6^{-1}$
$\qquad\qquad = (-1)\log_3 6$
$\qquad\qquad = -\dfrac{\ln 6}{\ln 3} \approx -1.6309$

59. $\log_8 6 = \dfrac{\ln 6}{\ln 8} \approx 0.8617$

61. $R = \log\left(\dfrac{a}{T}\right) + B,\ a = 200,\ T = 1.6$
$\qquad B = 2.1$
$\qquad R = \log\left(\dfrac{200}{1.6}\right) + 2.1 \approx 4.2$

The earthquake measures 4.2
on the Richter scale.

63. $R = \log\left(\dfrac{a}{T}\right) + B$, $a = 400$, $T = 2.6$

$B = 3.1$

$R = \log\left(\dfrac{400}{2.6}\right) + 3.1 \approx 5.3$

The earthquake measures 5.3 on the Richter scale.

65. $A = Pe^{rt}$, $t = 12$, $P = 1400$, $r = 0.08$

$A = 1400e^{(0.08)12} = 1400e^{0.96}$

≈ 3656.38

Dana has $3656.38 after 12 years.

67. $A = Pe^{rt}$, $t = 4$, $P = 2000$, $r = 0.06$

$A = 2000e^{(0.06)4} = 2000e^{0.24}$

≈ 2542.50

Barbara owes $2542.50 at the end of 4 years.

69. $\dfrac{4}{7}$

71. $x = \dfrac{3y}{4}$

73. $-6, -1$

75. $(2, -3)$

77. $\ln 50$ is larger

Answers may vary.

79. $f(x) = e^x$

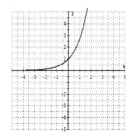

81. $f(x) = e^{-3x}$

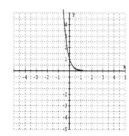

83. $f(x) = e^x + 2$

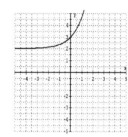

85. $f(x) = e^{x-1}$

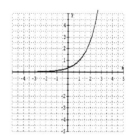

87. $f(x) = 3e^x$

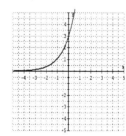

89. $f(x) = \ln x$

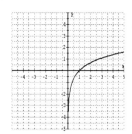

91. $f(x) = -2\log x$

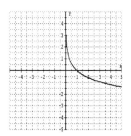

93. $f(x) = \log(x+2)$

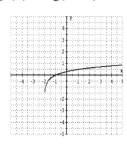

95. $f(x) = \ln x - 3$

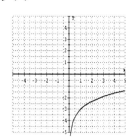

97. $f(x) = e^x$
$\quad\quad f(x) = e^x + 2$
$\quad\quad f(x) = e^x - 3$

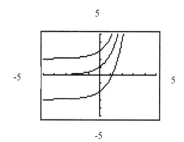

Section 12.6

Graphing Calculator Explorations

1. $Y_1 = 5000\left(1+\dfrac{0.05}{4}\right)^{4x}, Y_2 = 6000$

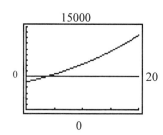

3. $Y_1 = 10000\left(1+\dfrac{0.06}{12}\right)^{12x}, Y_2 = 40000$

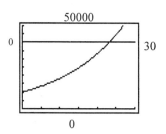

Exercise Set 12.6

1. $3^x = 6$
$\quad x = \log_3 6 = \dfrac{\log 6}{\log 3} \approx 1.6309$

3. $3^{2x} = 3.8$

$$2x = \log_3 3.8 = \frac{\log 3.8}{\log 3}$$

$$x = \frac{\log 3.8}{2\log 3} \approx 0.6076$$

5. $2^{x-3} = 5$

$x - 3 = \log_2 5$

$$x = 3 + \log_2 5 = 3 + \frac{\log 5}{\log 2}$$

$$\approx 5.3219$$

7. $9^x = 5$

$$x = \log_9 5 = \frac{\log 5}{\log 9} \approx 0.7325$$

9. $4^{x+7} = 3$

$x + 7 = \log_4 3$

$x = -7 + \log_4 3$

$$= -7 + \frac{\log 3}{\log 4}$$

$$\approx -6.2075$$

11. $7^{3x-4} = 11$

$3x - 4 = \log_7 11$

$3x = 4 + \log_7 11$

$$x = \frac{1}{3}\left(4 + \frac{\log 11}{\log 7}\right) \approx 1.7441$$

13. $e^{6x} = 5$

$6x = \ln 5$

$$x = \frac{\ln 5}{6} \approx 0.2682$$

15. $\log_2(x+5) = 4$

$x + 5 = 2^4$

$x + 5 = 16$

$x = 11$

17. $\log_3 x^2 = 4$

$x^2 = 3^4$

$x^2 = 81$

$x = \pm 9$

19. $\log_4 2 + \log_4 x = 0$

$\log_4(2x) = 0$

$2x = 4^0$

$2x = 1$

$$x = \frac{1}{2}$$

21. $\log_2 6 - \log_2 x = 3$

$$\log_2\left(\frac{6}{x}\right) = 3$$

$$\frac{6}{x} = 2^3$$

$$\frac{6}{x} = 8$$

$8x = 6$

$$x = \frac{3}{4}$$

23. $\log_4 x + \log_4(x+6) = 2$

$\log_4 x(x+6) = 2$

$x(x+6) = 4^2$

$x^2 + 6x = 16$

$x^2 + 6x - 16 = 0$

$(x+8)(x-2) = 0$

$x = -8$ or $x = 2$

We discard -8 as extraneous.

25. $\log_5(x+3) - \log_5 x = 2$

$$\log_5\left(\frac{x+3}{x}\right) = 2$$

$$\frac{x+3}{x} = 5^2$$

$$\frac{x+3}{x} = 25$$

$$x + 3 = 25x$$

$$3 = 24x$$

$$x = \frac{1}{8}$$

27. $\log_3(x-2) = 2$

$$x - 2 = 3^2$$

$$x - 2 = 9$$

$$x = 11$$

29. $\log_4(x^2 - 3x) = 1$

$$x^2 - 3x = 4$$

$$x^2 - 3x - 4 = 0$$

$$(x-4)(x+1) = 0$$

$$x = 4 \text{ or } x = -1$$

31. $\ln 5 + \ln x = 0$

$$\ln(5x) = 0$$

$$e^0 = 5x$$

$$1 = 5x$$

$$\frac{1}{5} = x$$

33. $3\log x - \log x^2 = 2$

$$3\log x - 2\log x = 2$$

$$\log x = 2$$

$$x = 10^2$$

$$x = 100$$

35. $\log_2 x + \log_2(x+5) = 1$

$$\log_2 x(x+5) = 1$$

$$x(x+5) = 2$$

$$x^2 + 5x - 2 = 0$$

$$a = 1, \, b = 5, \, c = -2$$

$$x = \frac{-5 \pm \sqrt{5^2 - 4(1)(-2)}}{2(1)}$$

$$x = \frac{-5 \pm \sqrt{33}}{2}$$

Discard $\dfrac{-5 - \sqrt{33}}{2}$

37. $\log_4 x - \log_4(2x-3) = 3$

$$\log_4\left(\frac{x}{2x-3}\right) = 3$$

$$\frac{x}{2x-3} = 64$$

$$x = 64(2x-3)$$

$$x = 128x - 192$$

$$192 = 127x$$

$$x = \frac{192}{127}$$

39. $\log_2 x + \log_2(3x+1) = 1$

$$\log_2 x(3x+1) = 1$$

$$x(3x+1) = 2$$

$$3x^2 + x - 2 = 0$$

$$(3x-2)(x+1) = 0$$

$$3x - 2 = 0 \text{ or } x + 1 = 0$$

$$x = \frac{2}{3} \text{ or } x = -1$$

We discard -1 as extraneous.

41. $y = y_0 e^{0.043t}$, $y_0 = 83$, $t = 5$

$y = 83e^{0.043(5)} = 83e^{0.215} \approx 103$

There should be 103 wolves in

5 years.

43. $y = y_0 e^{0.026t}$, $y_0 = 10,052,000$, $t = 6$

$y = 10,052,000e^{0.026(6)} \approx 11,750,000$

There will be approximately

11,750,000 inhabitants in 2005.

45. $y = y_0 e^{-0.005t}$

$y_0 = 146,394$, $y = 120,000$

$120,000 = 146,394e^{-0.005t}$

$\dfrac{120,000}{146,394} = e^{-0.005t}$

$t = \dfrac{\ln\left(\dfrac{120,000}{146,394}\right)}{-0.005} \approx 39.8$

It will take approximately 39.8

years to reach 120,000.

47. $A = P\left(1 + \dfrac{r}{n}\right)^{nt}$, $P = 600$,

$A = 2(600) = 1200$, $r = 0.07$,

$n = 12$

$1200 = 600\left(1 + \dfrac{0.07}{12}\right)^{12t}$

$2 = (1.0058\overline{3})^{12t}$

$12t = \log_{1.0058\overline{3}} 2$

$t \approx 10$

It would take approximately 10

years for the $600 to double.

49. $A = P\left(1 + \dfrac{r}{n}\right)^{nt}$, $P = 1200$,

$A = P + I = 1200 + 200 = 1400$

$r = 0.009$, $n = 4$

$1400 = 1200\left(1 + \dfrac{0.09}{4}\right)^{4t}$

$\dfrac{7}{6} = (1.0225)^{4t}$

$4t = \log_{1.0225}\left(\dfrac{7}{6}\right)$

$t \approx 1.73$

It would take the investment

approximately 1.7 years to earn

$200.

51. $A = P\left(1 + \dfrac{r}{n}\right)^{nt}$, $P = 1000$

$A = 2(1000) = 2000$, $r = 0.08$, $n = 2$

$2000 = 1000\left(1 + \dfrac{0.08}{2}\right)^{2t}$

$2 = (1.04)^{2t}$

$2t = \log_{1.04} 2 = \dfrac{\ln 2}{\ln 1.04}$

$t \approx 8.8$

53. $w = 0.00185h^{2.67}$, and $h = 35$

$w = 0.00185(35)^{2.67} \approx 24.5$

The expected weight of a boy 35

inches tall is 24.5 pounds.

55. $w = 0.00185h^{2.67}$, and $w = 85$

$$85 = 0.00185h^{2.67}$$

$$\frac{85}{0.00185} = h^{2.67}$$

$$h = \left(\frac{85}{0.00185}\right)^{1/2.67} \approx 55.7$$

The expected height of the boy is approximately 55.7 inches.

57. $P = 14.7e^{-0.21x}$, $x = 1$

$$= 14.7e^{-0.21(1)}$$

$$= 14.7e^{-0.21}$$

$$\approx 11.9$$

The average atmospheric pressure of Denver is approximately 11.9 pounds per square inch.

59. $P = 14.7e^{-0.21x}$, $P = 7.5$

$$7.5 = 14.7e^{-0.21x}$$

$$\frac{7.5}{14.7} = e^{-0.21x}$$

$$-0.21x = \ln\left(\frac{7.5}{14.7}\right)$$

$$x = -\frac{1}{0.21}\ln\left(\frac{7.5}{14.7}\right) \approx 3.2$$

The elevation of the jet is approximately 3.2 miles.

61. $t = \frac{1}{c}\ln\left(\frac{A}{A-N}\right)$

$$t = \frac{1}{0.09}\ln\left(\frac{75}{75-50}\right)$$

$$t = \frac{1}{0.09}\ln(3)$$

$$t \approx 12.21$$

It will take 12 weeks.

63. $t = \frac{1}{c}\ln\left(\frac{A}{A-N}\right)$

$$t = \frac{1}{0.07}\ln\left(\frac{210}{210-150}\right)$$

$$t = \frac{1}{0.07}\ln(3.5)$$

$$t \approx 17.9$$

It will take 18 weeks.

65. $-\dfrac{5}{3}$

67. $\dfrac{17}{4}$

69. $f^{-1}(x) = \dfrac{x-2}{5}$

71. $y = 5,130,632$, $y_0 = 3,665,228$

$$5,130,632 = 3,665,228e^{10k}$$

$$\frac{5,130,632}{3,665,228} = e^{10k}$$

$$\ln\left(\frac{5,130,632}{3,665,228}\right) = 10k$$

$$k \approx 0.0336 \text{ or } 3.4\%$$

73. Answers may vary.

75. $Y_1 = e^{0.3x}$, $Y_2 = 8$

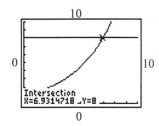

77. $Y_1 = 2\log(-5.6x + 1.3)$, $Y_2 = -x - 1$

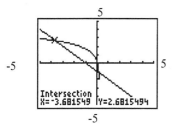

79. $Y_1 = 7^{3x-4}$, $Y_2 = 11$

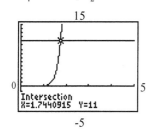

81. $Y_1 = \ln 5 + \ln x$, $Y_2 = 0$

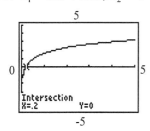

<u>**Chapter 12 Review**</u>

1. The function is one-to-one.
$$h^{-1} = \left\{ \begin{array}{l} (14,-9),(8,6),(12,-11) \\ ,(15,15) \end{array} \right\}$$

2. The function is not one-to-one.

3. The function is one-to-one.

Rank in Auto Thefts (Input)	2	4	1	3
U.S. Region (Output)	West	Midwest	South	Northeast

4. The function is not one-to-one.

5. $f(x) = \sqrt{x+2}$

   a. $f(7) = \sqrt{7+2} = \sqrt{9} = 3$

   b. $f^{-1}(3) = 7$

6. $f(x) = \sqrt{x+2}$

   a. $f(-1) = \sqrt{-1+2} = \sqrt{1} = 1$

   b. $f^{-1}(1) = -1$

7. The graphs does not represent a one-to-one function.

8. The graph is not a one-to-one.

9. The graph is not a one-to-one.

10. The graph is a one-to-one function.

11. $f(x) = x - 9$

$$y = x - 9$$
$$x = y - 9$$
$$y = x + 9$$
$$f^{-1}(x) = x + 9$$

12. $f(x) = x + 8$
$$y = x + 8$$
$$x = y + 8$$
$$y = x - 8$$
$$f^{-1}(x) = x - 8$$

13. $f(x) = 6x + 11$
$$y = 6x + 11$$
$$x = 6y + 11$$
$$6y = x - 11$$
$$y = \frac{x - 11}{6}$$
$$f^{-1}(x) = \frac{x - 11}{6}$$

14. $f(x) = 12x$
$$y = 12x$$
$$x = 12y$$
$$y = \frac{x}{12}$$
$$f^{-1}(x) = \frac{x}{12}$$

15. $f(x) = x^3 - 5$
$$y = x^3 - 5$$
$$x = y^3 - 5$$
$$y^3 = x + 5$$
$$y = \sqrt[3]{x + 5}$$
$$f^{-1}(x) = \sqrt[3]{x + 5}$$

16. $f(x) = \sqrt[3]{x + 2}$
$$y - \sqrt[3]{x + 2}$$
$$x = \sqrt[3]{y + 2}$$
$$x^3 = y + 2$$
$$y = x^3 - 2$$
$$f^{-1}(x) = x^3 - 2$$

17. $g(x) = \dfrac{12x - 7}{6}$
$$y = \frac{12x - 7}{6}$$
$$x = \frac{12y - 7}{6}$$
$$6x = 12y - 7$$
$$y = \frac{6x + 7}{12}$$
$$g^{-1}(x) = \frac{6x + 7}{12}$$

18. $r(x) = \dfrac{13}{2}x - 4$
$$y = \frac{13}{2}x - 4$$
$$x = \frac{13}{2}y - 4$$
$$x + 4 = \frac{13}{2}y$$
$$y = \frac{2(x + 4)}{13}$$
$$r^{-1}(x) = \frac{2(x + 4)}{13}$$

19. $y = g(x) = \sqrt{x}$
$$x = \sqrt{y}$$
$$x^2 = y = g^{-1}(x),\ x \geq 0$$

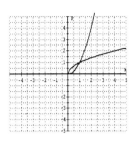

20. $h(x) = 5x - 5$

$y = 5x - 5$

$x = 5y - 5$

$5y = x + 5$

$y = \dfrac{x + 5}{5}$

$h^{-1}(x) = \dfrac{x + 5}{5}$

21. $f(x) = 2x - 3$

$x = 2y - 3$

$y = \dfrac{x + 3}{2}$

$f^{-1}(x) = \dfrac{x + 3}{2}$

22. $4^x = 64$

$4^x = 4^3$

$x = 3$

23. $3^x = \dfrac{1}{9}$

$3^x = 3^{-2}$

$x = -2$

24. $2^{3x} = \dfrac{1}{16}$

$2^{3x} = 2^{-4}$

$3x = -4$

$x = -\dfrac{4}{3}$

25. $5^{2x} = 125$

$5^{2x} = 5^3$

$2x = 3$

$x = \dfrac{3}{2}$

26. $9^{x+1} = 243$

$(3^2)^{x+1} = 3^5$

$3^{2x+2} = 3^5$

$2x + 2 = 5$

$2x = 3$

$x = \dfrac{3}{2}$

27. $8^{3x-2} = 4$

$(2^3)^{3x-2} = 2^2$

$2^{9x-6} = 2^2$

$9x - 6 = 2$

$9x = 8$

$x = \dfrac{8}{9}$

28.

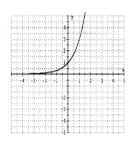

29.

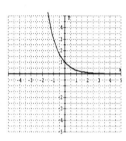

30.

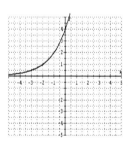

31.

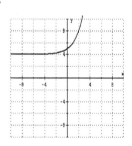

32. $A = P\left(1 + \dfrac{r}{n}\right)^{nt}$

$A = 1600\left(1 + \dfrac{0.09}{2}\right)^{(2)(7)}$

$A = \$2963.11$

33. $A = P\left(1 + \dfrac{r}{n}\right)^{nt}$

$A = 800\left(1 + \dfrac{0.07}{4}\right)^{(4)(5)}$

$A \approx 1131.82$

The certificate is worth \$1131.82 at the end of 5 years.

34.

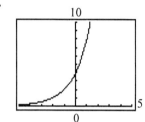

35. $7^2 = 49$

$\log_7 49 = 2$

36. $2^{-4} = \dfrac{1}{16}$

$\log_2 \dfrac{1}{16} = -4$

37. $\log_{1/2} 16 = -4$

$\left(\dfrac{1}{2}\right)^{-4} = 16$

38. $\log_{0.4} 0.064 = 3$

$0.4^3 = 0.064$

39. $\log_4 x = -3$

$x = 4^{-3} = \dfrac{1}{64}$

40. $\log_3 x = 2$

$x = 3^2 = 9$

41. $\log_3 x = 2$

$$3^x = 1$$
$$3^x = 3^0$$
$$x = 0$$

42. $\log_4 64 = x$

$$4^x = 64$$
$$4^x = 4^3$$
$$x = 3$$

43. $\log_x 64 = 2$

$$x^2 = 64$$
$$x = \pm\sqrt{64} = \pm 8$$
$$x = 8 \text{ since base} > 0$$

44. $\log_x 81 = 4$

$$x^4 = 81$$
$$x = \pm 3$$

We discard the negative base -3..

45. $\log_4 4^5 = x$

$$x = 5$$

46. $\log_7 7^{-2} = x$

$$x = -2$$

47. $5^{\log_5 4} = x$

$$x = 4$$

48. $2^{\log_2 9} = x$

$$9 = x$$

49. $\log_2(3x - 1) = 4$

$$3x - 1 = 2^4 = 16$$
$$3x = 17$$
$$x = \frac{17}{3}$$

50. $\log_3(2x + 5) = 2$

$$2x + 5 = 3^2$$
$$2x + 5 = 9$$
$$2x = 4$$
$$x = 2$$

51. $\log_4(x^2 - 3x) = 1$

$$x^2 - 3x = 4$$
$$x^2 - 3x - 4 = 0$$
$$(x + 1)(x - 4) = 0$$
$$x = -1 \text{ or } x = 4$$

52. $\log_8(x^2 + 7x) = 1$

$$x^2 + 7x = 8$$
$$x^2 + 7x - 8 = 0$$
$$(x + 8)(x - 1) = 0$$
$$x = -8 \text{ or } x = 1$$

53.

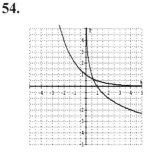

54.

55. $\log_3 8 + \log_3 4 = \log_3(8)(4) = \log_3 32$

56. $\log_2 6 + \log_2 3 = \log_2(6 \cdot 3) = \log_2 18$

57. $\log_7 15 - \log_7 20 = \log_7 \dfrac{3}{4}$

58. $\log 18 - \log 12 = \log \dfrac{3}{2}$

59. $\log_{11} 8 + \log_{11} 3 - \log_{11} 6 = \log_{11} \dfrac{(8)(3)}{6}$
$$= \log_{11} 4$$

60. $\log_5 14 + \log_5 3 - \log_5 21$
$$= \log_5 (14 \cdot 3) - \log_5 21$$
$$= \log_5 \left(\dfrac{42}{21} \right) = \log_5 2$$

61. $2\log_5 x - 2\log_5 (x+1) + \log_5 x$
$$= \log_5 x^2 - \log_5 (x+1)^2 + \log_5 x$$
$$= \log_5 \dfrac{(x^2)(x)}{(x+1)^2}$$
$$= \log_5 \dfrac{x^3}{(x+1)^2}$$

62. $4\log_3 x - \log_3 x + \log_3 (x+2)$
$$= 3\log_3 x + \log_3 (x+2)$$
$$= \log_3 x^3 + \log_3 (x+2)$$
$$= \log_3 \left[x^3 (x+2) \right]$$
$$= \log_3 (x^4 + 2x^3)$$

63. $\log_3 \dfrac{x^3}{x+2} = \log_3 x^3 - \log_3 (x+2)$
$$= 3\log_3 x - \log_3 (x+2)$$

64. $\log_4 \dfrac{x+5}{x^2} = \log_4 (x+5) - \log_4 x^2$
$$= \log_4 (x+5) - 2\log_4 x$$

65. $\log_2 \dfrac{3x^2 y}{z}$

$$= \log_2 3 + \log_2 x^2 + \log_2 y - \log_2 z$$
$$= \log_2 3 + 2\log_2 x + \log_2 y - \log_2 z$$

66. $\log_7 \dfrac{yz^3}{x} = \log_7 (yz^3) - \log_7 x$
$$= \log_7 y + \log_7 z^3 - \log_7 x$$
$$= \log_7 y + 3\log_7 z - \log_7 x$$

67. $\log_6 50 = \log_6 (5)(5)(2)$
$$= \log_6 (5) + \log_6 (5) + \log_6 (2)$$
$$= 0.83 + 0.83 + 0.36$$
$$= 2.02$$

68. $\log_b \dfrac{4}{5} = \log_b 4 - \log_b 5$
$$= \log_b 2^2 - \log_b 5$$
$$= 2\log_b 2 - \log_b 5$$
$$= 2(0.36) - 0.83$$
$$= 0.72 - 0.83$$
$$= -0.11$$

69. $\log 3.6 \approx 0.5563$

70. $\log 0.15 \approx -0.8239$

71. $\ln 1.25 \approx 0.2231$

72. $\ln 4.63 \approx 1.5326$

73. $\log 1000 = 3$

74. $\log \dfrac{1}{10} = \log 10^{-1} = -1$

75. $\ln \dfrac{1}{e} = \ln 1 - \ln e = 0 - 1 = -1$

76. $\ln(e^4) = 4$

77. $\ln(2x) = 2$

$$2x = e^2$$

$$x = \frac{e^2}{2}$$

78. $\ln(3x) = 1.6$

$$3x = e^{1.6}$$

$$x = \frac{e^{1.6}}{3}$$

79. $\ln(2x - 3) = -1$

$$2x - 3 = e^{-1}$$

$$x = \frac{e^{-1} + 3}{2}$$

80. $\ln(3x + 1) = 2$

$$3x + 1 = e^2$$

$$3x = e^2 - 1$$

$$x = \frac{e^2 - 1}{3}$$

81. $\ln \dfrac{I}{I_0} = -kx$

$$\ln \frac{0.03 I_0}{I_0} = -2.1x$$

$$\ln 0.03 = -2.1x$$

$$\frac{\ln 0.03}{-2.1} = x$$

$$x \approx 1.67 \, mm$$

82. $\ln \dfrac{I}{I_0} = -kx$

$$\ln \frac{0.02 I_0}{I_0} = -3.2x$$

$$\ln 0.02 = -3.2x$$

$$\frac{\ln 0.02}{-3.2} = x$$

$$x \approx 1.22$$

2% of the original radioactivity will penetrate at a depth of approximately 1.22 millimeters.

83. $\log_5 1.6 = \dfrac{\log 1.6}{\log 5} = 0.2920$

84. $\log_3 4 = \dfrac{\log 4}{\log 3} \approx 1.2619$

85. $A = Pe^{rt}$

$$A = 1450 e^{(0.06)(5)}$$

$$A = \$1957.30$$

86. $A = Pe^{rt}$

$$A = 940 e^{0.11(3)} = 940 e^{0.33} \approx 1307.51$$

87. $3^{2x} = 7$

$$2x \log 3 = \log 7$$

$$x = \frac{\log 7}{2 \log 3} \approx 0.8856$$

88. $6^{3x} = 5$

$$3x = \log_6 5$$

$$x = \frac{1}{3} \log_6 5 = \frac{\log 5}{3 \log 6} \approx 0.2994$$

89. $3^{2x+1} = 6$

$$(2x+1)\log 3 = \log 6$$

$$2x = \frac{\log 6}{\log 3} - 1$$

$$x = \frac{1}{2}\left(\frac{\log 6}{\log 3} - 1\right) \approx 0.315$$

90. $4^{3x+2} = 9$

$$3x + 2 = \log_4 9$$

$$3x = \log_4 9 - 2$$

$$x = \frac{1}{3}\left(\frac{\log 9}{\log 4} - 2\right) \approx -0.1383$$

91. $5^{3x-5} = 4$

$$(3x - 5)\log 5 = \log 4$$

$$3x = \frac{\log 4}{\log 5} + 5$$

$$x = \frac{1}{3}\left(\frac{\log 4}{\log 5} + 5\right)$$

$$x \approx 1.9538$$

92. $8^{4x-2} = 3$

$$4x - 2 = \log_8 3$$

$$4x = \log_8 3 + 2$$

$$4x = \frac{\log 3}{\log 8} + 2$$

$$x = \frac{1}{4}\left(\frac{\log 3}{\log 8} + 2\right)$$

$$x \approx 0.6321$$

93. $2 \cdot 5^{x-1} = 1$

$$\log 2 + (x-1)\log 5 = \log 1$$

$$(x-1)\log 5 = -\log 2$$

$$x = -\frac{\log 2}{\log 5} + 1$$

$$x \approx 0.5693$$

94. $3 \cdot 4^{x+5} = 2$

$$4^{x+5} = \frac{2}{3}$$

$$x + 5 = \log_4\left(\frac{2}{3}\right)$$

$$x = \log_4\left(\frac{2}{3}\right) - 5$$

$$x = \frac{\log\left(\frac{2}{3}\right)}{\log 4} - 5$$

$$x \approx -5.2925$$

95. $\log_5 2 + \log_5 x = 2$

$$\log_5 2x = 2$$

$$2x = 5^2 = 25$$

$$x = \frac{25}{2}$$

96. $\log_3 x + \log_3 10 = 2$

$$\log_3(10x) = 2$$

$$10x = 3^2$$

$$10x = 9$$

$$x = \frac{9}{10}$$

97. $\log(5x) - \log(x+1) = 4$

$$\log\frac{5x}{x+1} = 4$$

$$\frac{5x}{x+1} = 10^4 = 10{,}000$$

$$5x = 10000x + 10000$$

$$x = -1.0005$$

no solution

98. $\ln(3x) - \ln(x-3) = 2$

$$\ln\left(\frac{3x}{x-3}\right) = 2$$

$$\frac{3x}{x-3} = e^2$$

$$3x = e^2 x - 3e^2$$

$$3x - e^2 x = -3e^2$$

$$(3 - e^2)x = -3e^2$$

$$x = \frac{3e^2}{e^2 - 3}$$

99. $\log_2 x + \log_2 2x - 3 = 1$

$$\log_2 (x)(2x) = 4$$

$$2x^2 = 16$$

$$x^2 = 8$$

$$x = \pm 2\sqrt{2}$$

100. $-\log_6(4x+7) + \log_6 x = 1$

$$\log_6 \frac{x}{4x+7} = 1$$

$$\frac{x}{4x+7} = 6$$

$$x = 6(4x+7)$$

$$x = 24x + 42$$

$$x = -\frac{42}{23}$$

101. $y = y_0 e^{kt}$

$$y = 155,000 e^{0.06(4)}$$

$$= 197,044 \text{ ducks}$$

102. $y = y_0 e^{kt}$

$$y = 212,942,000 e^{0.015(8)}$$

$$= 212,942,000 e^{0.12}$$

$$= 240,091,435$$

The expected population of
Indonesia by the year 2006
is approximately 240,091,435.

103. $y = y_0 e^{kt}$

$$130,000,000 = 126,975,000 e^{0.001t}$$

$$t = \frac{\ln \dfrac{130,000,000}{126,975,000}}{0.001}$$

$$t \approx 24$$

It will take approximately 24 years.

104. $y = y_0 e^{kt}$

$$2(31,902,268) = 31,902,268 e^{0.08t}$$

$$2 = e^{0.08t}$$

$$t = \frac{\ln 2}{0.008}$$

$$t \approx 87$$

It will take approximately 87 years.

105. $y = y_0 e^{kt}$

$$2(70,712,345) = 70,712,345 e^{0.016t}$$

$$2 = e^{0.016t}$$

$$t = \frac{\ln 2}{0.016}$$

$$t \approx 43$$

It will take approximately 43 years.

106. $A = P\left(1 + \dfrac{r}{n}\right)^{nt}$

$$10{,}000 = 5{,}000\left(1 + \dfrac{0.08}{4}\right)^{4t}$$

$$2 = (1.02)^{4t}$$

$$\log 2 = 4t \log 1.02$$

$$t = \dfrac{\log 2}{4 \log 1.02} \approx 8.8 \text{ years}$$

107. $A = P\left(1 + \dfrac{r}{n}\right)^{nt}$

$$10{,}000 = 6{,}000\left(1 + \dfrac{0.06}{12}\right)^{12t}$$

$$\dfrac{5}{3} = (1.005)^{12t}$$

$$12t = \log_{1.005}\left(\dfrac{5}{3}\right)$$

$$t = \dfrac{1}{12}\left(\dfrac{\log\left(\dfrac{5}{3}\right)}{\log(1.005)}\right) \approx 8.5 \text{ years}$$

It was invested for approximately 8.5 years.

108. $x \approx 0.69$

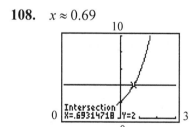

109. $x \approx 2.82$

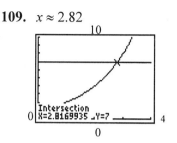

Chapter 12 Test

1. $f(x) = 7x - 14$, $f^{-1}(x) = \dfrac{x + 14}{7}$

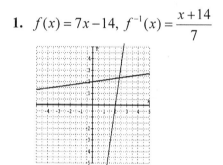

2. The graph is one-to-one.

3. The graph is not one-to-one.

4. $y = 6 - x$ is one-to-one.

$$x = 6 - 2y$$

$$2y = -x + 6$$

$$y = \dfrac{-x + 6}{2}$$

$$f^{-1}(x) = \dfrac{-x + 6}{2}$$

5. $f = \{(0,0),(2,3),(-1,5)\}$ is one-to-one.

$f^{-1} = \{(0,0),(3,2),(5,-1)\}$

6. The function is not one-to-one.

7. $\log_3 6 + \log_3 4 = \log_3(6 \cdot 4) = \log_3 24$

8. $\log_5 x + 3\log_5 x - \log_5(x+1)$

$$= 4\log_5 x - \log_5(x+1)$$
$$= \log_5 x^4 - \log_5(x+1)$$
$$= \log_5 \frac{x^4}{x+1}$$

9. $\log_6 \dfrac{2x}{y^3} = \log_6 2x - \log_6 y^3$
$$= \log_6 2 + \log_6 x - 3\log_6 y$$

10. $\log_b\left(\dfrac{3}{25}\right) = \log_b 3 - \log_b 25$
$$= \log_b 3 - \log_b 5^2$$
$$= \log_b 3 - 2\log_b 5$$
$$= 0.79 - 2(1.16)$$
$$= -1.53$$

11. $\log_7 8 = \dfrac{\ln 8}{\ln 7} \approx 1.0686$

12. $8^{x-1} = \dfrac{1}{64}$
$$8^{x-1} = 8^{-2}$$
$$x - 1 = -2$$
$$x = -1$$

13. $3^{2x+5} = 4$
$$2x + 5 = \log_3 4$$
$$2x = \frac{\log 4}{\log 3} - 5$$
$$x = \frac{1}{2}\left(\frac{\log 4}{\log 3} - 5\right)$$
$$x \approx -1.8691$$

14. $\log_3 x = -2$
$$x = 3^{-2}$$
$$x = \frac{1}{9}$$

15. $\ln \sqrt{e} = x$
$$\ln e^{1/2} = x$$
$$\frac{1}{2} = x$$

16. $\log_8(3x-2) = 2$
$$3x - 2 = 8^2$$
$$3x - 2 = 64$$
$$3x = 66$$
$$x = \frac{66}{3} = 22$$

17. $\log_5 x + \log_5 3 = 2$
$$\log_5(3x) = 2$$
$$3x = 5^2$$
$$3x = 25$$
$$x = \frac{25}{3}$$

18. $\log_4(x+1) - \log_4(x-2) = 3$
$$\log_4 \frac{x+1}{x-2} = 3$$
$$\frac{x+1}{x-2} = 4^3$$
$$x + 1 = 64x - 128$$
$$x = \frac{43}{21}$$

19. $\ln(3x+7) = 1.31$
$$3x + 7 = e^{1.31}$$
$$3x = e^{1.31} - 7$$
$$x = \frac{e^{1.31} - 7}{3} \approx -1.0979$$

20. $y = \left(\dfrac{1}{2}\right)^x + 1$

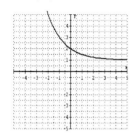

21. $y = 3^x$ and $y = \log_3 x$

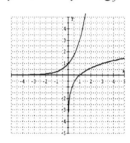

22. $A = \left(1 + \dfrac{r}{n}\right)^{nt}, P = 4000, t = 3, r = 0.09$

and $n = 12$

$$A = 4000\left(1 + \dfrac{0.09}{12}\right)^{12(3)}$$
$$= 4000(1.0075)^{36}$$
$$\approx 5234.58$$

$5234.58 will be in the account.

23. $A = \left(1 + \dfrac{r}{n}\right)^{nt}, P = 2000, A = 3000$

$r = 0.07, \ n = 2$

$$3000 = 2000\left(1 + \dfrac{0.07}{2}\right)^{2t}$$
$$1.5 = (1.035)^{2t}$$
$$2t = \log_{1.035} 1.5$$
$$t = \dfrac{1}{2}\dfrac{\ln 1.5}{\ln 1.035}$$
$$\approx 5.9$$

24. $y = y_0 e^{kt}$

$$y = 57{,}000 e^{0.026(5)}$$
$$= 57{,}000 e^{0.13}$$
$$\approx 64{,}913$$

There will be approximately 64,913 prairie dogs 5 years from now.

25. $y = y_0 e^{kt}$

$$1000 = 400 e^{0.062(t)}$$
$$2.5 = e^{0.062t}$$
$$0.062t = \ln 2.5$$
$$t = \dfrac{\ln 2.5}{0.062}$$
$$\approx 14.8$$

It will take the naturalists approximately 15 years to reach their goal.

26. $\log(1 + k) = \dfrac{0.3}{D}, D = 56$

$$\log(1 + k) = \dfrac{0.3}{56}$$
$$1 + k = 10^{0.3/56}$$
$$k = -1 + 10^{0.3/56}$$
$$k \approx 0.012$$

The rate of population increase is approximately 1.2%.

27. $Y_1 = e^{0.2x}, \ Y_2 = e^{-0.4x} + 2$

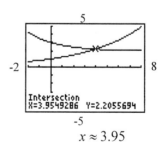

$x \approx 3.95$

Chapter 12 Cumulative Review

1. **a.** $\dfrac{4}{5} \div \dfrac{5}{16} = \dfrac{4}{5} \cdot \dfrac{16}{5} = \dfrac{64}{25}$

   b. $\dfrac{7}{10} \div 14 = \dfrac{7}{10} \cdot \dfrac{1}{14} = \dfrac{1}{20}$

   c. $\dfrac{3}{8} \div \dfrac{3}{10} = \dfrac{3}{8} \cdot \dfrac{10}{3} = \dfrac{5}{4}$

2. $\dfrac{1}{3}(x-2) = \dfrac{1}{4}(x+1)$

   $4(x-2) = 3(x+1)$

   $4x - 8 = 3x + 3$

   $\qquad x = 11$

3. $y = x^2$ with $f(x) = x^2$

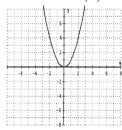

4. $f(x) - 6 = \dfrac{1}{3}(x+2)$

   $f(x) = \dfrac{1}{3}x + \dfrac{20}{3}$

5. Equation 2 is twice the opposite of equation 1. Therefore, the system is dependent. Thus,
   $\{(x, y, z) \mid x - 5y - 2z = 6\}$.

6. $x = 110°$, $y = 70°$

7. **a.** x^3

   b. 5^6

8. **a.** $16a^6$

   b. $\dfrac{8}{27}$

   c. $\dfrac{64a^{15}}{b^9}$

   d. $729x^3$

   e. $\dfrac{a^4 c^8}{b^6}$

9. **a.** $C(100) = \dfrac{2.6(100) + 10,000}{100}$

   $\qquad\quad = \$102.60$

   b. $C(1000) = \dfrac{2.6(1000) + 10,000}{1000}$

   $\qquad\quad\ \ = \$12.60$

10. **a.** $(3x-1)^2 = (3x-1)(3x-1)$

    $\qquad\qquad\quad = 9x^2 - 6x + 1$

    b. $\left(\dfrac{1}{2}x + 3\right)\left(\dfrac{1}{2}x - 3\right)$

    $\qquad = \dfrac{1}{4}x^2 + \dfrac{3}{2}x - \dfrac{3}{2}x - 9$

    $\qquad = \dfrac{1}{4}x^2 - 9$

    c. $(2x-5)(6x+7)$

    $\qquad = 12x^2 + 14x - 30x - 35$

    $\qquad = 12x^2 - 16x - 35$

11. $12a - 8a = 10 + 20 - 13 - 7$

    $\qquad 4a = 2a - 10$

    $\qquad 2a = -10$

    $\qquad\ a = -5$

12. $\dfrac{5}{x-2} + \dfrac{3}{x^2+4x+4} - \dfrac{6}{x+2}$

$= \dfrac{5(x+2)^2 + 3(x-2) - 6(x-2)(x+2)}{(x-2)(x+2)(x+2)}$

$= \dfrac{-x^2 - 23x + 38}{(x-2)(x+2)^2}$

13. $\dfrac{8x^2y^2 - 16xy + 2x}{4xy}$

$= \dfrac{8x^2y^2}{4xy} - \dfrac{16xy}{4xy} + \dfrac{2x}{4xy}$

$= 2xy - 4 + \dfrac{1}{2y}$

14. a. $\dfrac{\dfrac{5}{a-1}}{\dfrac{a}{10}} = \dfrac{a}{5} \cdot \dfrac{10}{a-1} = \dfrac{2a}{a-1}$

b. $\dfrac{-3a-18}{4a-12}$

c. $\left(\dfrac{\dfrac{1}{x} + \dfrac{1}{y}}{xy}\right) = \dfrac{\left(\dfrac{1}{x} + \dfrac{1}{y}\right)xy}{(xy)(xy)} = \dfrac{y+x}{x^2y^2}$

15. $3m^2 - 24m - 60$

$= 3\left(m^2 - 8m - 20\right)$

$= 3\left(m - 10\right)\left(m + 2\right)$

16. $5x^2 - 85x + 350$

$= 5\left(x^2 - 17x + 70\right)$

$= 5\left(x - 10\right)\left(x - 7\right)$

17. $\dfrac{3x^2 + 2x}{x-1} - \dfrac{10x-5}{x-1}$

$= \dfrac{3x^2 + 2x - \left(10x - 5\right)}{x-1}$

$= \dfrac{3x^2 + 2x - 10x + 5}{x-1}$

$= \dfrac{3x^2 - 8x + 5}{x-1}$

$= \dfrac{\left(3x - 5\right)\left(x - 1\right)}{x-1}$

$= 3x - 5$

18. $8x + 4 + \dfrac{1}{x-2}$

19. a. 3

b. -3

c. -5

d. not a real number

e. $4x$

20. $\dfrac{1}{a+5} = \dfrac{1}{3a+6} - \dfrac{a+2}{a^2+7a+10}$

$3(a+2) = a + 5 - (3a+6)$

$a = -\dfrac{3}{5}$

21. a. $\sqrt[4]{x^3}$

b. $\sqrt[6]{x}$

c. $\sqrt[6]{72}$

22. $\dfrac{1}{2} = 12k$

$k = \dfrac{1}{24}, \; y = \dfrac{1}{24}x$

23. a. $5\sqrt{3} + 3\sqrt{10}$

    b. $\sqrt{35} + \sqrt{5} + \sqrt{42} + \sqrt{6}$

    c. $7\sqrt{5x} + 15\sqrt{x} - 5\sqrt{5}$

    d. $49 - 8\sqrt{3}$

    e. $2x - 25$

    f. $x + 22 + 10\sqrt{x-3}$

24. a. 3

    b. -3

    c. $\dfrac{3}{8}$

    d. x^3

    e. $-5y^2$

25. $\dfrac{\sqrt[4]{x}}{\sqrt[4]{81y^5}} = \dfrac{\sqrt[4]{x}}{\sqrt[4]{81y^5}} \cdot \dfrac{\sqrt[4]{y^3}}{\sqrt[4]{y^3}} = \dfrac{\sqrt[4]{xy^3}}{3y^2}$

26. a. $a - a^2$

    b. $x + 3x^{\frac{1}{2}} - 15$

27. $\sqrt{4-x} = x - 2$

    $\left(\sqrt{4-x}\right)^2 = (x-2)^2$

    $4 - x = x - 2$

    $x = 3$

28. a. $\sqrt{\dfrac{54}{6}} = \sqrt{9} = 3$

    b. $\dfrac{\sqrt{108a^2}}{3\sqrt{3}} = \dfrac{1}{3}\sqrt{\dfrac{108a^2}{3}} = 2a$

    c. $3\sqrt[3]{\dfrac{81a^5b^{10}}{3b^4}} = 3\sqrt[3]{27a^5b^6} = 8ab^2\sqrt[3]{a^2}$

29. $3x^2 - 9x + 8 = 0$

    $3(x^2 - 3x) = -8$

    $3(x^2 - 3x + \dfrac{9}{4}) = -8 + \dfrac{9}{4}$

    $3(x - \dfrac{3}{2})^2 + \dfrac{5}{4} = 0$

    $x = \dfrac{9 \pm i\sqrt{15}}{6}$

30. a. $\dfrac{11\sqrt{15}}{12}$

    b. $\dfrac{\sqrt[3]{3x}}{6}$

31. $\dfrac{3x}{x-2} - \dfrac{x+1}{x} = \dfrac{6}{x(x-2)}$

    $3x(x) - (x+1)(x-2) = 6$

    $x = \dfrac{-1 \pm \sqrt{33}}{4}$

32. $\dfrac{3\sqrt[3]{m^2n}}{m^2n^2}$

33. $x^2 - 4x \le 0$

    $x(x-4) = 0$

    $x = 0,\ x = 4$

    See graph

    $[0, 4]$

34. $4\sqrt{3}$ in.

35. $F(x) = (x-3)^2 + 1$

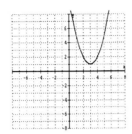

36. a. 1

b. i

c. -1

d. $-i$

37. $\dfrac{45}{x} = \dfrac{5}{7}$

$\quad 5x = 7(45)$

$\quad\; x = 7(9)$

$\quad\; x = 63$

38. $4x^2 + 8x - 1 = 0$

$\quad 4(x+1)^2 - 5 = 0$

$\qquad\qquad x = \dfrac{-2 \pm \sqrt{5}}{2}$

39. $\quad f(x) = x + 3$

$\qquad\quad x = y + 3$

$\qquad\quad y = x - 3$

$\quad f^{-1}(x) = x - 3$

40. $\left(x - \dfrac{1}{2}\right)^2 = \dfrac{1}{2}x$

$x^2 - x + \dfrac{1}{4} = \dfrac{1}{2}x$

$x^2 - \dfrac{3}{2}x + \dfrac{1}{4} = 0$

$a = 1,\, b = -\dfrac{3}{2},\, c = \dfrac{1}{4}$

$x = \dfrac{-b \pm \sqrt{b^2 - 4ac}}{2a}$

$x = \dfrac{3 \pm \sqrt{5}}{4}$

41. a. 2

b. -1

c. $\dfrac{1}{2}$

42. $f(x) = -(x+1)^2 + 1$

Vertex: $(-1, 1)$

Axis of symmetry: $x = -1$

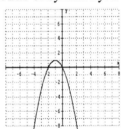

Chapter 13

Section 13.1

Graphing Calculator Explorations

1. $x^2 + y^2 = 55$
$$y^2 = 55 - x^2$$
$$y = \pm\sqrt{55 - x^2}$$

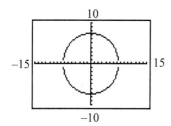

3. $5x^2 + 5y^2 = 50$
$$5y^2 = 50 - 5x^2$$
$$y^2 = 10 - x^2$$
$$y = \pm\sqrt{10 - x^2}$$

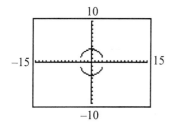

5. $2x^2 + 2y^2 - 34 = 0$
$$2y^2 = 34 - 2x^2$$
$$y^2 = 17 - x^2$$
$$y = \pm\sqrt{17 - x^2}$$

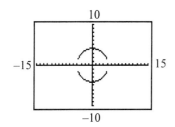

7. $7x^2 + 7y^2 - 89 = 0$
$$7y^2 = 89 - 7x^2$$
$$y^2 = \frac{89 - 7x^2}{7}$$
$$y = \pm\sqrt{\frac{89 - 7x^2}{7}}$$

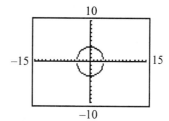

Mental Math

1. $y = x^2 - 7x + 5$; upward

2. $y = -x^2 + 16$; downward

3. $x = -y^2 - y + 2$; to the left

4. $x = 3y^2 + 2y - 5$; to the right

5. $y = -x^2 + 2x + 1$; downward

6. $x = -y^2 + 2y - 6$; to the left

Exercise Set 13.1

1. $x = 3y^2$
$$x = 3(y - 0)^2 + 0$$
Vertex: (0, 0)

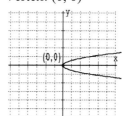

3. $x = (y-2)^2 + 3$
Vertex: (3, 2)

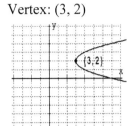

5. $y = 3(x-1)^2 + 5$
Vertex: (1, 5)

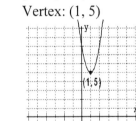

7.
$$x = y^2 + 6y + 8$$
$$x - 8 = y^2 + 6y$$
$$x - 8 + 9 = y^2 + 6y + 9$$
$$x + 1 = (y+3)^2$$
$$x = (y+3)^2 - 1$$
Vertex: (−1, −3)

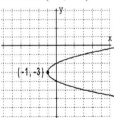

9.
$$y = x^2 + 10x + 20$$
$$y - 20 = x^2 + 10x$$
$$y - 20 + 25 = x^2 + 10x + 25$$
$$y + 5 = (x+5)^2$$
$$y = (x+5)^2 - 5$$

Vertex: (−5, −5)

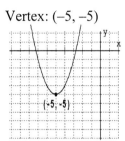

11.
$$x = -2y^2 + 4y + 6$$
$$x - 6 = -2(y^2 - 2y)$$
$$x - 6 + [-2(1)] = -2(y^2 - 2y + 1)$$
$$x - 8 = -2(y-1)^2$$
$$x = -2(y-1)^2 + 8$$
Vertex: (8, 1)

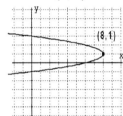

13. (5, 1), (8, 5)
$$d = \sqrt{(8-5)^2 + (5-1)^2}$$
$$= \sqrt{9+16}$$
$$= \sqrt{25}$$
$$= 5 \text{ units}$$

15. (−3, 2), (1, −3)
$$d = \sqrt{[1-(-3)]^2 + (-3-2)^2}$$
$$= \sqrt{4^2 + (-5)^2}$$
$$= \sqrt{16+25}$$
$$= \sqrt{41} \text{ units}$$

17. (−9, 4), (−8, 1)
$$d = \sqrt{[-8-(-9)]^2 + (1-4)^2}$$
$$= \sqrt{1^2 + (-3)^2}$$
$$= \sqrt{1+9}$$
$$= \sqrt{10} \text{ units}$$

19. $\left(0, -\sqrt{2}\right), \left(\sqrt{3}, 0\right)$

$$d = \sqrt{\left(\sqrt{3} - 0\right)^2 + \left[0 - \left(-\sqrt{2}\right)\right]^2}$$
$$= \sqrt{\left(\sqrt{3}\right)^2 + \left(\sqrt{2}\right)^2}$$
$$= \sqrt{3 + 5}$$
$$= \sqrt{5} \text{ units}$$

21. $(1.7, -3.6), (-8.6, 5.7)$

$$d = \sqrt{(-8.6 - 1.7)^2 + [5.7 - (-3.6)]^2}$$
$$= \sqrt{(-10.3)^2 + (9.3)^2}$$
$$= \sqrt{192.58}$$
$$= 13.88 \text{ units}$$

23. $\left(2\sqrt{3}, \sqrt{6}\right), \left(-\sqrt{3}, 4\sqrt{6}\right)$

$$d = \sqrt{\left(-\sqrt{3} - 2\sqrt{3}\right)^2 + \left(4\sqrt{6} - \sqrt{6}\right)^2}$$
$$= \sqrt{\left(-3\sqrt{3}\right)^2 + \left(3\sqrt{6}\right)^2}$$
$$= \sqrt{27 + 54}$$
$$= \sqrt{81}$$
$$= 9 \text{ units}$$

25. $(6, -8), (2, 4)$

$$\left(\frac{6+2}{2}, \frac{-8+4}{2}\right) = \left(\frac{8}{2}, \frac{-4}{2}\right) = (4, 2)$$

The midpoint of the segment is (4, 2).

27. $(-2, -1), (-8, 6)$

$$\left(\frac{-2+(-8)}{2}, \frac{-1+6}{2}\right) = \left(\frac{-10}{2}, \frac{5}{2}\right) = \left(-5, \frac{5}{2}\right)$$

The midpoint of the segment is $\left(-5, \frac{5}{2}\right)$.

29. $(7, 3), (-1, -3)$

$$\left(\frac{7+(-1)}{2}, \frac{3+(-3)}{2}\right) = \left(\frac{6}{2}, \frac{0}{2}\right) = (3, 0)$$

The midpoint of the segment is (3, 0).

31. $\left(\frac{1}{2}, \frac{3}{8}\right), \left(-\frac{3}{2}, \frac{5}{8}\right)$

$$\left(\frac{\frac{1}{2} + \left(-\frac{3}{2}\right)}{2}, \frac{\frac{3}{8} + \frac{5}{8}}{2}\right) = \left(\frac{-1}{2}, \frac{1}{2}\right)$$

The midpoint of the segment is $\left(-\frac{1}{2}, \frac{1}{2}\right)$.

33. $\left(\sqrt{2}, 3\sqrt{5}\right), \left(\sqrt{2}, -2\sqrt{5}\right)$

$$\left(\frac{\sqrt{2} + \sqrt{2}}{2}, \frac{3\sqrt{5} + \left(-2\sqrt{5}\right)}{2}\right) = \left(\frac{2\sqrt{2}}{2}, \frac{\sqrt{5}}{2}\right)$$
$$= \left(\sqrt{2}, \frac{\sqrt{5}}{2}\right)$$

The midpoint of the segment is $\left(\sqrt{2}, \frac{\sqrt{5}}{2}\right)$.

35. $(4.6, -3.5), (7.8, -9.8)$

$$\left(\frac{4.6 + 7.8}{2}, \frac{-3.5 + (-9.8)}{2}\right) = \left(\frac{12.4}{2}, \frac{-13.2}{2}\right)$$
$$= (6.2, -6.65)$$

The midpoint of the segment is (6.2, –6.65).

37. $x^2 + y^2 = 9$

$(x - 0)^2 + (y - 0)^2 = 3^2$

Center: (0, 0), radius $r = 3$.

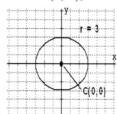

39. $x^2 + (y-2)^2 = 1$

$(x-0)^2 + (y-2)^2 = 1^2$

Center: (0, –2), radius $r = 1$.

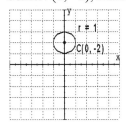

41. $(x-5)^2 + (y+2)^2 = 1$

$(x-5)^2 + (y+2)^2 = 1^2$

Center: (5, –2), radius $r = 1$.

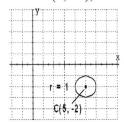

43. $x^2 + y^2 + 6y = 0$

$x^2 + (y^2 + 6y) = 0$

$x^2 + (y^2 + 6y + 9) = 9$

$(x-0)^2 + (y+3)^2 = 9$

Center: (0, –3), radius $r = 3$.

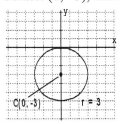

45. $x^2 + y^2 + 2x - 4y = 4$

$(x^2 + 2x) + (y^2 - 4y) = 4$

$(x^2 + 2x + 1) + (y^2 - 4y + 4) = 4 + 1 + 4$

$(x+1)^2 + (y-2)^2 = 9$

Center: (–1, 2), radius $r = 3$.

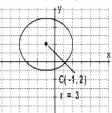

47. $x^2 + y^2 - 4x - 8y - 2 = 0$

$(x^2 - 4x) + (y^2 - 8y) = 2$

$(x^2 - 4x + 4) + (y^2 - 8y + 16) = 2 + 4 + 16$

$(x-2)^2 + (y-4)^2 = 22$

Center: (2, 4), radius $r = \sqrt{22}$.

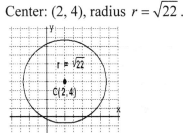

49. Center $(h, k) = (2, 3)$ and radius $r = 6$.

$(x-h)^2 + (y-k)^2 = r^2$

$(x-2)^2 + (y-3)^2 = 6^2$

$(x-2)^2 + (y-3)^2 = 36$

51. Center $(h, k) = (0, 0)$ and radius $r = \sqrt{3}$.

$(x-h)^2 + (y-k)^2 = r^2$

$(x-0)^2 + (y-0)^2 = \left(\sqrt{3}\right)^2$

$x^2 + y^2 = 3$

53. Center $(h, k) = (-5, 4)$ and radius $r = 3\sqrt{5}$.

$(x-h)^2 + (y-k)^2 = r^2$

$[x-(-5)]^2 + (y-4)^2 = \left(3\sqrt{5}\right)^2$

$(x+2)^2 + (y-4)^2 = 45$

55. Answers may vary.

57. $x = y^2 - 3$

$x = (y - 0)^2 - 3$

Vertex: $(-3, 0)$

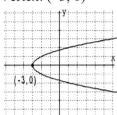

59. $y = (x - 2)^2 - 2$

Vertex: $(2, -2)$

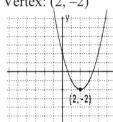

61. $x^2 + y^2 = 1$

Center: $(0, 0)$, radius $r = \sqrt{1} = 1$

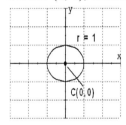

63. $x = (y + 3)^2 - 1$

Vertex: $(-1, -3)$

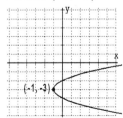

65. $(x - 2)^2 + (y - 2)^2 = 16$

Center: $(2, 2)$, radius $r = \sqrt{16} = 4$

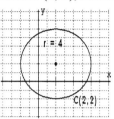

67. $x = -(y - 1)^2$

Vertex: $(0, 1)$

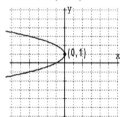

69. $(x - 4)^2 + y^2 = 7$

Center: $(4, 0)$, radius $r = \sqrt{7}$

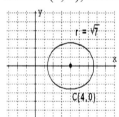

71. $y = 5(x + 5)^2 + 3$

Vertex: $(-5, 3)$

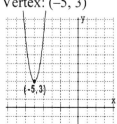

73.
$$\frac{x^2}{8} + \frac{y^2}{8} = 2$$
$$8\left(\frac{x^2}{8} + \frac{y^2}{8}\right) = 8(2)$$
$$x^2 + y^2 = 16$$

Center: $(0, 0)$, radius $r = \sqrt{16} = 4$

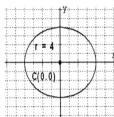

75.
$$y = x^2 + 7x + 6$$
$$y - 6 = x^2 + 7x$$
$$y - 6 + \frac{49}{4} = x^2 + 7x + \frac{49}{4}$$
$$y + \frac{25}{4} = \left(x + \frac{7}{2}\right)^2$$
$$y = \left(x + \frac{7}{2}\right)^2 - \frac{25}{4}$$

Vertex: $\left(-\frac{7}{2}, -\frac{25}{4}\right)$

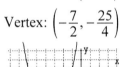
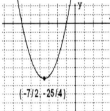

77.
$$x^2 + y^2 + 2x + 12y - 12 = 0$$
$$(x^2 + 2x) + (y^2 + 12y) = 12$$
$$(x^2 + 2x + 1) + (y^2 + 12y + 36) = 12 + 1 + 36$$
$$(x + 1)^2 + (y + 6)^2 = 49$$

Center: $(-1, -6)$, radius $r = \sqrt{49} = 7$

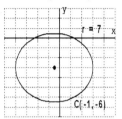

79.
$$x = y^2 + 8y - 4$$
$$x + 4 = y^2 + 8y$$
$$x + 4 + 16 = y^2 + 8y + 16$$
$$x + 20 = (y + 4)^2$$
$$x = (y + 4)^2 - 20$$

Vertex: $(-20, -4)$

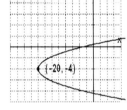

81.
$$x^2 - 10y + y^2 + 4 = 0$$
$$x^2 + (y^2 - 10y) = -4$$
$$x^2 + (y^2 - 10y + 25) = -4 + 25$$
$$x^2 + (y - 5)^2 = 21$$

Center: $(-1, -6)$, radius $r = \sqrt{21}$

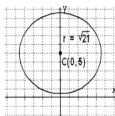

83.
$$x = -3y^2 + 30y$$
$$x = -3(y^2 - 10y)$$
$$x + [-3(25)] = -3(y^2 - 10y + 25)$$
$$x - 75 = -3(y - 5)^2$$
$$x = -3(y - 5)^2 + 75$$

Vertex: (75, 5)

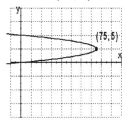

85. $5x^2 + 5y^2 = 25$
$$x^2 + y^2 = 5$$

Center: (0, 0), radius $r = \sqrt{5}$

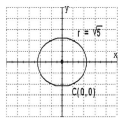

87.
$$y = 5x^2 - 20x + 16$$
$$y - 16 = 5(x^2 - 4x)$$
$$y - 16 + 5(4) = 5(x^2 - 4x + 4)$$
$$y - 4 = 4(x - 2)^2$$
$$y = 4(x - 2)^2 + 4$$

Vertex: (2, –4)

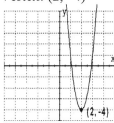

89. $y = -3x + 3$

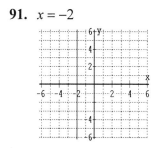

91. $x = -2$

93. $\dfrac{\sqrt{5}}{\sqrt{8}} = \dfrac{\sqrt{5}}{2\sqrt{2}} = \dfrac{\sqrt{5} \cdot \sqrt{2}}{2\sqrt{2} \cdot \sqrt{2}} = \dfrac{\sqrt{10}}{2 \cdot 2} = \dfrac{\sqrt{10}}{4}$

95. $\dfrac{10}{\sqrt{5}} = \dfrac{10 \cdot \sqrt{5}}{\sqrt{5} \cdot \sqrt{5}} = \dfrac{10\sqrt{5}}{5} = 2\sqrt{5}$

97. Height = 264 ft and diameter d = 250 ft

   a. radius $= \dfrac{1}{2} d = \dfrac{1}{2}(250) = 125$ ft

   b. 264 – 250 = 14 ft from the ground

   c. 125 + 14 = 139 ft from the ground

   d. center: (0, 139)

   e. $(x - h)^2 + (y - k)^2 = r^2$
$$(x - 0)^2 + (y - 139)^2 = 125^2$$
$$x^2 + (y - 139)^2 = 15,625$$

99. *B*: (3, 1) and *C*: (19, 13)
$$d = \sqrt{(13 - 1)^2 + (19 - 3)^2}$$
$$= \sqrt{12^2 + 16^2}$$
$$= \sqrt{144 + 256}$$
$$= \sqrt{400} = 20 \text{ meters}$$

101. $y = a(x - h)^2 + k$

Vertex: (0, 40)

$y = a(x - 0)^2 + 40$

$y = ax^2 + 40$

The parabola passes through (50, 0).

$0 = a(50)^2 + 40$

$-40 = 2500a$

$a = \dfrac{-40}{2500} = -\dfrac{2}{125}$

Thus, the equation is $y = -\dfrac{2}{125}x^2 + 40$.

103. $5x^2 + 5y^2 = 25$

$5y^2 = 25 - 5x^2$

$y^2 = 5 - x^2$

$y = \pm\sqrt{5 - x^2}$

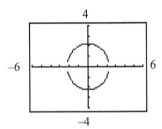

Answers are the same.

105. $y = 5x^2 - 20x + 16$

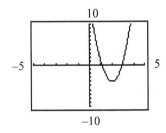

Answer are the same.

Section 13.2

Graphing Calculator Explorations

1. $10x^2 + y^2 = 32$

$y^2 = 32 - 10x^2$

$y = \pm\sqrt{32 - 10x^2}$

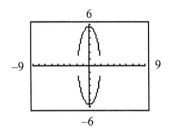

3. $20x^2 + 5y^2 = 100$

$5y^2 = 100 - 20x^2$

$y^2 = 20 - 4x^2$

$y = \pm\sqrt{20 - 4x^2}$

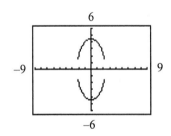

5. $7.3x^2 + 15.5y^2 = 95.2$

$15.5y^2 = 95.2 - 7.3x^2$

$y^2 = \dfrac{95.2 - 7.3x^2}{15.5}$

$y = \pm\sqrt{\dfrac{95.2 - 7.3x^2}{15.5}}$

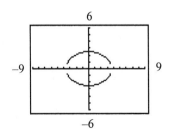

Mental Math

1. $\dfrac{x^2}{16} + \dfrac{y^2}{4} = 1$; Ellipse

2. $\dfrac{x^2}{16} - \dfrac{y^2}{4} = 1$; Hyperbola

3. $x^2 - 5y^2 = 3$; Hyperbola

4. $-x^2 + 5y^2 = 3$
   $5y^2 - x^2 = 3$; Hyperbola

5. $-\dfrac{y^2}{25} + \dfrac{x^2}{36} = 1$
   $\dfrac{x^2}{36} - \dfrac{y^2}{25} = 1$; Hyperbola

6. $\dfrac{y^2}{25} + \dfrac{x^2}{36} = 1$; Ellipse

Exercise Set 13.2

1. $\dfrac{x^2}{4} + \dfrac{y^2}{25} = 1$
   $\dfrac{x^2}{2^2} + \dfrac{y^2}{5^2} = 1$
   Center: $(0, 0)$
   x-intercepts: $(-2, 0)$, $(2, 0)$
   y-intercepts: $(0, -5)$, $(0, 5)$

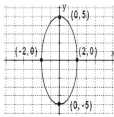

3. $\dfrac{x^2}{16} + \dfrac{y^2}{9} = 1$
   $\dfrac{x^2}{4^2} + \dfrac{y^2}{3^2} = 1$

Center: $(0, 0)$
x-intercepts: $(-4, 0)$, $(4, 0)$
y-intercepts: $(0, -3)$, $(0, 3)$

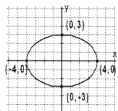

5. $9x^2 + 4y^2 = 36$
   $\dfrac{x^2}{4} + \dfrac{y^2}{9} = 1$
   $\dfrac{x^2}{2^2} + \dfrac{y^2}{3^2} = 1$
   Center: $(0, 0)$
   x-intercepts: $(-2, 0)$, $(2, 0)$
   y-intercepts: $(0, -3)$, $(0, 3)$

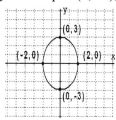

7. $4x^2 + 25y^2 = 100$
   $\dfrac{x^2}{25} + \dfrac{y^2}{4} = 1$
   $\dfrac{x^2}{5^2} + \dfrac{y^2}{2^2} = 1$
   Center: $(0, 0)$
   x-intercepts: $(-5, 0)$, $(5, 0)$
   y-intercepts: $(0, -2)$, $(0, 2)$

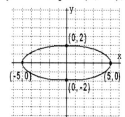

9. $\dfrac{(x+1)^2}{36} + \dfrac{(y-2)^2}{49} = 1$

$\dfrac{(x+1)^2}{6^2} + \dfrac{(y-2)^2}{7^2} = 1$

Center: $(-1, 2)$
Other points:
$(-1-6, 2) = (-7, 2)$
$(-1+6, 2) = (5, 2)$
$(-1, 2-7) = (-1, -5)$
$(-1, 2+7) = (-1, 9)$

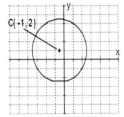

11. $\dfrac{(x-1)^2}{4} + \dfrac{(y-1)^2}{25} = 1$

$\dfrac{(x-1)^2}{2^2} + \dfrac{(y-1)^2}{5^2} = 1$

Center: $(1, 1)$
Other points:
$(1-2, 2) = (-1, 2)$
$(1+2, 2) = (3, 2)$
$(1, 1-5) = (1, -4)$
$(1, 1+5) = (1, 6)$

13. $\dfrac{x^2}{4} - \dfrac{y^2}{9} = 1$

$\dfrac{x^2}{2^2} - \dfrac{y^2}{3^2} = 1$

$a = 2, b = 3$

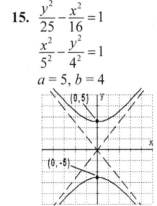

15. $\dfrac{y^2}{25} - \dfrac{x^2}{16} = 1$

$\dfrac{x^2}{5^2} - \dfrac{y^2}{4^2} = 1$

$a = 5, b = 4$

17. $x^2 - 4y^2 = 16$

$\dfrac{x^2}{16} - \dfrac{y^2}{4} = 1$

$\dfrac{x^2}{4^2} - \dfrac{y^2}{2^2} = 1$

$a = 4, b = 2$

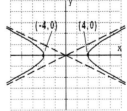

19. $16y^2 - x^2 = 16$

$$\frac{y^2}{1} - \frac{x^2}{16} = 1$$

$$\frac{y^2}{1^2} - \frac{x^2}{4^2} = 1$$

$a = 4$, $b = 1$

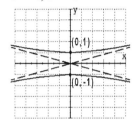

21. Answers may vary.

23. $y = x^2 + 4$

Parabola; vertex (0, 4), opens upward

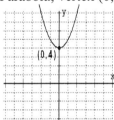

25. $\frac{x^2}{4} + \frac{y^2}{9} = 1$

$$\frac{x^2}{2^2} + \frac{y^2}{3^2} = 1$$

Ellipse; center: (0, 0)
x-intercepts: (−2, 0), (2, 0)
y-intercepts: (0, −3), (0, 3)

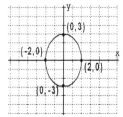

27. $\frac{x^2}{16} - \frac{y^2}{4} = 1$

$$\frac{x^2}{4^2} - \frac{y^2}{2^2} = 1$$

Hyperbola; center: (0, 0)
$a = 4$, $b = 2$

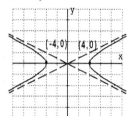

29. $x^2 + y^2 = 16$

Circle; center: (0, 0), radius: $r = \sqrt{16} = 4$

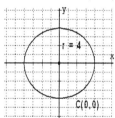

31. $x = -y^2 + 6y$

Parabola: $y = -\dfrac{b}{2a} = \dfrac{-6}{2(-1)} = 3$

$x = -(3)^2 + 6(3) = -9 + 18 = 9$

Vertex: (9, 3), opens to the left

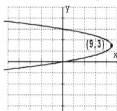

33. $9x^2 + 4y^2 = 36$

$$\frac{x^2}{4} + \frac{y^2}{9} = 1$$

$$\frac{x^2}{2^2} - \frac{y^2}{3^2} = 1$$

Ellipse; center: $(0, 0)$
$a = 2$, $b = 3$

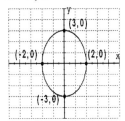

35. $y^2 = x^2 + 16$

$y^2 - x^2 = 16$

$$\frac{y^2}{16} - \frac{x^2}{16} = 1$$

$$\frac{y^2}{4^2} - \frac{x^2}{4^2} = 1$$

Hyperbola; center: $(0, 0)$
$a = 4$, $b = 4$

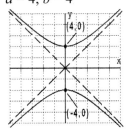

37. $y = -2x^2 + 4x - 3$

Parabola; $x = -\dfrac{b}{2a} = \dfrac{-4}{2(-2)} = 1$

$y = -2(1)^2 + 4(1) - 3 = -1$

Vertex: $(1, -1)$

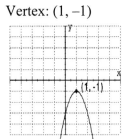

39. $x < 5$ or $x < 1$
$x < 5$
$(-\infty, 5)$

41. $2x - 1 \geq 7$ and $-3x \leq -6$
      $2x \geq 8$ and $x \geq 2$
        $x \geq 4$
$x \geq 4$
$[4, \infty)$

43. $2x^3 - 4x^3 = -2x^3$

45. $(-5x^2)(x^2) = -5x^{2+2} = -5x^4$

47. Circles: B, F
Ellipses: C, E, H
Hyperbolas: A, D, G

49. A: $c^2 = 36 + 13 = 49$; $c = \sqrt{49} = 7$
B: $c^2 = 4 - 4 = 0$; $c = \sqrt{0} = 0$
C: $c^2 = |25 - 16| = 9$; $c = \sqrt{9} = 3$
D: $c^2 = 39 + 25 = 64$; $c = \sqrt{64} = 8$
E: $c^2 = |81 - 17| = 64$; $c = \sqrt{64} = 8$
F: $c^2 = |36 - 36| = 0$; $c = \sqrt{0} = 0$
G: $c^2 = 65 + 16 = 81$; $c = \sqrt{81} = 9$
H: $c^2 = |144 - 140| = 4$; $c = \sqrt{4} = 2$

51. A: $e = \dfrac{7}{6}$

B: $e = \dfrac{0}{2} = 0$

C: $e = \dfrac{3}{5}$

D: $e = \dfrac{8}{5}$

E: $e = \dfrac{8}{9}$

F: $e = \dfrac{0}{6} = 0$

G: $e = \dfrac{9}{4}$

H: $e = \dfrac{2}{12} = \dfrac{1}{6}$

53. They are equal to 0.

55. Answers may vary.

57. $a = 130,000,000 \Rightarrow a^2 = (130,000,000)^2$
$$= 1.69 \times 10^{16}$$

$b = 125,000,000 \Rightarrow b^2 = (125,000,000)^2$
$$= 1.5625 \times 10^{16}$$

Thus, the equation is

$$\dfrac{x^2}{1.69 \times 10^{16}} + \dfrac{y^2}{1.5625 \times 10^{16}} = 1.$$

59. $9x^2 + 4y^2 = 36$
$$4y^2 = 36 - 9x^2$$
$$y^2 = \dfrac{36 - 9x^2}{4}$$
$$y = \pm\sqrt{\dfrac{36 - 9x^2}{4}} = \pm\dfrac{\sqrt{36 - 9x^2}}{2}$$

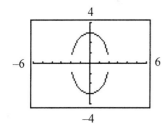

61. $\dfrac{(x-1)^2}{4} - \dfrac{(y+1)^2}{25} = 1$

Center: $(1, -1)$
$a = 2, b = 5$

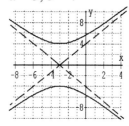

63. $\dfrac{y^2}{16} - \dfrac{(x+3)^2}{9} = 1$

Center: $(-3, 0)$
$a = 3, b = 4$

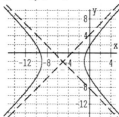

65. $\dfrac{(x+5)^2}{16} - \dfrac{(y+2)^2}{25} = 1$

Center: $(-5, -2)$
$a = 4, b = 5$

Integrated Review

1. $(x-7)^2 + (y-2)^2 = 4$
Circle; center: (7, 2),
radius: $r = \sqrt{4} = 2$

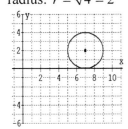

2. $y = x^2 + 4$
Parabola; vertex: (0, 4)

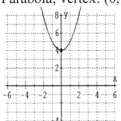

3. $y = x^2 + 12x + 36$

Parabola; $x = -\dfrac{b}{2a} = \dfrac{-12}{2(1)} = -6$

$y = (-6)^2 + 12(-6) + 36 = 0$
Vertex: (−6, 0)

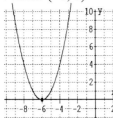

4. $\dfrac{x^2}{4} + \dfrac{y^2}{9} = 1$
Ellipse; center: (0, 0)
$a = 2, b = 3$

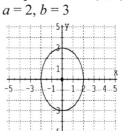

5. $\dfrac{y^2}{9} - \dfrac{x^2}{9} = 1$
Hyperbola; center: (0, 0)
$a = 3, b = 3$

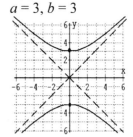

6. $\dfrac{x^2}{16} - \dfrac{y^2}{4} = 1$
Hyperbola; center: (0, 0)
$a = 4, b = 2$

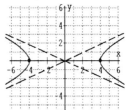

7. $\dfrac{x^2}{16} + \dfrac{y^2}{4} = 1$

Ellipse; center: $(0, 0)$
$a = 4$, $b = 2$

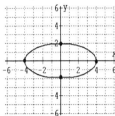

8. $x^2 + y^2 = 16$

Circle; center: $(0, 0)$
radius: $r = \sqrt{16} = 4$

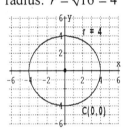

9. $x = y^2 + 4y - 1$

Parabola; $y = -\dfrac{b}{2a} = \dfrac{-4}{2(1)} = -2$

$x = (-2)^2 + 4(-2) - 1 = -5$
Vertex: $(-5, -2)$

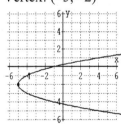

10. $x = -y^2 + 6y$

Parabola; $y = -\dfrac{b}{2a} = \dfrac{-6}{2(-1)} = 3$

$x = -(3)^2 + 6(3) = 9$
Vertex: $(9, 3)$

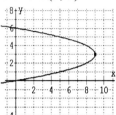

11. $9x^2 - 4y^2 = 36$

$\dfrac{x^2}{4} - \dfrac{y^2}{9} = 1$

Hyperbola; center: $(0, 0)$
$a = 2$, $b = 3$

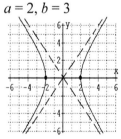

13. $\dfrac{(x-1)^2}{49} + \dfrac{(y+2)^2}{25} = 1$

Ellipse; center: $(1, -2)$,
$a = 7$, $b = 5$

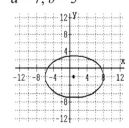

14.
$$y^2 = x^2 + 16$$
$$y^2 - x^2 = 16$$
$$\frac{y^2}{16} - \frac{x^2}{16} = 1$$

Hyperbola; center: $(0, 0)$
$a = 4, b = 4$

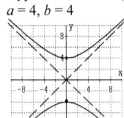

15. $\left(x + \frac{1}{2}\right)^2 + \left(y - \frac{1}{2}\right)^2 = 1$

Circle; center: $\left(-\frac{1}{2}, \frac{1}{2}\right)$,

radius: $r = \sqrt{1} = 1$

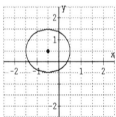

Exercise Set 13.3

1. $\begin{cases} x^2 + y^2 = 25 & (1) \\ 4x + 3y = 0 & (2) \end{cases}$

Solve E2 for y.
$$3y = -4x$$
$$y = -\frac{4x}{3}$$

Substitute into E1.

$$x^2 + \left(-\frac{4x}{3}\right)^2 = 25$$
$$x^2 + \frac{16x^2}{9} = 25$$
$$9\left(x^2 + \frac{16x^2}{9}\right) = 9(25)$$
$$9x^2 + 16x^2 = 225$$
$$25x^2 = 225$$
$$x^2 = 9$$
$$x = \pm\sqrt{9} = \pm 3$$

$x = 3 : y = -\frac{4(3)}{3} = -4$

$x = -3 : y = -\frac{4(-3)}{3} = 4$

The solutions are $(3, -4)$ and $(-3, 4)$.

3. $\begin{cases} x^2 + 4y^2 = 10 & (1) \\ y = x & (2) \end{cases}$

Substitute x for y in E1.
$$x^2 + 4x^2 = 10$$
$$5x^2 = 10$$
$$x^2 = 2$$
$$x = \pm\sqrt{2}$$

Replace these values into E2.
$$x = \sqrt{2} : y = x = \sqrt{2}$$
$$x = -\sqrt{2} : y = x = -\sqrt{2}$$

The solutions are $\left(\sqrt{2}, \sqrt{2}\right)$ and $\left(-\sqrt{2}, -\sqrt{2}\right)$.

5. $\begin{cases} y^2 = 4 - x & (1) \\ x - 2y = 4 & (2) \end{cases}$

Solve E2 for x.
$$x = 2y + 4$$

Substitute into E1.
$$y^2 = 4 - (2y + 4)$$
$$y^2 = -2y$$
$$y^2 + 2y = 0$$
$$y(y + 2) = 0$$

$y = 0$ or $y + 2 = 0$
$$y = -2$$

Replace these values into the equation
$x = 2y + 4$.
$y = 0 : x = 2(0) + 4 = 4$
$y = -2 : x = 2(-2) + 4 = 0$
The solutions are $(4, 0)$ and $(0, -2)$.

7. $\begin{cases} x^2 + y^2 = 9 & (1) \\ 16x^2 - 4y^2 = 64 & (2) \end{cases}$

Multiply E1 by 4 and add to E2.
$4x^2 + 4y^2 = 36$
$\underline{16x^2 + 4y^2 = 64}$
$20x^2 \qquad = 100$
$x^2 = 5$
$x = \pm\sqrt{5}$

Substitute 5 for x^2 into E1.
$5 + y^2 = 9$
$y^2 = 4$
$y = \pm 2$

The solutions are $\left(-\sqrt{5}, -2\right)$, $\left(-\sqrt{5}, 2\right)$,
$\left(\sqrt{5}, -2\right)$, and $\left(\sqrt{5}, 2\right)$.

9. $\begin{cases} x^2 + 2y^2 = 2 & (1) \\ x - y = 2 & (2) \end{cases}$

Solve E2 for x: $x = y + 2$
Substitute into E1.
$(y + 2)^2 + 2y^2 = 2$
$y^2 + 4y + 4 + 2y^2 = 2$
$3y^2 + 4y + 2 = 0$

$y = \dfrac{-4 \pm \sqrt{(4)^2 - 4(3)(2)}}{2(3)} = \dfrac{-4 \pm \sqrt{-8}}{6}$

which yields no real solutions.
The solution is $\varnothing$.

11. $\begin{cases} y = x^2 - 3 & (1) \\ 4x - y = 6 & (2) \end{cases}$

Substitute $x^2 - 3$ for y in E2.
$4x - (x^2 - 3) = 6$
$4x - x^2 + 3 = 6$
$0 = x^2 - 4x + 3$
$0 = (x - 3)(x - 1)$
$x - 3 = 0$ or $x - 1 = 0$
$x = 3$ or $x = 1$
Substitute these values into E1.
$x = 3 : y = (3)^2 - 3 = 6$
$x = 1 : y = (1)^2 - 3 = -2$
The solutions are $(3, 6)$ and $(1, -2)$.

13. $\begin{cases} y = x^2 & (1) \\ 3x + y = 10 & (2) \end{cases}$

Substitute x^2 for y in E2.
$3x + x^2 = 10$
$x^2 + 3x - 10 = 0$
$(x + 5)(x - 2) = 0$
$x + 5 = 0$ or $x - 2 = 0$
$x = -5$ or $x = 2$
Substitute these values into E1.
$x = -5 : y = (-5)^2 = 25$
$x = 2 : y = (2)^2 = 4$
The solutions are $(-5, 25)$ and $(2, 4)$.

15. $\begin{cases} y = 2x^2 + 1 & (1) \\ x + y = -1 & (2) \end{cases}$

Substitute $2x^2 + 1$ for y in E2.
$x + 2x^2 + 1 = -1$
$2x^2 + x + 2 = 0$

$x = \dfrac{-1 \pm \sqrt{(1)^2 - 4(2)(2)}}{2(2)} = \dfrac{-1 \pm \sqrt{-15}}{4}$

which yields no real solutions.
The solution is $\varnothing$.

17. $\begin{cases} y = x^2 - 4 & (1) \\ y = x^2 - 4x & (2) \end{cases}$

Substitute $x^2 - 4$ for y in E2.

$x^2 - 4 = x^2 - 4x$

$\quad -4 = -4x$

$\quad\quad 1 = x$

Substitute this value into E1.

$y = (1)^2 - 4 = -3$

The solution is (1, –3).

19. $\begin{cases} 2x^2 + 3y^2 = 14 & (1) \\ -x^2 + y^2 = 3 & (2) \end{cases}$

Multiply E2 by 2 and add to E1.

$-2x^2 + 2y^2 = 6$

$\underline{2x^2 + 3y^2 = 14}$

$5y^2 = 20$

$y^2 = 4$

$y = \pm 2$

Substitute 4 for y^2 into E2.

$-x^2 + 4 = 3$

$-x^2 = -1$

$x^2 = 1$

$x = \pm 1$

The solutions are (–1, –2), (–1, 2), (1, –2), and (1, 2).

21. $\begin{cases} x^2 + y^2 = 1 & (1) \\ x^2 + (y+3)^2 = 4 & (2) \end{cases}$

Multiply E1 by –1 and add to E2.

$-x^2 - y^2 = -1$

$\underline{x^2 + (y+3)^2 = 4}$

$(y+3)^3 - y^2 = 3$

$y^2 + 6y + 9 - y^2 = 3$

$6y = -6$

$y = -1$

Replace y with –1 in E1.

$x^2 + (1)^2 = 1$

$x^2 = 0$

$x = 0$

The solution is (0, –1).

23. $\begin{cases} y = x^2 + 2 & (1) \\ y = -x^2 + 4 & (2) \end{cases}$

Add E1 and E2.

$2y = 6$

$y = 3$

Substitute this value into E1.

$3 = x^2 + 2$

$1 = x^2$

$\pm 1 = x$

The solutions are (–1, 3) and (1, 3).

25. $\begin{cases} 3x^2 + y^2 = 9 & (1) \\ 3x^2 - y^2 = 9 & (2) \end{cases}$

Add E1 and E2.

$6x^2 = 18$

$x^2 = 3$

$x = \pm\sqrt{3}$

Substitute 3 for x^2 into E1.

$3(3) + y^2 = 9$

$y^2 = 0$

$y = 0$

The solutions are $\left(-\sqrt{3}, 0\right)$, $\left(\sqrt{3}, 0\right)$.

27. $\begin{cases} x^2 + 3y^2 = 6 & (1) \\ x^2 - 3y^2 = 10 & (2) \end{cases}$

Solve E2 for x^2: $x^2 = 3y^2 + 10$.

Substitute into E1.

$(3y^2 + 10) + 3y^2 = 6$

$6y^2 = -4$

$y^2 = -\dfrac{2}{3}$

which yields no real solutions.

The solution is $\varnothing$.

29. $\begin{cases} x^2 + y^2 = 36 & (1) \\ y = \frac{1}{6}x^2 - 6 & (2) \end{cases}$

Solve E1 for x^2: $x^2 = 36 - y^2$.
Substitute into E2.

$$y = \frac{1}{6}(36 - y^2) - 6$$
$$y = 6 - \frac{1}{6}y^2 - 6$$
$$6y = -y^2$$
$$y^2 + 6y = 0$$
$$y(y + 6) = 0$$
$$y = 0 \text{ or } y = -6$$

Replace these values into the equation
$x^2 = 36 - y^2$.

$$y = 0 : x^2 = 36 - (0)^2$$
$$x^2 = 36$$
$$x = \pm 6$$
$$y = -6 : x^2 = 36 - (6)^2$$
$$x^2 = 0$$
$$x = 0$$

The solutions are (–6, 0), (6, 0) and
(–6, 0).

31. $x > -3$

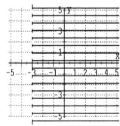

33. $y < 2x - 1$

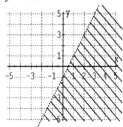

35. $P = x + (2x - 5) + (5x - 20)$
$ = (8x - 25)$ inches

37. $P = 2(x^2 + 3x + 1) + 2(x^2)$
$ = 2x^2 + 6x + 2 + 2x^2$
$ = (4x^2 + 6x + 2)$ meters

39. There are 0, 1, 2, 3, or 4 possible real
solutions. Answers may vary.

41. Let x and y represent the numbers.
$$\begin{cases} x^2 + y^2 = 130 \\ x^2 - y^2 = 32 \end{cases}$$
Add the equations.
$$2x^2 = 162$$
$$x^2 = 81$$
$$x = \pm 9$$
Replace x^2 with 81 in the first equation.
$$81 + y^2 = 130$$
$$y^2 = 49$$
$$y = \pm 7$$
The numbers are –9 and –7, –9 and 7, 9
and –7, and 9 and 7.

43. Let x and y be the length and width.
$$\begin{cases} xy = 285 \\ 2x + 2y = 68 \end{cases}$$

Solve the first equation for y: $y = \frac{285}{x}$.
Substitute into the second equation.
$$2x + 2\left(\frac{285}{x}\right) = 68$$
$$x + \frac{285}{x} = 34$$
$$x^2 + 285 = 34x$$
$$x^2 - 34x + 285 = 0$$
$$(x - 19)(x - 15) = 0$$
$$x = 19 \text{ or } x = 15$$
Using $x = 19$, $y = \frac{285}{x} = \frac{285}{19} = 15$.
The dimensions are 19 cm by 15 cm.

45. $\begin{cases} p = -0.01x^2 - 0.2x + 9 \\ p = 0.01x^2 - 0.1x + 3 \end{cases}$

Substitute.

$-0.01x^2 - 0.2x + 9 = 0.01x^2 - 0.1x + 3$

$0 = 0.02x^2 + 0.1x - 6$

$0 = x^2 + 5x - 300$

$0 = (x + 20)(x - 15)$

$x + 20 = 0 \quad$ or $\quad x - 15 = 0$

$\quad x = -20 \quad$ or $\quad\quad x = 15$

Disregard the negative.

$p = -0.01(15)^2 - 0.2(15) + 9$

$p = 3.75$

The equilibrium quantity is 15,000 compact discs, and the corresponding price is $3.75.

47.

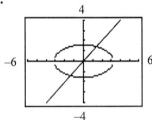

49.

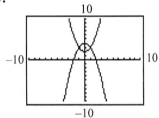

Exercise Set 13.4

1. $y < x^2$

First graph the parabola with dashes.

Test Point	$y < x^2$; Result
(0, 1)	$1 < 0^2$; False

Shade the portion of the graph which does not contain (0, 1).

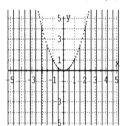

3. $x^2 + y^2 \geq 16$

First graph the circle with a solid curve.

Test Point	$x^2 + y^2 \geq 16$; Result
(0, 0)	$0^2 + 0^2 \geq 16$; False

Shade the portion of the graph which does not contain (0, 0).

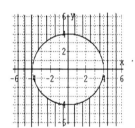

5. $\dfrac{x^2}{4} - y^2 < 1$

First graph the hyperbola with a dashed curve.

Test Points	$\dfrac{x^2}{4} - y^2 < 1$; Result
(–4, 0)	$\dfrac{(-4)^2}{4} - 0^2 < 1$; False
(0, 0)	$\dfrac{(0)^2}{4} - 0^2 < 1$; True
(4, 0)	$\dfrac{(4)^2}{4} - 0^2 < 1$; False

Shade the portion of the graph that contains (0, 0).

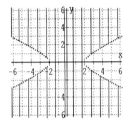

7. $y > (x-1)^2 - 3$

First graph the parabola with a dashed curve.

Test Point	$y > (x-1)^2 - 3$; Result
(0, 0)	$0 > (0-1)^2 - 3$; True

Shade the portion of the graph which contains (0, 0).

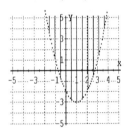

9. $x^2 + y^2 \leq 9$

First graph the circle with a solid curve.

Test Point	$x^2 + y^2 \leq 9$; Result
(0, 0)	$0^2 + 0^2 \leq 9$; True

Shade the portion of the graph which contains (0, 0).

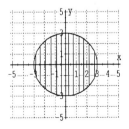

11. $y > -x^2 + 5$

First graph the parabola with a dashed curve.

Test Point	$y > -x^2 + 5$; Result
(0, 0)	$0 > -(0)^2 + 5$; False

Shade the portion of the graph which does not contain (0, 0).

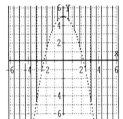

13. $\dfrac{x^2}{4} + \dfrac{y^2}{9} \leq 1$

First graph the ellipse with a solid curve.

Test Point	$\dfrac{x^2}{4} + \dfrac{y^2}{9} \leq 1$; Result
(0, 0)	$\dfrac{(0)^2}{4} + \dfrac{(0)^2}{9} \leq 1$; True

Shade the portion of the graph that contains (0, 0).

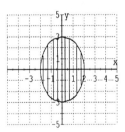

15. $\dfrac{y^2}{4} - x^2 \leq 1$

First graph the hyperbola with solid curves.

Test Points	$\dfrac{y^2}{4} - x^2 \leq 1$; Result
(0, 4)	$\dfrac{(-4)^2}{4} - 0^2 \leq 1$; False
(0, 0)	$\dfrac{(0)^2}{4} - 0^2 \leq 1$; True
(0, 4)	$\dfrac{(4)^2}{4} - 0^2 \leq 1$; False

Shade the portion of the graph that contains (0, 0).

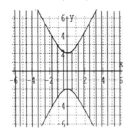

17. $y < (x-2)^2 + 1$

First graph the parabola with a dashed curve.

Test Point	$y < (x-2)^2 + 1$; Result
(0, 0)	$0 < (0-2)^2 + 1$; True

Shade the portion of the graph which contains (0, 0).

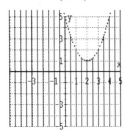

19. $y \leq x^2 + x - 2$

First graph the parabola with a solid curve.

Test Point	$y \leq x^2 + x - 2$; Result
(0, 0)	$0 \leq (0)^2 + (0) - 2$; False

Shade the portion of the graph which contains (0, 0).

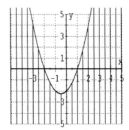

21. $\begin{cases} 2x - y < 2 \\ \quad y \leq -x^2 \end{cases}$

First graph $2x - y = 2$ with a dashed line.

Test Point	$2x - y < 2$; Result
(0, 0)	$2(0) - 0 < 2$; True

Shade the portion of the graph which contains (0, 0).

Next, graph the parabola $y = -x^2$ with a solid curve.

Test Point	$y \leq -x^2$; Result
(0, 1)	$1 \leq -(0)^2$; False

Shade the portion of the graph which does not contain (0, 1).

The solution to the system is the overlapping region.

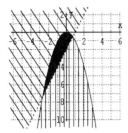

23. $\begin{cases} 4x + 3y \geq 12 \\ x^2 + y^2 < 16 \end{cases}$

First graph $4x + 3y = 12$ with a solid line.

Test Point	$4x + 3y \geq 12$; Result
(0, 0)	$4(0) + 3(0) \geq 12$; False

Shade the portion of the graph which does not contain (0, 0).

Next, graph the circle $x^2 + y^2 = 16$ with a dashed curve.

Test Point	$x^2 + y^2 < 16$; Result
(0, 0)	$0^2 + 0^2 < 16$; True

Shade the portion of the graph which contains (0, 0).

The solution to the system is the overlapping region.

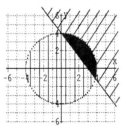

25. $\begin{cases} x^2 + y^2 \leq 9 \\ x^2 + y^2 \geq 1 \end{cases}$

First graph the circle with radius 3 with a solid curve.

Test Point	$x^2 + y^2 \leq 9$; Result
(0, 0)	$0^2 + 0^2 \leq 9$; True

Shade the portion of the graph which contains (0, 0).

Next, graph the circle with 1 with a dashed curve.

Test Point	$x^2 + y^2 \geq 1$; Result
(0, 0)	$0^2 + 0^2 \geq 1$; False

Shade the portion of the graph which does not contain (0, 0).

The solution to the system is the overlapping region.

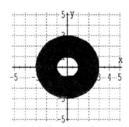

27. $\begin{cases} y > x^2 \\ y \geq 2x + 1 \end{cases}$

First graph the parabola with a dashed curve.

Test Point	$y > x^2$; Result
(0, 1)	$1 > 0^2$; True

Shade the portion of the graph which contains (0, 1).

Next, graph $y = 2x + 1$ with a solid line.

Test Point	$y \geq 2x + 1$; Result
(0, 0)	$0 \geq 2(0) + 1$; False

Shade the portion of the graph which does not contain (0, 0).

The solution to the system is the overlapping region.

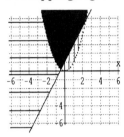

29. $\begin{cases} x > y^2 \\ y > 0 \end{cases}$

First graph the parabola with a dashed curve.

Test Point	$x > y^2$; Result
(1, 0)	$1 > 0^2$; True

Shade the portion of the graph which contains (1, 0).

Next, graph $y = 0$ with a dashed line.

Test Point	$y > x^2$; Result
(0, 1)	$1 \geq 0^2$; True

Shade the portion of the graph which contains (0, 1).

The solution to the system is the overlapping region.

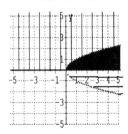

31. $\begin{cases} x^2 + y^2 > 9 \\ y > x^2 \end{cases}$

First graph the circle with a dashed curve.

Test Point	$x^2 + y^2 > 9$; Result
(0, 0)	$0^2 + 0^2 > 9$; False

Shade the portion of the graph which does not contain (0, 0).

Next, graph the parabola with a dashed curve.

Test Point	$y > x^2$; Result
(0, 1)	$1 > 0^2$; True

Shade the portion of the graph which contains (0, 1).

The solution to the system is the overlapping region.

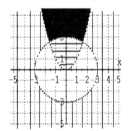

33. $\begin{cases} \dfrac{x^2}{4} + \dfrac{y^2}{9} \geq 1 \\ x^2 + y^2 \geq 4 \end{cases}$

First graph the ellipse with a solid curve.

Test Point	$\dfrac{x^2}{4} + \dfrac{y^2}{9} \geq 1$; Result
(0, 0)	$\dfrac{0^2}{4} + \dfrac{0^2}{9} \geq 1$; False

Shade the portion of the graph which does not contain (0, 0).

Next, graph the circle with a solid curve.

Test Point	$x^2 + y^2 \geq 4$; Result
(0, 0)	$0^2 + 0^2 \geq 4$; False

Shade the portion of the graph which does not contain (0, 0).

The solution to the system is the overlapping region.

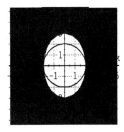

35. $\begin{cases} x^2 - y^2 \geq 1 \\ \quad\quad y \geq 0 \end{cases}$

First graph the hyperbola with solid curves.

Test Point	$x^2 - y^2 \geq 1$; Result
(0, 0)	$0^2 - 0^2 \geq 1$; False

Shade the portion of the graph which does not contain (0, 0).

Next, graph $y = 0$ with a solid line.

Test Point	$y > 0$; Result
(0, 1)	$1 \geq 0$; True

Shade the portion of the graph which contains (0, 1).

The solution to the system is the overlapping region.

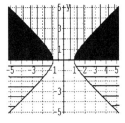

37. $\begin{cases} x + y \geq 1 \\ 2x + 3y < 1 \\ \quad\quad x > -3 \end{cases}$

First graph $x + y = 1$ with a solid line.

Test Point	$x + y \geq 1$; Result
(0, 0)	$0 + 0 \geq 1$; False

Shade the portion of the graph which does not contain (0, 0).

Next, graph $2x + 3y = 1$ with a dashed line.

Test Point	$2x + 3y < 1$; Result
(0, 0)	$2(0) + 3(0) < 1$; True

Shade the portion of the graph which contains (0, 0).

Now graph the line $x = -3$ with a dashed line.

Test Point	$x > -3$; Result
(0, 0)	$0 > -3$; True

Shade the portion of the graph which contains (0, 0).

The solution to the system is the overlapping region.

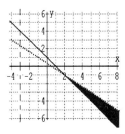

39. $\begin{cases} x^2 - y^2 < 1 \\ \dfrac{x^2}{16} + y^2 \leq 1 \\ \quad\quad x \geq -2 \end{cases}$

First graph the hyperbola with dashed curves.

Test Point	$x^2 - y^2 < 1$; Result
(0, 0)	$0^2 - 0^2 < 1$; True

Shade the portion of the graph which contains (0, 0).

Next, graph the ellipse with a solid curve.

Test Point	$\dfrac{x^2}{16} + y^2 \leq 1$; Result
(0, 0)	$\dfrac{0^2}{16} + 0^2 \leq 1$; True

Shade the portion of the graph which contains (0, 0).

Now graph the line $x = -2$ with a solid line.

Test Point	$x \geq -2$; Result
(0, 0)	$0 \geq -2$; True

Shade the portion of the graph which contains (0, 0).

The solution to the system is the overlapping region.

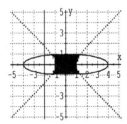

41. This is not a function because a vertical line can cross the graph in more than one place.

43. This is a function because a vertical line can cross the graph no more than one place.

45. $f(x) = 3x^2 - 2$
$f(-1) = 3(-1)^2 - 2 = 3 - 2 = 1$

47. $f(x) = 3x^2 - 2$
$f(a) = 3(a)^2 - 2 = 3a^2 - 2$

49. Answers may vary.

51. $\begin{cases} y \leq x^2 \\ y \geq x + 2 \\ x \geq 0 \\ y \geq 0 \end{cases}$

First graph $y = x^2$ with a solid curve.

Test Point	$y \leq x^2$; Result
(0, 1)	$1 \leq 0^2$; False

Shade the portion of the graph which does not contain (0, 1).

Next, graph $y = x + 2$ with a solid line.

Test Point	$y \geq x + 2$; Result
(0, 0)	$0 \geq 0 + 2$; False

Shade the portion of the graph which does not contain (0, 0).

Next graph the line $x = 0$ with a solid line, and shade to the right.

Now graph the line $y = 0$ with a solid line, and shade above the line.

The solution to the system is the overlapping region.

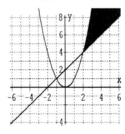

Chapter 13 Review

1. $(-6, 3), (8, 4)$

$$d = \sqrt{(4-3)^2 + [8-(-6)]^2}$$
$$= \sqrt{1^2 + 14^2}$$
$$= \sqrt{1+196}$$
$$= \sqrt{197} \text{ units}$$

2. $(3, 5), (8, 9)$

$$d = \sqrt{(8-3)^2 + (9-5)^2}$$
$$= \sqrt{5^2 + 4^2}$$
$$= \sqrt{25+16}$$
$$= \sqrt{41} \text{ units}$$

3. $(-4, -6), (-1, 5)$

$$d = \sqrt{[-1-(-4)]^2 + [5-(-6)]^2}$$
$$= \sqrt{3^2 + 11^2}$$
$$= \sqrt{9+121}$$
$$= \sqrt{130} \text{ units}$$

4. $(-1, 5), (2, -3)$

$$d = \sqrt{[2-(-1)]^2 + (-3-5)^2}$$
$$= \sqrt{3^2 + (-8)^2}$$
$$= \sqrt{9+64}$$
$$= \sqrt{73} \text{ units}$$

5. $\left(-\sqrt{2}, 0\right), \left(0, -4\sqrt{6}\right)$

$$d = \sqrt{\left[0-\left(-\sqrt{2}\right)\right]^2 + \left(-4\sqrt{6}-0\right)^2}$$
$$= \sqrt{\left(\sqrt{2}\right)^2 + \left(-4\sqrt{6}\right)^2}$$
$$= \sqrt{2 + 16\cdot 6}$$
$$= \sqrt{98}$$
$$= 7\sqrt{2} \text{ units}$$

6. $\left(-\sqrt{5}, -\sqrt{11}\right), \left(-\sqrt{5}, -3\sqrt{11}\right)$

$$d = \sqrt{\left[-\sqrt{5}-\left(-\sqrt{5}\right)\right]^2 + \left[-3\sqrt{11}-\left(-\sqrt{11}\right)\right]^2}$$
$$= \sqrt{0^2 + \left(-2\sqrt{11}\right)^2}$$
$$= \sqrt{4\cdot 11}$$
$$= 2\sqrt{11} \text{ units}$$

7. $(7.4, -8.6), (-1.2, 5.6)$

$$d = \sqrt{(-1.2-7.4)^2 + [5.6-(-8.6)]^2}$$
$$= \sqrt{(-8.6)^2 + (14.2)^2}$$
$$= \sqrt{275.6} \approx 16.60 \text{ units}$$

8. $(2.3, 1.8), (10.7, -9.2)$

$$d = \sqrt{(10.7-2.3)^2 + (-9.2-1.8)^2}$$
$$= \sqrt{(8.4)^2 + (-11)^2}$$
$$= \sqrt{191.56} \approx 13.84 \text{ units}$$

9. $(2, 6), (-12, 4)$

$$\left(\frac{2+(-12)}{2}, \frac{6+4}{2}\right) = \left(\frac{-10}{2}, \frac{10}{2}\right) = (-5, 5)$$

The midpoint of is $(-5, 5)$.

10. $(-3, 8), (11, 24)$

$$\left(\frac{-3+11}{2}, \frac{8+24}{2}\right) = \left(\frac{8}{2}, \frac{32}{2}\right) = (4, 16)$$

The midpoint of is $(4, 16)$.

11. $(-6, -5), (-9, 7)$

$$\left(\frac{-6+(-9)}{2}, \frac{-5+7}{2}\right) = \left(\frac{-15}{2}, \frac{2}{2}\right) = -\left(\frac{15}{2}, 1\right)$$

The midpoint of is $\left(-\frac{15}{2}, 1\right)$.

12. $(4, -6), (-15, 2)$

$$\left(\frac{4+(-15)}{2}, \frac{-6+2}{2}\right) = \left(-\frac{11}{2}, -2\right)$$

The midpoint of is $\left(-\frac{11}{2}, -2\right)$.

13. $\left(0, -\frac{3}{8}\right), \left(\frac{1}{10}, 0\right)$

$$\left(\frac{0 + \left(\frac{1}{10}\right)}{2}, \frac{-\frac{3}{8} + 0}{2}\right) = \left(\frac{1}{20}, -\frac{3}{16}\right)$$

The midpoint of is $\left(\frac{1}{20}, -\frac{3}{16}\right)$.

14. $\left(\frac{3}{4}, -\frac{1}{7}\right), \left(-\frac{1}{4}, -\frac{3}{7}\right)$

$$\left(\frac{\frac{3}{4} + \left(-\frac{1}{4}\right)}{2}, \frac{-\frac{1}{7} + \left(-\frac{3}{7}\right)}{2}\right) = \left(\frac{\frac{1}{2}}{2}, -\frac{\frac{4}{7}}{2}\right)$$

$$= \left(\frac{1}{4}, -\frac{2}{7}\right)$$

The midpoint of is $\left(\frac{1}{4}, -\frac{2}{7}\right)$.

15. $\left(\sqrt{3}, -2\sqrt{6}\right), \left(\sqrt{3}, -4\sqrt{6}\right)$

$$\left(\frac{\sqrt{3} + \sqrt{3}}{2}, \frac{-2\sqrt{6} + \left(-4\sqrt{6}\right)}{2}\right)$$

$$= \left(\frac{2\sqrt{3}}{2}, -\frac{6\sqrt{6}}{2}\right) = \left(\sqrt{3}, -3\sqrt{6}\right)$$

The midpoint of is $\left(\sqrt{3}, -3\sqrt{6}\right)$.

16. $\left(-5\sqrt{3}, 2\sqrt{7}\right), \left(-3\sqrt{3}, 10\sqrt{7}\right)$

$$\left(\frac{-5\sqrt{3} + \left(-3\sqrt{3}\right)}{2}, \frac{2\sqrt{7} + 10\sqrt{7}}{2}\right)$$

$$= \left(\frac{-8\sqrt{3}}{2}, \frac{12\sqrt{7}}{2}\right) = \left(-4\sqrt{3}, 6\sqrt{7}\right)$$

The midpoint of is $\left(-4\sqrt{3}, 6\sqrt{7}\right)$.

17. center (–4, 4), radius 3
$$[x - (-4)]^2 + (y - 4)^2 = 3^2$$
$$(x + 4)^2 + (y - 4)^2 = 9$$

18. center (5, 0), radius 5
$$(x - 5)^2 + (y - 0)^2 = 5^2$$
$$(x - 5)^2 + y^2 = 25$$

19. center (–7, –9), radius $\sqrt{11}$
$$[x - (-7)]^2 + [y - (-9)]^2 = \left(\sqrt{11}\right)^2$$
$$(x + 7)^2 + (y + 9)^2 = 11$$

20. center (0, 0), radius $\frac{7}{2}$
$$(x - 0)^2 + (y - 0)^2 = \left(\frac{7}{2}\right)^2$$
$$x^2 + y^2 = \frac{49}{4}$$

21. $x^2 + y^2 = 7$

Circle; center (0, 0), radius $r = \sqrt{7}$

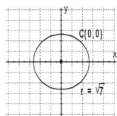
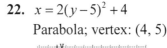

22. $x = 2(y - 5)^2 + 4$

Parabola; vertex: (4, 5)

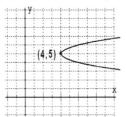

23. $x = -(y+2)^2 + 3$

Parabola; vertex: $(3, -2)$

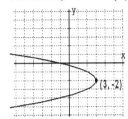

24. $(x-1)^2 + (y-2)^2 = 4$

Circle; center $(1, 2)$, radius $r = \sqrt{4} = 2$

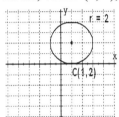

25. $y = -x^2 + 4x + 10$

Parabola; $x = -\dfrac{b}{2a} = \dfrac{-4}{2(-1)} = 2$

$y = -(2)^2 + 4(2) + 10 = 14$

Vertex: $(2, 14)$

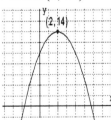

26. $x = -y^2 - 4y + 6$

Parabola; $y = -\dfrac{b}{2a} = \dfrac{-(-4)}{2(-1)} = -2$

$x = -(-2)^2 - 4(-2) + 6 = 10$

Vertex: $(10, -2)$

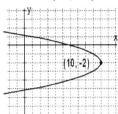

27. $x = \dfrac{1}{2}y^2 + 2y + 1$

Parabola; $y = -\dfrac{b}{2a} = \dfrac{-2}{2\left(\dfrac{1}{2}\right)} = -2$

$x = \dfrac{1}{2}(-2)^2 + 2(-2) + 1 = -1$

Vertex: $(-1, -2)$

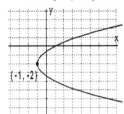

28. $y = -3x^2 + \dfrac{1}{2}x + 4$

Parabola; $x = -\dfrac{b}{2a} = \dfrac{-\dfrac{1}{2}}{2(-3)} = \dfrac{1}{12}$

$y = -3\left(\dfrac{1}{12}\right)^2 + \dfrac{1}{2}\left(\dfrac{1}{12}\right) + 4 = \dfrac{193}{48}$

Vertex: $\left(\dfrac{1}{12}, \dfrac{193}{48}\right)$

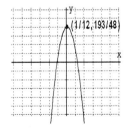

29.
$$x^2 + y^2 + 2x + y = \frac{3}{4}$$
$$(x^2 + 2x) + (y^2 + y) = \frac{3}{4}$$
$$(x^2 + 2x + 1) + \left(y^2 + y + \frac{1}{4}\right) = \frac{3}{4} + 1 + \frac{1}{4}$$
$$(x+1)^2 + \left(y + \frac{1}{2}\right)^2 = 2$$

Circle; center $\left(-1, -\frac{1}{2}\right)$, radius $r = \sqrt{2}$

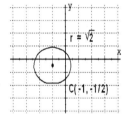

30.
$$x^2 + y^2 + 3y = \frac{7}{4}$$
$$x^2 + \left(y^2 + 3y + \frac{9}{4}\right) = \frac{3}{4} + \frac{9}{4}$$
$$x^2 + \left(y + \frac{3}{2}\right)^2 = 3$$

Circle; center $\left(0, -\frac{3}{2}\right)$, radius $r = \sqrt{3}$

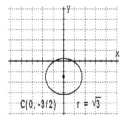

31.
$$4x^2 + 4y^2 + 16x + 8y = 1$$
$$(x^2 + 4x) + (y^2 + 2y) = \frac{1}{4}$$
$$(x^2 + 4x + 4) + (y^2 + 2y + 1) = \frac{1}{4} + 4 + 1$$
$$(x+2)^2 + (y+1)^2 = \frac{21}{4}$$

Circle; center $(-2, -1)$,

radius $r = \sqrt{\frac{21}{4}} = \frac{\sqrt{21}}{2}$

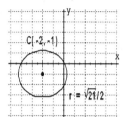

32.
$$3x^2 + 6x + 3y^2 = 9$$
$$x^2 + 2x + y^2 = 3$$
$$(x^2 + 2x + 1) + y^2 = 3 + 1$$
$$(x+1)^2 + y^2 = 4$$

Circle; center $(-1, 0)$, radius $r = \sqrt{4} = 2$

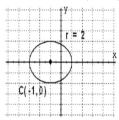

33. $y = x^2 + 6x + 9$
$$= (x+3)^2$$

Parabola; vertex: $(-3, 0)$

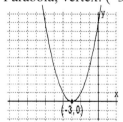

34. $x = y^2 + 6y + 9$
$\quad = (y + 3)^2$
Parabola; vertex: $(0, -3)$

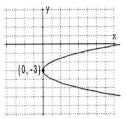

35. Center: $(5.6, -2.4)$, radius $\dfrac{6.2}{2} = 3.1$

$(x - 5.6)^2 + [y - (-2.4)]^2 = (3.1)^2$
$(x - 5.6)^2 + (y + 2.4)^2 = 9.61$

36. $x^2 + \dfrac{y^2}{4} = 1$

Center: $(0, 0)$; $a = 1$, $b = 2$

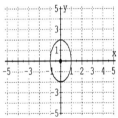

37. $x^2 - \dfrac{y^2}{4} = 1$

Center: $(0, 0)$; $a = 1$, $b = 2$

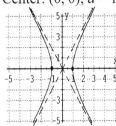

38. $\dfrac{y^2}{4} - \dfrac{x^2}{16} = 1$

Center: $(0, 0)$; $a = 4$, $b = 2$

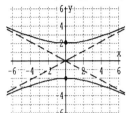

39. $\dfrac{y^2}{4} + \dfrac{x^2}{16} = 1$

Center: $(0, 0)$; $a = 4$, $b = 2$

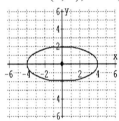

40. $\dfrac{x^2}{5} + \dfrac{y^2}{5} = 1$

$x^2 + y^2 = 5$

Center: $(0, 0)$; radius $r = \sqrt{5}$

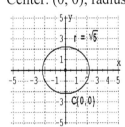

41. $\dfrac{x^2}{5} - \dfrac{y^2}{5} = 1$

Center: (0, 0); $a = \sqrt{5}$, $b = \sqrt{5}$

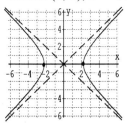

42. $-5x^2 + 25y^2 = 125$

$\dfrac{y^2}{5} - \dfrac{x^2}{25} = 1$

Center: (0, 0); $a = 5$, $b = \sqrt{5}$

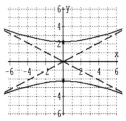

43. $4y^2 + 9x^2 = 36$

$\dfrac{y^2}{9} + \dfrac{x^2}{4} = 1$

Center: (0, 0); $a = 2$, $b = 3$

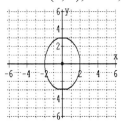

44. $\dfrac{(x-2)^2}{4} + (y-1)^2 = 1$

Center: (2, 1); $a = 2$, $b = 1$

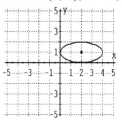

45. $\dfrac{(x+3)^2}{9} + \dfrac{(y-4)^2}{25} = 1$

Center: (–3, 4); $a = 3$, $b = 5$

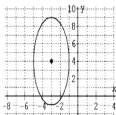

46. $x^2 - y^2 = 1$

Center: (0, 0); $a = 1$, $b = 1$

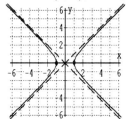

47. $36y^2 - 49x^2 = 1764$

$\dfrac{y^2}{49} - \dfrac{x^2}{36} = 1$

Center: (0, 0); $a = 6$, $b = 7$

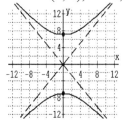

48. $\quad\quad y^2 = x^2 + 9$

$\quad\quad\quad y^2 - x^2 = 9$

$\quad\quad\quad \dfrac{y^2}{9} - \dfrac{x^2}{9} = 1$

Center: (0, 0); $a = 3$, $b = 3$

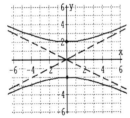

49. $x^2 = 4y^2 - 16$

$\quad\quad 16 = 4y^2 - x^2$

$\quad\quad 1 = \dfrac{y^2}{4} - \dfrac{x^2}{16}$

Center: (0, 0); $a = 4$, $b = 2$

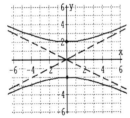

50. $100 - 25x^2 = 4y^2$

$\quad\quad 100 = 25x^2 + 4y^2$

$\quad\quad 1 = \dfrac{x^2}{4} + \dfrac{y^2}{25}$

Center: (0, 0); $a = 2$, $b = 5$

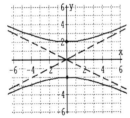

51. $\quad y = x^2 + 4x + 6$

Parabola; $x = -\dfrac{b}{2a} = \dfrac{-4}{2(1)} = -2$

$y = (-2)^2 + 4(-2) + 6 = 2$

Vertex: (−2, 2)

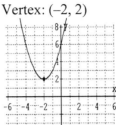

52. $\quad\quad y^2 = x^2 + 6$

$\quad\quad\quad y^2 - x^2 = 6$

$\quad\quad\quad \dfrac{y^2}{6} - \dfrac{x^2}{6} = 1$

Hyperbola; center: (0, 0),

$a = \sqrt{6}$, $b = \sqrt{6}$

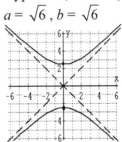

53. $\quad\quad\quad y^2 + x^2 = 4x + 6$

$\quad\quad (x^2 - 4x) + y^2 = 6$

$\quad (x^2 - 4x + 4) + y^2 = 6 + 4$

$\quad\quad\quad (x - 2)^2 + y^2 = 10$

Circle; center: (2, 0), radius $r = \sqrt{10}$

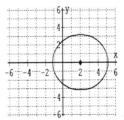

54.
$$y^2 + 2x^2 = 4x + 6$$
$$(2x^2 - 4x) + y^2 = 6$$
$$2(x^2 - 2x + 1) + y^2 = 6 + 2$$
$$2(x-1)^2 + y^2 = 8$$
$$\frac{(x-1)^2}{4} + \frac{y^2}{8} = 1$$
Ellipse; Center: (1, 0);
$a = 2$, $b = \sqrt{8} = 2\sqrt{2}$

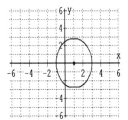

55.
$$x^2 + y^2 - 8y = 0$$
$$x^2 + (y^2 - 8y + 16) = 0 + 16$$
$$x^2 + (y-4)^2 = 16$$

Circle; center: (0, 4), radius $r = \sqrt{16} = 4$

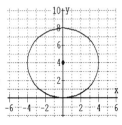

56. $x - 4y = y^2$
$$x = y^2 + 4y$$
$$x + 4 = y^2 + 4y + 4$$
$$x = (y+2)^2 - 4$$
Parabola; vertex: (–4, –2)

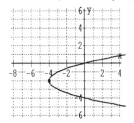

57. $x^2 - 4 = y^2$
$$x^2 - y^2 = 4$$
$$\frac{x^2}{4} - \frac{y^2}{4} = 1$$
Hyperbola; center: (0, 0), $a = 2$, $b = 2$

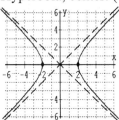

58. $x^2 = 4 - y^2$
$$x^2 + y^2 = 4$$
Circle; center: (0, 0), radius $r = \sqrt{4} = 2$

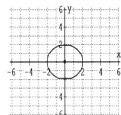

59. $6(x-2)^2 + 9(y+5)^2 = 36$
$$\frac{(x-2)^2}{6} + \frac{(y+5)^2}{4} = 1$$
Ellipse; Center: (2, –5); $a = \sqrt{6}$, $b = 2$

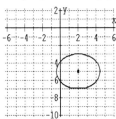

60.
$$36y^2 = 576 + 16x^2$$
$$36y^2 - 16x^2 = 576$$
$$\frac{y^2}{16} - \frac{x^2}{36} = 1$$

Hyperbola: center: (0, 0); $a = 6$, $b = 4$

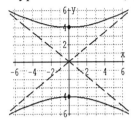

61. $\dfrac{x^2}{16} - \dfrac{y^2}{25} = 1$

Hyperbola; center: (0, 0), $a = 4$, $b = 5$

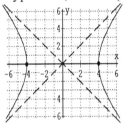

62. $3(x-7)^2 + 3(y+4)^2 = 1$

$\qquad (x-7)^2 + (y+4)^2 = \dfrac{1}{3}$

Circle; center: (7, –4);

radius $r = \sqrt{\dfrac{1}{3}} = \dfrac{1}{\sqrt{3}} = \dfrac{\sqrt{3}}{3}$

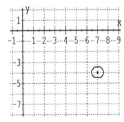

63. $\qquad \dfrac{y^2}{4} + \dfrac{x^2}{16} = 1$

$\qquad 16\left(\dfrac{y^2}{4} + \dfrac{x^2}{16}\right) = 16(1)$

$\qquad\qquad 4y^2 + x^2 = 16$

$$4y^2 = 16 - x^2$$
$$y^2 = \dfrac{16 - x^2}{4}$$
$$y = \pm\sqrt{\dfrac{16 - x^2}{4}}$$

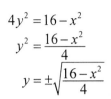

Answers are the same.

64. $\dfrac{x^2}{5} + \dfrac{y^2}{5} = 1$

$\qquad x^2 + y^2 = 5$

$\qquad\quad y^2 = 5 - x^2$

$\qquad\quad y = \pm\sqrt{5 - x^2}$

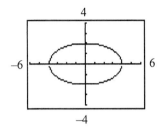

Answers are the same.

65. $y = x^2 + 4x + 6$

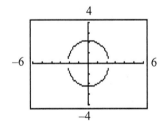

Answers are the same.

66. $x^2 = 4 - y^2$

$y^2 = 4 - x^2$

$y = \pm\sqrt{4 - x^2}$

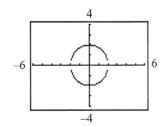

Answers are the same.

67. $\begin{cases} y = 2x - 4 & (1) \\ y^2 = 4x & (2) \end{cases}$

Substitute $2x - 4$ for y into E2.

$(2x - 4)^2 = 4x$

$4x^2 - 16x + 16 = 4x$

$4x^2 - 20x + 16 = 0$

$x^2 - 5x + 4 = 0$

$(x - 4)(x - 1) = 0$

$x = 4$ or $x = 1$

Use these values in E1.

$x = 4 : y = 2(4) - 4 = 4$

$x = 1 : y = 2(1) - 4 = -2$

The solutions are (4, 4) and (1, –2).

68. $\begin{cases} x^2 + y^2 = 2 & (1) \\ x - y = 4 & (2) \end{cases}$

Solve E2 for x: $x = y + 4$.

Substitute into E1.

$(y + 4)^2 + y^2 = 4$

$(y^2 + 8y + 16) + y^2 = 4$

$2y^2 + 8y + 12 = 0$

$y^2 + 4y + 6 = 0$

$y = \dfrac{-4 \pm \sqrt{(4)^2 - 4(1)(6)}}{2(1)} = \dfrac{-4 \pm \sqrt{-8}}{2}$

which yields no real solutions.

The solution is $\varnothing$.

69. $\begin{cases} y = x + 2 & (1) \\ y = x^2 & (2) \end{cases}$

Substitute $x + 2$ for y into E2.

$x + 2 = x^2$

$0 = x^2 - x - 2$

$0 = (x - 2)(x + 1)$

$x = 2$ or $x = -1$

Use these values in E1.

$x = 2 : y = x + 2 = 4$

$x = -1 : y = -1 + 2 = 1$

The solutions are (2, 4) and (–1, 1).

70. $\begin{cases} y = x^2 - 5x + 1 & (1) \\ y = -x + 6 & (2) \end{cases}$

Substitute $-x + 6$ for y into E2.

$-x + 6 = x^2 - 5x + 1$

$0 = x^2 - 4x - 5$

$0 = (x - 5)(x + 1)$

$x = 5$ or $x = -1$

Use these values in E2.

$x = 5 : y = -(5) + 6 = 1$

$x = -1 : y = -(-1) + 6 = 7$

The solutions are (5, 1) and (–1, 7).

71. $\begin{cases} 4x - y^2 = 0 & (1) \\ 2x^2 + y^2 = 16 & (2) \end{cases}$

Solve E1 for y^2: $y^2 = 4x$.

Substitute into E2.

$2x^2 + 4x = 16$

$2x^2 + 4x - 16 = 0$

$x^2 + 2x - 8 = 0$

$(x + 4)(x - 2) = 0$

$x = -4$ or $x = 2$

Use these values in the equation

$y^2 = 4x$.

$x = -4 : y^2 = 4(-4)$

$\quad\quad\quad y^2 = -16$ (no real solutions)

$x = 2 : y^2 = 4(2)$
$$y^2 = 8$$
$$y = \pm\sqrt{8} = \pm 2\sqrt{2}$$

The solutions are $\left(2, -2\sqrt{2}\right)$ and $\left(2, 2\sqrt{2}\right)$.

72. $\begin{cases} x^2 + 4y^2 = 16 & (1) \\ x^2 + y^2 = 4 & (2) \end{cases}$

Multiply E2 by -1 and add to E1.

$-x^2 - y^2 = -4$
$\underline{x^2 + 4y^2 = 16}$
$3y^2 = 12$
$y^2 = 4$
$y = \pm 2$

Replace y^2 with 4 into E2.

$x^2 + 4 = 4$
$x^2 = 0$
$x = 0$

The solutions are (0, 2) and (0, –2).

73. $\begin{cases} x^2 + y^2 = 10 & (1) \\ 9x^2 + y^2 = 18 & (2) \end{cases}$

Multiply E1 by -1 and add to E2.

$-x^2 - y^2 = -10$
$\underline{9x^2 + y^2 = 18}$
$8x^2 \quad\quad = 8$
$x^2 = 1$
$x = \pm 1$

Replace x^2 with 1 into E1.

$1 + y^2 = 10$
$y^2 = 9$
$y = \pm 3$

The solutions are (–1, –3), (–1, 3), (1, –3) and (1, 3).

74. $\begin{cases} x^2 + 2y = 9 & (1) \\ 5x - 2y = 5 & (2) \end{cases}$

Add E1 and E2.

$x^2 + 5x = 14$
$x^2 + 5x - 14 = 0$
$(x + 7)(x - 2) = 0$
$x = -7 \text{ or } x = 2$

Use these values into E1.

$x = -7 : (-7)^2 + 2y = 9$
$49 + 2y = 9$
$2y = -40$
$y = -20$

$x = 2 : (2)^2 + 2y = 9$
$4 + 2y = 9$
$2y = 5$
$y = \dfrac{5}{2}$

The solutions are (–7, –20) and $\left(2, \dfrac{5}{2}\right)$.

75. $\begin{cases} y = 3x^2 + 5x - 4 & (1) \\ y = 3x^2 - x + 2 & (2) \end{cases}$

Multiply E1 by -1 and add to E2.

$0 = -6x + 6$
$6x = 6$
$x = 1$

Use this value in E1.

$y = 3(1)^2 + 5(1) - 4 = 4$

The solution is (1, 4).

76. $\begin{cases} x^2 - 3y^2 = 1 & (1) \\ 4x^2 + 5y^2 = 21 & (2) \end{cases}$

Multiply E1 by -4 and add to E2.

$-4x^2 + 12y^2 = -4$
$\underline{4x^2 + 5y^2 = 21}$
$17y^2 = 17$
$y^2 = 1$
$y = \pm 1$

Replace y^2 with 1 into E1.

$$x^2 - 3(1) = 1$$
$$x^2 = 4$$
$$x = \pm 2$$

The solutions are $(-2, -1)$, $(-2, 1)$, $(2, -1)$ and $(2, 1)$.

77. Let x and y be the length and width.
$$\begin{cases} xy = 150 \\ 2x + 2y = 50 \end{cases}$$

Solve the first equation for y: $y = \dfrac{150}{x}$.

Substitute into E2.
$$2x + 2\left(\frac{150}{x}\right) = 50$$
$$x + \frac{150}{x} = 25$$
$$x^2 + 150 = 25x$$
$$x^2 - 25x + 150 = 0$$
$$(x - 15)(x - 10) = 0$$
$$x = 15 \text{ or } x = 10$$

Substitute these values into E1.

$$\begin{aligned} 15y &= 150 & 10y &= 150 \\ y &= 10 & y &= 15 \end{aligned}$$

The room is 15 feet by 10 feet.

78. Four real solutions.

79. $y \leq -x^2 + 3$

Graph $y = -x^2 + 3$ with a solid curve.

Test Point	$y \leq -x^2 + 3$; Result
$(0, 0)$	$0 \leq -(0)^2 + 3$; True

Shade the portion of the graph which contains $(0, 0)$.

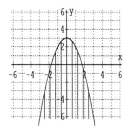

80. $x^2 + y^2 < 9$

First graph the circle with a dashed curve.

Test Point	$x^2 + y^2 < 9$; Result
$(0, 0)$	$0^2 + 0^2 < 9$; True

Shade the portion of the graph which contains $(0, 0)$.

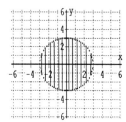

81. $x^2 - y^2 < 1$

First graph the hyperbola with a dashed curve.

Test Points	$x^2 - y^2 < 1$; Result
$(-2, 0)$	$(-2)^2 - 0^2 < 1$; False
$(0, 0)$	$0^2 - 0^2 < 1$; True
$(2, 0)$	$2^2 - 0^2 < 1$; False

Shade the portion of the graph that contains $(0, 0)$.

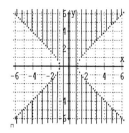

82. $\dfrac{x^2}{4} + \dfrac{y^2}{9} \geq 1$

First graph the ellipse with a solid curve.

Test Point	$\dfrac{x^2}{4} + \dfrac{y^2}{9} \geq 1$; Result
(0, 0)	$\dfrac{(0)^2}{4} + \dfrac{(0)^2}{9} \geq 1$; False

Shade the portion of the graph that does not contain (0, 0).

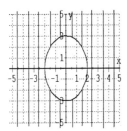

83. $\begin{cases} 2x \leq 4 \\ x + y \geq 1 \end{cases}$

First graph $2x = 4$, or $x = 2$, with a solid line, and shade to the left of the line.

Next, graph $x + y = 1$ with a solid line.

Test Point	$x + y \geq 1$; Result
(0, 0)	$0 + 0 \geq 1$; False

Shade the portion of the graph which does not contain (0, 0).

The solution to the system is the overlapping region.

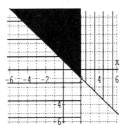

84. $\begin{cases} 3x + 4y \leq 12 \\ x - 2y > 6 \end{cases}$

First graph $3x + 4y = 12$ with a solid line.

Test Point	$3x + 4y \leq 12$; Result
(0, 0)	$3(0) + 4(0) \leq 12$; True

Shade the portion of the graph which contains (0, 0).

Next, graph $x - 2y = 6$ with a dashed line.

Test Point	$x - 2y > 6$; Result
(0, 0)	$0 - 2(0) > 6$; False

Shade the portion of the graph which does not contain (0, 0).

The solution to the system is the overlapping region.

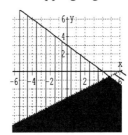

85. $\begin{cases} y > x^2 \\ x + y \geq 3 \end{cases}$

First graph the parabola with a dashed curve.

Test Point	$y > x^2$; Result
(0, 1)	$1 > 0^2$; True

Shade the portion of the graph which contains (0, 1).

Next, graph $x + y = 3$ with a solid line.

Test Point	$x + y \geq 3$; Result
(0, 0)	$0 + 0 \geq 3$; False

Shade the portion of the graph which does not contain (0, 0).

The solution to the system is the overlapping region.

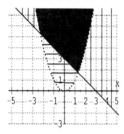

86. $\begin{cases} x^2 + y^2 \leq 16 \\ x^2 + y^2 \geq 4 \end{cases}$

First graph the first circle with a solid curve.

Test Point	$x^2 + y^2 \leq 16$; Result
(0, 0)	$0^2 + 0^2 \leq 16$; True

Shade the portion of the graph which contains (0, 0).

Next, graph the second circle with a solid curve.

Test Point	$x^2 + y^2 \geq 4$; Result
(0, 0)	$0^2 + 0^2 \geq 4$; False

Shade the portion of the graph which does not contain (0, 0).

The solution to the system is the overlapping region.

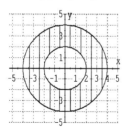

87. $\begin{cases} x^2 + y^2 < 4 \\ x^2 - y^2 \leq 1 \end{cases}$

First graph the first circle with a dashed curve.

Test Point	$x^2 + y^2 < 4$; Result
(0, 0)	$0^2 + 0^2 < 4$; True

Shade the portion of the graph which contains (0, 0).

Next, graph the hyperbola with a solid curve.

Test Point	$x^2 - y^2 \leq 1$; Result
(0, 0)	$0^2 - 0^2 \leq 1$; True

Shade the portion of the graph which contains (0, 0).

The solution to the system is the overlapping region.

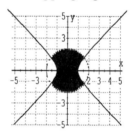

88. $\begin{cases} x^2 + y^2 < 4 \\ \quad y \geq x^2 - 1 \\ \quad x \geq 0 \end{cases}$

First graph the first circle with a dashed curve.

Test Point	$x^2 + y^2 < 4$; Result
(0, 0)	$0^2 + 0^2 < 4$; True

Shade the portion of the graph which contains (0, 0).

Next, graph the parabola with a solid curve.

Test Point	$y^2 \geq x^2 - 1$; Result
$(0, 0)$	$0 \geq 0^2 - 1$; True

Shade the portion of the graph which contains $(0, 0)$.

Now graph the line $x = 0$ with a solid line, and shade to the right.

The solution to the system is the overlapping region.

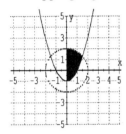

Chapter 13 Test

1. $(-6, 3), (-8, -7)$

$$d = \sqrt{[-8 - (-6)]^2 + (-7 - 3)^2}$$
$$= \sqrt{(-2)^2 + (-10)^2}$$
$$= \sqrt{4 + 100}$$
$$= \sqrt{104} = 2\sqrt{26} \text{ units}$$

2. $\left(-2\sqrt{5}, \sqrt{10}\right), \left(-\sqrt{5}, 4\sqrt{10}\right)$

$$d = \sqrt{\left[-\sqrt{5} - \left(-2\sqrt{5}\right)\right]^2 + \left(4\sqrt{10} - \sqrt{10}\right)^2}$$
$$= \sqrt{\left(\sqrt{5}\right)^2 + \left(3\sqrt{10}\right)^2}$$
$$= \sqrt{5 + 9 \cdot 10}$$
$$= \sqrt{95} \text{ units}$$

3. $(-2, -5), (-6, 12)$

$$\left(\frac{-2 + (-6)}{2}, \frac{-5 + 12}{2}\right) = \left(\frac{-8}{2}, \frac{7}{2}\right) = \left(-4, \frac{7}{2}\right)$$

The midpoint is $\left(-4, \frac{7}{2}\right)$.

4. $x^2 + y^2 = 36$

Circle; center: $(0, 0)$, radius $r = \sqrt{36} = 6$

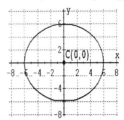

5. $x^2 - y^2 = 36$

$$\frac{x^2}{36} - \frac{y^2}{36} = 1$$

Hyperbola; center: $(0, 0)$, $a = 6$, $b = 6$

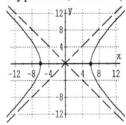

6. $16x^2 + 9y^2 = 144$

$$\frac{x^2}{9} + \frac{y^2}{16} = 1$$

Ellipse; center: $(0, 0)$, $a = 3$, $b = 4$

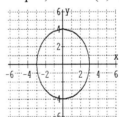

7. $y = x^2 - 8x + 16$
$$= (x - 4)^2$$

Parabola; vertex: (4, 0)

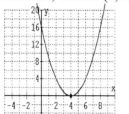

8.
$$x^2 + y^2 + 6x = 16$$
$$(x^2 + 6x) + y^2 = 16$$
$$(x^2 + 6x + 9) + y^2 = 16 + 9$$
$$(x + 3)^2 + y^2 = 25$$

Circle; center: (−3, 0),

radius $r = \sqrt{25} = 5$

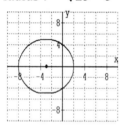

9.
$$x = y^2 + 8y - 3$$
$$x + 16 = (y^2 + 8y + 16) - 3$$
$$x = (y + 4)^2 - 19$$

Parabola; vertex: (−4, −19)

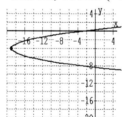

10. $\dfrac{(x-4)^2}{16} + \dfrac{(y-3)^2}{9} = 1$

Ellipse: center: (4, 3), $a = 4$, $b = 3$

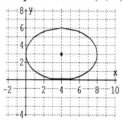

11. $y^2 - x^2 = 1$

Hyperbola: center: (0, 0), $a = 1$, $b = 1$

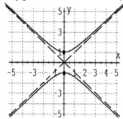

12. $\begin{cases} x^2 + y^2 = 26 & (1) \\ x^2 - 2y^2 = 23 & (2) \end{cases}$

Solve E1 for x^2: $x^2 = 26 - y^2$.

Substitute into E2.
$$(26 - y^2) - 2y^2 = 23$$
$$-3y^2 = -3$$
$$y^2 = 1$$
$$y = \pm 1$$

Replace y^2 with 1 in E1.
$$x^2 + 1 = 26$$
$$x^2 = 25$$
$$x = \pm 5$$

The solutions are (−5, −1), (−5, 1), (5, −1), and (5, 1).

13. $\begin{cases} y = x^2 - 5x + 6 & (1) \\ y = 2x & (2) \end{cases}$

Substitute 2x for y in E1.

$$2x = x^2 - 5x + 6$$
$$0 = x^2 - 7x + 6$$
$$0 = (x - 6)(x - 1)$$

$x = 6$ or $x = 1$

Use these values in E2.

$x = 6 : y = 2(6) = 12$

$x = 1 : y = 2(1) = 2$

The solutions are $(1, 2)$ and $(6, 12)$.

14. $\begin{cases} 2x + 5y \geq 10 \\ \quad\quad y \geq x^2 + 1 \end{cases}$

First graph $2x + 5y = 10$ with a solid line.

Test Point	$2x + 5y \geq 10$; Result
$(0, 0)$	$2(0) + 5(0) \geq 10$; False

Shade the portion of the graph which does not contain $(0, 0)$.

Next, graph $y = x^2 + 1$ with a solid curve.

Test Point	$y \geq x^2 + 1$; Result
$(0, 0)$	$0 \geq 0^2 + 1$; False

Shade the portion of the graph which does not contain $(0, 0)$.

The solution to the system is the overlapping region.

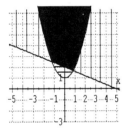

15. $\begin{cases} \dfrac{x^2}{4} + y^2 \leq 1 \\ \quad x + y > 1 \end{cases}$

First graph the ellipse with a solid curve.

Test Point	$\dfrac{x^2}{4} + y^2 \leq 1$; Result
$(0, 0)$	$\dfrac{0^2}{4} + 0^2 \leq 1$; True

Shade the portion of the graph which contains $(0, 0)$.

Next, graph $x + y = 1$ with a solid line.

Test Point	$x + y > 1$; Result
$(0, 0)$	$0 + 0 > 1$; False

Shade the portion of the graph which does not contain $(0, 0)$.

The solution to the system is the overlapping region.

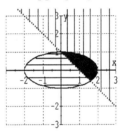

16. $\begin{cases} x^2 + y^2 \geq 4 \\ x^2 + y^2 < 16 \\ \quad\quad y \geq 0 \end{cases}$

First graph the circle $x^2 + y^2 = 4$ with a solid curve.

Test Point	$x^2 + y^2 \geq 4$; Result
$(0, 0)$	$0^2 + 0^2 \geq 4$; False

Shade the portion of the graph which does not contain $(0, 0)$.

Next graph the circle $x^2 + y^2 = 16$ with a dashed curve.

Test Point	$x^2 + y^2 < 16$; Result
$(0, 0)$	$0^2 + 0^2 < 16$; True

Shade the portion of the graph which contains (0, 0).

Now graph the inequality $y = 0$ by shading the region above the x-axis.

The solution to the system is the overlapping region.

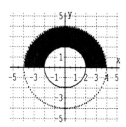

17. $100x^2 + 225y^2 = 22,500$

$$\frac{x^2}{225} + \frac{y^2}{100} = 1$$

$a = \sqrt{225} = 15$

$b = \sqrt{100} = 10$

Width = 15 + 15 = 30 feet

Height = 10 feet

Chapter 13 Cumulative Review

1. $2x \geq 0$ and $4x - 1 \leq -9$

$\quad x \geq 0$ and $\quad 4x \leq -8$

$\quad x \geq 0$ and $\quad\quad x \leq -2$

$\varnothing$

2. $3x + 4 > 1$ and $2x - 5 \leq 9$

$\quad 3x > -3$ and $\quad 2x \leq 14$

$\quad\quad x > -1$ and $\quad\quad x \leq 7$

$-1 < x \leq 7$

$(-1, 7]$

3. $5x - 3 \leq 10$ or $x + 1 \geq 5$

$\quad 5x \leq 13$ or $\quad x \geq 4$

$\quad x \leq \dfrac{13}{5}$ or $\quad x \geq 4$

$\left(-\infty, \dfrac{13}{5}\right] \cup [4, \infty)$

4. $(3, 2), (1, -4)$

$$m = \frac{-4 - 2}{1 - 3} = \frac{-6}{-2} = 3$$

5. $|5w + 3| = 7$

$\quad 5w + 3 = 7$ or $5w + 3 = -7$

$\quad\quad 5w = 4$ $\quad\quad\quad 5w = -10$

$\quad\quad w = \dfrac{4}{5}$ $\quad\quad\quad w = -2$

6. Let x = speed of one plane. Then $x + 25$ = speed of the other plane.

$d_{\text{plane 1}} + d_{\text{plane 2}} = 650$ miles

$2x + 2(x + 25) = 650$

$2x + 2x + 50 = 650$

$4x = 600$

$x = 150$

$x + 25 = 150 + 25 = 175$

The planes are traveling at 150 mph and 175 mph.

7. $\left|\dfrac{x}{2} - 1\right| = 11$

$\dfrac{x}{2} - 1 = 11$ or $\dfrac{x}{2} - 1 = -11$

$x - 2 = 22$ $\quad\quad x - 2 = -22$

$x = 24$ $\quad\quad\quad x = -20$

c. $\left(\dfrac{3p^4}{q^5}\right)^2 = \dfrac{3^2(p^4)^2}{(q^5)^2} = \dfrac{9p^8}{q^{10}}$

d. $\left(\dfrac{2^{-3}}{y}\right)^{-2} = \left(\dfrac{1}{2^3 y}\right)^{-2}$

$\qquad = \dfrac{1^{-2}}{2^{-6}y^{-2}} = \dfrac{2^6 y^2}{1} = 64y^2$

e. $(x^{-5}y^2 z^{-1})^7 = x^{-35}y^{14}z^{-7} = \dfrac{y^{14}}{x^{35}z^7}$

8. a. $\dfrac{4^8}{4^3} = 4^{8-3} = 4^5$

b. $\dfrac{y^{11}}{y^5} = y^{11-5} = y^6$

c. $\dfrac{32x^7}{4x^6} = 8x^{7-6} = 8x$

d. $\dfrac{18a^{12}b^6}{12a^8 b^6} = \dfrac{3a^{12-8}b^{6-6}}{2} = \dfrac{3a^4 b^0}{2} = \dfrac{3a^4}{2}$

9. $|3x+2| = |5x-8|$

$\quad 3x+2 = 5x-8 \quad$ or $\quad 3x-2 = -(5x-8)$

$\quad -2x+2 = -8 \qquad\qquad 3x-2 = -5x+8$

$\quad\; -2x = -10 \qquad\qquad\;\; 8x-2 = 8$

$\qquad\quad x = 5 \qquad\qquad\qquad\;\; 8x = 10$

$\qquad\qquad\qquad\qquad\qquad\qquad\;\; x = \dfrac{5}{4}$

10. a. $3y^2 + 14y + 15 = (3y+5)(y+3)$

b. $20a^5 + 54a^4 + 10a^3$

$\quad = 2a^3(10a^2 + 27a + 5)$

$\quad = 2a^3(2a+5)(5a+1)$

c. $(y-3)^2 - 2(y-3) - 8$

Let $u = y-3$. Then $u^2 = (y-3)^2$ and

$u^2 - 2u - 8 = (u-4)(u+2)$

$\qquad\qquad\quad = [(y-3)-4][(y-3)+2]$

$\qquad\qquad\quad = (y-7)(y-1)$

11. $|m-6| < 2$

$\quad -2 < m-6 < 2$

$\quad\; 4 < m < 8$

$\quad (4, 8)$

12. $\dfrac{2}{3a-15} - \dfrac{a}{25-a^2}$

$= \dfrac{2}{3(a-5)} + \dfrac{a}{a^2-25}$

$= \dfrac{2}{3(a-5)} + \dfrac{a}{(a+5)(a-5)}$

$= \dfrac{2(a+5)+3a}{3(a+5)(a-5)}$

$= \dfrac{2a+10+3a}{3(a+5)(a-5)}$

$= \dfrac{5a+10}{3(a+5)(a-5)}$

13. $\dfrac{x^{-1}+2xy^{-1}}{x^{-2}-x^{-2}y^{-1}} = \dfrac{\frac{1}{x}+\frac{2x}{y}}{\frac{1}{x^2}-\frac{1}{x^2 y}}$

$\dfrac{\left(\frac{1}{x}+\frac{2x}{y}\right)\left(x^2 y\right)}{\left(\frac{1}{x^2}-\frac{1}{x^2 y}\right)\left(x^2 y\right)} = \dfrac{xy+2x^3}{y-1}$

14. a. $(a^{-1} - b^{-1})^{-1} = \left(\dfrac{1}{a} - \dfrac{1}{b}\right)^{-1}$

$$= \left(\dfrac{b-a}{ab}\right)^{-1}$$

$$= \dfrac{ab}{b-a}$$

b. $\dfrac{2 - \dfrac{1}{x}}{4x - \dfrac{1}{x}} = \dfrac{\left(2 - \dfrac{1}{x}\right)x}{\left(4x - \dfrac{1}{x}\right)x}$

$$= \dfrac{2x - 1}{4x^2 - 1}$$

$$= \dfrac{2x - 1}{(2x+1)(2x-1)}$$

$$= \dfrac{1}{2x + 1}$$

15. $|2x + 9| + 5 > 3$

$$|2x + 9| > -2$$

$$(-\infty, \infty)$$

16. $\dfrac{2}{x+3} = \dfrac{1}{x^2 - 9} - \dfrac{1}{x - 3}$

$$\dfrac{2}{x+3} = \dfrac{1}{(x+3)(x-3)} - \dfrac{1}{x-3}$$

$$2(x - 3) = 1 - 1(x + 3)$$

$$2x - 6 = 1 - x - 3$$

$$2x - 6 = -x - 2$$

$$3x = 4$$

$$x = \dfrac{4}{3}$$

17.

$$\underline{4}\big|\;\begin{array}{ccccccc} 4 & -25 & 35 & 0 & 17 & 0 & 0 \\ & 16 & -36 & -4 & -16 & 4 & 16 \\ \hline 4 & -9 & -1 & -4 & 1 & 4 & 16 \end{array}$$

Thus, $P(4) = 16$.

18. $y = \dfrac{k}{x}$

$$3 = \dfrac{k}{\dfrac{2}{3}}$$

$$k = 3\left(\dfrac{2}{3}\right) = 2$$

Thus, the equation is $y = \dfrac{2}{x}$.

It will take them $2\dfrac{2}{9}$ hours. No, they can not finish before the movie starts.

19. a. $\sqrt[3]{1} = 1$

b. $\sqrt[3]{-64} = -4$

c. $\sqrt[3]{\dfrac{8}{125}} = \dfrac{\sqrt[3]{8}}{\sqrt[3]{125}} = \dfrac{2}{5}$

d. $\sqrt[3]{x^6} = x^2$

e. $\sqrt[3]{-27x^9} = -3x^3$

20. a. $\sqrt{5}\left(2+\sqrt{15}\right) = 2\sqrt{5} + \sqrt{5}\cdot\sqrt{15}$
$$= 2\sqrt{5} + \sqrt{75}$$
$$= 2\sqrt{5} + 5\sqrt{3}$$

b. $\left(\sqrt{3}-\sqrt{5}\right)\left(\sqrt{7}-1\right)$
$$= \sqrt{3}\cdot\sqrt{7} - \sqrt{3}\cdot 1 - \sqrt{5}\cdot\sqrt{7} + \sqrt{5}\cdot 1$$
$$= \sqrt{21} - \sqrt{3} - \sqrt{35} + \sqrt{5}$$

c. $\left(2\sqrt{5}-1\right)^2 = \left(2\sqrt{5}\right)^2 - 2\cdot 2\sqrt{5}\cdot 1 + 1^2$
$$= 4(5) - 4\sqrt{5} + 1$$
$$= 21 - 4\sqrt{5}$$

d. $\left(3\sqrt{2}+5\right)\left(3\sqrt{2}-5\right) = \left(3\sqrt{2}\right)^2 - 5^2$
$$= 9(2) - 25$$
$$= 18 - 25$$
$$= -7$$

21. a. $z^{2/3}\left(z^{1/3}-z^5\right) = z^{2/3+1/3} - z^{2/3+5}$
$$= z^{3/3} - z^{2/3+15/3}$$
$$= z - z^{17/3}$$

b. $(x^{1/3}-5)(x^{1/3}+2)$
$$= x^{1/3}\cdot x^{1/3} + 2x^{1/3} - 5x^{1/3} - 5(2)$$
$$= x^{2/3} - 3x^{1/3} - 10$$

22. $\dfrac{-2}{\sqrt{3}+3} = \dfrac{-2\left(\sqrt{3}-3\right)}{\left(\sqrt{3}+3\right)\left(\sqrt{3}-3\right)}$

$$= \dfrac{-2\left(\sqrt{3}-3\right)}{\left(\sqrt{3}\right)^2 - 3^2}$$

$$= \dfrac{-2\left(\sqrt{3}-3\right)}{3-9}$$

$$= \dfrac{-2\left(\sqrt{3}-3\right)}{-6} = \dfrac{\sqrt{3}-3}{3}$$

23. a. $\dfrac{\sqrt{20}}{\sqrt{5}} = \sqrt{\dfrac{20}{5}} = \sqrt{4} = 2$

b. $\dfrac{\sqrt{50x}}{2\sqrt{2}} = \dfrac{1}{2}\sqrt{\dfrac{50x}{2}} = \dfrac{1}{2}\sqrt{25x} = \dfrac{5\sqrt{x}}{2}$

c. $\dfrac{7\sqrt[3]{48x^4 y^8}}{\sqrt[3]{6y^2}} = 7\sqrt[3]{\dfrac{48x^4 y^8}{6y^2}}$
$$= 7\sqrt[3]{8x^4 y^6}$$
$$= 7\sqrt[3]{8x^3 y^6 \cdot x}$$
$$= 7\cdot 2xy^2 \sqrt[3]{x}$$
$$= 14xy^2 \sqrt[3]{x}$$

d. $\dfrac{2\sqrt[4]{32a^8 b^6}}{\sqrt[4]{a^{-1} b^2}} = 2\sqrt[4]{\dfrac{32a^8 b^6}{a^{-1} b^2}}$
$$= 2\sqrt[4]{32a^8 b^4}$$
$$= 2\sqrt[4]{16a^8 b^4 \cdot 2}$$
$$= 2\cdot 2a^2 b\sqrt[4]{2}$$
$$= 4a^2 b\sqrt[4]{2}$$

24. $\sqrt{2x-3} = x-3$
$$2x-3 = (x-3)^2$$
$$2x-3 = x^2 - 6x + 9$$
$$0 = x^2 - 8x + 12$$
$$0 = (x-6)(x-2)$$

$x - 6 = 0$ or $x - 2 = 0$

$x = 6$ or $x = 2$

Discard 2 as an extraneous solution.

The solution is 6.

25. a. $\dfrac{\sqrt{45}}{4} - \dfrac{\sqrt{5}}{3} = \dfrac{3\sqrt{5}}{4} - \dfrac{\sqrt{5}}{3}$

$= \dfrac{9\sqrt{5} - 4\sqrt{5}}{12}$

$= \dfrac{5\sqrt{5}}{12}$

b. $\sqrt[3]{\dfrac{7x}{8}} + 2\sqrt[3]{7x} = \dfrac{\sqrt[3]{7x}}{2} + 2\sqrt[3]{7x}$

$= \dfrac{\sqrt[3]{7x}}{2} + \dfrac{4\sqrt[3]{7x}}{2}$

$= \dfrac{5\sqrt[3]{7x}}{2}$

26. $9x^2 - 6x = -4$

$9x^2 - 6x + 4 = 0$

$a = 9,\ b = -6,\ c = 4$

$b^2 - 4ac = (-6)^2 - 4(9)(4)$

$= 36 - 144$

$= -108$

Two complex but not real solutions

27. $\sqrt{\dfrac{7x}{3y}} = \dfrac{\sqrt{7x}}{\sqrt{3y}} = \dfrac{\sqrt{7x} \cdot \sqrt{3y}}{\sqrt{3y} \cdot \sqrt{3y}} = \dfrac{\sqrt{21xy}}{3y}$

28. $\dfrac{4}{x - 2} - \dfrac{x}{x + 2} = \dfrac{16}{x^2 - 4}$

$\dfrac{4}{x - 2} - \dfrac{x}{x + 2} = \dfrac{16}{(x + 2)(x - 2)}$

$4(x + 2) - x(x - 2) = 16$

$4x + 8 - x^2 + 2x = 16$

$0 = x^2 - 6x + 8$

$0 = (x - 4)(x - 2)$

$x - 4 = 0$ or $x - 2 = 0$

$x = 4$ or $x = 2$

Discard the solutions 2 as extraneous.

The solution is 4.

29. $\sqrt{2x - 3} = 9$

$2x - 3 = 9^2$

$2x - 3 = 81$

$2x = 84$

$x = 42$

The solution is 42.

30. $x^3 + 2x^2 - 4x \geq 8$

$x^3 + 2x^2 - 4x - 8 \geq 0$

$x^2(x + 2) - 4(x + 2) \geq 0$

$(x + 2)(x^2 - 4) \geq 0$

$(x + 2)(x + 2)(x - 2) \geq 0$

$(x + 2)^2(x - 2) \geq 0$

$(x + 2)^2 = 0$ or $x - 2 = 0$

$x + 2 = 0$ or $x = 2$

$x = -2$

Region	Test Point	$(x + 2)^2(x - 2) \geq 0$ Result
A: $(-\infty, -2)$	-3	$(-1)^2(-5) \geq 0$ False
B: $(-2, 2)$	0	$(2)^2(-2) \geq 0$ False
C: $(2, \infty)$	3	$(5)^2(1) \geq 0$ True

Solution: $(2, \infty)$

31. a. $i^7 = i^4 \cdot i^3 = 1 \cdot (-i) = -i$

b. $i^{20} = (i^4)^5 = 1^5 = 1$

c. $i^{46} = i^{44} \cdot i^2 = (i^4)^{11} \cdot (-1) = 1^{11}(-1) = -1$

d. $i^{-12} = \dfrac{1}{i^{12}} = \dfrac{1}{(i^4)^3} = \dfrac{1}{1^3} = 1$

32. $f(x) = (x+2)^2 - 1$

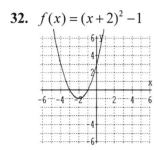

33.
$$p^2 + 2p = 4$$
$$p^2 + 2p + \left(\frac{2}{2}\right)^2 = 4 + 1$$
$$p^2 + 2p + 1 = 5$$
$$(p+1)^2 = 5$$
$$p + 1 = \pm\sqrt{5}$$
$$p = -1 \pm \sqrt{5}$$

The solutions are $-1 + \sqrt{5}$ and $-1 - \sqrt{5}$.

34. $f(x) = -x^2 - 6x + 4$

The maximum will occur at the vertex.
$$x = -\frac{b}{2a} = \frac{-(-6)}{2(-1)} = -3$$
$$f(-3) = -(-3)^2 - 6(-3) + 4 = 13$$

The maximum value is 13.

35.
$$\frac{1}{4}m^2 - m + \frac{1}{2} = 0$$
$$4\left(\frac{1}{4}m^2 - m + \frac{1}{2}\right) = 4(0)$$
$$m^2 - 4m + 2 = 0$$
$$a = 1, b = -4, c = 2$$
$$x = \frac{4 \pm \sqrt{(-4)^2 - 4(1)(2)}}{2(1)}$$
$$= \frac{4 \pm \sqrt{16 - 8}}{2}$$
$$= \frac{4 \pm \sqrt{8}}{2}$$
$$= \frac{4 \pm 2\sqrt{2}}{2} = 2 \pm \sqrt{2}$$

The solutions are $2 + \sqrt{2}$ and $2 - \sqrt{2}$.

36.
$$f(x) = \frac{x+1}{2}$$
$$y = \frac{x+1}{2}$$
$$x = \frac{y+1}{2}$$
$$2x = y + 1$$
$$2x - 1 = y$$
$$f^{-1}(x) = 2x - 1$$

37.
$$p^4 - 3p^2 - 4 = 0$$
$$(p^2 - 4)(p^2 + 1) = 0$$
$$(p+2)(p-2)(p^2 + 1) = 0$$
$$p + 2 = 0 \quad \text{or} \quad p - 2 = 0 \quad \text{or} \quad p^2 + 1 = 0$$
$$p = -2 \quad \text{or} \quad p = 2 \quad \text{or} \quad p^2 = -1$$
$$p = \pm i$$

The solutions are $-2, 2, -i, i$.

38. $f(x) = x^2 - 3x + 2$
$g(x) = -3x + 5$

a. $(f \circ g)(x) = f[g(x)]$
$$= f(-3x + 5)$$
$$= (-3x + 5)^2 - 3(-3x + 5) + 2$$
$$= 9x^2 - 30x + 25 + 9x - 15 + 2$$
$$= 9x^2 - 21x + 12$$

b. $(f \circ g)(-2) = f[g(-2)]$
$$= f[-3(-2) + 5]$$
$$= f(11)$$
$$= (11)^2 - 3(11) + 2$$
$$= 121 - 33 + 2$$
$$= 90$$

c. $(g \circ f)(x) = g[f(x)]$
$$= g(x^2 - 3x + 2)$$
$$= -3(x^2 - 3x + 2) + 5$$
$$= -3x^2 + 9x - 6 + 5$$
$$= -3x^2 + 9x - 1$$

d. $(g \circ f)(5) = g[f(5)]$
$= g[(5)^2 - 3(5) + 2)]$
$= g(12)$
$= -3(12) + 5$
$= -36 + 5$
$= -31$

39. $\dfrac{x+2}{x-3} \le 0$
$x + 2 = 0 \quad$ or $\quad x - 3 = 0$
$x = -2 \quad$ or $\qquad x = 3$

Region	Test Point	$\dfrac{x+2}{x-3} \le 0$ Result
A: $(-\infty, -2)$	-3	$\dfrac{-1}{-6} \le 0$; False
B: $(-2, 3)$	0	$\dfrac{2}{-3} \le 0$; True
C: $(3, \infty)$	4	$\dfrac{6}{1} \le 0$; False

Solution: $[-2, 3)$

40. $4x^2 + 9y^2 = 36$
$\dfrac{x^2}{9} + \dfrac{y^2}{4} = 1$
Ellipse: center $(0, 0)$, $a = 3$, $b = 2$

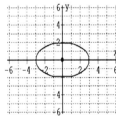

41. $g(x) = \dfrac{1}{2}(x+2)^2 + 5$
Vertex: $(-2, 5)$, axis: $x = -2$

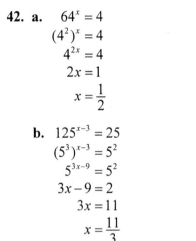

42. a. $64^x = 4$
$(4^2)^x = 4$
$4^{2x} = 4$
$2x = 1$
$x = \dfrac{1}{2}$

b. $125^{x-3} = 25$
$(5^3)^{x-3} = 5^2$
$5^{3x-9} = 5^2$
$3x - 9 = 2$
$3x = 11$
$x = \dfrac{11}{3}$

c. $\dfrac{1}{81} = 3^{2x}$
$81^{-1} = 3^{2x}$
$(3^4)^{-1} = 3^{2x}$
$3^{-4} = 3^{2x}$
$-4 = 2x$
$-\dfrac{4}{2} = x$
$-2 = x$

43. $f(x) = x^2 - 4x - 12$
$x = -\dfrac{b}{2a} = \dfrac{-(-4)}{2(1)} = 2$
$f(2) = (2)^2 - 4(2) - 12 = -16$
Vertex: $(2, -16)$

44. $\begin{cases} x + 2y < 8 \\ y \geq x^2 \end{cases}$

First, graph $x + 2y = 8$ with a dashed line.

Test Point	$x + 2y > 8$; Result
(0, 0)	$0 + 2(0) > 8$; False

Shade the portion of the graph which does not contain (0, 0).

Next, graph the parabola $y = x^2$ with a solid curve.

Test Point	$y \geq x^2$; Result
(0, 1)	$1 \geq 0^2$; True

Shade the portion of the graph which contains (0, 1).

The solution to the system is the overlapping region.

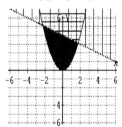

45. (2, –5), (1, –4)

$$d = \sqrt{[-4 - (-5)]^2 + (1 - 2)^2}$$
$$= \sqrt{1^2 + (-1)^2}$$
$$= \sqrt{2} \approx 1.414$$

46. $\begin{cases} x^2 + y^2 = 36 & (1) \\ y = x + 6 & (2) \end{cases}$

Substitute $x + 6$ for y in E1.
$$x^2 + (x + 6)^2 = 36$$
$$x^2 + (x^2 + 12x + 36) = 36$$
$$2x^2 + 12x = 0$$
$$2x(x + 6) = 0$$
$$2x = 0 \text{ or } x + 6 = 0$$
$$x = 0 \text{ or } x = -6$$

Use these values in E2 to find y.
$$x = 0 : y = 0 + 6 = 6$$
$$x = -6 : y = -6 + 6 = 0$$

The solutions are (0, 6) and (–6, 0).

Chapter 14

Exercise Set 14.1

1. $a_n = n + 4$
$a_1 = 1 + 4 = 5$
$a_2 = 2 + 4 = 6$
$a_3 = 3 + 4 = 7$
$a_4 = 4 + 4 = 8$
$a_5 = 5 + 4 = 9$

Thus, the first five terms of the sequence $a_n = n + 4$ are 5, 6, 7, 8, 9.

3. $a_n = (-1)^n$

$a_1 = (-1)^1 = -1$
$a_2 = (-1)^2 = 1$
$a_3 = (-1)^3 = -1$
$a_4 = (-1)^4 = 1$
$a_5 = (-1)^5 = -1$

Thus, the first five terms of the sequence $a_n = (-1)^n$ are $-1, 1, -1, 1, -1$.

5. $a_n = \dfrac{1}{n+3}$

$a_1 = \dfrac{1}{1+3} = \dfrac{1}{4}$

$a_2 = \dfrac{1}{2+3} = \dfrac{1}{5}$

$a_3 = \dfrac{1}{3+3} = \dfrac{1}{6}$

$a_4 = \dfrac{1}{4+3} = \dfrac{1}{7}$

$a_5 = \dfrac{1}{5+3} = \dfrac{1}{8}$

Thus, the first five terms of the sequence $a_n = \dfrac{1}{n+3}$ are $\dfrac{1}{4}, \dfrac{1}{5}, \dfrac{1}{6}, \dfrac{1}{7}, \dfrac{1}{8}$.

7. $a_n = 2n$
$a_1 = 2(1) = 2$
$a_2 = 2(2) = 4$
$a_3 = 2(3) = 6$
$a_4 = 2(4) = 8$
$a_5 = 2(5) = 10$
or 2, 4, 6, 8, 10.

9. $a_n = -n^2$
$a_1 = -1^2 = -1$
$a_2 = -2^2 = -4$
$a_3 = -3^2 = -9$
$a_4 = -4^2 = -16$
$a_5 = -5^2 = -25$

Thus, the first five terms of the sequence $a_n = -n^2$ are -1, -4, -8, -16, -25.

11. $a_n = 2^n$
$a_1 = 2^1 = 2$
$a_2 = 2^2 = 4$
$a_3 = 2^3 = 8$
$a_4 = 2^4 = 16$
$a_5 = 2^5 = 32$

Thus, the first five terms of the sequence $a_n = 2^n$ are 2, 4, 8, 16, 32.

13. $a_n = 2n + 5$
$a_1 = 2(1) + 5 = 2 + 5 = 7$
$a_2 = 2(2) + 5 = 4 + 5 = 9$
$a_3 = 2(3) + 5 = 6 + 5 = 11$

$$a_4 = 2(4) + 5 = 8 + 5 = 13$$
$$a_5 = 2(5) + 5 = 10 + 5 = 15$$

Thus, the first five teerms of the sequence $a_n = 2n + 5$ are 7, 9, 11, 13, 15.

15. $a_n = (-1)^n n^2$

$$a_1 = (-1)^1 (1)^1 = -1(1) = -1$$
$$a_2 = (-1)^2 (2)^2 = 1(4) = 4$$
$$a_3 = (-1)^3 (3)^2 = -1(9) = -9$$
$$a_4 = (-1)^4 (4)^2 = 1(16) = 16$$
$$a_5 = (-1)^5 (5)^2 = -1(25) = -25$$

Thus, the first five terms of the sequence $a_n = (-1)^n n^2$ are -1, 4, -9, 16, -25.

17. $a_n = 3n^2$
$$a_5 = 3(5)^2 = 3(25) = 75$$

19. $a_n = 6n - 2$
$$a_{20} = 6(20) - 2 = 120 - 2 = 118$$

21. $a_n = \dfrac{n+3}{n}$
$$a_{15} = \dfrac{15+3}{15} = \dfrac{18}{15} = \dfrac{6}{5}$$

23. $a_n = (-3)^n$
$$a_6 = (-3)^6 = 729$$

25. $a_n = \dfrac{n-2}{n+1}$
$$a_6 = \dfrac{6-2}{6+1} = \dfrac{4}{7}$$

27. $a_n = \dfrac{(-1)^n}{n}$
$$a_8 = \dfrac{(-1)^8}{8} = \dfrac{1}{8}$$

29. $a_n = -n^2 + 5$
$$a_{10} = -10^2 + 5 = -100 + 5 = -95$$

31. $a_n = \dfrac{(-1)^n}{n+6}$
$$a_{19} = \dfrac{(-1)^{19}}{19+6} = -\dfrac{1}{25}$$

33. 3, 7, 11, 15, or
$4(1) - 1$, $4(2) - 1$, $4(3) - 1$,
$4(4) - 1$. In general, $a_n = 4n - 1$

35. -2, -4, -8, -16, or -2, -2^2, -2^3, -2^4
In general, $a_n = -2^n$

37. $\dfrac{1}{3}, \dfrac{1}{9}, \dfrac{1}{27}, \dfrac{1}{81}$, or
$\dfrac{1}{3}, \dfrac{1}{3^2}, \dfrac{1}{3^3}, \dfrac{1}{3^4}$
In general, $a_n = \dfrac{1}{3^n}$

39. $a_n = 32n - 16$
$$a_2 = 32(2) - 16 = 64 - 16 = 48 \text{ ft}$$
$$a_3 = 32(3) - 16 = 96 - 16 = 80 \text{ ft}$$
$$a_4 = 32(4) - 16 = 128 - 16 = 112 \text{ ft}$$

41. 0.10, 0.20, 0.40, or
0.10, 0.10(2), $0.10(2)^2$
In general, $a_n = 0.10(2)^{n-1}$
$$a_{14} = 0.10(2)^{13} = \$819.20$$

43. $a_n = 75(2)^{n-1}$

$a_6 = 75(2)^5 = 75(32) = 2400$ cases

$a_1 = 75(2)^0 = 75(1) = 75$ cases

45. $a_n = \frac{1}{2}a_{n-1}$ for $n > 1, a_1 = 800$

In 2000, $n = 1$ and $a_1 = 800$.

In 2001, $n = 2$ and $a_2 = \frac{1}{2}(800) = 400$.

In 2002, $n = 3$ and $a_3 = \frac{1}{2}(400) = 200$.

In 2003, $n = 4$ and $a_4 = \frac{1}{2}(200) = 100$.

In 2004, $n = 5$ and $a_5 = \frac{1}{2}(100) = 50$.

The population estimaate for 2004 is 50 sparrows.

Continuing the sequence:

in 2005, $n = 6$ and $a_6 = \frac{1}{2}(50) = 25$;

in 2006, $n = 7$ and $a_7 = \frac{1}{2}(25) \approx 12$;

in 2007, $n = 8$ and $a_8 = \frac{1}{2}(12) = 6$;

in 2008, $n = 9$ and $a_9 = \frac{1}{2}(6) = 3$;

in 2009, $n = 10$ and $a_{10} = \frac{1}{2}(3) \approx 1$;

in 2010, $n = 11$ and $a_{11} = \frac{1}{2}(1) \approx 0$.

The population is estimated to become extinct in 2010.

47. $f(x) = (x-1)^2 + 3$

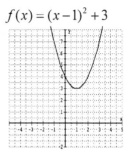

49. $f(x) = 2(x+4)^2 + 2$

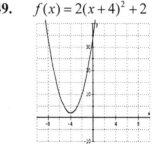

51. $(-4,-1)$ and $(-7,-3)$

$d = \sqrt{[-7-(-4)]^2 + [-3-(-1)]^2}$

$d = \sqrt{(-7+4)^2 + (-3+1)^2}$

$d = \sqrt{(-3)^2 + (-2)^2}$

$d = \sqrt{9+4} = \sqrt{13}$ units

53. $(2,-7)$ and $(-3,-3)$

$d = \sqrt{(-3-2)^2 + [-3-(-7)]^2}$

$d = \sqrt{(-5)^2 + (-3+7)^2}$

$d = \sqrt{(-5)^2 + (4)^2}$

$d = \sqrt{25+16} = \sqrt{41}$ units

55. $a_n = \dfrac{1}{\sqrt{n}}$

$a_1 = \dfrac{1}{\sqrt{1}} = \dfrac{1}{1} = 1$

$a_2 = \dfrac{1}{\sqrt{2}} \approx 0.7071$

$a_3 = \dfrac{1}{\sqrt{3}} \approx 0.5774$

$a_4 = \dfrac{1}{\sqrt{4}} = \dfrac{1}{2} = 0.5$

$a_5 = \dfrac{1}{\sqrt{5}} \approx 0.4472$

Thus, the first five terms of the sequence $a_n = \dfrac{1}{\sqrt{n}}$ are 1, 0.7071, 0.5774, 0.5, 0.4472.

57. $a_n = \left(1 + \dfrac{1}{n}\right)^n$

$a_1 = \left(1 + \dfrac{1}{1}\right)^1 = (2)^1 = 2$

$a_2 = \left(1 + \dfrac{1}{2}\right)^2 = \left(\dfrac{3}{2}\right)^2 = 2.25$

$a_3 = \left(1 + \dfrac{1}{3}\right)^3 = \left(\dfrac{4}{3}\right)^3 \approx 2.3704$

$a_4 = \left(1 + \dfrac{1}{4}\right)^4 = \left(\dfrac{5}{4}\right)^4 \approx 2.4414$

$a_5 = \left(1 + \dfrac{1}{5}\right)^5 = \left(\dfrac{6}{5}\right)^5 \approx 2.4883$

Thus, the first five terms of the sequence $a_n = \left(1 + \dfrac{1}{n}\right)^n$ are

Exercise Set 14.2

1. $a_n = a_1 + (n-1)d$

$a_1 = 4; \ d = 2$

$a_1 = 4$

$a_2 = 4 + (2-1)2 = 6$

$a_3 = 4 + (3-1)2 = 8$

$a_4 = 4 + (4-1)2 = 10$

$a_5 = 4 + (5-1)2 = 12$

The first five terms are 4, 6, 8, 10, 12.

3. $a_n = a_1 + (n-1)d$

$a_1 = 6, \ d = -2$

$a_1 = 6$

$a_2 = 6 + (2-1)(-2) = 4$

$a_3 = 6 + (3-1)(-2) = 2$

$a_4 = 6 + (4-1)(-2) = 0$

$a_5 = 6 + (5-1)(-2) = -2$

The first five terms are $6, 4, 2, 0, -2$.

5. $a_n = a_1 r^{n-1}$

$a_1 = 1, \ r = 3$

$a_1 = 1(3)^{1-1} = 1$

$a_2 = 1(3)^{2-1} = 3$

$a_3 = 1(3)^{3-1} = 9$

$a_4 = 1(3)^{4-1} = 27$

$a_5 = 1(3)^{5-1} = 81$

The first five terms are 1, 3, 9, 27, 81.

7. $a_n = a_1 r^{n-1}$

$a_1 = 48,\ r = \dfrac{1}{2}$

$a_1 = 48\left(\dfrac{1}{2}\right)^{1-1} = 48$

$a_2 = 48\left(\dfrac{1}{2}\right)^{2-1} = 24$

$a_3 = 48\left(\dfrac{1}{2}\right)^{3-1} = 12$

$a_4 = 48\left(\dfrac{1}{2}\right)^{4-1} = 6$

$a_5 = 48\left(\dfrac{1}{2}\right)^{5-1} = 3$

The first five terms are 48, 24, 12, 6, 3.

9. $a_n = a_1 + (n-1)d$

$a_1 = 12,\ d = 3$

$a_n = 12 + (n-1)3$

$a_8 = 12 + 7(3) = 12 + 21 = 33$

11. $a_n = a_1 r^{n-1}$

$a_1 = 7,\ d = 3$

$a_n = a_1 r^{n-1}$

$a_4 = 7(-5)^3 = 7(-125) = -875$

13. $a_n = a_1 + (n-1)d$

$a_1 = -4,\ d = -4$

$a_n = -4 + (n-1)(-4)$

$a_{15} = -4 + 14(-4) = -4 - 56 = -60$

15. 0, 12, 24

$a_1 = 0$ and $d = 12$

$a_n = 0 + (n-1)12$

$a_9 = 8(12) = 96$

17. 20, 18, 16

$a_1 = 20$ and $d = -2$

$a_n = 20 + (n-1)(-2)$

$a_{25} = 20 + 24(-2) = 20 - 48 = -28$

19. 2, −10, 50

$a_1 = 2$ and $r = -5$

$a_n = 2(-5)^{n-1}$

$a_5 = 2(-5)^4 = 2(625) = 1250$

21. $a_4 = 19,\ a_{15} = 52$

$\begin{cases} a_4 = a_1 + (4-1)d \\ a_{15} = a_1 + (15-1)d \end{cases}$ or

$\begin{cases} 19 = a_1 + 3d \\ 52 = a_1 + 14d \end{cases}$

Solving the system gives $d = 3$ and $a_1 = 10$.

$a_n = 10 + (n-1)3$

$= 10 + 3n - 3$

$= 7 + 3n$

and $a_8 = 7 + 3(8)$

$= 7 + 24$

$= 31$

23. $a_2 = -1, a_4 = 5$

$$\begin{cases} a_2 = a_1 + (2-1)d \\ a_4 = a_1 + (4-1)d \end{cases} \text{ or}$$

$$\begin{cases} -1 = a_1 + d \\ 5 = a_1 + 3d \end{cases}$$

Solving the system gives $d = 3$

and $a_1 = -4$.

$a_n = -4(n-1)3$

$\quad = -4 + 3n - 3$

$\quad = -7 + 3n$

and $a_9 = -7 + 3(9)$

$\qquad = -7 + 27$

$\qquad = 20$

25. $a_2 = -\dfrac{4}{3}$ and $a_3 = \dfrac{8}{3}$

Notice that $\dfrac{8}{3} \div \dfrac{-4}{3} = \dfrac{8}{3} \cdot -\dfrac{3}{4} = -2,$

so $r = -2$. Then

$a_2 = a_1(-2)^{2-1}$

$-\dfrac{4}{3} = a_1(-2)$

$\dfrac{2}{3} = a_1.$

The first term is $\dfrac{2}{3}$ and the

common ratio is -2.

27. Answers may vary.

29. $2, 4, 6$ is an arithmetic sequence.

$a_1 = 2$ and $d = 2$

31. $5, 10, 20$ is a geometric sequence.

$a_1 = 5$ and $r = 2$

33. $\dfrac{1}{2}, \dfrac{1}{10}, \dfrac{1}{50}$ is a geometric sequence.

35. $x, 5x, 25x$ is a geometric sequence.

$a_1 = x$ and $r = 5$

37. $p, p+4, p+8$ is an arithmetic sequence.

$a_1 = p$ and $d = 4$

39. $a_1 = 14$ and $d = \dfrac{1}{4}$

$a_n = 14 + (n-1)\dfrac{1}{4}$

$a_{21} = 14 + 20\left(\dfrac{1}{4}\right) = 14 + 5 = 19$

41. $a_1 = 3$ and $r = -\dfrac{2}{3}$

$a_n = 3\left(-\dfrac{2}{3}\right)^{n-1}$

$a_4 = 3\left(-\dfrac{2}{3}\right)^3 = 3\left(-\dfrac{8}{27}\right) = -\dfrac{8}{9}$

43. $\dfrac{3}{2}, 2, \dfrac{5}{2}, \ldots$

$a_1 = \dfrac{3}{2}$ and $d = \dfrac{1}{2}$

$a_n = \dfrac{3}{2} + (n-1)\dfrac{1}{2}$

$a_{15} = \dfrac{3}{2} + 14\left(\dfrac{1}{2}\right) = \dfrac{17}{2}$

45. $24, 8, \dfrac{8}{3}, \ldots$

$a_1 = 24$ and $r = \dfrac{1}{3}$

$a_n = 24\left(\dfrac{1}{3}\right)^{n-1}$

$a_6 = 24\left(\dfrac{1}{3}\right)^{5} = 24\left(\dfrac{1}{243}\right) = \dfrac{8}{81}$

47. $a_3 = 2$, $a_{17} = -40$

$\begin{cases} a_3 = a_1 + (3-1)d \\ a_{17} = a_1 + (17-1)d \end{cases}$ or

$\begin{cases} 2 = a_1 + 2d \\ -40 = a_1 + 16d \end{cases}$

Solving the system gives $d = -3$
and $a_1 = 8$.

$a_n = 8 + (n-1)(-3)$

$\quad = 8 - 3n + 3$

$\quad = 11 - 3n$

and $a_8 = 11 - 3(10)$

$\quad = 11 - 30$

$\quad = -19$

49. $54, 58, 62$

$a_1 = 54$ and $d = 4$

$a_n = 54 + (n-1)4$

$a_{20} = 54 + 19(4) = 54 + 76 = 130$

The general term of the sequence is

$a_n = 4n + 50$.

There are 130 seats in the twentieth
row.

51. $a_1 = 6$ and $r = 3$

$a_n = 6(3)^{n-1} = 2 \cdot 3 \cdot (3)^{n-1} = 2(3)^{n}$

The general term of the sequence
is $a_n = 6(3)^{n-1}$ or $a_n = 2(3)^{n}$.

53. $a_1 - 486$ and $r = \dfrac{1}{3}$

Initial Height $= a_1$

$\quad = 486\left(\dfrac{1}{3}\right)^{1-1} = 486$

Rebound 1 $= a_2$

$\quad = 486\left(\dfrac{1}{3}\right)^{2-1} = 162$

Rebound 2 $= a_3$

$\quad = 486\left(\dfrac{1}{3}\right)^{3-1} = 54$

Rebound 3 $= a_4$

$\quad = 486\left(\dfrac{1}{3}\right)^{4-1} = 18$

Rebound 4 $= a_5$

$\quad = 486\left(\dfrac{1}{3}\right)^{5-1} = 6$

The first five terms of the
sequence are 486, 162, 54,
18, 6.

55. $a_1 = 4000$ and $d = 125$

$a_n = 4000 + (n-1)125$ or

$a_n = 3875 + 125n$

$a_{12} = 4000 + 11(125)$

$a_{12} = 5375$

His salary for his last month of
training is \$5375.

57. $a_1 = 400$ and $r = \dfrac{1}{2}$

12 hrs = 4(3 hrs), so we seek the
fourth term after a_1, namely a_5.

$a_n = a_1 r^{n-1}$

$a_5 = 400\left(\dfrac{1}{2}\right)^4 = \dfrac{400}{16} = 25$

25 grams of the radioactive
material remains after 12 hours.

59.   50

61. $3^0 + 3^1 + 3^2 + 3^3 = 40$

63. $\dfrac{8-1}{8+1} + \dfrac{8-2}{8+2} + \dfrac{8-3}{8+3}$

$= \dfrac{7}{9} + \dfrac{6}{10} + \dfrac{5}{11}$

$= \dfrac{770}{990} + \dfrac{594}{990} + \dfrac{450}{990}$

$= \dfrac{1814}{990} = \dfrac{907}{495}$

65. $a_1 = \$11,782.40$

$r = 0.5$

$a_2 = (11,782.40)(0.5) = \$5891.20$

$a_3 = (5891.20)(0.5) = \$2945.60$

$a_4 = (2945.60)(0.5) = \$1472.80$

The first four terms of the
sequence are $11,782.40,
$5891.20, $2945.60, $1472.80.

67. $a_1 = 19.652$ and $d = -0.034$

$a_2 = 19.652 - 0.034 = 19.618$

$a_3 = 19.618 - 0.034 = 19.584$

$a_4 = 19.584 - 0.034 = 19.550.$

69.   Answers may vary.

<u>**Exercise Set 14.3**</u>

1. $\displaystyle\sum_{i=1}^{4}(i-3)$

$= (1-3) + (2-3) + (3-3) + (4-3)$

$= -2 + (-1) + 0 + 1$

$= -2$

3. $\displaystyle\sum_{i=4}^{7}(2i+4)$

$= \big[2(4)+4\big] + \big[2(5)+4\big] + \big[2(6)+4\big]$

$\qquad\qquad + \big[2(7)+4\big]$

$= 12 + 14 + 16 + 18 = 60$

5. $\displaystyle\sum_{i=2}^{4}(i^2-3)$

$= (2^2-3) + (3^2-3) + (4^2-3)$

$= 1 + 6 + 13$

$= 20$

7. $\displaystyle\sum_{i=1}^{3}\dfrac{1}{i+5} = \dfrac{1}{1+5} + \dfrac{1}{2+5} + \dfrac{1}{3+5}$

$= \dfrac{1}{6} + \dfrac{1}{7} + \dfrac{1}{8} = \dfrac{28}{168} + \dfrac{24}{168} + \dfrac{21}{168}$

$= \dfrac{73}{168}$

9. $\displaystyle\sum_{i=1}^{3}\dfrac{1}{6i} = \dfrac{1}{6(1)} + \dfrac{1}{6(2)} + \dfrac{1}{6(3)}$

$= \dfrac{1}{6} + \dfrac{1}{12} + \dfrac{1}{18}$

$= \dfrac{6+3+2}{36}$

$= \dfrac{11}{36}$

11. $\displaystyle\sum_{i=2}^{6} 3i$

$= 3(2) + 3(3) + 3(4) + 3(5) + 3(6)$

$= 6 + 9 + 12 + 15 + 18$

$= 60$

13. $\displaystyle\sum_{i=3}^{5} i(i+2)$

$= 3(3+2) + 4(4+2) + 5(5+2)$

$= 15 + 24 + 35$

$= 74$

15. $\displaystyle\sum_{i=1}^{5} 2^i = 2^1 + 2^2 + 2^3 + 2^4 + 2^5$

$= 2 + 4 + 8 + 16 + 32$

$= 62$

17. $\displaystyle\sum_{i=1}^{4} \frac{4i}{i+3} = \frac{4(1)}{1+3} + \frac{4(2)}{2+3} + \frac{4(3)}{3+3} + \frac{4(4)}{4+3}$

$= 1 + \dfrac{8}{5} + 2 + \dfrac{16}{7} = \dfrac{105}{35} + \dfrac{56}{35} + \dfrac{80}{35}$

$= \dfrac{241}{35}$

19. $1 + 3 + 5 + 7 + 9$

$[(2)-1] + [2(2)-1] + [2(3)-1]$

$\qquad\qquad + [2(4)-1] + [2(5)-1]$

$= \displaystyle\sum_{i=1}^{5} (2i-1)$

21. $4 + 12 + 36 + 108$

$= 4 + 4(3) + 4(3)^2 + 4(3)^3$

$= \displaystyle\sum_{i=1}^{4} 4(3)^{i-1}$

23. $12 + 9 + 6 + 3 + 0 + (-3)$

$= [-3(1) + 15] + [-3(2) + 15]$

$\qquad + [-3(3) + 15] + [-3(4) + 15]$

$\qquad + [-3(5) + 15] + [-3(6) + 15]$

$= \displaystyle\sum_{i=1}^{6} (-3i + 15)$

25. $12 + 4 + \dfrac{4}{3} + \dfrac{4}{9} = \dfrac{4}{3^{-1}} + \dfrac{4}{3^0} + \dfrac{4}{3} + \dfrac{4}{3^2}$

$= \displaystyle\sum_{i=1}^{4} \dfrac{4}{3^{i-2}}$

27. $1 + 4 + 9 + 16 + 25 + 36 + 49$

$= 1^2 + 2^2 + 3^2 + 4^2 + 5^2 + 6^2 + 7^2$

$= \displaystyle\sum_{i=1}^{7} i^2$

29. $a_n = (n+2)(n-5)$

$a_1 = (1+2)(1-5) = 3(-4) = -12$

$a_2 = (2+2)(2-5) = 4(-3) = -12$

$a_1 + a_2 = -12 + (-12) = -24$

31. $a_n = n(n-6)$

$= a_1 + a_2 = 1(1-6) + 2(2-6)$

$= 1(-5) + 2(-4) = -13$

33. $a_n = (n+3)(n+1)$

$a_1 = (1+3)(1+1) = 4(2) = 8$

$a_2 = (2+3)(2+1) = 5(3) = 15$

$a_3 = (3+3)(3+1) = 6(4) = 24$

$a_4 = (4+3)(4+1) = 7(5) = 35$

$\displaystyle\sum_{i=1}^{4} a_i = 8 + 15 + 24 + 35 = 82$

35. $a_n = -2n$

$$\sum_{i=1}^{4} (-2i) =$$

$$= -2(1) + (-2)(2) + (-2)(3) + (-2)(4)$$

$$= -2 - 4 - 6 - 8$$

$$= -20$$

37. $a_n = -\dfrac{n}{3}$

$$a_1 + a_2 + a_3 = -\frac{1}{3} - \frac{2}{3} - \frac{3}{3} = -2$$

39. $1, 2, 3, \ldots, 10$

$$a_n = n$$

$$\sum_{i=1}^{10} i = 1 + 2 + 3 + \ldots + 10$$

$$= \frac{10(11)}{2} = 55$$

A total of 55 trees were planted.

41. $a_1 = 6$ and $r = 2$

$$a_n = 6 \cdot 2^{n-1}$$

$$a_5 = 6 \cdot 2^4 = 6 \cdot 16 = 96$$

There will be 96 fungus units at the beginning of the 5th day.

43. $a_1 = 50$ and $r = 2$

Since $48 = 4(12)$, we seek the fourth term after a_1, namely a_5.

The general term of the sequence is

$a_n = 50(2)^{n-1}$, where n represents the number of 12-hr periods.

$$a_5 = 50(2)^4 = 50(16) = 800$$

There are 800 bacteria after 48 hours.

45. $a_n = (n+1)(n+2)$

$$a_4 = (4+1)(4+2)$$

$$= 5(6) = 30 \text{ oppossums}$$

$$a_1 = (1+1)(1+2) = 2(3) = 6$$

$$a_2 = (2+1)(2+2) = 3(4) = 12$$

$$a_3 = (3+1)(3+2) = 4(5) = 20$$

$$\sum_{i=1}^{4} a_i = 6 + 12 + 20 + 30$$

$$= 68 \text{ oppossums}$$

47. $a_n = 100(0.5)^n$

$$a_4 = 100(0.5)^4 = 6.25 \text{ lbs of decay.}$$

$$a_1 = 100(0.5)^1 = 50$$

$$a_2 = 100(0.5)^2 = 25$$

$$a_3 = 100(0.5)^3 = 12.5$$

$$\sum_{i=1}^{4} a_i = 50 + 25 + 12.5 + 6.25$$

$$= 93.75 \text{ lbs of decay}$$

49. $a_1 = 40$ and $r = \dfrac{4}{5}$

$$a_5 = 40 \left(\frac{4}{5}\right)^4 = 16.384 \text{ or } 16.4 \text{ in.}$$

$$a_2 = 40 \left(\frac{4}{5}\right)^1 = 32$$

$$a_3 = 40 \left(\frac{4}{5}\right)^2 = 25.6$$

$$a_4 = 40 \left(\frac{4}{5}\right)^3 = 20.48$$

$$\sum_{i=1}^{5} a_i = 40 + 32 + 25.6 + 20.48$$

$$+ 16.384$$

$$= 134.464 \text{ or } 134.5 \text{ in.}$$

51. $\dfrac{5}{1-\dfrac{1}{2}} - \dfrac{5}{\dfrac{1}{2}} - 5 \cdot \dfrac{2}{1} - 10$

53. $\dfrac{\dfrac{1}{3}}{1-\dfrac{1}{10}} = \dfrac{\dfrac{1}{3}}{\dfrac{9}{10}} = \dfrac{1}{3} \cdot \dfrac{10}{9} = \dfrac{10}{27}$

55. $\dfrac{3(1-2^4)}{1-2} = \dfrac{3(1-16)}{-1}$

$= \dfrac{3(-15)}{-1} = \dfrac{-45}{-1} = 45$

57. $\dfrac{10}{2}(3+15) = \dfrac{10}{2}(18) = \dfrac{180}{2} = 90$

59. a. $\displaystyle\sum_{i=1}^{7} i + i^2$

$= (1+1^2) + (2+2^2) + (3+3^2)$

$\quad + (4+4^2) + (5+5^2) + (6+6^2)$

$\quad + (7+7^2)$

$= 2 + 6 + 12 + 20 + 30 + 42 + 56$

b. $\displaystyle\sum_{i=1}^{7} i + \sum_{i=1}^{7} i^2$

$= (1+2+3+4+5+6+7) + (1+4$

$\quad + 9 + 16 + 25 + 36 + 49)$

c. They are equal; 168

d. True

Exercise Set 14.4

1. $1, 3, 5, 7, \ldots$

The first term is 1 and the sixth term is 11.

$S_6 = \dfrac{6}{2}(1+11) = 3(12) = 36$

3. $4, 12, 36, \ldots$

$a_1 = 4,\ r = 3,\ n = 5$

$S_5 = \dfrac{4(1-3^5)}{1-3} = 484$

5. $3, 6, 9, \ldots$

The first term is 3 and the sixth term is 18.

$S_6 = \dfrac{6}{2}(3+18)$

$= 3(21)$

$= 63$

7. $2, \dfrac{2}{5}, \dfrac{2}{25}, \ldots$

$a_1 = 2,\ r = \dfrac{1}{5},\ n = 4$

$S_4 = \dfrac{2\left[1 - \left(\dfrac{1}{5}\right)^4\right]}{1 - \dfrac{1}{5}} = 2.496$

9. $1, 2, 3, \ldots, 10$

The first term is 1 and the tenth term is 10.

$$S_{10} = \frac{10}{2}(1+10)$$
$$= 5(11)$$
$$= 55$$

11. $1, 2, 3, 7$

The first term is 1 and the fourth term is 7.

$$S_4 = \frac{4}{2}(1+7)$$
$$= 2(8)$$
$$= 16$$

13. $12, 6, 3, \ldots$

$$a_1 = 12, \quad r = \frac{1}{2}$$
$$S_\infty = \frac{12}{1-\frac{1}{2}} = \frac{12}{\frac{1}{2}} = 24$$

15. $\frac{1}{10}, \frac{1}{100}, \frac{1}{1000}, \ldots$

$$a_1 = \frac{1}{10}, \quad r = \frac{1}{10}$$
$$S_\infty = \frac{\frac{1}{10}}{1-\frac{1}{10}} = \frac{1}{9}$$

17. $-10, -5, -\frac{5}{2}, \ldots$

$$a_1 = -10, \quad r = \frac{1}{2}$$
$$S_\infty = \frac{-10}{1-\frac{1}{2}} = -20$$

19. $2, -\frac{1}{4}, \frac{1}{32}, \ldots$

$$a_1 = 2, \quad r = -\frac{1}{8}$$
$$S_\infty = \frac{2}{1-\left(-\frac{1}{8}\right)} = \frac{16}{9}$$

21. $\frac{2}{3}, -\frac{1}{3}, \frac{1}{6}, \ldots$

$$a_1 = \frac{2}{3}, \quad r = -\frac{1}{2}$$
$$S_\infty = \frac{\frac{2}{3}}{1-\left(-\frac{1}{2}\right)} = \frac{4}{9}$$

23. $-4, 1, 6, \ldots, 41$

The first term is -4 and the tenth term is 41.

$$S_{10} = \frac{10}{2}(-4+41)$$
$$= 5(37)$$
$$= 185$$

25. $3, \dfrac{3}{2}, \dfrac{3}{4}, \ldots$

$$a_1 = 3, \; r = \dfrac{1}{2}, \; n = 7$$

$$S_7 = \dfrac{3\left[1-\left(\dfrac{1}{2}\right)^7\right]}{1-\dfrac{1}{2}} = \dfrac{381}{64}$$

27. $-12, 6, -3, \ldots$

$$a_1 = -12, \; r = -\dfrac{1}{2}, \; n = 5$$

$$S_5 = \dfrac{-12\left[1-\left(-\dfrac{1}{2}\right)^5\right]}{1-\left(-\dfrac{1}{2}\right)} = -\dfrac{33}{4}$$

29. $\dfrac{1}{2}, \dfrac{1}{4}, 0, \ldots, -\dfrac{17}{4}$

The first term is $\dfrac{1}{2}$ and the

twentieth term is $-\dfrac{17}{4}$.

$$S_{20} = \dfrac{20}{2}\left(\dfrac{1}{2} - \dfrac{17}{4}\right)$$

$$= 10\left(\dfrac{-15}{4}\right)$$

$$= -\dfrac{75}{2}$$

31. $a_1 = 8, \; r = -\dfrac{2}{3}, \; n = 3$

$$S_3 = \dfrac{8\left[1-\left(-\dfrac{2}{3}\right)^3\right]}{1-\left(-\dfrac{2}{3}\right)} = \dfrac{56}{9}$$

33. The first five terms are 4000, 3950, 3900, 3850, 3800

$$a_1 = 4000, \; d = -50, \; n = 12$$

$$a_{12} = 4000 + 11(-50)$$

$$= 3450 \text{ cars sold in month 12.}$$

$$S_{12} = \dfrac{12}{2}(4000 + 3450)$$

$$= 44{,}700 \text{ cars sold in the first}$$

12 months.

35. Firm A:

The first term is 22,000 and the tenth term is 31,000.

$$S_{10} = \dfrac{10}{2}(22000 + 31000)$$

$$= \$265{,}000$$

Firm B:

The first term is 20,000 and the tenth term is 30,800.

$$S_{10} = \dfrac{10}{2}(20000 + 30800)$$

$$= \$254{,}000$$

Thus, Firm A is making the better offer.

37. $a_1 = 30,000, r = 1.10, n = 4$

$a_4 = 30000(1.10)^{4-1}$

$a_4 = \$39,930$ made during her

fourth year of business

$S_4 = \dfrac{30000(1 - 1.10^4)}{1 - 1.10}$

    $= \$139,230$ made during the

first four years of business.

39. $a_1 = 30, r = 0.9, n = 5$

$a_5 = 30(0.9)^{5-1} = 19.63$

Approximately 20 minutes to

assemble the first computer.

$S_5 = \dfrac{30(1 - 0.9^5)}{1 - 0.9} = 122.853$

Approximately 123 minutes to

assemble the first 5 computers.

41. $a_1 = 20, r = \dfrac{4}{5}$

$S_\infty = \dfrac{20}{1 - \dfrac{4}{5}} = 100$

We double the number (to account

for the flight up as well as down)

and subtract 20 (since the first

bounce was preceded by only a

downward flight). Thus, the ball

$2(100) - 20 = 180$ feet

43. Player A:

The first term is 1 and the

ninth term is 9.

$S_9 = \dfrac{9}{2}(1 + 9)$

   $= 45$ points

Player B:

The first term is 10 and the

sixth term is 15.

$S_6 = \dfrac{6}{2}(10 + 15)$

   $= 75$ points

45. The first term is 200 and the

twentieth is 105.

$S_{20} = \dfrac{20}{2}(200 + 105)$

   $= 3050$

Thus, \$3050 rent is paid for 20 days

during the holiday rush.

47. $a_1 = 0.01, r = 2, n = 30$

$S_3 = \dfrac{0.01\left[1 - 2^{30}\right]}{1 - 2} = 10,737,418.23$

He would pay \$10,737,418.23

in room and board for the 30 days.

49. 720

51. 3

53. $x^2 + 10x + 25$

55. $8x^3 - 12x^2 + 6x - 1$

57. $0.8\overline{88} = 0.8 + 0.08 + 0.008 + \cdots$

$$= \frac{8}{10} + \frac{8}{100} + \frac{8}{1000} + \cdots$$

This is a geometric series with

$$a_1 = \frac{8}{10}, \ r = \frac{1}{10}$$

$$S_\infty = \frac{\dfrac{8}{10}}{1 - \dfrac{1}{10}} = \frac{8}{9}$$

59. Answers may vary.

Exercise Set 14.5

1. $(m+n)^3$

$\quad = m^3 + 3m^2n + 3mn^2 + n^3$

3. $(c+d)^5$

$\quad = c^5 + 5c^4d + 10c^3d^2 + 10c^2d^3 + 5cd^4$

$\quad + d^4$

5. $(y-x)^5 = \left[y + (-x)\right]^5$

$\quad = y^5 - 5y^4x + 10y^3x^2 - 10y^2x^3 + 5yx^4$

$\quad - x^5$

7. Answers may vary.

9. $\dfrac{8!}{7!} = \dfrac{8 \cdot 7!}{7!} = 8$

11. $\dfrac{7!}{5!} = \dfrac{7 \cdot 6 \cdot 5!}{5!} = 7 \cdot 6 = 42$

13. $\dfrac{10!}{7!2!} = \dfrac{10 \cdot 9 \cdot 8 \cdot 7!}{7!2!} = \dfrac{10 \cdot 9 \cdot 8}{2 \cdot 1} = 360$

15. $\dfrac{8!}{6!0!} = \dfrac{8 \cdot 7 \cdot 6!}{6! \cdot 1} = 56$

17. $(a+b)^7$

$$= a^7 + 7a^6b + \frac{7\cdot6}{2!}a^5b^2 + \frac{7\cdot6\cdot5}{3!}a^4b^3 + \frac{7\cdot6\cdot5\cdot4}{4!}a^3b^4 + \frac{7\cdot6\cdot5\cdot4\cdot3}{5!}a^2b^5$$

$$+ \frac{7\cdot6\cdot5\cdot4\cdot3\cdot2}{6!}ab^6 + b^7$$

$$= a^7 + 7a^6b + 21a^5b^2 + 35a^3b^4 + 21a^2b^5 + 7ab^6 + b^7$$

19. $(a+2b)^5 = a^5 + 10a^4b + 40a^3b^2 + 80a^2b^3 + 80ab^4 + 32b^5$

21. $(q+r)^2 = q^9 + 9q^8r + 36q^7r^2 + 84q^6r^3 + 126q^4r^5 + 84q^3r^6 + 36q^2r^7 + 9qr^8 + r^9$

23. $(4a+b)^5 = 1024a^5 + 1280a^4b + 640a^3b^2 + 160a^2b^3 + 20ab^4 + b^5$

25. $(5a-2b)^4 = 625a^4 - 1000a^3b + 600a^2b^2 - 160ab^3 + 16b^4$

27. $(2a+3b)^3 = 8a^3 + 36a^2b + 54ab^2 + 27b^3$

29. $(x+2)^5 = x^5 + 10x^4 + 40x^3 + 80x^2 + 80x + 32$

31. 5th term of $(c-d)^5$ corresponds to
$r = 4$:

$$\frac{5!}{4!(5-4)!}c^{5-4}(-d)^4 = 5cd^4$$

33. 8th term of $(2c+d)^7$ corresponds to
$r = 7$:

$$\frac{7!}{7!(7-7)!}(2c)^{7-7}(d)^7 = d^7$$

35. 4th term of $(2r-s)^5$ corresponds to
$r = 3$:

$$\frac{5!}{3!(5-3)!}(2r)^{5-3}(-s)^3 = -40r^2s^3$$

37. 3rd term of $(x+y)^4$ corresponds to
$r = 2$:

$$\frac{4!}{2!(4-2)!}(x)^{4-2}(y)^2 = 6x^2y^2$$

39. 2nd term of $(a+3b)^{10}$ corresponds to $r = 1$:

$$\frac{10!}{1!(10-1)!}(a)^{10-1}(3b)^1 = 30a^9b$$

41. $f(x) = |x|$

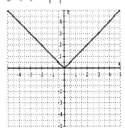

Not one-to-one

43. $H(x) = 2x + 3$

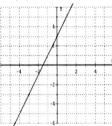

One-to-one

45. $f(x) = x^2 + 3$

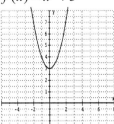

Not one-to-one

Chapter 14 Review

1. $a_n = -3n^2$

$a_1 = -3(1)^2 = -3$

$a_2 = -3(2)^2 = -12$

$a_3 = -3(3)^2 = -27$

$a_4 = -3(4)^2 = -48$

$a_5 = -3(5)^2 = -75$

2. $a_n = n^2 + 2n$

$a_1 = 1^2 + 2(1) = 3$

$a_2 = 2^2 + 2(2) = 8$

$a_3 = 3^2 + 2(3) = 15$

$a_4 = 4^2 + 2(4) = 24$

$a_5 = 5^2 + 2(5) = 35$

3. $a_n = \dfrac{(-1)^n}{100}$

$a_{100} = \dfrac{(-1)^{100}}{100} = \dfrac{1}{100}$

4. $a_n = \dfrac{2n}{(-1)^2}$

$a_{50} = \dfrac{2(50)}{(-1)^2} = 100$

5. $\dfrac{1}{6 \cdot 1}, \dfrac{1}{6 \cdot 2}, \dfrac{1}{6 \cdot 3}, \cdots$

In general, $a_n = \dfrac{1}{6n}$

6. $-1, 4, -9, 16, \ldots$

$a_n = (-1)^n n^2$

7. $a_n = 32n - 16$

$a_5 = 32(5) - 16 = 144$ ft

$a_6 = 32(6) - 16 = 176$ ft

$a_7 = 32(7) - 16 = 208$ ft

8. $a_n = 100(2)^{n-1}$

$10,000 = 100(2)^{n-1}$

$100 = 2^{n-1}$

$\log 100 = (n-1)\log 2$

$n = \dfrac{\log 100}{\log 2} + 1 \approx 7.6$

Eigthth day culture will be at least 10,000. Originally, 100.

9. $a_1 = 450$

$a_2 = 3(450) = 1350$

$a_3 = 3(1350) = 4050$

$a_4 = 3(4050) = 12,150$

$a_5 = 3(12,150) = 36,450$

In 2003, the number of infected people should be 36,450.

10. $a_n = 50 + (n-1)8$

$a_1 = 50$

$a_2 = 50 + 8 = 58$

$a_3 = 50 + 2(8) = 66$

$a_4 = 50 + 4(8) = 74$

$a_5 = 50 + 4(8) = 82$

$a_6 = 50 + 5(8) = 90$

$a_7 = 50 + 6(8) = 98$

$a_8 = 50 + 7(8) = 106$

$a_9 = 50 + 8(8) = 114$

$a_{10} = 50 + 9(8) = 122$

There are 122 seats in the tenth row.

11. $a_1 = -2,\ r = \dfrac{2}{3}$

The first 5 terms of the sequence are

$-2,\ -\dfrac{4}{3},\ -\dfrac{8}{9},\ -\dfrac{16}{27},\ -\dfrac{32}{81}.$

12. $a_n = 12 + (n-1)(-1.5)$

$a_1 = 12$

$a_2 = 12 + (1)(-1.5) = 10.5$

$a_3 = 12 + 2(-1.5) = 9$

$a_4 = 12 + 3(-1.5) = 7.5$

$a_5 = 12 + 4(-1.5) = 6$

13. $a_1 = -5,\ d = 4,\ n = 30$

$a_{30} = 5 + (30-1)4 = 111$

14. $a_n = 2 + (n-1)\dfrac{3}{4}$

$a_{11} = 2 + 10\left(\dfrac{3}{4}\right) = \dfrac{19}{2}$

15. $12,\ 7,\ 2,\ldots$

$a_1 = 12,\ d = -5,\ n = 20$

$a_{20} = 12 + (20-1)(-5) = -83$

16. $a_n = a_1 r^{n-1}$

$a_6 = 4\left(\dfrac{3}{2}\right)^{6-1} = \dfrac{243}{8}$

17. $a_4 = 18, a_{20} = 98$

Use the relationship:

$a_4 + 16d = a_{20}$

$18 + 16d = 98$

$d = 5$

Now use the relationship:

$a_4 = a_1 + 3d$

$18 = a_1 + 3(5)$

$a_1 = 3$

18. $-48 = a_3 = a_1 r^{3-1}$

$192 = a_4 = a_1 r^{4-1}$

$-48 = a_1 r^2$

$192 = a_1 r^3$

$r = -4, a_1 = -3.$

19. $\dfrac{3}{10}, \dfrac{3}{10^2}, \dfrac{3}{10^3}, \ldots$

In general, $a_n = \dfrac{3}{10^n}$

20. 50, 58, 66, ...

$a_n = 50 + (n-1)8$

21. $\dfrac{8}{3}$, 4, 6, ...

Geometric, $a_1 = \dfrac{8}{3}, r = \dfrac{3}{2}$

22. arithmetic; $a_1 = -10.5, d = 4.4$

23. $7x, -14x, 28x$

Geometric: $a_1 = 7x, r = -2$

24. neither

25. $a_1 = 8, r = 0.75$

8, 6, 4.5, 3.4, 2.5, 1.9

Yes, a ball that rebounds to a height of 2.5 feet after the fifth bounce is good, since $2.5 \geq 1.9$.

26. $a_n = 25 + (n-1)(-4)$

$a_n = 25 + 6(-4) = 1$

Continuing the progression as far as possible leaves 1 can in the top row.

27. $a_1 = 1, r = 2$

$a_n = 2^{n-1}$

$a_{10} = 2^9 = \$512$

$a_{30} = 2^{29} = \$536,870,912$

28. $a_n = a_1 r^{n-1}$

$a_5 = 30(0.7)^4 = 7.203$ in.

29. $a_1 = 900, d = 150$

$a_n = 900 + (n-1)150$

$a_6 = 900 + (6-1)150$

$\quad = \$1650 / \text{month}$

30. $\dfrac{1}{512}, \dfrac{1}{256}, \dfrac{1}{128}, \ldots$

first fold: $a_1 = \dfrac{1}{256}, r = 2$

$a_{15} = \dfrac{1}{256}(2)^{15-1} = 64$ inches

31. $\displaystyle\sum_{i=1}^{5}(2i-1)$

$= [2(1)-1] + [2(2)-1] + [2(3)-1]$

$\quad + [2(4)-1] + [2(5)-1]$

$= 25$

32. $\displaystyle\sum_{i=1}^{5} i(i+2)$

$= 1(1+2) + 2(2+2) + 3(3+2)$

$\qquad + 4(4+2) + 5(5+2)$

$= 85$

33. $\displaystyle\sum_{i=2}^{4} \frac{(-1)^i}{2i} = \frac{(-1)^2}{2(2)} + \frac{(-1)^3}{2(3)} + \frac{(-1)^4}{2(4)}$

$\qquad = \dfrac{5}{24}$

34. $\displaystyle\sum_{i=3}^{5} 5(-1)^{i-1}$

$= 5(-1)^{3-1} + 5(-1)^{4-1} + 5(-1)^{5-1}$

$= 5(1) + 5(-1) + 5(1) = 5$

35. $a_n = (n-3)(n+2)$

$S_4 = (1-3)(1+2) + (2-3)(2+2)$

$\qquad = -4$

36. $a_n = n^2$

$S_6 = (1)^2 + (2)^2 + (3)^2 + (4)^2 + (5)^2$

$\qquad + (6)^2$

$\qquad = 91$

37. $a_n = -8 + (n-1)3$

$a_1 = -8 + (1-1)3 = -8$

$a_2 = -8 + (2-1)3 = -5$

$a_3 = -8 + (3-1)3 = -2$

$a_4 = -8 + (4-1)3 = 1$

$a_5 = -8 + (5-1)3 = 4$

So $S_5 = -10$

38. $a_n = 5(4)^{n-1}$

$S_3 = 5(4)^0 + 5(4)^1 + 5(4)^2 = 105$

39. $1 + 3 + 9 + 27 + 81 + 243$

$= 3^0 + 3^1 + 3^2 + 3^3 + 3^4 + 3^5$

$= \displaystyle\sum_{i=1}^{6} 3^{i-1}$

40. $6 + 2 + (-2) + (-6) + (-10) + (-14)$

$+(-18)$

$a_n = 6 + (n-1)(-4)$

$\displaystyle\sum_{i=1}^{7} 6 + (i-1)(-4)$

41. $\dfrac{1}{4} + \dfrac{1}{16} + \dfrac{1}{64} + \dfrac{1}{256} = \displaystyle\sum_{i=1}^{4} \dfrac{1}{4^i}$

42. $1 + \left(-\dfrac{3}{2}\right) + \dfrac{9}{4} = \displaystyle\sum_{i=1}^{3} \left(-\dfrac{3}{2}\right)^{i-1}$

43. $a_1 = 20, \; r = 2$

$a_n = 20(2)^n$ represents the number of yeast, where n represents the number of 8-hr periods. Since $48 = 6(8)$ here, $n = 6$.

$a_6 = 20(2)^6 = 1280$ yeast

44. $a_n = n^2 + 2n - 1$

$a_4 = (4)^2 + 2(4) - 1 = 23$ cranes

$\displaystyle\sum_{i=1}^{4} i^2 + 2i - 1$

$= (1+2-1) + (4+4-1) +$

$\qquad + (9+6-1) + (16+8-1)$

$= 46$ cranes

45. For Job A: $a_1 = 39{,}500, \ d = 2200;$

$a_5 = 39{,}500 + (5-1)2200 = \$48{,}330$

For Job B: $a_1 = 41{,}000, \ d = 1400$

$a_5 = 41{,}000 + (5-1)1400 = \$46{,}600$

For the fifth year, Job A has a higher salary.

46. $a_n = 200(0.5)^n$

$a_3 = 200(0.5)^3 = 25 \text{ kg}$

$\displaystyle\sum_{i=1}^{3} 200(0.5)^i$

$\qquad = 200(0.5) + 200(0.5)^2 + 200(0.5)^3$

$\qquad = 175 \text{ kg}$

47. 15, 19, 23, ...

$a_1 = 15, \ d = 4$

$S_6 = \dfrac{6}{2}\big[2(15) + (6-1)4\big] = 150$

48. 5, −10, 20, ...

$a_1 = 5, \ r = -2$

$S_n = \dfrac{a_1(1 - r^n)}{1 - r}$

$S_9 = \dfrac{5(1 - (-2)^9)}{1 - (-2)} = 855$

49. $a_1 = 1, \ d = 2, \ n = 30$

$S_{30} = \dfrac{30}{2}\big[2(1) + (30-1)2\big] = 900$

50. 7, 14, 21, 28, ...

$a_n = 7 + (n-1)7$

$a_{20} = 7 + (20-1)7 = 140$

$S_{20} = \dfrac{20}{2}(7 + 140) = 1470$

51. 8, 5, 2, ...

$a_1 = 8, \ d = -3, \ n = 20$

$S_{20} = \dfrac{20}{2}\big[2(8) + (20-1)(-3)\big]$

$\qquad = -410$

52. $\dfrac{3}{4}, \dfrac{9}{4}, \dfrac{27}{4}, \ldots$

$a_1 = \dfrac{3}{4}, \ r = 3$

$S_8 = \dfrac{\dfrac{3}{4}(1 - 3^8)}{1 - 3} = 2460$

53. $a_1 = 6, \ r = 5$

$S_4 = \dfrac{6(1 - 5^4)}{1 - 5} = 936$

54. $a_1 = -3, \ d = -6$

$a_n = -3 + (n-1)(-6)$

$a_{100} = -3 + (100-1)(-6) = -597$

$S_{100} = \dfrac{100}{2}(-3 + (-597)) = -30{,}000$

55. $5, \dfrac{5}{2}, \dfrac{5}{4}, \ldots$

$a_1 = 5, \ r = \dfrac{1}{2}$

$S_\infty = \dfrac{5}{1 - \dfrac{1}{2}} = 10$

56. $18, -2, \dfrac{2}{9}, \ldots$

$a_1 = 18, \ r = -\dfrac{1}{9}$

$S_\infty = \dfrac{18}{1 + \dfrac{1}{9}} = \dfrac{81}{5}$

57. $-20, -4, -\dfrac{4}{5}, \ldots$

$$a_1 = -20, \ r = \dfrac{1}{5}$$

$$S_\infty = \dfrac{-20}{1 - \dfrac{1}{5}} = -25$$

58. $0.2, 0.02, 0.002, \ldots$

$$a_1 = 0.2, \ r = \dfrac{1}{10}$$

$$S_\infty = \dfrac{0.2}{1 - \dfrac{1}{10}} = \dfrac{2}{9}$$

59. $a_1 = 20{,}000, \ r = 1.15, \ n = 4$

$$a_4 = 20{,}000(1.15)^{4-1} = 30{,}418$$

Earned in his fourth year.

$$S_4 = \dfrac{20{,}000(1 - 1.15^4)}{1 - 1.15} = \$99{,}868$$

earned in his first four years.

60. $a_n = 40(0.8)^{n-1}$

$$a_4 = 40(0.8)^{4-1} = 20.48 \text{ min}$$

$$S_4 = \dfrac{40(1 - 0.8^4)}{1 - 0.8} = 118 \text{ min}$$

61. $a_1 = 100, d = -7, n = 7$

$$a_7 = 100 + (7-1)(-7)$$

$\quad = \$58$ rent paid for the seventh day.

$$S_7 = \dfrac{7}{2}\big[2(100) + (7-1)(-7)\big]$$

$\quad = \$553$ rent paid for the first seven days.

62. $a_1 = 15, r = 0.8$

$$S_\infty = \dfrac{15}{1 - 0.8} = 75 \text{ feet downward}$$

$$a_1 = 12, r = 0.8$$

$$S_\infty = \dfrac{12}{1 - 0.8} = 60 \text{ feet upward}$$

The total is 135 feet.

63. $1800, 600, 200, \ldots$

$$a_1 = 1800, \ r = \dfrac{1}{3}, \ n = 6$$

$$S_6 = 1800 \dfrac{\left(1 - \left(\dfrac{1}{3}\right)^6\right)}{1 - \dfrac{1}{3}}$$

≈ 2696 mosquitoes killed during the first six days after the spraying.

64. $1800, 600, 200, \ldots$
For which n is $a_n > 1$?

$$a_n = 1800\left(\dfrac{1}{3}\right)^{n-1} > 1$$

$$(n-1)\log\left(\dfrac{1}{3}\right) > \log\dfrac{1}{1800}$$

$$n < 7.8$$

No longer effective on the 8th day.
About 2700 mosquitoes were killed.

65. $0.5\overline{55} = 0.5 + 0.05 + 0.005 + \cdots$

$$a_1 = 0.5, r = 0.1$$

$$S_\infty = \dfrac{0.5}{1 - 0.1} = \dfrac{5}{9}$$

66. 27, 30, 33, ...

$$a_n = 27 + (n-1)(3)$$

$$a_{20} = 27 + (20-1)(3) = 84$$

$$S_{20} = \frac{20}{2}(27+84) = 1110$$

1110 seats

67. $(x+z)^5 = x^5 + 5x^4z + 10x^3z^2 + 10x^2z^3 + 5xz^4 + z^5$

68. $(y-r)^6 = y^6 - 6y^5r + 15y^4r^2 - 20y^3r^3 + 15y^2r^4 - 6yr^5 + r^6$

69. $(2x+y)^4 = 16x^4 + 32x^3y + 24x^2y^2 + 8xy^3 + y^4$

70. $(3y-z)^4 = 81y^4 - 108y^3z + 54y^2z^2 - 12yz^3 + z^4$

71. $(b+c)^8 = b^8 + 8b^7c + 28b^6c^2 + 56b^5c^3 + 70b^4c^4 + 56b^3c^5 + 28b^2c^6 + 8bc^7 + c^8$

72. $(x-w)^7 = x^7 - 7x^6w + 21x^5w^2 - 35x^4w^3 + 35x^3w^4 - 21x^2w^5 + 7xw^6 - w^7$

73. $(4m-n)^4 = (4m+(-n))^4$
$$= 256m^4 - 256m^3n + 96m^2n^2 - 16mn^3 + n^4$$

74. $(p-2r)^5 = p^5 - 10p^4r + 40p^3r^2 - 80p^2r^3 + 80pr^4 - 32r^5$

75. The 4th term corresponds to $r = 3$.

$$\frac{7!}{3!(7-3)!}a^{7-3}b^3 = 35a^4b^3$$

76. The 11th term is $\dfrac{10!}{10!0!}y^{10-10}(2z)^{10} = 1024z^{10}$

Chapter 14 Test

1. $a_n = \dfrac{(-1)^n}{n+4}$

$a_1 = \dfrac{(-1)^1}{1+4} = -\dfrac{1}{5}$

$a_2 = \dfrac{(-1)^2}{2+4} = \dfrac{1}{6}$

$a_3 = \dfrac{(-1)^3}{3+4} = -\dfrac{1}{7}$

$a_4 = \dfrac{(-1)^4}{4+4} = \dfrac{1}{8}$

$a_5 = \dfrac{(-1)^5}{5+4} = -\dfrac{1}{9}$

2. $a_n = \dfrac{3}{(-1)^n}$

$a_1 = \dfrac{3}{(-1)^1} = -3$

$a_2 = \dfrac{3}{(-1)^2} = 3$

$a_3 = \dfrac{3}{(-1)^3} = -3$

$a_4 = \dfrac{3}{(-1)^4} = 3$

$a_5 = \dfrac{3}{(-1)^5} = -3$

3. $\dfrac{2}{5}, \dfrac{2}{25}, \dfrac{2}{125}, \ldots$

In general, $a_n = \dfrac{2}{5}\left(\dfrac{1}{5}\right)^{n-1}$

4. $a_n = (n+1)(n-1)(-1)^n$

$a_{200} = (200+1)(200-1)(-1)^{200}$

$\qquad = 200^2 - 1^2 = 39{,}999$

5. $a_n = 5(2)^{n-1}$,

$S_5 = \dfrac{5(1-2^5)}{1-2} = 155$

6. $(-1)^1 9\cdot 1, \ (-1)^2 9\cdot 2, \ldots, a_n = (-1)^n 9n$

7. $a_1 = 24, \ r = \dfrac{1}{6}, \ S_\infty = \dfrac{24}{1-\frac{1}{6}} = \dfrac{144}{5}$

8. $a_n = 18 + (n-1)(-2)$

$a_1 = 18, \ d = -2$

$S_{30} = \dfrac{30}{2}\big[2(18) + (30-1)(-2)\big] = -330$

9. $a_1 = 24, \ r = \dfrac{1}{6}$

$S_\infty = \dfrac{24}{1-\dfrac{1}{6}} = \dfrac{144}{5}$

10. $\dfrac{3}{2}, -\dfrac{3}{4}, \dfrac{3}{8}, \ldots$

$a_1 = \dfrac{3}{2}, \ r = -\dfrac{1}{2}$

$S_\infty = \dfrac{\dfrac{3}{2}}{1-\left(-\dfrac{1}{2}\right)} = 1$

11. $\displaystyle\sum_{i=1}^{4} i(i-2)$

$= 1(1-2) + 2(2-2) + 3(3-2) +$

$\quad 4(4-2)$

$= 10$

12. $\displaystyle\sum_{i=2}^{4} 5(2)^i(-1)^{i-1}$

$= 5(2)^2(-1)^{2-1} + 5(2)^3(-1)^{3-1} + 5(2)^4(-1)^{4-1} = -60$

13. $(a-b)^6 = a^6 - 6a^5b + 15a^4b^2 - 20a^3b^3 + 15a^2b^4 - 6ab^5 + b^6$

14. $(2x+y)^5 = 32x^5 + 80x^4y + 80x^3y^2 + 40x^2y^3 + 10xy^4 + y^5$

15. $(y+z)^8 = y^8 + 8y^7z + 28y^6z^2 + 56y^5z^3 + 70y^4z + 56y^3z + 28y^2z^6 + 8yz^7 + z^8$

16. $(2p+r)^7 = 128p^7 + 448p^6r + 672p^5r^2 + 560p^4r^3 + 280p^3r^4 + 84p^2r^5 + 14pr^6 + r^7$

17. $a_n = 250 + 75(n-1)$

$a_{10} = 250 + 75(10-1) = 925$

There were 925 people in the town at the beginning of the tenth year.

$a_1 = 250 + 75(1-1) = 250$

There were 250 people in the town at the beginning of the first year.

18. $1, 3, 5, \cdots$

$a_1 = 1,\ d = 2,\ n = 8$

$a_8 = 1 + (8-1)2 = 15$

We want $1+3+5+...+15$

$S_8 = \dfrac{8}{2}[1+15] = 64$

There were 64 shrubs planted in the 8 rows.

19. $a_1 = 80,\ r = \dfrac{3}{4},\ n = 4$

$a_4 = 80\left(\dfrac{3}{4}\right)^{4-1} = 33.75$

The arc length is 33.75 cm on the 4th swing.

$S_4 = \dfrac{80\left(1-\left(\dfrac{3}{4}\right)^4\right)}{1-\dfrac{3}{4}} = 218.75$

The total of the arc lengths is 218.75 cm for the first 4 swings.

20. $a_1 = 80$, $r = \dfrac{3}{4}$

$$S_\infty = \frac{80}{1-\dfrac{3}{4}} = 320$$

The total of the arc lengths is 320 cm before the pendulum comes to rest.

21. 16, 48, 80,...

$a_{10} = 16 + (10-1)32 = 304$

He falls 304 feet during the 10th second.

$$S_{10} = \frac{10}{2}[16+304] = 1600$$

He falls 1600 feet during the first 10 seconds.

22. $0.42\overline{42} = 0.42 + 0.0042 + 0.000042$

$$S_\infty = \frac{0.42}{1-0.01} = \frac{14}{33}$$

Thus, $0.42\overline{42} = \dfrac{14}{33}$

Chapter 14 Cumulative Review

1. a. $(-2)^3 = -8$

b. $-2^3 = -8$

c. $(-3)^2 = 9$

d. $-3^2 = -9$

2. a. $3a - (4a+3) =$
$= 3a - 4a - 3$
$= -a - 3$

b. $(5x-3)+(2x+6) =$
$= 7x + 3$

c. $4(2x-5)-3(5x+1) =$
$= 8x - 20 - 15x - 3$
$= -7x - 23$

3. $(2x-3)-(4x-2)$
$= 2x - 3 - 4x + 2$
$= -2x - 1$

4. $x + 0.06x = 344.50$
$x = \$325$

5. $m = \dfrac{1}{4}$, $b = -3$

$y = mx + b$

$y = \dfrac{1}{4}x - 3$

6. $f(x) + 2 = \dfrac{3}{2}(x-3)$

$f(x) = \dfrac{3}{2}x - \dfrac{13}{2}$

7. $(2, 5)$ and $(-3, 4)$

$m = \dfrac{y_2 - y_1}{x_2 - x_1} = \dfrac{4-5}{-3-2} = \dfrac{1}{5}$

$y - y_1 = m(x - x_1)$

$y - 5 = \dfrac{1}{5}(x-2)$

$5y - 25 = x - 2$

$$-25 = x - 5y - 2$$
$$-23 = x - 5y$$
$$x - 5y = 23$$

8. $y^3 + 5y^2 - y - 5 = 0$

$$(y^3 + 5y^2) + (-y - 5) = 0$$
$$y^2(y + 5) - 1(y + 5) = 0$$
$$(y^2 - 1)(y + 5) = 0$$
$$y = -5, -1, 1$$

9. $x^3 - 4x^2 - 3x + 11 + \dfrac{12}{x+2}$

10. $\dfrac{5}{3a-6} - \dfrac{a}{a-2} + \dfrac{3+2a}{5a-10}$

$$= 5a^2 + 25a - 30 + (3 + 2a)(3a - 6)$$
$$= \dfrac{34 - 9a}{15(a - 2)}$$

11. a. $5\sqrt{2}$

    b. $2\sqrt[3]{3}$

    c. $\sqrt{26}$

    d. $2\sqrt[4]{2}$

12. $\sqrt{3x+6} - \sqrt{7x-6} = 0$

$$\left(\sqrt{3x+6}\right)^2 = \left(\sqrt{7x-6}\right)^2$$
$$3x + 6 = 7x - 6$$
$$x = 3$$

13. $2420 = 2000(1+r)^2$

$$\sqrt{1.21} = \sqrt{(1+r)^2}$$
$$r = 10\%$$

14. a. $\dfrac{\sqrt[3]{36x^2}}{3x}$

    b. $\dfrac{\sqrt{2}+1}{\sqrt{2}-1} =$

$$= \dfrac{\sqrt{2}+1}{\sqrt{2}-1} \cdot \left(\dfrac{\sqrt{2}+1}{\sqrt{2}+1}\right)$$
$$= 3 + 2\sqrt{2}$$

15. $(x-3)^2 - 3(x-3) - 4 = 0$

$$x^2 - 6x + 9 - 3x + 9 - 4 = 0$$
$$x^2 - 9x + 14 = 0$$
$$x = 2, 7$$

16. $\dfrac{10}{(2x+4)^2} - \dfrac{1}{2x+4} = 3$

$10 - (2x+4) = 3(2x+4)^2$

$x = \dfrac{-7}{6}, -3$

17. $\dfrac{5}{x+1} < -2$

$5 = -2(x+1)$

$x = -\dfrac{7}{2}, -1$

$\left(\dfrac{-7}{2}, -1\right)$ See graph.

18. Axis of symmetry: $x = -2$
vertex: $(-2, -6)$

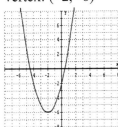

19. $-16t^2 + 20t = 0$

$-16(t - \dfrac{5}{8})^2 + \dfrac{25}{4} = 0$

$t = \dfrac{25}{4}$ feet, $\dfrac{5}{2}$ sec

20. $f(x) = x^2 + 3x - 18$

$f(x) = \left(x - \dfrac{3}{2}\right)^2 - \dfrac{81}{4}$

$\left(-\dfrac{3}{2}, \dfrac{-81}{4}\right)$

21. a. $25; 7$

b. $x^2 + 6x + 9; x^2 + 3$

22. $f(x) = -2x + 3$

$x = -2y + 3$

$x - 3 = -2y$

$f^{-1}(x) = -\dfrac{x-3}{2}$

23. $f^{-1} = \{(1,0), (7,-2), (-6,3), (4,4)\}$

24. a. $7; 3$

b. $x^2 + 2x; -x^2 - 1$

25. a. $x = 4$

b. $\dfrac{3}{2}$

c. 6

26. a. 5

b. -3

c. $\dfrac{1}{32}$

27. a. 2

b. -1

c. 3

d. 6

28. a. $4^x = 64$

$\ln 4^x = \ln 64$

$x \ln 4 = \ln 64$

$x = 3$

b. $8^x = 32$

$\ln 8^x = \ln 32$

$x \ln 8 = \ln 32$

$x = \dfrac{5}{3}$

c. $9^{x+4} = 243^x$

$3^{2(x+4)} = 3^{5x}$

$2x + 8 = 5x$

$x = \dfrac{8}{3}$

29. a. $\log_{11} 30$

   b. $\log_3 6$

   c. $\log_2(x^2 + 2x)$

30. a. 5

   b. -3

   c. $\dfrac{1}{5}$

   d. 4

31. $A = 1600e^{0.9(5)}$

$A = \$2509.31$

32 a. $\log_6 20$

   b. $\log_8 3$

   c. $\log_2 \dfrac{x^5}{(-x-1)^2}$

33. $\quad 3^x = 7$

$x \log 3 = \log 7$

$x = \dfrac{\log 7}{\log 3} \approx 1.7712$

34. $10000 = 5000\left(1 + \dfrac{0.02}{4}\right)^{4t}$

$\ln 2 = 4t \ln\left(1 + \dfrac{0.02}{4}\right)$

$t \approx 34.7$ years

35. $\log_4(x - 2) = 2$

$4^{\log_4(x-2)} = 4^2$

$x - 2 = 16$

$x = 18$

36. $\log_4 \dfrac{10}{x} = 2$

$4^{\log_4 \frac{10}{x}} = 4^2$

$\dfrac{10}{x} = 16$

$x = \dfrac{5}{8}$

37.

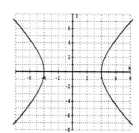

38. $\sqrt{101}$

39. $\begin{cases} y = \sqrt{x} \\ x^2 + y^2 = 6 \end{cases}$

$(x)^2 + \left(\sqrt{x}\right)^2 = 6$

$x^2 + x - 6 = 0$

$\left(2, \sqrt{2}\right)$

40. $\begin{cases} x^2 + y^2 = 36 \\ x - y = 6 \end{cases}$

$(y+6)^2 + y^2 = 36$

$2y^2 + 12y = 0$

$(0, -6); (6, 0)$

41.

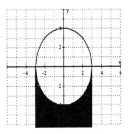

42.

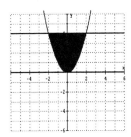

43. 0, 3, 8, 15, 24

44. $\dfrac{2}{3}$

45. 72

46. 6250

47. a. $\displaystyle\sum_{i=0}^{6}\frac{i-2}{2} = -1 - \frac{1}{2} + 0 + \cdots + \frac{3}{2} = \frac{7}{2}$

    b. $\displaystyle\sum_{i=3}^{5} 2^{i} = 8 + 16 + 32 = 56$

48. a. $\displaystyle\sum_{i=0}^{4} i^{2} + i = 0 + 2 + 6 + \cdots + 20 = 40$

    b. $\displaystyle\sum_{i=0}^{3} 2^{i} = 1 + 2 + 4 + 8 = 15$

49. 465

50. $15x^{4}y^{2}$

Appendix A Exercise Set

Section 2.3

1. $3x - 4 = 3(2x - 1) + 7$

$3x - 4 = 6x - 3 + 7$

$3x - 4 = 6x + 4$

$-3x = 8$

$x = -\dfrac{8}{3}$

The solution is $-\dfrac{8}{3}$.

3. $5 + 2x = 5(x + 1)$

$5 + 2x = 5x + 5$

$-3x = 0$

$x = 0$

The solution is 0.

Section 2.7

1. $\dfrac{x + 3}{2} > 1$

$x + 2 > 2$

$x > -1$

The solution is $(-1, \infty)$.

3. $\dfrac{x - 2}{2} - \dfrac{x - 4}{3} = \dfrac{5}{6}$

$6\left(\dfrac{x - 2}{2} - \dfrac{x - 4}{3}\right) = 6\left(\dfrac{5}{6}\right)$

$3(x - 2) - 2(x - 4) = 5$

$3x - 6 - 2x + 8 = 5$

$x + 2 = 5$

$x = 2$

The solution is 3.

Section 6.5

1. $2x^2 - 17x = 9$

$2x^2 - 17x - 9 = 0$

$(2x + 1)(x - 9) = 0$

$2x + 1 = 0 \quad \text{or} \quad x - 9 = 0$

$2x = -1 \qquad\qquad x = 9$

$x = -\dfrac{1}{2}$

The solutions are $-\dfrac{1}{2}$ and 9.

3. $x(x + 5) = 36$

$x^2 + 5x = 36$

$x^2 + 5x - 36 = 0$

$(x + 9)(x - 4) = 0$

$x + 9 = 0 \quad \text{or} \quad x - 4 = 0$

$x = -9 \qquad\qquad x = 4$

5. $3(2x-1) < 9$ and $-4x > -12$

$\qquad 6x - 3 < 9 \qquad\qquad x < 3$

$\qquad\quad 6x < 12$

$\qquad\qquad x < 2$

$x < 2$

The solution is $(-\infty, 2)$.

Section 7.5

1. $\qquad \dfrac{x}{10} - \dfrac{1}{2} = \dfrac{7}{5x}$

$\qquad 10x\left(\dfrac{x}{10} - \dfrac{1}{2}\right) = 10x\left(\dfrac{7}{5x}\right)$

$\qquad\quad x^2 - 5x = 2(7)$

$\qquad x^2 - 5x - 14 = 0$

$\qquad (x+1)(x-7) = 0$

$\qquad x + 2 = 0$ or $x - 7 = 0$

$\qquad\quad x = -2 \qquad\quad x = 7$

The solutions are -2 and 7.

3. $\qquad x(2x - 11) = 21$

$\qquad\quad 2x^2 - 11x = 21$

$\quad 2x^2 - 11x - 21 = 0$

$\quad (2x + 3)(x - 7) = 0$

$\quad 2x + 3 = 0$ or $x - 7 = 0$

$\qquad 2x = -3 \qquad\quad x = 7$

$\qquad\quad x = -\dfrac{3}{2}$

5. $-8 + 2x - 4 \le -2$

$\qquad\quad 2x - 12 \le -2$

$\qquad\qquad 2x \le 10$

$\qquad\qquad x \le 5$

The solution is $(-\infty, 5]$.

Section 9.1

1. $x - 2 \le 1$ and $3x - 1 \ge -4$

$\qquad x \le 3 \qquad\qquad 3x \ge -3$

$\qquad\qquad\qquad\qquad\quad x \ge -1$

$-1 \le x \le 3$

The solution is $[-1, 3]$.

3. $-2x + 2.5 = -7.7$

$\qquad -2x = -10.2$

$\qquad\quad x = \dfrac{-10.2}{-2} = 5.1$

The solution is 5.1.

5. $x \le -3$ or $x \le -5$

$\quad x \le -3$

The solution is $(-\infty, -3]$.

Section 9.2

1. $|2 + 3x| = 7$

$2 + 3x = 7$ or $2 + 3x = -7$

$\quad 3x = 5 \qquad\qquad 3x = -9$

$\quad\; x = \dfrac{5}{3} \qquad\qquad x = -3$

The solutions are -3 and $\dfrac{5}{3}$.

3. $\dfrac{5t}{2} - \dfrac{3t}{4} = 7$

$4\left(\dfrac{5t}{2} - \dfrac{3t}{4}\right) = 4(7)$

$2(5t) - 3t = 28$

$10t - 3t = 28$

$7t = 28$

$t = 4$

The solution is 4.

5. $\dfrac{4x}{5} - 1 = \dfrac{x}{2} + 2$

$10\left(\dfrac{4x}{5} - 1\right) = 10\left(\dfrac{x}{2} + 2\right)$

$2(4x) - 10 = 5x + 20$

$8x - 10 = 5x + 20$

$3x = 30$

$x = 10$

The solution is 10.

5. $5(x - 3) + x + 2 \geq 3(x + 2) + 2x$

$5x - 15 + x + 2 \geq 3x + 6 + 2x$

$6x - 13 \geq 5x + 6$

$x \geq 9$

The solution is $[19, \infty)$.

Section 9.3

1. $|x - 11| \geq 7$

$x - 11 \leq 7 \quad \text{or} \quad x - 11 \geq 7$

$x \leq 4 \qquad\qquad x \geq 18$

The solution is $(-\infty, 4] \cup [18, \infty)$.

3. $-5 < x - (2x + 3) < 0$

$-5 < x - 2x - 3 < 0$

$-5 < -x - 3 < 0$

$-2 < -x < 3$

$2 > x > -3$

$-3 < x < 2$

The solution is $(-3, 2)$.

Section 9.6

1. $4^x = 8^{x-1}$

$\left(2^2\right)^x = \left(2^3\right)^{x-1}$

$2^{2x} = 2^{3x-3}$

$2x = 3x - 3$

$-x = -3$

$x = 3$

3. $x(x - 9) > 0$

$x = 0 \quad \text{or} \quad x - 9 = 0$

$x = 9$

The solution is $(-\infty, 0) \cup (9, \infty)$.

5. $\log_4\left(x^2 - 3x\right) = 1$

$x^2 - 3x = 4^1$

$x^2 - 3x - 4 = 0$

$(x - 4)(x + 1) = 0$

$x - 4 = 0 \quad \text{or} \quad x + 1 = 0$

$x = 4 \qquad\qquad x = -1$

The solutions are -1 and 4.

7. $\dfrac{6}{x-2} \geq 3$

The denominator is 0 when

$x - 2 = 0$, or $x = 2$.

$\dfrac{6}{x-2} = 3$

$6 = 3(x-2)$

$6 = 3x - 6$

$12 = 3x$

$4 = x$

The solution is $(2, 4]$.

9. $\log_3 (2x+1) - \log_3 x = 1$

$\log_3 \dfrac{2x+1}{x} = 1$

$\dfrac{2x+1}{x} = 3^1$

$2x + 1 = 3x$

$1 = x$

$x - 4 = 0$ or $x + 1 = 0$

$x = 4$ $x = -1$

The solution is 1.

Section 10.6

1. $x(3x+14) = 5$

$3x^2 + 14x = 5$

$3x^2 + 14x - 5 = 0$

$(3x-1)(x+5) = 0$

$3x - 1 = 0$ or $x + 5 = 0$

$3x = 1$ $x = -5$

$x = \dfrac{1}{3}$

The solutions are -5 and $\dfrac{1}{3}$.

3. $|5x-4| = |4x+1|$

$5x - 4 = 4x + 1$ or $5x - 4 = -(4x+1)$

$x = 5$ $5x - 4 = -4x - 1$

$9x = 3$

$x = \dfrac{1}{3}$

The solutions are 5 and $\dfrac{1}{3}$.

5. $-2(x-4) + 3x \leq -3(x+2) - 2$

$-2x + 8 + 3x \leq -3x - 6 - 2$

$x + 8 \leq -3x - 8$

$4x \leq -16$

$x \leq -4$

The solution is $(-\infty, -4]$.

Section 11.2

1. $(x-2)^2 = 17$

$$x - 2 = \pm\sqrt{17}$$
$$x = 2 \pm \sqrt{17}$$

3. $x^2 - 5x + 6 = 0$

$$(x-2)(x-3) = 0$$
$$x - 2 = 0 \quad \text{or} \quad x - 3 = 0$$
$$x = 2 \qquad\qquad x = 3$$

The solutions are 2 and 3.

5. $\sqrt{2x+30} = x+3$

$$2x + 30 = (x+3)^2$$
$$2x + 30 = x^2 + 6x + 9$$
$$0 = x^2 + 4x - 21$$
$$0 = (x+7)(x-3)$$
$$x + 7 = 0 \quad \text{or} \quad x - 3 = 0$$
$$x = -7 \qquad\qquad x = 3$$

The solution are is 3.

7. $\dfrac{3x^2-7}{3x^2-8x-3} = \dfrac{1}{x-3} + \dfrac{2}{3x+1}$

$$\frac{3x^2-7}{(3x+1)(x-3)} = \frac{1}{x-3} + \frac{2}{3x+1}$$
$$3x^2 - 7 = 1(3x+1) + 2(x-3)$$
$$3x^2 - 7 = 3x + 1 + 2x - 6$$
$$3x^2 - 7 = 5x - 5$$
$$3x^2 - 5x - 2 = 0$$

$$(3x+1)(x-2) = 0$$
$$3x + 1 = 0 \quad \text{or} \quad x - 2 = 0$$
$$3x = -1 \qquad\qquad x = 2$$
$$x = -\frac{1}{3}$$

The solution is 2.

Section 11.4

1. $x^2 - 3x - 10 = 0$

$$(x-5)(x+2) = 0$$
$$x - 5 = 0 \quad \text{or} \quad x + 2 = 0$$
$$x = 5 \qquad\qquad x = -2$$

The solutions are -2 and 5.

3. $\dfrac{x+4}{x-10} = 0$

$$x + 4 = 0$$
$$x = -4$$

The solution is -4.

5. $\sqrt{x-7} - 12 = -8$

$$\sqrt{x-7} = 4$$
$$x - 7 = 4^2$$
$$x - 7 = 16$$
$$x = 23$$

The solution is 23.

7. $\left|\dfrac{3x+5}{2}\right| = -9$ is impossible

There is no solution, or $\varnothing$.

9. $-4(x-3)+2x<6x+4$

$\quad -4x+12+2x<6x+4$

$\quad\quad -2x+12<6x+4$

$\quad\quad\quad -8x<-8$

$\quad\quad\quad\quad x>1$

The solution is $(1,\infty)$.

13.
$$3\overline{)8.1} = 2.7$$

$$\begin{array}{r} 2.7 \\ 3\overline{)8.1} \\ \underline{6} \\ 21 \\ \underline{21} \\ 0 \end{array}$$

15.
$$\begin{array}{r} 55.4050 \\ -6.1711 \\ \hline 49.2339 \end{array}$$

Appendix B Exercise Set

1.
$$\begin{array}{r} 9.076 \\ +\ 8.004 \\ \hline 17.080 \end{array}$$

17.
$$\begin{array}{r} 80. \\ 75.\overline{)6000.} \\ \underline{600} \\ 00 \\ \underline{00} \end{array}$$

3.
$$\begin{array}{r} 27.004 \\ -\ 14.200 \\ \hline 12.804 \end{array}$$

19.
$$\begin{array}{r} 0.07612 \\ 100.\overline{)7.61200} \\ \underline{700} \\ 612 \\ \underline{600} \\ 120 \\ \underline{100} \\ 200 \\ \underline{200} \\ 0 \end{array}$$

5.
$$\begin{array}{r} 107.92 \\ +\ 3.04 \\ \hline 110.96 \end{array}$$

7.
$$\begin{array}{r} 10.0 \\ -\ 7.6 \\ \hline 2.4 \end{array}$$

9.
$$\begin{array}{r} 126.32 \\ -\ 97.89 \\ \hline 28.43 \end{array}$$

11.
$$\begin{array}{r} 3.25 \\ \times\ \ \ 70 \\ \hline 227.50 \end{array}$$

21.

$$\begin{array}{r} 4.56 \\ 27.\overline{)123.12} \\ \underline{108} \\ 151 \\ \underline{135} \\ 162 \\ \underline{162} \\ 0 \end{array}$$

23.

$$\begin{array}{r} 569.20 \\ 71.25 \\ +\ \ 8.01 \\ \hline 648.46 \end{array}$$

25.

$$\begin{array}{r} 768.00 \\ -\ \ 0.17 \\ \hline 767.83 \end{array}$$

27.

$$\begin{array}{r} 12.00 \\ +\ \ 0.062 \\ \hline 12.062 \end{array}$$

29.

$$\begin{array}{r} 76.00 \\ -\ 14.52 \\ \hline 61.48 \end{array}$$

31.

$$\begin{array}{r} 7.7 \\ 43.\overline{)331.1} \\ \underline{301} \\ 30\ 1 \\ \underline{30\ 1} \\ 0 \end{array}$$

33.

$$\begin{array}{r} 762.12 \\ 89.70 \\ +\ \ 11.55 \\ \hline 863.37 \end{array}$$

35.

$$\begin{array}{r} 23.400 \\ -\ \ 0.821 \\ \hline 22.579 \end{array}$$

37.

$$\begin{array}{r} 476.12 \\ -\ 112.97 \\ \hline 363.15 \end{array}$$

39.

$$\begin{array}{r} 0.007 \\ +\ 7.000 \\ \hline 7.007 \end{array}$$

Appendix C-1 Exercise Set

1. $x^2 + 11x + 24 = 0$

$(x+3)(x+8) = 0$

$x+3 = 0 \quad \text{or} \quad x+8 = 0$

$x = -3 \qquad\qquad x = -8$

3. $3x - 4 - 5x = x + 4 + x$

$-2x - 4 = 2x + 4$

$-4x - 4 = 4$

$-4x = 8$

$x = -2$

5. $12x^2 + 5x - 2 = 0$

$(3x + 2)(4x - 1) = 0$

$3x + 2 = 0$ or $4x - 1 = 0$

$3x = -2$ $4x = 1$

$x = -\dfrac{2}{3}$ $x = \dfrac{1}{4}$

7. $z^2 + 9 = 10z$

$z^2 - 10z + 9 = 0$

$(z - 1)(z - 9) = 0$

$z - 1 = 0$ or $z - 9 = 0$

$z = 1$ $z = 9$

9. $5(y + 4) = 4(y + 5)$

$5y + 20 = 4y + 20$

$y + 20 = 20$

$y = 0$

11. $0.6x - 10 = 1.4x - 14$

$-0.8x - 10 = -14$

$-0.8x = -4$

$x = 5$

13. $x(5x + 2) = 3$

$5x^2 + 2x = 3$

$5x^2 + 2x - 3 = 0$

$(5x - 3)(x + 1) = 0$

$5x - 3 = 0$ or $x + 1 = 0$

$5x = 3$ $x = -1$

$x = \dfrac{3}{5}$

15. $6x - 2(x - 3) = 4(x + 1) + 4$

$6x - 2x + 6 = 4x + 4 + 4$

$4x + 6 = 4x + 8$

$6 = 8$

$x = -2$

There is no solution.

17. $\dfrac{3}{8} + \dfrac{b}{3} = \dfrac{5}{12}$

$24\left(\dfrac{3}{8}\right) + 24\left(\dfrac{b}{3}\right) = 24\left(\dfrac{5}{12}\right)$

$9 + 8b = 10$

$8b = 1$

$b = \dfrac{1}{8}$

19. $x^2 - 6x = x(8 + x)$

$x^2 - 6x = 8x + x^2$

$-6x = 8x$

$-14x = 0$

$x = 0$

21. $\dfrac{z^2}{6} - \dfrac{z}{2} - 3 = 0$

$6\left(\dfrac{z^2}{6}\right) - 6\left(\dfrac{z}{2}\right) - 6(3) = 6(0)$

$z^2 - 3z - 18 = 0$

$(z - 6)(z + 3) = 0$

$z - 6 = 0$ or $z + 3 = 0$

$z = 6$ $z = -3$

23. $z + 3(2 + 4z) = 6(z + 1) + 5z$
$z + 6 + 12z = 6z + 6 + 5z$
$6 + 13z = 11z + 6$
$6 + 2z = 6$
$2z = 0$
$z = 0$

25. $\dfrac{x^2}{2} + \dfrac{x}{20} = \dfrac{1}{10}$

$20\left(\dfrac{x^2}{2}\right) + 20\left(\dfrac{x}{20}\right) = 20\left(\dfrac{1}{10}\right)$

$10x^2 + x = 2$
$10x^2 + x - 2 = 0$
$(5x - 2)(2x + 1) = 0$
$5x - 2 = 0 \quad \text{or} \quad 2x + 1 = 0$
$5x = 2 \qquad\qquad 2x = -1$
$x = \dfrac{2}{5} \qquad\qquad x = -\dfrac{1}{2}$

27. $\dfrac{4t^2}{5} = \dfrac{t}{5} + \dfrac{3}{10}$

$10\left(\dfrac{4t^2}{\cdot\,5}\right) = 10\left(\dfrac{t}{5}\right) + 10\left(\dfrac{3}{10}\right)$

$8t^2 = 2t + 3$
$8t^2 - 2t - 3 = 0$
$(4t - 3)(2t + 1) = 0$
$4t - 3 = 0 \qquad 2t + 1 = 0$
$4t = 3 \qquad\qquad 2t = -1$
$t = \dfrac{3}{4} \qquad\qquad t = -\dfrac{1}{2}$

29. $\dfrac{3t + 1}{8} = \dfrac{5 + 2t}{7} + 2$

$56\left(\dfrac{3t + 1}{8}\right) = 56\left(\dfrac{5 + 2t}{7}\right) + 56(2)$

$7(3t + 1) = 8(5 + 2t) + 112$
$21t + 7 = 40 + 16t + 112$
$21t + 7 = 16t + 152$
$5t + 7 = 152$
$5t = 145$
$t = 29$

31. $\dfrac{m - 4}{3} - \dfrac{3m - 1}{5} = 1$

$15\left(\dfrac{m - 4}{3}\right) - 15\left(\dfrac{3m - 1}{5}\right) = 15(1)$

$5(m - 4) - 3(3m - 1) = 15$
$5m - 20 - 9m + 3 = 15$
$-4m - 20 - 9m + 3 = 15$
$-4m - 17 = 15$
$-4m = 32$
$m = -8$

33. $3x^2 = -x$
$3x^2 + x = 0$
$x(3x + 1) = 0$
$x = 0 \quad \text{or} \quad 3x + 1 = 0$
$\qquad\qquad\qquad 3x = -1$
$\qquad\qquad\qquad x = -\dfrac{1}{3}$

35. $x(x-3) = x^2 + 5x + 7$

$x^2 - 3x = x^2 + 5x + 7$

$-3x = 5x + 7$

$-8x = 7$

$x = -\dfrac{7}{8}$

37. $3(t-8) + 2t = 7 + t$

$3t - 24 + 2t = 7 + t$

$5t - 24 = 7 + t$

$4t - 24 = 7$

$4t = 31$

$t = \dfrac{31}{4}$

39. $-3(x-4) + x = 5(3-x)$

$-3x + 12 + x = 15 - 5x$

$-2x + 12 = 15 - 5x$

$3x + 12 = 15$

$3x = 3$

$x = 1$

41. $(x-1)(x+4) = 24$

$x^2 + 3x - 4 = 24$

$x^2 + 3x - 28 = 0$

$(x+7)(x-4) = 0$

$x + 7 = 0 \quad \text{or} \quad x - 4 = 0$

$x = -7 \qquad\qquad x = 4$

43. $\dfrac{x^2}{4} - \dfrac{5}{2}x + 6 = 0$

$4\left(\dfrac{x^2}{4}\right) - 4\left(\dfrac{5}{2}x\right) + 4(6) = 4(0)$

$x^2 - 10x + 24 = 0$

$(x-6)(x-4) = 0$

$x - 6 = 0 \quad \text{or} \quad x - 4 = 0$

$x = 6 \qquad\qquad x = 4$

45. $y^2 + \dfrac{1}{4} = -y$

$4(y^2) + 4\left(\dfrac{1}{4}\right) = -4y$

$4y^2 + 1 = -4y$

$4y^2 + 4y + 1 = 0$

$(2y+1)(2y+1) = 0$

$2y + 1 = 0$

$2y = -1$

$y = -\dfrac{1}{2}$

47. a. incorrect: answers may vary

b. correct

c. correct

d. incorrect: answers may vary

49. $3.2x + 4 = 5.4x - 7$

$3.2x = 5.4x - 11$

$k = -11$

51. $\dfrac{x}{6} + 4 = \dfrac{x}{3}$

$$6\left(\dfrac{x}{6}\right) + 6(4) = 6\left(\dfrac{x}{3}\right)$$

$$x + 24 = 2x$$

$$k = 24$$

53. $2.569x = -12.48534$

$$x = -4.86$$

Check:

$$2.569(-4.86) \overset{?}{=} -12.48534$$

$$-12.48534 = -12.48534$$

The solution is $x = -4.86$.

55. $2.86z - 8.1258 = -3.75$

$$2.86z = 4.3758$$

$$z = 1.53$$

Check:

$$2.86(1.53) - 8.1258 \overset{?}{=} -3.75$$

$$4.3758 - 8.1258 \overset{?}{=} -3.75$$

$$-3.75 = -3.75$$

The solution is $z = 1.53$.

57. $\dfrac{8}{x}$

59. $8x$

61. $3x + 2$

Appendix C-2 Exercise Set

1. perimeter $= 4y$

3. $z + (z+1) + z + 2 = z + 3$

5. $5x + 10(x+3) = (15x + 30)$ cents

7. $4x + 3(2x+1) = 4x + 6x + 3$
$$= 10x + 3$$

9. $4(x-2) = 6x + 2$
$$4x - 8 = 6x + 2$$
$$-8 = 2x + 2$$
$$10 = 2x$$
$$5 = x$$

The number is 5.

11. Let $x =$ one number and
$5x =$ another number

$$x + 5x = 270$$
$$6x = 270$$
$$x = 45$$
$$5x = 225$$

The numbers are 45 and 225.

13. $30\%(260) = 0.3(260) = 78$

15. $12\%(16) = 0.12(16) = 1.92$

17. $100\% - 29\% = 71\%$ not federally owned
$$71\%(2271) = 0.71(2271) = 1612.41$$
1612.41 million acres are not federally owned.

21. $100\% - 31\% = 66\dfrac{2}{3}\% = \dfrac{200}{3}$

$\dfrac{200}{3}\%(1290) = \dfrac{200}{300}(1290) = 86$

860 online shoppers do not spend more than they intended.

23. 17%

25. $6\%(112,500) = 0.06(112,500) = 67$

6750 e-mail users check their e-mail about once a week.

27. $3x + 21.1 = 205.9$

$\qquad 3x = 184.8$

$\qquad x = 61.6$

$\quad x + 5.8 = 61.6 + 5.8 = 67.4$

$x + 15.3 = 61.6 + 15.3 = 76.9$

Number of arrivals and departures: Los Angeles, 61.6 million. Chicago, 67.4 million. Atlanta, 76.9 million.

29. Let x = the number of seats in the 737-200;

$21 + x$ = the number in the 737-300 and

$2x - 36$ = the number in the 757-200.

$x + (21 + x) + (2x - 36) = 437$

$\quad x + 21 + 2x - 36 = 437$

$\qquad\qquad 4x - 15 = 437$

$\qquad\qquad\qquad 4x = 452$

$\qquad\qquad\qquad x = 113$

$21 + x = 21 + 113 = 134$

$2x - 36 = 2(113) - 36 = 190$

The 737-200 has 113 seats.

The 737-300 has 134 seats.

The 757-200 has 190 seats.

31. Let x = the price before taxes

$108\%x = 464.40$

$1.08x = 464.40$

$\quad x = 430$

The cost before taxes was $430.

33. Let x = the number of seats at Heinz Field and $x + 11,675$ = the number at INVESCO.

$x + x + 11,675 = 140,575$

$\quad 2x + 11,675 = 140,575$

$\quad 2x + 11,675 = 140,575$

$\qquad\qquad 2x = 128,900$

$\qquad\qquad x = 64,450$

$x + 11,675 = 64,450 + 11,675$

$\qquad\qquad = 76,125$

There are 64,450 seats at Heinz Field and 76,125 seats at INVESCO.

35. Let x = the number of subscribers to MSN,

   $x + 700,000$ = the number to Earthlink and

   $5x + 3,700,000$ = the number to AOL.

   $x + (x + 700,000) + (5x + 3,700,000)$

   $= 32,400,000$

   $x + x + 700,000 + 5x + 3,700,000$

   $= 32,400,000$

   $7x + 4,400,000 = 32,400,000$

   $\qquad\qquad 7x = 28,000,000$

   $\qquad\qquad\quad x = 4,000,000$

   4,000,000 subscribers to MSN.

   4,700,000 subscribers to Earthlink.

   23,700,000 subscribers to AOL.

37. Let x = the population of Morocco

   in the year 2000.

   $109.1\% x = 31.2$

   $\quad 1.091x = 31.2$

   $\qquad\quad x = 28.6$

   The population was 28.6 million.

39. a. Let x = the number of switchboard

   operators in the year 1998.

   $(100\% - 13.9\%)x = 185,000$

   $\qquad\quad 86.1\% x = 185,000$

   $\qquad\quad 0.861x = 185,000$

   $\qquad\qquad\quad x = 214,866$

   There were 214,866 switchboard

   operators in 1998.

   b. Answes may vary

41. Let x = the first integer,

   $x + 1$ = the next consecutive integer and

   $x + 2$ = the next.

   $x + (x + 1) + (x + 2) = 228$

   $\qquad\qquad\quad 3x + 3 = 228$

   $\qquad\qquad\qquad 3x = 225$

   $\qquad\qquad\qquad\; x = 75$

   $x + 1 = 75 + 1 = 76$

   $x + 2 = 75 + 2 = 77$

   The integers are 75, 76, 77.

43. Let x = the second angle,

   $2x$ = the first angle

   $3x - 12$ = the third.

   $x + 2x + 3x - 12 = 180$

   $\qquad\quad 6x - 12 = 180$

   $\qquad\qquad\; 6x = 192$

   $\qquad\qquad\quad x = 32$

   $2x = 2(32) = 64$

   $3x - 12 = 3(32) - 12 = 84$

   The angles are 32°, 64°, 84°.

45. Let x = the height and

   $2x + 12$ = the length.

   $P = 2L + 2h$

   $312 = 2(2x + 12) + 2x$

   $312 = 4x + 14 + 2x$

   $312 = 6x + 24$

   $288 = 6x$

   $\;\, 48 = x$

   $2x + 12 = 2(48) + 12 = 108$

   The height is 48 inches and the

   width is 108 inches.

47. Let x = the width and
$2x + 12$ = the length.
$P = 2L + 2W$
$40 = 2(2x + 2) + 2x$
$40 = 4x + 4 + 2x$
$40 = 6x + 4$
$36 = 6x$
$6 = x$
$2x + 2 = 2(6) + 2 = 14$
The width is 14 cm and the
length is 14 cm.

49. $x + (x + 20) = 180$
$2x + 20 = 180$
$2x = 160$
$x = 80$
$x + 20 = 80 + 20 = 100$
The angles are $80°$ and $100°$.

51. $5x + x = 90$
$6x = 90$
$x = 15$
$5x = 5(15) = 75$
The angles are $15°$ and $75°$.

53. Let x = the supplement of an angle,
$3x + 20$ = the angle.
$x + 3x + 20 = 180$
$4x + 20 = 180$
$4x = 160$
$x = 40$
$3x + 20 = 3(40) + 20$
$= 120 + 20$
$= 140$
The angles are $40°$, $140°$.

55. Let x = the width and
$5(x + 1)$ = the height
$x + 5(x + 1) = 55.4$
$6x + 5 = 55.4$
$6x = 50.4$
$x = 8.4$
$5(x + 1) = 5(8.4 + 1) = 5(9.4) = 47$
The width is 8.4 m and the height is 47 m.

57. Let x = the number of hours for a halogen bulb,

$25x$ = the number of hours for a fluorescent bulb and

$x - 2500$ = the number for an incandescent bulb.

$$x + 25x + x - 2500 = 105,000$$
$$27x - 2500 = 105,500$$
$$27x = 108,000$$
$$x = 4000$$
$$25x = 25(4000) = 100,000$$
$$x - 2500 = 4000 - 2500 = 1500$$

The number of bulb hours for each type is,

Incandescent: 1500

Fluorescent: 100,000

Halogen: 4000

Appendix C-3 Exercise Set

1. Point A: $(5, 2)$

3. Point C: $(3, 0)$

5. Point D: $(-1, 3)$

7. Point G: $(-1, 0)$

9. $(2, 3)$, QI

11. $(-2, 7)$, QII

13. $(-1, -4)$, QIII

15. $(0, -100)$, y-axis

17. $(-10, -30)$, QIII

19. $(-87, 0)$, x-axis

21. $(x, -y)$, QIV

23. $(x, 0)$, x-axis

25. $(-y, -y)$, QIII

27. $y = -x - 2$

x	y
0	-2
-2	0

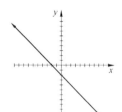

29. $3x - 4y = 8$

x	y
0	-2
$\frac{8}{3}$	0

31. $y = \dfrac{1}{3}x$ **See Example 3 in Text.**

33. $y + 4 = 0$
      $y = -4$

35. $f(4) = 1$

37. $g(0) = -4$

39. $f(x) = 0;\ x = 1,\ 3$

41. $(-1, -2)$ means $g(-1) = -2$

Appendix C-4 Exercise Set

1. $\left(-y^2 + 6y - 1\right) + \left(3y^2 - 4y - 10\right)$
   $= -y^2 + 6y - 1 + 3y^2 - 4y - 10$
   $= 2y^2 + 2y - 11$

3. $\left(x^2 - 6x + 2\right) - (x - 5)$
   $= x^2 - 6x + 2 - x + 5$
   $= x^2 - 7x + 7$

5. $(5x - 3)^2 = (5x)^2 - 2(3)(5x) + 3^2$
              $= 25x^2 - 30x + 9$

7.
$$
\begin{array}{r}
2x^3 - 4x^2 + 5x - 5 \\
x + 2\overline{\smash{\big)}\,2x^4 + 0x^3 - 3x^2 + 5x - 2} \\
\underline{2x^4 + 4x^3} \\
-4x^3 - 3x^2 \\
\underline{-4x^3 - 8x^2} \\
5x^2 + 5x \\
\underline{5x^2 + 10x} \\
-5x - 2 \\
\underline{-5x - 10} \\
8
\end{array}
$$

$\left(2x^4 - 3x^2 + 5x - 2\right) \div (x + 2)$

$= 2x^3 - 4x^2 + 5x - 5 + \dfrac{8}{x + 2}$

9. $x^2 - 8x + 16 - y^2$
   $= (x - 4)^2 - y^2$
   $= (x - 4 + y)(x - 4 - y)$

11. $x^4 - x = x\left(x^3 - 1\right)$
           $= x(x - 1)\left(x^2 + x + 1\right)$

13. $14x^2y - 2xy = 2xy(7x - 1)$

15. $4x^2 - 16 = (2x + 4)(2x - 4)$

17. $3x^2 - 8x - 11 = (3x - 11)(x + 1)$

19. $4x^2 + 8x - 12 = 4\left(x^2 + 2x - 3\right)$
                      $= 4(x + 3)(x - 1)$

21. $4x^2 + 36x + 81 = (2x+9)^2$

23. $8x^3 + 125y^3$
$$= (2x+5y)(4x^2 - 10xy + 25y^2)$$

25. $64x^2y^3 - 8x^2$
$$= 8x^2(8y^3 - 1)$$
$$= 8x^2(2y-1)(4y^2 + 2y + 1)$$

27. $(x+5)^3 + y^3$
$$= (x+5+y)\left[(x+5)^2 - (x+5)(y) + y^2\right]$$
$$= (x+y+5)(x^2 + 10x + 25 - xy - 5y - y^2)$$

29. $(5a-3)^2 - 6(5a-3) + 9$
$$= \left[(5aa-3)-3\right]^2$$
$$= (5a-6)^2$$

31. $7x^2 - 63x = (x-9)$

33. $ab - 6a + 7b - 42$
$$= a(b-6) + 7(b-6)$$
$$= (a+7)(b-6)$$

35. $x^4 - 1 = (x^2+1)(x^2-1)$
$$= (x^2+1)(x+1)(x-1)$$

37. $10x^2 - 7x - 33 = (5x-11)(21x+3)$

39. $5a^3b^3 - 50a^3b$
$$= 5a^3b(b^2 - 10)$$

41. $16x^2 + 25$ is prime

43. $10x^3 - 210x^2 + 1100x$
$$= 10x(x^2 - 21x + 110)$$
$$= 10x(x-11)(x-10)$$

45. $64a^3b^4 - 27a^3b$
$$= a^3b(64b^3 - 27)$$
$$= a^3b(4b-3)(16b^2 + 12b + 9)$$

47. $2x^3 - 54 = 2(x^3 - 27)$
$$= 2(x-3)(x^2 + 3x + 9)$$

49. $3y^5 - 5y^4 + 6y - 10$
$$= y^4(3y-5) + 2(3y-5)$$
$$= (y^4 + 2)(3y-5)$$

51. $100z^3 + 100$
$$= 100(z^3 + 1)$$
$$= 100(z+1)(z^2 - z + 1)$$

53. $4b^2 - 36b + 81 = (2b-9)^2$

55. $(y-6)^2 + 3(y-6) + 2$
$$= (y-6+2)(y-6+1)$$
$$= (y-4)(y-5)$$

Appendix C-5 Exercise Set

1. $\dfrac{x}{2} = \dfrac{1}{8} + \dfrac{x}{4}$

$8\left(\dfrac{x}{2}\right) = 8\left(\dfrac{1}{8}\right) + 8\left(\dfrac{x}{4}\right)$

$\quad 4x = 1 + 2x$

$\quad 2x = 1$

$\qquad x = \dfrac{1}{2}$

3. $\dfrac{1}{8} + \dfrac{x}{4} = \dfrac{1}{8} + \dfrac{x(2)}{x(2)}$

$\qquad = \dfrac{1}{8} + \dfrac{2x}{8}$

$\qquad = \dfrac{2x+1}{8}$

5. $\dfrac{4}{x+2} - \dfrac{2}{x-1}$

$= \dfrac{4(x-1)}{(x+2)(x-1)} - \dfrac{2(x+2)}{(x+2)(x-1)}$

$= \dfrac{4(x-1) - 2(x+2)}{(x+2)(x-1)}$

$= \dfrac{4x - 4 - 2x - 4}{(x+2)(x-1)}$

$= \dfrac{2x - 8}{(x+2)(x-1)}$

$= \dfrac{2(x-4)}{(x+2)(x-1)}$

7. $\dfrac{4}{x+2} = \dfrac{2}{x-1}$

$(x+2)(x-1)\left(\dfrac{4}{x+2}\right) = (x+2)(x-1)\left(\dfrac{2}{x-1}\right)$

$\qquad 4(x-1) = 2(x+2)$

$\qquad 4x - 4 = 2x + 4$

$\qquad 2x - 4 = 4$

$\qquad 2x = 8$

$\qquad x = 4$

9. $\dfrac{2}{x^2-4} = \dfrac{1}{x+2} - \dfrac{3}{x-2}$

$(x+2)(x-2)\left(\dfrac{2}{x^2-4}\right)(x+2)(x-2)\left(\dfrac{1}{x+2}\right)$

$\quad -(x+2)(x-2)\left(\dfrac{3}{x-2}\right)$

$2 = x - 2 - 3(x+2)$

$2 = x - 2 - 3x - 6$

$2 = -2x - 8$

$10 = -2x$

$-5 = x$

11. $\dfrac{5}{x^2-3x} + \dfrac{4}{2x-6}$

$= \dfrac{5(2)}{x(x-3)(2)} + \dfrac{4(x)}{2(x-3)(x)}$

$= \dfrac{10 + 4x}{2x(x-3)}$

$= \dfrac{2(2x+5)}{2x(x-3)}$

$= \dfrac{2x+5}{x(x-3)}$

13. $\dfrac{x-1}{x+1}+\dfrac{x+7}{x-1}=\dfrac{4}{x^2-1}$

$(x+1)(x-1)\left(\dfrac{x-1}{x+1}\right)+(x+1)(x-1)\left(\dfrac{x+7}{x-1}\right)$

$=(x+1)(x-1)\left(\dfrac{4}{x^2-1}\right)$

$(x-1)(x-1)+(x+1)(x+7)=4$

$x^2-2x+1+x^2+8x+7=4$

$2x^2+6x+8=4$

$2x^2+6x+4=0$

$2\left(x^2+3x+2\right)=0$

$2(x+2)(x+1)=0$

$x+2=0 \quad \text{or} \quad x+1=0$

$x=-2 \qquad\quad x=-1$

$x=-1$ is not allowed

The solution is $x=-2$.

15. $\dfrac{a^2-9}{a-6}\cdot\dfrac{a^2-5a-6}{a^2-a-6}$

$=\dfrac{(a+3)(a-3)}{a-6}\cdot\dfrac{(a-6)(a+1)}{(a-3)(a+2)}$

$=\dfrac{(a+3)(a+1)}{a+2}$

17. $\dfrac{2x+3}{3x-2}=\dfrac{4x+1}{6x+1}$

$(3x-2)(6x+1)\left(\dfrac{2x+3}{3x-2}\right)$

$=(3x-2)(6x+1)\left(\dfrac{4x+1}{6x+1}\right)$

$(6x+1)(2x+3)=(3x-2)(4x+1)$

$12x^2+20x+3=12x^2-5x-2$

$20x+3=-5x-2$

$25x+3=-2$

$25x=-5$

$x=-\dfrac{1}{5}$

19. $\dfrac{a}{9a^2-1}+\dfrac{2}{6a-2}$

$=\dfrac{a}{(3a+1)(3a-1)(2)}+\dfrac{2(3a+1)}{2(3a+1)(3a-1)}$

$=\dfrac{2a+2(3a+1)}{2(3a+1)(3a-1)}$

$=\dfrac{2a+6a+2}{2(3a+1)(3a-1)}$

$=\dfrac{8a+2}{2(3a+1)(3a-1)}$

$=\dfrac{2(4a+1)}{2(3a+1)(3a-1)}$

$=\dfrac{4a+1}{(3a+1)(3a-1)}$

21. $-\dfrac{3}{x^2} - \dfrac{1}{x} + 2 = 0$

$$x^2\left(-\dfrac{3}{x^2}\right) - x^2\left(\dfrac{1}{x}\right) + 2x^2 = 0$$

$$-3 - x + 2x^2 = 0$$

$$2x^2 - x - 3 = 0$$

$$(2x - 3)(x + 1) = 0$$

$$2x - 3 = 0 \quad \text{or} \quad x + 1 = 0$$

$$2x = 3 \qquad\qquad x = -1$$

$$x = \dfrac{3}{2}$$

23. $\dfrac{x-8}{x^2-x-2} + \dfrac{2}{x-2}$

$$= \dfrac{x-8}{(x-2)(x+1)} + \dfrac{2}{x-2}$$

$$= \dfrac{x-8}{(x-2)(x+1)} + \dfrac{2(x+1)}{(x-2)(x+1)}$$

$$= \dfrac{x-8+2(x+1)}{(x-2)(x+1)}$$

$$= \dfrac{x-8+2x+2}{(x-2)(x+1)}$$

$$= \dfrac{3x-6}{(x-2)(x+1)}$$

$$= \dfrac{3}{x+1}$$

25. $\qquad\dfrac{3}{a} - 5 = \dfrac{7}{a} - 1$

$$a\left(\dfrac{3}{a}\right) - a(5) = a\left(\dfrac{7}{a}\right) - a(1)$$

$$3 - 5a = 7 - a$$

$$3 - 4a = 7$$

$$-4a = 4$$

$$a = -1$$

27. a. $\dfrac{x}{5} - \dfrac{x}{4} + \dfrac{1}{10}$

    b. Write each term so that the denominator is the LCD of 20.

    c. $\dfrac{x(4)}{5(4)} - \dfrac{x(5)}{4(5)} + \dfrac{1(2)}{10(2)}$

$$= \dfrac{4x}{20} - \dfrac{5x}{20} + \dfrac{2}{20}$$

$$= \dfrac{4x - 5x + 2}{20}$$

$$= \dfrac{-x + 2}{20}$$

29. B

31. D

33. D

Appendix D Exercise Set

Viewing Window and Interpreting Window Exercise Set

1. Yes, since every coordinate is between −10 and 10.

3. No, since −11 is less than −10.

5. Answers may vary. Any values such that Xmin < −90, Ymin < −80, Xmax > 55, and Ymax > 80.

7. Answers may vary. Any values such that Xmin < −11, Ymin < −5, Xmax > 7, and Ymax > 2.

9. Answers may vary. Any values such that Xmin < 50, Ymin < −50, Xmax > 200, and Ymax > 200.

11. Xmax = −12   Ymax = −12
    Xmin = 12   Ymin = 12
    Xscl = 3   Yscl = 3

13. Xmax = −9   Ymax = −12
    Xmin = 9   Ymin = 12
    Xscl = 1   Yscl = 2

15. Xmax = −10   Ymax = −25
    Xmin = 10   Ymin = 25
    Xscl = 2   Yscl = 5

17. Xmax = −10   Ymax = −30
    Xmin = 10   Ymin = 30
    Xscl = 1   Yscl = 3

19. Xmax = −20   Ymax = −30
    Xmin = 30   Ymin = 50
    Xscl = 5   Yscl = 10

Graphing Equations and Square Viewing Window Exercise Set

1. Setting A:

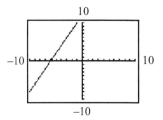

   Setting B:

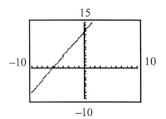

   Setting B shows all intercepts.

3. Setting A:

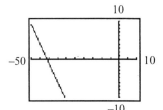

   Setting B:

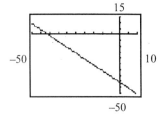

   Setting B shows all intercepts.

5. Setting A:

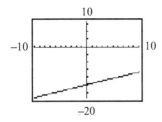

Setting B:

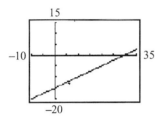

Setting B shows all intercepts.

7. $3x = 5y$

$y = \dfrac{3}{5} x$

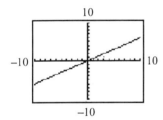

9. $9x - 5y = 30$

$-5y = -9x + 30$

$y = \dfrac{9}{5} x - 6$

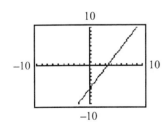

11. $y = -7$

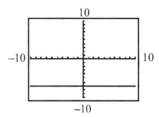

13. $x + 10y = -5$

$10y = -x - 5$

$y = -\dfrac{1}{10} x - \dfrac{1}{2}$

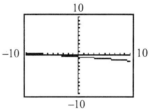

15. $y = \sqrt{x}$

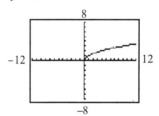

17. $y = x^2 + 2x + 1$

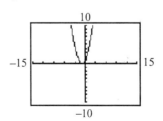

19. $y = |x|$

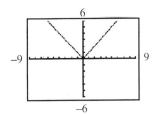

21. $x + 2y = 30$

$2y = -x + 30$

$y = -\dfrac{1}{2}x + 15$

Standard window:

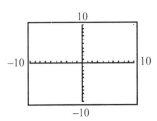

Adjusted window:

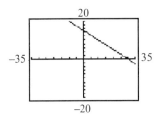

Appendix E Exercise Set

1. $\begin{cases} x + y = 1 \\ x - 2y = 4 \end{cases}$

$\begin{bmatrix} 1 & 1 & | & 1 \\ 1 & -2 & | & 4 \end{bmatrix}$

Multiply R1 by −1 and add to R2.

$\begin{bmatrix} 1 & 1 & | & 1 \\ 0 & -3 & | & 3 \end{bmatrix}$

Divide R2 by −3.

$\begin{bmatrix} 1 & 1 & | & 1 \\ 0 & 1 & | & -1 \end{bmatrix}$

This corresponds to $\begin{cases} x + y = 1 \\ y = -1 \end{cases}$.

$x + (-1) = 1$

$x - 1 = 1$

$x = 2$

The solution is $(2, -1)$.

3. $\begin{cases} x + 3y = 2 \\ x + 2y = 0 \end{cases}$

$\begin{bmatrix} 1 & 3 & | & 2 \\ 1 & 2 & | & 0 \end{bmatrix}$

Multiply R1 by −1 and add to R2.

$\begin{bmatrix} 1 & 3 & | & 2 \\ 0 & -1 & | & -2 \end{bmatrix}$

Multiply R2 by −1.

$\begin{bmatrix} 1 & 3 & | & 2 \\ 0 & 1 & | & 2 \end{bmatrix}$

This corresponds to $\begin{cases} x + 3y = 2 \\ y = 2 \end{cases}$.

$x + 3(2) = 6$

$x + 6 = 2$

$x = -4$

The solution is $(-4, 2)$.

5. $\begin{cases} x - 2y = 4 \\ 2x - 4y = 4 \end{cases}$

$\begin{bmatrix} 1 & -2 & | & 4 \\ 2 & -4 & | & 4 \end{bmatrix}$

Multiply R1 by −2 and add to R2.

$\begin{bmatrix} 1 & -2 & | & 4 \\ 0 & 0 & | & -4 \end{bmatrix}$

This corresponds to $\begin{cases} x - 2y = 4 \\ 0 = -4 \end{cases}$.

This is an inconsistent system.

The solution is $\varnothing$.

7. $\begin{cases} 3x - 3y = 9 \\ 2x - 2y = 6 \end{cases}$

$\begin{bmatrix} 3 & -3 & | & 9 \\ 2 & -2 & | & 6 \end{bmatrix}$

Divide R1 by 3.

$\begin{bmatrix} 1 & -1 & | & 3 \\ 2 & -2 & | & 6 \end{bmatrix}$

Multiply R1 by -2 and add to R2.

$\begin{bmatrix} 1 & -1 & | & 3 \\ 0 & 0 & | & 0 \end{bmatrix}$

This corresponds to $\begin{cases} x - y = 3 \\ 0 = 0 \end{cases}$.

This is a dependent system.
The solution is $\{(x, y) \mid x - y = 3\}$.

9. $\begin{cases} x + y \quad\;\; = 3 \\ \quad\;\; 2y \quad\;\; = 10 \\ 3x + 2y - 4z = 12 \end{cases}$

$\begin{bmatrix} 1 & 1 & 0 & | & 3 \\ 0 & 2 & 0 & | & 10 \\ 3 & 2 & -4 & | & 12 \end{bmatrix}$

Multiply R1 by -3 and add to R3.

$\begin{bmatrix} 1 & 1 & 0 & | & 3 \\ 0 & 2 & 0 & | & 10 \\ 0 & -1 & -4 & | & 3 \end{bmatrix}$

Divide R2 by 2.

$\begin{bmatrix} 1 & 1 & 0 & | & 3 \\ 0 & 1 & 0 & | & 5 \\ 0 & -1 & -4 & | & 3 \end{bmatrix}$

Add R2 to R3.

$\begin{bmatrix} 1 & 1 & 0 & | & 3 \\ 0 & 1 & 0 & | & 5 \\ 0 & 0 & -4 & | & 8 \end{bmatrix}$

Divide R3 by -4.

$\begin{bmatrix} 1 & 1 & 0 & | & 3 \\ 0 & 1 & 0 & | & 5 \\ 0 & 0 & 1 & | & -2 \end{bmatrix}$

This corresponds to $\begin{cases} x + y = 3 \\ y = 5 \\ z = -2 \end{cases}$

$x + 5 = 3$
$\quad x = -2$
The solution is $(-2, 5, -2)$.

11. $\begin{cases} 2y - z = -7 \\ x + 4y + z = -4 \\ 5x - y + 2z = 13 \end{cases}$

$\begin{bmatrix} 0 & 2 & -1 & | & -7 \\ 1 & 4 & 1 & | & -4 \\ 5 & -1 & 2 & | & 13 \end{bmatrix}$

Interchange R1 and R2.

$\begin{bmatrix} 1 & 4 & 1 & | & -4 \\ 0 & 2 & -1 & | & -7 \\ 5 & -1 & 2 & | & 13 \end{bmatrix}$

Multiply R1 by -5 and add to R3.

$\begin{bmatrix} 1 & 4 & 1 & | & -4 \\ 0 & 2 & -1 & | & -7 \\ 0 & -21 & -3 & | & 33 \end{bmatrix}$

Divide R2 by 2.

$\begin{bmatrix} 1 & 4 & 1 & | & -4 \\ 0 & 1 & -\frac{1}{2} & | & -\frac{7}{2} \\ 0 & -21 & -3 & | & 33 \end{bmatrix}$

Multiply R2 by 21 and add to R3.

$\begin{bmatrix} 1 & 4 & 1 & | & -4 \\ 0 & 1 & -\frac{1}{2} & | & -\frac{7}{2} \\ 0 & 0 & -\frac{27}{2} & | & -\frac{81}{2} \end{bmatrix}$

Multiply R2 by $-\frac{2}{27}$.

$$\begin{bmatrix} 1 & 4 & 1 & | & -4 \\ 0 & 1 & -\frac{1}{2} & | & -\frac{7}{2} \\ 0 & 0 & 1 & | & 3 \end{bmatrix}$$

This corresponds to $\begin{cases} x + 4y + z = 4 \\ y - \frac{1}{2}z = -\frac{7}{2} \\ z = 3 \end{cases}$.

$y - \frac{1}{2}(3) = -\frac{7}{2}$

$y - \frac{3}{2} = -\frac{7}{2}$

$y = -2$

$x + 4(-2) + 3 = -4$

$x - 8 + 3 = -4$

$x = 1$

The solution is $(1, -2, 3)$.

13. $\begin{cases} x - 4 = 0 \\ x + y = 1 \end{cases}$ or $\begin{cases} x = 4 \\ x + y = 1 \end{cases}$

$$\begin{bmatrix} 1 & 0 & | & 4 \\ 1 & 1 & | & 1 \end{bmatrix}$$

Multiply R1 by –1 and add to R2.

$$\begin{bmatrix} 1 & 0 & | & 4 \\ 0 & 1 & | & -3 \end{bmatrix}$$

This corresponds to $\begin{cases} x = 4 \\ y = -3 \end{cases}$

The solution is $(4, -3)$.

15. $\begin{cases} x + y + z = 2 \\ 2x \quad - z = 5 \\ 3y + z = 2 \end{cases}$

$$\begin{bmatrix} 1 & 1 & 1 & | & 2 \\ 2 & 0 & -1 & | & 5 \\ 0 & 3 & 1 & | & 2 \end{bmatrix}$$

Multiply R1 by –2 and add to R2.

$$\begin{bmatrix} 1 & 1 & 1 & | & 2 \\ 0 & -2 & -3 & | & 1 \\ 0 & 3 & 1 & | & 2 \end{bmatrix}$$

Divide R2 by –2.

$$\begin{bmatrix} 1 & 1 & 1 & | & 2 \\ 0 & 1 & \frac{3}{2} & | & -\frac{1}{2} \\ 0 & 3 & 1 & | & 2 \end{bmatrix}$$

Multiply R2 by –3 and add to R3.

$$\begin{bmatrix} 1 & 1 & 1 & | & 2 \\ 0 & 1 & \frac{3}{2} & | & -\frac{1}{2} \\ 0 & 0 & -\frac{7}{2} & | & \frac{7}{2} \end{bmatrix}$$

Multiply R3 by $-\frac{2}{7}$.

$$\begin{bmatrix} 1 & 1 & 1 & | & 2 \\ 0 & 1 & \frac{3}{2} & | & -\frac{1}{2} \\ 0 & 0 & 1 & | & -1 \end{bmatrix}$$

This corresponds to $\begin{cases} x + y + z = 2 \\ y + \frac{3}{2}z = -\frac{1}{2} \\ z = -1 \end{cases}$.

$y + \frac{3}{2}(-1) = -\frac{1}{2}$

$y - \frac{3}{2} = -\frac{1}{2}$

$y = 1$

$x + 1 + (-1) = 2$

$x = 2$

The solution is $(2, 1, -1)$.

17. $\begin{cases} 5x - 2y = 27 \\ -3x + 5y = 18 \end{cases}$

$$\begin{bmatrix} 5 & -2 & | & 27 \\ -3 & 5 & | & 18 \end{bmatrix}$$

Divide R1 by 5.

$$\begin{bmatrix} 1 & -\frac{2}{5} & | & \frac{27}{5} \\ -3 & 5 & | & 18 \end{bmatrix}$$

Multiply R1 by 3 and add to R2.

$$\begin{bmatrix} 1 & -\dfrac{2}{5} & \bigg| & \dfrac{27}{5} \\ 0 & \dfrac{19}{5} & \bigg| & \dfrac{171}{5} \end{bmatrix}$$

Multiply R2 by $\dfrac{5}{19}$.

$$\begin{bmatrix} 1 & -\dfrac{2}{5} & \bigg| & \dfrac{27}{5} \\ 0 & 1 & \bigg| & 9 \end{bmatrix}$$

This corresponds to $\begin{cases} x - \dfrac{2}{5}y = \dfrac{27}{5} \\ \qquad\quad y = 9 \end{cases}$.

$x - \dfrac{2}{5}(9) = \dfrac{27}{5}$

$x - \dfrac{18}{5} = \dfrac{27}{5}$

$x = 9$

The solution is (9, 9).

19. $\begin{cases} 4x - 7y = 7 \\ 12x - 21y = 24 \end{cases}$

$$\begin{bmatrix} 4 & -7 & | & 7 \\ 12 & -21 & | & 24 \end{bmatrix}$$

Divide R1 by 4.

$$\begin{bmatrix} 1 & -\dfrac{7}{4} & \bigg| & \dfrac{7}{4} \\ 12 & -21 & | & 24 \end{bmatrix}$$

Multiply R1 by −12 and add to R2.

$$\begin{bmatrix} 1 & -\dfrac{7}{4} & \bigg| & \dfrac{7}{4} \\ 0 & 0 & | & 3 \end{bmatrix}$$.

This corresponds to $\begin{cases} x - \dfrac{7}{4}y = \dfrac{7}{4} \\ \qquad\quad 0 = 3 \end{cases}$.

This is an inconsistent system.
The solution set is ∅.

21. $\begin{cases} 4x - y + 2z = 5 \\ \quad 2y + z = 4 \\ 4x + y + 3z = 10 \end{cases}$

$$\begin{bmatrix} 4 & -1 & 2 & | & 5 \\ 0 & 2 & 1 & | & 4 \\ 4 & 1 & 3 & | & 10 \end{bmatrix}$$

Divide R1 by 4.

$$\begin{bmatrix} 1 & -\dfrac{1}{4} & \dfrac{1}{2} & \bigg| & \dfrac{5}{4} \\ 0 & 2 & 1 & | & 4 \\ 0 & 2 & 1 & | & 5 \end{bmatrix}$$

Divide R2 by 2.

$$\begin{bmatrix} 1 & -\dfrac{1}{4} & \dfrac{1}{2} & \bigg| & \dfrac{5}{4} \\ 0 & 1 & \dfrac{1}{2} & | & 2 \\ 0 & 2 & 1 & | & 5 \end{bmatrix}$$

Multiply R2 by −2 and add to R3.

$$\begin{bmatrix} 1 & -\dfrac{1}{4} & \dfrac{1}{2} & \bigg| & \dfrac{5}{4} \\ 0 & 1 & \dfrac{1}{2} & | & 2 \\ 0 & 0 & 0 & | & 1 \end{bmatrix}$$

This corresponds to $\begin{cases} x - \dfrac{1}{4}y + \dfrac{1}{2}z = \dfrac{5}{4} \\ \qquad\quad y + \dfrac{1}{2}z = 2 \\ \qquad\qquad\qquad 0 = 1 \end{cases}$.

This is an inconsistent system.
The solution set is ∅.

23. $\begin{cases} 4x + y + \ z = 3 \\ -x + y - 2z = -11 \\ x + 2y + 2z = -1 \end{cases}$

$$\begin{bmatrix} 4 & 1 & 1 & | & 3 \\ -1 & 1 & -2 & | & -11 \\ 1 & 2 & 2 & | & -1 \end{bmatrix}$$

Interchange R1 and R3.

$$\begin{bmatrix} 1 & 2 & 2 & | & -1 \\ -1 & 1 & -2 & | & -11 \\ 4 & 1 & 1 & | & 3 \end{bmatrix}$$

Add R1 to R2.
Multiply R1 by –4 and add to R3.

$$\begin{bmatrix} 1 & 2 & 2 & | & -1 \\ 0 & 3 & 0 & | & -12 \\ 0 & -7 & -7 & | & 7 \end{bmatrix}$$

Divide R2 by 3.

$$\begin{bmatrix} 1 & 2 & 2 & | & -1 \\ 0 & 1 & 0 & | & -4 \\ 0 & -7 & -7 & | & 7 \end{bmatrix}$$

Multiply R2 by 7 and add to R3.

$$\begin{bmatrix} 1 & 2 & 2 & | & -1 \\ 0 & 1 & 0 & | & -4 \\ 0 & 0 & -7 & | & -21 \end{bmatrix}$$

Divide R3 by –7.

$$\begin{bmatrix} 1 & 2 & 2 & | & -1 \\ 0 & 1 & 0 & | & -4 \\ 0 & 0 & 1 & | & 3 \end{bmatrix}$$

This corresponds to $\begin{cases} x + 2y + 2z = -1 \\ \quad\quad y \quad\quad = -4 \\ \quad\quad\quad\quad z = 3 \end{cases}$.

$$x + 2(-4) + 2(3) = -1$$
$$x - 8 + 6 = -1$$
$$x = 1$$

The solution is $(1, -4, 3)$.

25. a. Solve the system $\begin{cases} 2.3x + y = 52 \\ -5.4x + y = 14 \end{cases}$.

$$\begin{bmatrix} 2.3 & 1 & | & 52 \\ -5.4 & 1 & | & 14 \end{bmatrix}$$

Since getting 1 in the first column would lead to repeating decimals, we multiply R1 by –1 and add to R2.

$$\begin{bmatrix} 2.3 & 1 & | & 52 \\ -7.7 & 0 & | & -38 \end{bmatrix}$$

This corresponds to $\begin{cases} 2.3x + y = 52 \\ -7.7x \quad\quad = -38 \end{cases}$.

From the second equation,

$$x = \frac{-3.8}{-7.7} \approx 4.935.$$

Thus, the percent of U.S. households owning black-and-white television sets was the same as the percent of U.S. households owning a microwave oven in the end of 1984 (about 4.9 years after 1980).

b. Solve the television equation for y:
$y = -2.3x + 52$. Thus, for 1980,
$y = -2.3(0) + 52 = 52$, and for 1993,
$y = -2.3(13) + 52 = 22.1$.
Solve the microwave equation for y:
$y = 5.4x + 14$. Thus, for 1980
$y = 5.4(0) + 14 = 14$, and for 1993,
$y = 5.4(13) + 14 = 84.2$.

In 1980, a greater percent of (and hence more) U.S. households owned black-and-white television sets. In 1993, more households owned a microwave oven. The percent owning black-and-white television sets is decreasing and the percent owning a microwave oven is increasing. Answers may vary.

c. Let $y = 0$ in the television equation.
$$2.3x + y = 52$$
$$2.3x + 0 = 52$$
$$x = \frac{52}{2.3} \approx 22.6$$

According to this model, the percent of U.S. households owning a black-and-white television set will be 0% about 22.6 years after 1980, or sometime in 2002.

27. Answers may vary.

Appendix F Exercise Set

1. $\begin{vmatrix} 3 & 5 \\ -1 & 7 \end{vmatrix} = 3(7) - 5(-1) = 21 + 5 = 26$

3. $\begin{vmatrix} 9 & -2 \\ 4 & -3 \end{vmatrix} = 39(-3) - 4(-2) = -27 + 8 = -19$

5. $\begin{vmatrix} -2 & 9 \\ 4 & -18 \end{vmatrix} = -2(-18) - 9(4) = 36 - 36 = 0$

7. $\begin{cases} 2y - 4 = 0 \\ x + 2y = 5 \end{cases}$ or $\begin{cases} 2y = 4 \\ x + 2y = 5 \end{cases}$

$D = \begin{vmatrix} 0 & 2 \\ 1 & 2 \end{vmatrix} = 0(2) - 2(1) = 0 - 2 = -2$

$D_x = \begin{vmatrix} 4 & 2 \\ 5 & 2 \end{vmatrix} = 4(2) - 2(5) = 8 - 10 = -2$

$D_y = \begin{vmatrix} 0 & 4 \\ 1 & 5 \end{vmatrix} = 0(5) - 4(1) = 0 - 4 = -4$

$x = \dfrac{-2}{-2} = 1$ and $y = \dfrac{-4}{-2} = 2$

The solution is (1, 2).

9. $\begin{cases} 3x + y = 1 \\ 2y = 2 - 6x \end{cases}$ or $\begin{cases} 3x + y = 1 \\ 6x + 2y = 2 \end{cases}$

$D = \begin{vmatrix} 3 & 1 \\ 6 & 2 \end{vmatrix} = 3(2) - 1(6) = 6 - 6 = 0$

Thus, the system cannot be solved by Cramer's rule. Since E2 is 2 times E1, the system is dependent.
The solution is $\{(x, y) \mid 3x + y = 1\}$.

11. $\begin{cases} 5x - 2y = 27 \\ -3x + 5y = 5 \end{cases}$

$D = \begin{vmatrix} 5 & -2 \\ -3 & 5 \end{vmatrix}$
$= 5(5) - (-2)(-3)$
$= 25 - 6$
$= 19$

$D_x = \begin{vmatrix} 27 & -2 \\ 18 & 5 \end{vmatrix}$
$= 27(5) - (-2)(18)$
$= 135 + 36$
$= 171$

$D_y = \begin{vmatrix} 5 & 27 \\ -3 & 18 \end{vmatrix}$
$= 5(18) - 27(-3)$
$= 90 + 81$
$= 171$

$x = \dfrac{D_x}{D} = \dfrac{171}{19} = 9$ and $y = \dfrac{D_y}{D} = \dfrac{171}{19} = 9$
The solution is (9, 9).

13. $\begin{vmatrix} 2 & 1 & 0 \\ 0 & 5 & -3 \\ 4 & 0 & 2 \end{vmatrix}$
$= 2 \begin{vmatrix} 5 & -3 \\ 0 & 2 \end{vmatrix} - 1 \begin{vmatrix} 0 & -3 \\ 4 & 2 \end{vmatrix} + 0 \begin{vmatrix} 0 & 5 \\ 4 & 0 \end{vmatrix}$
$= 2[5(2) - (-3)(0)] - [0(2) - 4(-3)] - 0$
$= 2(10) - 12$
$= 8$

15. $\begin{vmatrix} 4 & -6 & 0 \\ -2 & 3 & 0 \\ 4 & -6 & 1 \end{vmatrix}$
$= 0 \begin{vmatrix} -2 & 3 \\ 4 & -6 \end{vmatrix} - 0 \begin{vmatrix} 4 & -6 \\ 4 & -6 \end{vmatrix} + 1 - 2 \begin{vmatrix} 4 & -6 \\ 4 & 3 \end{vmatrix}$
$= 0 - 0 + [4(3) - (-6)(-2)]$
$= 0$

17. $\begin{vmatrix} 3 & 6 & -3 \\ -1 & -2 & 3 \\ 4 & -1 & 6 \end{vmatrix}$

$= 3\begin{vmatrix} -2 & 3 \\ -1 & 6 \end{vmatrix} - 6\begin{vmatrix} -1 & 3 \\ 4 & 6 \end{vmatrix} + (-3)\begin{vmatrix} -1 & -2 \\ 4 & -1 \end{vmatrix}$

$= 3[-2(-6) - 3(-1)] - 6[-1(6) - 3(4)]$
$\quad - 3[(-1)(-1) - (-2)(4)]$

$= 3(-9) - 6(-18) - 3(9)$

$= -27 + 108 - 27$

$= 54$

19. $\begin{cases} 3x \quad\ + z = -1 \\ -x - 3y + z = 7 \\ \quad\ \ 3y + z = 5 \end{cases}$

$D = \begin{vmatrix} 3 & 0 & 1 \\ -1 & -3 & 1 \\ 0 & 3 & 1 \end{vmatrix}$

$= 3\begin{vmatrix} -3 & 1 \\ 3 & 1 \end{vmatrix} - 0\begin{vmatrix} -1 & 1 \\ 0 & 1 \end{vmatrix} + 1\begin{vmatrix} -1 & -3 \\ 0 & 3 \end{vmatrix}$

$= 3[(-3)(1) - 1(3)] - 0$
$\quad + [(-1)(3) - (-3)(0)]$

$= 3(-6) - 3$

$= -21$

$D_x = \begin{vmatrix} -1 & 0 & 1 \\ 7 & -3 & 1 \\ 5 & 3 & 1 \end{vmatrix}$

$= -1\begin{vmatrix} -3 & 1 \\ 3 & 1 \end{vmatrix} - 0\begin{vmatrix} 7 & 1 \\ 5 & 1 \end{vmatrix} + 1\begin{vmatrix} 7 & -3 \\ 5 & 3 \end{vmatrix}$

$= -[(-3)(1) - 1(3)] - 0$
$\quad + [(7)(3) - (-3)(5)]$

$= 6 + 36$

$= 42$

$D_y = \begin{vmatrix} 3 & -1 & 1 \\ -1 & 7 & 1 \\ 0 & 5 & 1 \end{vmatrix}$

$= 3\begin{vmatrix} 7 & 1 \\ 5 & 1 \end{vmatrix} - (-1)\begin{vmatrix} -1 & 1 \\ 0 & 1 \end{vmatrix} + 1\begin{vmatrix} -1 & 7 \\ 0 & 5 \end{vmatrix}$

$= 3[7(1) - 1(5)] + 1[(-1)(1) - 1(0)]$
$\quad + [(-1)(5) - 7(0)]$

$= 3(2) + (-1) + (-5)$

$= 0$

$D_z = \begin{vmatrix} 3 & 0 & -1 \\ -1 & -3 & 7 \\ 0 & 3 & 5 \end{vmatrix}$

$= 3\begin{vmatrix} -3 & 7 \\ 3 & 5 \end{vmatrix} - 0\begin{vmatrix} -1 & 7 \\ 0 & 5 \end{vmatrix} + 1\begin{vmatrix} -1 & -3 \\ 0 & 3 \end{vmatrix}$

$= 3[(-3)(5) - 7(3)] - 0$
$\quad - [(-1)(3) - (-3)(0)]$

$= 3(-36) - (-3)$

$= -105$

$x = \dfrac{D_x}{D} = \dfrac{42}{-21} = -2, \quad y = \dfrac{D_y}{D} = \dfrac{0}{-21} = 0,$

$z = \dfrac{D_z}{D} = \dfrac{-105}{-21} = 5$

The solution is $(-2, 0, 5)$.

21. $\begin{cases} x + \ y + \ z = 8 \\ 2x - \ y - \ z = 10 \\ x - 2y + 3z = 22 \end{cases}$

$D = \begin{vmatrix} 1 & 1 & 1 \\ 2 & -1 & -1 \\ 1 & -2 & 3 \end{vmatrix}$

$= 1\begin{vmatrix} -1 & -1 \\ -2 & 3 \end{vmatrix} - 1\begin{vmatrix} 2 & -1 \\ 1 & 3 \end{vmatrix} + 1\begin{vmatrix} 2 & -1 \\ 1 & -2 \end{vmatrix}$

$= (-3 - 2) - [6 - (-1)] + [-4 - (-1)]$

$= -5 - 7 - 3$

$= -15$

$D_x = \begin{vmatrix} 8 & 1 & 1 \\ 10 & -1 & -1 \\ 22 & -2 & 3 \end{vmatrix}$

$= 8\begin{vmatrix} -1 & -1 \\ -2 & 3 \end{vmatrix} - 1\begin{vmatrix} 10 & -1 \\ 22 & 3 \end{vmatrix} + 1\begin{vmatrix} 10 & -1 \\ 22 & -2 \end{vmatrix}$

$= 8(-3 - 2) - [30 - (-22)]$
$\quad + [-20 - (-22)]$

$= 8(-5) - 52 + 2$

$\quad -40 - 52 + 2$

$= -90$

$$D_y = \begin{vmatrix} 1 & 8 & 1 \\ 2 & 10 & -1 \\ 1 & 22 & 3 \end{vmatrix}$$

$$= 1\begin{vmatrix} 10 & -1 \\ 22 & 3 \end{vmatrix} - 8\begin{vmatrix} 2 & -1 \\ 1 & 3 \end{vmatrix} + 1\begin{vmatrix} 2 & 10 \\ 1 & 22 \end{vmatrix}$$

$$= [30 - (-22)] - 8[6 - (-1)] + (44 - 10)$$

$$= 52 - 8(7) + 34$$

$$= 52 - 56 + 34$$

$$= 30$$

$$D_z = \begin{vmatrix} 1 & 1 & 8 \\ 2 & -1 & 10 \\ 1 & -2 & 22 \end{vmatrix}$$

$$= 1\begin{vmatrix} -1 & 10 \\ -2 & 22 \end{vmatrix} - 1\begin{vmatrix} 2 & 10 \\ 1 & 22 \end{vmatrix} + 8\begin{vmatrix} 2 & -1 \\ 1 & -2 \end{vmatrix}$$

$$= [-22 - (20)] - (44 - 10)$$
$$\quad + 8[-4 - (-1)]$$

$$= -2 - 34 + 8(-3)$$

$$= -36 - 24$$

$$= -60$$

$$x = \frac{D_x}{D} = \frac{-90}{-15} = 6, \quad y = \frac{D_y}{D} = \frac{30}{-15} = -2,$$

$$z = \frac{D_z}{D} = \frac{-60}{-15} = 4$$

The solution is $(6, -2, 4)$.

23. $\begin{vmatrix} 10 & -1 \\ -4 & 2 \end{vmatrix} = 10(2) - (-1)(-4) = 20 - 4 = 16$

25. $\begin{vmatrix} 1 & 0 & 4 \\ 1 & -1 & 2 \\ 3 & 2 & 1 \end{vmatrix}$

$$= 1\begin{vmatrix} -1 & 2 \\ 2 & 1 \end{vmatrix} - 0\begin{vmatrix} 1 & 2 \\ 3 & 1 \end{vmatrix} + 4\begin{vmatrix} 1 & -1 \\ 3 & 2 \end{vmatrix}$$

$$= 1(-1 - 4) - 1 + 4[2 - (-3)]$$

$$= -5 + 4(5)$$

$$= -5 + 20$$

$$= 15$$

27. $\begin{vmatrix} \frac{3}{4} & \frac{5}{2} \\ -\frac{1}{6} & \frac{7}{3} \end{vmatrix} = \frac{3}{4}\left(\frac{7}{3}\right) - \frac{5}{2}\left(-\frac{1}{6}\right)$

$$= \frac{21}{12} + \frac{5}{12}$$

$$= \frac{26}{12}$$

$$= \frac{13}{6}$$

29. $\begin{vmatrix} 4 & -2 & 2 \\ 6 & -1 & 3 \\ 2 & 1 & 1 \end{vmatrix}$

$$= 4\begin{vmatrix} -1 & 3 \\ 1 & 1 \end{vmatrix} - (-2)\begin{vmatrix} 6 & 3 \\ 2 & 1 \end{vmatrix} + 2\begin{vmatrix} 6 & -1 \\ 2 & 1 \end{vmatrix}$$

$$= 4(-1 - 3) + 2(6 - 6) + 2[6 - (-2)]$$

$$= 4(-4) + 2(0) + 2(8)$$

$$= -16 + 0 + 16$$

$$= 0$$

31. $\begin{vmatrix} -2 & 5 & 4 \\ 5 & -1 & 3 \\ 4 & 1 & 2 \end{vmatrix}$

$$= -2\begin{vmatrix} -1 & 3 \\ 1 & 2 \end{vmatrix} - 5\begin{vmatrix} 5 & 3 \\ 4 & 2 \end{vmatrix} + 4\begin{vmatrix} 5 & -1 \\ 4 & 1 \end{vmatrix}$$

$$= -2(-2 - 3) - 5(10 - 12) + 4[5 - (-4)]$$

$$= -2(-5) - 5(-2) + 4(9)$$

$$= 10 + 10 + 36$$

$$= 56$$

33. $\begin{cases} 2x - 5y = 4 \\ x + 2y = -7 \end{cases}$

$D = \begin{vmatrix} 2 & -5 \\ 1 & 2 \end{vmatrix}$

$= 2(2) - (-5)(1) = 4 + 5 = 9$

$D_x = \begin{vmatrix} 4 & -5 \\ -7 & 2 \end{vmatrix}$

$= 4(2) - (-5)(-7)$

$= 8 - 35$

$= -27$

$D_y = \begin{vmatrix} 2 & 4 \\ 1 & -7 \end{vmatrix}$

$= 2(-7) - 4(1)$

$= -14 - 4$

$= -18$

$x = \dfrac{D_x}{D} = \dfrac{-27}{9} = -3$

$y = \dfrac{D_y}{D} = \dfrac{-18}{9} = -2$

The solution is $(-3, -2)$.

35. $\begin{cases} 4x + 2y = 5 \\ 2x + y = -1 \end{cases}$

$D = \begin{vmatrix} 4 & 2 \\ 2 & 1 \end{vmatrix}$

$= 4(1) - (2)(2) = 4 + 4 = 0$

Thus, the system cannot be solved by Cramer's rule. Multiply E2 by 2 yielding the new system:

$\begin{cases} 4x + 2y = 5 \\ 4x + 2y = -2 \end{cases}$

Therefore, the system is inconsistent. The solution is $\varnothing$.

37. $\begin{cases} 2x + 2y + z = 1 \\ -x + y + 2z = 3 \\ x + 2y + 4z = 0 \end{cases}$

$D = \begin{vmatrix} 2 & 2 & 1 \\ -1 & 1 & 2 \\ 1 & 2 & 4 \end{vmatrix}$

$= 2\begin{vmatrix} 1 & 2 \\ 2 & 4 \end{vmatrix} - 2\begin{vmatrix} -1 & 2 \\ 1 & 4 \end{vmatrix} + 1\begin{vmatrix} -1 & 1 \\ 1 & 2 \end{vmatrix}$

$= 2(4 - 4) - 2(-4 - 2) + (-2 - 1)$

$= 2(0) - 2(-6) + (-3)$

$= 0 + 12 - 3$

$= 9$

$D_x = \begin{vmatrix} 1 & 2 & 1 \\ 3 & 1 & 2 \\ 0 & 2 & 4 \end{vmatrix}$

$= 1\begin{vmatrix} 1 & 2 \\ 2 & 4 \end{vmatrix} - 3\begin{vmatrix} 2 & 1 \\ 2 & 4 \end{vmatrix} + 0\begin{vmatrix} 2 & 1 \\ 1 & 2 \end{vmatrix}$

$= (4 - 4) - 3(8 - 2) + 0$

$= 0 - 3(6)$

$= -18$

$D_y = \begin{vmatrix} 2 & 1 & 1 \\ -1 & 3 & 2 \\ 1 & 0 & 4 \end{vmatrix}$

$= 1\begin{vmatrix} 1 & 1 \\ 3 & 2 \end{vmatrix} - 0\begin{vmatrix} 2 & 1 \\ -1 & 2 \end{vmatrix} + 4\begin{vmatrix} 2 & 1 \\ -1 & 3 \end{vmatrix}$

$= (2 - 3) - 0 + 4[6 - (-1)]$

$= -1 + 4(7)$

$= -1 + 28$

$= -27$

$D_z = \begin{vmatrix} 2 & 2 & 1 \\ -1 & 1 & 3 \\ 1 & 2 & 0 \end{vmatrix}$

$= 1\begin{vmatrix} 2 & 1 \\ 1 & 3 \end{vmatrix} - 2\begin{vmatrix} 2 & 1 \\ -1 & 4 \end{vmatrix} + 0\begin{vmatrix} 2 & 2 \\ -1 & 1 \end{vmatrix}$

$= (6 - 1) - 2[6 - (-1)] + 0$

$= 5 - 2(7)$

$= 5 - 14$

$= -9$

$x = \dfrac{D_x}{D} = \dfrac{-18}{9} = -2, \quad y = \dfrac{D_y}{D} = \dfrac{27}{9} = 3,$

$z = \dfrac{D_z}{D} = \dfrac{-9}{9} = -1$

The solution is $(-2, 3, -1)$.

39. $\begin{cases} \dfrac{2}{3}x - \dfrac{3}{4}y = -1 \\ -\dfrac{1}{6}x + \dfrac{3}{4}y = \dfrac{5}{2} \end{cases}$

$D = \begin{vmatrix} \dfrac{2}{3} & -\dfrac{3}{4} \\ -\dfrac{1}{6} & \dfrac{3}{4} \end{vmatrix}$

$= \dfrac{2}{3}\left(\dfrac{3}{4}\right) - \left(-\dfrac{3}{4}\right)\left(-\dfrac{1}{6}\right)$

$= \dfrac{1}{2} - \dfrac{1}{8}$

$= \dfrac{3}{8}$

$D_x = \begin{vmatrix} -1 & -\dfrac{3}{4} \\ \dfrac{5}{2} & \dfrac{3}{4} \end{vmatrix}$

$= (-1)\left(\dfrac{3}{4}\right) - \left(-\dfrac{3}{4}\right)\left(\dfrac{5}{2}\right)$

$= -\dfrac{3}{4} + \dfrac{15}{8}$

$= \dfrac{9}{8}$

$D_y = \begin{vmatrix} \dfrac{2}{3} & -1 \\ -\dfrac{1}{6} & \dfrac{5}{2} \end{vmatrix}$

$= \dfrac{2}{3}\left(\dfrac{5}{2}\right) - (-1)\left(-\dfrac{1}{6}\right)$

$= \dfrac{5}{3} - \dfrac{1}{6}$

$= \dfrac{3}{2}$

$x = \dfrac{D_x}{D} = \dfrac{\frac{9}{8}}{\frac{3}{8}} = 3$ and $y = \dfrac{D_y}{D} = \dfrac{\frac{3}{2}}{\frac{3}{8}} = 4$

The solution is (3, 4).

41. $\begin{cases} 0.7x - 0.2y = -1.6 \\ 0.2x - y = -1.4 \end{cases}$

$D = \begin{vmatrix} 0.7 & -0.2 \\ 0.2 & -1 \end{vmatrix}$

$= 0.7(-1) - (-0.2)(0.2)$

$= -0.7 + 0.04$

$= -0.66$

$D_x = \begin{vmatrix} -1.6 & -0.2 \\ -1.4 & -1 \end{vmatrix}$

$= -1.6(-1) - (-0.2)(-1.4)$

$= 1.6 - 2.8$

$= 1.32$

$D_y = \begin{vmatrix} 0.7 & -1.6 \\ 0.2 & -1.4 \end{vmatrix}$

$= 0.7(-1.4) - (-1.6)(0.2)$

$= -0.98 + 0.32$

$= -0.66$

$x = \dfrac{D_x}{D} = \dfrac{1.32}{-0.66} = -2$ and

$y = \dfrac{D_y}{D} = \dfrac{-0.66}{-0.66} = 1$

The solution is (–2, 1).

43. $\begin{cases} -2x + 4y - 2z = 6 \\ x - 2y + z = -3 \\ 3x - 6y + 3z = -9 \end{cases}$

$D = \begin{vmatrix} -2 & 4 & -2 \\ 1 & -2 & 1 \\ 3 & -6 & 3 \end{vmatrix}$

$= -2\begin{vmatrix} -2 & 1 \\ -6 & 3 \end{vmatrix} - 4\begin{vmatrix} 1 & 1 \\ 3 & 3 \end{vmatrix} + (-2)\begin{vmatrix} 1 & -2 \\ 3 & -6 \end{vmatrix}$

$= -2[-6 - (-6)] - 4(3 - 3) - 2[-6 - (-6)]$

$= 2(0) - 4(0) - 2(0)$

$= 0$

Therefore, Cramer's rule will not provide the solution. Note that E1 is –2 times E2 and that E3 is 3 times E2. Thus, the system is dependent. The solution is $\{(x, y, z) \mid x - 2y + z = -3\}$.

45. $\begin{cases} x - 2y + z = -5 \\ \quad\; -3y + 2z = 4 \\ 3x - y \quad\;\; = -2 \end{cases}$

$D = \begin{vmatrix} 1 & -2 & 1 \\ 0 & 3 & 2 \\ 3 & -1 & 0 \end{vmatrix}$

$= 1\begin{vmatrix} 3 & 2 \\ -1 & 0 \end{vmatrix} - 0\begin{vmatrix} -2 & 1 \\ -1 & 0 \end{vmatrix} + 3\begin{vmatrix} -2 & 1 \\ 3 & 2 \end{vmatrix}$

$= [0 - (-2)] - 0 + 3(-4 - 3)$

$= 2 + 3(-7)$

$= -19$

$D_x = \begin{vmatrix} -5 & -2 & 1 \\ 4 & 3 & 2 \\ -2 & -1 & 0 \end{vmatrix}$

$= 1\begin{vmatrix} 4 & 3 \\ -2 & -1 \end{vmatrix} - 2\begin{vmatrix} -5 & -2 \\ -2 & -1 \end{vmatrix} + 0\begin{vmatrix} 5 & -2 \\ 4 & 3 \end{vmatrix}$

$= [4 - (-6)] - 2(5 - 4) + 0$

$= 2 - 2(1)$

$= 0$

$D_y = \begin{vmatrix} 1 & -5 & 1 \\ 0 & 4 & 2 \\ 3 & -2 & 0 \end{vmatrix}$

$= 1\begin{vmatrix} 4 & 2 \\ -2 & 0 \end{vmatrix} - 0\begin{vmatrix} -5 & 1 \\ -2 & 0 \end{vmatrix} + 3\begin{vmatrix} -5 & 1 \\ 4 & 2 \end{vmatrix}$

$= [0 - (-4)] - 0 + 3(-10 - 4)$

$= 4 + 3(-14)$

$= 4 - 42$

$= -38$

$D_z = \begin{vmatrix} 1 & -2 & -5 \\ 0 & 3 & 4 \\ 3 & -1 & -2 \end{vmatrix}$

$= 1\begin{vmatrix} 3 & 4 \\ -1 & -2 \end{vmatrix} - 0\begin{vmatrix} -2 & -5 \\ -1 & -2 \end{vmatrix} + 3\begin{vmatrix} -2 & -5 \\ 3 & 4 \end{vmatrix}$

$= [-6 - (-4)] - 0 + 3[-8 - (-15)]$

$= -2 + 3(7)$

$= 19$

$x = \dfrac{D_x}{D} = \dfrac{0}{-19} = 0, \quad y = \dfrac{D_y}{D} = \dfrac{-38}{-19} = 2,$

$z = \dfrac{D_z}{D} = \dfrac{19}{-19} = -1$

The solution is $(0, 2, -1)$.

47. $\begin{vmatrix} 1 & x \\ 2 & 7 \end{vmatrix} = -3$

$1(7) - 2x = -3$

$7 - 2x = -3$

$-2x = -10$

$x = 5$

49. If the elements of a single row (or column) of a determinant are all zero, the value of the determinant will be zero. To see this, consider expanding on that row (or column) containing all zeros.

51. The array of signs for use with a 4×4 matrix is

$$\begin{matrix} + & - & + & - \\ - & + & - & + \\ + & - & + & - \\ - & + & - & + \end{matrix}$$

53. $\begin{vmatrix} 5 & 0 & 0 & 0 \\ 0 & 4 & 2 & -1 \\ 1 & 3 & -2 & 0 \\ 0 & -3 & 1 & 2 \end{vmatrix}$

$= 5\begin{vmatrix} 4 & 2 & -1 \\ 3 & -2 & 0 \\ -3 & 1 & 2 \end{vmatrix} - 0\begin{vmatrix} 0 & 2 & -1 \\ 1 & -2 & 0 \\ 0 & 1 & 2 \end{vmatrix}$

$\quad + 0\begin{vmatrix} 0 & 4 & -1 \\ 1 & 3 & 0 \\ 0 & -3 & 2 \end{vmatrix} - 0\begin{vmatrix} 0 & 4 & 2 \\ 1 & 3 & -2 \\ 0 & -3 & 1 \end{vmatrix}$

$= 5\left[(-1)\begin{vmatrix} 3 & -2 \\ -3 & 1 \end{vmatrix} - 0\begin{vmatrix} 4 & 2 \\ -3 & 1 \end{vmatrix} + 2\begin{vmatrix} 4 & 2 \\ 3 & -2 \end{vmatrix} \right]$

$= 5[-(3 - 6) - 0 + 2(-8 - 6)]$

$= 5[3 + 2(-14)]$

$= 5(3 - 28)$

$= 5(-25)$

$= -125$

55.

$$\begin{vmatrix} 4 & 0 & 2 & 5 \\ 0 & 3 & -1 & 1 \\ 0 & 0 & 2 & 0 \\ 0 & 0 & 0 & 1 \end{vmatrix} = 4\begin{vmatrix} 3 & -1 & 1 \\ 0 & 2 & 0 \\ 0 & 0 & 1 \end{vmatrix} - 0\begin{vmatrix} 0 & 2 & 5 \\ 0 & 2 & 0 \\ 0 & 0 & 1 \end{vmatrix}$$

$$+ 0\begin{vmatrix} 0 & 2 & 5 \\ 3 & -1 & 1 \\ 0 & 0 & 1 \end{vmatrix} - 0\begin{vmatrix} 0 & 2 & 5 \\ 3 & -1 & 1 \\ 0 & 2 & 0 \end{vmatrix}$$

$$= 4\left[3\begin{vmatrix} 2 & 0 \\ 0 & 1 \end{vmatrix} - 0\begin{vmatrix} -1 & 1 \\ 0 & 1 \end{vmatrix} + 0\begin{vmatrix} -1 & 1 \\ 2 & 0 \end{vmatrix} \right]$$

$$= 4[3(2-0) - 0 + 0]$$
$$= 4(6)$$
$$= 24$$

Appendix G Exercise Set

1. 21, 28, 16, 42, 38

$$\overline{x} = \frac{21 + 28 + 16 + 42 + 38}{5} = \frac{145}{5} = 29$$

16, 21, 28, 38, 42

median $= 28$

no mode

3. 7.6, 8.2, 8.2, 9.6, 5.7, 9.1

$$\overline{x} = \frac{7.6 + 8.2 + 8.2 + 9.6 + 5.7 + 9.1}{6}$$

$$= \frac{48.4}{6} = 8.1$$

5.7, 7.6, 8.2, 8.2, 9.1, 9.6

$$\text{median} = \frac{8.2 + 8.2}{2} = 8.2$$

mode $= 8.2$

5. 0.2, 0.3, 0.5, 0.6, 0.6, 0.9, 0.2, 0.7, 1.1

$$\overline{x} = \frac{0.2 + 0.3 + 0.5 + 0.6 + 0.6 + 0.9 + 0.2 + 0.7 + 1.1}{9}$$

$$= \frac{5.1}{9}$$

$$= 0.6$$

0.2, 0.2, 0.3, 0.5, 0.6, 0.6, 0.7, 0.9, 1.1

median $= 0.6$

mode $= 0.2$ and 0.6

7. 231, 543, 601, 293, 588, 109, 334, 268

$$\overline{x} = \frac{231 + 543 + 601 + 293 + 588 + 109 + 334 + 268}{8}$$

$$= \frac{2967}{8}$$

$$= 370.9$$

109, 231, 268, 293, 334, 543, 588, 601

$$\text{median} = \frac{293 + 334}{2} = 313.5$$

no mode

9. 1454, 1250, 1136, 1127, 1107

$$\overline{x} = \frac{1454 + 1250 + 1136 + 1127 + 1107}{5}$$

$$= \frac{6074}{5}$$

$$= 1214.8 \text{ feet}$$

11. 1454, 1250, 1136, 1127,

1107, 1046, 1023, 1002

$$\text{median} = \frac{1127 + 1107}{2} = 1117 \text{ feet}$$

13. $$\overline{x} = \frac{7.8 + 6.9 + 7.5 + 4.7 + 6.9 + 7.0}{6}$$

$$= \frac{40.8}{6} = 6.8 \text{ seconds}$$

15. 4.7, 6.9, 6.9, 7.0, 7.5, 7.8

mode = 6.9

17. 74, 77, 85, 86, 91, 95

median $= \dfrac{85+86}{2} = 85.5$

19. Sum $= 78 + 80 + 66 + 68 + 71$
$+ 64 + 82 + 71 + 70 + 65$
$+ 70 + 75 + 77 + 86 + 72$
$= 1095$

$\overline{x} = \dfrac{1095}{15} = 73$

21. 64, 65, 66, 68, 70, 70, 71, 71, 72, 75,
77, 78, 80, 82, 86

mode = 70 and 71

23. 64, 65, 66, 68,
70, 70, 71, 71, 72, 75, 77, 78, 80, 82, 86
$\uparrow$
mean = 73

9 rates were lower than the mean.

25. _, _, 16, 18, _;

Since the mode is 21, at least two of
the missing numbers must be 21. The
mean is 20. Let the one unknown
number be x.

$\overline{x} = \dfrac{21 + 21 + 16 + 18 + 24}{5} = 20$

$\dfrac{76 + x}{5} = 20$

$76 + x = 100$

$x = 24$

The missing numbers are 21, 21, 24.

Appendix H Exercise Set

1. $90° - 19° = 71°$

3. $90° - 70.8° = 19.2°$

5. $90° - 11\dfrac{1}{4}° = 78\dfrac{3}{4}°$

7. $180° - 150° = 30°$

9. $180° - 30.2° = 149.8°$

11. $180° - 79\dfrac{1}{2}° = 100\dfrac{1}{2}°$

13. $m\angle 1 = 110°$

$m\angle 2 = 180° - 110° = 70°$

$m\angle 3 = m\angle 2 = 70°$

$m\angle 4 = m\angle 2 = 70°$

$m\angle 5 = m\angle 1 = 110°$

$m\angle 6 = m\angle 4 = 70°$

$m\angle 7 = m\angle 5 = 110°$

15. $180° - 11° - 79° = 90°$

17. $180° - 25° - 65° = 90°$

19. $180° - 30° - 60° = 90°$

21. $90° - 45° = 45°$
$45°, 90°$

23. $90° - 17° = 73°$
$73°, 90°$

25. $90° - 39\frac{3}{4}° = 50\frac{1}{4}°$

$\quad 50\frac{1}{4}°$, $90°$

27. $\dfrac{12}{4} = \dfrac{18}{x}$

$\quad 4x\left(\dfrac{12}{4}\right) = 4x\left(\dfrac{18}{x}\right)$

$\quad\quad\quad 12x = 72$

$\quad\quad\quad\quad x = 6$

29. $\dfrac{6}{9} = \dfrac{3}{x}$

$\quad 9x\left(\dfrac{6}{9}\right) = 9x\left(\dfrac{3}{x}\right)$

$\quad\quad\quad 6x = 27$

$\quad\quad\quad\quad x = 4.5$

31. $a^2 + b^2 = c^2$

$\quad 6^2 + 8^2 = c^2$

$\quad 36 + 64 = c^2$

$\quad\quad\quad 100 = c^2$

$\quad\quad\quad\; 10 = c$

33. $a^2 + b^2 = c^2$

$\quad 5^2 + b^2 = 13^2$

$\quad 25 + b^2 = 169$

$\quad\quad\quad\; b^2 = 144$

$\quad\quad\quad\;\; b = 12$